FOUNDATIONS OF
EARTH SCIENCE

EIGHTH EDITION

FOUNDATIONS OF EARTH SCIENCE

EIGHTH EDITION

FREDERICK K. LUTGENS

EDWARD J. TARBUCK

ILLUSTRATED BY DENNIS TASA

PEARSON

Editor-in-Chief: Beth Wilbur
Senior Marketing Manager: Mary Salzman
Senior Acquisitions Editor: Andrew Dunaway
Executive Marketing Manager: Neena Bali
Senior Project Manager: Nicole Antonio
Director of Development: Jennifer Hart
Development Editor: Margot Otway
Senior Content Producer: Timothy Hainley
Program Manager: Sarah Shefveland
Editorial Assistant: Michelle Koski
Marketing Assistant: Ami Sampat
Team Lead, Project Management: David Zielonka
Team Lead, Program Management: Kristen Flathman

Project Manager, Instructor Media: Kyle Doctor
Full Service/Composition: Cenveo® Publisher Services
Full Service Project Manager: Heidi Allgair/Cenveo®
  Publisher Services
Design Manager: Derek Bacchus
Cover and Interior Design: Jeff Puda Design
Photo and Illustration Support: International Mapping
Permissions Project Manager: Maya Gómez
Photo Research: Kristin Piljay
Photo Permissions: Jillian Santos, QBS Learning
Text Permissions Research: Jillian Santos, QBS Learning
Procurement Specialist: Maura Zaldivar-Garcia
Cover Image Credit: © Michael Collier

Credits and acknowledgments for materials borrowed from other sources and reproduced, with permission, in this textbook appear on the appropriate page within the book.

Many of the designations used by manufacturers and sellers to distinguish their products are claimed as trademarks. Where those designations appear in this book, and the publisher was aware of a trademark claim, the designations have been printed in initial caps or all caps.

Library of Congress Cataloging-in-Publication Data
Names: Lutgens, Frederick K., author. | Tarbuck, Edward J., author.
Title: Foundations of earth science/Frederick K. Lutgens, Edward J. Tarbuck; illustrated by
    Dennis Tasa.
Description: Eighth edition. | Boston: Pearson, 2016. | Includes index.
Identifiers: LCCN 2015043673 | ISBN 9780134184814 | ISBN 0134184815
Subjects: LCSH: Earth sciences--Textbooks.
Classification: LCC QE28 .L96 2016 | DDC 550--dc23
LC record available at hgp://lccn.loc.gov/2015043673

2   16

Student Edition
ISBN-10: 0-134-18481-5
ISBN-13: 978-0-134-18481-4

Instructor's Review Copy
ISBN-10: 0-134-24076-6
ISBN-13: 978-0-134-24076-3

www.pearsonhighered.com

To Our Grandchildren

Allison and Lauren

Shannon, Amy, Andy, Ali, and Michael

Each is a bright promise for the future.

BRIEF TABLE OF CONTENTS

CONTENTS

UNIT I EARTH MATERIALS   22

1

Matter and Minerals   22

viii   Contents

UNIT II   SCULPTING EARTH'S SURFACE   76

3

Landscapes Fashioned by Water   76

4

Glacial and Arid Landscapes   114

UNIT III FORCES WITHIN 142

5

Plate Tectonics: A Scientific Revolution Unfolds   142

6

Restless Earth: Earthquakes and Mountain Building   172

7

Volcanoes and Other Igneous Activity   210

UNIT
IV
DECIPHERING EARTH'S HISTORY **248**

8

Geologic Time 248

UNIT V
THE GLOBAL OCEAN **272**

UNIT **VI** EARTH'S DYNAMIC ATMOSPHERE **328**

11

Heating the Atmosphere   328

12

Moisture, Clouds, and Precipitation   360

13

The Atmosphere
in Motion   394

15
The Nature of the Solar System   444

16
Beyond Our Solar System   482

SMARTFIGURES

Using a cutting-edge technology called augmented reality, Pearson's BouncePages app launches engaging, interactive video and animations that bring textbook pages to life. Use your mobile device to scan a SmartFigure identified by the BouncePage icon, and an animation/video/interactive illustrating the SmartFigure's concept launches immediately. No slow websites or hard-to-remember logins required. BouncePages' augmented reality technology transforms textbooks into convenient digital platforms, breathes life into your learning experience, and helps you grasp difficult academic concepts. Learning geology from a textbook will never be the same.

Preface

Foundations of Earth Science, eighth edition, is a college-level text designed for an introductory course in Earth science. It consists of seven units that emphasize broad and up-to-date coverage of basic topics and principles in geology, oceanography, meteorology, and astronomy. The book is intended to be a meaningful, nontechnical survey for undergraduate students who may have a modest science background. Usually these students are taking an Earth science class to meet a portion of their college's or university's general requirements.

In addition to being informative and up-to-date, *Foundations of Earth Science,* eighth edition, strives to meet the need of beginning students for a readable and user-friendly text and a highly usable tool for learning basic Earth science principles and concepts.

New and Important Features

This eighth edition is an extensive and thorough revision of *Foundations of Earth Science* that integrates improved textbook resources with new online features to enhance the learning experience:

- **Significant updating and revision of content.** A basic function of a college science textbook is to present material in a clear, understandable way that is accurate, engaging, and up-to-date. In the long history of this textbook, our number-one goal has always been to keep *Foundations of Earth Science* current, relevant, and highly readable for beginning students. To that end, every part of this text has been examined carefully. Many discussions, case studies, examples, and illustrations have been updated and revised.

- **SmartFigures that make *Foundations* much more than a traditional textbook.** Through its many editions, an important strength of *Foundations of Earth Science* has always been clear, logically organized, and well-illustrated explanations. Now, complementing and reinforcing this strength are a series of SmartFigures. Simply by scanning a SmartFigure with a mobile device and **Pearson's BouncePages Augmented Reality app** (available for iOS and Android), students can follow hundreds of unique and innovative avenues that will increase their insight and understanding of important ideas and concepts. SmartFigures are truly art that teaches! This eighth edition of *Foundations* has more than 200 SmartFigures, of five different types:

  1. **SmartFigure Tutorials.** Each of these 2- to 4-minute features, prepared and narrated by Professor Callan Bentley, is a mini-lesson that examines and explains the concepts illustrated by the figure.

  2. **SmartFigure Mobile Field Trips.** Scattered throughout this new edition are 24 video field trips that explore classic sites from Iceland to Hawaii. On each trip you will accompany geologist-pilot-photographer Michael Collier in the air and on the ground to see and learn about landscapes that relate to discussions in the chapter.

  3. **SmartFigure Condor Videos.** The 10 *Condor* videos take you to locations in the American West. By coupling aerial footage acquired by a drone aircraft with ground-level views, effective narratives, and helpful animations, these videos will engage you in real-life case studies.

  4. **SmartFigure Animations.** Scanning the many figures with this designation brings art to life. These animations and accompanying narrations illustrate and explain many difficult-to-visualize topics and ideas more effectively than static art alone.

  5. **SmartFigure Videos.** Rather than provide a single image to illustrate an idea, these figures include short video clips that help illustrate such diverse subjects as mineral properties and the structure of ice sheets.

- **Revised active learning path.** *Foundations of Earth Science* is designed for learning. Every chapter begins with *Focus on Concepts.* Each numbered learning objective corresponds to a major section in the chapter. The statements identify the knowledge and skills students should master by the end of the chapter and help students prioritize key concepts. Within the chapter, each major section concludes with *Concept Checks* that allow students to check their understanding and comprehension of important ideas and terms before moving on to the next section. Two end-of-chapter features complete the learning path. *Concepts in Review* coordinates with the *Focus on Concepts* at the start of the chapter and with the numbered sections within the chapter. It is a readable and concise overview of key ideas, with photos, diagrams, and questions that also help students focus on important ideas and test their understanding of key concepts. Chapters conclude with *Give It Some Thought.* The questions and problems in this section challenge learners by involving them in activities that require higher-order thinking skills, such as application, analysis, and synthesis of chapter material.

- **An unparalleled visual program.** In addition to more than 100 new, high-quality photos and satellite images, dozens of figures are new or have been redrawn by the gifted and highly respected geoscience illustrator Dennis Tasa. Maps and diagrams are frequently paired with photographs for greater effectiveness. Further, many new and revised figures have additional labels that narrate the process being illustrated and guide students as they examine the figures. Overall, the visual program of this text is clear and easy to understand.

- **MasteringGeology™.** MasteringGeology delivers engaging, dynamic learning opportunities—focused on course objectives and responsive to each student's progress—that have been proven to help students learn course material and understand difficult concepts. Assignable activities in MasteringGeology include SmartFigure (Tutorial, Condor, Animation, Mobile Field Trip, and Video) activities, GigaPan® activities, Encounter Earth activities using Google Earth™, GeoTutor activities, Geoscience Animation activities, GEODe tutorials, and more. MasteringGeology also includes all instructor resources and a robust Study Area with resources for students.

The Teaching and Learning Package

MasteringGeology™ with Pearson eText

Used by more than 1 million science students, the Mastering platform is the most effective and widely used online tutorial, homework, and assessment system for the sciences. Now available with *Foundations of Earth Science,* eighth edition, **MasteringGeology**™ offers tools for use before, during, and after class:

- **Before class:** Assign adaptive Dynamic Study Modules and reading assignments from the eText with Reading Quizzes to ensure that students come prepared to class, having done the reading.
- **During class:** Learning Catalytics, a "bring your own device" student engagement, assessment, and classroom intelligence system, allows students to use a smartphone, tablet, or laptop to respond to questions in class. With Learning Catalytics, you can assess students in real-time, using open-ended question formats to determine student misconceptions, and adjust lectures accordingly.
- **After class:** Assign an array of assessment resources such as Mobile Field Trips, Project Condor videos, Interactive Simulations, GeoDrone activities, Google Earth Encounter Activities, and much more. Students receive wrong-answer feedback personalized to their answers, which will help them get back on track.

MasteringGeology Student Study Area also provides students with self-study materials including all of the SmartFigures, geoscience animations, *GEODe: Earth Science* tutorials, *In the News* RSS feeds, Self Study Quizzes, Web Links, Glossary, and Flashcards.

For more information or access to MasteringGeology, please visit www.masteringgeology.com.

Instructor's Resource Materials (Download Only)

The authors and publisher have been pleased to work with a number of talented people who have produced an excellent supplements package.

Instructor's Resource Materials (IRM) The IRM puts all your lecture resources in one easy-to-reach place:

- The IRM provides all of the line art, tables, and photos from the text in .jpg files.
- The IRM provides three PowerPoint files for each chapter. Cut down on your preparation time, no matter what your lecture needs, by taking advantage of these components of the PowerPoint files:
  - **Exclusive art.** All of the photos, art, and tables from the text, in order, loaded into PowerPoint slides.
  - **Lecture outlines.** This set averages 70 slides per chapter and includes customizable lecture outlines with supporting art.
  - **Classroom Response System (CRS) questions.** Authored for use in conjunction with classroom response systems, these PowerPoints allow you to electronically poll your class for responses to questions, pop quizzes, attendance, and more.

Instructor Manual (Download Only)

The Instructor Manual has been designed to help seasoned and new professors alike, and it offers the following for each chapter: an introduction to the chapter, an outline, and learning objectives/Focus on Concepts; teaching strategies; teacher resources; and answers to *Concept Checks, Concepts in Review,* and *Give It Some Thought* questions from the textbook.

TestGen Computerized Test Bank (Download Only)

TestGen is a computerized test generator that lets instructors view and edit Test Bank questions, transfer questions to tests, and print tests in a variety of customized formats. The Test Bank includes approximately 1,200 multiple-choice, matching, and essay questions. Questions are correlated to Bloom's Taxonomy, each chapter's learning objectives, the Earth Science Literacy Initiative Big Ideas, and the Pearson Science Global Outcomes to help instructors better map the assessments against both broad and specific teaching and learning objectives. The Test Bank is also available in Microsoft Word and can be imported into Blackboard. www.pearsonhighered.com/irc

Blackboard Already have your own website set up? We will provide a Test Bank in Blackboard or formats for importation upon request. Additional course resources are available on the IRC and are available for use with permission.

Acknowledgments

Writing a college textbook requires the talents and cooperation of many people. It is truly a team effort, and the authors are fortunate to be part of an extraordinary team at Pearson Education. In addition to being great people to work with, all are committed to producing the best textbooks possible. Special thanks to our geoscience editor, Andy Dunaway, who invested a great deal of time, energy, and effort in this project. We appreciate his enthusiasm, hard work, and quest for excellence. We also appreciate our conscientious project manager, Nicole Antonio, whose job it was to keep track of all that was going on—and a lot was going on. As always, our marketing managers, Neena Bali and Mary Salzman, who talk with faculty daily, provide us with helpful input. The eighth edition of *Foundations of Earth Science* was certainly improved by the talents of our developmental editor, Margot Otway. Many thanks. It was the job of the production team, led by Heidi Allgair at Cenveo® Publisher Services, to turn our manuscript into a finished product. The team also included copyeditor Kitty Wilson, compositor Annamarie Boley, proofreader Heather Mann, and photo researcher Kristin Piljay. We think these talented people did great work. All are true professionals, with whom we are very fortunate to be associated.

The authors owe special thanks to three people who were very important contributors to this project:

- Working with Dennis Tasa, who is responsible for all of the text's outstanding illustrations and some excellent animations, is always special for us. He has been part of our team for more than 30 years. We value not only his artistic talents, hard work, patience, and imagination but his friendship as well.

- As you read this text, you will see dozens of extraordinary photographs by Michael Collier. Most are aerial shots taken from his 60-year-old Cessna 180. Michael was also responsible for preparing the 24 remarkable Mobile Field Trips that are scattered through the text. Among his many awards is the American Geological Institute Award for Outstanding Contribution to the Public Understanding of Geosciences. We think that Michael's photographs and field trips are the next best thing to being there. We were very fortunate to have had Michael's assistance on *Foundations of Earth Science*, eighth edition. Thanks, Michael.

- Callan Bentley has been an important addition to the *Foundations of Earth Science* team. Callan is a professor of geology at Northern Virginia Community College in Annandale, where he has been honored many times as an outstanding teacher. He is a frequent contributor to *Earth* magazine and is author of the popular geology blog *Mountain Beltway*. Callan was responsible for preparing the SmartFigure Tutorials that appear throughout the text. As you take advantage of these outstanding learning aids, you will hear his voice explaining the ideas. We appreciate Callan's contributions to this new edition of *Foundations*.

Great thanks also go to those colleagues who prepared in-depth reviews. Their critical comments and thoughtful input helped guide our work and clearly strengthened the text. Special thanks to:

Larry Braile, *Purdue University*
Mitch Chapura, *Piedmont College*
Kathy DeBusk, *Longwood University*
Stewart Farrar, *Eastern Kentucky University*
Gary Johnpeer, *Cerritos College*
Garrett Love, *North Carolina Central University*
Rick Lynn, *Grayson County College*
Jill Murray, *Pennsylvania State University, Abington*
Marcus Ross, *Liberty University*
Janelle Sikorski, *Miami University–Oxford*
Jana Svec, *Moraine Valley Community College*
Scott Vetter, *Centenary College of Louisiana*
Maggie Zimmerman, *Saint Paul College*

Last but certainly not least, we gratefully acknowledge the support and encouragement of our wives, Nancy Lutgens and Joanne Bannon. Preparation of *Foundations of Earth Science*, eighth edition, would have been far more difficult without their patience and understanding.

Fred Lutgens
Ed Tarbuck

AUGMENTED REALITY:
Bringing the Textbook to Life

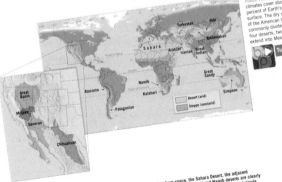

SmartFigure 4.26 Dry climates Arid and semiarid climates cover about 30 percent of Earth's land surface. The dry region of the American West is commonly divided into four deserts, two of which extend into Mexico.

Tutorial

Augmented Reality Enhances the Reading Experience, Bringing the Textbook to Life

Using a cutting-edge technology called augmented reality, Pearson's BouncePages app launches engaging, interactive videos and animations that bring textbook pages to life. Use your mobile device to scan a SmartFigure identified by the BouncePages icon, and an animation or video illustrating the SmartFigure's concept launches immediately. No slow websites or hard-to-remember logins required.

BouncePages' augmented reality technology transforms textbooks into convenient digital platforms, breathes life into your learning experience, and helps you grasp difficult academic concepts. Learning geology from a textbook will never be the same.

Download the FREE BP App for Android

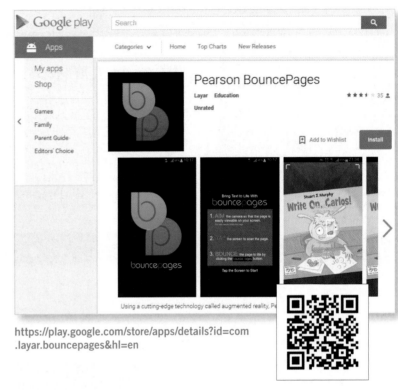

https://play.google.com/store/apps/details?id=com
.layar.bouncepages&hl=en

Download the FREE BP App for iOS

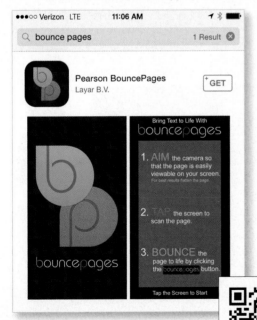

https://itunes.apple.com/us/app/pearson
-bouncepages/id659370955?mt=8

By scanning figures associated with the BouncePages icon, students will be immediately connected to the digital world and will deepen their learning experience with the printed text.

SmartFigure 2.31 Confining pressure and differential stress

Tutorial

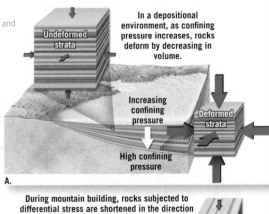

In a depositional environment, as confining pressure increases, rocks deform by decreasing in volume.

Undeformed strata

Increasing confining pressure

Deformed strata

High confining pressure

A.

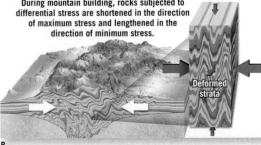

During mountain building, rocks subjected to differential stress are shortened in the direction of maximum stress and lengthened in the direction of minimum stress.

Deformed strata

B.

Bring the Field to YOUR Teaching and Learning Experience

NEW! **SmartFigure: Condor Videos.** Bringing Physical Geology to life for GenEd students, three geologists, using a GoPro camera mounted to a quadcopter, have ventured out into the field to film **10 key geologic locations**. These process-oriented videos, accessed through BouncePages technology, are designed to bring the field to the classroom or dorm room and enhance the learning experience in our texts.

NEW! **SmartFigure: Mobile Field Trips.** Scattered throughout this new edition of *Foundations of Earth Science* are **24 video field trips**. On each trip, you will accompany geologist-pilot-photographer Michael Collier in the air and on the ground to see and learn about iconic landscapes that relate to discussions in the chapter. These extraordinary field trips are accessed by using the BouncePages app to scan the figure in the chapter—usually one of Michael's outstanding photos.

Visualize Processes and Tough Topics

Dip-slip fault   Dip-slip fault   Dip-slip fault

NEW! **SmartFigure: Animations** are brief videos, many created by text illustrator Dennis Tasa, that animate a process or concept depicted in the textbook's figures. This technology allows students to view moving figures rather than static art to depict how a geologic process actually changes through time. The videos can be accessed using Pearson's BouncePages app for use on mobile devices, and will also be available via MasteringGeology.

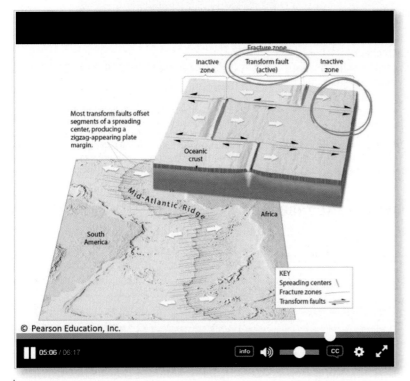

Callan Bentley, SmartFigure Tutorial author, is a Chancellor's Commonwealth Professor of Geology at Northern Virginia Community College (NOVA) in Annandale, Virginia. Trained as a structural geologist, Callan teaches introductory level geology at NOVA, including field-based and hybrid courses. Callan writes a popular geology blog called *Mountain Beltway*, contributes cartoons, travel articles, and book reviews to *EARTH* magazine, and is a digital education leader in the two-year college geoscience community.

SmartFigure: Tutorials bring key chapter illustrations to life! Found throughout the book, these Tutorials are sophisticated, annotated illustrations that are also narrated videos. They are accessible on mobile devices via scannable BouncePages printed in the text and through the Study Area in MasteringGeology.

Modular Approach Driven by Learning Objectives

The new edition is designed to support a four-part learning path, an innovative structure that facilitates active learning and allows students to focus on important ideas as they pause to assess their progress at frequent intervals.

The chapter-opening **Focus on Concepts** lists the learning objectives for each chapter. Each section of the chapter is tied to a specific learning objective, providing students with a clear learning path to the chapter content.

Each chapter section concludes with **Concept Checks,** a feature that lists questions tied to the section's learning objective, allowing students to monitor their grasp of significant facts and ideas.

8.2 CONCEPT CHECKS

1. Distinguish between numerical dates and relative dates.
2. Sketch and label five simple diagrams that illustrate each of the following: superposition, original horizontality, lateral continuity, cross-cutting relationships, and inclusions.
3. What is the significance of an unconformity?
4. Distinguish among angular unconformity, disconformity, and nonconformity.

Concepts in Review, a fresh approach to the typical end-of-chapter material, provides students with a structured and highly visual review of each chapter. Consistent with the Focus on Concepts and Concept Checks, the Concepts in Review is structured around the section title and the corresponding learning objective for each section.

CONCEPTS IN REVIEW
Geologic Time

8.1 A Brief History of Geology
Explain the principle of uniformitarianism and discuss how it differs from catastrophism.

KEY TERMS: catastrophism, uniformitarianism

- Early ideas about the nature of Earth were based on religious traditions and notions of great catastrophes.
- In the late 1700s, James Hutton emphasized that the same slow processes have acted over great spans of time and are responsible for Earth's rocks, mountains, and landforms. This similarity of processes over vast spans of time led to this principle being called uniformitarianism.

8.2 Creating a Time Scale—Relative Dating Principles
Distinguish between numerical and relative dating and apply relative dating principles to determine a time sequence of geologic events.

KEY TERMS: numerical date, relative date, principle of superposition, principle of original horizontality, principle of lateral continuity, principle of cross-cutting relationships, inclusion, conformable, unconformity, angular unconformity, disconformity, nonconformity

- The two types of dates that geologists use to interpret Earth history are (1) relative dates, which put events in their proper sequence of formation, and (2) numerical dates, which pinpoint the time in years when an event took place.
- Relative dates can be established using the principles of superposition, original horizontality, lateral continuity, cross-cutting relationships, and inclusions. Unconformities, gaps in the geologic record, may be identified during the relative dating process.

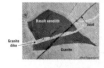

? The accompanying photo shows four features. Place the features in the proper sequence, from oldest to youngest. Explain your reasoning.

8.3 Fossils: Evidence of Past Life
Define fossil and discuss the conditions that favor the preservation of organisms as fossils. List and describe various fossil types.

KEY TERMS: fossil, paleontology

- Fossils are remains or traces of ancient life. Paleontology is the branch of science that studies fossils.
- Fossils can form through many processes. For an organism to be preserved as a fossil, it usually needs to be buried rapidly. Also, an organism's hard parts are most likely to be preserved because soft tissue decomposes rapidly in most circumstances.

? What term is used to describe the type of fossil that is shown here? Briefly describe how it formed.

8.4 Correlation of Rock Layers
Explain how rocks of similar age that are in different places can be matched up.

KEY TERMS: correlation, principle of fossil succession, index fossil, fossil assemblage

- Matching up exposures of rock that are the same age but are in different places is called correlation. By correlating rocks from around the world, geologists developed the geologic time scale and obtained a fuller perspective on Earth history.
- Fossils can be used to correlate sedimentary rocks in widely separated places by using the rocks' distinctive fossil content and applying the principle of fossil succession. The principle states that fossil organisms succeed one another in a definite and determinable order, and, therefore, a time period can be recognized by examining its fossil content.
- Index fossils are particularly useful in correlation because they are widespread and associated with a relatively narrow time span. The overlapping ranges of fossils in an assemblage may be used to establish an age for a rock layer that contains multiple fossils.
- Fossils may be used to establish ancient environmental conditions that existed when sediment was deposited.

8.5 Determining Numerical Dates with Radioactivity
Discuss three types of radioactive decay and explain how radioactive isotopes are used to determine numerical dates.

KEY TERMS: radioactivity, radioactive decay, radiometric dating, half-life, radiocarbon dating

GIVE IT SOME THOUGHT

1. The accompanying image shows the metamorphic rock gneiss, a basaltic dike, and a fault. Place these three features in their proper sequence (which came first, second, and third) and explain your logic.

2. A mass of granite is in contact with a layer of sandstone. Using a principle described in this chapter, explain how you might determine whether the sandstone was deposited on top of the granite or whether the magma that formed the granite was intruded after the sandstone was deposited.

3. This scenic image is from Monument Valley in the northeastern corner of Arizona. The bedrock in this region consists of layers of sedimentary rocks. Although the prominent rock exposures ("monuments") in this photo are widely separated, we can infer that they represent a once-continuous layer. Discuss the principle that allows us to make this inference.

4. The accompanying photo shows two layers of sedimentary rock. The lower layer is shale from the late Mesozoic era. Note the old river channel that was carved into the shale after it was deposited. Above is a younger layer of boulder-rich breccia. Are these layers conformable? Explain why or why not. What term from relative dating applies to the line separating the two layers?

5. Refer to Figure 8.9, which shows the historic angular unconformity at Scotland's Siccar Point that James Hutton studied in the late 1700s. Refer to this photo for the following exercises.
   a. Describe in general what occurred to produce this feature.
   b. Suggest ways in which at least three of Earth's four spheres could have been involved.
   c. The Earth system is powered by energy from two sources. How are both sources represented in the Siccar Point unconformity?

6. These polished stones are called *gastroliths*. Explain how such objects can be considered fossils. What category of fossil are they? Name another example of a fossil in this category.

7. If a radioactive isotope of thorium (atomic number 90, mass number 232) emits 6 alpha particles and 4 beta particles during the course of radioactive decay, what are the atomic number and mass number of the stable daughter product?

8. A hypothetical radioactive isotope has a half-life of 10,000 years. If the ratio of radioactive parent to stable daughter product is 1:3, how old is the rock that contains the radioactive material?

9. Solve the problems below that relate to the magnitude of Earth history. To make calculations easier, round Earth's age to 5 billion years.
   a. What percentage of geologic time is represented by recorded history? (Assume 5000 years for the length of recorded history.)
   b. Humans and their close relatives (hominins) have been around for roughly 5 million years. What percentage of geologic time is represented by the history of this group?
   c. The first abundant fossil evidence does not appear until the beginning of the Cambrian period, about 540 million years ago. What percentage of geologic time is represented by abundant fossil evidence?

10. A portion of a popular college text in historical geology includes 10 chapters (281 pages) in a unit titled "The Story of Earth." Two chapters (49 pages) are devoted to Precambrian time. By contrast, the last two chapters (67 pages) focus on the most recent 23 million years, with 25 of those pages devoted to the Holocene Epoch, which began 10,000 years ago.
    a. Compare the percentage of pages devoted to the Precambrian to the actual percentage of geologic time that this span represents.
    b. How does the number of pages about the Holocene compare to its actual percentage of geologic time?
    c. Suggest some reasons why the text seems to have such an unequal treatment of Earth history.

11. The accompanying diagram is a cross section of a hypothetical area. Place the lettered features in the proper sequence, from oldest to youngest. Where in the sequence can you identify an unconformity?

Give It Some Thought (GIST) is found at the end of each chapter and consists of questions and problems asking students to analyze, synthesize, and think critically about Geology. GIST questions relate back to the chapter's learning objectives, and can easily be assigned using MasteringGeology.

Continuous Learning Before, During, and After Class with MasteringGeology™

MasteringGeology delivers engaging, dynamic learning opportunities—focusing on course objectives responsive to each student's progress—that are proven to help students learn geology course material and understand challenging concepts.

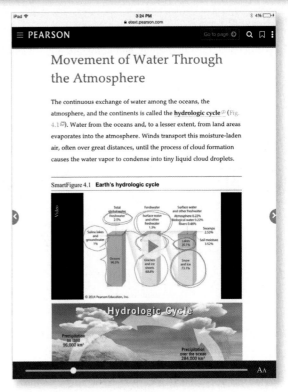

Movement of Water Through the Atmosphere

The continuous exchange of water among the oceans, the atmosphere, and the continents is called the **hydrologic cycle** (Fig. 4.1). Water from the oceans and, to a lesser extent, from land areas evaporates into the atmosphere. Winds transport this moisture-laden air, often over great distances, until the process of cloud formation causes the water vapor to condense into tiny liquid cloud droplets.

SmartFigure 4.1   **Earth's hydrologic cycle**

Before Class

Dynamic Study Modules and eText 2.0 provide students with a preview of what's to come.

Dynamic Study Modules enable students to study effectively on their own in an adaptive format. Students receive an initial set of questions with a unique answer format asking them to indicate their confidence.

Once completed, Dynamic Study Modules include explanations using material taken directly from the text.

NEW! **Interactive eText 2.0** complete with embedded media. eText 2.0 is mobile friendly and ADA accessible.

- Now available on smartphones and tablets.
- Seamlessly integrated videos and other rich media.
- Accessible (screen-reader ready).
- Configurable reading settings, including resizable type and night reading mode.
- Instructor and student note-taking, highlighting, bookmarking, and search.

During Class

Engage Students with Learning Catalytics

Learning Catalytics, a "bring your own device" student engagement, assessment, and classroom intelligence system, allows students to use their smartphone, tablet, or laptop to respond to questions in class.

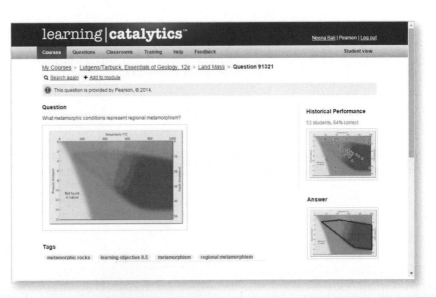

MasteringGeology™

After Class

Easy-to-Assign, Customizeable, and Automatically Graded Assignments

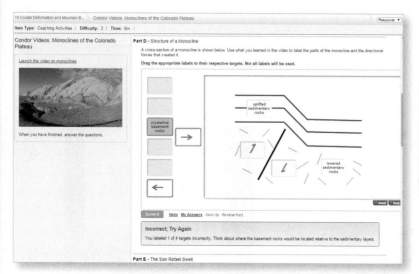

NEW! Project Condor Videos capture stunning footage of the Mountain West region with a quadcopter and a GoPro camera. A series of videos have been created with annotations, sketching, and narration to improve the way students learn about faults and folds, streams, volcanoes, and so much more. In Mastering, these videos are accompanied by questions designed to assess students on the main takeaways from each video.

NEW! Mobile Field Trips take students to classic geologic locations as they accompany geologist–pilot–photographer–author Michael Collier in the air and on the ground to see and learn about landscapes that relate to concepts in the chapter. In Mastering, these videos will be accompanied by auto-gradable assessments that will track what students have learned.

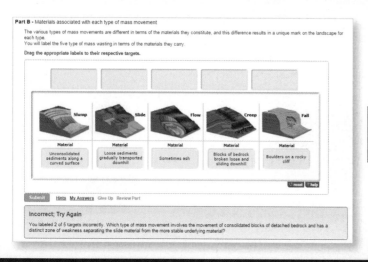

GeoTutor coaching activities help students master important geologic concepts with highly visual, kinesthetic activities focused on critical thinking and application of core geoscience concepts.

MasteringGeology™

Encounter Activities provide rich, interactive explorations of geology and earth science concepts using the dynamic features of Google Earth™ to visualize and explore earth's physical landscape. Dynamic assessment includes questions related to core concepts. All explorations include corresponding Google Earth KMZ media files, and questions include hints and specific wrong-answer feedback to help coach students toward mastery of the concepts while improving students' geospatial skills.

NEW! GigaPan Activities allow students to take advantage of a virtual field experience with high-resolution picture technology that has been developed by Carnegie Mellon University in conjunction with NASA.

Part D - Making Observations

After exploring the Gigapan field site, arrange the following observations/inferences by their respective rock unit. These observations/inferences describe the material, appearance and weathering pattern of the respective rock units.

Drag the appropriate items into their respective bins. Each item may be used only once.

Rock Unit 1
- Red and white in color
- Appears to be made up of many thin layers
- Weathers in small irregular shapes
- Weathers in large blocks
- Appears to be massive (no layers)
- Sediments too small to see

Rock Unit 2
- Black and dark gray in color
- Crystals too small to see

reset   help

Submit   Hints My Answers Give Up Review Part

Incorrect; Try Again

You sorted 2 out of 8 items incorrectly. Compare the weathering pattern of rock unit #2 to the weathering pattern of rock unit #1. Which rock unit produces large blocks?

Additional MasteringGeology assignments available:

- SmartFigures
- Interactive Animations
- Give It Some Thought Activities
- Reading Quizzes
- MapMaster Interactive Maps

FOCUS ON CONCEPTS

Each statement represents the primary learning objective for the corresponding major heading within the chapter. After you complete the chapter, you should be able to:

I.1 List and describe the sciences that collectively make up Earth science. Discuss the scales of space and time in Earth science.

I.2 Describe the four "spheres" that comprise Earth's natural environment.

I.3 Define *system* and explain why Earth is considered to be a system.

I.4 Summarize some important connections between people and the physical environment.

I.5 Discuss the nature of scientific inquiry and distinguish between a hypothesis and a theory.

Earth's four spheres—the geosphere (solid Earth), atmosphere (air), hydrosphere (water), and biosphere (life)—are represented in this scene in the Canadian Rockies. (Photo by John E. Marriott/Glow Images)

INTRODUCTION
TO EARTH SCIENCE

The spectacular eruption of a volcano, the magnificent scenery of a rocky coast, and the destruction created by a hurricane are all subjects for the Earth scientist. The study of Earth science deals with many fascinating and practical questions about our environment. What forces produce mountains? Why is our daily weather so variable? Is climate really changing? How old is Earth, and how is it related to the other planets in the solar system? What causes ocean tides? What was the Ice Age like? Will there be another? Can a successful well be located at a particular site?

The subject of this text is *Earth science*. To understand Earth is not an easy task because our planet is not a static and unchanging mass. Rather, it is a dynamic body with many interacting parts and a long and complex history.

I.1 What Is Earth Science?

List and describe the sciences that collectively make up Earth science. Discuss the scales of space and time in Earth science.

Earth science is the name for all the sciences that collectively seek to understand Earth and its neighbors in space. It includes geology, oceanography, meteorology, and astronomy. Throughout its long existence, Earth has been changing. In fact, it is changing as you read this page and will continue to do so into the foreseeable future. Sometimes the changes are rapid and violent, as when severe storms, landslides, and volcanic eruptions occur. Conversely, many changes take place so gradually that they go unnoticed during a lifetime. Scales of size and space also vary greatly among the phenomena studied in Earth science.

Earth science is often perceived as science that is performed in the out of doors—and rightly so. A great deal of an Earth scientist's study is based on observations and experiments conducted in the field. But Earth science is also conducted in the laboratory, where, for example, the study of various Earth materials provides insights into many basic processes, and the creation of complex computer models allows for the simulation of our planet's complicated climate system. Frequently, Earth scientists require an understanding and application of knowledge and principles from physics, chemistry, and biology. Geology, oceanography, meteorology, and astronomy are sciences that seek to expand our knowledge of the natural world and our place in it.

Geology

In this text, Units 1–4 focus on the science of **geology**, a word that literally means "study of Earth." Geology is traditionally divided into two broad areas: physical and historical.

Physical geology examines the materials composing Earth and seeks to understand the many processes that operate beneath and upon its surface. Earth is a dynamic, ever-changing planet. *Internal processes* create earthquakes, build mountains, and produce volcanic structures (**Figure I.1**). At the surface, *external processes* break rock apart and sculpt a broad array of landforms. The erosional effects of water, wind, and ice result in a great diversity of landscapes. Because rocks and minerals form in response to Earth's internal and external processes, their interpretation is basic to an understanding of our planet.

In contrast to physical geology, the aim of *historical geology* is to understand the origin of Earth and the development of the planet through its 4.6-billion-year history (**Figure I.2**). It strives to establish an orderly chronological arrangement of the multitude of physical and biological changes that have occurred in the geologic past. The study of physical geology logically precedes the study of Earth history because we must first understand how Earth works before we attempt to unravel its past.

Oceanography

Earth is often called the "water planet" or the "blue planet." Such terms relate to the fact that more than 70 percent of Earth's surface is covered by the global ocean. If we are to understand Earth, we must learn about its oceans. Unit 5, *The Global Ocean*, is devoted to **oceanography**.

Oceanography is actually not a separate and distinct science. Rather, it involves the application of all sciences in a comprehensive and interrelated study of the oceans in all their aspects and relationships. Oceanography integrates chemistry, physics, geology, and biology. It includes the study of the composition and movements of seawater, as well as coastal processes, seafloor topography, and marine life.

Meteorology

The continents and oceans are surrounded by an atmosphere. Unit 6, *Earth's Dynamic Atmosphere*, examines the mixture of gases that is held to the planet by gravity and thins rapidly with altitude. Acted on by the combined effects of Earth's motions and energy from the Sun, and influenced by Earth's land and sea surface, the formless and invisible atmosphere reacts by producing an infinite variety of weather, which in turn creates the basic pattern of global climates. **Meteorology** is the study of the atmosphere and the processes that produce weather and climate. Like oceanography, meteorology involves the application of other sciences in an integrated study of the thin layer of air that surrounds Earth.

Astronomy

Unit 7, *Earth's Place in the Universe*, demonstrates that an understanding of Earth requires that we relate our planet to the larger universe. Because Earth is related to all the other objects in space, the science of **astronomy**—the study of the universe—is very useful in probing the origins of our own environment. Because we are so closely acquainted with the planet on which we live, it is easy to forget that Earth is just a tiny object in a vast universe. Indeed, Earth is subject to the same physical laws that govern the many other objects populating the great expanses of space. Thus, to understand explanations of our planet's origin, it is useful to learn something

Figure I.1 Volcanic eruption Molten lava from Hawaii's Kilauea volcano is spilling into the Pacific Ocean. Internal processes are those that occur beneath Earth's surface. Sometimes they lead to the formation of major features at the surface. (Photo by Stuart Westmoreland/ Cultura/ Getty Images)

SmartFigure I.2 Arizona's Grand Canyon The erosional work of the Colorado River along with other external processes created this natural wonder. For someone studying historical geology, hiking down the South Kaibab Trail in Grand Canyon National Park is a trip through time. These rock layers hold clues to millions of years of Earth history. (Photo by Michael Collier) (http://goo.gl/7KwQLk)

Mobile Field Trip

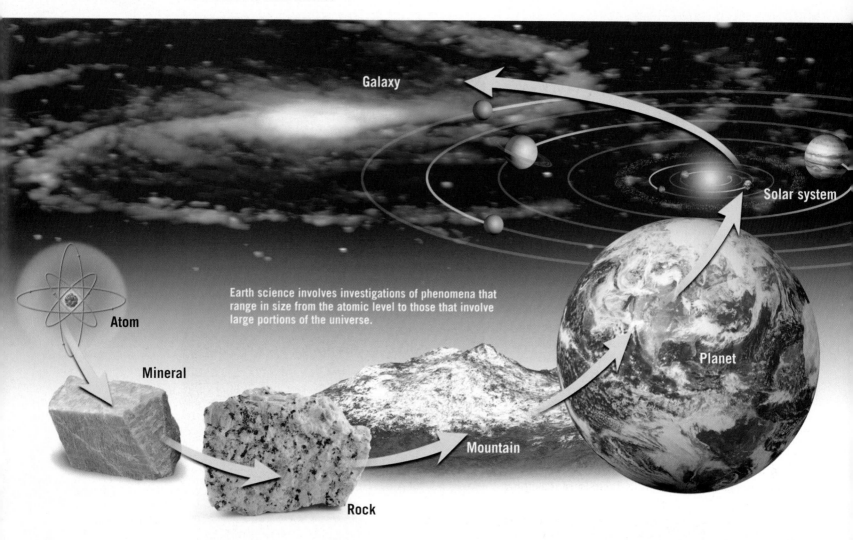

Earth science involves investigations of phenomena that range in size from the atomic level to those that involve large portions of the universe.

Galaxy

Solar system

Planet

Mountain

Atom

Mineral

Rock

Figure I.3 From atoms to galaxies Earth science involves investigations of phenomena that range in size from atoms to galaxies and beyond.

about the other members of our solar system. Moreover, it is helpful to view the solar system as a part of the great assemblage of stars that comprise our galaxy, which is but one of many galaxies.

Scales of Space and Time in Earth Science

When we study Earth, we must contend with a broad array of space and time scales (**Figure I.3**). Some phenomena are relatively easy for us to imagine, such as the size and duration of an afternoon thunderstorm or the dimensions of a sand dune. Other phenomena are so vast or so small that they are difficult to imagine. The number of stars and distances in our galaxy (and beyond!) or the internal arrangement of atoms in a mineral crystal are examples of such phenomena.

Some of the events we study occur in fractions of a second. Lightning is an example. Other processes extend over spans of tens or hundreds of millions of years.

The lofty Himalaya Mountains began forming about 45 million years ago, and they continue to develop today.

The concept of **geologic time**, the span of time since the formation of Earth, is new to many nonscientists. People are accustomed to dealing with increments of time that are measured in hours, days, weeks, and years. Our history books often examine events over spans of centuries, but even a century is difficult to appreciate fully. For most of us, someone or something that is 90 years old is *very old*, and a 1000-year-old artifact is *ancient*.

By contrast, those who study Earth science must routinely deal with vast time periods—millions or billions (thousands of millions) of years. When viewed in the context of Earth's nearly 4.6-billion-year history, an event that occurred 100 million years ago may be characterized as "recent" by a geologist, and a rock sample that has been dated at 10 million years may be called "young."

An appreciation for the *magnitude of geologic time* is important in the study of our planet because many processes are so gradual that vast spans of time are needed

What if we compress the 4.6 billion years of Earth history into a single year?

SmartFigure I.4
Magnitude of geologic time
(https://goo.gl/odwyUE)

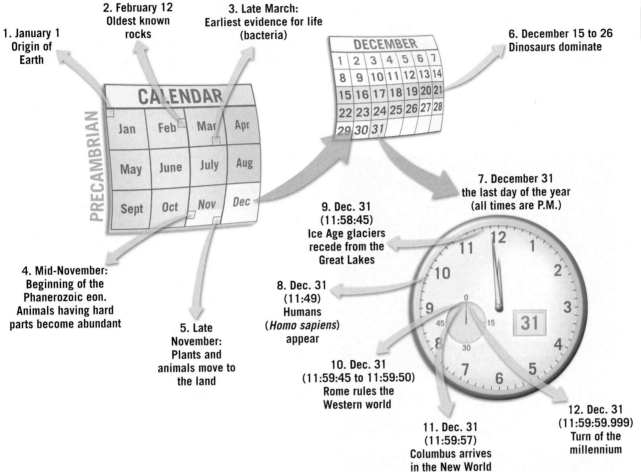

1. January 1 Origin of Earth

2. February 12 Oldest known rocks

3. Late March: Earliest evidence for life (bacteria)

6. December 15 to 26 Dinosaurs dominate

4. Mid-November: Beginning of the Phanerozoic eon. Animals having hard parts become abundant

5. Late November: Plants and animals move to the land

7. December 31 the last day of the year (all times are P.M.)

8. Dec. 31 (11:49) Humans (*Homo sapiens*) appear

9. Dec. 31 (11:58:45) Ice Age glaciers recede from the Great Lakes

10. Dec. 31 (11:59:45 to 11:59:50) Rome rules the Western world

11. Dec. 31 (11:59:57) Columbus arrives in the New World

12. Dec. 31 (11:59:59.999) Turn of the millennium

before significant changes occur. How long is 4.6 billion years? If you were to begin counting at the rate of one number per second and continued 24 hours a day, 7 days a week and never stopped, it would take about two lifetimes (150 years) to reach 4.6 billion!

The previous is just one of many analogies that have been conceived in an attempt to convey the magnitude of geologic time. Although helpful, all of them, no matter how clever, only begin to help us comprehend the vast expanse of Earth history. **Figure I.4** provides another interesting way of viewing the age of Earth.

Over the past 200 years or so, Earth scientists have developed a *geologic time scale* of Earth history. It subdivides the 4.6-billion-year history of Earth into many different units and provides a meaningful time frame within which the events of the geologic

past are arranged (see Figure 8.23, page 266). The geologic time scale and the principles used to develop it are examined in Chapter 8.

I.1 CONCEPT CHECKS

1. List and briefly describe the sciences that collectively make up Earth science.
2. Name the two broad subdivisions of geology and distinguish between them.
3. List two examples of size/space scales in Earth science that are at opposite ends of the spectrum.
4. How old is Earth?
5. If you compress geologic time into a single year, how much time has elapsed since Columbus arrived in the New World?

Did You Know?

The Sun contains 99.86 percent of the mass of the solar system, and its circumference is 109 times that of Earth. A jet plane traveling at 1000 km/hr (620 mi/hr) would require nearly 182 days to circle the Sun.

I.2 Earth's Spheres

Describe the four "spheres" that comprise Earth's natural environment.

The images in **Figure I.5** are considered classics because they let humanity see Earth differently than ever before. These early views profoundly altered our conceptualizations of Earth and remain powerful images decades after they were first viewed. Such images remind us that our home is, after all, a planet—small, self-contained, and in some ways even fragile. Bill Anders, the *Apollo 8*

astronaut who took the "Earthrise" photo, expressed it this way: "We came all this way to explore the Moon, and the most important thing is that we discovered the Earth."

As we look closely at our planet from space, it becomes clear that Earth is much more than rock and soil. In fact, the most conspicuous features in Figure I.5A are not continents but swirling clouds suspended above the surface and the vast global ocean. These features emphasize the importance of air and water to our planet.

The closer view of Earth from space shown in Figure I.5B helps us appreciate why traditionally the physical environment is divided into three major parts: the water portion of our planet, called the hydrosphere; Earth's gaseous envelope, called the atmosphere; and, of course, the solid Earth, or geosphere.

It should be emphasized that our environment is highly integrated and not dominated by water, air, or rock alone. Rather, it is characterized by continuous interactions as air comes in contact with rock, rock with water, and water with air. Moreover, the biosphere, the totality of life-forms on our planet, extends into each of the three physical realms and is an equally integral part of the planet. Thus, Earth can be thought of as consisting of four major spheres: the hydrosphere, atmosphere, geosphere, and biosphere.

The interactions among Earth's four spheres are incalculably complex. **Figure I.6** provides an easy-to-visualize example. The shoreline is an obvious meeting place for rock, water, and air. In this scene, ocean waves that were created by the drag of air moving across the water are breaking against the rocky shore. The force of the water can be powerful, and the erosional work that is accomplished can be great.

Hydrosphere

Earth is sometimes called the *blue planet*. Water, more than anything else, makes Earth unique. The **hydrosphere** is a dynamic mass of water that is continually moving, evaporating from the oceans to the atmosphere, precipitating to the land, and flowing back to the ocean. The global ocean is certainly the most prominent feature of the hydrosphere, blanketing nearly 71 percent of Earth's surface to an average depth of about 3800 meters (12,500 feet). It accounts for more than 96 percent of Earth's water (**Figure I.7**). The hydrosphere also includes the freshwater found underground and in streams, lakes, and glaciers. Moreover, water is an important component of all living things.

Although freshwater accounts for just a tiny fraction of the total, its importance goes beyond supporting life

Figure I.5 Two classic views of Earth from space (Johnson Space Center/NASA)

View called "Earthrise" that greeted *Apollo 8* astronauts as their spacecraft emerged from behind the Moon in December 1968. This classic image let people see Earth differently than ever before.

A.

This image taken from *Apollo 17* in December 1972 is perhaps the first to be called "The Blue Marble." The dark blue ocean and swirling cloud patterns remind us of the importance of the oceans and atmosphere.

B.

Did You Know?
The volume of ocean water is so large that if Earth's solid mass were perfectly smooth (level) and spherical, the oceans would cover Earth's entire surface to a uniform depth of more than 2000 m (1.2 mi)!

on land. Streams, glaciers, and groundwater are responsible for sculpturing and creating many of our planet's varied landforms. Water in the atmosphere, in the form of clouds and water vapor, plays a critical role in weather and climate processes.

Atmosphere

Earth is surrounded by a life-giving gaseous envelope called the **atmosphere** (**Figure I.8**). When we watch a high-flying jet plane cross the sky, it seems that the atmosphere extends upward for a great distance. However, when compared to the thickness (radius) of the solid Earth (about 6400 kilometers [4000 miles]), the atmosphere is a very shallow layer. Despite its modest dimensions, this thin blanket of air is nevertheless an integral part of the planet. It not only provides the air we breathe but also acts to protect us from the dangerous radiation emitted by the Sun. The energy exchanges that continually occur between the atmosphere and Earth's surface, as well as between the atmosphere and space, produce the effects we call *weather* and *climate*. Climate has a strong influence on

Figure I.6 Interactions among Earth's spheres The shoreline is one obvious interface—a common boundary where different parts of a system interact. In this scene, ocean waves (hydrosphere) that were created by the force of moving air (atmosphere) break against a rocky shore (geosphere). The force of the water can be powerful, and the erosional work that is accomplished can be great. (Photo by Michael Collier)

Figure I.7 The water planet Distribution of water in the hydrosphere.

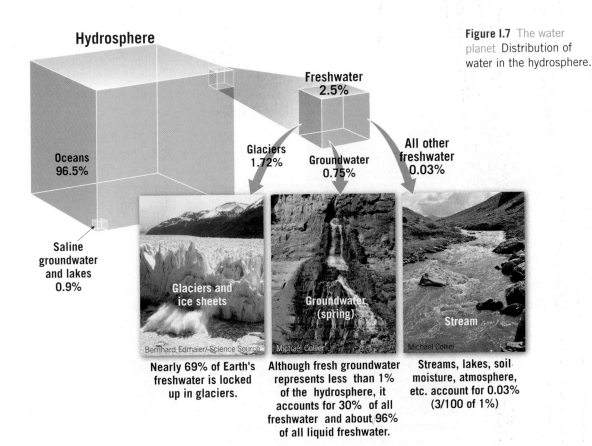

Hydrosphere

Oceans
96.5%

Saline groundwater and lakes 0.9%

Freshwater 2.5%

Glaciers 1.72%

Groundwater 0.75%

All other freshwater 0.03%

Glaciers and ice sheets
Bernhard Edmaier/Science Source
Nearly 69% of Earth's freshwater is locked up in glaciers.

Groundwater (spring)
Michael Collier
Although fresh groundwater represents less than 1% of the hydrosphere, it accounts for 30% of all freshwater and about 96% of all liquid freshwater.

Stream
Michael Collier
Streams, lakes, soil moisture, atmosphere, etc. account for 0.03% (3/100 of 1%)

Did You Know?
Since the mid-1970s, the global average surface temperature has increased by about 0.6°C (1°F). By the end of the twenty-first century, the global average surface temperature may increase by an additional 2° to 4.5°C (3.5° to 8.1°F).

Figure I.8 A shallow layer The atmosphere is an integral part of the planet. (NASA)

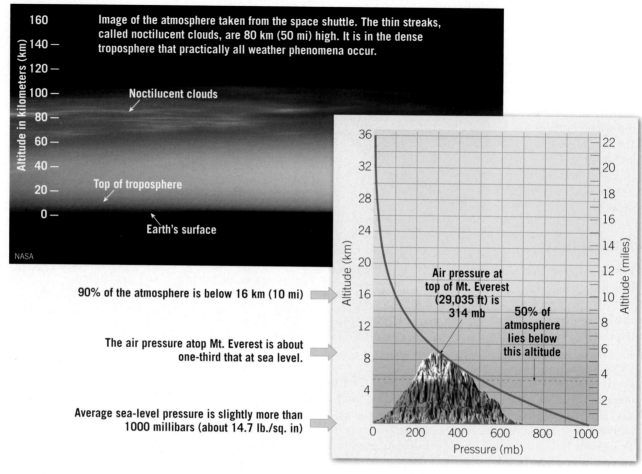

Image of the atmosphere taken from the space shuttle. The thin streaks, called noctilucent clouds, are 80 km (50 mi) high. It is in the dense troposphere that practically all weather phenomena occur.

Noctilucent clouds

Top of troposphere

Earth's surface

NASA

90% of the atmosphere is below 16 km (10 mi)

The air pressure atop Mt. Everest is about one-third that at sea level.

Average sea-level pressure is slightly more than 1000 millibars (about 14.7 lb./sq. in)

Air pressure at top of Mt. Everest (29,035 ft) is 314 mb

50% of atmosphere lies below this altitude

Pressure (mb)

the nature and intensity of Earth's surface processes. When climate changes, these processes respond.

If, like the Moon, Earth had no atmosphere, our planet would be lifeless because many of the processes and interactions that make Earth's surface such a dynamic place could not operate. Without weathering and erosion, the face of our planet might more closely resemble the surface of the Moon, which has not changed appreciably in nearly 3 billion years.

Biosphere

The **biosphere** includes all life on Earth (**Figure I.9**). Ocean life is concentrated in the sunlit upper

Figure I.9 The biosphere The biosphere, one of Earth's four spheres, includes all life. **A.** Tropical rain forests are teeming with life and occur in the vicinity of the equator. (Photo by Korup Cameroon/AGE Fotostock/SuperStock) **B.** Some life occurs in extreme environments such as the absolute darkness of the deep ocean. (Photo by Verena Tunnicliffe/Uvic/Fisheries and Ocean Canada/Newscom)

A. Tropical rain forests are characterized by hundreds of different species per square kilometer.

Tubeworms

Deep-sea vent

B. Microorganisms are nourished by hot, mineral-rich fluids spewing from vents on the deep-ocean floor. The microbes support larger organisms such as tube worms.

waters. Most life on land is also concentrated near the surface; tree roots and burrowing animals reach a few meters underground, and flying insects and birds mostly stay within a kilometer or so above the surface (**Figure I.9A**). A surprising variety of life-forms are also adapted to extreme environments. For example, on the deep ocean floor, where pressures are extreme and no light penetrates, there are places where vents spew hot, mineral-rich fluids that support communities of exotic life-forms (**Figure I.9B**). On land, some bacteria thrive in rocks as deep as 4 kilometers (2.5 miles) and in boiling hot springs. Moreover, air currents can carry microorganisms many kilometers into the atmosphere. But even when we consider these extremes, life still must be thought of as being confined to a narrow band very near Earth's surface.

Plants and animals depend on the physical environment for the basics of life. However, organisms do more than just respond to this environment. Through countless interactions, life-forms help maintain and alter their physical environment. Without life, the makeup and nature of the geosphere, hydrosphere, and atmosphere would be very different.

Geosphere

Lying beneath the atmosphere and the ocean is the solid Earth, or **geosphere**, extending from the surface to the center of the planet, at a depth of nearly 6400 kilometers (4000 miles)—by far the largest of Earth's spheres. Much of our study of the solid Earth focuses on the more accessible surface and near-surface features, but it is worth noting that many of these features are linked to the dynamic behavior of Earth's interior. As you can see from **Figure I.10**, Earth's interior is layered. The figure shows that we can think of this layering as being due to differences in both *chemical composition* and *physical properties*. On the basis of chemical composition, the Earth has three layers: a dense inner sphere called the **core**, the

SmartFigure I.10
Earth's layers Structure of Earth's interior, based on chemical composition and physical properties. (https://goo.gl/70au1N)

Tutorial

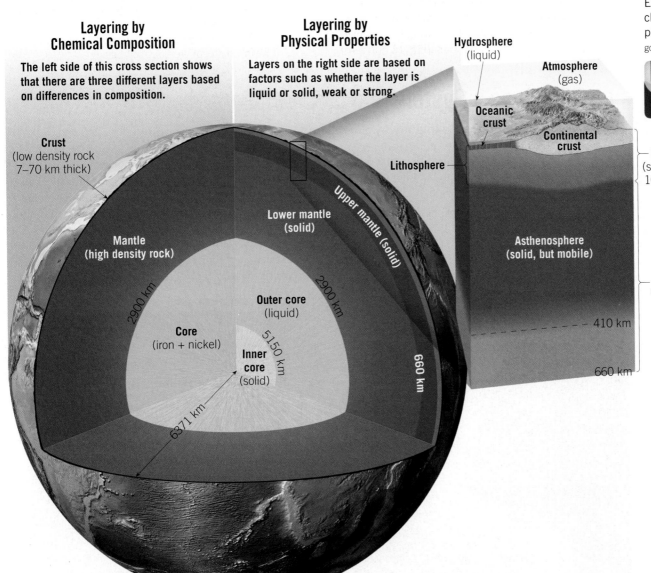

Layering by Chemical Composition

The left side of this cross section shows that there are three different layers based on differences in composition.

Crust (low density rock 7–70 km thick)

Mantle (high density rock)

2900 km

Core (iron + nickel)

Outer core (liquid)

Inner core (solid)

5150 km

6371 km

Layering by Physical Properties

Layers on the right side are based on factors such as whether the layer is liquid or solid, weak or strong.

Upper mantle (solid)

Lower mantle (solid)

2900 km

660 km

Hydrosphere (liquid)

Atmosphere (gas)

Oceanic crust

Continental crust

Lithosphere

Asthenosphere (solid, but mobile)

Lithosphere (solid and rigid 100 km thick)

Upper mantle

410 km

660 km

less dense **mantle**, and the **crust**, which is the light and very thin outer skin of Earth. The crust is not a layer of uniform thickness. It is thinnest beneath the oceans and thickest where continents exist. Although the crust may seem insignificant when compared with the other layers of the geosphere, which are much thicker, it was created by the same general processes that formed Earth's present structure. Thus, the crust is important in understanding the history and nature of our planet.

The layering of Earth in terms of physical properties reflects the way the Earth's materials behave when various forces and stresses are applied. The term **lithosphere** refers to the rigid outer layer that includes the crust and uppermost mantle. Beneath the rigid rocks that compose the lithosphere, the rocks of the **asthenosphere** are weak and are able to slowly flow in response to the uneven distribution of heat deep within Earth.

The two principal divisions of Earth's surface are the continents and the ocean basins. The most obvious difference between these two provinces is their relative levels. The average elevation of the continents above sea level is about 840 meters (2750 feet), whereas the average depth of the oceans is about 3800 meters (12,500 feet).

Thus, the continents stand an average of 4640 meters (about 4.6 kilometers [nearly 3 miles]) above the level of the ocean floor.

Soil, the thin veneer of material at Earth's surface that supports the growth of plants, may be thought of as part of all four spheres. The solid portion is a mixture of weathered rock debris (geosphere) and organic matter from decayed plant and animal life (biosphere). The decomposed and disintegrated rock debris is the product of weathering processes that require air (atmosphere) and water (hydrosphere). Air and water also occupy the open spaces between solid particles.

I.2 CONCEPT CHECKS

1. List and briefly define the four "spheres" that constitute our environment.
2. Compare the height of the atmosphere to the thickness of the geosphere.
3. How much of Earth's surface do oceans cover? How much of the planet's total water supply do oceans represent?
4. Briefly summarize Earth's layered structure.
5. To which sphere does soil belong?

I.3 Earth as a System

Define *system* and explain why Earth is considered to be a system.

Anyone who studies Earth soon learns that our planet is a dynamic body with many separate but interacting parts or spheres. The hydrosphere, atmosphere, biosphere, geosphere, and all of their components can be studied separately. However, the parts are not isolated. Each is related in many ways to the others, producing a complex and continuously interacting whole that we call the **Earth system**.

A simple example of the interactions among different parts of the Earth system occurs every winter and spring as moisture evaporates from the Pacific Ocean and subsequently falls as rain in coastal hills and mountains, triggering destructive debris flows (**Figure I.11**). The processes that move water from the hydrosphere to the atmosphere and then to the geosphere have a profound impact on the physical landscape and on the plants and

Figure I.11 Deadly debris flow This image provides an example of interactions among different parts of the Earth system. Extraordinary rains triggered this debris flow (popularly called a mudslide) on March 22, 2014, near Oso, Washington. The mass of mud and debris blocked the North Fork of the Stillaguamish River and engulfed an area of about 2.6 square kilometers (1 square mile). Forty-three people perished.
(Photo by Michael Collier)

animals (including humans) that inhabit the affected regions.

Earth System Science

Scientists have recognized that in order to more fully understand our planet, they must learn how its individual components (land, water, air, and life-forms) are interconnected. This endeavor, called **Earth system science**, aims to study Earth as a *system* composed of numerous interacting parts, or *subsystems*. Rather than looking through the limited lens of only one of the traditional sciences—geology, atmospheric science, chemistry, biology, and so on—Earth system science attempts to integrate the knowledge of several academic fields. Using an interdisciplinary approach, those engaged in Earth system science attempt to achieve the level of understanding necessary to comprehend and solve many of our global environmental problems.

A **system** is a group of interacting, or interdependent, parts that form a complex whole. Most of us hear and use the term *system* frequently. We may service our car's cooling *system*, make use of the city's transportation *system*, and be a participant in the political *system*. A news report might inform us of an approaching weather *system*. Further, we know that Earth is just a small part of a larger system known as the *solar system*, which in turn is a subsystem of an even larger system called the Milky Way Galaxy.

The Earth System

The Earth system has a nearly endless array of subsystems in which matter is recycled over and over. One familiar loop, or subsystem, is the *hydrologic cycle*. It represents the unending circulation of Earth's water among the hydrosphere, atmosphere, biosphere, and geosphere (**Figure I.12**). Water enters the atmosphere through evaporation from Earth's surface and transpiration from plants. Water vapor condenses in the atmosphere to form clouds, which in turn produce precipitation that falls back to Earth's surface. Some of the rain that falls onto the land sinks in and then is taken up by plants or becomes groundwater, and some flows across the surface toward the ocean.

Viewed over long time spans, the rocks of the geosphere are constantly forming, changing, and re-forming. The loop that involves the processes by which one rock

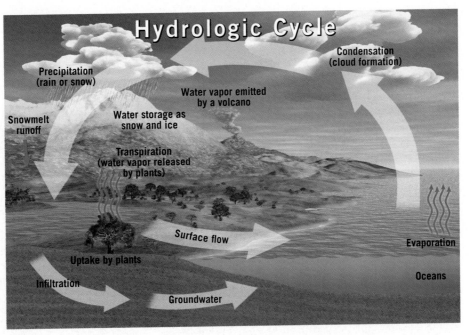

Hydrologic Cycle

Precipitation (rain or snow)

Condensation (cloud formation)

Water vapor emitted by a volcano

Snowmelt runoff

Water storage as snow and ice

Transpiration (water vapor released by plants)

Surface flow

Uptake by plants

Evaporation

Infiltration

Oceans

Groundwater

Figure I.12 The hydrologic cycle Earth's surface offers temperatures and pressures that allow water to change readily among three states—liquid, gas (water vapor), and solid (ice). This cycle traces the movements of water among Earth's four spheres. It is one of many subsystems that collectively make up the Earth system.

changes to another is called the *rock cycle* and is discussed at some length in Chapter 2. The cycles of the Earth system are not independent of one another. To the contrary, there are many places where the cycles come in contact and interact.

The Parts Are Linked The parts of the Earth system are linked so that a change in one part can produce changes in any or all of the other parts. For example, when a volcano erupts, lava from Earth's interior may flow out at the surface and block a nearby valley. This new obstruction influences the region's drainage system by creating a lake or causing streams to change course. The large quantities of volcanic ash and gases that can be emitted during an eruption might be blown high into the atmosphere and influence the amount of solar energy that can reach Earth's surface. The result could be a drop in air temperatures over the entire hemisphere.

Where the surface is covered by lava flows or a thick layer of volcanic ash, existing soils are buried. This causes the soil-forming processes to begin anew to transform the new surface material into soil (**Figure I.13**). The soil that eventually forms will reflect the interactions among many parts of the Earth system—the volcanic parent material, the climate, and the impact of biological activity. Of course, there would also be significant changes in the biosphere. Some organisms and their habitats would be eliminated by the lava and ash, and new settings for life, such as a lake formed by a lava dam, would be created. The potential climate change could also impact sensitive life-forms.

Time and Space Scales The Earth system is characterized by processes that vary on spatial scales from fractions of millimeters to thousands of kilometers. Time scales for Earth's processes range from milliseconds to billions of

Figure I.13 Change is a constant When Mount St. Helens erupted in May 1980, the area shown here was buried by a volcanic mudflow. Now plants are reestablished, and new soil is forming. (Photo by Terry Donnelly/Alamy)

years. As we learn about Earth, it becomes increasingly clear that despite significant separations in distance or time, many processes are connected, and a change in one component can influence the entire system.

Energy for the Earth System The Earth system is powered by energy from two sources. The Sun drives external processes that occur in the atmosphere, in the hydrosphere, and at Earth's surface. Weather and climate, ocean circulation, and erosional processes are driven by energy from the Sun. Earth's interior is the second source of energy. Heat remaining from when our planet formed and heat that is continuously generated by radioactive decay power the internal processes that produce volcanoes, earthquakes, and mountains.

People and the Earth System Humans are *part of* the Earth system, a system in which the living and nonliving components are entwined and interconnected. Therefore,

our actions produce changes in all the other parts. When we burn gasoline and coal, dispose of our wastes, and clear the land, we cause other parts of the system to respond, often in unforeseen ways. Throughout this text, you will learn about many of Earth's subsystems, including the hydrologic system, the tectonic (mountain-building) system, the rock cycle, and the climate system. Remember that these components *and we humans* are all part of the complex interacting whole we call the Earth system.

The organization of this text involves traditional groupings of chapters that focus on closely related topics. Nevertheless, the theme of *Earth as a system* keeps recurring through *all* major units of this text. It is a thread that weaves through the chapters and helps tie them together.

 I.3 **CONCEPT CHECKS**

1. What is a system? List three examples of systems.
2. What are the two sources of energy for the Earth system?
3. Predict how a change in the hydrologic cycle, such as increased rainfall in an area, might influence the biosphere and geosphere in that area.
4. Are humans part of the Earth system? Briefly explain.

I.4 Resources and Environmental Issues

Summarize some important connections between people and the physical environment.

Environment refers to everything that surrounds and influences an organism. Some of these things are biological and social, but others are nonliving. The factors in this latter category are collectively called our physical environment. The **physical environment** encompasses water, air, soil, and rock, as well as conditions such as temperature, humidity, and sunlight. The phenomena and processes studied by the Earth sciences are basic to an understanding of the physical environment. In this

sense, most of Earth science may be characterized as environmental science.

However, when the term *environmental* is applied to Earth science today, it often refers to relationships between people and the physical environment. Application of the Earth sciences is necessary to understand and solve problems that arise from these interactions.

We can dramatically influence natural processes. For example, river flooding is natural, but the magnitude and frequency of flooding can be changed significantly by human activities such as clearing forests, building cities, and constructing dams. Unfortunately, natural systems do not always adjust to artificial changes in ways we can anticipate. Thus, an alteration to the environment that was intended to benefit society may have the opposite effect.

Resources

Resources are an important focus of the Earth sciences that is of great value to people. Resources include water and soil, a great variety of metallic and nonmetallic minerals, and energy. Together, they form the very foundation of modern civilization. The Earth sciences deal not only with the formation and occurrence of these vital resources but also with maintaining supplies and the environmental impact of their extraction and use.

Few people who live in highly industrialized nations realize the quantity of resources needed to maintain their present standard of living. **Figure I.14** shows the annual per capita consumption of energy resources and several important metallic and nonmetallic mineral

Figure I.15 Wind energy is renewable These wind turbines are operating near Tehachapi Pass in Kern County, California. California was the first state to develop significant wind power, but today Texas is the state with the greatest installed capacity. (Photo by Dazzo/Radius Images/ Getty Images)

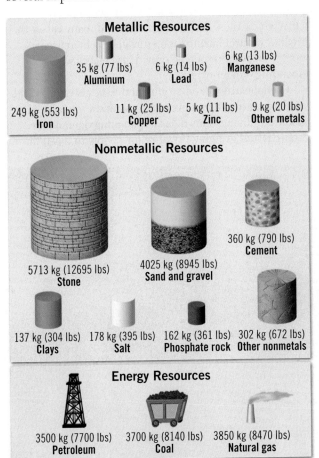

Metallic Resources

35 kg (77 lbs) Aluminum   6 kg (14 lbs) Lead   6 kg (13 lbs) Manganese

249 kg (553 lbs) Iron   11 kg (25 lbs) Copper   5 kg (11 lbs) Zinc   9 kg (20 lbs) Other metals

Nonmetallic Resources

5713 kg (12695 lbs) Stone   4025 kg (8945 lbs) Sand and gravel   360 kg (790 lbs) Cement

137 kg (304 lbs) Clays   178 kg (395 lbs) Salt   162 kg (361 lbs) Phosphate rock   302 kg (672 lbs) Other nonmetals

Energy Resources

3500 kg (7700 lbs) Petroleum   3700 kg (8140 lbs) Coal   3850 kg (8470 lbs) Natural gas

Figure I.14 How much do each of us use? The annual per capita consumption of nonmetallic and metallic resources for the United States is about 11,000 kilograms (12 tons). About 97 percent of the materials used are nonmetallic. The per capita use of oil, coal, and natural gas exceeds 11,000 kilograms. (U.S. Geological Survey)

resources for the United States. This is each person's prorated share of the materials required by industry to provide the vast array of products modern society demands. Figures for other highly industrialized countries are comparable.

Resources are commonly divided into two broad categories. Some are classified as **renewable**, which means that they can be replenished over relatively short time spans. Common examples are plants and animals for food, natural fibers for clothing, and forest products for lumber and paper. Energy from flowing water, wind, and the Sun is also considered renewable (**Figure I.15**).

By contrast, many other basic resources are classified as **nonrenewable**. Important metals such as iron, aluminum, and copper fall into this category, as do our most important fuels: oil, natural gas, and coal. Although these and other resources continue to form, the processes that create them are so slow that significant deposits take millions of years to accumulate. In essence, Earth contains *fixed* quantities of these substances. When the present supplies are mined or pumped from the ground, there will be no more. Although some nonrenewable resources, such as aluminum, can be used over and over again, others, such as oil, cannot be recycled.

How long will the remaining supplies of basic resources last? How long can we sustain the rising standard of living in today's industrial countries and still provide for the growing needs of developing regions? How much environmental deterioration are we willing to accept in pursuit of basic resources? Can alternatives be found? If we are to cope with an increasing demand and a growing

Figure I.16 People influence the atmosphere Smoke bellows from a coal-fired electricity-generating plant in New Delhi, India, in June 2008. In addition to smoke, this power plant also emits gases such as sulfur dioxide and carbon dioxide that contribute to air pollution and global climate change. (Photo by Gurinder Osan/AP Images)

world population, it is important that we have some understanding of our present and potential resources.

Environmental Problems

In addition to the quest for adequate mineral and energy resources, the Earth sciences must also deal with a broad array of environmental problems. Some are local, some are regional, and still others are global in extent. Serious difficulties face developed and developing nations alike. Urban air pollution, the disposal of toxic wastes, and global climate change are just a few that pose significant threats (**Figure I.16**). Other problems involve the loss of fertile soils to erosion and the contamination and depletion of water resources. The list continues to grow.

In addition to human-induced and human-accentuated problems, people must also cope with the many **natural hazards** posed by the physical environment. Earthquakes, landslides, volcanic eruptions, floods, and hurricanes are just five of the many risks (**Figure I.17**). Others, such as drought, although not as spectacular, are nevertheless equally important environmental concerns. It should be emphasized that *natural hazards* are *natural phenomena* that only become hazards when people try to live where these phenomena occur. In many cases, the threat of natural hazards is aggravated by increases in population as more people crowd into places where an impending danger exists.

Complicating all environmental issues is rapid world population growth and everyone's aspiration to a better standard of living. This means ballooning demand for resources *and* growing pressure for people to dwell in

Figure I.17 Tsunami destruction Undersea earthquakes sometimes create large, fast-moving ocean waves that can cause significant death and destruction in coastal areas. Natural hazards are *natural* processes. They become hazards only when people try to live where these processes occur. (Photo by Mainichi Newspaper/ Aflo/Newscom)

environments having significant natural hazards. Therefore, an understanding of Earth is essential not only for the location and recovery of basic resources but also for dealing with the human impact on the environment and minimizing the effects of natural hazards. Knowledge about our planet and how it works is necessary to our survival and well-being. Earth is the only suitable habitat we have, and its resources are limited.

I.4 CONCEPT CHECKS

1. Define *physical environment*.
2. Distinguish between renewable and nonrenewable resources. What are some examples of each?
3. List at least four phenomena that can be regarded as natural hazards.

I.5 The Nature of Scientific Inquiry

Discuss the nature of scientific inquiry and distinguish between a hypothesis and a theory.

As members of a modern society, we are constantly reminded of the benefits derived from science. But what exactly is the nature of scientific inquiry? Science is a process of producing knowledge. The process depends both on making careful observations and on creating explanations that make sense of the observations. Developing an understanding of how science is done and how scientists work is an important theme that appears throughout this text. You will explore the difficulties in gathering data and some of the ingenious methods that have been developed to overcome these difficulties. You will also see examples of how hypotheses are formulated and tested, as well as learn about the evolution and development of some major scientific theories.

All science is based on the assumption that the natural world behaves in a consistent and predictable manner that is comprehensible through careful, systematic study. The overall goal of science is to discover the underlying patterns in nature and then to use this knowledge to make predictions about what should or should not be expected, given certain facts or circumstances. For example, by understanding the processes and conditions that produce certain cloud types, meteorologists are often able to predict the approximate time and place of their formation.

The development of new scientific knowledge involves some basic logical processes that are universally accepted. To determine what is occurring in the natural world, scientists collect scientific data through observation and measurement (**Figure I.18**). The types of data that are collected often seek to answer a well-defined question about the natural world. Because some error is inevitable, the accuracy of a particular measurement or observation is always open to question. Nevertheless, these data are essential to science and serve as the springboard for the development of scientific theories.

Hypothesis

Once data have been gathered and principles have been formulated to describe a natural phenomenon, investigators try to explain how or why things happen in the manner observed. They often do this by constructing a tentative (or untested) explanation, which is called a scientific

Did You Know?
It took until about the year 1800 for the world population to reach 1 billion people. By 1927, the number doubled to 2 billion. According to U.N. estimates, world population reached 7 billion in late October 2011 and stood at nearly 7.4 billion in 2015. We are currently adding about 80 million people per year to the planet.

Figure I.18 Observation and measurement Gathering data and making careful observations are basic parts of scientific inquiry. Scientific data are gathered in many ways. **A.** Instruments onboard satellites provide data that often is not available from any other source. (NASA) **B.** Earth science involves not only outdoor fieldwork but work in the laboratory. This researcher is using a special microscope to study the mineral composition of rock samples that were collected by researchers in the field. (Photo by Jon Wilson/Science Source)

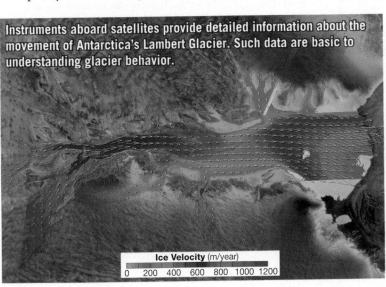

Instruments aboard satellites provide detailed information about the movement of Antarctica's Lambert Glacier. Such data are basic to understanding glacier behavior.

Ice Velocity (m/year)
0   200   400   600   800   1000   1200

A.

B.

hypothesis. It is best if an investigator can formulate more than one hypothesis to explain a given set of observations. If an individual scientist is unable to devise multiple hypotheses, others in the scientific community will almost always develop alternative explanations. A spirited debate frequently ensues. As a result, extensive research is conducted by proponents of opposing hypotheses, and the results are made available to the wider scientific community in scientific journals.

Before a hypothesis can become an accepted part of scientific knowledge, it must pass objective testing and analysis. If a hypothesis cannot be tested, it is not scientifically useful, no matter how interesting it might seem. The verification process requires that *predictions* be made based on the hypothesis being considered and that the predictions be tested by comparing them against objective observations of nature. Put another way, hypotheses must fit observations other than those used to formulate them in the first place. Those hypotheses that fail rigorous testing are ultimately discarded. The history of science is littered with discarded hypotheses. One of the best known is the Earth-centered model of the universe—a proposal that was supported by the apparent daily motion of the Sun, Moon, and stars around Earth. As the mathematician Jacob Bronowski so ably stated, "Science is a great many things, but in the end they all return to this: Science is the acceptance of what works and the rejection of what does not."

Theory

When a hypothesis has survived extensive scrutiny and when competing ones have been eliminated, a hypothesis

may be elevated to the status of a scientific **theory**. In everyday language we may say, "That's only a theory." But a scientific theory is a well-tested and widely accepted view that the scientific community agrees best explains certain observable facts. Some theories that are extensively documented and extremely well supported are comprehensive in scope. One example is the *nebular theory* discussed in Chapter 15, which explains the formation of our solar system. Another, called the *theory of plate tectonics*, provides the framework for understanding the origin of mountains, earthquakes, and volcanic activity—ideas that are explored in some detail in Chapters 5, 6, and 7.

Scientific Methods

The process just described, in which researchers gather data through observations and formulate scientific hypotheses and theories, is called the **scientific method**. Contrary to popular belief, the scientific method is not a standard recipe that scientists apply in a routine manner to unravel the secrets of our natural world but an endeavor that involves creativity and insight. Rutherford and Ahlgren put it this way: "Inventing hypotheses or theories to imagine how the world works and then figuring out how they can be put to the test of reality is as creative as writing poetry, composing music, or designing skyscrapers."[*]

There is no fixed path that scientists always follow that leads unerringly to scientific knowledge. However, many scientific investigations involve the steps outlined in **Figure I.19**. In addition, some scientific discoveries result from purely theoretical ideas that stand up to extensive examination. Some researchers use high-speed computers to create models that simulate what is happening in the "real" world. These models are useful when dealing with natural processes that occur on very long time scales or that take place in extreme or inaccessible locations. Still other scientific advancements are made when a totally unexpected happening occurs during an experiment. These serendipitous discoveries are more than pure luck, for as the nineteenth-century French scientist Louis Pasteur said, "In the field of observation, chance favors only the prepared mind."

Scientific knowledge is acquired through several avenues, so it might be best to describe the nature of scientific inquiry as the methods of science rather than as the scientific method. In addition, we should always remember that even the most compelling scientific theories are still simplified explanations of the natural world.

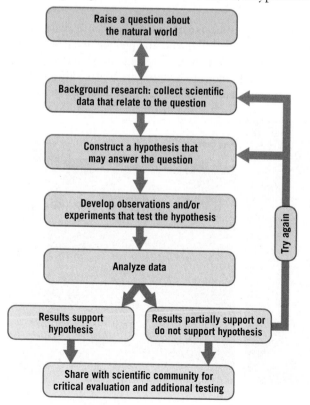

Figure I.19 Steps frequently followed in scientific investigations The diagram depicts the steps involved in the process many refer to as the *scientific method*.

Raise a question about the natural world

Background research: collect scientific data that relate to the question

Construct a hypothesis that may answer the question

Develop observations and/or experiments that test the hypothesis

Analyze data

Results support hypothesis

Results partially support or do not support hypothesis

Try again

Share with scientific community for critical evaluation and additional testing

I.5 **CONCEPT CHECKS**

1. How is a scientific hypothesis different from a scientific theory?
2. Summarize the basic steps followed in many scientific investigations.

*F. James Rutherford and Andrew Ahlgren, *Science for All Americans* (New York: Oxford University Press, 1990), p. 7.

CONCEPTS IN REVIEW
Introduction to Earth Science

I.1 What Is Earth Science?

List and describe the sciences that collectively make up Earth science. Discuss the scales of space and time in Earth science.

KEY TERMS: Earth science, geology, oceanography, meteorology, astronomy, geologic time

- Earth science includes geology, oceanography, meteorology, and astronomy.
- There are two broad subdivisions of geology. Physical geology studies Earth materials and the internal and external processes that create and shape Earth's landscape. Historical geology examines Earth's history.
- The other Earth sciences seek to understand the oceans, the atmosphere's weather and climate, and Earth's place in the universe.
- Earth science must deal with processes and phenomena that vary from the subatomic scale of matter to the nearly infinite scale of the universe.
- The time scales of phenomena studied in Earth science range from tiny fractions of a second to many billions of years.
- Geologic time, the span of time since the formation of Earth, is about 4.6 billion years, a number that is difficult to comprehend.

I.2 Earth's Spheres

Describe the four "spheres" that comprise Earth's natural environment.

KEY TERMS: hydrosphere, atmosphere, biosphere, geosphere, core, mantle, crust, lithosphere, asthenosphere

- Earth's physical environment is traditionally divided into three major parts: the solid Earth, called the geosphere; the water portion of our planet, called the hydrosphere; and Earth's gaseous envelope, called the atmosphere.
- A fourth Earth sphere is the biosphere, the totality of life on Earth. It is concentrated in a relatively thin zone that extends a few kilometers into the hydrosphere and geosphere and a few kilometers up into the atmosphere.
- Of all the water on Earth, more than 96 percent is in the oceans, which cover nearly 71 percent of the planet's surface.

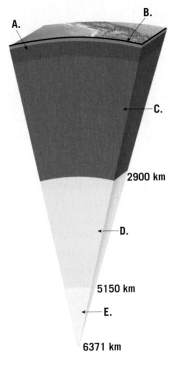

A.
B.
C.
2900 km
D.
5150 km
E.
6371 km

(?) The diagram represents Earth's layered structure. Is it showing layering based on physical properties or layering based on composition? Identify the lettered layers.

I.3 Earth as a System

Define *system* and explain why Earth is considered to be a system.

KEY TERMS: Earth system, Earth system science, system

- Although each of Earth's four spheres can be studied separately, they are all related in a complex and continuously interacting whole that is called the Earth system.
- Earth system science uses an interdisciplinary approach to integrate the knowledge of several academic fields in the study of our planet and its global environmental problems.
- The two sources of energy that power the Earth system are (1) the Sun, which drives the external processes that occur in the atmosphere, hydrosphere, and at Earth's surface, and (2) heat from Earth's interior, which powers the internal processes that produce volcanoes, earthquakes, and mountains.
- People are part of the Earth system; our actions can cause changes in all the other parts.

(?) Give a specific example of how people are affected by the Earth system and another example of how humans affect the Earth system.

I.4 Resources and Environmental Issues

Summarize some important connections between people and the physical environment.

KEY TERMS: physical environment, renewable resource, nonrenewable resource, natural hazard

- Environment refers to everything that surrounds and influences an organism. Nonliving components, such as air, water, soil, and rock are collectively called the physical environment.
- Important relationships between people and the environment include the quest for resources, the impact of people on the natural environment, and the effects of natural hazards.
- Two broad categories of resources are (1) renewable, which means that they can be replenished over relatively short time spans, and (2) nonrenewable.

(?) As the photo reminds us, large quantities of aluminum are recycled. Does this mean that aluminum is a renewable resource? Explain.

ALUMINUM CANS ONLY!

Courtney Weber/Alamy

I.5 The Nature of Scientific Inquiry

Discuss the nature of scientific inquiry and distinguish between a hypothesis and a theory.

KEY TERMS: hypothesis, theory, scientific method

- Scientists make careful observations, construct tentative explanations for those observations (hypotheses), and then test those hypotheses with field investigations and laboratory work. In science, a theory is a well-tested and widely accepted explanation that the scientific community agrees best fits certain observable facts.

- As failed hypotheses are discarded, scientific knowledge moves closer to a correct understanding, but we can never be fully confident that we know all the answers. Scientists must always be open to new information that forces change in our model of the world.

GIVE IT SOME THOUGHT

1. Is glacial ice part of the geosphere, or does it belong to the hydrosphere? Explain your answer.

Michael Collier

2. The average distance between Earth and the Sun is 150 million kilometers (93 million miles). About how long would it take a jet plane traveling from Earth at 1000 kilometers (620 miles) per hour to reach the Sun?

3. This scene is in British Columbia's Mount Robson Provincial Park. The park is named for the highest peak in the Canadian Rockies. List examples of features associated with each of Earth's four spheres. Which, if any, of these features was created by internal processes? Describe the role of external processes in this scene.

Michael Wheatley/AGE Fotostock

4. Humans are part of the Earth system. List at least three examples of how you, in particular, influence one or more of Earth's major spheres.

5. Examine Figure I.7 to answer these questions.
   a. Where is most of Earth's freshwater stored?
   b. Where is most of Earth's liquid freshwater found?

6. Refer to the graph in Figure I.8 to answer the following questions.
   a. If you were to climb to the top of Mount Everest, how many breaths of air would you have to take at that altitude to equal one breath at sea level?
   b. If you were flying in a commercial jet at an altitude of 12 kilometers (about 39,000 feet), about what percentage of the atmosphere would be below the plane?

7. The accompanying photo provides an example of interactions among different parts of the Earth system. It is a view of a mudflow that was triggered by extraordinary rains. Which of Earth's four "spheres" were involved in this natural disaster that buried a small town on the Philippine island of Leyte? Describe how each contributed to the mudflow.

Pat Roque/AP Images

8. As you enter a dark room, you turn on a wall switch, but the ceiling light does not come on. Formulate at least three hypotheses that might explain this observation. Once you have formulated your hypotheses, what should be your next step?

MasteringGeology™

1

FOCUS ON CONCEPTS

Each statement represents the primary learning objective for the corresponding major heading within the chapter. After you complete the chapter, you should be able to:

1.1 List the main characteristics that an Earth material must possess to be considered a mineral and describe each characteristic.

1.2 Compare and contrast the three primary particles contained in atoms.

1.3 Distinguish among ionic bonds, covalent bonds, and metallic bonds.

1.4 List and describe the properties used in mineral identification.

1.5 List the common silicate and nonsilicate minerals and describe what characterizes each group.

The Cave of Crystals, Chihuahua, Mexico, contains giant gypsum crystals, some of the largest natural crystals ever found. (Photo by Carsten Pete/Speleoresearch & Films/Getty Images)

MATTER AND MINERALS

Earth's crust and oceans are home to a wide variety of useful and essential minerals. Most people are familiar with the common uses of many basic metals, including aluminum in beverage cans, copper in electrical wiring, and gold and silver in jewelry. But some people are not aware that pencil "lead" contains the greasy-feeling mineral graphite and that bath powders and many cosmetics contain the mineral talc. Moreover, many do not know that dentists use drill bits impregnated with diamonds to drill through tooth enamel. In fact, practically every manufactured product contains materials obtained from minerals.

In addition to the economic uses of rocks and minerals, every geologic process in some way depends on the properties of these basic Earth materials. Events such as volcanic eruptions, mountain building, weathering and erosion, and even earthquakes involve rocks and minerals. Consequently, a basic knowledge of Earth materials is essential to understanding all geologic phenomena.

Did You Know?
Archaeologists discovered that more than 2000 years ago, the Romans transported water in lead pipes. In fact, Roman smelting of lead and copper ores between 500 B.C. and A.D. 300 caused a small rise in atmospheric pollution that was recorded in Greenland ice cores.

1.1 Minerals: Building Blocks of Rocks

List the main characteristics that an Earth material must possess to be considered a mineral and describe each characteristic.

We begin our discussion of Earth materials with an overview of **mineralogy** (*mineral* = mineral, *ology* = study of) because minerals are the building blocks of rocks. In addition, humans have used minerals for both practical and decorative purposes for thousands of years. For example, the common mineral quartz is the source of silicon for computer chips. The first Earth materials mined were flint and chert, which humans fashioned into weapons and cutting tools. As early as 3700 B.C.E., Egyptians began mining gold, silver, and copper. By 2200 B.C.E., humans had discovered how to combine copper with tin to make bronze, a strong, hard alloy. Later, a process was developed to extract iron from minerals such as hematite—a discovery that marked the decline of the Bronze Age. During the Middle Ages, mining of a variety of minerals became common, and the impetus for the formal study of minerals was in place.

The term *mineral* is used in several different ways. For example, those concerned with health and fitness extol the benefits of vitamins and minerals. The mining industry typically uses the word *mineral* to refer to anything extracted from Earth, such as coal, iron ore, or sand and gravel. The guessing game *Twenty Questions* usually begins with the question *Is it animal, vegetable, or mineral?* What criteria do geologists use to determine whether something is a mineral (**Figure 1.1**)?

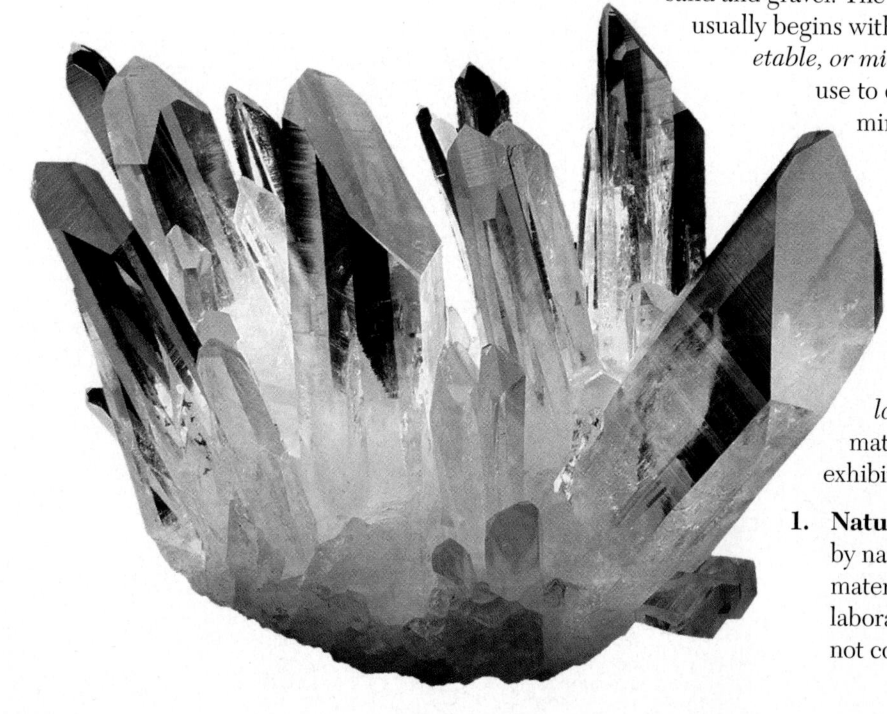

Figure 1.1 Quartz crystals A collection of well-developed quartz crystals found near Hot Springs, Arkansas. (Photo by Jeffrey A. Scovil)

Defining a Mineral

Geologists define **mineral** as *any naturally occurring inorganic solid that possesses an orderly crystalline structure and a definite chemical composition that allows for some variation.* Thus, Earth materials that are classified as minerals exhibit the following characteristics:

1. **Naturally occurring.** Minerals form by natural geologic processes. Synthetic materials, meaning those produced in a laboratory or by human intervention, are not considered minerals.

2. **Generally inorganic.** Inorganic crystalline solids, such as ordinary table salt (halite), that are found naturally in the ground are considered minerals. (Organic compounds, on the other hand, are generally not. Sugar, a crystalline solid like salt but extracted from sugarcane or sugar beets, is a common example of such an organic compound.) Many marine animals secrete inorganic compounds, such as calcium carbonate (calcite), in the form of shells and coral reefs. If these materials are buried and become part of the rock record, geologists consider them minerals.

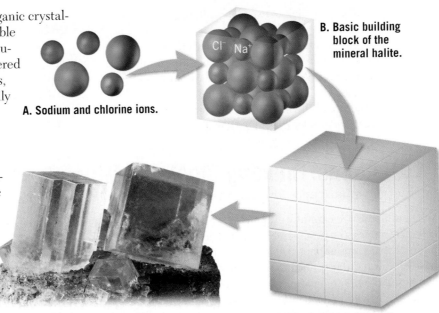

A. Sodium and chlorine ions.

B. Basic building block of the mineral halite.

C. Collection of basic building blocks (crystal).

D. Crystals of the mineral halite.

Figure 1.2 Arrangement of sodium and chloride ions in the mineral halite The arrangement of atoms (ions) into basic building blocks that have a cubic shape results in regularly shaped cubic crystals. (Photo by Dennis Tasa)

3. **Solid substance.** Only solid crystalline substances are considered minerals. Ice (frozen water) fits this criterion and is considered a mineral, whereas liquid water and water vapor do not.

4. **Orderly crystalline structure.** Minerals are crystalline substances, made up of atoms (or ions) that are arranged in an orderly, repetitive manner (**Figure 1.2**). This orderly packing of atoms is reflected in regularly shaped objects called *crystals*. Some naturally occurring solids, such as volcanic glass (obsidian), lack a repetitive atomic structure and are not considered minerals.

5. **Definite chemical composition that allows for some variation.** Most minerals are chemical compounds having compositions that can be expressed by a chemical formula. For example, the common mineral quartz has the formula SiO_2, which indicates that quartz consists of silicon (Si) and oxygen (O) atoms, in a 1:2 ratio. This proportion of silicon to oxygen is true for any sample of pure quartz, regardless of its origin. However, the compositions of some minerals vary *within specific, well-defined limits*. This occurs because certain elements can substitute for others of similar size without changing the mineral's internal structure.

What Is a Rock?

In contrast to minerals, rocks are more loosely defined. Simply, a **rock** is any solid mass of mineral, or mineral-like, matter that occurs naturally as part of our planet. Most rocks, like the sample of granite shown in **Figure 1.3**, occur as aggregates of several different

minerals. The term *aggregate* implies that the minerals are joined in such a way that their individual properties are retained. Note that the different minerals that make up granite can be easily identified. However, some rocks are composed almost entirely of one mineral. A common example is the sedimentary rock *limestone*, which is an impure mass of the mineral calcite.

In addition, some rocks are composed of nonmineral matter. These include the volcanic rocks *obsidian* and *pumice*, which are noncrystalline glassy

Granite (Rock)

Quartz (Mineral)

Hornblende (Mineral)

Feldspar (Mineral)

SmartFigure 1.3 Most rocks are aggregates of minerals Shown here is a hand sample of the igneous rock granite and three of its major constituent minerals. (Photos by E. J. Tarbuck) (https://goo.gl/7cZXyr)

Tutorial

substances, and *coal*, which consists of solid organic debris.

Although this chapter deals primarily with the nature of minerals, keep in mind that most rocks are simply aggregates of minerals. Because the properties of rocks are determined largely by the chemical composition and crystalline structure of the minerals contained within them, we will first consider these Earth materials.

1.1 **CONCEPT CHECKS**

1. List five characteristics of a mineral.
2. Based on the definition of mineral, which of the following—gold, liquid water, synthetic diamonds, ice, and wood—are not classified as minerals?
3. Define the term *rock*. How do rocks differ from minerals?

1.2 Atoms: Building Blocks of Minerals

Compare and contrast the three primary particles contained in atoms.

When minerals are carefully examined, even under optical microscopes, the innumerable tiny particles of their internal structures are not visible. Nevertheless, scientists have discovered that all matter, including minerals, is composed of minute building blocks called **atoms**—the smallest particles that cannot be chemically split. Atoms, in turn, contain even smaller particles—*protons* and *neutrons* located in a central **nucleus** that is surrounded by *electrons* (**Figure 1.4**).

Properties of Protons, Neutrons, and Electrons

Protons and **neutrons** are very dense particles with almost identical masses. By contrast, **electrons** have a negligible mass, about 1/2000 that of a proton. To visualize this difference, imagine a scale where a proton or neutron has the mass of a baseball compared to an electron that has the mass of a single grain of rice.

Both protons and electrons share a fundamental property called *electrical charge*. Protons have an electrical charge of +1, and electrons have a charge of –1. Neutrons, as the name suggests, have no charge. The charges of protons and electrons are equal in magnitude but opposite in polarity, so when these two particles are paired, the charges cancel each other out. Since matter typically contains equal numbers of positively charged protons and

negatively charged electrons, most substances are electrically neutral.

Illustrations sometimes show electrons orbiting the nucleus in a manner that resembles the planets of our solar system orbiting the Sun (see Figure 1.4A). However, electrons do not actually behave this way. A more realistic depiction would show electrons as a cloud of negative charges surrounding the nucleus (see Figure 1.4B). Studies of the arrangements of electrons show that they move about the nucleus in regions called *principal shells*, each with an associated energy level. In addition, each shell can hold a specific number of electrons, with the outermost shell generally containing **valence electrons**. These electrons can be transferred or shared with other atoms to form chemical bonds.

Most of the atoms in the universe (except hydrogen and helium) were created inside massive stars by nuclear fusion and released into interstellar space during hot, fiery supernova explosions. As this ejected material cooled, the newly formed nuclei attracted electrons to complete their atomic structure. At the temperatures found at Earth's surface, free atoms (those not bonded to other atoms) generally have a full complement of electrons—one for each proton in the nucleus.

Elements: Defined by Their Number of Protons

The simplest atoms have only 1 proton in their nuclei, whereas others have more than 100. The number of protons in the nucleus of an atom, called the **atomic number**, determines its chemical nature. All atoms with the same

Figure 1.4 Two models of an atom **A.** Simplified view of an atom having a central nucleus composed of protons and neutrons, encircled by high-speed electrons. **B.** This model of an atom shows spherically shaped electron clouds (shells) surrounding a central nucleus. The nucleus contains virtually all of the mass of the atom. The remainder of the atom is the space occupied by negatively charged electrons. (The relative sizes of the nuclei shown are greatly exaggerated.)

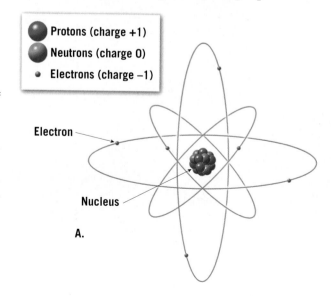

- Protons (charge +1)
- Neutrons (charge 0)
- Electrons (charge –1)

Electron

Nucleus

A.

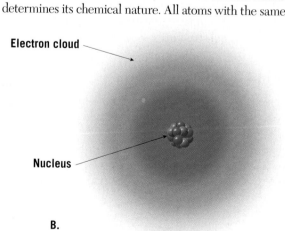

Electron cloud

Nucleus

B.

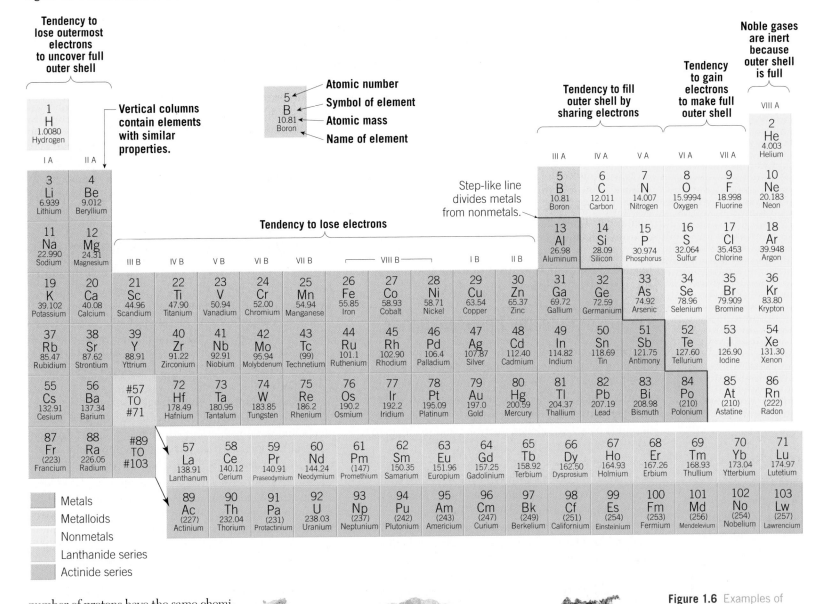

Tendency to lose outermost electrons to uncover full outer shell

Vertical columns contain elements with similar properties.

Atomic number
Symbol of element
Atomic mass
Name of element

Step-like line divides metals from nonmetals.

Tendency to fill outer shell by sharing electrons

Tendency to gain electrons to make full outer shell

Noble gases are inert because outer shell is full

Tendency to lose electrons

Metals
Metalloids
Nonmetals
Lanthanide series
Actinide series

number of protons have the same chemical and physical properties; collectively they constitute an **element**. There are about 90 naturally occurring elements, and several more have been synthesized in the laboratory. You are probably familiar with the names of many elements, including carbon, nitrogen, and oxygen. All carbon atoms have six protons, whereas all nitrogen atoms have seven protons, and all oxygen atoms have eight.

The **periodic table**, shown in **Figure 1.5**, organizes elements so those with similar properties line up in columns, referred to as groups. Each element is assigned a one- or two-letter symbol. The atomic number and atomic mass for each element are also included in the periodic table.

Atoms of the naturally occurring elements are the basic building blocks of Earth's minerals. Most elements join with atoms of other elements to form **chemical compounds**. Therefore, most minerals are chemical compounds composed of atoms of two or more elements.

A. Gold on quartz

B. Sulfur

C. Copper

Figure 1.6 Examples of minerals composed of a single element (Photos by Dennis Tasa)

These include the minerals quartz (SiO_2), halite (NaCl), and calcite ($CaCO_3$). However, a few minerals, such as diamonds, sulfur, gold, and native copper, are made entirely of atoms of only one element (**Figure 1.6**). (Copper is called "native" when it is found in its pure form in nature.)

1.2 CONCEPT CHECKS

1. Make a simple sketch of an atom and label its three main particles. Explain how these particles differ from one another.
2. What is the significance of valence electrons?

1.3 Why Atoms Bond

Distinguish among ionic bonds, covalent bonds, and metallic bonds.

Under the temperature and pressure conditions found on Earth, most elements do not occur in the form of individual atoms—instead, their atoms bond with other atoms. (A group of elements known as the noble gases are an exception.) Some atoms bond to form *ionic compounds*, some form *molecules*, and still others form *metallic substances*. Why does this happen? Experiments show that electrical forces hold atoms together and bond them to each other. These electrical attractions lower the total energy of the bonded atoms, and this, in turn, generally makes them more stable. Consequently, atoms that are bonded in compounds tend to be more stable than atoms that are free (not bonded).

The Octet Rule and Chemical Bonds

As noted earlier, valence (outer shell) electrons are generally involved in chemical bonding. **Figure 1.7** shows a shorthand way of representing the number of valence electrons for some selected elements. Notice that the elements in Group I have one valence electron, those in Group II have two valence electrons, and so on, up to eight valence electrons in Group VIII.

The noble gases (except helium) have very stable electron arrangements with eight valence electrons and, therefore, tend to lack chemical reactivity. Many other atoms gain, lose, or share electrons during chemical reactions, ending up with electron arrangements of the noble gases. This observation led to a chemical guideline known as the **octet rule**: *Atoms tend to gain, lose, or share electrons until they are surrounded by eight valence electrons.* Although there are exceptions to the octet rule, it is a useful *rule of thumb* for understanding chemical bonding.

When an atom's outer shell does not contain eight electrons, it is likely to chemically bond to other atoms to achieve an octet in its outer shell. A **chemical bond** is a transfer or sharing of electrons that allows each atom to attain a full valence shell of electrons. Some atoms do this by transferring all their valence electrons to other atoms, so that an inner shell becomes the full valence shell.

When the valence electrons are transferred between the elements to form ions, the bond is an *ionic bond*. When the electrons are shared between the atoms, the bond is a *covalent bond*. When the valence electrons are shared among all the atoms in a substance, the bonding is *metallic*.

Ionic Bonds: Electrons Transferred

Perhaps the easiest type of bond to visualize is the **ionic bond**, in which one atom gives up one or more valence electrons to another atom to form **ions**—*positively and negatively charged atoms*. The atom that loses electrons becomes a positive ion, and the atom that gains electrons becomes a negative ion. Oppositely charged ions are strongly attracted to one another and join to form ionic compounds.

Consider the ionic bonding that occurs between sodium (Na) and chlorine (Cl) to produce the solid ionic compound sodium chloride—the mineral halite (common table salt). Notice in **Figure 1.8A** that a sodium atom gives up its single valence electron to chlorine and, as a result, becomes a positively charged sodium ion (Na^+). Chlorine, on the other hand, gains one electron and becomes a negatively charged chloride ion (Cl^-). We know that ions having unlike charges attract. Thus, an ionic bond is an attraction of oppositely charged ions to one another that produces an electrically neutral ionic compound.

Figure 1.8B illustrates the arrangement of sodium and chlorine ions in ordinary table salt. Notice that salt consists of alternating sodium and chlorine ions, positioned in such a manner that each positive ion is attracted to and surrounded on all sides by negative ions and vice versa. This arrangement maximizes the attraction between ions with opposite charges while minimizing the repulsion between ions with identical charges. Thus, ionic compounds consist of an orderly arrangement of oppositely charged ions assembled in a definite ratio that provides overall electrical neutrality.

The properties of a chemical compound are dramatically different from the properties of the various elements comprising it. For example, sodium is a soft silvery metal that is extremely reactive and poisonous. If you were to consume even a small amount of elemental sodium, you would need immediate medical attention. Chlorine, a green poisonous gas, is so toxic that it was used as a chemical weapon during World War I. Together, however, these elements produce sodium chloride, a harmless flavor enhancer that we call table salt.

Did You Know?
One of the world's heaviest cut and polished gemstones is a 22,892.5-carat golden-yellow topaz. Currently housed in the Smithsonian Institution, this roughly 10-lb gem is about the size of an automobile headlight and could hardly be used as a piece of jewelry, except perhaps by an elephant.

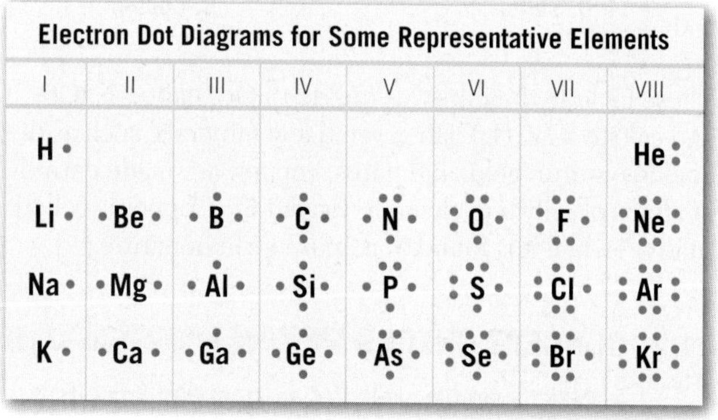

Figure 1.7 Dot diagrams for certain elements Each dot represents a valence electron found in the outermost principal shell.

Electron Dot Diagrams for Some Representative Elements							
I	II	III	IV	V	VI	VII	VIII
H							He
Li	Be	B	C	N	O	F	Ne
Na	Mg	Al	Si	P	S	Cl	Ar
K	Ca	Ga	Ge	As	Se	Br	Kr

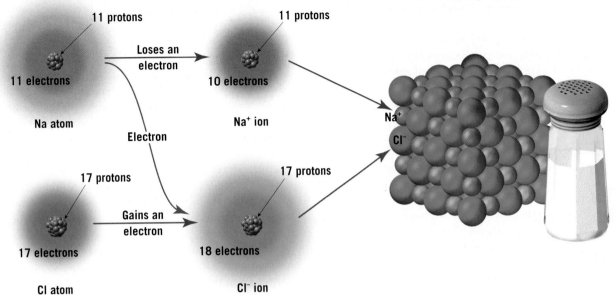

A. The transfer of an electron from a sodium (Na) atom to a chlorine (Cl) atom leads to the formation of a Na⁺ ion and a Cl⁻ ion.

B. The arrangement of Na⁺ and Cl⁻ in the solid ionic compound sodium chloride (NaCl), table salt.

Figure 1.8 Formation of the ionic compound sodium chloride

11 protons
11 electrons
Na atom

Loses an electron

11 protons
10 electrons
Na⁺ ion

Electron

17 protons
17 electrons
Cl atom

Gains an electron

17 protons
18 electrons
Cl⁻ ion

Na⁺
Cl⁻

Thus, when elements combine to form compounds, their properties change significantly.

Covalent Bonds: Electron Sharing

Sometimes the forces that hold atoms together cannot be understood on the basis of the attraction of oppositely charged ions. One example is the hydrogen molecule (H_2), in which the two hydrogen atoms are held together tightly, and no ions are present. The strong attractive force that holds two hydrogen atoms together results from a **covalent bond**, a chemical bond formed by the *sharing* of one or more valence electrons between a pair of atoms. (Hydrogen is one of the exceptions to the octet rule: Its single shell is full with just two electrons.)

Imagine two hydrogen atoms (each with one proton and one electron) approaching one another, as shown in **Figure 1.9**. Once they meet, the electron configuration will change so that both electrons will primarily occupy the space between the atoms. In other words, the two electrons are shared by both hydrogen atoms and are attracted simultaneously by the positive charge of the proton in the nucleus of each atom. Although hydrogen atoms do not form ions, the force that holds these atoms together arises from the attraction of oppositely charged particles—positively charged protons in the nuclei and negatively charged electrons that surround these nuclei.

Metallic Bonds: Electrons Free to Move

In **metallic bonds**, the valence electrons are free to move from one atom to another so that all atoms share the available valence electrons. This type of bonding is found in metals such as copper, gold, aluminum, and silver and in alloys such as brass and bronze. Metallic

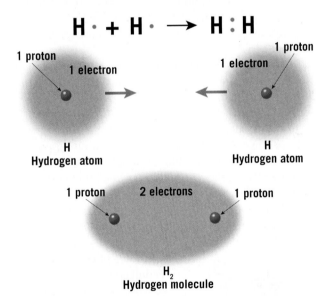

Two hydrogen atoms combine to form a hydrogen molecule, held together by the attraction of oppositely charged particles—positively charged protons in each nucleus and negatively charged electrons that surround these nuclei.

H · + H · → H : H

1 proton
1 electron
H
Hydrogen atom

1 proton
1 electron
H
Hydrogen atom

1 proton
2 electrons
1 proton
H_2
Hydrogen molecule

Figure 1.9 Formation of a covalent bond When hydrogen atoms bond, the negatively charged electrons are shared by both hydrogen atoms and attracted simultaneously by the positive charge of the proton in the nucleus of each atom.

bonding accounts for the high electrical conductivity of metals, the ease with which metals can be shaped, and numerous other special properties.

1.3 CONCEPT CHECKS

1. What is the difference between an atom and an ion?
2. How does an atom become a positive ion? A negative ion?
3. Briefly distinguish between ionic and covalent bonding and discuss the role that electrons play in both.

1.4 Properties of Minerals

List and describe the properties used in mineral identification.

A. This freshly broken sample of galena displays a metallic luster.

Metallic

B. This sample of galena is tarnished and has a submetallic luster.

Submetallic

Figure 1.10 Metallic versus submetallic luster (Photo courtesy of E. J. Tarbuck)

SmartFigure 1.11 Color variations in minerals Some minerals, such as fluorite and quartz, exhibit a variety of colors. (Photo by E. J. Tarbuck) (https://goo.gl/wznOpk)

Tutorial

Minerals have definite crystalline structures and chemical compositions that give them unique sets of physical and chemical properties shared by all specimens of that mineral, regardless of when or where they formed. For example, two samples of the mineral quartz will be equally hard and equally dense, and they will break in a similar manner. However, the physical properties of individual samples may vary within specific limits due to ionic substitutions, inclusions of foreign elements (impurities), and defects in the crystalline structure.

Some mineral properties, called **diagnostic properties**, are particularly useful in identifying an unknown mineral. The mineral halite, for example has a salty taste. Because so few minerals share this property, a salty taste is considered a diagnostic property of halite. Other properties of certain minerals vary among different specimens of the same mineral. These properties are referred to as **ambiguous properties**.

Optical Properties

Of the many optical properties of minerals, their luster, their ability to transmit light, their color, and their streak are most frequently used for mineral identification.

Luster The appearance or quality of light reflected from the surface of a mineral is known as **luster**. Minerals that have the appearance of a metal, regardless of color, are said to have a *metallic luster* (**Figure 1.10A**). Some metallic minerals, such as native copper and galena, develop a dull coating or tarnish when exposed to the atmosphere. Because they are not as shiny as samples with freshly broken surfaces, these samples are often said to exhibit a *submetallic luster* (**Figure 1.10B**).

Most minerals have a *nonmetallic luster* and are described using various adjectives such as *vitreous*, or *glassy*. Other nonmetallic minerals are described as having a *dull*, or *earthy*, *luster* (a dull appearance like soil) or a *pearly luster* (such as a pearl or the inside of a clamshell). Still others exhibit a *silky luster* (like satin cloth) or a *greasy luster* (as though coated in oil).

A. Fluorite

B. Quartz

Color Although **color** is generally the most conspicuous characteristic of any mineral, it is considered a diagnostic property of only a few minerals. Slight impurities in the common minerals fluorite and quartz, for example, give them a variety of tints, including pink, purple, yellow, white, gray, and even black (**Figure 1.11**). Other minerals, such as tourmaline, also exhibit a variety of hues, with multiple colors sometimes occurring in the same sample. Thus, the use of color as a means of identification is often ambiguous or even misleading.

Streak The color of a mineral in powdered form, called **streak**, is often useful in identification. A mineral's streak is obtained by rubbing it across a *streak plate* (a piece of unglazed porcelain) and observing the color of the mark it leaves (**Figure 1.12**). Although a mineral's color may vary from sample to sample, its streak is usually consistent in color. (Note that not all minerals produce a streak when rubbed across a streak plate. Quartz, for example, is harder than a porcelain streak plate and therefore leaves no streak.)

Streak can also help distinguish between minerals with metallic luster and those with nonmetallic luster. Metallic minerals generally have a dense, dark streak, whereas minerals with nonmetallic luster typically have a light-colored streak.

Ability to Transmit Light Another optical property used to identify minerals is the ability to transmit light. When no light is transmitted through a mineral sample, that mineral is described as *opaque*; when light, but not an image, is transmitted, the mineral is said to be *translucent*. When both light and an image are visible through the sample, the mineral is described as *transparent*.

Mineral (Pyrite)

Color (Brass yellow)

Streak (Black)

Although the color of a mineral is not always helpful in identification, the streak, which is the color of the powdered mineral, can be very useful.

SmartFigure 1.12 Streak (https://goo.gl/ULVLM4)

Animation

Crystal Shape, or Habit

Mineralogists use the term **crystal shape**, or **habit**, to refer to the common or characteristic shape of individual crystals or aggregates of crystals. Some minerals tend to grow equally in all three dimensions, whereas others tend to be elongated in one direction or flattened if growth in one dimension is suppressed. A few minerals have crystals that exhibit regular polygons that are helpful in their identification. For example, magnetite crystals sometimes occur as octahedrons, garnets often form dodecahedrons, and halite and fluorite crystals tend to grow as cubes or near-cubes. Most minerals have just one common crystal shape, but a few, such as the pyrite samples shown in **Figure 1.13**, have two or more characteristic crystal shapes.

In addition, some mineral samples consist of numerous intergrown crystals exhibiting characteristic shapes that are useful for identification. Terms commonly used to describe these and other crystal habits include *equant* (equidimensional), *bladed, fibrous, tabular, prismatic, platy, blocky, cubic,* and *banded*. Some of these habits are pictured in **Figure 1.14**.

Mineral Strength

How easily minerals break or deform under stress is determined by the type and strength of the chemical bonds that hold the crystals together. Mineralogists use terms including *hardness, cleavage, fracture,* and *tenacity* to describe mineral strength and how minerals break when stress is applied.

Hardness One of the most useful diagnostic properties is **hardness**, a measure of the resistance of a mineral to abrasion or scratching. This property is determined by rubbing a mineral of unknown hardness against one of known hardness or vice versa. A numerical value of hardness can be obtained by using the **Mohs scale** of hardness, which consists of 10 minerals arranged in order from 1 (softest) to 10 (hardest), as shown in **Figure 1.15A**. It should be noted that the Mohs scale is a relative ranking and does not imply that a mineral with a hardness of 2, such as gypsum, is twice as hard as mineral with

a hardness of 1, like talc. In fact, gypsum is only slightly harder than talc, as **Figure 1.15B** indicates.

In the laboratory, common objects used to determine the hardness of a mineral can include a human fingernail, which has a hardness of about 2.5, a copper penny (3.5), and a piece of glass (5.5). The mineral gypsum, which has a hardness of 2, can be easily scratched with a fingernail. On the other hand, the mineral calcite, which has a hardness of 3, will scratch a fingernail but will not scratch glass. Quartz, one of the hardest common minerals, will easily scratch glass. Diamonds, hardest of all, scratch anything, including other diamonds.

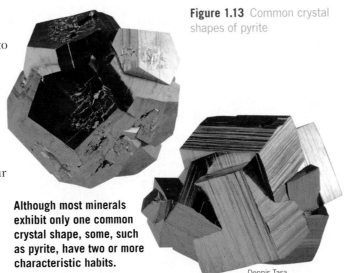
Figure 1.13 Common crystal shapes of pyrite

Although most minerals exhibit only one common crystal shape, some, such as pyrite, have two or more characteristic habits.
Dennis Tasa

Did You Know?
The name *crystal* is derived from the Greek (*krystallos* = ice) and was originally applied to quartz crystals. The ancient Greeks thought quartz was water that had crystallized at high pressures deep inside Earth.

SmartFigure 1.14
Some common crystal habits **A.** Thin, rounded crystals that break into fibers. **B.** Elongated crystals that are flattened in one direction. **C.** Minerals that have stripes or bands of different color or texture. **D.** Groups of crystals that are cube shaped.
(https://goo.gl/IBw4OJ)

 **Tutorial**

A. Fibrous E.J. Tarbuck

B. Bladed Dennis Tasa

C. Banded Dennis Tasa

D. Cubic crystals Dennis Tasa

SmartFigure 1.15 Hardness scales **A.** The Mohs scale of hardness, with the hardnesses of some common objects. **B.** Relationship between the Mohs relative hardness scale and an absolute hardness scale. (https://goo.gl/ZNODAG)

▶ **Tutorial**

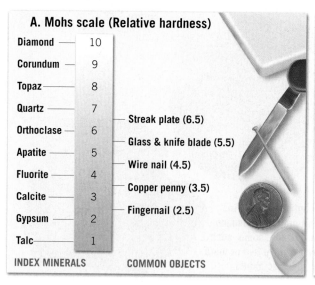

A. Mohs scale (Relative hardness)

INDEX MINERALS		COMMON OBJECTS
Diamond	10	
Corundum	9	
Topaz	8	
Quartz	7	
Orthoclase	6	Streak plate (6.5)
Apatite	5	Glass & knife blade (5.5)
Fluorite	4	Wire nail (4.5)
Calcite	3	Copper penny (3.5)
Gypsum	2	Fingernail (2.5)
Talc	1	

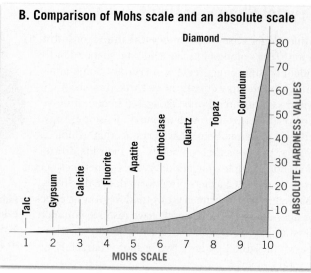

B. Comparison of Mohs scale and an absolute scale

(Graph plotting ABSOLUTE HARDNESS VALUES against MOHS SCALE with index minerals: Talc, Gypsum, Calcite, Fluorite, Apatite, Orthoclase, Quartz, Topaz, Corundum, Diamond)

Did You Know?

Clay minerals are sometimes used as an additive to thicken milkshakes in fast-food restaurants.

SmartFigure 1.16 Micas exhibit perfect cleavage The thin sheets shown here exhibit one plane of cleavage. (Photo by Chip Clark/Fundamental Photographs) (https://goo.gl/JYClSi)

▶ **Animation**

Knife blade

Weak bonds

Strong bonds

Cleavage In the crystal structure of many minerals, some atomic bonds are weaker than others. It is along these weak bonds that minerals tend to break when they are stressed. **Cleavage** (*kleiben* = carve) is the tendency of a mineral to break (cleave) along planes of weak bonding. Not all minerals have cleavage, but those that do can be identified by the relatively smooth, flat surfaces that are produced when the mineral is broken.

The simplest type of cleavage is exhibited by the micas (**Figure 1.16**). Because these minerals have very weak bonds in one direction, they cleave to form thin, flat sheets. Some minerals have excellent cleavage in one, two, three, or more directions, whereas others exhibit fair or poor cleavage, and still others have no cleavage at all. When minerals break evenly in more than one direction, cleavage is described by *the number of cleavage directions and the angle(s) at which they meet* (**Figure 1.17**).

Each cleavage surface that has a different orientation is counted as a different direction of cleavage. For example, some minerals, such as halite, cleave to form six-sided cubes. Because cubes are defined by three different sets of parallel planes that intersect at 90-degree angles, cleavage for the mineral halite is described as *three directions of cleavage that meet at 90 degrees.*

Do not confuse cleavage with crystal shape. When a mineral exhibits cleavage, it will break into pieces that all have the same geometry. By contrast, the smooth-sided quartz crystals shown in Figure 1.1 do not have cleavage. If broken, they fracture into shapes that do not resemble one another or the original crystals.

Fracture Minerals having chemical bonds that are equally, or nearly equally, strong in all directions exhibit a property called **fracture** (**Figure 1.18A**). When minerals fracture, most produce uneven surfaces and are described as exhibiting *irregular fracture*. However, some minerals, including quartz, sometimes break into smooth, curved surfaces resembling broken glass. Such breaks are called *conchoidal fractures* (**Figure 1.18B**). Still other minerals exhibit fractures that produce splinters or fibers referred to as *splintery fracture* and *fibrous fracture*, respectively.

Tenacity The term **tenacity** describes a mineral's resistance to breaking, bending, cutting, or other forms of deformation. As mentioned earlier, nonmetallic minerals such as quartz and halite tend to be *brittle* and will fracture or exhibit cleavage when struck. Minerals that are ionically bonded, such as fluorite and halite, tend to be *brittle* and shatter into small pieces when struck. By contrast, native metals, such as copper and gold, are *malleable*, which means they can be hammered without breaking. In addition, minerals that can be cut into thin shavings, including gypsum and talc, are described as *sectile*. Still others, notably the micas, are *elastic* and will

A. Cleavage in one direction.
Example: Muscovite

B. Cleavage in two directions at 90° angles.
Example: Feldspar

Fracture not cleavage

SmartFigure 1.17 Cleavage directions exhibited by minerals (Photos by E. J. Tarbuck and Dennis Tasa) (https://goo.gl/5lkkd6)

Tutorial

Did You Know?
The mineral pyrite is commonly called "fool's gold" because its golden-yellow color closely resembles gold. The name *pyrite* is derived from the Greek *pyros* ("fire") because it gives off sparks when struck sharply.

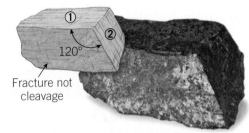

Fracture not cleavage

C. Cleavage in two directions not at 90° angles. Example: Hornblende

D. Cleavage in three directions at 90° angles. Example: Halite

E. Cleavage in three directions not at 90° angles.
Example: Calcite

F. Cleavage in four directions.
Example: Fluorite

bend and snap back to their original shape after stress is released.

Density and Specific Gravity

Density, an important property of matter, is defined as mass per unit volume. Mineralogists often use a related measure called **specific gravity** to describe the density of minerals. Specific gravity is a number representing the ratio of a mineral's weight to the weight of an equal volume of water.

Most common minerals have a specific gravity between 2 and 3. For example, quartz has a specific gravity of 2.65. By contrast, some metallic minerals, such as pyrite, native copper, and magnetite, are more than twice as dense and thus have more than twice the specific gravity of quartz. Galena, an ore from which lead is extracted, has a specific gravity of roughly 7.5, whereas 24-karat gold has a specific gravity of approximately 20.

With a little practice, you can estimate the specific gravity of a mineral by hefting it in your hand.

Does this mineral feel about as "heavy" as similarly sized rocks you have handled? If the answer is "yes," the specific gravity of the sample will likely be between 2.5 and 3.

Other Properties of Minerals

In addition to the properties discussed thus far, some minerals can be recognized by other distinctive properties. For example, halite is ordinary salt, so it can be

A. Irregular fracture (Quartz)

B. Conchoidal fracture (Quartz)

Figure 1.18 Irregular versus conchoidal fracture (Photo by E. J. Tarbuck)

quickly identified through taste. Talc and graphite both have distinctive feels: Talc feels soapy, and graphite feels greasy. Further, the streaks of many sulfur-bearing minerals smell like rotten eggs. A few minerals, such as magnetite, have high iron content and can be picked up with a magnet, while some varieties (such as lodestone) are themselves natural magnets and will pick up small iron-based objects such as pins and paper clips (see Figure 1.30F, page 39).

Moreover, some minerals exhibit special optical properties. For example, when a transparent piece of

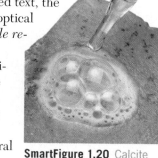

calcite is placed over printed text, the letters appear twice. This optical property is known as *double refraction* (**Figure 1.19**).

One very simple chemical test to detect carbonate mineral involves placing a drop of dilute hydrochloric acid from a dropper bottle onto a freshly broken mineral surface. Samples containing carbonate minerals will effervesce (fizz) as carbon dioxide gas is released (**Figure 1.20**). This test is especially useful in identifying calcite, a common carbonate mineral.

SmartFigure 1.20 Calcite reacting with a weak acid (Photo by Chip Clark/ Fundamental Photographs) (https://goo.gl/pnGkML)

Figure 1.19 Double refraction This sample of calcite exhibits double refraction. (Photo by Chip Clark/Fundamental Photographs)

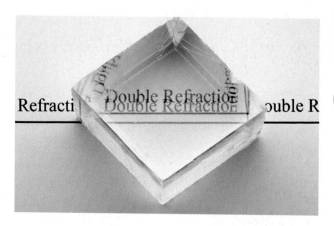

Refracti Double Refraction ouble R

1.4 CONCEPT CHECKS

1. Define *luster*.
2. Why is color not always a useful property in mineral identification? Give an example of a mineral that supports your answer.
3. What differentiates cleavage from fracture?
4. What is meant by a mineral's *tenacity*? List three terms that describe tenacity.
5. Describe a simple chemical test that is useful in identifying the mineral calcite.

1.5 Mineral Groups

List the common silicate and nonsilicate minerals and describe what characterizes each group.

More than 4000 minerals have been named, and several new ones are identified each year. Fortunately for students who are beginning to study minerals, no more than a few dozen are abundant. Collectively, these few make

up most of the rocks of Earth's crust and are therefore generally known as the **rock-forming minerals**.

Although less abundant, many other minerals are used extensively in the manufacture of products; these are called **economic minerals**. However, rock-forming minerals and economic minerals are not mutually exclusive groups. When found in large deposits, some rock-forming minerals are economically significant. One example is calcite, a mineral that is the primary component of the sedimentary rock limestone. Calcite's many uses include cement production.

It is worth noting that *only eight elements* make up the vast majority of the rock-forming minerals and represent more than 98 percent (by weight) of the continental crust (**Figure 1.21**). These elements, in order of most to least abundant, are oxygen (O), silicon (Si), aluminum (Al), iron (Fe), calcium (Ca), sodium (Na), potassium (K), and magnesium (Mg). As shown in Figure 1.21, oxygen and silicon are by far the most common elements in Earth's crust. Furthermore, these two elements readily combine to form the basic "building block" for the most common mineral group,

Figure 1.21 The eight most abundant elements in the continental crust The numbers represent percentages by weight.

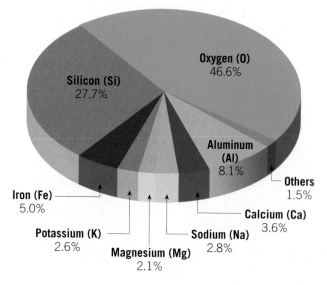

Oxygen (O) 46.6%

Silicon (Si) 27.7%

Aluminum (Al) 8.1%

Others 1.5%

Iron (Fe) 5.0%

Potassium (K) 2.6%

Magnesium (Mg) 2.1%

Sodium (Na) 2.8%

Calcium (Ca) 3.6%

the **silicates**. More than 800 silicate minerals are known, and they account for more than 90 percent of Earth's crust.

Because other mineral groups are far less abundant in Earth's crust than the silicates, they are often grouped together under the heading **nonsilicates**. Although not as common as silicates, some nonsilicate minerals are very important economically. They provide us with iron and aluminum to build automobiles, gypsum for plaster and drywall for home construction, and copper wire that carries electricity and connects us to the Internet. Common nonsilicate mineral groups include the carbonates, sulfates, and halides. In addition to their economic importance, these groups include minerals that are major constituents in sediments and sedimentary rocks.

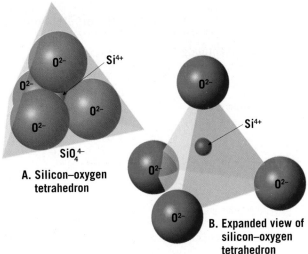

Figure 1.22 Two representations of the silicon–oxygen tetrahedron

SiO_4^{4-}

A. Silicon–oxygen tetrahedron

Si^{4+}

B. Expanded view of silicon–oxygen tetrahedron

Silicate Minerals

Every silicate mineral contains oxygen and silicon atoms. Except for a few silicate minerals such as quartz, most silicate minerals also contain one or more additional elements in their crystalline structure. These elements give rise to the great variety of silicate minerals and their varied properties.

All silicates have the same fundamental building block, the **silicon–oxygen tetrahedron** (SiO_4^{4-}). This structure consists of four oxygen ions that are covalently bonded to a comparatively smaller silicon ion, forming a tetrahedron—a pyramid shape with four identical faces (**Figure 1.22**). In some minerals, the tetrahedra are joined into chains, sheets, or three-dimensional networks by sharing oxygen atoms (**Figure 1.23**). These larger silicate structures are then connected to one another by other elements. The primary elements that join silicate structures are iron (Fe), magnesium (Mg), potassium (K), sodium (Na), and calcium (Ca).

Major groups of silicate minerals and common examples are given in Figure 1.23. The *feldspars* are by far the most plentiful group, comprising over 50 percent of Earth's crust. *Quartz*, the second-most-abundant mineral in the continental crust, is the only common mineral made completely of silicon and oxygen.

Notice in Figure 1.23 that each mineral *group* has a particular silicate *structure*. A relationship exists between this internal structure of a mineral and the *cleavage* it exhibits. Because the silicon–oxygen bonds are strong, silicate minerals tend to cleave between the silicon–oxygen structures rather than across them. For example, the micas have a sheet structure and thus tend to cleave into flat plates (see muscovite in Figure 1.17A). Quartz has equally strong silicon–oxygen bonds in all directions; therefore, it has no cleavage but fractures instead.

How do silicate minerals form? Most of them crystallize from molten rock as it cools. This cooling can occur at or near Earth's surface (low temperature and pressure) or at great depths (high temperature and pressure). The *environment* during crystallization and the *chemical composition of the molten rock* mainly determine which minerals are produced. For example, the silicate mineral olivine crystallizes at high temperatures (about 1200°C [2200°F]), whereas quartz crystallizes at much lower temperatures (about 700°C [1300°F]).

In addition, some silicate minerals form at Earth's surface from the weathered (disintegrated) products of other silicate minerals. Clay minerals are an example. Still other silicate minerals are formed under the extreme pressures associated with mountain building. Each silicate mineral, therefore, has a structure and a chemical composition that *indicate the conditions under which it formed*. Thus, by carefully examining the mineral makeup of rocks, geologists can often determine the circumstances under which the rocks formed.

We will now examine some of the most common silicate minerals, which are divided into two major groups based on their chemical makeup.

Common Light Silicate Minerals The most common light silicates are the feldspars, quartz, muscovite, and the clay minerals. Generally light in color and having a specific gravity of about 2.7, **light silicate minerals** contain varying amounts of aluminum, potassium, calcium, and sodium.

The most abundant mineral group, the *feldspars*, are found in many igneous, sedimentary, and metamorphic rocks. One group of feldspar minerals containing potassium ions in its crystalline structure is called **potassium feldspar** (**Figure 1.24A, B**). The other group, called **plagioclase feldspar**, contains calcium and/or sodium ions (**Figure 1.24C, D**). All feldspar minerals have two directions of cleavage that meet at 90-degree angles and are

SmartFigure 1.23 Common silicate minerals Note that the complexity of the silicate structure increases from the top of the chart to the bottom. (Photos by Dennis Tasa and E. J. Tarbuck) (https://goo.gl/xpaEPC)

Tutorial

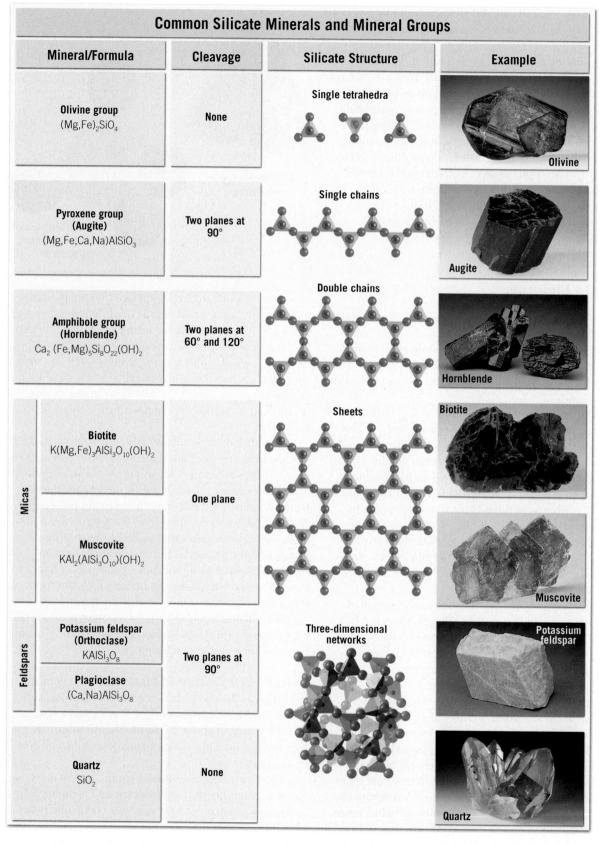

Common Silicate Minerals and Mineral Groups

Mineral/Formula	Cleavage	Silicate Structure	Example
Olivine group $(Mg,Fe)_2SiO_4$	None	Single tetrahedra	Olivine
Pyroxene group (Augite) $(Mg,Fe,Ca,Na)AlSiO_3$	Two planes at 90°	Single chains	Augite
Amphibole group (Hornblende) $Ca_2(Fe,Mg)_5Si_8O_{22}(OH)_2$	Two planes at 60° and 120°	Double chains	Hornblende
Micas — **Biotite** $K(Mg,Fe)_3AlSi_3O_{10}(OH)_2$	One plane	Sheets	Biotite
Micas — **Muscovite** $KAl_2(AlSi_3O_{10})(OH)_2$	One plane	Sheets	Muscovite
Feldspars — **Potassium feldspar (Orthoclase)** $KAlSi_3O_8$	Two planes at 90°	Three-dimensional networks	Potassium feldspar
Feldspars — **Plagioclase** $(Ca,Na)AlSi_3O_8$	Two planes at 90°	Three-dimensional networks	
Quartz SiO_2	None	Three-dimensional networks	Quartz

relatively hard (6 on the Mohs scale). The only reliable way to physically differentiate the feldspars is to look for striations (lines) that are present on some cleavage surfaces of plagioclase feldspar (see Figure 1.24D) but do not appear in potassium feldspar.

Quartz is a major constituent of many igneous, sedimentary, and metamorphic rocks. Found in a wide variety of colors (caused by impurities), quartz is quite hard (7 on the Mohs scale) and exhibits conchoidal fracture when broken (**Figure 1.25**). Pure quartz is clear,

Potassium Feldspar

A. Potassium feldspar crystal (orthoclase)

B. Potassium feldspar showing cleavage (orthoclase)

Plagioclase Feldspar

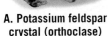

C. Sodium-rich plagioclase feldspar (albite)

D. Plagioclase feldspar showing striations (labradorite)

Figure 1.24 Feldspar minerals **A.** Characteristic crystal form of potassium feldspar. **B.** Most salmon-colored feldspar belongs to the potassium feldspar subgroup. **C.** Sodium-rich plagioclase feldspar tends to be light in color with a pearly luster. **D.** Calcium-rich plagioclase feldspar tends to be gray, blue-gray, or black in color. Labradorite, the variety shown here, exhibits striations on one of its crystal faces. (Photos by Dennis Tasa and E. J. Tarbuck)

A. Smoky quartz

B. Rose quartz

C. Milky quartz

D. Jasper

Figure 1.25 Quartz is one of the most common minerals and has many varieties **A.** Smoky quartz is commonly found in coarse-grained igneous rocks. **B.** Rose quartz owes its color to small amounts of titanium. **C.** Milky quartz often occurs in veins that occasionally contain gold. **D.** Jasper is a variety of quartz composed of minute crystals. (Photos by Dennis Tasa and E. J. Tarbuck)

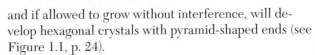

Dennis Tasa

Figure 1.26 Kaolinite
Kaolinite is a common clay mineral formed by weathering of feldspar minerals.

and if allowed to grow without interference, will develop hexagonal crystals with pyramid-shaped ends (see Figure 1.1, p. 24).

Another light silicate mineral, **muscovite**, is an abundant member of the mica family and has excellent cleavage in one direction. Muscovite is relatively soft (2.5 to 3 on the Mohs scale).

Clay minerals are light silicates that typically form as products of chemical weathering of igneous rocks. They make up much of the surface material we call soil, and nearly half of the volume of sedimentary rocks is composed of clay minerals. Kaolinite is a common clay mineral formed from the weathering of feldspar (**Figure 1.26**).

Common Dark Silicate Minerals
The **dark silicate minerals** contain iron and magnesium in their crystalline structures and include the pyroxenes, amphiboles, olivine, biotite, and garnet. Iron gives the dark silicates their color and contributes to their high specific gravity, which is between 3.2 and 3.6, significantly greater than the specific gravity of the light silicate minerals.

Olivine is an important group of dark silicate minerals that are major constituents of dark-colored igneous rocks. Abundant in Earth's upper mantle, olivine is black to olive green in color, has a glassy luster, and often forms small crystals, which give it a granular appearance (**Figure 1.27**).

Olivine-rich peridotite (variety dunite)

Dennis Tasa

Figure 1.27 Olivine
Commonly black to olive green in color, olivine has a glassy luster and is often granular in appearance. Olivine is commonly found in the igneous rock basalt.

Figure 1.28 Augite and hornblende These dark-colored silicate minerals are common constituents of a variety of igneous rocks. (Photos by E. J. Tarbuck)

A. Augite

B. Hornblende

The *pyroxenes* are a group of dark silicate minerals that are also important components of dark-colored igneous rocks. The most common member, **augite**, is a black, opaque mineral with two directions of cleavage that meet at nearly 90-degree angles (**Figure 1.28A**).

The *amphibole group*, the most common of which is **hornblende**, is usually dark green to black in color (**Figure 1.28B**). Except for its cleavage angles, which are about 60 degrees and 120 degrees, hornblende is very similar in appearance to augite. Hornblende is common in igneous rocks, where it often forms elongated black crystals.

Biotite is the dark, iron-rich member of the mica family. Like other micas, biotite possesses a sheet structure that gives it excellent cleavage in one direction. Biotite's shiny appearance helps distinguish it from other dark silicate minerals. Like hornblende, biotite is a common constituent of most light-colored igneous rocks, including granite.

Another common dark silicate is **garnet**, which has a glassy luster, lacks cleavage, and exhibits conchoidal fracture. Although garnet colors vary, the mineral is most often brown to deep red. Well-developed garnet crystals have 12 diamond-shaped faces and are most commonly found in metamorphic rocks (**Figure 1.29**).

Figure 1.29 Well-formed garnet crystal Garnets come in a variety of colors and are commonly found in mica-rich metamorphic rocks. (Photo by E. J. Tarbuck)

← 2 cm →

Important Nonsilicate Minerals

Although the nonsilicates make up only about 8 percent of Earth's crust, some nonsilicate minerals, such as gypsum, calcite, and halite, occur as constituents in sedimentary rocks in significant amounts. Many nonsilicates are also economically important.

Nonsilicate minerals are typically divided into groups based on the negatively charged ion or complex ion that the members have in common. For example, the *oxides* contain negative oxygen ions (O^{2-}), which bond to one or more kinds of positive ions. Thus, within each mineral group, the basic structure and type of bonding is similar. As a result, the minerals in each group have similar physical properties that are useful in mineral identification. **Figure 1.30** lists some of the major nonsilicate mineral groups and includes a few examples of each.

Some of the most common nonsilicate minerals belong to one of three classes of minerals: the carbonates (CO_3^{2-}), the sulfates (SO_4^{2-}), and the halides (Cl^{1-}, F^{1-}, Br^{1-}). The carbonate minerals are much simpler structurally than the silicates. This mineral group is composed of the carbonate ion (CO_3^{2-}) and one or more kinds of positive ions. The two most common carbonate minerals are **calcite**, $CaCO_3$ (calcium carbonate), and **dolomite**, $CaMg(CO_3)_2$ (calcium/magnesium carbonate) (see Figure 1.30A,B). Calcite and dolomite are usually found together as the primary constituents in the sedimentary rocks limestone and dolostone. When calcite is the dominant mineral, the rock is called *limestone*, whereas *dolostone* results from a predominance of dolomite. Limestone is used in many ways; for example, it is used as road aggregate and building stone, and it is the main ingredient in Portland cement.

Two other nonsilicate minerals frequently found in sedimentary rocks are **halite** and **gypsum** (see Figure 1.30C,I). Both of these minerals are commonly found in thick layers that are the last vestiges of ancient seas that have long since evaporated (**Figure 1.31**). Like limestone, both halite and gypsum are important nonmetallic resources. Halite is the mineral name for common table salt (NaCl). Gypsum ($CaSO_4 \cdot 2H_2O$), which is calcium sulfate with water bound into the structure, is the mineral from which plaster and other similar building materials are composed.

Most nonsilicate mineral classes contain members that are prized for their economic value. This includes the oxides, whose members *hematite* and *magnetite* are important ores of iron (see Figure 1.30E,F). Also significant are the sulfides, which are basically compounds of sulfur (S) and one or more metals. Important sulfide minerals include galena (lead), sphalerite (zinc), and chalcopyrite (copper). In addition, native elements—including gold, silver, and carbon (diamonds)—plus a host of other nonsilicate minerals—fluorite (flux in making steel), corundum (gemstone, abrasive), and uraninite (a uranium source)—are economically important.

Common Nonsilicate Mineral Groups

Mineral Group (key ion(s) or element(s))	Mineral Name	Chemical Formula	Economic Use	Examples
Carbonates (CO_3^{2-})	Calcite Dolomite	$CaCO_3$ $CaMg(CO_3)_2$	Portland cement, lime Portland cement, lime	 B. Dolomite A. Calcite
Halides (Cl^{1-}, F^{1-}, Br^{1-})	Halite Fluorite Sylvite	$NaCl$ CaF_2 KCl	Common salt Used in steelmaking Used as fertilizer	 C. Halite D. Fluorite
Oxides (O^{2-})	Hematite Magnetite Corundum Ice	Fe_2O_3 Fe_3O_4 Al_2O_3 H_2O	Ore of iron, pigment Ore of iron Gemstone, abrasive Solid form of water	 E. Hematite F. Magnetite
Sulfides (S^{2-})	Galena Sphalerite Pyrite Chalcopyrite Cinnabar	PbS ZnS FeS_2 $CuFeS_2$ HgS	Ore of lead Ore of zinc Sulfuric acid production Ore of copper Ore of mercury	 H. Chalcopyrite G. Galena
Sulfates (SO_4^{2-})	Gypsum Anhydrite Barite	$CaSO_4 \cdot 2H_2O$ $CaSO_4$ $BaSO_4$	Plaster Plaster Drilling mud	 J. Anhydrite I. Gypsum
Native elements (single elements)	Gold Copper Diamond Graphite Sulfur Silver	Au Cu C C S Ag	Trade, jewelry Electrical conductor Gemstone, abrasive Pencil lead Sulfadrugs, chemicals Jewelry, photography	 K. Copper L. Sulfur

Figure 1.30 Important nonsilicate minerals (Photos by Dennis Tasa and E. J. Tarbuck)

Figure 1.31 Thick bed of halite exposed in an underground mine Halite (salt) mine in Grand Saline, Texas. Note the person for scale. (Photo by Tom Bochsler /Pearson Education, Inc.)

1.5 CONCEPT CHECKS

1. List the eight most common elements in Earth's crust.
2. Explain the difference between the terms *silicon* and *silicate*.
3. Sketch the silicon–oxygen tetrahedron and label its parts.
4. What is the most abundant mineral in Earth's crust?
5. List six common nonsilicate mineral groups and the key ion(s) or element(s) in each group.
6. What is the most common carbonate mineral?
7. List eight common nonsilicate minerals and their economic uses.

CONCEPTS IN REVIEW
Matter and Minerals

1.1 Minerals: Building Blocks of Rocks

List the main characteristics that an Earth material must possess to be considered a mineral and describe each characteristic.

KEY TERMS: mineralogy, mineral, rock

- In Earth science, the word *mineral* refers to naturally occurring inorganic solids that possess an orderly crystalline structure and a characteristic chemical composition. The study of minerals is called *mineralogy*.
- Minerals are the building blocks of rocks. Rocks are naturally occurring masses of minerals or mineral-like matter, such as natural glass or organic material.

1.2 Atoms: Building Blocks of Minerals

Compare and contrast the three primary particles contained in atoms.

KEY TERMS: atom, nucleus, proton, neutron, electron, valence electron, atomic number, element, periodic table, chemical compound

- Minerals are composed of atoms of one or more elements. The atoms of any element consist of the same three basic ingredients: protons, neutrons, and electrons.
- The atomic number represents the number of protons found in the nucleus of an atom of a particular element. For example, an oxygen atom has eight protons, so its atomic number is eight. Protons and neutrons

are approximately the same size and mass, but protons are positively charged, and by contrast, neutrons have no charge.
- Electrons are much smaller than both protons and neutrons; an electron weighs about 2000 times less than a proton. Each electron has a negative charge, equal in magnitude to a proton's positive charge. Electrons swarm around an atom's nucleus at a distance, in several distinctive energy levels called principal shells. The electrons in the outermost principal shell, called valence electrons, are important when one atom bonds with other atoms to form chemical compounds.
- Elements with similar numbers of valence electrons tend to behave in similar ways. The periodic table displays these similarities in its graphical arrangement of the elements.

(?) Use the periodic table (see Figure 1.5) to identify these geologically important elements by their number of protons: (A) 14, (B) 6, (C) 13, (D) 17, and (E) 26.

1.3 Why Atoms Bond

Distinguish among ionic bonds, covalent bonds, and metallic bonds.

KEY TERMS: octet rule, chemical bond, ionic bond, ion, covalent bond, metallic bond

- When atoms are attracted to other atoms, they can form chemical bonds, which generally involve the transfer or sharing of valence electrons. The most stable arrangement for most atoms is to have eight electrons in the outermost principal shell. This idea is called the *octet rule*.

(1.3 continued)

- Ionic bonds involve atoms of one element giving up electrons to atoms of another element, forming positively and negatively charged atoms called *ions.* Positively charged ions bond with negative ions to form ionic bonds.
- Covalent bonds involve the sharing of one or more electrons between two adjacent atoms.
- In metallic bonds, the sharing is more extensive: Electrons can freely move from one atom to the next throughout the entire mass.

Cloud of electrons

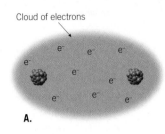

A.

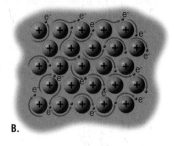

B.

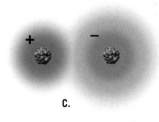

C.

(?) Which of the accompanying diagrams (A, B, or C) best illustrates ionic bonding? What are the distinguishing characteristics of ionic versus covalent bonding?

1.4 Properties of Minerals

List and describe the properties used in mineral identification.

KEY TERMS: diagnostic property, ambiguous property, luster, color, streak, crystal shape (habit), hardness, Mohs scale, cleavage, fracture, tenacity, density, specific gravity

- The composition and internal crystalline structure of a mineral give it specific physical properties. Mineral properties are useful in the identification of minerals.
- Luster is a mineral's ability to reflect light. The terms *transparent, translucent,* and *opaque* describe the degree to which a mineral can transmit light. Color can be unreliable for mineral identification, as impurities can "stain" minerals with misleading colors. A more reliable identifier is streak, the color of the powder generated by scraping a mineral against a porcelain streak plate.
- Crystal shape, also referred to as crystal habit, is often useful for mineral identification.
- Variations in the strength of chemical bonds give minerals properties such as hardness (resistance to being scratched) and tenacity (tendency to break in a brittle fashion or bend when stressed). Cleavage, the preferential breakage of a mineral along planes of weakly bonded atoms, is very useful in identifying minerals.
- The amount of matter packed into a given volume determines a mineral's density. To compare the densities of minerals, mineralogists use a related quantity, known as specific gravity, which is the ratio between a mineral's density and the density of water.
- Other properties are diagnostic for certain minerals but rare in most others—examples include smell, taste, feel, reaction to hydrochloric acid, magnetism, and double refraction.

(?) Research the minerals *quartz* and *calcite*. List five physical characteristics that may be used to distinguish one from the other.

Dennis Tasa

Quartz

Dennis Tasa

Calcite

1.5 Mineral Groups

List the common silicate and nonsilicate minerals and describe what characterizes each group.

KEY TERMS: rock-forming mineral, economic mineral, silicate, nonsilicate, silicon–oxygen tetrahedron, light silicate mineral, potassium feldspar, plagioclase feldspar, quartz, muscovite, clay mineral, dark silicate mineral, olivine, augite, hornblende, biotite, garnet, calcite, dolomite, halite, gypsum

- Silicate minerals have a basic building block in common: a small pyramid-shaped structure consisting of one silicon atom surrounded by four oxygen atoms called the *silicon—oxygen tetrahedron.* Neighboring tetrahedra can share some of their oxygen atoms, causing them to develop long chains, sheet structures, and three-dimensional networks.
- Silicate minerals are the most common mineral class on Earth. They are subdivided into minerals that contain iron and/or magnesium (dark silicates) and those that do not (light silicates). The light silicate minerals are generally light in color and have relatively low specific gravities. Feldspar, quartz, muscovite, and clay minerals are examples. The dark silicate minerals are generally dark in color and relatively dense. Olivine, pyroxene, amphibole, biotite, and garnet are examples.
- Nonsilicate minerals include oxides, which contain oxygen ions that bond to other elements (usually metals); carbonates, which have CO_3^{2-} as a critical part of their crystal structure; sulfates, which have SO_4^{2-} as their basic building block; and halides, which contain a nonmetal ion such as chlorine, bromine, or fluorine that bonds to a metal ion such as sodium or calcium.
- Nonsilicate minerals are often economically important. Hematite is an important source of industrial iron, while calcite is an essential component of cement.

GIVE IT SOME THOUGHT

1. Using the geologic definition of *mineral* as your guide, determine which of the items in this list are minerals and which are not. If something in this list not a mineral, explain.
   a. Gold nugget
   b. Seawater
   c. Quartz
   d. Cubic zirconia
   e. Obsidian
   f. Ruby
   g. Glacial ice
   h. Amber

2. Assume that the number of protons in a neutral atom is 92 and its atomic mass is 238.03. (*Hint:* Refer to the periodic table in Figure 1.5 to answer this question.)
   a. What is the name of the element?
   b. How many electrons does it have?
   c. Given its atomic mass, how many neutrons must it have?

3. Referring to the accompanying photos of five minerals, determine which of these specimens exhibit a metallic luster and which have a nonmetallic luster.

A. B. C.

D. E.

(Photos by Dennis Tasa)

4. Gold has a specific gravity of almost 20. A 5-gallon bucket of water weighs 40 pounds. How much would a 5-gallon bucket of gold weigh?

5. Examine the accompanying photo of a mineral that has several smooth, flat surfaces that resulted when the specimen was broken.
   a. How many flat surfaces are present on this specimen?
   b. How many different directions of cleavage does this specimen have?
   c. Do the cleavage directions meet at 90-degree angles?

Cleaved sample

6. Each of the following statements describes a silicate mineral or mineral group. In each case, provide the appropriate name:
   a. The most common member of the amphibole group
   b. The most common light-colored member of the mica family
   c. The only common silicate mineral made entirely of silicon and oxygen
   d. A silicate mineral with a name based on its color
   e. A silicate mineral characterized by striations
   f. A silicate mineral that originates as a product of chemical weathering

7. What mineral property is illustrated in the accompanying photo?

Dennis Tasa

8. Do an Internet search to determine which minerals are used to manufacture the following products:
   a. Stainless steel utensils
   b. Cat litter
   c. Tums brand antacid tablets
   d. Lithium batteries
   e. Aluminum beverage cans

9. Most states have designated a state mineral, rock, or gemstone to promote interest in that state's natural resources. Describe your state mineral, rock, or gemstone and explain why it was selected. If your state does not have a state mineral, rock, or gemstone, describe one from a state adjacent to yours.

MasteringGeology™

www.masteringgeology.com Looking for additional review and test prep materials? With individualized coaching on the toughest topics of the course, MasteringGeology offers a wide variety of ways for you to move beyond memorization to begin thinking like a geologist. Visit the Study Area in www.masteringgeology.com to find practice quizzes, study tools, and multimedia that will improve your understanding of this chapter's content. Sign in today to enjoy the following features: **Self Study Quizzes, SmartFigure: Tutorials/Animations/Condor Videos/Mobile Field Trips, Geoscience Animation Library, GEODe, RSS Feeds, Digital Study Modules,** and an optional **Pearson eText.**

2

FOCUS ON CONCEPTS

Each statement represents the primary learning objective for the corresponding major heading within the chapter. After you complete the chapter, you should be able to:

2.1 Sketch, label, and explain the rock cycle.

2.2 Describe the two criteria used to classify igneous rocks and explain how the rate of cooling influences the crystal size of minerals.

2.3 Define *weathering* and distinguish between the two main categories of weathering.

2.4 List and describe the different categories of sedimentary rocks and discuss the processes that change sediment into sedimentary rock.

2.5 Define *metamorphism*, explain how metamorphic rocks form, and describe the agents of metamorphism.

Sedimentary strata, Castle Rock, Capitol Reef National Park, Utah. (Photo by Michael Weber/Image BROKER/AGE Fotostock)

ROCKS
MATERIALS OF THE SOLID EARTH

Why study rocks? You have already learned that some rocks and minerals have great economic value. In addition, all Earth processes depend in some way on the properties of these basic Earth materials. Events such as volcanic eruptions, mountain building, weathering, erosion, and even earthquakes involve rocks and minerals. Consequently, a basic knowledge of Earth materials is essential to understanding most geologic phenomena.

Every rock contains clues about the environment in which it formed. For example, some rocks are composed entirely of small shell fragments. This tells Earth scientists that the rock likely originated in a shallow marine environment. Other rocks contain clues indicating that they formed from a volcanic eruption or deep in the Earth during mountain building. Thus, rocks contain a wealth of information about events that have occurred over Earth's long history.

2.1 Earth as a System: The Rock Cycle

Sketch, label, and explain the rock cycle.

Earth as a system is illustrated most vividly when we examine the rock cycle. The **rock cycle** allows us to see many of the interactions among the many components and processes of the Earth system (**Figure 2.1**). It helps us understand the origins of igneous, sedimentary, and metamorphic rocks and how these rocks are connected. In addition, the rock cycle demonstrates that any rock type, given the right sequence of events, can be transformed into any other type.

The Basic Cycle

We begin our discussion of the rock cycle with molten rock, called *magma*, which forms by melting that occurs primarily within Earth's crust and upper mantle (see Figure 2.1). Once formed, a magma body rises toward the surface because it is less dense than the surrounding rock. If magma reaches Earth's surface and erupts, we call it *lava*. Eventually, molten rock cools and solidifies, a process called *crystallization*, or *solidification*. Molten rock may solidify either beneath the surface or, following a volcanic eruption, at the surface. In either situation, the resulting rocks are called *igneous rocks*.

If igneous rocks are exposed at the surface, they undergo *weathering*, the slow disintegration and decomposition of rocks by the daily influences of the atmosphere. The loose materials that result are often moved downslope by gravity and then picked up and transported by one or more erosional agents—running water, glaciers, wind, or waves. These rock particles and dissolved substances, called *sediment*, are eventually deposited. Although most sediment ultimately comes to rest in the ocean, other sites of deposition include river floodplains, desert basins, lakes, inland seas, and sand dunes.

Next, the sediments undergo *lithification*, a term meaning "conversion into rock." Sediment is usually lithified into *sedimentary rock* when compacted by the weight of overlying materials or when cemented by percolating groundwater that fills the pores with mineral matter.

If the resulting sedimentary rock becomes deeply buried or is involved in the dynamics of mountain building, it will be subjected to great pressures and intense heat. The sedimentary rock may react to the changing environment by turning into the third rock type, *metamorphic rock*. If metamorphic rock is subjected to still higher temperatures, it may melt, creating magma, and the cycle begins again.

Although rocks may appear to be stable, unchanging masses, the rock cycle shows that they are not. The changes, however, take time—sometimes millions or even billions of years. In addition, different locations on Earth are at different stages of the rock cycle. Today, new magma is forming under the island of Hawaii, whereas the rocks that comprise the Colorado Rockies are slowly being worn down by weathering and erosion. Some of this weathered debris will eventually be carried to the Gulf of Mexico, where it will add to the already substantial mass of sediment that has accumulated there.

Alternative Paths

Rocks do not necessarily go through the cycle in the order just described. Other paths are also possible. For example, rather than being exposed to weathering and erosion at Earth's surface, igneous rocks may remain deeply buried (see Figure 2.1). Eventually, these masses may be subjected to the strong compressional forces and high temperatures associated with mountain building. When this occurs, they are transformed directly into metamorphic rocks.

Uplift and erosion may bring even deeply buried rocks of any type to the surface. When this happens, the material is attacked by weathering processes and turned into new raw materials for sedimentary rocks. Conversely,

ROCK CYCLE

Viewed over long time spans, rocks are constantly forming, changing, and re-forming.

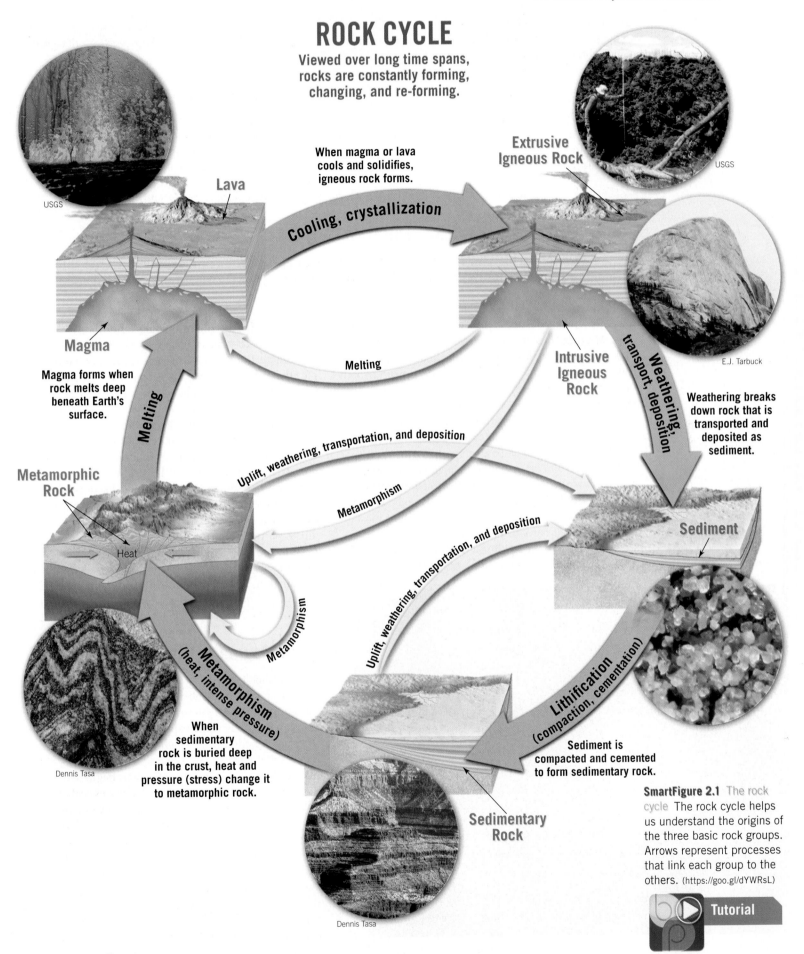

Lava

USGS

When magma or lava cools and solidifies, igneous rock forms.

Extrusive Igneous Rock

USGS

Cooling, crystallization

Magma

Magma forms when rock melts deep beneath Earth's surface.

Melting

Melting

Intrusive Igneous Rock

E.J. Tarbuck

Weathering; transport, deposition

Weathering breaks down rock that is transported and deposited as sediment.

Uplift, weathering, transportation, and deposition

Metamorphic Rock

Metamorphism

Heat

Uplift, weathering, transportation, and deposition

Sediment

Metamorphism

Dennis Tasa

Metamorphism (heat, intense pressure)

When sedimentary rock is buried deep in the crust, heat and pressure (stress) change it to metamorphic rock.

Lithification (compaction, cementation)

Sediment is compacted and cemented to form sedimentary rock.

Sedimentary Rock

SmartFigure 2.1 The rock cycle The rock cycle helps us understand the origins of the three basic rock groups. Arrows represent processes that link each group to the others. (https://goo.gl/dYWRsL)

Tutorial

Dennis Tasa

igneous or metamorphic rocks that form at depth (deep underground) may remain there, where the high temperatures and forces associated with mountain building may metamorphose or even melt them. Over time, rocks may be transformed into any other rock type, or even into a different form of the original type. Rocks may take many paths through the rock cycle.

What drives the rock cycle? Earth's internal heat is responsible for the processes that form igneous and metamorphic rocks. Weathering and the transport of weathered material are external processes, powered by energy from the Sun. External processes produce sedimentary rocks.

2.1 CONCEPT CHECKS

1. Sketch and label the rock cycle. Make sure your sketch includes alternative paths.
2. Use the rock cycle to explain the statement "One rock is the raw material for another."

2.2 Igneous Rocks: "Formed by Fire"

Describe the two criteria used to classify igneous rocks and explain how the rate of cooling influences the crystal size of minerals.

In the discussion of the rock cycle, we pointed out that **igneous rocks** form as *magma* or *lava* cools and crystallizes. **Magma** is most often generated by melting of rocks in Earth's mantle, although some magma originates from the melting of crustal rock. Once formed, a magma body buoyantly rises toward the surface because it is less dense than the surrounding rocks.

When magma reaches the surface, it is called **lava** (**Figure 2.2**). Sometimes, lava is emitted as fountains, which are produced when escaping gases propel molten rock skyward. On other occasions, magma is ejected explosively from vents, producing a spectacular eruption such as the 1980 eruption of Mount St. Helens. However, most eruptions are not violent; rather, volcanoes most often emit quiet outpourings of lava.

Molten rock may solidify at depth, or it may solidify at Earth's surface. When molten rock solidifies *at the* surface, the resulting igneous rocks are classified as **extrusive rocks**, or **volcanic rocks** (after the Roman fire god, Vulcan). Extrusive igneous rocks are abundant in western portions of the Americas, including the volcanic cones of the Cascade Range and the extensive lava flows of the Columbia Plateau. In addition, many oceanic islands, including the Hawaiian Islands, are composed almost entirely of volcanic igneous rocks.

Most magma loses its mobility before reaching Earth's surface and eventually solidifies deep below the surface. When magma solidifies at depth, it forms igneous rocks known as **intrusive rocks**, or **plutonic rocks** (after Pluto, the god of the underworld in classical mythology). Intrusive igneous rocks remain at depth unless portions of the crust are uplifted and the overlying rocks are stripped away by erosion. Exposures of intrusive igneous rocks occur in many places, including Mount Washington, New Hampshire; Stone Mountain, Georgia; Mount Rushmore in the Black Hills of South Dakota; and Yosemite National Park, California (see Figure 2.8).

From Magma to Crystalline Rock

Magma is molten rock (or *melt*) composed of ions of the elements found in silicate minerals, mainly silicon and oxygen, that move about freely. Magma also contains gases, particularly water vapor, that are confined within the magma body by the weight (pressure) of the overlying rocks, and it may contain some solids (mineral crystals). As magma cools, the once-mobile ions begin to arrange themselves into orderly patterns—a process called *crystallization*. As cooling continues, numerous small crystals develop, and ions are systematically added to these centers of crystal growth. When the crystals grow large enough for their edges to meet, their growth ceases. Eventually, all the liquid melt is transformed into a solid mass of interlocking crystals.

The rate of cooling strongly influences crystal size. If magma cools very slowly, ions can migrate over great distances. Consequently, *slow cooling results in the*

Figure 2.2 Fluid basaltic lava emitted from Mount Etna, Italy (Photo by Stockteck Images/Super Stock)

formation of fewer, larger crystals. On the other hand, if cooling occurs rapidly, the ions lose their motion and quickly combine. This results in a large number of tiny crystals that all compete for the available ions. Therefore, *rapid cooling results in the formation of a solid mass of small intergrown crystals.*

If the molten material is quenched, or cooled almost instantly, there is insufficient time for the ions to arrange themselves into a crystalline network. Solids produced in this manner consist of randomly distributed atoms. Such rocks are called *glass* and are quite similar to ordinary manufactured glass. "Instant" quenching often occurs during violent volcanic eruptions that produce tiny shards of glass called *volcanic ash.*

In addition to the rate of cooling, the composition of a magma and the amount of dissolved gases influence crystallization. Because magmas differ in each of these aspects, the physical appearance and mineral composition of igneous rocks vary widely.

Igneous Compositions

Igneous rocks are composed mainly of silicate minerals. Chemical analysis shows that silicon and oxygen—usually expressed as the silica (SiO_2) content of a magma—are by far the most abundant constituents of igneous rocks.

These two elements, plus ions of aluminum (Al), calcium (Ca), sodium (Na), potassium (K), magnesium (Mg), and iron (Fe), make up roughly 98 percent by weight of most magmas. Magma also contains small amounts of many other elements, including titanium and manganese, and trace amounts of much rarer elements, such as gold, silver, and uranium.

As magma cools and solidifies, these elements combine to form two major groups of silicate minerals. The *dark silicates* are rich in iron and/or magnesium and are comparatively low in silica (SiO_2). *Olivine, pyroxene, amphibole,* and *biotite mica* are the common dark silicate minerals of Earth's crust. By contrast, the *light silicates* contain greater amounts of potassium, sodium, and calcium and are richer in silica than dark silicates. Light silicate minerals include *quartz, muscovite mica,* and the most abundant mineral group, the *feldspars.* Feldspars make up at least 40 percent of most igneous rocks. Thus, in addition to feldspar, igneous rocks contain some combination of the other light and/or dark silicates listed earlier.

Granitic (Felsic) Versus Basaltic (Mafic) Compositions Despite their great compositional diversity, igneous rocks (and the magmas from which they form) can be divided into broad groups according to their proportions of light and dark minerals (**Figure 2.3**). Near one end

SmartFigure 2.3
Composition of common igneous rocks (After Dietrich, Daily, and Larsen) (https://goo.gl/vST3z8)

Tutorial

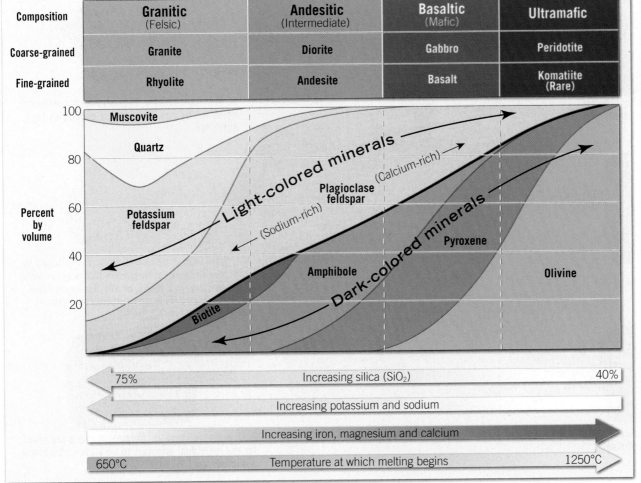

of the continuum are rocks composed almost entirely of light-colored silicates—quartz and potassium feldspar. Igneous rocks in which these are the dominant minerals have a **granitic composition**. Geologists refer to granitic rocks as being **felsic**, a term derived from *feld*spar and *si*lica (quartz). In addition to quartz and feldspar, most granitic rocks contain about 10 percent dark silicate minerals, usually biotite mica and amphibole. Granitic rocks are major constituents of the continental crust.

Rocks that contain at least 45 percent dark silicate minerals and calcium-rich plagioclase feldspar (but no quartz) are said to have a **basaltic composition** (see Figure 2.3). Basaltic rocks contain a high percentage of dark silicate minerals, so geologists refer to them as **mafic** (from *ma*gnesium and *fer*rum, the Latin word for iron). Because of their high iron content, mafic rocks are typically darker and denser than granitic rocks. Basaltic rocks make up the ocean floor as well as many of the volcanic islands located within the ocean basins.

Other Compositional Groups As you can see in Figure 2.3, rocks with a composition between granitic and basaltic rocks are said to have an **andesitic composition**, or **intermediate composition**, after the common volcanic rock *andesite*. Intermediate rocks contain at least 25

percent dark silicate minerals, mainly amphibole, pyroxene, and biotite mica, with the other dominant mineral being plagioclase feldspar. This important category of igneous rocks is associated with volcanic activity that is typically confined to the seaward margins of the continents and on volcanic island arcs such as the Aleutian chain.

At the far end of the compositional spectrum are the **ultramafic** igneous rocks, composed almost entirely of dense ferromagnesian minerals. Although ultramafic rocks are rare at Earth's surface, the rock **peridotite**, which is composed mostly of the minerals olivine and pyroxene, is the main constituent of the upper mantle.

What Can Igneous Textures Tell Us?

To describe the *size, shape,* and *arrangement* of the mineral grains that make up a rock, geologists use the word **texture**. Texture is an important property because it allows geologists to make inferences about a rock's origin, based on careful observations of crystal size and other characteristics (**Figure 2.4**). Rapid cooling produces small crystals, whereas very slow cooling produces much larger crystals. As you might expect, the rate of cooling is slow in magma chambers that lie deep within the crust, whereas a thin layer of lava extruded upon Earth's

SmartFigure 2.4 Igneous rock textures (Photos courtesy of E. J. Tarbuck and Dennis Tasa) ((https://goo.gl/U0OIx8))

Tutorial

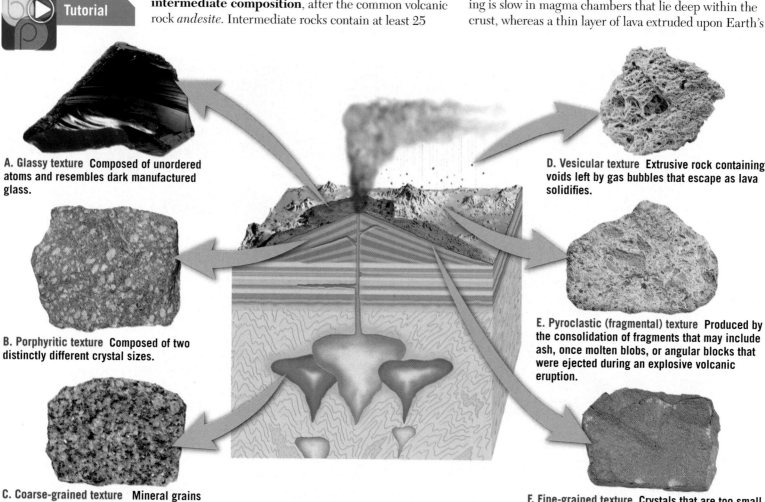

A. Glassy texture Composed of unordered atoms and resembles dark manufactured glass.

B. Porphyritic texture Composed of two distinctly different crystal sizes.

C. Coarse-grained texture Mineral grains that are large enough to be identified without a microscope.

D. Vesicular texture Extrusive rock containing voids left by gas bubbles that escape as lava solidifies.

E. Pyroclastic (fragmental) texture Produced by the consolidation of fragments that may include ash, once molten blobs, or angular blocks that were ejected during an explosive volcanic eruption.

F. Fine-grained texture Crystals that are too small for the individual minerals to be identified without a microscope.

surface may chill to form solid rock in a matter of hours. Small molten blobs ejected from a volcano during a violent eruption can solidify in mid-air.

Fine-Grained Texture Igneous rocks that form at Earth's surface or as small intrusive masses within the upper crust, where cooling is relatively rapid, exhibit a **fine-grained texture** (see Figure 2.4F). By definition, the crystals that make up fine-grained igneous rocks are so small that individual minerals can be distinguished only with the aid of a polarizing microscope or other sophisticated techniques. Therefore, we commonly characterize fine-grained rocks as being light, intermediate, or dark in color.

Coarse-Grained Texture When large masses of magma slowly crystallize at great depth, they form igneous rocks that exhibit a **coarse-grained texture**. Coarse-grained rocks consist of a mass of intergrown crystals that are roughly equal in size and large enough so that the individual minerals can be identified without the aid of a microscope (see Figure 2.4C). Geologists often use a small magnifying lens to aid in identifying minerals in coarse-grained igneous rocks.

Porphyritic Texture A large mass of magma may require thousands or even millions of years to solidify. Because different minerals crystallize under different conditions of temperature and pressure, it is possible for crystals of one mineral to become quite large before others even begin to form. If molten rock containing some large crystals erupts or otherwise moves to a cooler location, the remaining liquid portion of the lava will cool more quickly. The resulting rock, which has large crystals embedded in a matrix of smaller crystals, is said to have a **porphyritic texture** (**Figure 2.5**). The large crystals in porphyritic rocks are referred to as **phenocrysts**, whereas the matrix of smaller crystals is called **groundmass**.

Vesicular Texture Many extrusive rocks exhibit voids that represent gas bubbles that formed as the lava solidified. These nearly spherical openings are called *vesicles*, and the rocks that contain them are said to have a **vesicular texture**. Rocks that exhibit a vesicular texture often form in the upper zone of a lava flow, where cooling occurs rapidly enough to preserve the openings produced by

the expanding gas bubbles (**Figure 2.6**). Another common vesicular rock, called *pumice*, forms when silica-rich lava is ejected during an explosive eruption (see Figure 2.4D).

Glassy Texture During some volcanic eruptions, molten rock is ejected into the atmosphere, where it is quenched and becomes a solid. Rapid cooling of this type may generate rocks having a **glassy texture** (see Figure 2.4A). Glass results when unordered ions are "frozen in place" before they are able to unite into an orderly crystalline structure. *Obsidian*, a common type of natural glass, is similar in appearance to dark chunks of manufactured glass.

Pyroclastic (Fragmental) Texture Another group of igneous rocks is formed from the consolidation of individual rock fragments ejected during explosive volcanic eruptions. The ejected particles might be very fine ash, molten blobs, or large angular blocks torn from the walls of a vent during an eruption. Igneous rocks composed of these rock fragments are said to have a **pyroclastic texture**, or **fragmental texture** (see Figure 2.4E). A common type of pyroclastic rock, called *welded tuff*, is composed of fine fragments of glass that remained hot enough to eventually fuse together.

- Groundmass
- Phenocryst

1 cm

Figure 2.5 Porphyritic texture The large crystals in porphyritic rocks are called *phenocrysts*, and the small crystals constitute the *groundmass*. (Photos by Dennis Tasa)

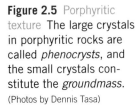

Did You Know?

A quartz watch actually contains a quartz crystal to keep time. Before quartz watches, timepieces used some sort of oscillating mass or tuning fork. Cogs and wheels converted this mechanical movement to the movement of the hand. It turns out that if voltage is applied to a quartz crystal, it will oscillate with a consistency that is hundreds of times better for timing than a tuning fork. Because of this property and modern integrated-circuit technology, quartz watches are now built so inexpensively that when they stop working, they are typically replaced rather than repaired. Modern watches that employ mechanical movements are very expensive indeed.

Scoria, an extrusive igneous rock with a vesicular texture,

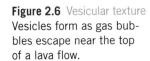

USGS

Figure 2.6 Vesicular texture Vesicles form as gas bubbles escape near the top of a lava flow.

SmartFigure 2.7
Classification of igneous rocks, based on their mineral composition and texture Coarse-grained rocks are plutonic, solidifying deep underground. Fine-grained rocks are volcanic, or solidify as shallow, thin rock bodies. Ultramafic rocks are dark, dense rocks, composed almost entirely of minerals containing iron and magnesium. Although relatively rare on Earth's surface, ultramafic rocks are major constituents of the upper mantle. (Photos by E. J. Tarbuck and Dennis Tasa) (https://goo.gl/WiyTul)

 Tutorial

IGNEOUS ROCK CLASSIFICATION CHART

		MINERAL COMPOSITION			
		Granitic (Felsic)	**Andesitic** (Intermediate)	**Basaltic** (Mafic)	**Ultramafic**
	Dominant Minerals	Quartz Potassium feldspar	Amphibole Plagioclase feldspar	Pyroxene Plagioclase feldspar	Olivine Pyroxene
	Accessory Minerals	Plagioclase feldspar Amphibole Muscovite Biotite	Pyroxene Biotite	Amphibole Olivine	Plagioclase feldspar
TEXTURE	**Coarse-grained**	Granite	Diorite	Gabbro	Peridotite
	Fine-grained	Rhyolite	Andesite	Basalt	Komatiite (rare)
	Porphyritic (two distinct grain sizes)	Granite porphyry	Andesite porphyry	Basalt porphyry	Uncommon
	Glassy	Obsidian	Less common	Less common	Uncommon
	Vesicular (contains voids)	Pumice (also glassy)		Scoria	Uncommon
	Pyroclastic (fragmental)	Most fragments < 4mm — Tuff or welded tuff		Most fragments > 4mm — Volcanic breccia	Uncommon
	Rock Color (based on % of dark minerals)	0% to 25%	25% to 45%	45% to 85%	85% to 100%

Cory Rich/Aurora/Getty Images

Michael Collier

Granite

Dennis Tasa

SmartFigure 2.8
Rocks contain information about the processes that formed them This massive granitic monolith (El Capitan) located in Yosemite National Park was once a molten mass deep within Earth. (http://goo.gl/z8XJ1i)

Mobile Field Trip

Common Igneous Rocks

Geologists classify igneous rocks by their texture and mineral composition. The texture of an igneous rock is mainly a result of its cooling history, whereas its mineral composition is largely a result of the chemical makeup of the parent magma (**Figure 2.7**). Because igneous rocks are classified on the basis of both mineral composition and texture, some rocks having similar mineral constituents but exhibiting different textures are given different names.

Granitic (Felsic) Rocks **Granite** is a coarse-grained igneous rock that forms where large masses of granitic magma slowly solidify at depth. Episodes of mountain building may uplift granite and related crystalline rocks, with the processes of weathering and erosion stripping away the overlying crust. Areas where large quantities of granite are exposed at the surface include Pikes Peak in the Rockies, Mount Rushmore in the Black Hills, Stone Mountain in Georgia, and Yosemite National Park in the Sierra Nevada (**Figure 2.8**).

Granite is perhaps the best-known igneous rock, in part because of its natural beauty, which is enhanced when polished, and partly because of its abundance. Slabs of polished granite are commonly used for tombstones, monuments, and countertops.

Rhyolite is the extrusive equivalent of granite (same chemical composition but different texture) and, likewise, is composed essentially of light-colored silicates (see Figure 2.7). This fact accounts for its color, which is usually buff to pink or light gray. Rhyolite is fine grained and frequently contains glass fragments and voids, indicating rapid cooling in a surface environment. In contrast to granite, which is widely distributed on continents as large intrusive masses, rhyolite deposits are less common and generally less voluminous. Yellowstone Park is one well-known exception where extensive rhyolite lava flows are found (along with thick ash deposits of rhyolitic composition).

Obsidian is a common type of volcanic glass. Although dark in color, obsidian usually has a felsic composition; its color results from small amounts of metallic ions in an otherwise clear, glassy substance. Because of its excellent conchoidal fracture and ability to hold a sharp, hard edge, obsidian was a prized material from which Native Americans chipped arrowheads and cutting tools (**Figure 2.9**).

Another silica-rich volcanic rock that exhibits a glassy and also vesicular texture is **pumice**. Often found with obsidian, pumice forms

Figure 2.9 Obsidian, a natural glass Native Americans used obsidian to make arrowheads and cutting tools. (Photo by Jeffrey Scovil)

Figure 2.10 Pumice, a vesicular glassy rock Pumice is very light-weight because it contains numerous vesicles. (Photo by E. J. Tarbuck; inset photo by Chip Clark/Fundamental Photographs)

← 2 cm →

when large amounts of gas escape from molten rock to generate a gray, frothy mass (**Figure 2.10**). In some samples, the vesicles are quite noticeable, whereas in others, the pumice resembles fine shards of intertwined glass. Because of the large volume of air-filled voids, many samples of pumice float in water (see Figure 2.10).

Andesitic (Intermediate) Rocks **Andesite** is a medium-gray extrusive rock. It may be fine-grained, or it may have a porphyritic texture (see Figure 2.7), often with phenocrysts of plagioclase feldspar (pale and rectangular) or amphibole (black and elongated). Andesite is a major constituent of many of the volcanoes that are found around the Pacific Rim, including those of the Andes Mountains (after which it is named) and the Cascade Range.

Diorite, the intrusive equivalent of andesite, is a coarse-grained rock that resembles gray granite. However, it

Figure 2.11 Fluid basaltic lava flowing from Kilauea Volcano, Hawaii (Photo by David Reggie/Perspectives/Getty Images)

can be distinguished from granite because it contains few or no visible quartz crystals and has a higher percentage of dark silicate minerals.

Basaltic (Mafic) Rocks The most common extrusive igneous rock is **basalt**, a very dark green to black, fine-grained volcanic rock composed primarily of pyroxene, olivine, and plagioclase feldspar. Many volcanic islands, such as the Hawaiian Islands and Iceland, are composed mainly of basalt (**Figure 2.11**). Furthermore, the upper layers of the oceanic crust consist of basalt. In the United States, large portions of central Oregon and Washington were the sites of extensive basaltic outpourings.

The coarse-grained, intrusive equivalent of basalt is **gabbro** (see Figure 2.7). Gabbro is not commonly exposed at the surface, but it makes up a significant percentage of the oceanic crust.

How Different Igneous Rocks Form

Because a large variety of igneous rocks exist, it is logical to assume that an equally large variety of magmas also exist. However, geologists have observed that a single volcano may extrude lavas exhibiting quite different compositions, rather than just one kind. Data of this type led them to examine the possibility that magma might change (evolve) and thus become the parent to a variety of igneous rocks. To explore this idea, a pioneering investigation into the crystallization of magma was carried out by N. L. Bowen in the first quarter of the twentieth century.

Bowen's Reaction Series In a laboratory setting, Bowen demonstrated that magma, with its diverse chemistry, crystallizes over a temperature range of at least 200°C (360°F), unlike simple compounds (such as water), which solidify at specific temperatures. As magma cools, certain minerals crystallize first at relatively high temperatures. At successively lower temperatures, other minerals begin to crystallize. This arrangement of minerals, shown in **Figure 2.12**, became known as **Bowen's reaction series**.

Bowen discovered that the first mineral to crystallize from a body of magma is *olivine*. Further cooling results in the formation of *pyroxene*, as well as *plagioclase feldspar*. At intermediate temperatures, the minerals *amphibole* and *biotite* begin to crystallize.

During the last stage of crystallization, after most of the magma has solidified, the minerals *muscovite* and *potassium feldspar* may form (see Figure 2.12). Finally, *quartz* crystallizes from any remaining liquid. Olivine and quartz are seldom found in the same igneous rock because quartz crystallizes at much lower temperatures than olivine.

Analysis of igneous rocks provides evidence that this crystallization model approximates what can happen in nature. In particular, we find that minerals that form in the same general range on Bowen's reaction series are

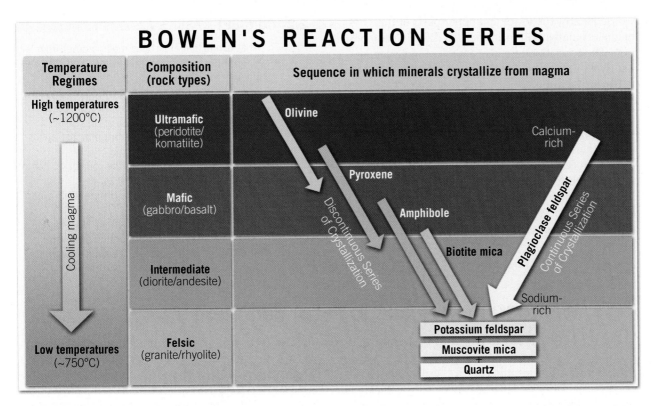

BOWEN'S REACTION SERIES

Temperature Regimes	Composition (rock types)	Sequence in which minerals crystallize from magma
High temperatures (~1200°C)	**Ultramafic** (peridotite/komatiite)	Olivine
Cooling magma ↓	**Mafic** (gabbro/basalt)	Pyroxene — Amphibole — Plagioclase feldspar (Calcium-rich / Sodium-rich) — Discontinuous Series of Crystallization / Continuous Series of Crystallization
	Intermediate (diorite/andesite)	Biotite mica
Low temperatures (~750°C)	**Felsic** (granite/rhyolite)	Potassium feldspar + Muscovite mica + Quartz

Figure 2.12 Bowen's reaction series This diagram shows the sequence in which minerals crystallize from a magma. Compare this figure to the mineral composition of the rock groups in Figure 2.7. Note that each rock group consists of minerals that crystallize in the same temperature range.

found together in the same igneous rocks. For example, notice in Figure 2.12 that the minerals quartz, potassium feldspar, and muscovite, located in the same region of Bowen's diagram, are typically found together as major constituents of the igneous rock *granite*.

Magmatic Differentiation We said earlier that the composition of an igneous rock depends on the composition of the magma from which it forms. However, a single volcano, fed from a single magma chamber, can form rocks of quite different composition over its lifetime. How does this work? First, as Bowen demonstrated, different

minerals crystallize from magma according to a predictable pattern. As each mineral forms, it selectively removes certain elements from the melt. For example, crystallization of olivine and pyroxene will selectively remove iron and magnesium, leaving the remaining melt more felsic. If some mechanism then physically separates these crystals from the remaining melt, the remaining melt will crystallize to form a different, more felsic rock type.

One such mechanism is **crystal settling**. If early-formed crystals are more dense (heavier) than the remaining melt, then they will tend to sink toward the bottom of the magma chamber, as shown in **Figure 2.13**.

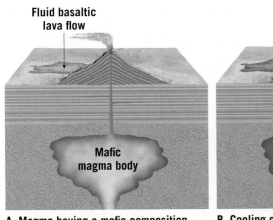

A. Magma having a mafic composition erupts fluid basaltic lavas.

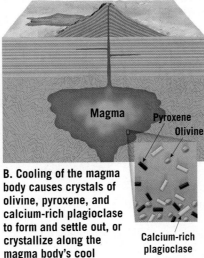

B. Cooling of the magma body causes crystals of olivine, pyroxene, and calcium-rich plagioclase to form and settle out, or crystallize along the magma body's cool margins.

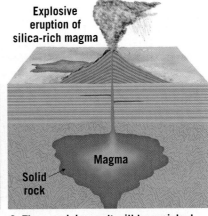

C. The remaining melt will be enriched with silica, and should a subsequent eruption occur, the rocks generated will be more silica-rich and closer to the granitic end of the compositional range than the initial magma.

Figure 2.13 Magmatic differentiation and crystal settling Illustration of how a magma evolves as the earliest-formed minerals (those richer in iron, magnesium, and calcium) crystallize and settle to the bottom of the magma chamber, leaving the remaining melt richer in sodium, potassium, and silica (SiO_2).

Consequently, the lower and upper parts of the magma chamber will form rocks of differing composition. The formation of one or more secondary magmas from a single parent magma is called **magmatic differentiation**.

At any stage in the evolution of a magma, the solid and liquid components can separate into two chemically distinct units. Furthermore, magmatic differentiation within the secondary magma can generate other chemically distinct masses of molten rock. Consequently, magmatic differentiation and separation of the solid and liquid components at various stages of crystallization can produce several chemically diverse magmas and, ultimately, a variety of igneous rocks.

 2.2 CONCEPT CHECKS

1. What is magma? How does magma differ from lava?
2. In what basic settings do intrusive and extrusive igneous rocks originate?
3. How does the rate of cooling influence crystal size? What other factors influence the texture of igneous rocks?
4. What does a porphyritic texture indicate about the history of an igneous rock?
5. List and distinguish among the four basic compositional groups of igneous rocks.
6. How are granite and rhyolite different? In what way are they similar?
7. What is magmatic differentiation? How might this process lead to the formation of several different igneous rocks from a single magma?

2.3 Weathering of Rocks to Form Sediment

Define *weathering* and distinguish between the two main categories of weathering.

All materials are susceptible to weathering. Consider, for example, the fabricated product concrete, which closely resembles the sedimentary rock conglomerate. A newly poured concrete sidewalk has a smooth, unweathered look. However, after only a few years, the same sidewalk will appear chipped, cracked, and rough, with pebbles exposed at the surface. If a tree is nearby, its roots may grow under the concrete, heaving and buckling it. The same natural processes that eventually break apart a concrete sidewalk also act to disintegrate rocks, regardless of their type or strength.

In the following sections, we discuss the various modes of mechanical and chemical weathering. Although we consider these two categories separately, keep in mind that mechanical and chemical weathering processes usually work simultaneously in nature and reinforce each other.

Mechanical Weathering

When a rock undergoes **mechanical weathering**, it is broken into smaller and smaller pieces, each of which retains the characteristics of the original material. The end result is many small pieces from a single large one. **Figure 2.14** shows that breaking a rock into smaller pieces increases the surface area available for chemical attack. Hence, by breaking rocks into smaller pieces, mechanical weathering increases the amount of surface area available for chemical weathering.

In nature, four important physical processes lead to the fragmentation of rock: frost wedging, salt crystal growth, expansion resulting from unloading (sheeting), and biological activity. In addition, as erosional agents such as waves, wind, glacial ice, and running water move rock debris, these particles are further broken and abraded.

Frost Wedging If you leave a glass bottle filled with water in the freezer too long, you will find the bottle fractured, as in **Figure 2.15**. The bottle breaks because water has the unique property of expanding about 9 percent when it freezes.

You might also expect this same process to fracture rocks in nature. This is, in fact, the basis for the traditional explanation of **frost wedging**. After water works its way into the cracks in rock, the freezing water enlarges the cracks, and angular fragments are eventually produced (**Figure 2.16**).

SmartFigure 2.14
Mechanical weathering increases surface area
Mechanical weathering adds to the effectiveness of chemical weathering because chemical weathering occurs mainly on exposed surfaces. (https://goo.gl/XkBfd2)

 **Tutorial**

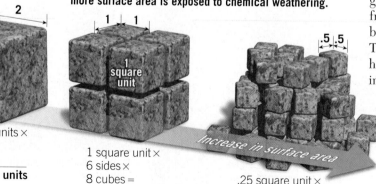

As mechanical weathering breaks rock into smaller pieces, more surface area is exposed to chemical weathering.

4 square units × 6 sides × 1 cube = **24 square units**

1 square unit × 6 sides × 8 cubes = **48 square units**

.25 square unit × 6 sides × 64 cubes = **96 square units**

Increase in surface area

More recent research has shown that frost wedging can also occur in another way. It has long been known that when moist soils freeze, they expand, or *frost heave*, due to the growth of ice lenses. These masses of ice grow larger because they are supplied with water migrating from unfrozen areas as thin liquid films. As more water accumulates and freezes, the soil is heaved upward. A similar process occurs within the cracks and pore spaces of rocks. Lenses of ice grow larger as they attract liquid water from surrounding pores. The growth of these ice masses gradually weakens the rock, causing it to fracture.

Salt Crystal Growth Another expansive force that can split rocks is generated by the growth of salt crystals. Rocky shorelines and arid regions are common settings for this process. It begins when sea spray from breaking waves or salty groundwater penetrates crevices and pore spaces in rock. As this water evaporates, salt crystals form. As these crystals gradually grow larger, they weaken the rock by pushing apart the surrounding grains or enlarging tiny cracks.

Figure 2.15 Ice breaks bottle The bottle broke because water expands about 9 percent when it freezes. (Photo by Martyn F. Chillmaid/Science Source)

Sheeting When large masses of igneous rock, particularly those composed of granite, are exposed by erosion, concentric slabs begin to break loose. The process generating these onion-like layers is called **sheeting**. It takes place, at least in part, due to the great reduction in pressure that occurs when the overlying rock is eroded away, a process called *unloading*. **Figure 2.17** illustrates what happens: As the overburden is removed, the outer parts of the granitic mass expand more than the rock below and separate from the rock body. Continued weathering eventually causes the slabs to separate and fall off, creating **exfoliation domes** (*ex* = off, *folium* = leaf). Excellent examples of exfoliation domes include

Frost wedging

Slightly tilted sedimentary beds

Falling rock debris

Falling rock debris

Patches of snow

Talus slope composed of angular rock fragments

SmartFigure 2.16 Frost wedging In mountainous areas frost wedging creates angular rock fragments that accumulate to form steep rocky slopes. (Photo by Marli Miller) (https://goo.gl/5uqnS1)

Tutorial

SmartFigure 2.17
Unloading leads to sheeting Sheeting leads to the formation of an exfoliation dome. (Photo by Gary Moon/ AGE Fotostock) (https://goo.gl/ k4N2o3)

A. This large igneous mass formed deep beneath the surface, where confining pressure is great.

Confining pressure

Deep pluton

B. As erosion removes the overlying bedrock (unloading), the outer parts of the igneous mass expand. Joints form parallel to the surface. Continued weathering causes thin slabs to separate and fall off.

Joints

Expansion and sheeting

Uplift

C. The summit of Half Dome in California's Yosemite National Park is an exfoliation dome and illustrates the onion-like layers created by sheeting.

Stone Mountain in Georgia and Half Dome in Yosemite National Park.

Biological Activity The activities of organisms, including plants, burrowing animals, and humans, can cause weathering. Plant roots in search of minerals and water grow into fractures, and as the roots grow, they wedge apart the rock (**Figure 2.18**). Burrowing animals further break down the rock by moving fresh material to the surface, where physical and chemical processes can more effectively attack it. Decaying organisms also produce acids, which contribute to chemical weathering. Where rock has been blasted in search of minerals or for road construction, the impact of humans is particularly noticeable.

Chemical Weathering

In the preceding discussion you learned that breaking rock into smaller pieces aids chemical weathering by increasing the surface area available for chemical attack. It should also be pointed out that chemical weathering contributes to mechanical weathering. It does so by weakening the outer portions of some rocks, which, in turn, makes them more susceptible to being broken by mechanical weathering processes.

Chemical weathering involves the complex processes that alter the internal structures of minerals by removing and/or adding elements. During this transformation, the original rock decomposes into substances that are stable in the surface environment. Consequently, the products of chemical weathering will remain essentially unchanged as long as they remain in an environment similar to the one in which they formed.

Water and Carbonic Acid Water is by far the most important agent of chemical weathering. Although pure water is nonreactive, a small amount of dissolved material is generally all that is needed to activate it. Oxygen dissolved in water will *oxidize* some materials. For example, when an iron nail is found in moist soil, it will have a coating of rust (iron oxide), and if the time of exposure has been long, the nail will be so weak that it can be broken as easily as a toothpick. When rocks containing iron-rich minerals oxidize, a yellow to reddish-brown rust will appear on the surface.

Carbon dioxide (CO_2) dissolved in water (H_2O) forms **carbonic acid** (H_2CO_3)—the same weak acid produced when soft drinks are carbonated. Rain dissolves some carbon dioxide as it falls through the atmosphere, and additional amounts released by decaying organic matter are acquired as the water percolates through the soil. Carbonic acid ionizes to form the very reactive hydrogen ion (H^+) and the bicarbonate ion (HCO_3^-). Acids such as carbonic acid readily decompose many rocks and produce certain products that are water soluble. For example, the mineral calcite ($CaCO_3$), which is in the common building stones marble and limestone, is easily attacked by even a weakly acidic solution. In nature, over spans of thousands of years, large quantities of limestone are dissolved and carried away by groundwater. This activity is largely responsible for the formation of limestone caverns.

Products of Chemical Weathering

To illustrate how a rock composed of silicate minerals chemically weathers when attacked by carbonic acid, we will consider the weathering of granite, a common constituent of Earth's continental crust. Recall that granite consists mainly of quartz and potassium feldspar. The weathering of the potassium feldspar component of granite involves a chemical reaction in which the hydrogen ions (H^+) in carbonic acid replace the potassium ions (K^+) in the feldspar structure. Once removed, the potassium is available as a nutrient for plants or becomes the soluble salt potassium bicarbonate, which may be incorporated into other minerals or carried to the ocean by groundwater and streams.

The remaining elements in feldspar reorganize to form clay minerals. Because clay minerals are the end product of chemical weathering, they are very stable under surface conditions. Clay minerals, which make up a high percentage of the inorganic material found in soils, are also found in shale, a common sedimentary rock.

In addition to forming clay minerals, this chemical reaction also removes some silica from the feldspar structure, which tends to be carried away by groundwater. This dissolved silica will eventually precipitate (settle out of the solution as a solid) to produce nodules of chert, or fill in the pore spaces between sediment grains, or be carried to the ocean, where microscopic animals will remove it to build hard silica shells.

Quartz, the other main component of granite, is *very resistant* to chemical weathering; it remains substantially unaltered when attacked by weakly acidic solutions. As a result, when granite weathers, the feldspar crystals dull and slowly turn to clay, releasing the once-interlocked quartz grains, which still retain their fresh, glassy appearance. Although some quartz remains in the soil, much is transported to the sea or to other sites of deposition, where it becomes the main constituent of landforms such as sandy beaches and sand dunes. In time it may become lithified to form the sedimentary rock *sandstone*.

Table 2.1 lists the weathered products of some of the most common silicate minerals. Remember that silicate minerals make up most of Earth's crust and that these minerals are composed essentially of only eight elements. When chemically weathered, silicate minerals yield

Figure 2.18 Plants can break rock Root wedging near Boulder, Colorado. (Photo by Kristin Piljay)

Plant roots can extend into joints and grow in diameter and length. This process enlarges fractures and breaks rock.

sodium, calcium, potassium, and magnesium ions, which form soluble products that may be removed by groundwater. The element iron combines with oxygen, producing relatively insoluble iron oxides, which give soil a reddish-brown or yellowish color. Under most conditions, the three remaining elements—aluminum, silicon, and oxygen—join with water to produce residual clay minerals.

2.3 CONCEPT CHECKS

1. When a rock is mechanically weathered, how does its surface area change? How does this influence chemical weathering?
2. Explain how water can cause mechanical weathering.
3. How does an exfoliation dome form?
4. How does biological activity contribute to weathering?
5. How is carbonic acid formed in nature? What products result when carbonic acid reacts with potassium feldspar?

Table 2.1 Products of Weathering

Mineral	Residual Products	Material in Solution
Quartz	Quartz grains	Silica
Feldspars	Clay minerals	Silica, K^+, Na^+, Ca^{2+}
Amphibole	Clay minerals	Silica, Ca^{2+}, Mg^{2+}
	Iron oxides	
Olivine	Iron oxides	Silica, Mg^{2+}

2.4 Sedimentary Rocks: Compacted and Cemented Sediment

List and describe the different categories of sedimentary rocks and discuss the processes that change sediment into sedimentary rock.

Recall the rock cycle, which shows the origin of **sedimentary rocks**. Weathering begins the process. Next, gravity and erosional agents remove the products of weathering and carry them to a new location, where they are deposited. Usually, the particles are broken down further during this transport phase. Following deposition, this **sediment** may become lithified, or "turned to rock."

The word *sedimentary* indicates the nature of these rocks, for it is derived from the Latin *sedimentum*, which means "settling," a reference to a solid material settling out of a fluid. Most sediment is deposited in this fashion.

Weathered debris is constantly being swept from bedrock and carried away by running water, waves, glacial ice, or wind. Eventually, the material is deposited in lakes, river valleys, seas, and countless other places. The particles in a desert sand dune, the mud on the floor of a swamp, the gravels in a streambed, and even household dust are examples of sediment produced by this never-ending process.

The weathering of bedrock and the transport and deposition of this weathered rock material are continuous processes. Therefore, sediment is found almost everywhere. As piles of sediment accumulate, the materials near the bottom are *compacted* by the weight of the overlying layers. Over long periods, these sediments are *cemented* together by mineral matter deposited from water into the spaces between particles. This forms solid sedimentary rock.

Geologists estimate that sedimentary rocks account for only about 5 percent (by volume) of Earth's outer 16 kilometers (10 miles). However, the importance of this group of rocks is far greater than this percentage implies. If you sampled the rocks exposed at Earth's surface, you would find that the great majority are sedimentary (**Figure 2.19**). Indeed,

SmartFigure 2.19
Sedimentary rocks exposed along the Waterpocket Fold, Capitol Reef National Park, Utah About 75 percent of all rock exposures on continents consist of sedimentary rocks. (Photo by Michael Collier) (http://goo.gl/eeryMM)

Mobile Field Trip

about 75 percent of all rock outcrops on the continents are sedimentary. Therefore, we can think of sedimentary rocks as comprising a relatively thin and somewhat discontinuous layer in the uppermost portion of the crust. This makes sense because sediment accumulates at the surface.

It is from sedimentary rocks that geologists reconstruct many details of Earth's history. Because sediments are deposited in a variety of different settings at the surface, the rock layers that they eventually form hold many clues to past surface environments. They may also exhibit characteristics that allow geologists to decipher information about the method and distance of sediment transport. Furthermore, sedimentary rocks contain fossils, which are vital evidence in the study of the geologic past.

Finally, many sedimentary rocks are important economically. Coal, which is burned to provide a significant portion of U.S. electrical energy, is classified as a sedimentary rock. Other major energy resources (such as petroleum and natural gas) occur in pores within sedimentary rocks. Other sedimentary rocks are major sources of iron, aluminum, manganese, and fertilizer, plus numerous materials essential to the construction industry.

Types of Sedimentary Rocks

Materials that accumulate as sediment have two principal sources. First, sediments may originate as solid particles from weathered rocks, such as the igneous rocks described earlier. These particles are called *detritus*, and the sedimentary rocks they form are called **detrital sedimentary rocks** (Figure 2.20).

Detrital Sedimentary Rocks Though a wide variety of minerals and rock fragments may be found in detrital rocks, clay minerals and quartz dominate. As you learned earlier, clay minerals are the most abundant product of the chemical weathering of silicate minerals, especially the feldspars. Quartz, on the other hand, is abundant because it is extremely durable and very resistant to chemical weathering. Thus, when igneous rocks such as granite are weathered, individual quartz grains are set free.

Geologists use particle size to distinguish among detrital sedimentary rocks. Figure 2.20 presents the four size categories for particles making up detrital rocks. When gravel-size particles predominate, the rock is called **conglomerate** if the sediment is rounded and **breccia** if the pieces are angular (see Figure 2.20). Angular fragments indicate that the particles were not transported very far from their source prior to deposition and so have not had corners and rough edges abraded. **Sandstone** is the name given to rocks when sand-size grains prevail. **Shale** (mudstone), the most common sedimentary rock, is made of very fine-grained sediment and composed mainly of clay minerals (see Figure 2.20). **Siltstone**, another rather fine-grained rock, is composed

Figure 2.20 Detrital sedimentary rocks (Photos by Dennis Tasa and E. J. Tarbuck)

Detrital Sedimentary Rocks

Particle Size	Sediment Name	Rock Name
Coarse (over 2 mm)	Gravel (Rounded particles)	Conglomerate
Coarse (over 2 mm)	Gravel (Angular particles)	Breccia
Medium (1/16 to 2 mm)	Sand	Sandstone
Medium (1/16 to 2 mm)	Sand	Arkose*
Fine (1/16 to 1/256 mm)	Silt	Siltstone
Very fine (less than 1/256 mm)	Clay	Shale or Mudstone

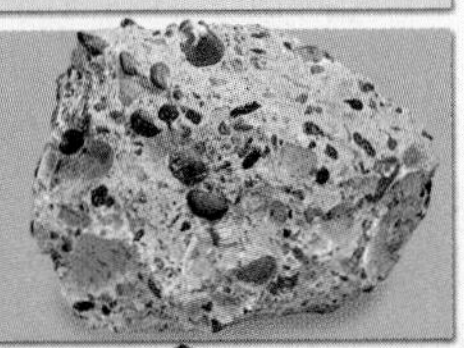

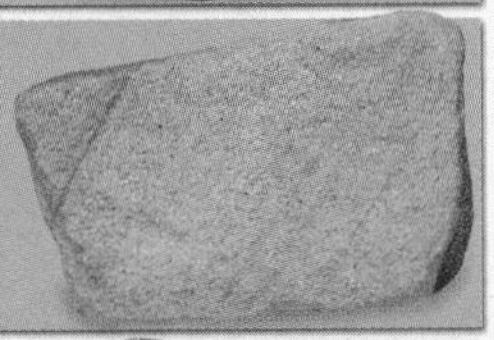

*If abundant feldspar is present the rock is called arkose.

Figure 2.21 Chemical, biochemical, and organic sedimentary rocks (Photos by Dennis Tasa and E. J. Tarbuck)

Chemical, Biochemical, and Organic Sedimentary Rocks

Composition	Texture	Rock Name
Calcite, $CaCO_3$	Fine to coarse crystalline	Crystalline Limestone 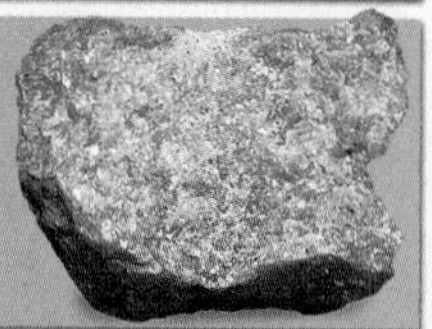
	Very fine-grained crystals	Microcrystalline Limestone 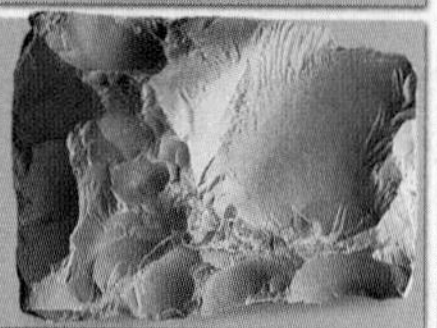
	Fine to coarse crystalline	Travertine 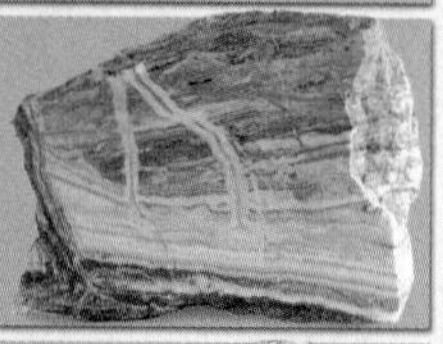
Biochemical Limestone	Visible shells and shell fragments loosely cemented	Coquina 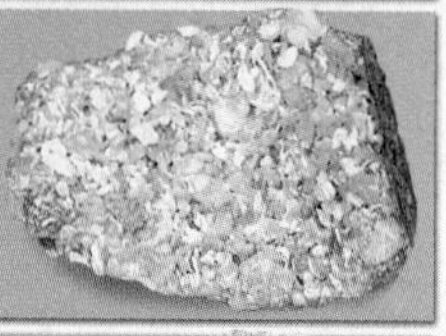
	Various size shells cemented with calcite cement	Fossiliferous Limestone 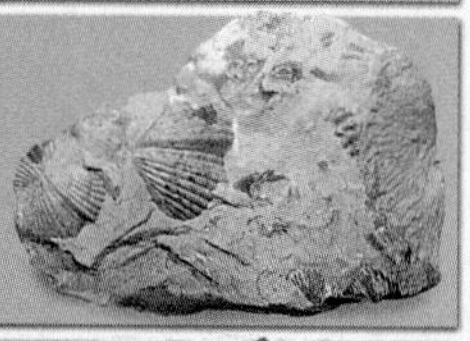
	Microscopic shells and clay	Chalk 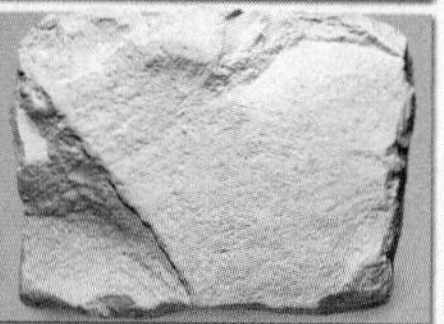
Quartz, SiO_2	Very fine crystalline	Chert (light colored) 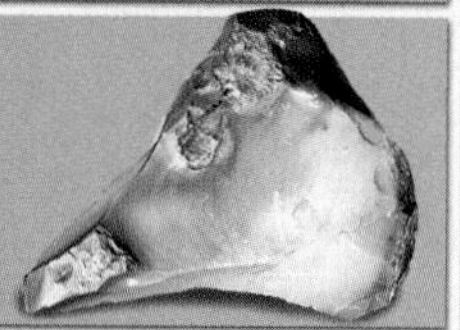
Gypsum $CaSO_4 \cdot 2H_2O$	Fine to coarse crystalline	Rock Gypsum 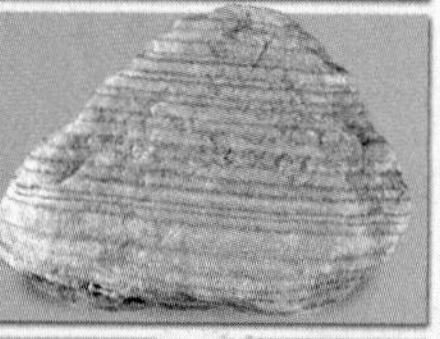
Halite, NaCl	Fine to coarse crystalline	Rock Salt 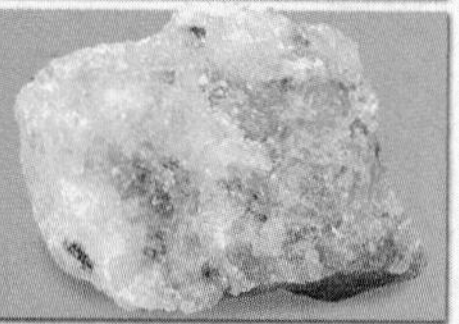
Altered plant fragments (organic matter)	Fine-grained	Bituminous Coal 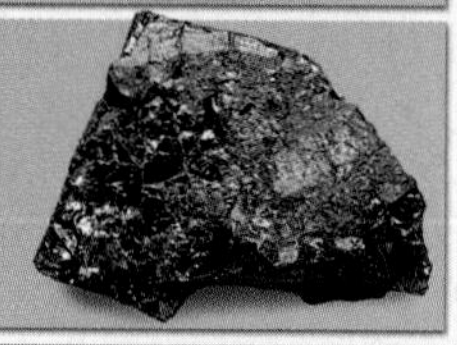

of clay-sized sediment intermixed with slightly larger silt-sized grains.

Particle size also provides useful information about the environment in which the sediment was deposited. Currents of water or air sort the particles by size. The stronger the current, the larger the particle size carried. Gravels, for example, are moved by swiftly flowing rivers, rockslides, and glaciers. Less energy is required to transport sand; thus, it is common in windblown dunes, river deposits, and beaches. Because silts and clays settle very slowly, accumulations of these materials are generally associated with the quiet waters of a lake, lagoon, swamp, or marine environment.

Although detrital sedimentary rocks are classified by particle size, in certain cases, the mineral composition is also part of naming a rock. For example, most sandstones are rich in quartz, so they are referred to as quartz sandstone. In addition, rocks consisting of detrital sediments are rarely composed of grains of just one size. Consequently, a rock containing quantities of both sand and silt can be correctly classified as sandy siltstone or silty sandstone, depending on which particle size dominates.

Chemical, Biochemical, and Organic Sedimentary Rocks In contrast to detrital rocks, which are the solid products of weathering, **chemical sedimentary rocks** and **biochemical sedimentary rocks** are derived from material (ions) carried in solution to lakes and seas (**Figure 2.21**). This

Did You Know?
The most important and most common material used for making glass is silica, which is usually obtained from the quartz in "clean," well-sorted sandstones.

material does not remain dissolved in water indefinitely. Under certain conditions and due to physical processes, the dissolved matter precipitates to form *chemical sediments*. An example of chemical sediments resulting from physical processes is the salt left behind as a body of saltwater evaporates.

Precipitation may also occur indirectly through life processes of water-dwelling organisms that form materials called *biochemical sediments*. Many aquatic animals and plants extract the mineral matter dissolved in the water to form shells and other hard parts. After the organisms die, their skeletons may accumulate on the floor of a lake or an ocean.

Limestone, an abundant sedimentary rock, is composed chiefly of the mineral calcite ($CaCO_3$). Nearly 90 percent of limestone is formed from biochemical sediments secreted by marine organisms, and the remaining amount consists of chemical sediments that precipitated directly from seawater.

One easily identified biochemical limestone is **coquina**, a coarse rock composed of loosely cemented shells and shell fragments (**Figure 2.22**). Another less obvious but familiar example is *chalk*, a soft, porous rock made up almost entirely of the hard parts of microscopic organisms that are no larger than the head of a pin. Among the most famous chalk deposits are the White Chalk Cliffs exposed along the southeast coast of England (**Figure 2.23**).

Inorganic limestone forms when chemical changes or high water temperatures cause calcium carbonate (calcite) to precipitate. **Travertine**, the type of limestone that decorates caverns, is one example. Groundwater is the source of travertine that is deposited in caves. As water drops reach the air in a cavern, some of the carbon dioxide dissolved in the water escapes, causing calcium carbonate to precipitate.

Dissolved silica (SiO_2) precipitates to form varieties of *microcrystalline rocks*—rocks consisting of very fine-grained quartz crystals (**Figure 2.24**). Sedimentary rocks composed of microcrystalline quartz include *chert* (light color), *flint* (dark), *jasper* (red), and *petrified wood* (multi-colored). These chemical sedimentary rocks may have either an inorganic or biochemical origin, but the mode of origin is usually difficult to determine.

Figure 2.22 Coquina This variety of limestone consists of shell fragments; therefore, it has a biochemical origin. (Rock sample photo by E. J. Tarbuck; beach photo by Donald R. Frazier Photolibrary, Inc./Alamy Images)

Close up

The massive White Chalk Cliffs. Chalk is a biochemical limestone made up almost entirely of the tiny hard parts of microscopic marine organisms, mainly plankton.

View of a group of plankton called *coccolithophores* from a scanning electron microscope. Individual plates shaped like hubcaps are only three one-thousandths of a millimeter in diameter; so tiny they could pass through the eye of a needle.

Figure 2.23 The White Chalk Cliffs This prominent deposit underlies large portions of southern England as well as parts of northern France. (Photo by David Wall/Alamy; coccolithophores photo by Steve Gschmeissner/SPL/Alamy

Figure 2.24 Colorful varieties of chert Chert is the name applied to a number of dense, hard chemical sedimentary rocks made of microcrystalline quartz. (**A.** and **B.** by E. J. Tarbuck; **C**, by Daniel Sambraus/Science Source; **D**, by gracious_tiger/Shutterstock)

A. Flint

B. Jasper

C. Chert arrowhead

D. Petrified wood

Very often, evaporation causes minerals to precipitate from water. Such minerals include halite, the chief component of *rock salt*, and gypsum, the main ingredient of *rock gypsum*. Both materials have significant commercial importance. Halite is familiar to everyone as the common salt used in cooking and seasoning foods. Of course, it has many other uses and has been considered important enough that people have sought, traded, and fought over it for much of human history. Gypsum is the basic ingredient in plaster of Paris. This material is used most extensively in the construction industry for drywall and plaster.

In the geologic past, many areas that are now dry land were covered by shallow arms of the sea that had only narrow connections to the open ocean. Under these conditions, water continually moved into the bay to replace water lost by evaporation. Eventually, the waters of the bay became saturated, and salt deposition began. Today, these arms of the sea are gone, and the remaining deposits are called **evaporite deposits**.

Evaporite deposits are also found in enclosed basins on land. One example is Death Valley, California, where following periods of rainfall or snowmelt in the surrounding mountains, streams carry mineral-rich water into the lowest parts of the valley. As the water in these desert basins evaporates, materials that were dissolved in the water are left behind, forming a white, salt-rich crust on the ground that grows in thickness to generate a *salt flat* (**Figure 2.25**).

Coal—An Organic Sedimentary Rock In contrast to sedimentary rocks that are calcite or silica rich, **coal** consists mostly of *organic matter*. Because coal is produced by biochemical activity and contains organic matter, it is often classified as a *biochemical*, or *organic*, rock. Examining a piece of *lignite*, also called *brown coal*, under a magnifying glass reveals plant structures such as leaves, bark, and wood that have been chemically altered but remain identifiable (**Figure 2.26**). This observation supports the conclusion that coal is the end product of the burial of large amounts of plant material over extended periods.

The initial stage in coal formation is the accumulation of large quantities of plant remains. However, special conditions are required for such accumulations because dead plants normally decompose when exposed to the atmosphere. Ideal environments that allow plant matter to accumulate are swamps. Stagnant swamp water is oxygen deficient, which makes it impossible for plant material to

SmartFigure 2.25
Bonneville Salt Flats This well-known Utah site was once a large salt lake. (Photo by Jupiter Images/Getty Images; satellite image by NASA) (https://goo.gl/mhgwN3)

Tutorial

This extensive evaporite deposit is a 30,000-acre expanse of hard white salt that in places is nearly 2 meters thick.

completely decay (oxidize). At various times during Earth history, such environments have been common.

Coal forms in a series of stages. With each successive stage, higher temperatures and pressures drive off impurities and volatiles, as shown in Figure 2.26. Lignite and bituminous coals are sedimentary rocks, but anthracite is a metamorphic rock. Anthracite forms when sedimentary layers are subjected to the folding and deformation associated with mountain building, discussed in the final section of this chapter.

Lithification of Sediment

Lithification refers to the processes by which sediments are transformed into solid sedimentary rocks. One of the most common processes is **compaction** (Figure 2.27A). As sediments accumulate through time, the weight of overlying material compresses the deeper sediments. As the grains are pressed closer and closer, pore space is greatly reduced. For example, when clays are buried beneath several thousand meters of material, the volume of the clay may be reduced as much as 40 percent. Compaction is most effective in converting very fine-grained sediments, such as clay-size particles, into sedimentary rocks.

Because sand and coarse sediments (gravel) are not easily compressed, they are generally transformed into sedimentary rocks by the process of **cementation** (Figure 2.27B). The cementing materials are carried in solution by water that percolates through the pore spaces between particles. Over time, the cement precipitates onto the sediment grains, fills the open spaces, and acts like glue to join the particles together. Calcite, silica, and iron oxide are the most common cements. The cementing material is relatively easy to identify. Calcite cement will effervesce (fizz) in contact with dilute hydrochloric acid. Silica is the hardest cement and thus produces the hardest sedimentary rocks. An orange or red color in a sedimentary rock usually means iron oxide is present.

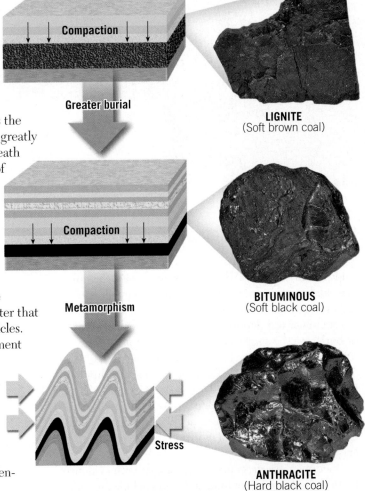

SWAMP ENVIRONMENT

Burial

PEAT (Partially altered plant material)

Compaction

Greater burial

LIGNITE (Soft brown coal)

Compaction

Metamorphism

BITUMINOUS (Soft black coal)

Stress

ANTHRACITE (Hard black coal)

SmartFigure 2.26 From plants to coal Successive stages in the formation of coal. (Photos by E. J. Tarbuck) (https://goo.gl/EHKyKU)

Tutorial

Figure 2.27 Compaction and cementation

A. COMPACTION

Water filled pore spaces

Pressure

Loosely packed clay size particles (magnified)

Compacted sediment (sedimentary rock)

B. CEMENTATION

Circulation of mineral-bearing groundwater

Cement

Loosely packed sand or gravel size particles (magnified)

Gradually the cementing material fills much of the pore space and "glues" the grains together

Figure 2.28 Sedimentary environments **A.** Ripple marks preserved in sedimentary rocks may indicate a beach or stream channel environment. (Photo by Tim Graham/Alamy) **B.** Mud cracks form when wet mud or clay dries and shrinks, perhaps signifying a tidal flat or desert basin. (Photo by Marli Miller)

A.

B.

example, indicates a high-energy environment, such as a rushing stream, where only the coarse materials can settle out. By contrast, both coal and black shale (with its high carbon content) are associated with a low-energy, organic-rich environment, such as a swamp or lagoon. Other features found in some sedimentary rocks also give clues to past environments (**Figure 2.28**).

Fossils, the traces or remains of prehistoric life, are perhaps the most important inclusions found in some sedimentary rock. Knowing the nature of the life-forms that existed at a particular time may help answer many questions about the environment. Was it land or ocean, lake or swamp? Was the climate hot or cold, rainy or dry? Was the ocean water shallow or deep, turbid or clear? Furthermore, fossils are important time indicators and play a key role in matching up rocks from different places that are the same age. Fossils are important tools used in interpreting the geologic past and will be examined in some detail in Chapter 8.

Features of Sedimentary Rocks

Sedimentary rocks are particularly important in the study of Earth history. These rocks form at Earth's surface, and as layer upon layer of sediment accumulates, each records the nature of the environment at the time the sediment was deposited. These layers, called **strata**, or **beds**, are the *single most characteristic feature of sedimentary rocks* (see the chapter-opening photo).

Bed thickness ranges from microscopically thin to tens of meters thick. Separating the strata are *bedding planes*, flat surfaces along which rocks tend to separate or break. Generally, each bedding plane marks the end of one episode of sedimentation and the beginning of another.

Sedimentary rocks provide geologists with evidence for deciphering past environments. A conglomerate, for

2.4 CONCEPT CHECKS

1. Why are sedimentary rocks important?
2. What minerals are most abundant in detrital sedimentary rocks? In which rocks do these sediments predominate?
3. Distinguish between conglomerate and breccia.
4. What are the two groups of chemical sedimentary rock? Give an example of a rock that belongs to each group.
5. How do evaporites form? Give an example.
6. Describe the two processes by which sediments are transformed into sedimentary rocks. Which is the more effective process in the lithification of sand- and gravel-sized sediments?
7. List three common cements. How might each be identified?
8. What is the single most characteristic feature of sedimentary rocks?

2.5 Metamorphic Rocks: New Rock from Old

Define metamorphism, explain how metamorphic rocks form, and describe the agents of metamorphism.

Recall from the discussion of the rock cycle that metamorphism is the transformation of one rock type into another. **Metamorphic rocks** are produced from preexisting igneous, sedimentary, or even other metamorphic rocks (**Figure 2.29**). Thus, every metamorphic rock has a *parent rock*—the rock from which it was formed.

Metamorphism, which means "to change form," is a process that leads to changes in the mineralogy, texture (for example, grain size), and sometimes chemical composition of rocks. Metamorphism occurs most often when rock is subjected to a significant increase in temperature and/or pressure. In response to these new conditions, the

SmartFigure 2.29 Folded and metamorphosed rocks This rock outcrop is in Anza-Borrego Desert State Park, California. (Photo by Pearson Education, Inc.) (https://goo.gl/kkykbZ)

Animation

rock gradually changes until it reaches a state of equilibrium with the new environment. Most metamorphic changes occur at the elevated temperatures and pressures that exist in the zone beginning a few kilometers below Earth's surface and extending into the mantle.

Metamorphism often progresses incrementally, from slight changes (*low-grade metamorphism*) to substantial changes (*high-grade metamorphism*). For example, under low-grade metamorphism, the common sedimentary rock *shale* becomes the more compact metamorphic rock *slate* (**Figure 2.30A**). Hand samples of shale and slate are sometimes difficult to distinguish, illustrating that the transition from sedimentary to metamorphic rock is often gradual, and the changes can be subtle.

In more extreme environments, metamorphism causes a transformation so complete that the identity of the parent rock cannot be determined. In high-grade metamorphism, such features

A. Parent rock (Shale) → Low-grade metamorphism / Low temperatures and pressures → Metamorphic rock (Slate)

Loosely packed clay minerals

Tightly packed chlorite and mica minerals

B. Parent rock (Granodiorite) → High-grade metamorphism / Strong compressional forces, high temperatures and pressures → Metamorphic rock (Folded gneiss)

Randomly oriented minerals

Deformed layers of segregated minerals

Figure 2.30 Metamorphic grade **A.** Low-grade metamorphism illustrated by the transformation of the common sedimentary rock shale to the more compact metamorphic rock slate. **B.** High-grade metamorphic environments obliterate the existing texture and often change the mineralogy of the parent rock. High-grade metamorphism occurs at temperatures that approach those at which rocks melt. (Photos by Dennis Tasa)

as bedding planes, fossils, and vesicles that may have existed in the parent rock are obliterated. Furthermore, when rocks deep in the crust (where temperatures are high) are subjected to directed pressure, the entire mass may deform, producing large-scale structures such as folds (**Figure 2.30B**).

By definition, rock undergoing metamorphism remains essentially solid. In the most extreme metamorphic environments, the temperatures approach those at which rocks melt. However, if appreciable melting occurs, the rocks have entered the realm of igneous activity.

Most metamorphism occurs in one of two settings:

1. When rock is intruded by magma, **contact meta-morphism** may take place, as the magma heats the adjacent rock to temperatures that cause metamor-phic changes.
2. During mountain building, great quantities of rock are subjected to pressures and high temperatures associated with large-scale deformation called **re-gional metamorphism**.

Extensive areas of metamorphic rocks are exposed on every continent. Metamorphic rocks are an impor-tant component of many mountain belts, where they make up a large portion of a mountain's crystalline core. Even the stable continental interiors, which are gener-ally covered by sedimentary rocks, are underlain by metamorphic basement rocks. In each of these settings,

the metamorphic rocks are usually highly deformed and intruded by igneous masses. Consequently, significant parts of Earth's continental crust are composed of meta-morphic and associated igneous rocks.

What Drives Metamorphism?

The agents of metamorphism include *heat, confining pressure, differential stress,* and *chemically active fluids.* During metamorphism, rocks are often subjected to all four metamorphic agents simultaneously. However, the degree of metamorphism and the contribution of each agent vary greatly from one environment to another.

Heat as a Metamorphic Agent *Thermal energy*, com-monly referred to as heat, is the most important factor driving metamorphism. It triggers chemical reactions that result in the recrystallization of existing minerals and the formation of new minerals. Thermal energy for metamor-phism comes mainly from two sources. Rocks experience a rise in temperature when they are intruded by magma ris-ing from below (contact metamorphism). In this situation, the adjacent host rock is "baked" by the emplaced magma.

By contrast, rocks that formed at Earth's surface will experience a gradual increase in temperature and pres-sure as they are taken to greater depths. In the upper crust, this increase in temperature averages about 25°C per kilometer. When buried to a depth of about 8 kilome-ters (5 miles), where temperatures are between 150° and 200°C, clay minerals tend to become unstable and begin to recrystallize into other minerals, such as chlorite and muscovite, that are stable in this environment. (Chlorite is a mica-like mineral formed by the metamorphism of iron- and magnesium-rich silicates.) However, many silicate minerals, particularly those found in crystalline igneous rocks—quartz and feldspar, for example—re-main stable at these temperatures. Thus, these minerals require much higher temperatures in order to meta-morphose and recrystallize.

Confining Pressure and Differential Stress as Metamorphic Agents Pressure, like tempera-ture, increases with depth as the thickness of the overly-ing rock increases. Buried rocks are subjected to **confin-ing pressure**—similar to water pressure in that the forces are equally applied in all directions (**Figure 2.31A**). The deeper you go in the ocean, the greater the confin-ing pressure. The same is true for buried rock. Confin-ing pressure causes the spaces between mineral grains to close, producing a more compact rock that has greater density. Further, at great depths, confining pressure may cause minerals to recrys-tallize into new minerals that display more com-pact crystalline forms.

During episodes of mountain building, large rock bodies become highly crumpled and metamor-phosed (**Figure 2.31B**). Unlike confining pressure, which

SmartFigure 2.31 Confining pressure and differential stress (https://goo.gl/GbqzMz)

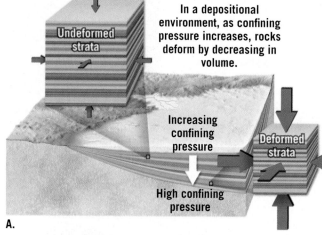

In a depositional environment, as confining pressure increases, rocks deform by decreasing in volume.

Undeformed strata

Increasing confining pressure

High confining pressure

Deformed strata

A.

During mountain building, rocks subjected to differential stress are shortened in the direction of maximum stress and lengthened in the direction of minimum stress.

Deformed strata

B.

"squeezes" rock equally in all directions, the forces that generate mountains are unequal in different directions and are called **differential stress**. As shown in Figure 2.31B, rocks subjected to differential stress are shortened in the direction of greatest stress, and they are elongated, or lengthened, in the direction perpendicular to that stress. The deformation caused by differential stresses plays a major role in developing metamorphic textures.

In surface environments where temperatures are relatively low, rocks are *brittle* and tend to fracture when subjected to differential stress. (Think of a heavy boot crushing a piece of fine crystal.) Continued deformation grinds and pulverizes the mineral grains into small fragments. By contrast, in high-temperature, high-pressure environments deep in Earth's crust, rocks are *ductile* and tend to flow rather than break. (Think of a heavy boot crushing a soda can.) When rocks exhibit ductile behavior, their mineral grains tend to flatten and elongate when subjected to differential stress. This accounts for their ability to generate intricate folds (see Figure 2.29).

Chemically Active Fluids as Metamorphic Agents

Ion-rich fluids composed mainly of water and other volatiles (materials that readily change to gases at surface conditions) are believed to play an important role in some types of metamorphism. Fluids that surround mineral grains act as catalysts that promote recrystallization by enhancing ion migration. In progressively hotter environments, these ion-rich fluids become correspondingly more reactive. Chemically active fluids can produce two types of metamorphism, explained below. The first type changes the arrangement and shape of mineral grains within a rock; the second type changes the rock's chemical composition.

When two mineral grains are squeezed together, the parts of their crystalline structures that touch are the most highly stressed. Atoms at these sites are readily dissolved by the hot fluids and move to fill the voids between individual grains. Thus, hot fluids aid in the recrystallization of mineral grains by dissolving material from regions of high stress and then precipitating (depositing) this material in areas of low stress. As a result, *minerals tend to recrystallize and grow longer in a direction perpendicular to compressional stresses.*

When hot fluids circulate freely through rocks, ionic exchange may occur between adjacent rock layers, or ions may migrate great distances before they are finally deposited. The latter situation is particularly common when we consider hot fluids that escape during the crystallization of an intrusive mass of magma. If the rocks surrounding the magma differ markedly in composition from the invading fluids, there may be a substantial exchange of ions between the fluids and host rocks. When this occurs, the overall composition of the surrounding rock changes.

Metamorphic Textures

The degree of metamorphism is reflected in a rock's *texture* and *mineralogy*. (Recall that the term *texture* is used to describe the size, shape, and arrangement of grains within a rock.) When rocks are subjected to low-grade metamorphism, they become more compact and thus denser. A common example is the metamorphic rock slate, which forms when shale is subjected to temperatures and pressures only slightly greater than those associated with the compaction that lithifies sediment. In this case, differential stress causes the microscopic clay minerals in shale to align into the more compact arrangement found in slate.

Under more extreme temperature and pressure, stress causes certain minerals to recrystallize. In general, recrystallization encourages the growth of larger crystals. Consequently, many metamorphic rocks consist of visible crystals, much like coarse-grained igneous rocks.

Foliation The term **foliation** refers to any nearly flat arrangement of mineral grains or structural features within a rock. Although foliation may occur in some sedimentary and even a few types of igneous rocks, it is a fundamental characteristic of regionally metamorphosed rocks—that is, rock units that have been strongly deformed, mainly during folding. As we see in **Figure 2.32**, foliation in metamorphic environments is ultimately driven by

SmartFigure 2.32 Rotation of platy and elongated mineral grains to produce foliated texture When subjected to differential stress during metamorphism, some mineral grains become reoriented and aligned at right angles to the stress. The resulting orientation of mineral grains gives the rock a foliated (layered) texture. If the coarse-grained igneous rock (granite) on the left underwent intense metamorphism, it could end up closely resembling the metamorphic rock on the right (gneiss). (Photos by E. J. Tarbuck) (https://goo.gl/vaWv6l)

Platy and elongated mineral grains having random orientation.

When differential stress causes rocks to flatten, the mineral grains rotate and align roughly perpendicular to the direction of maximum differential stress.

Animation

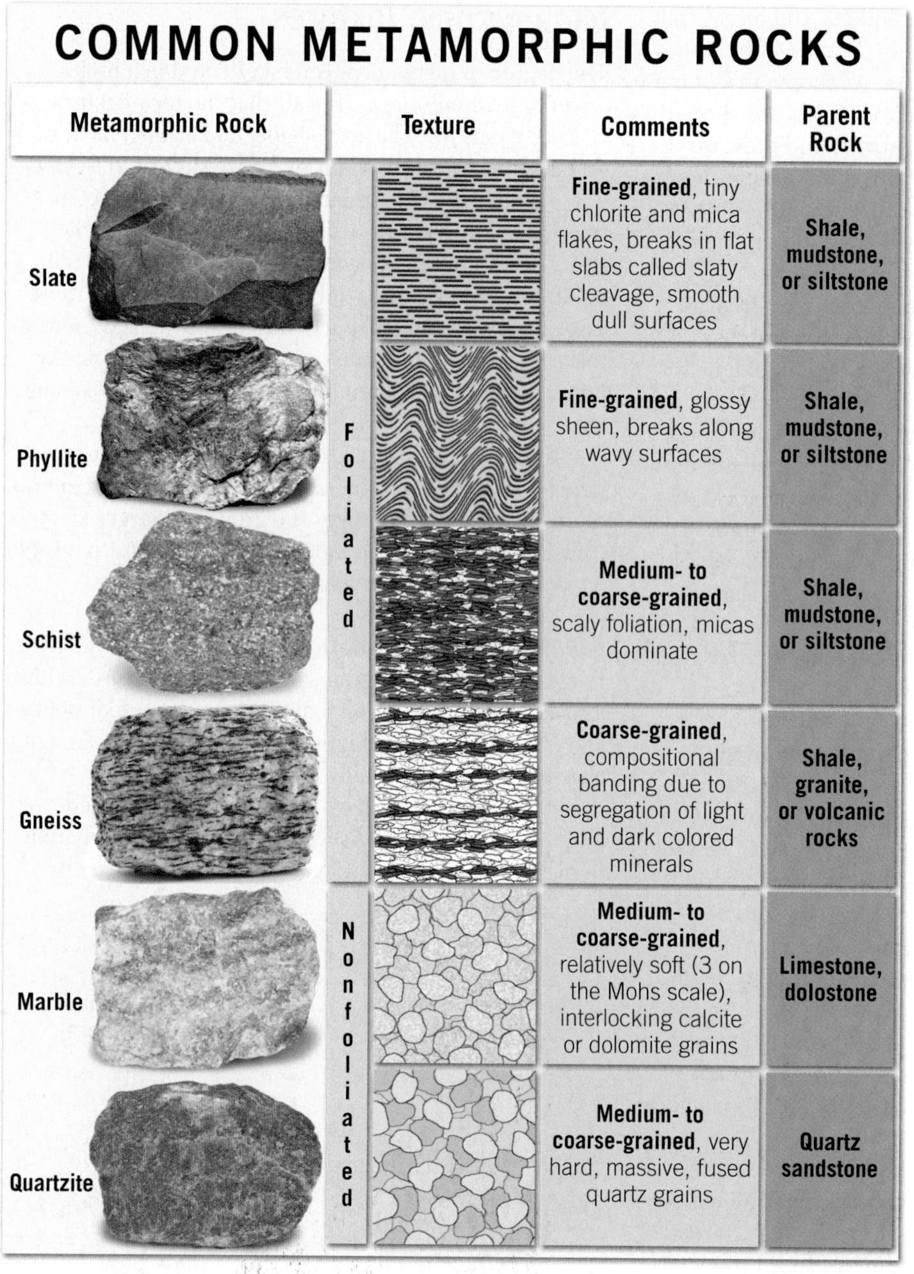

Figure 2.33 Classification of common metamorphic rocks (Photos by E. J. Tarbuck)

COMMON METAMORPHIC ROCKS

Metamorphic Rock	Texture	Comments	Parent Rock
Slate	Foliated	**Fine-grained**, tiny chlorite and mica flakes, breaks in flat slabs called slaty cleavage, smooth dull surfaces	**Shale, mudstone, or siltstone**
Phyllite	Foliated	**Fine-grained**, glossy sheen, breaks along wavy surfaces	**Shale, mudstone, or siltstone**
Schist	Foliated	**Medium- to coarse-grained**, scaly foliation, micas dominate	**Shale, mudstone, or siltstone**
Gneiss	Foliated	**Coarse-grained**, compositional banding due to segregation of light and dark colored minerals	**Shale, granite, or volcanic rocks**
Marble	Nonfoliated	**Medium- to coarse-grained**, relatively soft (3 on the Mohs scale), interlocking calcite or dolomite grains	**Limestone, dolostone**
Quartzite	Nonfoliated	**Medium- to coarse-grained**, very hard, massive, fused quartz grains	**Quartz sandstone**

environments where deformation is minimal and the parent rocks are composed of minerals that have a relatively simple chemical composition, such as quartz or calcite. For example, when a fine-grained limestone (made of calcite) is metamorphosed by the intrusion of a hot magma body (contact metamorphism), the small calcite grains recrystallize and form larger interlocking crystals. The resulting rock, *marble*, exhibits large, equidimensional grains that are randomly oriented, similar to those in a coarse-grained igneous rock.

Common Metamorphic Rocks

Figure 2.33 depicts the common rocks produced by metamorphic processes, which are described below.

Foliated Rocks **Slate** is a very fine-grained foliated rock composed of minute mica flakes that are too small to be visible (see Figure 2.33). A noteworthy characteristic of slate is its excellent rock cleavage, or tendency to break into flat slabs. This property has made slate a useful rock for roof and floor tile, as well as billiard tables (**Figure 2.34**). Slate is usually generated by the low-grade metamorphism of shale. Less frequently, it is produced when volcanic ash is metamorphosed. Slate's color is variable. Black slate contains organic material, red slate gets its color from iron oxide, and green slate is usually composed of chlorite, a greenish mica-like mineral.

Phyllite represents a degree of metamorphism between slate and schist. Its constituent platy minerals, mainly muscovite and chlorite, are larger than those in slate but not large enough to be readily identifiable with the unaided eye. Although phyllite appears similar to slate, it can be easily distinguished from slate by its glossy sheen and wavy surface (see Figure 2.33).

compressional stresses that shorten rock units, causing mineral grains in preexisting rocks to develop parallel, or nearly parrllel, alignments. Examples of foliation include the *parallel alignment of platy (flat and disk-like) minerals* such as the micas; *elongated* or *flattened pebbles* that are characteristic of metaconglomerates; *compositional banding*, in which dark and light minerals separate, generating a layered appearance; and *rock cleavage*, in which rocks can be easily split into tabular slabs. It is important to note that rock cleavage is not related to the mineral cleavage discussed in Chapter 1.

Nonfoliated Textures Not all metamorphic rocks exhibit a foliated texture. Those that do not are referred to as **nonfoliated** and typically develop in

Figure 2.34 Slate exhibits rock cleavage Because slate breaks into flat slabs, it has many uses. The larger image shows a quarry near Alta, Norway. (Photo by Fred Bruemmer/Getty Images) In the inset photo, slate is used to roof a house in Switzerland. (Photo by E. J. Tarbuck)

Schists are moderately to strongly foliated rocks formed by regional metamorphism (see Figure 2.33). They are platy and can be readily split into thin flakes or slabs. Many schists originate from shale parent rock. The term *schist* describes the *texture* of a rock regardless of composition. For example, schists composed primarily of muscovite and biotite are called *mica schists*.

Gneiss (pronounced "nice") is the term applied to banded metamorphic rocks in which elongated and granular (as opposed to platy) minerals predominate (see Figure 2.33). The most common minerals in gneisses are quartz and feldspar, with lesser amounts of muscovite, biotite, and hornblende. Gneisses exhibit strong segregation of light and dark silicates, giving them a characteristic banded texture. While still deep below the surface where temperatures and pressures are great, banded gneisses can be deformed into intricate folds.

Nonfoliated Rocks **Marble**

is a coarse, crystalline rock whose parent rock is limestone (see Figure 2.33). Marble is composed of large interlocking calcite crystals formed from the recrystallization of smaller grains in the parent rock. Because of its color and relative softness (hardness of only 3 on the Mohs scale), marble is a popular building stone. White marble is particularly prized as a stone from which to carve monuments and statues, such as the Lincoln Memorial in Washington, DC, and the Taj Mahal in India (**Figure 2.35**). Marble can also be colored—pink, gray, green, or even black—if the parent rocks from which it formed contain impurities that color the stone.

Quartzite is a very hard metamorphic rock most often formed from quartz sandstone (see Figure 2.33).

Figure 2.35 Marble, because of its workability, is a widely used building stone The exterior of India's Taj Mahal is constructed primarily of the metamorphic rock marble. (Photo by Sam Dcruz/Shutterstock)

SmartFigure 2.36 Garnet-mica schist The dark red garnet crystals are embedded in a matrix of fine-grained micas. (Photo by E. J. Tarbuck) (http://goo.gl/KrkufS)

▶ **Mobile Field Trip**

Garnet crystals

Close-up

newly formed minerals, commonly referred to as *accessory minerals*, tend to form large crystals that are surrounded by smaller crystals of other minerals, such as muscovite and biotite. When naming a metamorphic rock that contains one or more easily recognizable accessory minerals, geologists add a prefix to the appropriate rock name. For example, **Figure 2.36** shows a mica schist that contains large dark red garnet crystals embedded in a matrix of fine-grained micas; consequently this rock is called a *garnet-mica schist*. The metamorphic rock gneiss also frequently contains accessory minerals, including garnet and staurolite. These rocks would be called *garnet gneiss* and *staurolite gneiss*, respectively.

Under moderate- to high-grade metamorphism, the quartz grains in sandstone fuse. Pure quartzite is white, but iron oxide may produce reddish or pinkish stains, and dark minerals may impart a gray color.

Naming Metamorphic Rocks

During intermediate- to high-grade metamorphism, recrystallization of existing minerals often produces new minerals that are mainly associated with metamorphic rocks, for example the mineral *garnet*. These

2.5 CONCEPT CHECKS

1. Metamorphism means "to change form." Describe how a rock may change during metamorphism.
2. Explain what is meant by the statement "every metamorphic rock has a parent rock."
3. List the four agents of metamorphism and describe the role of each.
4. Distinguish between regional and contact metamorphism.
5. What feature would easily distinguish schist and gneiss from quartzite and marble?
6. In what ways do metamorphic rocks differ from the igneous and sedimentary rocks from which they formed?

CONCEPTS IN REVIEW
Rocks: Materials of the Solid Earth

2.1 Earth as a System: The Rock Cycle

Sketch, label, and explain the rock cycle.

KEY TERM: rock cycle

• The rock cycle is a good model for thinking about the transformation of one rock to another due to Earth processes. Igneous rocks form when molten rock solidifies. Sedimentary rocks are made from weathered products of other rocks. Metamorphic rocks are the products of preexisting rocks subjected to conditions of high temperatures and/or pressures. Given the right sequence of conditions, any rock type can be transformed into any other type of rock.

❓ **Name the processes that are represented by each of the letters (A-E) in this rock cycle diagram.**

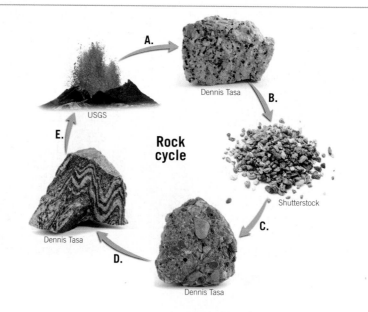

A.

Dennis Tasa

B.

USGS

E.

Rock cycle

Shutterstock

C.

Dennis Tasa

D.

Dennis Tasa

2.2 Igneous Rocks: "Formed by Fire"

Describe the two criteria used to classify igneous rocks and explain how the rate of cooling influences the crystal size of minerals.

KEY TERMS: igneous rock, magma, lava, extrusive (volcanic) rock, intrusive (plutonic) rock, granitic (felsic) composition, basaltic (mafic) composition, andesitic (intermediate) composition, ultramafic, peridotite, texture, fine-grained texture, coarse-grained texture, porphyritic texture, phenocryst, groundmass, vesicular texture, glassy texture, pyroclastic (fragmental) texture, granite, rhyolite, obsidian, pumice, andesite, diorite, basalt, gabbro, Bowen's reaction series, crystal settling, magmatic differentiation

- Completely or partly molten rock is called magma if it is below Earth's surface and lava if it has erupted onto the surface. It consists of a liquid melt that contains gases (volatiles) such as water vapor, and it may contain solids (mineral crystals).
- Magmas that cool at depth produce intrusive igneous rocks, whereas those that erupt onto Earth's surface produce extrusive igneous rocks.
- In geology, texture refers to the size, shape, and arrangement of mineral grains in a rock. Careful observation of the texture of igneous rocks can lead to insights about the conditions under which they formed. Lava on or near the surface cools rapidly, resulting in a large number of very small crystals that gives the rock a fine-grained texture. Magma at depth is insulated by the surrounding rock and cools very slowly. This allows sufficient time for the magma's ions to organize into larger crystals, resulting in a rock with a coarse-grained texture. If crystals begin to form at depth and then the magma rises to a shallow depth or erupts at the surface, it will have a two-stage cooling history. The result is a rock with a porphyritic texture.
- Pioneering experimentation by N. L. Bowen revealed that as magma cools, minerals crystallize in a specific order. The dark-colored silicate minerals, such as olivine, crystallize first at the highest temperatures (1250°C [2300°F]), whereas the light silicates, such as quartz, crystallize last at the lowest temperatures (650°C [1200°F]). Separation of minerals by mechanisms such as crystal settling results in igneous rocks having a wide variety of chemical compositions.
- Igneous rocks are classified into compositional groups based on the percentage of dark and light silicate minerals they contain. Granitic (or felsic) rocks such as granite and rhyolite are composed mostly of the light-colored silicate minerals potassium feldspar and quartz. Rocks of andesitic (or intermediate) composition such as andesite contain plagioclase feldspar and amphibole. Basaltic (or mafic) rocks such as basalt contain abundant pyroxene and calcium-rich plagioclase feldspar.

2.3 Weathering of Rocks to Form Sediment

Define *weathering* and distinguish between the two main categories of weathering.

KEY TERMS: mechanical weathering, frost wedging, sheeting, exfoliation dome, chemical weathering, carbonic acid

- Mechanical weathering is the physical breaking up of rock into smaller pieces. Rocks can be broken into smaller fragments by frost wedging, salt crystal growth, unloading, and biological activity. In addition, rocks that form under high pressure deep in Earth will expand when exposed at the surface. This can cause the rock to fracture in onion-like layers, a process called sheeting, which can generate broad, dome-shaped exposures of rock called exfoliation domes.
- Chemical weathering alters a rock's chemistry, changing it into different substances. Water is by far the most important agent of chemical weathering. Oxygen in water can oxidize some materials, while carbon dioxide (CO_2) dissolved in water forms carbonic acid. The chemical weathering of silicate minerals produces soluble products containing sodium, calcium, potassium, and magnesium; insoluble iron oxides; and clay minerals.

(?) **Which category of weathering is represented by the broken glass in this image? What about the rusty cans?**

Michael Collier

2.4 Sedimentary Rocks: Compacted and Cemented Sediment

List and describe the different categories of sedimentary rocks and discuss the processes that change sediment into sedimentary rock.

KEY TERMS: sedimentary rock, sediment, detrital sedimentary rock, conglomerate, breccia, sandstone, shale, siltstone, chemical sedimentary rock, biochemical sedimentary rock, limestone, coquina, travertine, evaporite deposit, coal, lithification, compaction, cementation, strata (beds), fossil

- Although igneous and metamorphic rocks make up most of Earth's crust by volume, sediment and sedimentary rocks are concentrated near the surface.
- Detrital sedimentary rocks are made of solid particles, mostly quartz grains and microscopic clay minerals. Common detrital sedimentary rocks include shale (the most abundant sedimentary rock), sandstone, and conglomerate.
- Chemical and biochemical sedimentary rocks are derived from mineral matter (ions) that is carried in solution to lakes and seas. Under certain conditions, ions in solution precipitate (settle out) to form chemical sediments as a result of physical processes, such as evaporation. Precipitation may also occur indirectly through life processes of water-dwelling organisms that form materials called biochemical sediments. Many water-dwelling animals and plants extract dissolved mineral matter to form shells and other hard parts. After the organisms die, their skeletons may accumulate on the floor of a lake or an ocean.
- Limestone, an abundant sedimentary rock, is composed chiefly of the mineral calcite ($CaCO_3$). Rock gypsum and rock salt are chemical rocks that form as water evaporates.

- Coal forms from the burial of large amounts of plant matter in low-oxygen depositional environments such as swamps and bogs.
- The transformation of sediment into sedimentary rock is called lithification. The two main processes that contribute to lithification are compaction (a reduction in pore space by packing grains more tightly together) and cementation (a reduction in pore space by adding new mineral material that acts as a "glue" to bind the grains to each other).

Dennis Tasa

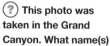 **This photo was taken in the Grand Canyon. What name(s) do geologists use for the characteristic features found in the sedimentary rocks shown in this image?**

2.5 Metamorphic Rocks: New Rock from Old

Define *metamorphism*, explain how metamorphic rocks form, and describe the agents of metamorphism.

KEY TERMS: metamorphic rock, metamorphism, contact metamorphism, regional metamorphism, confining pressure, differential stress, foliation, nonfoliated, slate, phyllite, schist, gneiss, marble, quartzite

- When rocks are subjected to elevated temperatures and pressures, they can change form, producing metamorphic rocks. Every metamorphic rock has a parent rock—the rock it used to be prior to metamorphism. When the minerals in parent rocks are subjected to heat and pressure, new minerals can form. Depending on the intensity of alteration, metamorphism ranges from low grade to high grade.
- Heat, confining pressure, differential stress, and chemically active fluids are four agents that drive metamorphic reactions. Any one alone may trigger metamorphism, or all four may act simultaneously.
- Confining pressure results from burial. The force it exerts is the same in all directions, like the pressure exerted by water on a diver. An increase in confining pressure causes rocks to compact into more dense configurations.
- Differential stresses, which occur during mountain building, are greater in one direction than in others. Rocks subjected to differential stress under ductile conditions deep in the crust tend to shorten in the direction of greatest stress and elongate in the direction(s) of least stress, producing flattened or stretched grains. In the shallow crust, most rocks respond to differential stress with brittle deformation, breaking into pieces.
- A common kind of texture is foliation, the planar arrangement of mineral grains. Common foliated metamorphic rocks include (in order of increasing metamorphic grade) slate, phyllite, schist, and gneiss.
- Common nonfoliated metamorphic rocks include quartzite and marble, recrystallized rocks that form from quartz sandstone and limestone respectively.

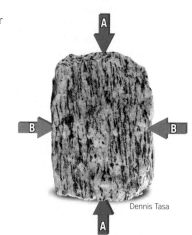

Dennis Tasa

Examine the photograph. Determine whether this rock is foliated or nonfoliated and then determine whether it formed under confining pressure or from differential stress. Which of the pairs of arrows shows the direction of maximum stress?

GIVE IT SOME THOUGHT

1. Refer to Figure 2.1. How does the rock cycle diagram—in particular, the labeled arrows—support the fact that sedimentary rocks are the most abundant rock type on Earth's surface?

2. Would you expect all the crystals in an intrusive igneous rock to be the same size? Explain why or why not.

3. Is it possible for two igneous rocks to have the same mineral composition but be different rocks? Support your answer with an example.

4. Use your understanding of magmatic differentiation to explain how magmas of different composition can be generated in a cooling magma chamber.

5. Give two reasons sedimentary rocks are more likely to contain fossils than igneous rocks.

6. If you hiked to a mountain peak and found limestone at the top, what would that indicate about the likely geologic history of the rock there?

7. Apply your understanding of igneous rock textures to describe the cooling history of each of the igneous rocks labeled A–D.

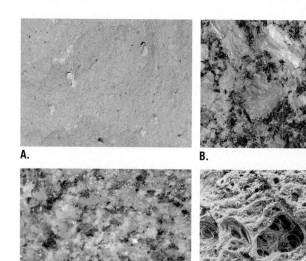

8. One of the accompanying photos shows an outcrop of metamorphic rock; the other two show igneous and sedimentary outcrops, respectively. Which do you think is the metamorphic rock? Explain why you ruled out the other rock bodies. (Photos by E. J. Tarbuck)

A.

B.

C.

9. Dust collecting on furniture is an everyday example of a sedimentary process. Provide another example of a sedimentary process that might be observed in or around where you live.

10. Examine the accompanying photos, which show the geology of the Grand Canyon. Notice that most of the walls of the canyon consist of layers of sedimentary rocks, but if you were to hike down into what is known as the Inner Gorge, you would encounter the Vishnu Schist, which is metamorphic rock.

  a. What process might have been responsible for the formation of the Vishnu Schist? How does this process differ from the processes that formed the sedimentary rocks that are atop the Vishnu Schist?

  b. What does the Vishnu Schist tell you about the history of the Grand Canyon prior to the formation of the canyon itself?

  c. Why is the Vishnu Schist visible at Earth's surface?

  d. Is it likely that rocks similar to the Vishnu Schist exist elsewhere but are not exposed at Earth's surface? Explain.

A. Inner Gorge of the Grand Canyon

B. Close up of Vishnu Schist (dark color)

MasteringGeology™

www.masteringgeology.com Looking for additional review and test prep materials? With individualized coaching on the toughest topics of the course, MasteringGeology offers a wide variety of ways for you to move beyond memorization to begin thinking like a geologist. Visit the Study Area in www.masteringgeology.com to find practice quizzes, study tools, and multimedia that will improve your understanding of this chapter's content. Sign in today to enjoy the following features: **Self Study Quizzes, SmartFigure: Tutorials/Animations/Condor Videos/Mobile Field Trips, Geoscience Animation Library, GEODe, RSS Feeds, Digital Study Modules,** and an optional **Pearson eText.**

3

FOCUS ON CONCEPTS

Each statement represents the primary learning objective for the corresponding major heading within the chapter. After you complete the chapter, you should be able to:

3.1 List three important external processes and describe where they fit into the rock cycle.

3.2 Explain the role of mass wasting in the development of valleys and discuss the factors that trigger and influence mass-wasting processes.

3.3 List the hydrosphere's major reservoirs and describe the different paths that water takes through the hydrologic cycle.

3.4 Describe the nature of drainage basins and river systems.

3.5 Discuss streamflow and the factors that cause it to change.

3.6 Outline the ways in which streams erode, transport, and deposit sediment.

3.7 Contrast bedrock and alluvial stream channels. Distinguish between two types of alluvial channels.

3.8 Contrast narrow V-shaped valleys, broad valleys with floodplains, and valleys that display incised meanders.

3.9 Discuss the formation of deltas and natural levees.

3.10 Discuss the causes of floods and some common flood control measures.

3.11 Discuss the importance of groundwater and describe its distribution and movement.

3.12 Compare and contrast springs, wells, and artesian systems.

3.13 List and discuss three important environmental problems associated with groundwater.

3.14 Explain the formation of caverns and the development of karst topography.

The Colorado River winding through Canyonlands National Park in southern Utah. When this meandering path was established, the river flowed across a relatively flat landscape. Subsequently, the region was gradually lifted upward while downward erosion lowered the riverbed. The meandering pattern persists, but the loops, locked within confining walls, are now referred to as incised meanders. (Photo by Michael Collier)

LANDSCAPES FASHIONED BY WATER

E arth is a dynamic planet. Internal forces such as those that create mountains elevate the land, while opposing external processes continually wear it down. The Sun and gravity drive external processes occurring at Earth's surface. Rock is disintegrated and decomposed, moved to lower elevations by gravity, and carried away by water, wind, or glacial ice. Collectively these processes sculpt the physical landscape. This chapter deals with some of these external processes. After a brief examination of mass wasting, we focus on the part of the hydrologic cycle in which water moves from the land back to the sea. Some water travels quickly via rushing streams, and some moves much more slowly, beneath the surface. When viewed as part of the Earth system, streams and groundwater are basic links in the constant cycling of the planet's water.

3.1 Earth's External Processes

List three important external processes and describe where they fit into the rock cycle.

Weathering, mass wasting, and erosion are called **external processes** because they occur at or near Earth's surface and are powered by energy from the Sun. External processes are a basic part of the rock cycle because they are responsible for transforming solid rock into sediment.

To the casual observer, Earth's surface may appear to be without change, unaffected by time. In fact, 200 years ago, most people believed that mountains, lakes, and deserts were permanent features of an Earth that was thought to be no more than a few thousand years old. Today we know Earth is about 4.6 billion years old and that mountains eventually succumb to weathering and erosion, lakes fill with sediment or are drained by streams, and deserts come and go with changes in climate.

Earth is a dynamic body. Some parts of Earth's surface are gradually elevated by mountain building and volcanic activity. These **internal processes** derive their energy from Earth's interior. Meanwhile, opposing external processes are continually breaking rock apart and moving the debris to lower elevations. The latter processes include:

1. **Weathering**—the physical breakdown (disintegration) and chemical alteration (decomposition) of rock at or near Earth's surface
2. **Mass wasting**—the transfer of rock and soil downslope, under the influence of gravity
3. **Erosion**—the physical removal of material by a mobile agent such as flowing water, waves, wind, or glacial ice

Weathering processes were treated in Chapter 2. In this chapter, we will focus on mass wasting and on two important erosional processes—erosion by running water and by groundwater. Chapter 4 will examine the geologic work of two other significant erosional processes—erosion by glacial ice and by wind.

3.1 CONCEPT CHECKS

1. Distinguish between internal and external processes.
2. Contrast weathering, mass wasting, and erosion.

3.2 Mass Wasting: The Work of Gravity

Explain the role of mass wasting in the development of valleys and discuss the factors that trigger and influence mass-wasting processes.

Earth's surface is never perfectly flat but instead consists of slopes. Some are steep and precipitous; others are moderate or gentle. Some are long and gradual; others are short and abrupt. Some slopes are mantled with soil and covered by vegetation; others consist of barren rock and rubble. Their form and variety are great.

Although most slopes appear to be stable and unchanging, they are not static features because the force of gravity causes material to move downslope. At one extreme, the movement may be gradual and practically imperceptible. At the other extreme, it may consist of a roaring debris flow or a thundering rock avalanche. Landslides are a worldwide natural hazard. When these natural processes lead to loss of life and property, they become natural disasters (**Figure 3.1**).

Figure 3.1 Watch out for falling rocks! Road signs in mountain areas often warn of this mass-wasting hazard. On February 22, 2011, rock weakened by weathering, combined with precipitous slopes, and the shock of an earthquake to produce the rockfall pictured here near Christchurch, New Zealand. (Photo by Marty Melville/AFP/Getty Images)

Mass Wasting and Landform Development

Events such as the one that created the scene in Figure 3.1 are spectacular examples of a common geologic process called mass wasting. *Mass wasting* is the downslope movement of rock and soil under the direct influence of gravity. It is distinct from erosional processes because mass wasting does not require a transporting medium, such as water, wind, or glacial ice. There is a broad array of mass-wasting processes. Four of them are illustrated in **Figure 3.2**.

The Role of Mass Wasting Mass wasting is the step that follows weathering in the evolution of most landforms. Once weathering weakens and breaks apart rock, mass wasting transfers the debris downslope, where a stream, acting as a conveyor belt, usually carries it away

A. Slump: Downward sliding of a mass of rock or unconsolidated material moving as a unit along a curved surface.

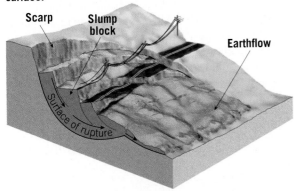

B. Rockslide: Blocks of bedrock break loose and slide very rapidly downslope.

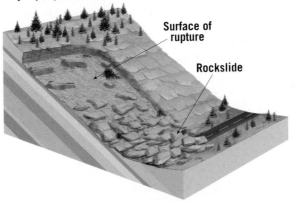

C. Debris flow: A flow of weathered debris containing a large amount of water. Often confined to channels. Sometimes called a mudflow.

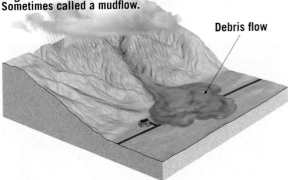

D. Earthflow: The tongue-like flow of water-saturated clay-rich soil on a hillside that breaks away and moves downslope.

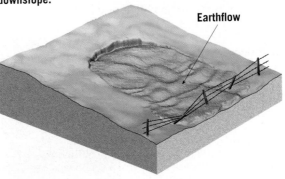

SmartFigure 3.2 Types of mass wasting The four processes illustrated here are all considered to be relatively rapid forms of mass wasting. Because the material in slumps and rockslides moves along well-defined surfaces, it is said to move by sliding. The animation illustrates the sliding movement of a slump. By contrast, when material moves like a thick fluid, it is described as a flow. Debris flow and earthflow are examples. (https://goo.gl/Y604zh)

Animation

SmartFigure 3.3
Excavating the Grand Canyon The walls of the canyon extend far from the channel of the Colorado River. This results primarily from the transfer of weathered debris downslope to the river and its tributaries by mass-wasting processes. (Photo by Bryan Brazil/Shutterstock) (https://goo.gl/9geHci)

Tutorial

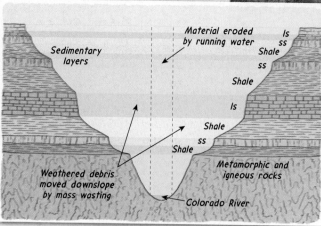
Geologist's Sketch

Did You Know?
The cost of damages by all types of mass wasting in the United States is conservatively estimated to exceed $2 billion in an average year.

(**Figure 3.3**). Although there may be many intermediate stops along the way, most of the sediment is eventually transported to the sea. The combined effects of mass wasting and running water produce stream valleys, which are the most common and conspicuous landforms at Earth's surface and the focus of a later part of this chapter.

If streams alone were responsible for creating the valleys in which they flow, valleys would be very narrow features. However, the fact that most river valleys are much wider than they are deep is a strong indication of the significance of mass-wasting processes in supplying material to streams. The walls of a canyon extend far from the river because of the transfer of weathered debris downslope to the river and its tributaries by mass-wasting processes. In this manner, streams and mass wasting combine to modify and sculpt the surface. Of course, glaciers, groundwater, waves, and wind are also important agents in shaping landforms and developing landscapes.

Slopes Change Through Time Most rapid and spectacular mass-wasting events occur in areas of rugged, geologically young mountains. Newly formed mountains are rapidly eroded by rivers and glaciers into regions characterized by steep and unstable slopes. It is in such settings that massive destructive landslides occur. As mountain

building subsides, mass-wasting and erosional processes lower the land. Over time, steep and rugged mountain slopes give way to gentler, more subdued terrain. Thus, as a landscape ages, massive and rapid mass-wasting processes give way to smaller, less dramatic downslope movements.

Controls and Triggers of Mass Wasting

Gravity is the controlling force of mass wasting, but several factors play important roles in overcoming inertia and creating downslope movements. Long before a landslide occurs, various processes work to weaken slope material, gradually making it more and more susceptible to the pull of gravity. During this span, the slope remains stable but gets closer and closer to being unstable. Eventually, the strength of the slope is weakened to the point that something causes it to cross the threshold from stability to instability. Such an event that initiates downslope movement is called a **trigger**. Remember that the trigger is not the sole cause of a mass-wasting event; it is just the last of many causes. Among the common factors that trigger mass-wasting processes are saturation of material with water, oversteepening of slopes, removal of anchoring vegetation, and ground vibrations from earthquakes.

The Role of Water

Mass wasting is sometimes triggered when heavy rains or periods of snowmelt saturate surface materials. Such a situation is shown in **Figure 3.4**. When the pores in soil and sediment become filled with water, two things

SmartFigure 3.4 Debris flow in the Colorado Front Range During the week of September 9–13, 2013, residents in and near Boulder, Colorado, received a harsh reminder of the dangers posed by debris flows. During that 5-day span, nearly continuous rainfall triggered numerous flash floods and more than 1100 debris flows in an area covering more than 3400 square kilometers (1300 square miles). Most occurred on slopes steeper than 25 degrees. (Photo by Rick Wilking/Reuters) (http://goo.gl/M7j5H6)

Mobile Field Trip

happen. First, the water reduces the cohesion among particles, allowing them to slide past each other more easily. For example, slightly moist sand sticks together well, allowing you to build a sand castle, but if you add enough water to fill the pore spaces between grains, the sand will ooze like a fluid. Similarly, wet clay is notoriously slick. Second, water adds weight, making the material likelier to slide or flow downslope.

Oversteepened Slopes If you pile up dry sand, you will find that it forms a slope with a specific incline called the **angle of repose** (Figure 3.5). Any kind of unconsolidated granular material (sand size or coarser) will behave similarly. Depending on the size and shape of the particles, the angle of repose varies from 25 to 40 degrees, with larger or more angular particles supporting steeper slopes. If you try to make such a slope steeper than its angle of repose, material will eventually move downslope until the angle is reestablished. A slope that is steeper than its stable angle is said to be *oversteepened*.

Although materials such as cohesive soils and bedrock do not have a specific angle of repose, slopes made of such materials can also be oversteepened, and they will eventually respond through mass wasting. In fact, oversteepening is a common trigger for mass wasting in nature, as when a stream undercuts a valley wall or waves erode the base of a cliff. Human activities can also create oversteepened slopes.

Removal of Vegetation Plants protect against erosion and contribute to the stability of slopes because their root systems bind soil particles together. Where plants are lacking, mass wasting is enhanced, especially if slopes are steep and water is plentiful. When anchoring vegetation is removed by forest fires or by people (for timber, farming, or development), surface materials frequently move downslope.

Earthquakes as Triggers Among the most important and dramatic mass-wasting triggers are earthquakes. An

Angle of repose

earthquake and its aftershocks can dislodge enormous volumes of rock and unconsolidated material. In many areas that are jolted by earthquakes, it is not ground vibrations directly but landslides and ground subsidence triggered by the vibrations that cause the greatest damage. The scene in Figure 3.1 was triggered by an earthquake.

Landslides Without Triggers? Do rapid mass-wasting events always require some sort of trigger, such as heavy rains or an earthquake? The answer is "no." Many rapid mass-wasting events occur without discernible triggers. Slope materials gradually weaken over time, under the influence of long-term weathering, infiltration of water, and other physical processes. Eventually, if the strength falls below what is necessary to maintain slope stability, a landslide will occur. The timing of such events is random, and accurate prediction is not possible.

3.2 CONCEPT CHECKS

1. How does water affect mass-wasting processes?
2. Describe the significance of the angle of repose.
3. How might a forest fire influence mass wasting?
4. Link earthquakes and landslides.

3.3 The Hydrologic Cycle

List the hydrosphere's major reservoirs and describe the different paths that water takes through the hydrologic cycle.

Water is continually on the move, from the ocean to the atmosphere to the land and back again, in an endless cycle called the **hydrologic cycle**. The remainder of this chapter deals with the part of the cycle that returns water to the sea. Some water travels by way of rivers, and some moves more slowly, beneath the surface. We will examine the factors that influence the distribution and movement of water, as well as look at how water sculpts the landscape. The Grand Canyon, Niagara Falls, the Old Faithful geyser, and Mammoth Cave all owe their existence to water making its way to the sea.

Earth's Water Water is almost everywhere on Earth—in the oceans, glaciers, rivers, lakes, air, rock, soil, and living tissue. All of these "reservoirs" constitute Earth's hydrosphere. In all, the water content of the hydrosphere is an estimated 1.36 billion cubic kilometers (326 million cubic miles). The vast bulk of it, about 96.5 percent, is stored in the global ocean. Ice sheets and glaciers account for an additional 1.76 percent, leaving just slightly more than 2 percent to be divided among lakes, streams, groundwater, and the atmosphere (see Figure I.7). Although the percentage of Earth's total water found in each of the latter

Did You Know?
Each year, a field of crops may transpire the equivalent of a water layer 60 cm (2 ft) deep over the entire field. The same area of trees may pump twice this amount into the atmosphere.

SmartFigure 3.6 The hydrologic cycle The primary movement of water through the cycle is shown by the large arrows. Numbers refer to the annual amount of water taking a particular path.
(https://goo.gl/8IRwFJ)

Tutorial

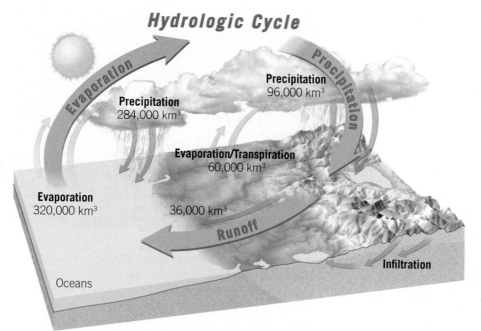

Hydrologic Cycle

Evaporation

Precipitation

Precipitation
284,000 km³

Precipitation
96,000 km³

Evaporation/Transpiration
60,000 km³

Evaporation
320,000 km³

36,000 km³

Runoff

Infiltration

Oceans

Storage in Glaciers When precipitation falls in very cold places—at high elevations or high latitudes—the water may not immediately soak in, run off, or evaporate. Instead, it may become part of a snowfield or a glacier. In this way, glaciers (especially the huge ice sheets covering Antarctica and Greenland) store large quantities of water on land. As you will see in Chapter 4, over the past 2 million years, huge ice sheets have formed and melted on several occasions, each time affecting the balance of the water cycle.

sources is just a small fraction of the total inventory, the absolute quantities are great.

Water's Paths The hydrologic cycle is a gigantic, worldwide system powered by energy from the Sun, in which the atmosphere provides a vital link between the oceans and continents (**Figure 3.6**). **Evaporation**, the process by which liquid water changes into water vapor (gas), is how water enters the atmosphere from the ocean and, to a much lesser extent, from the continents. Winds transport this moisture-laden air, often great distances. Complex processes of cloud formation eventually result in precipitation. The precipitation that falls into the ocean has completed its cycle and is ready to begin another. The water that falls on land, however, makes its way back to the ocean.

What happens to precipitation once it has fallen on land? A portion of the water soaks into the ground (called **infiltration**), slowly moving downward, then moving laterally, finally seeping into lakes, into streams, or directly into the ocean. When the rate of rainfall exceeds Earth's ability to absorb it, the surplus water flows over the surface into lakes and streams, a process called **runoff**. Much of the water that infiltrates or runs off eventually returns to the atmosphere because of evaporation from the soil, lakes, and streams. Also, some of the water that soaks into the ground is absorbed by plants, which then release it into the atmosphere. This process is called **transpiration**. Because we cannot clearly distinguish between the amount of water that is evaporated and the amount that is transpired by plants, the term **evapotranspiration** is often used for the combined effect.

Water Balance Figure 3.6 shows the volume of water that passes through each part of the cycle annually. The quantities that are cycled through the atmosphere over a 1-year period are immense—enough to cover Earth's entire surface to a depth of about 1 meter (39 inches).

The hydrologic cycle is balanced, which means the average annual precipitation worldwide equals the quantity of water that enters the atmosphere by evapotranspiration. However, notice that for all land areas taken together, precipitation exceeds evaporation, whereas over the oceans, evaporation exceeds precipitation. Because the level of the world ocean is not dropping, the system must be in balance. Balance is achieved because 36,000 cubic kilometers (8600 cubic miles) of water annually makes its way from the land back to the ocean.

About one-quarter of global precipitation falls on land and flows on and below the surface. This water is *the most important force sculpting Earth's land surface.* In the rest of this chapter, we will observe the work of water running over the surface, including floods, erosion, and the formation of valleys. Then we will look underground at the slow labors of groundwater as it forms springs and caverns and provides water for people on its long migration to the sea.

(3.3) CONCEPT CHECKS

1. Describe or sketch the movement of water through the hydrologic cycle. Once precipitation has fallen on land, what paths might the water take?
2. What is meant by the term *evapotranspiration*?
3. Over the oceans, evaporation exceeds precipitation, yet sea level does not drop. Explain this.

Did You Know?
According to the American Water Works Association, daily indoor per capita water use in an average American home is 69.3 gallons. Toilets (18.5 gallons), clothes washers (15 gallons), and showers and baths (13 gallons) are the top three uses. Leaks account for more than 9 gallons per home per day.

3.4 Running Water

Describe the nature of drainage basins and river systems.

Recall that most of the precipitation that falls on land either enters the soil (infiltration) or remains at the surface, moving downslope as runoff. The amount of water that runs off rather than soaking into the ground depends on several factors: (1) intensity and duration of rainfall, (2) amount of water already in the soil, (3) nature of the surface material, (4) slope of the land, and (5) extent and type of vegetation. When the surface material is highly impermeable or when it becomes saturated, runoff is the dominant process. Runoff is also high in urban areas because large areas are covered by impermeable buildings, roads, and parking lots.

Runoff initially flows in broad, thin sheets across hillslopes. This unconfined flow eventually develops threads of current that form tiny channels called *rills*. Where rills merge, the flowing water creates *gullies*, which join to form larger stream channels. At first streams are small, but as one intersects another, larger and larger streams form. Eventually they merge into rivers that carry water from a broad region.

Drainage Basins

The land area that contributes water to a river system is called a **drainage basin** or **watershed** (**Figure 3.7**). The drainage basin of one stream is separated from the drainage basin of another by an imaginary line called a **divide**. Divides range in scale from a ridge separating two small gullies on a hillside to a *continental divide*, which splits a whole continent into enormous drainage basins. The Mississippi River has the largest drainage

basin in North America (**Figure 3.8**). Extending between the Rocky Mountains in the West and the Appalachian Mountains in the East, the Mississippi River and its tributaries collect water from more than 3.2 million square kilometers (1.2 million square miles) of the continent.

Figure 3.7 Drainage basin and divide A drainage basin, also called a watershed, is the area drained by a stream and its tributaries. Boundaries between basins are called divides.

SmartFigure 3.8
Mississippi River drainage basin The drainage basin of the Mississippi River forms a funnel that stretches from Montana and southern Canada in the west to New York State in the east and runs down to a spout in Louisiana. It consists of many smaller drainage basins. The drainage basin of the Yellowstone River is one of many that contribute water to the Missouri River, which, in turn, is one of many that make up the drainage basin of the Mississippi River. (https://goo.gl/z6epSn)

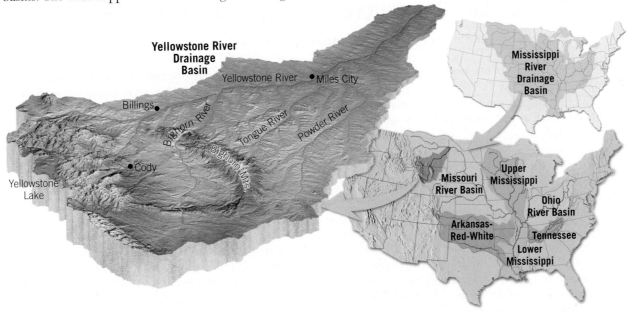

Tutorial

Figure 3.9 Zones of a river Each of the three zones is based on the dominant process that is operating in that part of the river system.

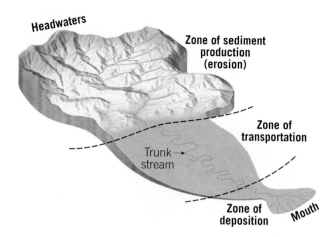

River Systems

River systems involve not only a network of stream channels but their entire drainage basin. Based on the dominant processes operating within them, river systems can be divided into three zones: *sediment production—* where erosion dominates—*sediment transport*, and *sediment deposition* (**Figure 3.9**). It is important to recognize that sediment is being eroded, transported, and deposited along the entire length of a stream, regardless of which process is dominant within each zone.

The zone of *sediment production*, where most of the water and sediment is derived, is located in the headwaters region of the river system. Much of the sediment carried by streams begins as bedrock that is subsequently broken down by weathering and then transported downslope by mass wasting and overland

flow. Bank erosion can also contribute significant amounts of sediment. In addition, scouring of the channel bed deepens the channel and adds to the stream's sediment load.

Sediment acquired by a stream is then transported through the channel network, along sections referred to as *trunk streams*. When trunk streams are in balance, the amount of sediment eroded from their banks equals the amount deposited elsewhere in the channel. Although trunk streams rework their channels over time, they are not a source of sediment.

When a river reaches the ocean or another large body of water, it slows, and the energy to transport sediment is greatly reduced. Most of the sediments accumulate at the mouth of the river to form a delta, are reconfigured by wave action to form a variety of coastal features, or are moved far offshore by ocean currents. Because coarse sediments tend to be deposited upstream, it is primarily the fine sediments (clay and fine sand) that eventually reach the ocean. Taken together, erosion, transportation, and deposition are the processes by which rivers move Earth's surface materials and sculpt landscapes.

3.4 CONCEPT CHECKS

1. List several factors that cause infiltration and runoff to vary from place to place and from time to time.
2. Draw a simple sketch of a drainage basin and a divide. Label each.
3. What are the three main parts (zones) of a river system?

3.5 Streamflow Characteristics

Discuss streamflow and the factors that cause it to change.

Water may flow in one of two ways—either as **laminar flow** or **turbulent flow**. In slow-moving streams, the flow is often laminar, meaning that the water moves in roughly straight-line paths that parallel the stream channel (**Figure 3.10A**). However, streamflow is usually turbulent, with the water moving in an erratic fashion that can be characterized as a swirling motion. Strong, turbulent flow may be seen in whirlpools and eddies, as well as rolling whitewater rapids (**Figure 3.10B**). Even streams that appear smooth on the surface often exhibit turbulent flow near the bottom and sides of the channel. Turbulence contributes to a stream's ability to erode its channel because it acts to lift sediment from the streambed.

An important factor influencing stream turbulence is the water's flow velocity. As the velocity of a stream increases, the flow becomes more turbulent. Flow

velocities can vary significantly from place to place along a stream, as well as over time, in response to variations in the amount and intensity of precipitation. If you have ever waded into a stream, you may have noticed that the strength of the current increased as you moved into deeper parts of the channel. This is related to the fact that frictional resistance is greatest near the banks and bed of the stream channel.

Factors Affecting Flow Velocity

The ability of a stream to erode and transport material is directly related to its flow velocity. Even slight variations in flow rate can lead to significant changes in the load of sediment that water can transport. Several factors influence flow velocity and, therefore, control a stream's

Running the rapids in the Grand Canyon— an extreme example of turbulent flow.

B.

This water is not standing still. It is moving slowly toward the bottom of the image. The flow in the foreground is primarily laminar.

A.

Figure 3.10 Laminar and turbulent flow Most often stream-flow is turbulent. (Photos by Michael Collier)

of several kilometers. The size of a stream channel is largely determined by the amount of water supplied from the drainage basin. The measure most often used to compare the sizes of streams is **discharge**—the volume of water flowing past a certain point in a given unit of time. Discharge, usually measured in cubic meters per second or cubic feet per second, is determined by multiplying a stream's cross-sectional area by its velocity.

The largest river in North America, the Mississippi, discharges an average of 16,800 cubic meters (nearly 600,000 cubic feet) per second (**Figure 3.11**). Although this is a huge quantity of water, it is dwarfed by the mighty Amazon in South America, the world's largest river. Fed by a vast rainy region that is nearly three-fourths the size of the conterminous United States, the Amazon discharges more than 12 times more water than the Mississippi.

The discharges of most rivers are far from constant. This is the case because of variables such as rainfall and snowmelt. In areas with seasonal variations in precipitation, streamflow tends to be highest during the wet season

potential to do "work." These factors include (1) channel slope, or gradient, (2) channel size and cross-sectional shape, (3) channel roughness, and (4) the amount of water flowing in the channel.

Gradient The slope of a stream channel expressed as the vertical drop of a stream over a specified distance is the **gradient**. Portions of the lower Mississippi River have very low gradients of 10 centimeters per kilometer or less. By contrast, some mountain stream channels decrease in elevation at a rate of more than 40 meters per kilometer—a gradient 400 times steeper than the lower Mississippi. Gradient varies not only among different streams but also over a particular stream's length. The steeper the gradient, the more energy available for streamflow. If two streams were identical in every respect except gradient, the stream with the higher gradient would have the greater velocity.

Channel Shape, Size, and Roughness A stream's channel is a conduit that guides the flow of water, but the water encounters friction as it flows. The shape, size, and roughness of the channel affect the amount of friction. Larger channels have more rapid flow because a smaller proportion of water is in contact with the channel. A smooth channel promotes a more uniform flow, whereas an irregular channel filled with boulders creates enough turbulence to slow the stream significantly.

Discharge Streams vary in size from small headwater creeks less than 1 meter wide to large rivers with widths

SmartFigure 3.11 The Mighty Mississippi near Helena, Arkansas The Mississippi is North America's largest river. From head to mouth, it is nearly 3900 kilometers (2400 miles) long. Its watershed encompasses about 40 percent of the lower 48 states and includes all or parts of 31 states and 2 Canadian provinces. Average discharge at its mouth is about 16,800 cubic meters (593,000 cubic feet) per second. (Photo by Michael Collier) (http://goo.gl/LkgZXX)

Mobile Field Trip

Figure 3.12 Longitudinal profile California's Kings River originates high in the Sierra Nevada and flows into the San Joaquin Valley.

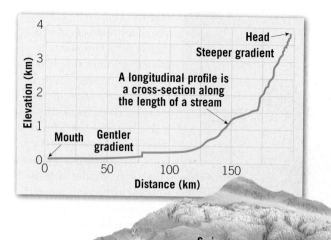

Changes from Upstream to Downstream

One useful way of studying a stream is to examine its **longitudinal profile**—a cross-sectional view of a stream from its source area (called the *head* or *headwaters*) to its *mouth*, the point downstream where the river empties into another water body—a river, a lake, or an ocean. As shown in **Figure 3.12**, the most obvious feature of a typical profile is a constantly decreasing gradient from the head to the mouth. Although many local irregularities may exist, the overall profile is a relatively smooth concave curve.

The change in slope observed on most stream profiles is usually accompanied by an increase in discharge and channel size, as well as a reduction in sediment particle size (**Figure 3.13**). Along most rivers in humid regions, discharge increases toward the mouth because as we move downstream, more and more tributaries contribute water to the main channel. In order to accommodate the growing volume of water, channel size typically increases downstream as well. Recall that flow velocities are higher in large channels than in small channels. Observations also show a general decline in sediment size downstream, making the channel smoother and more efficient (less friction).

Although the gradient decreases toward a stream's mouth, the flow velocity generally increases. This contradicts our intuitive assumptions of swift, narrow headwater streams and wide, placid rivers flowing across more subtle topography. Increases in channel size and discharge, as well as decreases in channel roughness that occur downstream compensate for the decrease in gradient, thereby making the stream more efficient. Thus the average flow velocity is typically lower in headwater streams than in wide, placid-appearing rivers.

or during spring snowmelt, and it is lowest during the dry season or during periods when high temperatures increase water losses through evaporation. Streams also sometimes dry up. Streams that exhibit flow only during wet periods are referred to as *intermittent streams*. In arid climates, many streams carry water only occasionally after a heavy rainstorm; these are called *ephemeral streams*.

SmartFigure 3.13 Channel changes from head to mouth Although the gradient decreases toward the mouth of a stream, increases in discharge and channel size and decreases in roughness more than offset the decrease in slope. Consequently, flow velocity usually increases toward the mouth. (https://goo.gl/6srX2s)

Tutorial

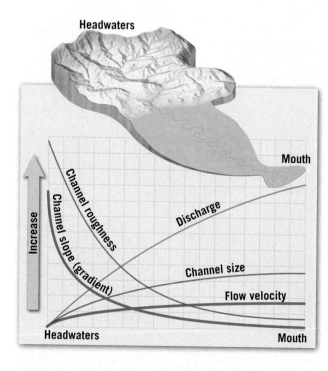

(3.5) CONCEPT CHECKS

1. Distinguish between laminar flow and turbulent flow.
2. Summarize the factors that influence flow velocity.
3. What is a longitudinal profile of a stream?
4. How do gradient, discharge, channel size, and channel roughness typically change from the head to the mouth of a stream?
5. Is flow velocity usually greater at the head or at the mouth of a stream? Explain.

3.6 The Work of Running Water

Outline the ways in which streams erode, transport, and deposit sediment.

Streams are Earth's most important erosional agents. Not only do they have the ability to downcut and widen their channels, but streams also have the capacity to transport the enormous quantities of sediment that are delivered to them by overland flow, mass wasting, and groundwater. Eventually, much of this material is deposited to create a variety of depositional features.

Stream Erosion

A stream's ability to accumulate and transport soil and weathered rock is sometimes aided by the work of raindrops, which knock sediment particles loose (**Figure 3.14**). When the ground is saturated, rainwater begins to flow downslope, transporting some of the material it has dislodged. On barren slopes the flow of muddy water will often create small channels that grow larger as one merges with another.

Once surface flow reaches a stream, the water's ability to erode is greatly enhanced by the increase in volume. When the flow of water is sufficiently strong, *hydraulic lifting* occurs, in which particles are dislodged from the channel and incorporated into the moving water. In this manner, the force of running water swiftly erodes poorly consolidated materials on the bed and sides of a stream channel. On occasion, the banks of the channel may be undercut, dumping

Abrasion by long-lived whirlpools armed with sand and pebbles creates bowl-shaped bedrock depressions called potholes.

Figure 3.15 Potholes The rotational motion of swirling pebbles acts like a drill to create potholes. (Photo by StormStudio/ Alamy)

Figure 3.14 Raindrop impact Soil dislodged by raindrops is more easily moved by water flowing across the surface. (USDA)

Raindrops may strike the surface at velocities approaching 35 km per hour. When a drop strikes an exposed surface, soil particles may splash as high as one meter and land more than a meter away from the point of raindrop impact.

even more loose debris into the water to be carried downstream.

In addition to eroding unconsolidated materials, flowing water can also cut a channel into solid bedrock. The process by which the bed and banks of a bedrock channel are bombarded by particles carried in the stream is called *abrasion*. These particles vary in size, from gravel in fast-flowing waters to sand in somewhat slower flows. Just as the particles of grit on sandpaper can wear away a piece of wood, so too can the sand and gravel carried by a stream abrade a bedrock channel. Moreover, pebbles caught in swirling eddies can act like "drills" and bore circular **potholes** into the channel floor (**Figure 3.15**).

Bedrock channels formed in soluble rock such as limestone are susceptible to *corrosion*—a process in which rock is gradually dissolved by the flowing water. Corrosion is a type of chemical weathering that occurs between the solutions in the stream water and the mineral matter composing the bedrock.

SmartFigure 3.16
Transport of sediment
Streams transport their load of sediment in three ways. The dissolved and suspended loads are carried in the general flow. The bed load includes coarse sand, gravel, and boulders that move by rolling, sliding, and saltation. (https://goo.gl/F9TOv1)

Animation

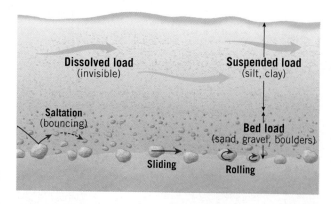

SmartFigure 3.16
Transport of sediment
Streams transport their load of sediment in three ways. The dissolved and suspended loads are carried in the general flow. The bed load includes coarse sand, gravel, and boulders that move by rolling, sliding, and saltation. (https://goo.gl/F9TOv1)

Transportation of Sediment by Streams

All streams, regardless of size, transport some rock material (**Figure 3.16**). Streams also sort the solid sediment they transport because finer, lighter material is carried more readily than larger, heavier particles. Streams transport their load of sediment in three ways: (1) in solution (**dissolved load**), (2) in suspension (**suspended load**), and (3) by sliding, skipping, or rolling along the bottom (**bed load**).

Dissolved Load Most of the *dissolved load* is brought to a stream by groundwater and is dispersed throughout the flow. When water percolates through the ground, it acquires soluble soil compounds. Then it seeps through cracks and pores in bedrock, dissolving additional mineral matter. Eventually, much of this mineral-rich water finds its way into streams.

Velocity has essentially no effect on a stream's ability to carry its dissolved load; material in solution goes wherever the water flows. Deposition (precipitation) of the dissolved mineral matter occurs when the chemistry of the water changes. In arid regions, the water may enter an inland lake or sea and evaporate, leaving behind the dissolved load.

Suspended Load Most streams carry the largest part of their load in *suspension* (**Figure 3.17**). The visible cloud of sediment suspended in the water is the most obvious portion of a stream's load. Usually, only fine particles consisting of silt and clay can be carried this way, but during a flood, larger particles can also be transported in suspension. Also during a flood, the total quantity of material carried in suspension increases dramatically, as can be verified by anyone whose home has been a site for the deposition of this material.

Bed Load A portion of a stream's load consists of sand, gravel, and occasionally large boulders. These coarser particles, which are too large to be carried in suspension, move along the bottom (bed) of the stream channel and constitute the *bed load*. Unlike the suspended and dissolved loads, which are constantly in motion, the bed load is in motion only intermittently, when the force of the water is sufficient

to move the larger particles. Many smaller particles, mainly sand and gravel, move by *saltation*, which resembles a series of jumps or skips. Larger particles either roll or slide along the bottom, depending on their shape.

Capacity and Competence A stream's ability to carry solid particles is described using two criteria: *capacity* and *competence*. **Capacity** is the maximum load of solid particles a stream can transport per unit of time. The greater the discharge, the greater the stream's capacity for hauling sediment. Consequently, large rivers with high flow velocities have large capacities.

Competence is a measure of a stream's ability to transport particles based on size rather than quantity. Flow velocity is the key: Swift streams have greater competencies than slow streams, regardless of channel size. A stream's competence increases proportionally to the square of its velocity. Thus, if the velocity doubles, the impact force of the water increases four times; if the velocity triples, the force increases nine times, and so forth. Consequently, large boulders that are often visible during low water and seem immovable can, in fact, be transported during exceptional floods because of the stream's increased competence.

By now it should be clear why the greatest erosion and transportation of sediment occur during floods. The increase in discharge results in a greater capacity, and the increase in velocity results in greater competence. With rising velocity, the water becomes more turbulent, and larger and larger particles are set in motion. In just a few

Figure 3.17 Suspended load An aerial view of the Colorado River in the Grand Canyon. Heavy rains washed sediment into the river. (Photo by Michael Collier)

This river's muddy appearance is a result of suspended sediment.

days or perhaps a few hours, a flooding stream can erode and transport more sediment than it does during months of normal flow.

Deposition of Sediment by Streams

When a stream slows down, the situation reverses: As its velocity decreases, the stream's competence is reduced, and sediment begins to drop out, largest particles first. Each particle size has a *critical settling velocity*. As streamflow drops below the critical settling velocity of a certain particle size, sediment in that category begins to settle out. Thus, stream transport provides a mechanism by which solid particles of various sizes are separated. This process, called **sorting**, explains why particles of similar size are deposited together.

The material deposited by a stream is called **alluvium**, the general term for *any* stream-deposited sediment. Many different depositional features are composed of alluvium. Some occur within stream channels, some occur on the valley floor adjacent to the channel, and some exist at the mouth of the stream. We will consider the nature of these features later in the chapter.

3.6 CONCEPT CHECKS

1. List two ways in which streams erode their channels.
2. In what three ways does a stream transport its load? Which part of the load moves most slowly?
3. What is the difference between capacity and competence?
4. What is alluvium?

3.7 Stream Channels

Contrast bedrock and alluvial stream channels. Distinguish between two types of alluvial channels.

A basic characteristic of streamflow that distinguishes it from overland flow is that it is confined to a channel. A stream channel can be thought of as an open conduit consisting of the streambed and banks that act to confine the flow except during floods.

Although this is somewhat oversimplified, we can divide stream channels into two types. A *Bedrock channel* is one in which the stream is actively cutting into solid rock. In contrast, when the bed and banks are composed mainly of unconsolidated sediment, the channel is called an *alluvial channel.*

Bedrock Channels

In their headwaters, where the gradient is steep, many rivers cut into bedrock. These streams typically transport coarse particles that actively abrade the bedrock channel. Potholes are often visible evidence of the erosional forces at work.

Bedrock channels typically alternate between relatively gently sloping segments where alluvium tends to accumulate and steeper segments where bedrock is exposed. These steeper areas may contain rapids or occasionally a waterfall. The channel pattern exhibited by streams cutting into bedrock is controlled by the underlying geologic structure. Even when flowing over rather uniform bedrock, streams tend to exhibit a winding or irregular pattern rather than flow in a straight channel. Anyone who has gone on a white-water rafting trip has observed the steep, winding nature of a stream flowing in a bedrock channel.

Alluvial Channels

Many stream channels are composed of loosely consolidated sediment (alluvium) and therefore can undergo significant changes in shape because the sediments are continually being eroded, transported, and redeposited. The major factors affecting the shapes of these channels are the average size of the sediment being transported and the channel's gradient and discharge.

Alluvial channel patterns reflect a stream's ability to transport its load at a uniform rate, while expending the least amount of energy. Thus, the size and type of sediment being carried help determine the nature of the stream channel. Two common types of alluvial channels are *meandering channels* and *braided channels.*

Meandering Channels Streams that transport much of their load in suspension generally move in sweeping bends called **meanders**. These streams flow in relatively deep, smooth channels and transport mainly mud (silt and clay). The lower Mississippi River exhibits a channel of this type.

Because of the cohesiveness of consolidated mud, the banks of stream channels carrying fine particles tend to resist erosion. As a consequence, most of the erosion in such channels occurs on the outside of the meander, where velocity and turbulence are greatest. In time, the outside bank is undermined, especially during periods of high water. Because the outside of a meander is a zone of active erosion, it is often referred

Did You Know?
The world's highest uninterrupted waterfall is Angel Falls on Venezuela's Churun River. Named for American aviator Jimmie Angel, who first sighted the falls from the air in 1933, the river plunges 979 m (3212 ft).

SmartFigure 3.18
Formation of cut banks and point bars By eroding its outer bank and depositing material on the inside of the bend, a stream is able to shift its channel. (https://goo.gl/4bXxsu)

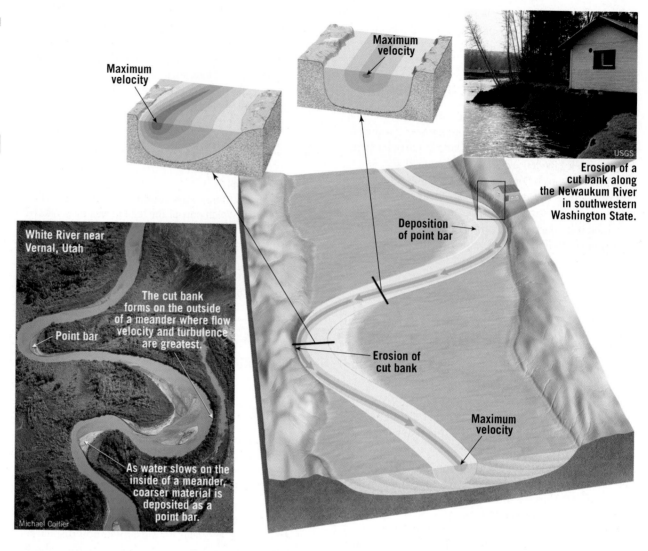

White River near Vernal, Utah

Point bar

The cut bank forms on the outside of a meander where flow velocity and turbulence are greatest.

As water slows on the inside of a meander, coarser material is deposited as a point bar.

Michael Collier

Maximum velocity

Maximum velocity

Deposition of point bar

Erosion of cut bank

Maximum velocity

USGS

Erosion of a cut bank along the Newaukum River in southwestern Washington State.

to as the **cut bank** (**Figure 3.18**). The debris acquired by the stream at a cut bank moves downstream, and the coarser material is generally deposited as **point bars** in zones of decreased velocity on the insides of meanders. In this manner, meanders migrate laterally by eroding the outside of the bends and depositing on the inside.

In addition to migrating laterally, the bends in a channel also migrate down the valley. This occurs because erosion is more effective on the downstream (downslope) side of the meander. The downstream migration of a meander is sometimes slowed when it reaches a more resistant material. This allows the next meander upstream to "catch up" and overtake it, as shown in **Figure 3.19**. The neck of land between the meanders is gradually narrowed. Eventually, the river may erode through the narrow neck of land to the next loop. The new, shorter channel segment is called a **cutoff** and, because of its shape, the abandoned bend is called an **oxbow lake**.

Braided Channels A stream may consist of a complex network of converging and diverging channels that thread their way among numerous islands or gravel bars. Because these channels have an interwoven appearance, they are said to be **braided channels**. Braided channels form where a large proportion of the stream's load consists of coarse material (sand and gravel) and the stream has a highly variable discharge. Because the bank material is readily erodible, braided channels are wide and shallow.

One setting in which braided streams form is at the end of a glacier, where there is a large seasonal variation in discharge (**Figure 3.20**). During summer, large amounts of ice-eroded sediment are dumped into the meltwater streams flowing away from the glacier. However, when flow is sluggish, the stream is unable to move all the sediment and therefore deposits the coarsest material as bars in the channel that force the flow to split and follow several paths. Usually the laterally shifting channels completely rework most of the

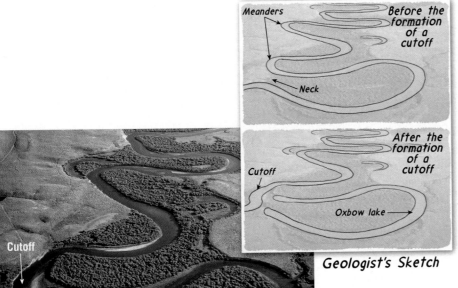

SmartFigure 3.19 Formation of an oxbow lake Oxbow lakes occupy abandoned meanders. Aerial view of an oxbow lake created by the meandering Green River near Bronx, Wyoming. (Photo by Michael Collier) (https://goo.gl/J1KomF)

Geologist's Sketch

In summary, meandering channels develop where the load consists largely of fine-grained particles that are transported as suspended load in a deep, smooth channel. By contrast, wide, shallow braided channels develop where coarse-grained alluvium is transported as bed load.

3.7 CONCEPT CHECKS

1. Are bedrock channels more likely to be found near the head or the mouth of a stream?
2. Describe or sketch the evolution of a meander, including how an oxbow lake forms.
3. Describe a situation that might cause a stream channel to become braided.

surface sediments each year, thereby transforming the entire streambed. In some braided streams, however, the bars have built up to form islands that are anchored by vegetation.

Figure 3.20 Braided stream The Knik River is a classic braided stream with multiple channels separated by migrating gravel bars. The Knik is choked with sediment from four melting glaciers in the Chugach Mountains north of Anchorage, Alaska. (Photo by Michael Collier)

3.8 Shaping Stream Valleys

Contrast narrow V-shaped valleys, broad valleys with floodplains, and valleys that display incised meanders.

Did You Know?
Not all rivers have deltas. Some streams that transport large sediment loads lack deltas because ocean waves and powerful currents redistribute the sediment. The Columbia River in the Pacific Northwest is one example. In other cases, rivers do not carry sufficient quantities of sediment to build a delta. The St. Lawrence River, for example, does not acquire much sediment between Lake Ontario and its mouth at the Gulf of St. Lawrence.

Streams, with the aid of weathering and mass wasting, shape the landscape through which they flow. As a result, streams continuously modify the valleys that they occupy.

A **stream valley** consists of not only the channel but also the surrounding terrain that directly contributes water to the stream. Thus, it includes the valley bottom, which is the lower, flatter area that is partially or totally occupied by the stream channel, and the sloping valley walls that rise above the valley bottom on both sides. Most stream valleys are much broader at the top than they are wide at their channel at the bottom. This would not be the case if the only agent responsible for eroding valleys were the streams flowing through them. The sides of most valleys are shaped by a combination of weathering, overland flow, and mass wasting. In some arid regions, where weathering is slow and where rock is particularly resistant, narrow valleys that have nearly vertical walls are common.

Stream valleys can be divided into two general types—narrow V-shaped valleys and wide valleys with flat floors—with many gradations between.

Base Level and Stream Erosion

Streams cannot endlessly erode their channels deeper and deeper. There is a lower limit to how deep a stream can erode, and that limit is called **base level**. Most often, a stream's base level occurs where a stream enters the ocean, a lake, or another stream.

Two general types of base level are recognized. Sea level is considered the *ultimate base level* because it is the lowest level to which stream erosion could lower the land. *Temporary*, or *local, base levels* include lakes, resistant layers of rock, and main streams that act as base level for their tributaries. For example, when a stream enters a lake, its velocity quickly approaches zero, and its ability to erode ceases. Thus, the lake prevents the stream from eroding below its level at any point upstream from the lake. However, because the outlet of the lake can cut downward and drain the lake, the lake is only a temporary hindrance to the stream's ability to lower its channel. In a similar manner, the layer of resistant rock at the lip of a waterfall acts as a temporary base level. Until the ledge of hard rock is eliminated, it will limit the amount of downcutting upstream.

Any change in base level will cause a corresponding readjustment of stream activities. When a dam is built along a stream, the reservoir that forms behind it raises the base level of the stream (**Figure 3.21**). Upstream from the dam, the gradient is reduced, lowering the stream's velocity and, hence, its sediment-transporting ability. The stream, now having too little energy to transport its entire load, will deposit sediment. This builds up its channel. Deposition will be the dominant process until the stream's gradient increases sufficiently to transport its load.

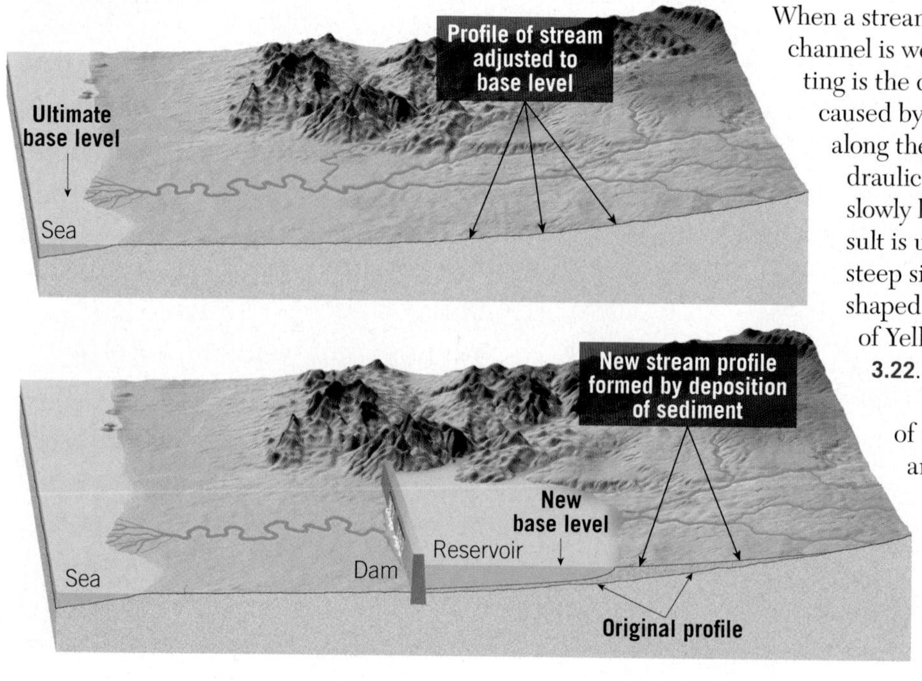

Figure 3.21 Building a dam The base level upstream from the reservoir is raised, which reduces the stream's flow velocity and leads to deposition and a reduced gradient.

Valley Deepening

When a stream's gradient is steep and the channel is well above base level, downcutting is the dominant activity. Abrasion caused by bed-load sliding and rolling along the bottom, along with the hydraulic power of fast-moving water, slowly lower the streambed. The result is usually a V-shaped valley with steep sides. A classic example of a V-shaped valley is located in the section of Yellowstone River shown in **Figure 3.22**.

The most prominent features of a V-shaped valley are *rapids* and *waterfalls*. Both occur where a stream's gradient increases significantly, a situation usually caused by variations in the erodibility of the bedrock into which a stream channel is cutting.

Figure 3.22 Yellowstone River The V-shaped valley, rapids, and waterfalls indicate that the river is vigorously downcutting. (Photo by Charles A. Blakeslee/AGE Fotostock)

Waterfall
of the
Yellowstone
River

Valley wall   Valley wall

Rapids   V-shaped valley
formed by vigorus
downcutting

Geologist's Sketch

Resistant beds create rapids by acting as a temporary base level upstream while allowing downcutting to continue downstream. In time, erosion usually eliminates the resistant rock. Waterfalls are places where the stream makes an abrupt vertical drop.

Valley Widening

Once a stream has cut its channel closer to base level, downward erosion becomes less dominant. At this point, the stream's channel takes on a meandering pattern, and more of the stream's energy is directed from side to side. The result is a widening of the valley as the river cuts away first at one bank and then at the other (**Figure 3.23**). The continuous lateral erosion caused by shifting of the stream's meanders produces an increasingly broad, flat valley floor covered with alluvium. This feature, called a **floodplain**, is appropriately named because when a river overflows its banks during flood stage, it inundates the floodplain.

Over time, the floodplain will widen to the point that the stream is actively eroding the valley walls in only a few places. In fact, in large rivers such as the lower Mississippi River valley, the distance from one valley wall to another can exceed 160 kilometers (100 miles).

Incised Meanders and Stream Terraces

We usually expect a stream with a highly meandering

course to be on a floodplain in a wide valley. However, certain rivers exhibit meandering channels that flow in steep, narrow valleys. Such meanders are called **incised** (*incisum* = to cut into) **meanders** (see the chapter-opening photo). How do such features form?

Originally, the meanders probably developed on the floodplain of a stream that was relatively near base level. Then, a change in base level caused the stream to begin downcutting. One of two events could have occurred. Either base level dropped, or the land on which the river flowed was uplifted.

An example of the first circumstance happened during the Ice Age, when large quantities of water were withdrawn from the ocean and locked up in glaciers on land. The result was that sea level (ultimate base level) dropped, causing meandering rivers flowing into the ocean to downcut.

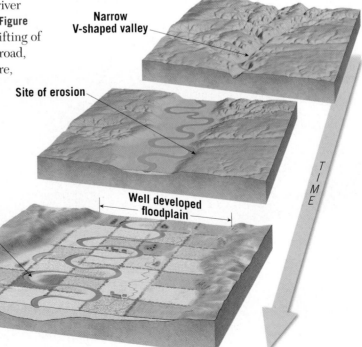

Narrow
V-shaped valley

Site of erosion

Site of
deposition

Well developed
floodplain

TIME

SmartFigure 3.23
Development of an erosional floodplain Continuous side-to-side erosion by shifting meanders gradually produces a broad, flat valley floor. Alluvium deposited during floods covers the valley floor. (https://goo.gl/WNfY8s)

Condor Video

SmartFigure 3.24 Stream terraces Terraces result when a stream adjusts to a relative drop in base level. (https://goo.gl/6cDynS)

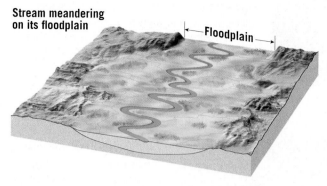

Stream meandering on its floodplain

← Floodplain →

Because of a relative drop in base level, the river erodes downward through previously deposited alluvium. Eventually a new floodplain forms. Terraces represent elevated remnants of the former floodplain.

Terrace   Terrace

Regional uplift of the land, the second cause of incised meanders, is exemplified by the Colorado Plateau in the southwestern United States. As the plateau was gradually uplifted, numerous meandering rivers adjusted to being higher above base level by downcutting. This is what created the incised meanders shown in the chapter opener.

Other features associated with a relative drop in base level are **stream terraces**. After a river that had been flowing on a floodplain has adjusted to a relative drop in base level, it may once again produce a floodplain at a level below the old one. As shown in **Figure 3.24**, the remnants of the former floodplain are present as relatively flat surfaces above the newly forming floodplain.

3.8 CONCEPT CHECKS

1. Define *base level* and distinguish between ultimate base level and temporary (local) base level.
2. Explain why V-shaped valleys often contain rapids and waterfalls.
3. Describe or sketch how an erosional floodplain develops.
4. Relate the formation of incised meanders and stream terraces to changes in base level.

3.9 Depositional Landforms

Discuss the formation of deltas and natural levees.

Recall that a stream continually picks up sediment in one part of its channel and deposits it downstream. Such channel deposits are most often composed of sand and gravel, and they are commonly referred to as **bars**. For example, in Figure 3.18, material acquired at the stream's cut bank is carried downstream and deposited as a point bar. These deposits are only temporary because the material will be picked up again and eventually carried to the ocean. In addition to sand and gravel bars, streams also create depositional features that have longer life spans. These include deltas and natural levees.

Figure 3.25 Formation of a simple delta Structure and growth of a simple delta that forms in relatively quiet waters.

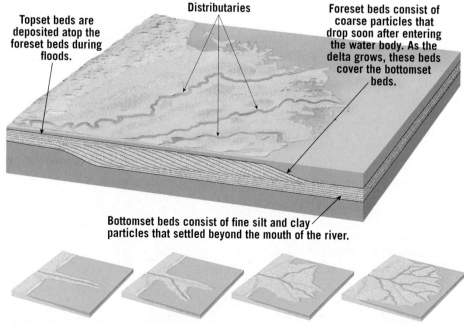

Distributaries

Topset beds are deposited atop the foreset beds during floods.

Foreset beds consist of coarse particles that drop soon after entering the water body. As the delta grows, these beds cover the bottomset beds.

Bottomset beds consist of fine silt and clay particles that settled beyond the mouth of the river.

As the stream extends its channel, the gradient is reduced. During flood stage some of the flow is diverted to a shorter, higher-gradient route forming a new distributary.

Deltas

A **delta** forms where a sediment-charged stream enters the relatively still waters of an ocean, a lake, or an inland sea (**Figure 3.25**). As the stream's forward motion slows, sediments are deposited by the dying current. As the delta grows outward, the stream's gradient continually lessens. This circumstance eventually causes the channel to become choked with sediment deposited from the slowing water. As a consequence, the river seeks a shorter, higher-gradient route to base level, as illustrated in Figure 3.25. This illustration shows the main channel dividing into several smaller ones, called **distributaries**. Most deltas are characterized by these shifting channels that act in an opposite way to that of tributaries.

Rather than carry water into the main channel, distributaries carry water *away* from the main channel. After numerous shifts of the channel, a delta may grow into a roughly triangular shape like the Greek letter delta (Δ), for which it is named. Note, however, that many deltas do not exhibit the idealized shape. Differences in the configurations of shorelines and variations in the nature and strength of wave activity result in many shapes. Many large rivers have deltas extending over thousands of square kilometers. The delta of the Mississippi River is one example. It resulted from the accumulation of huge quantities of sediment derived from the vast region drained by the river and its tributaries. Today, New Orleans rests where there was ocean less than 5000 years ago. **Figure 3.26** shows the portion of the Mississippi delta that has been built over the past 6000 years. As you can see, the delta is actually a series of seven coalescing subdeltas. Each formed when the river left its existing channel for a shorter, more direct path to the Gulf of Mexico. The individual subdeltas interfinger and partially cover one another, producing a very complex structure. The present subdelta, called a *bird-foot* delta because of the configuration of its distributaries, has been built by the Mississippi in the past 500 years.

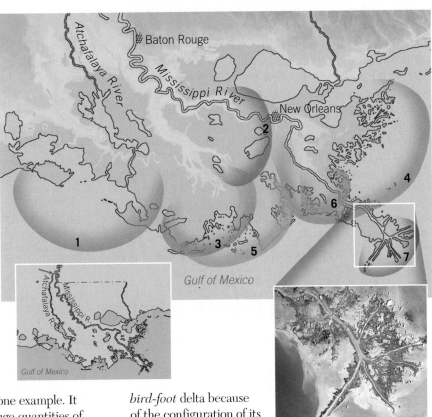

Figure 3.26 Growth of the Mississippi River delta During the past 6000 years, the river has built a series of seven coalescing subdeltas. The numbers indicate the order in which the subdeltas were deposited. The present bird-foot delta (number 7) represents the activity of the past 500 years. The left inset shows the point where the Mississippi may sometime break through (arrow) and the shorter path it would take to the Gulf of Mexico. (Image courtesy of JPL/Cal Tech/NASA)

Natural Levees

Meandering rivers that occupy valleys with broad floodplains tend to build **natural levees** that parallel their channels on both banks (**Figure 3.27**). Natural levees are built by years of successive floods. When a stream

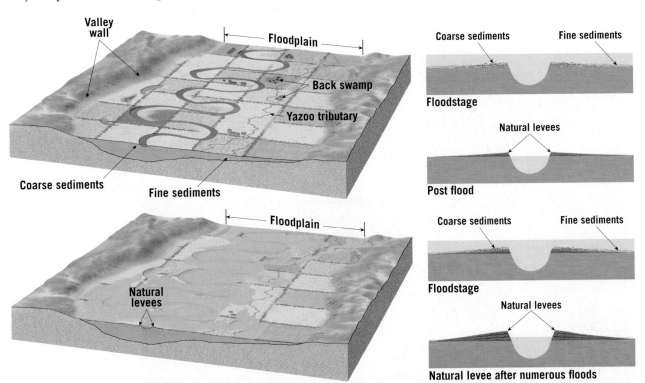

SmartFigure 3.27 Formation of a natural levee These gently sloping structures that parallel a river channel are created by repeated floods. Because the ground next to the channel is higher than the adjacent floodplain, back swamps and yazoo tributaries may develop. (https://goo.gl/7ZcsYJ)

Animation

overflows onto the floodplain, the water moves over the surface as a broad sheet. Because the flow velocity drops significantly, the coarser portion of the suspended load is immediately deposited adjacent to the channel. As the water spreads across the floodplain, a thin layer of fine sediment is laid down over the valley floor. This uneven distribution of material produces the gentle, almost imperceptible, slope of the natural levee.

The natural levees of the lower Mississippi rise 6 meters (20 feet) above the floodplain. The area behind a levee is characteristically poorly drained for the obvious reason that water cannot flow over the levee and into the river. Marshes called **back swamps** result. A tributary stream that cannot enter a river because levees block

the way often has to flow parallel to the river until it can breach the levee. Such streams are called **yazoo tributaries**, after the Yazoo River, which parallels the Mississippi for more than 300 kilometers (about 190 miles).

 3.9 CONCEPT CHECKS

1. What feature may form where a stream enters the relatively still waters of a lake, an inland sea, or an ocean?
2. What are distributaries, and why do they form?
3. Briefly describe the formation of a natural levee. How is this feature related to back swamps and yazoo tributaries?

3.10 Floods and Flood Control

Discuss the causes of floods and some common flood control measures.

When the discharge of a stream becomes so great that it exceeds the capacity of its channel, it overflows its banks as a **flood**. Floods are among the most common and most destructive of all natural hazards. They are, nevertheless, simply part of the *natural* behavior of streams.

Causes of Floods

Rivers flood because of the weather. Rapid melting of snow and/or major storms that bring heavy rains over a large areas cause most *regional floods*. In April 2011, unrelenting storms brought record rains to the Mississippi watershed. The Ohio Valley, which makes up the eastern portion of the Mississippi's drainage basin, received nearly

300 percent of its normal springtime precipitation. When that rainfall combined with water from the past winter's extensive and rapidly melting snowpack, the Mississippi River and many of its tributaries began to swell to record levels by early May. The resulting floods were among the largest and most damaging in nearly a century (**Figure 3.28**). Like most other regional floods, these were associated with weather phenomena that could be forecast with a good deal of accuracy. This allowed adequate time to warn and evacuate thousands of people who were in harm's way. Although economic losses approached $4 billion, loss of life was small.

Figure 3.28 Floods are dangerous natural hazards **A.** Extraordinary rains caused record flooding along the Mississippi River from Illinois to Louisiana in the spring of 2011. This scene is from Vicksburg, Mississippi. (Photo by Scott Olson/Getty Images) **B.** In most years floods are responsible for the greatest number of storm-related deaths. The average number of hurricane deaths was dramatically affected by Hurricane Katrina (more than 1000). For all other years on this graph, hurricane fatalities numbered fewer than 20.

Unlike extensive regional floods, *flash floods* are more limited in extent. Flash floods occur with little warning and can be deadly because they produce a rapid rise in water levels and can have a devastating flow velocity. Several factors influence flash flooding. Among them are rainfall intensity and duration, surface conditions, and topography. Urban areas are susceptible to flash floods because a high percentage of the surface area is composed of impervious roofs, streets, and parking lots, where runoff is very rapid (**Figure 3.29**). Mountainous areas are susceptible because steep slopes can quickly funnel runoff into narrow canyons.

Human interference with the stream system can worsen or even cause floods. A prime example is the failure of a dam or an artificial levee. These structures are built for flood protection. They are designed to contain floods of a certain magnitude. If a larger flood occurs, the dam or levee may be overtopped. If the dam or levee fails or is washed out, the water behind it is released and becomes a flash flood. The bursting of a dam in 1889 on the Little Conemaugh River caused the devastating Johnstown, Pennsylvania, flood that took some 3000 lives. A second dam failure occurred there in 1977, causing 77 fatalities.

Flood Control

Several strategies have been devised to eliminate or lessen the catastrophic effects of floods. Engineering efforts include the construction of artificial levees, the building of flood-control dams, and river channelization.

Artificial Levees Artificial levees are earthen mounds built on the banks of a river to increase the volume of water the channel can hold. These most common of stream-containment structures have been used since ancient times and continue to be used today. Artificial levees are usually easy to distinguish from natural levees because their slopes are much steeper. When exceptional floods threaten to overwhelm levees in densely populated areas, water is sometimes intentionally diverted from a river by creating openings in artificial levees. The purpose is to spare vulnerable urban areas by allowing water to flood sparsely populated rural areas. The areas that are intentionally flooded are called *floodways*. For example, to prevent the town of

Figure 3.29 Flash flood Although short-lived, flash floods can be powerful and often occur with little advance warning. More than half of U.S. flash-flood fatalities are automobile related. (Sue Ogrocki/AP Images)

Cairo, Illinois, from being inundated during the 2011 floods along the Mississippi River, an opening about 3 kilometers (2 miles) wide was blasted in a levee. This allowed water to spill into the 130,000-acre Birds Point–New Madrid Floodway. Similar steps were taken downstream in Louisiana to protect the cities of Baton Rouge and New Orleans.

Flood-Control Dams Flood-control dams are built to store floodwater and then let it out slowly. This lowers the flood crest by spreading it out over a longer time span. Since the 1920s, thousands of dams have been built on nearly every major river in the United States. Many dams have significant non-flood-related functions, such as providing water for irrigated agriculture and for hydroelectric power generation. Many reservoirs are also major regional recreational facilities.

Although dams may reduce flooding and provide other benefits, building these structures also involves significant costs and consequences. For example, reservoirs created by dams may cover fertile farmland, useful forests, historic sites, and scenic valleys. Of course, dams trap sediment. Therefore, deltas and floodplains downstream erode because they are no longer replenished with sediment during floods. Large dams can also cause significant ecologic damage to river environments that took thousands of years to establish.

Building a dam is not a permanent solution to flooding. Sedimentation behind a dam causes the volume of its reservoir to gradually diminish, reducing the effectiveness of this flood-control measure.

Channelization Channelization involves altering a stream channel to speed the flow of water to prevent it from reaching flood height. This may simply involve

Did You Know?
Urban development increases runoff. As a result, peak discharge and flood frequency increase in urban areas. Humans have covered an amazing amount of land with buildings, parking lots, and roads. A recent study indicated that the area of such impervious surfaces in the United States (excluding Alaska and Hawaii) amounts to more than 112,600 sq km (nearly 44,000 sq mi), which is slightly less than the area of the state of Ohio.

clearing a channel of obstructions or dredging a channel to make it wider and deeper.

A more radical alteration involves straightening a channel by creating *artificial cutoffs*. The idea is that by shortening the stream, the gradient and therefore also the velocity are increased. By increasing velocity, the larger discharge associated with flooding can be dispersed more rapidly.

Beginning in the early 1930s, the U.S. Army Corps of Engineers created many artificial cutoffs on the Mississippi for the purpose of increasing the efficiency of the channel and reducing the threat of flooding. In all, the river has been shortened more than 240 kilometers (150 miles). The program has been somewhat successful in reducing the height of the river in flood stage. However, because the river's tendency toward meandering still exists, preventing the river from returning to its previous course has been difficult.

A Nonstructural Approach All the flood-control measures described so far have involved structural solutions aimed at "controlling" a river. These solutions are expensive and often give people residing on the floodplain a false sense of security.

Today, many scientists and engineers advocate a nonstructural approach to flood control. They suggest that an alternative to artificial levees, dams, and channelization is sound floodplain management. By identifying high-risk areas, appropriate zoning regulations can be implemented to minimize development and promote more appropriate land use.

3.10 CONCEPT CHECKS

1. Contrast regional floods and flash floods.
2. Describe three basic flood-control strategies.
3. What is meant by a *nonstructural approach* to flood control?

3.11 Groundwater: Water Beneath the Surface

Discuss the importance of groundwater and describe its distribution and movement.

Groundwater is one of our most important and widely available resources. Yet people's perceptions of the subsurface environment from which it comes are often unclear and incorrect. The reason is that groundwater is hidden from view except in caves and mines, and the impressions people gain from these subsurface openings are often misleading. Observations on the land surface give an impression that Earth is "solid." This view is not changed very much when we enter a cave and see water flowing in a channel that appears to have been cut into solid rock.

Because of such observations, many people believe that groundwater occurs only in underground "rivers." But actual rivers underground are extremely rare. In reality, most of the subsurface environment is not "solid"

at all. Rather, it includes countless tiny *pore spaces* between grains of soil and sediment plus narrow joints and fractures in bedrock. Together, these spaces add up to an immense volume. Groundwater collects and moves in these tiny openings.

The Importance of Groundwater

Only a tiny percentage of Earth's water occurs underground. Nevertheless, this small percentage, stored in the rocks and sediments beneath Earth's surface, is a vast quantity. When the oceans are excluded and only sources of freshwater are considered, the significance of groundwater becomes more apparent.

Figure 3.30 shows estimates of the distribution of freshwater in the hydrosphere. Clearly, the largest volume occurs as glacial ice. Second in rank is groundwater, with about 30 percent of the total. However, when glacial ice is excluded and just liquid water is considered, about 96 percent is groundwater. Without question, *groundwater represents the largest reservoir of freshwater that is readily available to humans.* Its value in terms of economics and human well-being is incalculable.

Worldwide, wells and springs provide water for cities, crops, livestock, and industry. In the United States, groundwater is the source of about 40 percent of the water used for all purposes (except hydroelectric power generation and power plant cooling). Groundwater is the drinking water for about 44 percent of the population and provides 40 percent of the water used for irrigation. In some areas, however, overuse of this basic resource has caused serious problems, including streamflow

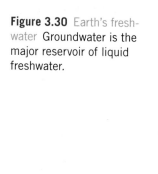

Figure 3.30 Earth's freshwater Groundwater is the major reservoir of liquid freshwater.

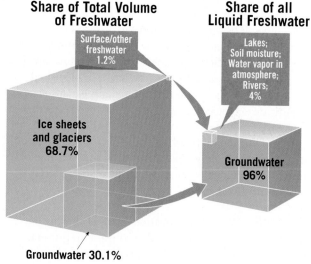

Share of Total Volume of Freshwater

Surface/other freshwater 1.2%

Ice sheets and glaciers 68.7%

Groundwater 30.1%

Share of all Liquid Freshwater

Lakes; Soil moisture; Water vapor in atmosphere; Rivers; 4%

Groundwater 96%

A.

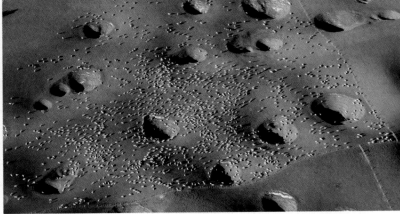

B.

Figure 3.31 Caverns and sinkholes **A.** A view of the interior of Kentucky's Mammoth Cave. The dissolving action of acidic groundwater created the caverns. Later, groundwater deposited the limestone decorations. (Photo by Clint Farlinger) **B.** Groundwater was responsible for creating these depressions, called sinkholes, west of Timaru on New Zealand's South Island. The white dots in this photo are grazing sheep. (Photo by David Wall/Alamy)

depletion, land subsidence, and increased pumping costs. In addition, groundwater contamination resulting from human activities is a real and growing threat in many places.

Groundwater's Geologic Roles

Geologically, groundwater is important as an erosional agent. The dissolving action of groundwater slowly removes soluble rock such as limestone, allowing surface depressions known as sinkholes to form and creating subterranean caverns (**Figure 3.31**). Groundwater is also an equalizer of streamflow. Much of the water that flows in rivers is not direct runoff from rain and snowmelt. Rather, a large percentage of precipitation soaks in and then moves slowly underground to stream channels. Groundwater is thus a form of storage that sustains streams during periods when rain does not fall. When we see water flowing in a river during a dry period, it is water from rain that fell at some earlier time and was stored underground.

Distribution of Groundwater

When rain falls, some of the water runs off, some returns to the atmosphere through evaporation and transpiration, and the remainder soaks into the ground. This last path is the primary source of practically all groundwater. The amount of water that takes each of these paths, however, varies greatly from time to time and place to place. Influential factors include the steepness of the slope, the nature of the surface material, the intensity of the rainfall, and the type and amount of vegetation. Heavy rains falling on steep slopes underlain by impervious materials will obviously result in a high percentage of the water running off. Conversely, if rain falls steadily and gently on more gradual slopes composed of materials that are more easily penetrated by water, a much larger percentage of the water soaks into the ground.

Underground Zones Some of the water that soaks in does not penetrate very far because

it is held by molecular attraction as a surface film on soil particles. This near-surface zone is called the *belt of soil moisture*. It is crisscrossed by roots, voids left by decayed roots, and animal and worm burrows that enhance the infiltration of rainwater into the soil. Soil water is used by plants for life functions and transpiration. Some of this water also evaporates back into the atmosphere.

Water that is not held as soil moisture moves downward until it reaches a zone where all the open spaces in sediment and rock are completely filled with water. This is the **zone of saturation**. Water within it is called **groundwater**. The top of this zone is known as the **water table**. The area above the water table where the soil, sediment, and rock are not saturated is called the **unsaturated zone** (**Figure 3.32**). Although a considerable amount of water can be present in the unsaturated zone, this water cannot be pumped by wells because it clings too tightly to rock and soil particles. By contrast, below the water table, the water pressure is great enough to allow water to enter wells, thus permitting groundwater to be withdrawn for use. We will examine wells more closely later in the chapter.

Water Table The water table is rarely level, as we might expect a table to be. Instead, its shape is usually

Figure 3.32 Water beneath the surface This diagram illustrates the relative positions of many features associated with subsurface water.

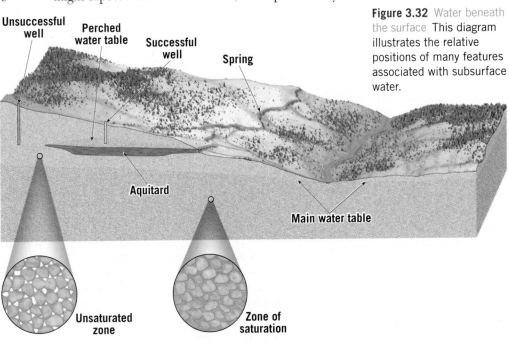

Unsuccessful well

Perched water table

Successful well

Spring

Aquitard

Main water table

Unsaturated zone

Zone of saturation

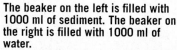

The beaker on the left is filled with 1000 ml of sediment. The beaker on the right is filled with 1000 ml of water.

The sediment-filled beaker now contains 500 ml of water. Pore spaces (porosity) must represent 50 percent of the volume of the sediment.

and, in sedimentary rocks, the amount of cementing material. Most igneous and metamorphic rocks, as well as some sedimentary rocks, are composed of tightly interlocking crystals, so the voids between grains may be negligible. In these rocks, fractures must provide the voids.

a subdued replica of the surface, reaching its highest elevations beneath hills and decreasing in height toward valleys (see Figure 3.32). The water table of a wetland (swamp) is right at the surface. Lakes and streams generally occupy areas low enough that the water table is above the land surface.

Several factors contribute to the irregular surface of the water table. One important influence is the fact that groundwater moves very slowly. Because of this, water tends to "pile up" beneath high areas between stream valleys. If rainfall were to cease completely, these water "hills" would slowly subside and gradually approach the level of the adjacent valleys. However, new supplies of rainwater are usually added often enough to prevent this. Nevertheless, in times of extended drought, the water table may drop enough to dry up shallow wells. Other causes for the uneven water table are variations in rainfall and in the permeability of Earth materials from place to place.

Factors Influencing the Storage and Movement of Groundwater

The nature of subsurface materials strongly influences the rate of groundwater movement and the amount of groundwater that can be stored. Two factors are especially important: porosity and permeability.

Porosity Water soaks into the ground because bedrock, sediment, and soil contain countless voids or openings. These openings are similar to those of a sponge and are often called *pore spaces*. The quantity of groundwater that can be stored depends on the **porosity** of the material, which is the percentage of the total volume of rock or sediment that consists of pore spaces (**Figure 3.33**). Voids most often are spaces between sedimentary particles, but also common are joints, faults, cavities formed by the dissolving of soluble rock such as limestone, and vesicles (voids left by gases escaping from lava).

Variations in porosity can be great. Sediment is commonly quite porous, and open spaces may occupy 10 percent to 50 percent of the sediment's total volume. Pore space depends on the size and shape of the grains; how they are packed together; the degree of sorting;

Permeability Porosity alone cannot measure a material's capacity to yield groundwater. Rock or sediment may be very porous and still prohibit water from moving through it. The **permeability** of a material indicates its ability to *transmit* a fluid. Groundwater moves by twisting and turning through interconnected small openings. The smaller the pore spaces, the slower the groundwater moves. If the spaces between particles are too small, water cannot move at all. For example, clay's ability to store water can be great, due to its high porosity, but its pore spaces are so small that water is unable to move through it. Thus, we say that clay is *impermeable*.

Aquitards and Aquifers Impermeable layers such as clay that hinder or prevent water movement are termed **aquitards** (*aqua* = water, *tard* = slow). In contrast, larger particles, such as sand or gravel, have larger pore spaces. Therefore, water moves with relative ease. Permeable rock strata or sediments that transmit groundwater freely are called **aquifers** ("water carriers"). Aquifers are important because they are the water-bearing layers sought after by well drillers.

Groundwater Movement

The movement of most groundwater is exceedingly slow, from pore to pore. A typical rate is a few centimeters per day. The energy that makes the water move is provided by the force of gravity. In response to gravity, water moves from areas where the water table is high to zones where the water table is lower. This means that water usually gravitates toward a stream channel, lake, or spring. Although some water takes the most direct path down the slope of the water table, much of the water follows long, curving paths toward the zone of discharge.

Figure 3.34 shows how water percolates into a stream from all possible directions. Some paths clearly turn upward, apparently against the force of gravity, and enter through the bottom of the channel. This is easily explained: The deeper you go into the zone of saturation, the greater the water pressure. Thus, the looping curves followed by water in the saturated zone may be thought of as a compromise between the downward pull of

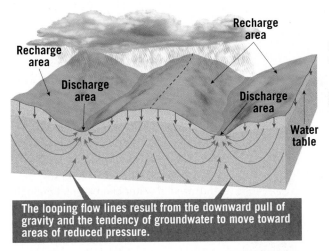

The looping flow lines result from the downward pull of gravity and the tendency of groundwater to move toward areas of reduced pressure.

Figure 3.34 Groundwater movement Arrows show paths of groundwater movement through uniformly permeable material.

gravity and the tendency of water to move toward areas of reduced pressure.

3.11 CONCEPT CHECKS

1. About what percentage of freshwater is groundwater? How does this change if glacial ice is excluded?
2. What are two geologic roles for groundwater?
3. When it rains, what factors influence the amount of water that soaks in?
4. Define *groundwater* and relate it to the water table.
5. Distinguish between porosity and permeability. Contrast aquifers and aquitards.
6. What factors cause water to follow the paths shown in Figure 3.34?

3.12 Springs, Wells, and Artesian Systems

Compare and contrast springs, wells, and artesian systems.

A great deal of groundwater eventually makes its way to the surface. Sometimes this occurs as a naturally flowing spring. Much of the groundwater that people use must be brought to the surface by being pumped from a well. To understand these phenomena, it is necessary to understand Earth's sometimes complex underground "plumbing."

Springs

Springs have aroused the curiosity and wonder of people for thousands of years. The fact that springs were (and to some people still are) rather mysterious phenomena is not difficult to understand because water is flowing freely from the ground in all kinds of weather in seemingly inexhaustible supply but with no obvious source. Today, we know that the source of springs is water from the zone of saturation and that the ultimate source of this water is precipitation.

Whenever the water table intersects Earth's surface, a natural outflow of groundwater, which we call a **spring**, results. Springs often form when an aquitard blocks the downward movement of groundwater and forces it to move laterally. Where the permeable bed (aquifer) outcrops at the surface, one or more springs result.

Another situation that can produce a spring is illustrated in Figure 3.32. Here an aquitard is situated above the main water table. As water percolates downward, a portion of it is intercepted by the aquitard, thereby creating a localized zone of saturation and a **perched water table**. Springs, however, are not confined to places where a perched water table creates a flow at the surface. Many geologic situations lead to the formation of springs because subsurface conditions vary greatly from place to

place. Even in areas underlain by impermeable crystalline rocks, permeable zones may exist in the form of fractures or solution channels. If these openings fill with water and intersect the ground surface along a slope, a spring results (**Figure 3.35**).

Wells

The most common method for removing groundwater is to use a **well**, a hole bored into the zone of saturation.

Figure 3.35 Vasey's Paradise A spring is a natural outflow of groundwater that occurs when the water table intersects the surface. This spring creates a waterfall after emerging from the steep rock wall of the Grand Canyon. (Photo by Michael Collier)

SmartFigure 3.36 Cone of depression For most small domestic wells, the cone of depression is negligible. When wells are heavily pumped, the cone of depression can be large and may lower the water table such that nearby shallower wells may be left dry. (https://goo.gl/nupO6m)

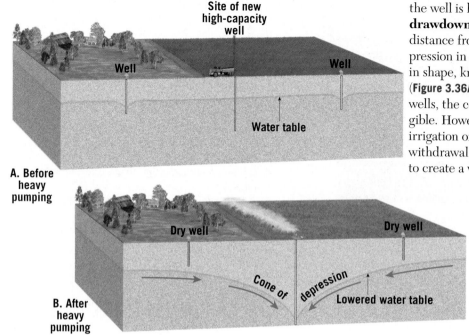

A. Before heavy pumping

B. After heavy pumping

the well is lowered. This effect, termed **drawdown**, decreases with increasing distance from the well. The result is a depression in the water table, roughly conical in shape, known as a **cone of depression** (**Figure 3.36A**). For most small domestic wells, the cone of depression is negligible. However, when wells are used for irrigation or for industrial purposes, the withdrawal of water can be great enough to create a very wide and steep cone of depression that may substantially lower the water table in an area and cause nearby shallow wells to become dry. **Figure 3.36B** illustrates this situation.

Artesian Systems

In most wells, water cannot rise on its own. If water is first encountered at 30 meters (100 feet) depth, it remains at that level, fluctuating perhaps 1 or 2 meters with seasonal wet and dry periods. However, in some wells, water rises and sometimes overflows at the surface. Such wells are abundant in the Artois region of northern France, and so we call these self-rising wells *artesian*.

The term **artesian** is applied to *any* situation in which groundwater under pressure rises above the level of the aquifer. For an artesian system to exist, two conditions usually are met (**Figure 3.37**): (1) Water is confined to an aquifer that is inclined so that one end can receive water,

Wells serve as small reservoirs into which groundwater migrates and from which it can be pumped to the surface. The use of wells dates back many centuries and continues to be an important method of obtaining water.

The water-table level may fluctuate considerably during the course of a year, dropping during dry seasons and rising following periods of precipitation. Therefore, to ensure a continuous supply of water, a well must penetrate below the water table. Whenever a substantial amount of water is withdrawn from a well, the water table around

SmartFigure 3.37 Artesian systems These groundwater systems occur where an inclined aquifer is surrounded by impermeable beds (aquitards). Such aquifers are called *confined aquifers*. The photo shows a flowing artesian well. (Photo by James E. Patterson, courtesy of Fred Lutgens) (https://goo.gl/yVJhLE)

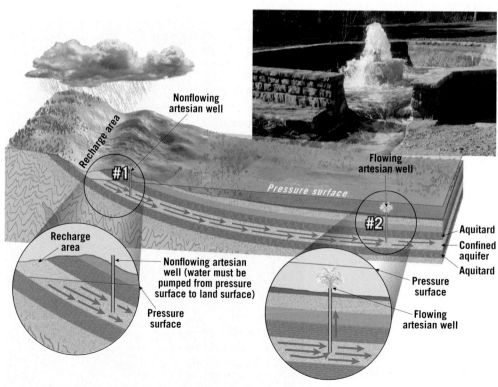

and (2) aquitards, both above and below the aquifer, must be present to prevent the water from escaping. Such an aquifer is called a **confined aquifer**. When such a layer is tapped, the pressure created by the weight of the water above forces the water to rise. If there were no friction, the water in the well would rise to the level of the water at the top of the aquifer. However, friction reduces the height of the pressure surface. The greater the distance from the recharge area (where water enters the inclined aquifer), the greater the friction and the less the rise of water.

In Figure 3.37, well 1 is a *nonflowing artesian well* because at this location, the pressure surface is below ground level. When the pressure surface is above the ground and a well is drilled into the aquifer, a *flowing artesian well* is created (well 2 in Figure 3.37). Not all artesian systems are wells. *Artesian springs* also exist. Here groundwater may reach the surface by rising along a natural fracture such as a fault rather than through an artificially produced hole. In deserts, artesian springs are sometimes responsible for creating oases.

Artesian systems act as "natural pipelines," transmitting water from remote areas of recharge great distances to the points of discharge. In this manner, water that fell in central Wisconsin years ago is now taken from the ground and used by communities many kilometers to the south, in Illinois. In South Dakota, such a system brings water from the western Black Hills eastward across the state.

On a different scale, city water systems may be considered examples of artificial artesian systems (**Figure 3.38**). A

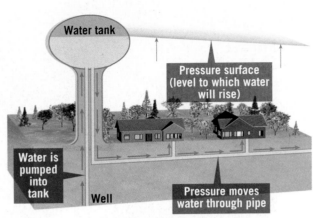

Figure 3.38 City water systems City water systems can be considered to be artificial artesian systems.

water tower, into which water is pumped, may be considered the area of recharge, the pipes the confined aquifer, and the faucets in homes the flowing artesian wells.

3.12 CONCEPT CHECKS

1. Describe the circumstances that created the springs in Figure 3.32.
2. Relate drawdown to cone of depression.
3. In Figure 3.32, two wells were drilled down to the same level. Why was one successful and the other not?
4. Sketch a simple cross section of an artesian system with a flowing artesian well. Label aquitards, the aquifer, and the pressure surface.

3.13 Environmental Problems of Groundwater

List and discuss three important environmental problems associated with groundwater.

As with many of our other valuable natural resources, groundwater is being exploited. In some regions overuse threatens the groundwater supply. In places, excessive groundwater withdrawal has caused the land surface and everything resting on it to sink. Still other localities are concerned with possible contamination of their groundwater supply.

Treating Groundwater as a Nonrenewable Resource

Many natural systems tend to establish a condition of equilibrium. The groundwater system is one of them. The water table's height reflects the rate at which water is added by precipitation and the rate at which water is removed by discharge and withdrawal. An imbalance will either raise or lower the water table. A long-term drop in the water table can occur if there is either a decrease in recharge due to prolonged drought or an increase in groundwater discharge or withdrawal.

For many people, groundwater appears to be an endlessly renewable resource, for it is continually replenished by rainfall and melting snow. But in some regions, groundwater has been and continues to be treated as a *nonrenewable* resource because the amount of water available to recharge the aquifer is significantly less than the amount being withdrawn.

The High Plains, a relatively dry region that extends from South Dakota to western Texas, provides one example of an extensive agricultural economy that is largely dependent on irrigation (**Figure 3.39**). Underlying about 111 million acres (450,000 square kilometers [174,000 square miles]) in parts of eight states, the High Plains aquifer is one of the largest and most agriculturally significant aquifers in the United States. It accounts for about 30 percent of all groundwater withdrawn for irrigation in the country. In the southern part of this region, which includes the Texas panhandle, the natural recharge of the aquifer is very slow, and the problem of declining groundwater levels is acute. In fact, in years of average or below-average precipitation, recharge is negligible

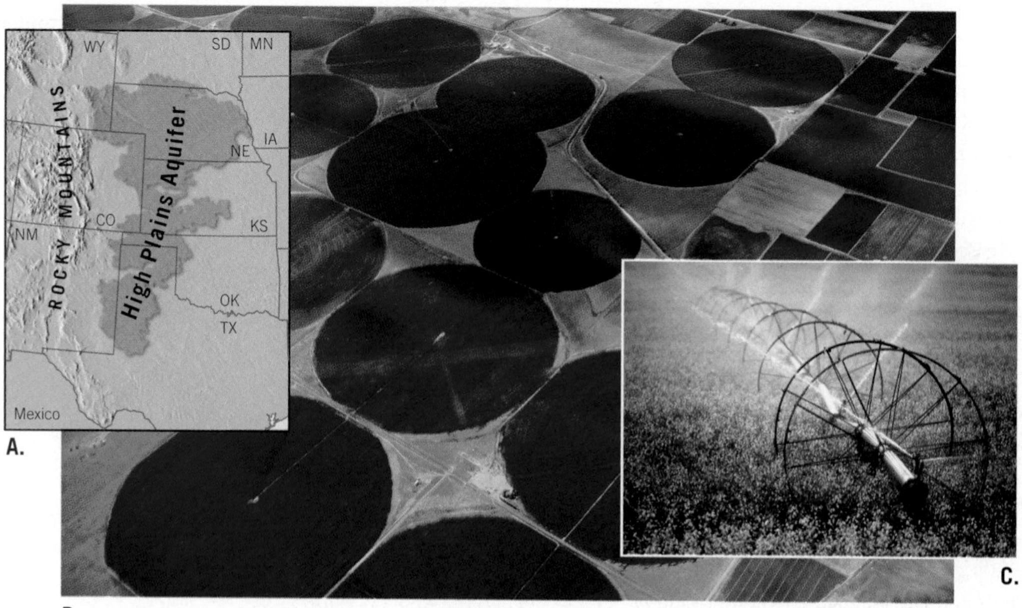

A. B. C.

Figure 3.39 Mining groundwater **A.** The High Plains aquifer is one of the largest aquifers in the United States. **B.** In parts of the High Plains aquifer, water is pumped from the ground faster than it is replenished. In such instances, groundwater is being treated as a nonrenewable resource. This aerial view shows circular crop fields irrigated by center-pivot irrigating systems in semiarid eastern Colorado. (Photo by James L. Amos/CORBIS). **C.** Groundwater provides more than 54 billion gallons per day in support of agriculture in the United States. (Photo by Michael Collier)

because all or nearly all of the meager rainfall is returned to the atmosphere by evaporation and transpiration.

Therefore, where intense irrigation has been practiced for an extended period, depletion of groundwater can be severe. Declines in the water table at rates as great as 1 meter per year have led to an overall drop in excess of 30 meters (100 feet) in some areas. Under these circumstances, it can be said that the groundwater is literally being "mined." Even if pumping were to cease immediately, it would take thousands of years for the groundwater to be fully replenished.

Groundwater depletion has been a concern in the High Plains and other areas of the West for many years, but it is worth pointing out that the problem is not confined to that part of the country. Increased demands on groundwater resources have overstressed aquifers in many areas, not just in arid and semiarid regions.

Land Subsidence Caused by Groundwater Withdrawal

As you will see later in this chapter, surface subsidence can result from natural processes related to groundwater. However, the ground may also sink when water is pumped from wells faster than natural recharge processes can replace it. This effect is particularly pronounced in areas underlain by thick layers of loose sediments. As water is withdrawn, the water pressure drops, and the weight of the overburden is transferred to the sediment. The greater pressure packs the sediment grains more tightly together, and the ground subsides.

Many areas can be used to illustrate such land subsidence. A classic example in the United States occurred in the San Joaquin Valley of California (**Figure 3.40**). Other well-known cases of land subsidence resulting from groundwater pumping in the United

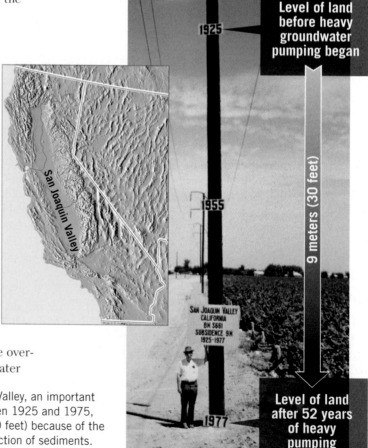

Figure 3.40 That sinking feeling! The San Joaquin Valley, an important agricultural area, relies heavily on irrigation. Between 1925 and 1975, this part of the valley subsided almost 9 meters (30 feet) because of the withdrawal of groundwater and the resulting compaction of sediments. (Photo courtesy of USGS)

Did You Know?
The U.S. Geological Survey estimates that during the past 60 years, water in storage in the High Plains aquifer declined by about 200 million acre feet (65 trillion gallons), with 62 percent of the total decline occurring in western Texas.

States include Las Vegas, Nevada; New Orleans and Baton Rouge, Louisiana; and the Houston–Galveston area of Texas. In the low-lying coastal area between Houston and Galveston, land subsidence ranges from 1.5 to 3 meters (5 to 10 feet). The result is that about 78 square kilometers (30 square miles) were permanently flooded.

Outside the United States, one of the most spectacular examples of subsidence occurred in Mexico City, a portion of which is built on a former lake bed. In the first half of the twentieth century, thousands of wells were sunk into the water-saturated sediments beneath the city. As water was withdrawn, portions of the city subsided by as much as 6 meters (20 feet) or more.

Groundwater Contamination

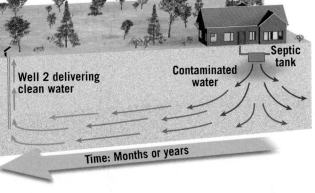

Figure 3.41
Comparing two aquifers In this example, the limestone aquifer allowed the contamination to reach a well, but the sandstone aquifer did not.

Although the contaminated water has traveled more than 100 meters before reaching Well 1, the water moves too rapidly through the cavernous limestone to be purified.

As the discharge from the septic tank percolates through the permeable sandstone, it moves more slowly and is purified in a relatively short distance.

The pollution of groundwater is a serious matter, particularly in areas where aquifers provide a large part of the water supply. One common source of groundwater pollution is sewage emanating from an ever-increasing number of septic tanks. Other sources are inadequate or broken sewer systems and farm wastes.

If sewage water, which is contaminated with bacteria, enters the groundwater system, it may become purified through natural processes. The harmful bacteria may be mechanically filtered by the sediment through which the water percolates, destroyed by chemical oxidation, and/or assimilated by other organisms. For purification to occur, however, the aquifer must be of the correct composition. For example, extremely permeable aquifers (such as highly fractured crystalline rock, coarse gravel, or cavernous

limestone) have such large openings that contaminated groundwater may travel long distances without being cleansed. In this case, the water flows too rapidly and is not in contact with the surrounding material long enough for purification to occur. This is the problem at well 1 in **Figure 3.41A**.

On the other hand, when an aquifer consists of sand or permeable sandstone, the water can sometimes be purified after traveling only a few dozen meters through it. The openings between sand grains are large enough to permit water movement, yet the movement of the water is slow enough to allow ample time for its purification (well 2 in **Figure 3.41B**).

Other sources and types of contamination also threaten groundwater supplies (**Figure 3.42**). These include widely used substances such as highway salt,

Figure 3.42 Potential sources of contamination Sometimes materials leached from landfills and leaking gasoline storage tanks contaminate an aquifer. (Landfill photo by Picsfive/Shutterstock; storage tank photo by Earth Gallery Environment/Alamy)

fertilizers that are spread across the land surface, and pesticides. In addition, a wide array of chemicals and industrial materials may leak from pipelines, storage tanks, landfills, and holding ponds. Some of these pollutants are classified as *hazardous*, meaning that they are flammable, corrosive, explosive, or toxic. As rainwater oozes through the refuse, it may dissolve a variety of potential contaminants. If the leached material reaches the water table, it will mix with the groundwater and contaminate the supply.

Because groundwater movement is usually slow, polluted water may go undetected for a long time. In fact, contamination is sometimes discovered only after drinking water has been affected and people become ill. By this time, the volume of polluted water could be quite large, and even if the source of contamination is removed immediately, the problem is not solved. Although the sources of groundwater contamination are numerous, the solutions are relatively few.

Once a source of the problem has been identified and eliminated, the most common practice is simply to abandon the water supply and allow the pollutants to be flushed away gradually. This is the least costly and easiest solution, but the aquifer must remain unused for many years. To accelerate this process, polluted water is sometimes pumped out and treated. Following removal of the tainted water, the aquifer is allowed to recharge naturally or, in some cases, the treated water or other freshwater is pumped back in. This process is costly, time-consuming, and possibly risky because there is no way to be certain that all of the contamination has been removed. Clearly, the most effective solution to groundwater contamination is prevention.

3.13 CONCEPT CHECKS

1. Describe the problem associated with pumping groundwater for irrigation in parts of the High Plains.
2. What happened in the San Joaquin Valley as a result of excessive groundwater pumping?
3. Which aquifer would be most effective in purifying polluted groundwater: one consisting of coarse gravel, sand, or cavernous limestone? Explain.

3.14 The Geologic Work of Groundwater

Explain the formation of caverns and the development of karst topography.

Groundwater can dissolve rock. This fact is key to understanding how caverns and sinkholes form. Because soluble rocks, especially limestone, underlie millions of square kilometers of Earth's surface, it is here that groundwater carries on its important role as an erosional agent (**Figure 3.43**). Limestone is nearly insoluble in pure water but is quite easily dissolved by water containing small quantities of carbonic acid, and most groundwater contains this acid. It forms because rainwater readily dissolves carbon dioxide from the air and from decaying plants. Therefore, when groundwater comes in contact with limestone, the carbonic acid reacts with calcite (calcium carbonate) in the rocks to form calcium bicarbonate, a soluble material that is then carried away in solution.

Caverns

The most spectacular results of groundwater's erosional handiwork are limestone **caverns**. In the United States alone, about 17,000 caves have been discovered. Although most are relatively small, some have spectacular dimensions. Carlsbad Caverns in southeastern New Mexico and Mammoth Cave in Kentucky are famous examples. One chamber in Carlsbad Caverns has an area equivalent to 14 football fields and enough height to accommodate the U.S. Capitol Building. At Mammoth Cave, the total length of interconnected caverns is over 540 kilometers (340 miles).

Most caverns are created at or just below the water table, in the zone of saturation. Acidic groundwater follows lines of weakness in the rock, such as joints and bedding planes. As time passes, the dissolving process slowly creates cavities and gradually enlarges them into caverns. Material that is dissolved by the groundwater is eventually discharged into streams and carried to the ocean.

Certainly, the features that arouse the greatest curiosity for most cavern visitors are the stone formations that give some caverns a wonderland

SmartFigure 3.43
Kentucky's Mammoth Cave area Portions of Kentucky are underlain by limestone. Dissolution by groundwater has created a landscape characterized by caves and sinkholes.
(Photo by Michael Collier) (http://goo.gl/jsqQfh)

Mobile Field Trip

A.

B.

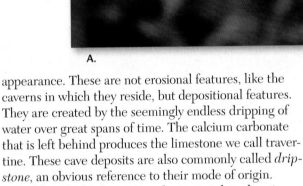

Figure 3.44 Cave decorations Speleothems are of many types, including stalactites, stalagmites, and columns. **A.** Close-up of a delicate live soda-straw stalactite in Chinn Springs Cave, Independence County, Arkansas. (Photo by Dante Fenolio/Science Source) **B.** Stalagmites and stalactites in New Mexico's Carlsbad Caverns National Park. (Photo by Deposit Photos/Glow Images)

appearance. These are not erosional features, like the caverns in which they reside, but depositional features. They are created by the seemingly endless dripping of water over great spans of time. The calcium carbonate that is left behind produces the limestone we call travertine. These cave deposits are also commonly called *dripstone*, an obvious reference to their mode of origin.

Although the formation of caverns takes place in the zone of saturation, the deposition of dripstone is not possible until the caverns are above the water table, in the unsaturated zone. This commonly occurs as nearby streams cut their valleys deeper, lowering the water table as the elevation of the rivers drops. As soon as the chamber is filled with air, the conditions are right for the decoration phase of cavern building to begin.

Of the various dripstone features found in caverns, perhaps the most familiar are **stalactites**. These icicle-like pendants hang from the ceiling of the cavern and form where water seeps through cracks above. When water reaches air in the cave, some of the dissolved carbon dioxide escapes from the drop, and calcite begins to precipitate. Deposition occurs as a ring around the edge of the water drop. As drop after drop follows, each leaves an infinitesimal trace of calcite behind, and a hollow limestone tube is created. Water then moves through the tube, remains suspended momentarily at the end, contributes a tiny ring of calcite, and falls to the cavern floor.

The stalactite just described is appropriately called a *soda straw* (**Figure 3.44A**). Often the hollow tube of the soda straw becomes plugged or its supply of water increases. In either case, the water is forced to flow and deposit along the outside of the tube. As deposition continues, the stalactite takes on the more common conical shape.

Formations that develop on the floor of a cavern and reach upward toward the ceiling are called **stalagmites** (**Figure 3.44B**). The water supplying the calcite for stalagmite growth falls from the ceiling and splatters over the surface. As a result, stalagmites do not have a central tube and are usually more massive in appearance and more rounded on their upper ends than are stalactites. Given enough time, a downward-growing stalactite and an upward-growing stalagmite may join to form a *column*.

Karst Topography

Many areas of the world have landscapes that, to a large extent, have been shaped by the dissolving power of groundwater. Such areas are said to exhibit **karst topography**, named for the *Krs* region in the border area between Slovenia and Italy, where this type of topography is strikingly developed. In the United States, karst landscapes occur in many areas that are underlain by limestone, including portions of Kentucky, Tennessee, Alabama, southern Indiana, and central and northern Florida (**Figure 3.45**). Generally, arid and semiarid areas do not develop karst topography because there is insufficient groundwater. When karst

Figure 3.45 Development of a karst landscape

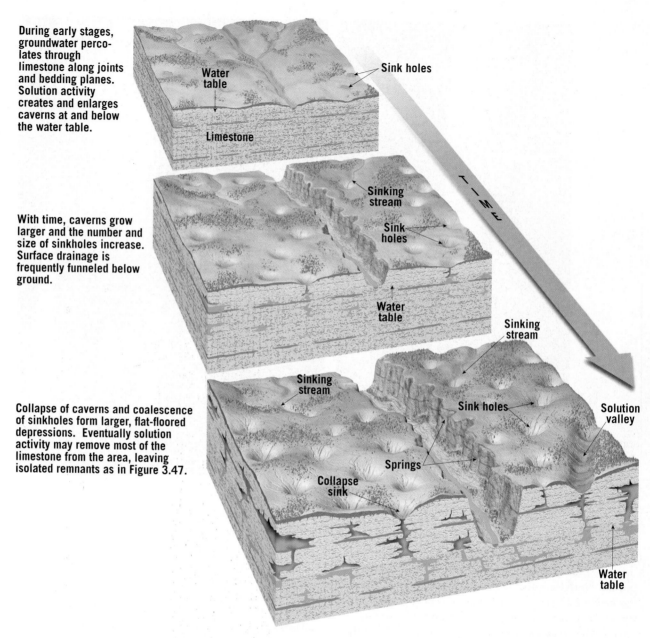

During early stages, groundwater percolates through limestone along joints and bedding planes. Solution activity creates and enlarges caverns at and below the water table.

With time, caverns grow larger and the number and size of sinkholes increase. Surface drainage is frequently funneled below ground.

Collapse of caverns and coalescence of sinkholes form larger, flat-floored depressions. Eventually solution activity may remove most of the limestone from the area, leaving isolated remnants as in Figure 3.47.

Did You Know?

Although most caves and sinkholes are associated with regions underlain by limestone, these features can also form in soluble rocks such as gypsum and rock salt (halite).

landscapes exist in dry regions, they are likely remnants of a time when rainier conditions prevailed.

Sinkholes Karst areas typically have irregular terrain punctuated with many depressions called **sinkholes** or, simply, **sinks** (see Figure 3.31B). In the limestone areas of Florida, Kentucky, and southern Indiana, literally tens of thousands of these depressions vary in depth from just 1 to 2 meters to a maximum of more than 50 meters (165 feet).

Sinkholes commonly form in one of two ways. Some develop gradually over many years, without any physical disturbance to the rock. In these situations, the limestone immediately below the soil is dissolved by downward-seeping rainwater that is freshly charged with carbon dioxide. These depressions are usually not deep and are characterized by relatively gentle slopes.

By contrast, sinkholes can also form suddenly and without warning when the roof of a cavern collapses under its own weight. Typically, the depressions created in this manner are steep sided and deep. When they form in populous areas, they may represent a serious geologic hazard. Such a situation is clearly the case in **Figure 3.46**.

In addition to a surface pockmarked by sinkholes, karst regions characteristically show a striking lack of surface drainage (streams). Following a rainfall, runoff is quickly funneled belowground, through sinks. It then flows through caverns until it finally reaches the water table. Where streams do exist at the surface, their paths are usually short. The names of such streams often give a clue to their fate. The Mammoth Cave area of Kentucky, for example, is home to Sinking Creek, Little Sinking Creek, and Sinking Branch. Some sinkholes become

Figure 3.46 Sinkholes can be geologic hazards This sinkhole formed suddenly in the backyard of a home in Lake City, Florida. Sinkholes such as this one form when the roof of a cavern collapses. (Jon M. Fletcher/The Florida Times-Union/AP Images)

plugged with clay and debris, creating small lakes or ponds.

Tower Karst Landscapes Some regions of karst development exhibit landscapes that look very different from the sinkhole-studded terrain depicted in Figure 3.46. One striking example is an extensive region in southern China that is described as exhibiting *tower karst*. As **Figure 3.47** shows, the term *tower* is appropriate because the landscape consists of a maze of isolated steep-sided hills that rise abruptly from the

Figure 3.47 Tower karst landscape in China One of the best-known and most distinctive regions of tower karst development is along the Li River in the Guilin District of southeastern China. (Photo by Philippe Michel/AGE Fotostock)

ground. Each is riddled with interconnected caves and passageways. This type of karst topography forms in wet tropical and subtropical regions having thick beds of highly jointed limestone. In such settings, groundwater dissolves large volumes of limestone, leaving only these residual towers. Karst development is more rapid in tropical climates due to the abundant rainfall and greater availability of carbon dioxide from the decay of lush tropical vegetation. The extra carbon dioxide in the soil means there is more carbonic acid for dissolving limestone. Other tropical areas of advanced karst development include portions of Puerto Rico, western Cuba, and northern Vietnam.

3.14 CONCEPT CHECKS

1. How does groundwater create caverns?
2. How do stalactites and stalagmites form?
3. Describe two ways in which sinkholes form.

CONCEPTS IN REVIEW
Landscapes Fashioned by Water

3.1 Earth's External Processes

List three important external processes and describe where they fit into the rock cycle.

KEY TERMS: external process, internal process, weathering, mass wasting, erosion

- Weathering, mass wasting, and erosion are responsible for creating, transporting, and depositing sediment. They are called external processes because they occur at or near Earth's surface and are powered by energy from the Sun.
- Internal processes result in such features as volcanoes and mountains and derive their energy from Earth's interior.

3.2 Mass Wasting: The Work of Gravity

Explain the role of mass wasting in the development of valleys and discuss the factors that trigger and influence mass-wasting processes.

KEY TERMS: trigger, angle of repose

- Mass wasting is the downslope movement of rock and soil under the direct influence of gravity.
- An event that initiates a mass-wasting process is referred to as a trigger. The addition of water, oversteepening of slopes, removal of vegetation, and ground shaking due to an earthquake are four examples of triggers. Not all landslides are initiated by one of these triggers, but many are.

3.3 The Hydrologic Cycle

List the hydrosphere's major reservoirs and describe the different paths that water takes through the hydrologic cycle.

KEY TERMS: hydrologic cycle, evaporation, infiltration, runoff, transpiration, evapotranspiration

- Water moves through the hydrosphere's many reservoirs by evaporating, condensing into clouds, and falling as precipitation. Once it reaches the ground, rain can either soak in, evaporate, be returned to the atmosphere by plant transpiration, or run off. Running water is the most important agent sculpting Earth's varied landscapes.

3.4 Running Water

Describe the nature of drainage basins and river systems.

KEY TERMS: drainage basin, watershed, divide

- The land area that contributes water to a stream is its drainage basin. Drainage basins are separated by imaginary lines called divides.
- As a generalization, river systems tend to erode at the upstream end, transport sediment through the middle section, and deposit sediment at the downstream end.

3.5 Streamflow Characteristics

Discuss streamflow and the factors that cause it to change.

KEY TERMS: laminar flow, turbulent flow, gradient, discharge, longitudinal profile

- The flow of water in a stream may be laminar or turbulent. A stream's flow velocity is influenced by the channel's gradient; the size, shape, and roughness of the channel; and the stream's discharge.
- A cross-sectional view of a stream from head to mouth is a longitudinal profile. Usually the gradient and roughness of the stream channel decrease going downstream, whereas the size of the channel, stream discharge, and flow velocity increase in the downstream direction.

(?) **Sketch a typical longitudinal profile. Where does most erosion happen? Where is sediment transport the dominant process?**

3.6 The Work of Running Water

Outline the ways in which streams erode, transport, and deposit sediment.

KEY TERMS: pothole, dissolved load, suspended load, bed load, capacity, competence, sorting, alluvium

- Streams erode when turbulent water lifts loose particles from the streambed. The focused "drilling" of the stream armed with swirling particles also creates potholes in solid rock.
- Streams transport their load of sediment dissolved in the water, in suspension, and along the bottom (bed) of the channel. A stream's ability to transport solid particles is described using two criteria. Capacity refers to how much sediment a stream is transporting, and competence refers to the particle sizes the stream is capable of moving.
- Streams deposit sediment when velocity slows and competence is reduced. This results in sorting, the process by which like-size particles are deposited together.

3.7 Stream Channels

Contrast bedrock and alluvial stream channels. Distinguish between two types of alluvial channels.

KEY TERMS: meander, cut bank, point bar, cutoff, oxbow lake, braided channel

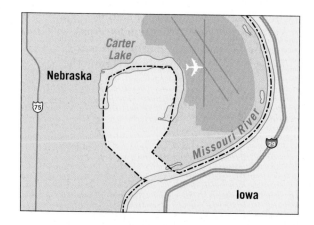

(3.7 continued)

- Bedrock channels are cut into solid rock and are most common in headwaters areas where gradients are steep. Rapids and waterfalls are common features.
- Alluvial channels are dominated by streamflow through alluvium previously deposited by the stream. A floodplain usually covers the valley floor, with the river meandering or moving through braided channels.
- Meanders change shape through erosion at the cut bank (the outer edge of the meander) and deposition of sediment on point bars (the inside of a meander). A meander may become cut off and form an oxbow lake.

(?) **The town of Carter Lake is the *only* portion of the state of Iowa that lies on the west side of the Missouri River. It is bounded on the north by its namesake, Carter Lake, on the south by the Missouri River, and on the east and west by Nebraska. After examining the map on the preceding page, prepare a hypothesis that explains how this unusual situation could have developed.**

3.8 Shaping Stream Valleys

Contrast narrow V-shaped valleys, broad valleys with floodplains, and valleys that display incised meanders.

KEY TERMS: stream valley, base level, floodplain, incised meander, stream terrace

- A stream valley includes the channel itself, the adjacent floodplain, and the relatively steep valley wall. Streams erode downward until they approach base level, the lowest point to which a stream can erode its channel. A river flowing toward the ocean (the ultimate base level) may encounter several local base levels along its route. These could be lakes or resistant rock layers that retard downcutting by the stream.
- A stream valley is widened through the meandering action of the stream, which erodes the valley walls and widens the floodplain. If base level were to drop or if the land were uplifted, a meandering stream might start downcutting and develop incised meanders.

Michael Collier

(?) **Meanders are associated with a river that is eroding from side to side, whereas narrow canyons are associated with rivers that are vigorously downcutting. The river in this image is confined to a narrow canyon but is also meandering. Explain.**

3.9 Depositional Landforms

Discuss the formation of deltas and natural levees.

KEY TERMS: bar, delta, distributary, natural levee, back swamp, yazoo tributary

- Deltas may form where a river deposits sediment in another water body at its mouth. The partitioning of streamflow into multiple distributaries spreads sediment in different directions.

- Natural levees result from sediment deposited along the margins of a stream channel by many flooding events. Because the levees slope gently away from the channel, the adjacent floodplain is poorly drained, resulting in back swamps and yazoo tributaries that flow parallel to the main river.

3.10 Floods and Flood Control

Discuss the causes of floods and some common flood control measures.

KEY TERMS: flood

- When a stream receives more discharge than its channel can hold, a flood occurs. Floods are triggered by heavy rains and/or snowmelt. Sometimes human interference can worsen or even cause floods.
- Three strategies for coping with floods are construction of artificial levees to constrain streamflow to the channel, alterations to make a stream channel's flow more efficient, and building of dams on a river's tributaries so that a sudden influx of water will be temporarily stored and released slowly to the river system. A nonstructural approach is sound floodplain management. Here, a solid scientific understanding of flood dynamics informs policy and regulation of areas that are subject to flooding.

(?) **Artificial levees are constructed to protect property from floods. Sometimes artificial levees in rural areas are intentionally opened up to protect a city from experiencing flooding. How does this work?**

3.11 Groundwater: Water Beneath the Surface

Discuss the importance of groundwater and describe its distribution and movement.

KEY TERMS: zone of saturation, groundwater, water table, unsaturated zone, porosity, permeability, aquitard, aquifer

- Groundwater represents the largest reservoir of freshwater that is readily available to humans. Geologically, groundwater is an equalizer of streamflow, and the dissolving action of groundwater produces caverns and sinkholes.

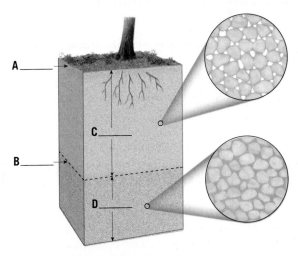

- Groundwater is water that occupies the pore spaces in sediment and rock in a zone beneath the surface called the zone of saturation. The upper limit of this zone is called the water table. The zone above the water table where the material is not saturated is the unsaturated zone.
- The quantity of water that can be stored in the open spaces in rock or sediment is termed porosity. Permeability, the ability of a material

(3.11 continued)

to transmit a fluid through interconnected pore spaces, is a key factor affecting the movement of groundwater. Aquifers are permeable materials that transmit groundwater freely, whereas aquitards are impermeable materials.

(?) **Examine this cross section that shows the vertical distribution of water in a mass of uniform sediments. Provide the correct term for each lettered feature.**

3.12 Springs, Wells, and Artesian Systems
Compare and contrast springs, wells, and artesian systems.

KEY TERMS: spring, perched water table, well, drawdown, cone of depression, artesian system, confined aquifer

- Springs occur where the water table intersects the land surface and a natural flow of groundwater results. They may be due to the intersection of a perched water table and the ground surface.
- Wells, which are openings bored into the zone of saturation, withdraw groundwater and may create roughly conical depressions in the water table known as cones of depression.
- Artesian wells tap into inclined aquifers bounded above and below by aquitards. For a system to qualify as artesian, the water in the well must be under sufficient pressure that it can rise above the top of the confined aquifer. Artesian wells may be flowing or nonflowing, depending on whether the pressure surface is above or below the ground surface.

ifong/Shutterstock

(?) **Relate this image to an artesian system.**

3.13 Environmental Problems of Groundwater
List and discuss three important environmental problems associated with groundwater.

- Groundwater can be "mined" by being extracted at a rate that is greater than the rate of recharge. When groundwater is treated as a

nonrenewable resource, as it is in parts of the High Plains aquifer, the water table drops.
- The extraction of groundwater can cause pore space to decrease in volume and the grains of loose Earth materials to pack more closely together. This overall compaction of sediment volume results in subsidence of the land surface.
- Contamination of groundwater with sewage, highway salt, fertilizer, or industrial chemicals is another issue of critical concern. Once groundwater is contaminated, the problem is very difficult to solve, requiring expensive remediation or abandonment of the aquifer.

3.14 The Geologic Work of Groundwater
Explain the formation of caverns and the development of karst topography.

KEY TERMS: cavern, stalactite, stalagmite, karst topography, sinkhole (sink)

- Groundwater dissolves rock, in particular limestone, leaving behind spaces in the rock. Caverns form at the zone of saturation, but later dropping of the water table may leave them open and dry—and available for people to explore.
- Dripstone is rock deposited by dripping of water containing dissolved calcium carbonate inside caverns. Features made of dripstone include stalactites, stalagmites, and columns.
- Karst topography develops in limestone regions and exhibits irregular terrain punctuated with many depressions called sinkholes. Some sinkholes form when the cavern roofs collapse.

(?) **Identify the three cavern deposits labeled in this photograph.**

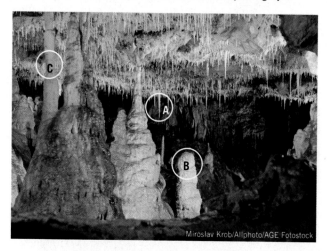

Miroslav Krob/Allphoto/AGE Fotostock

GIVE IT SOME THOUGHT

1. During summer, wildfires are relatively common occurrences in the mountains of the western United States. We have learned that Earth is a system in which various parts of the four major spheres interact in uncountable ways. Let's relate that idea to a wildfire in the western mountains.
   a. What atmospheric conditions might precede and thus set the stage for a wildfire?
   b. What might ignite the blaze? Suggest a natural possibility and a human possibility.
   c. Describe at least one way that wildfires might influence future mass-wasting processes.

2. Describe at least one situation in which an internal process might cause or contribute to a mass-wasting process.

3. If you collect a jar of water from a stream, what part of its load will settle to the bottom of the jar? What portion will remain in the water indefinitely? What part of the stream's load will probably not be represented in your sample?

4. A river system consists of three zones, based on the dominant process operating in each zone. On the accompanying diagram, match each process with one of the three zones: sediment production (erosion), sediment deposition, and sediment transportation.

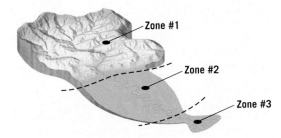

5. The meandering White River in Arkansas is a tributary of the Mississippi River.
   a. In this aerial view, the color of the White River is brown. What part of the stream's load gives it this color?
   b. If an artificial channel were created across the narrow neck of land shown by the arrow, how would the river's gradient change?
   c. How would the flow velocity be affected by the change in gradient?

Michael Collier

6. An acquaintance is considering purchasing a large tract of productive farmland in the Texas panhandle. His intention is to continue growing crops on the land for years to come. If he asked your opinion about the area he selected and his plans for the future, what advice would you give him?

7. During a trip to the grocery store, your friend wants to buy some bottled water. Some brands promote the fact that their product is artesian. Other brands boast that their water comes from a spring. Your friend asks, "Is artesian water or spring water necessarily better than water from other sources?" How would you reply?

8. Building a dam is one method of regulating the flow of a river to control flooding. Dams and their reservoirs may also provide recreational opportunities and water for irrigation and hydroelectric power generation. This image, from near Page, Arizona, shows Glen Canyon Dam on the Colorado River upstream from the Grand Canyon and a portion of Lake Powell, the reservoir it created.
   a. How did the behavior of the stream likely change upstream from Lake Powell?
   b. Given enough time, how might the reservoir change?
   c. Speculate on the possible environmental impacts of building a dam such as this one.

Michael Collier

9. A glance at the map shows that the drainage basin of the Republican River occupies portions of Colorado, Nebraska, and Kansas. A significant part of the basin is considered semiarid. In 1943, the three states made a legal agreement regarding sharing the river's water. In 1998, Kansas went to court to force farmers in southern Nebraska to substantially reduce the amount of groundwater they used for irrigation. Nebraska officials claimed that the farmers were not taking water from the Republican River and thus were not violating the 1943 agreement. The court ruled in favor of Kansas.
   a. Explain why the court ruled that groundwater in southern Nebraska should be considered part of the Republican River system.
   b. How might heavy irrigation in a drainage basin influence the flow of a river?

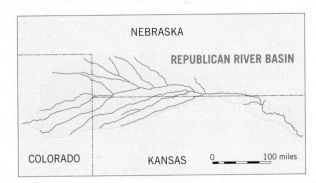

MasteringGeology™

www.masteringgeology.com Looking for additional review and test prep materials? With individualized coaching on the toughest topics of the course, MasteringGeology offers a wide variety of ways for you to move beyond memorization to begin thinking like a geologist. Visit the Study Area in www.masteringgeology.com to find practice quizzes, study tools, and multimedia that will improve your understanding of this chapter's content. Sign in today to enjoy the following features: **Self Study Quizzes, SmartFigure: Tutorials/Animations/Condor Videos/Mobile Field Trips, Geoscience Animation Library, GEODe, RSS Feeds, Digital Study Modules,** and an optional **Pearson eText.**

4

FOCUS ON CONCEPTS

Each statement represents the primary learning objective for the corresponding major heading within the chapter. After you complete the chapter, you should be able to:

4.1 Explain the role of glaciers in the hydrologic and rock cycles and describe the different types of glaciers and their present-day distribution.

4.2 Describe how glaciers move, the rates at which they move, and the significance of the glacial budget.

4.3 Discuss the processes of glacial erosion and the major features created by these processes.

4.4 Distinguish between the two basic types of glacial deposits and briefly describe the features associated with each type.

4.5 Describe and explain several important effects of Ice Age glaciers other than the formation of erosional and depositional landforms.

4.6 Discuss the extent of glaciation and climate variability during the Quaternary Ice Age.

4.7 Describe the general distribution and causes of Earth's dry lands and the role that water plays in modifying desert landscapes.

4.8 Discuss the stages of landscape evolution in the Basin and Range region of the western United States.

4.9 Describe the ways in which wind transports sediment and the features created by wind erosion. Distinguish between two basic types of wind deposits.

Salt flats cover a portion of the floor of California's Saline Valley. Death Valley is nearby. This desert area is part of the Basin and Range region. (Photo by Michael Collier)

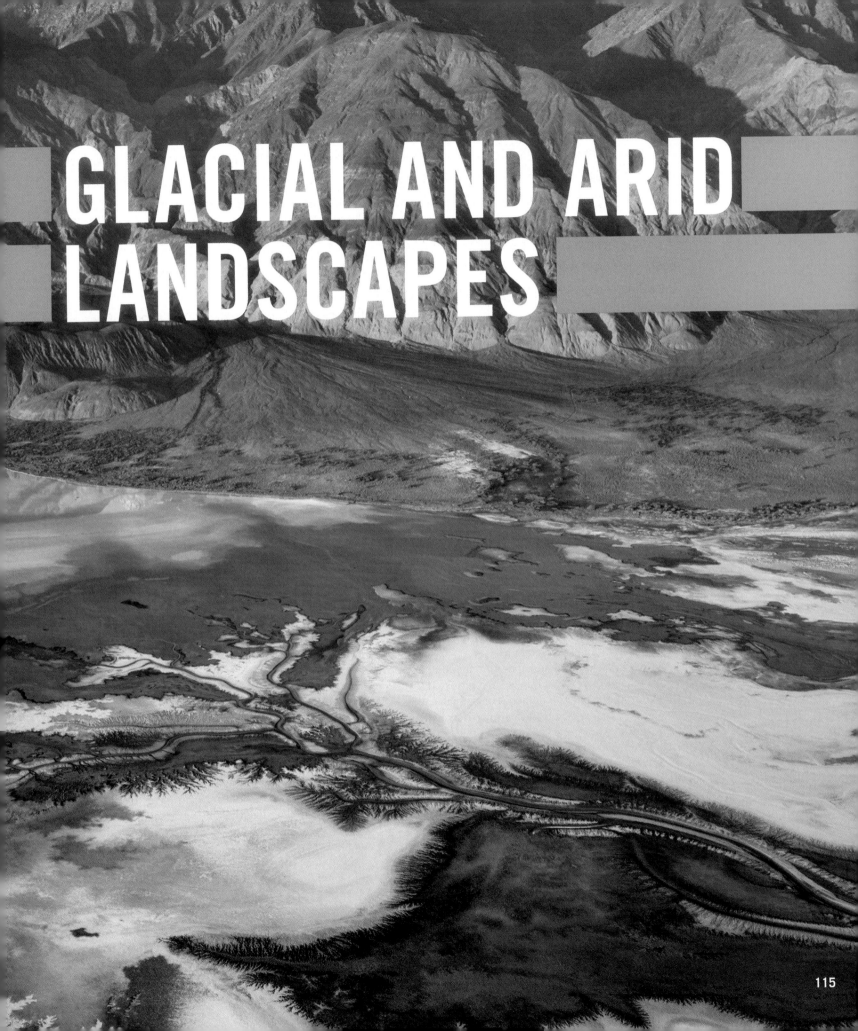

GLACIAL AND ARID LANDSCAPES

Like the running water and groundwater that were the focus of Chapter 3, glaciers and wind are significant agents of erosion. They are responsible for creating many different landforms and are part of an important link in the rock cycle, in which the products of weathering are transported and deposited as sediment.

Climate has a strong influence on the nature and intensity of Earth's external processes. This fact is dramatically illustrated in this chapter. The existence and extent of glaciers are largely controlled by Earth's changing climate. Another excellent example of the strong link between climate and geology is seen when we examine the development of arid landscapes.

Today, glaciers cover nearly 10 percent of Earth's land surface; however, in the recent geologic past, ice sheets were three times more extensive, covering vast areas with ice thousands of meters thick. Many regions still bear the mark of these glaciers. The first part of this chapter will examine glaciers and the erosional and depositional features they create. The second part will explore dry lands and the geologic work of wind. Because desert and near-desert conditions prevail over an area as large as that affected by massive Ice Age glaciers, the nature of such landscapes is indeed worth investigating.

4.1) Glaciers and the Earth System

Explain the role of glaciers in the hydrologic and rock cycles and describe the different types of glaciers and their present-day distribution.

A **glacier** is a thick ice mass that forms over hundreds or thousands of years. It originates on land from the accumulation, compaction, and recrystallization of snow. A glacier appears to be motionless, but it is not; glaciers move very slowly. Like running water, groundwater, wind, and waves, glaciers are dynamic erosional agents that accumulate, transport, and deposit sediment. Although glaciers are found in many parts of the world today, most are located in remote areas, either near Earth's poles or in high mountains.

Many present-day landscapes were modified by the widespread glaciers that advanced and retreated several times during the past 2.6 million years, a span called the *Quaternary period*. The basic features of such diverse places as the Alps, Cape Cod, and Yosemite Valley were fashioned by now-vanished masses of glacial ice. Moreover, Long Island, New York; the Great Lakes; and the fiords of Norway and Alaska all owe their existence to glacial ice. Glaciers, of course, are not just a phenomenon of the geologic past. They are still sculpting landscapes and depositing sediment in many regions today.

Glaciers are a part of two basic and important cycles in the Earth system: the hydrologic cycle and the rock cycle. In the hydrologic cycle, when precipitation falls at high elevations or high latitudes, the water may not immediately make its way back toward the sea. Instead, it may become part of a glacier. Although the ice will eventually melt, allowing the water to continue its path to the sea, water can be stored as glacial ice for tens, hundreds, or even thousands of years. During the time that the water is

part of a glacier, the moving mass of ice can do enormous amounts of work—scouring the land surface and acquiring, transporting, and depositing great quantities of sediment. This activity is a basic part of the rock cycle.

Valley (Alpine) Glaciers

Literally thousands of relatively small glaciers exist in lofty mountain areas, where they usually follow valleys originally occupied by streams. Unlike the rivers that previously flowed in these valleys, the glaciers advance slowly, perhaps only a few centimeters per day. Because of their location, these moving ice masses are termed **valley glaciers**, or **alpine glaciers** (**Figure 4.1**). Each glacier is a stream of ice, bounded by precipitous rock walls, that flows downvalley from a snow accumulation center near its head. Like rivers, valley glaciers can be long or short, wide or narrow, single or with branching tributaries. The widths of alpine glaciers are generally small compared to their lengths. In length, some extend for just a kilometer or less, whereas others go on for many tens of kilometers. The west branch of the Hubbard Glacier, for example, runs through 112 kilometers (70 miles) of mountainous terrain in Alaska and Canada's Yukon Territory.

Ice Sheets

Ice sheets exist on a much larger scale than valley glaciers. These enormous masses flow out in all directions

SmartFigure 4.1 Valley glacier Johns Hopkins Glacier in Alaska's Glacier Bay National Park is still eroding the Alaskan landscape. The dark stripes of sediment within this glacier are called *medial moraines*. This moving mass of ice. is one path through the hydrologic cycle. The transport and deposition of sediment make glaciers a part of the rock cycle. (Photo by Michael Collier) (http://goo.gl/jkqqPA)

Mobile Field Trip

Greenland's ice sheet occupies 1.7 million square kilometers (663,000 square miles), about 80 percent of the island.

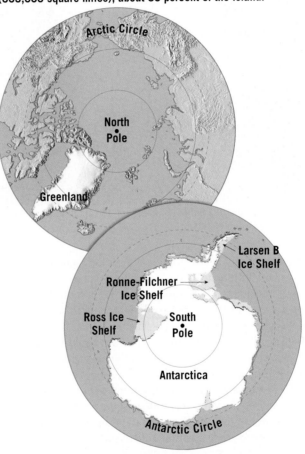

The area of the Antarctic Ice Sheet is almost 14 million square kilometers (5,460,000 square miles). Ice shelves occupy an additional 1.4 million square kilometers (546,000 square miles).

SmartFigure 4.2 Ice sheets The only present-day ice sheets are those covering Greenland and Antarctica. Their combined areas represent almost 10 percent of Earth's land area. (https://goo.gl/JdyThn)

Video

from one or more snow-accumulation centers and completely obscure all but the highest areas of underlying terrain. The low total annual solar energy reaching the poles makes these regions eligible for such great ice accumulations. Presently each of Earth's polar regions supports an ice sheet—Greenland in the Northern Hemisphere and Antarctica in the Southern Hemisphere (**Figure 4.2**).

Ice Age Ice Sheets About 18,000 years ago, glacial ice not only covered Greenland and Antarctica but also covered large portions of North America, Europe, and Siberia. This date in Earth history is appropriately known as the *Last Glacial Maximum*. The term implies that there were other glacial maximums, and this is indeed the case. Throughout the Quaternary period ice sheets formed, advanced over broad areas, and then melted away. These alternating glacial and interglacial periods occurred over and over again.

Greenland and Antarctica Some people mistakenly think that the North Pole is covered by glacial ice, but this is not the case. The ice that covers the Arctic Ocean

is **sea ice**—frozen seawater. Sea ice floats because ice is less dense than liquid water. Although sea ice never completely disappears from the Arctic, the area covered expands and contracts with the seasons. The thickness of sea ice ranges from a few centimeters for new ice to 4 meters for sea ice that has survived for years. By contrast, glaciers can be hundreds or thousands of meters thick.

Glaciers form on land, and in the Northern Hemisphere, Greenland supports an ice sheet. Greenland extends between about 60° and 80° north latitude. This largest island on Earth is covered by an imposing ice

Did You Know?
The annual mean temperature at Amundsen–Scott Station at the South Pole is –56.9°F. By contrast, the annual mean at McMurdo Station along the Antarctic coast adjacent to the Ronne–Filchner Ice Shelf is a "balmy" 1.6°F.

SmartFigure 4.3 Iceland's Vatnajökull ice cap In 1996 the Grímsvötn Volcano erupted beneath this ice cap, an event that triggered melting and floods. (NASA) (http://goo.gl/RsbHWM)

Mobile Field Trip

Ice caps completely bury the underlying terrain but are much smaller than ice sheets.

sheet that occupies 1.7 million square kilometers (more than 660,000 square miles), or about 80 percent of the island. Averaging nearly 1500 meters (5000 feet) thick, the ice extends 3000 meters (10,000 feet) above the island's bedrock floor in some places. The video that is linked to Figure 4.2 is an exploration of the Greenland Ice Sheet.

In the Southern Hemisphere, the huge Antarctic Ice Sheet attains a maximum thickness of about 4300 meters (14,000 feet) and covers an area of more than 13.9 million square kilometers (5.4 million square miles)—nearly the entire continent. Because of the proportions of these huge features, they are often called *continental ice sheets*.

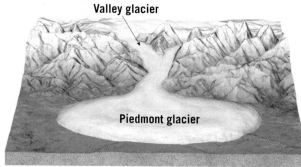

When a valley glacier is no longer confined, it spreads out to become a piedmont glacier.

Figure 4.4 Piedmont glacier Piedmont glaciers occur where valley glaciers exit a mountain range onto broad lowlands.

Indeed, the combined areas of present-day continental ice sheets represent almost 10 percent of Earth's land area.

Ice Shelves Along portions of the Antarctic coast, glacial ice flows into the adjacent ocean, creating features called **ice shelves**. These are large, relatively flat masses of floating ice that extend seaward from the coast but remain attached to the land along one or more sides. The shelves are thickest on their landward sides, and they become thinner seaward. They are sustained by ice from the adjacent ice sheet and are nourished by snowfall and the freezing of seawater to their bases. Antarctica's ice shelves extend over approximately 1.4 million square kilometers (0.6 million square miles). The Ross and Ronne-Filchner Ice Shelves are the largest, with the Ross Ice Shelf alone covering an area approximately the size of Texas (see Figure 4.2).

Other Types of Glaciers

In addition to valley glaciers and ice sheets, other types of glaciers are also recognized. Covering some uplands and plateaus are masses of glacial ice called **ice caps**. Like ice sheets, ice caps completely bury the underlying landscape, but they are much smaller than the continental-scale features. Ice caps occur in many places, including Iceland and several of the large islands in the Arctic Ocean (**Figure 4.3**).

Another type, known as **piedmont glaciers**, occupy broad lowlands at the bases of steep mountains and form when one or more valley glaciers emerge from the confining walls of mountain valleys. The advancing ice spreads out to form a broad sheet. The size of individual piedmont glaciers varies greatly. Among the largest is the broad Malaspina Glacier along the coast of southern Alaska. It covers more than 5000 square kilometers (2000 square miles) of the flat coastal plain at the foot of the lofty St. Elias Range (**Figure 4.4**).

Often ice caps and ice sheets feed **outlet glaciers**. These tongues of ice flow down valleys, extending outward from the margins of these larger ice masses. The tongues are essentially valley glaciers that are avenues for ice movement from an ice cap or ice sheet through mountainous terrain to the sea. Where they encounter the ocean, some outlet glaciers spread out as floating ice shelves. Often large numbers of icebergs are produced.

4.1 CONCEPT CHECKS

1. Where are glaciers found on Earth today, and what percentage of Earth's land surface do they cover?
2. Describe how glaciers fit into the hydrologic cycle. What role do they play in the rock cycle?
3. List and briefly distinguish among four types of glaciers.
4. What is the difference between an ice sheet, sea ice, and an ice shelf?

4.2 How Glaciers Move

Describe how glaciers move, the rates at which they move, and the significance of the glacial budget.

The way in which ice moves is complex and of two basic types. The first of these, *plastic flow*, involves movement *within* the ice. Ice behaves as a brittle solid until the pressure on it is equivalent to the weight of about 50 meters (165 feet) of ice. Once that load is surpassed, ice behaves as a plastic material and flow begins. Such flow occurs because of the molecular structure of ice. Glacial ice consists of layers of molecules stacked one upon the other. The bonds between layers are weaker than those within each layer. Therefore, when a stress exceeds the strength of the bonds between the layers, the layers remain intact and slide over one another.

A second and often equally important mechanism of glacial movement occurs when an entire ice mass slips along the ground. The lowest portions of most glaciers probably move by this sliding process.

The uppermost 50 meters (165 feet) of a glacier is appropriately referred to as the *zone of fracture*.

Figure 4.5 Crevasses As a glacier moves, internal stresses cause large cracks to develop in the brittle upper portion of the glacier, called the zone of fracture. Crevasses can extend to depths of 50 meters (165 feet) and can make travel across glaciers dangerous. (Photo by Wave/Glow Images)

Because there is not enough overlying ice to cause plastic flow, this upper part of the glacier consists of brittle ice. Consequently, the ice in this zone is carried along piggyback-style by the ice below. When the glacier moves over irregular terrain, the zone of fracture is subjected to tension, resulting in cracks called **crevasses** (**Figure 4.5**). These gaping cracks, which often make travel across glaciers dangerous, may extend to depths of 50 meters (165 feet). Below this depth, plastic flow seals them off.

Observing and Measuring Movement

Unlike the movement of water in streams, the movement of glacial ice is not obvious. If we could watch a valley glacier move, however, we would see that, as with the water in a river, the center of the glacier moves faster than the edges because of the drag created by the walls and floor of the valley.

Figure 4.6 shows how glacial movement was first investigated during the nineteenth century. Markers were placed in a straight line across an alpine glacier (in this example, Rhône Glacier in the Swiss Alps), and the original position of the line was marked on the valley walls. The positions of the markers were noted periodically revealing the nature of the movement described above. In this particular study, investigators also mapped the position of the glacier's terminus and demonstrated that the terminus could retreat even as the ice within the glacier moved forward.

Currently, satellite data help us track the movement of glaciers and observe glacial behavior, particularly in remote areas (see Figure I.18). Time-lapse photography can give a more detailed picture of a glacier's movement.

SmartFigure 4.6
Measuring the movement of a glacier Ice movement and changes in the terminus of Rhône Glacier, Switzerland. In this classic study of a valley glacier, the movement of stakes clearly shows that glacial ice moves and that movement along the sides of the glacier is slower than movement in the center. Also notice that even though the ice front was retreating, the ice within the glacier was advancing. (https://goo.gl/JoKPM3)

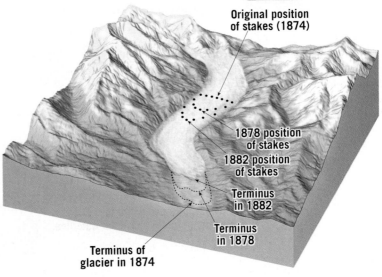

Original position of stakes (1874)

1878 position of stakes

1882 position of stakes

Terminus in 1882

Terminus in 1878

Terminus of glacier in 1874

SmartFigure 4.7 Zones of a glacier The snowline separates the zone of accumulation and the zone of wastage. Whether the ice front advances, retreats, or remains stationary depends on the balance or lack of balance between accumulation and wastage. (https://goo.gl/25XUcw)

Tutorial

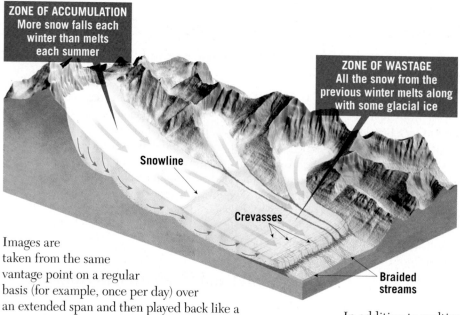

ZONE OF ACCUMULATION
More snow falls each winter than melts each summer

ZONE OF WASTAGE
All the snow from the previous winter melts along with some glacial ice

Snowline

Crevasses

Braided streams

Images are taken from the same vantage point on a regular basis (for example, once per day) over an extended span and then played back like a movie.

How rapidly does glacial ice move? Average rates vary considerably from one glacier to another. Some glaciers move so slowly that trees and other vegetation may become well established in the debris that accumulates on the glacier's surface. Others advance up to several meters each day. Recent satellite radar imaging provided insights into movements within the Antarctic Ice Sheet. This study showed that portions of some outlet glaciers move at rates greater than 800 meters (2600 feet) per year; on the other hand, ice in some interior regions creeps along at less than 2 meters (6.5 feet) per year.

Movement of some glaciers is characterized by occasional periods of extremely rapid advance called *surges*, followed by periods during which movement is much slower.

Budget of a Glacier: Accumulation Versus Wastage

Snow is the raw material from which glacial ice originates. Therefore, glaciers form in areas where more snow

falls in winter than can melt during the summer. Glaciers are constantly gaining and losing ice.

Glacial Zones Snow accumulation and ice formation occur in the **zone of accumulation** (**Figure 4.7**). The addition of snow thickens the glacier and promotes movement. Beyond this area of ice formation is the **zone of wastage**, where there is a net loss to the glacier as all the snow from the previous winter melts, as does some of the glacial ice (see Figure 4.7).

In addition to melting, glaciers also waste as large pieces of ice break off the front of a glacier, in a process called *calving*. Calving creates **icebergs** when glaciers reach the sea (**Figure 4.8**). Because icebergs are just slightly less dense than seawater, they float very low in the water, with 80 percent or more of their mass submerged. The margins of the Greenland Ice Sheet produce thousands of icebergs each year. Many drift southward and find their way into the North Atlantic, where they are hazardous to navigation.

Glacial Budget Whether the margin of a glacier is advancing, retreating, or remaining stationary depends on the *budget of the glacier*. The **glacial budget** refers to the balance or lack of balance between accumulation on the one hand and wastage on the other. If ice accumulation exceeds wastage, the glacial front advances until the two factors balance. At this point, the terminus of the glacier becomes stationary. If a warming trend increases wastage and/or if a drop in snowfall decreases accumulation, the ice front will retreat. As the terminus of the glacier retreats, the extent of the zone of wastage gets smaller. Therefore, in time a new balance will be reached between accumulation and wastage, and the ice front will again become stationary.

Whether the margin of a glacier is advancing, retreating, or stationary, the ice within the glacier continues to move forward. In the case of a receding glacier, the ice simply does not flow forward rapidly enough to offset wastage. This point is illustrated in Figure 4.6. While the line of stakes within the Rhône Glacier continued to advance downvalley, the terminus of the glacier slowly retreated upvalley.

Glaciers in Retreat: Unbalanced Glacial Budgets Because glaciers are sensitive to changes in temperature and precipitation, they provide clues

Figure 4.8 Icebergs Icebergs form when large masses of ice break off from the front of a glacier after it reaches a water body, in a process known as calving. (Photo by Radius/Getty Images)

Geologist's Sketch

Only about 20 percent or less of an iceberg protrudes above the waterline.

1935

2013

Figure 4.9 Retreating glaciers These two images were taken 78 years apart from about the same vantage point along the southwest coast of Greenland. Between 1935 and 2013 the outlet glacier that is the primary focus of these photos retreated about 3 kilometers (about 2 miles). (NASA)

about changes in climate. With few exceptions, glaciers around the world have been retreating and shrinking at unprecedented rates over the past century. The photos in **Figure 4.9** provide an example from the coast of Greenland. Many valley glaciers have disappeared altogether. For example, 150 years ago, there were 147 glaciers in Montana's Glacier National Park. Today only 37 remain.

4.2 CONCEPT CHECKS

1. Describe two components of glacial movement.
2. How rapidly does glacial ice move? Provide some examples.
3. What are crevasses, and where do they form?
4. Relate glacial budget to two zones of a glacier.
5. Under what circumstances will the front of a glacier advance? Retreat? Remain stationary?

4.3 Glacial Erosion

Discuss the processes of glacial erosion and the major features created by these processes.

Glaciers erode tremendous volumes of rock. For anyone who has observed the terminus of an alpine glacier, the evidence of its erosive force is plain. You can witness firsthand the release of rock fragments of various sizes from the ice as it melts (**Figure 4.10**). All signs lead to the conclusion that the ice has scraped, scoured, and torn rock debris from the floor and walls of the valley and carried it downvalley. In mountainous regions, mass-wasting processes can also make substantial contributions to the sediment load of a glacier.

Once rock debris is acquired by a glacier, it cannot settle out as does the load carried by a stream or by the wind. Consequently, glaciers can carry huge blocks that no other erosional agent could possibly budge. Although today's glaciers are of limited importance as erosional agents, many landscapes that were modified by the widespread glaciers of the Quaternary Ice Age still reflect to a high degree the work of ice.

How Glaciers Erode

Glaciers erode land primarily in two ways—by plucking and abrasion. First, as a glacier flows over a fractured bedrock surface, it loosens and lifts blocks of rock and incorporates them

into the ice. This process, known as **plucking**, occurs when meltwater penetrates the cracks and joints along the rock floor of the glacier and freezes. As the water freezes, it expands and exerts tremendous leverage that pries rock loose. In this manner, sediment of all sizes becomes part of the glacier's load.

The second major erosional process is **abrasion**. As the ice and its load of rock fragments move over bedrock, they function like sandpaper, smoothing and polishing the surface below. The pulverized rock produced by the glacial "grist mill" is appropriately called **rock flour**. So much rock flour may be produced that

Figure 4.10 Evidence of glacial erosion As the terminus of this glacier in Alaska wastes away, it deposits large quantities of sediment. The close-up view shows that the rock debris dropped by the melting ice is a jumbled mixture of different-size sediments. (Photos by Michael Collier)

Figure 4.11 Glacial abrasion Moving glacial ice, armed with sediment, acts like sandpaper, scratching and polishing rock. (Photos by Michael Collier)

Glacially polished granite in California's Yosemite National Park.

B.

Glacial abrasion created the scratches and grooves in this bedrock.

A.

meltwater streams leaving a glacier often have the grayish appearance of skim milk—visible evidence of the grinding power of the ice.

When ice at the bottom of a glacier contains larger rock fragments, long scratches and grooves called **glacial striations** may be gouged into the bedrock (**Figure 4.11A**). These linear scratches provide clues to the direction of ice flow. By mapping the striations over large areas, patterns of glacial flow can often be reconstructed.

Not all abrasive action produces striations. The rock surfaces over which a glacier moves may become highly polished by the ice and its load of finer particles. The broad expanses of smoothly polished granite in California's Yosemite National Park provide an excellent example (**Figure 4.11B**).

As is the case with other agents of erosion, the rate of glacial erosion is highly variable. This differential erosion by ice is largely controlled by four factors: (1) rate of glacial movement; (2) thickness of the ice; (3) shape, abundance, and hardness of the rock fragments contained in the ice at the base of the glacier; and (4) erodibility of the surface beneath the glacier. Variations in any or all of these factors from time to time and/or from place to place mean that the features, effects, and degree of landscape modification in glaciated regions can vary greatly.

Landforms Created by Glacial Erosion

Although the erosional accomplishments of ice sheets can be tremendous, landforms carved by these huge ice masses usually do not inspire the same awe as do the erosional features created by valley glaciers. In regions where the erosional effects of ice sheets are significant, glacially scoured surfaces and subdued terrain are the

rule. By contrast, in mountainous areas, erosion by valley glaciers produces many truly spectacular features. Much of the rugged mountain scenery so celebrated for its majestic beauty is the product of erosion by valley glaciers.

Take a moment to study **Figure 4.12**, which shows a mountain setting before, during, and after glaciation. You will refer to this figure often in the following discussion.

Glaciated Valleys Prior to glaciation, alpine valleys are characteristically V-shaped because streams are well above base level and are therefore downcutting (see Figure 4.12A). However, in mountainous regions that have been glaciated, the valleys are no longer narrow. As a glacier moves down a valley once occupied by a stream, the ice modifies it in three ways: The glacier widens, deepens, and straightens the valley so that what was once a narrow V-shaped valley is transformed into a U-shaped **glacial trough** (see Figure 4.12C).

The amount of glacial erosion depends in part on the thickness of the ice. Consequently, main glaciers, also called *trunk glaciers*, cut their valleys deeper than do their smaller tributary glaciers. Thus, after the ice has receded, the valleys of tributary glaciers are left standing above the main glacial trough and are termed **hanging valleys**. Rivers flowing through hanging valleys may produce spectacular waterfalls, such as those in Yosemite National Park, California (see Figure 4.12C).

Cirques At the head of a glacial valley is a characteristic and often imposing feature associated with an alpine glacier—a **cirque**. As Figure 4.12 illustrates, these hollowed-out, bowl-shaped depressions have precipitous walls on three sides and are open on the downvalley side. The cirque is the focal point of the glacier's growth because it is the area of snow accumulation and ice formation. Cirques begin as irregularities in the mountainside that are subsequently enlarged by frost wedging and plucking along the sides and bottom of the glacier. The glacier, in turn, acts as a conveyor belt that carries away

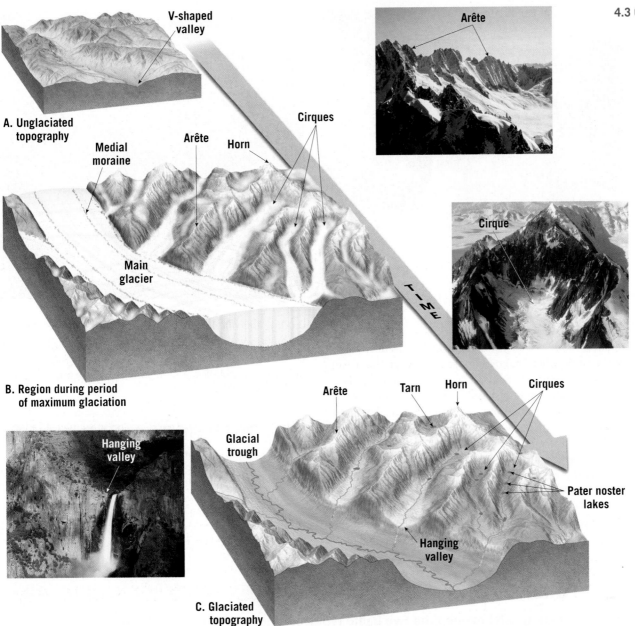

A. Unglaciated topography

V-shaped valley

B. Region during period of maximum glaciation

Medial moraine

Arête

Horn

Cirques

Main glacier

TIME

Hanging valley

Glacial trough

Arête

Tarn

Horn

Cirques

Pater noster lakes

Hanging valley

C. Glaciated topography

Arête

Cirque

SmartFigure 4.12 Erosional landforms created by alpine glaciers The unglaciated landscape in part A is modified by valley glaciers in part B. After the ice recedes, in part C, the terrain looks very different than it looked before glaciation. (Arête photo by James E. Patterson courtesy of F. Lutgens; cirque photo by Marli Miller; hanging valley photo by John Warden/SuperStock) (https://goo.gl/XPgbvY)

Tutorial

the debris. After the glacier has melted away, the cirque basin is often occupied by a small lake called a *tarn*.

Arêtes and Horns The Alps, Northern Rockies, and many other mountain landscapes carved by valley glaciers reveal more than glacial troughs and cirques. In addition, sinuous, sharp-edged ridges called **arêtes** and sharp, pyramid-like peaks called **horns** project above the surroundings (see Figure 4.12C). Both features can originate from the same basic process—the enlargement of cirques produced by plucking and frost action. Cirques around a single high mountain create the spires of rock called horns. As the cirques enlarge and converge, an isolated horn is produced. A famous example is the Matterhorn in the Swiss Alps (**Figure 4.13**).

Arêtes can form in a similar manner, except that the cirques are not clustered around a point but rather exist on opposite sides of a divide. As the cirques grow, the divide separating them is reduced to a very narrow,

Figure 4.13 The Matterhorn Horns are sharp, pyramid-like peaks that were shaped by alpine glaciers. The Matterhorn, in the Swiss Alps, is a famous example. (Photo by Andy Selinger/AGE Fotostock)

Figure 4.14 Fiords The coast of Norway is known for its many fiords. Frequently these ice-sculpted inlets of the sea are hundreds of meters deep. (Satellite images courtesy of NASA; photo by Yoshio Tomii/SuperStock)

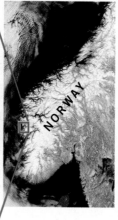

knifelike partition. An arête may also be created when the area separating two parallel glacial valleys is narrowed as the glaciers scour and widen their valleys.

Fiords **Fiords** are deep, often spectacular, steep-sided inlets of the sea that exist in many high-latitude areas of the world where mountains are adjacent to the ocean (**Figure 4.14**). Norway, British Columbia, Greenland, New Zealand, Chile, and Alaska all have coastlines characterized by fiords. They are glacial troughs that became submerged as the ice left the valley and sea level rose following the Ice Age.

The depth of some fiords exceeds 1000 meters (3300 feet). However, the great depths of these flooded troughs are only partly explained by the post–Ice Age rise in sea level. Unlike the situation governing the downward erosional work of rivers, sea level does not act as base level for glaciers. As a consequence, glaciers are capable of eroding their beds far below the surface of the sea. For example, a valley glacier 300 meters (1000 feet) thick can carve its valley floor more than 250 meters (800 feet) below sea level before downward erosion ceases and the ice begins to float.

 4.3 CONCEPT CHECKS

1. How do glaciers acquire their load of sediment?
2. How does a glaciated mountain valley differ in appearance from a mountain valley that was not glaciated?
3. Describe the features created by glacial erosion that you might see in an area where valley glaciers recently existed.

4.4 Glacial Deposits

Distinguish between the two basic types of glacial deposits and briefly describe the features associated with each type.

Glaciers pick up and transport a huge load of debris as they slowly advance across the land. These materials are ultimately deposited when the ice melts. Glacial sediment can play a truly significant role in forming the physical landscape in regions where it is deposited. For example, in many areas once covered by the ice sheets of the Quaternary Ice Age, the bedrock is rarely exposed because glacial deposits that are tens or even hundreds of meters thick completely bury the terrain. The general effect of these deposits is to level the topography. Indeed, many of today's familiar rural landscapes—rocky pastures in New England, wheat fields in the Dakotas, rolling farmland in the Midwest—resulted directly from glacial deposition.

Types of Glacial Drift

Long before the theory of an extensive Ice Age was proposed, much of the soil and rock debris covering portions of Europe was recognized as coming from elsewhere. At the time, these foreign materials were believed to have "drifted" into their present positions via floating ice during an ancient flood. As a consequence, the term *drift* was applied to this sediment. Although rooted in a concept that was not correct, this term was so well established by the time the true glacial origin of the debris became widely recognized that it remained in the glacial vocabulary. **Glacial drift** is an all-embracing term for sediments of glacial origin, no matter how, where, or in what form they were deposited.

Glacial drift is divided into two distinct types: (1) materials deposited directly by a glacier, which are known as *till*, and (2) sediments laid down by glacial meltwater, called *stratified drift*. Here is the difference: **Till** is deposited as glacial ice melts and drops its load of rock debris. Unlike moving water and wind, ice cannot sort the sediment it carries; therefore, deposits

Glacial till is an unsorted mixture of many different sediment sizes.

A close examination of glacial till often reveals cobbles that have been scratched as they were dragged along by the ice.

Figure 4.15 Glacial till Unlike sediment deposited by running water and wind, material deposited directly by a glacier is not sorted. Figure 4.10 provides an example of till being deposited at the terminus of a glacier. (Top photo by Michael Collier; lower photo by E.J. Tarbuck)

of till are characteristically unsorted mixtures of many particle sizes (**Figure 4.15**). **Stratified drift** is sorted according to the size and weight of the fragments. Because ice is not capable of such sorting activity, these sediments are not deposited directly by the glacier. Rather, they reflect the sorting action of glacial meltwater.

Some deposits of stratified drift are made by streams coming directly from a glacier. Other stratified deposits involve sediment that was originally laid down as till and later picked up, transported, and redeposited by meltwater beyond the margin of the ice. Accumulations of stratified drift often consist largely of sand and gravel because the meltwater is not capable of moving larger material and because the finer rock flour remains suspended and is commonly carried far from the glacier. An indication that stratified drift consists primarily of sand and gravel can be seen in many areas where these deposits are actively mined as aggregate for roadwork and other construction projects.

Boulders found in the till or lying free on the surface are called **glacial**

erratics if they are different from the bedrock below (**Figure 4.16**). Of course, this means that they must have been derived from a source outside the area where they are found. Although the locality of origin for most erratics is unknown, the origin of some can be determined. Therefore, by studying glacial erratics as well as the mineral composition of till, geologists can sometimes trace the path of a lobe of ice. In portions of New England as well as other areas, erratics can be seen dotting pastures and farm fields. In some places, these rocks are cleared from fields and piled to make stone walls and fences.

Moraines, Outwash Plains, and Kettles

Perhaps the most widespread features created by glacial deposition are *moraines*, which are simply layers or ridges of till. Several types of moraines are identified; some are common only to mountain valleys, and others are associated with areas affected by either ice sheets or valley glaciers. Lateral and medial moraines fall in the first category, whereas end moraines and ground moraines are in the second.

Lateral and Medial Moraines The sides of a valley glacier accumulate large quantities of debris from the valley walls. When the glacier wastes away, these materials are left as ridges, called **lateral moraines**, along the sides of the valley. **Medial moraines** form when two advancing valley glaciers come together, forming a single ice stream. The till that was once carried along the margins of each glacier joins to form a single dark stripe of debris within the newly enlarged glacier. The creation of these dark stripes within the ice stream is one obvious proof that glacial ice moves because the medial moraine could not form if the ice did not flow downvalley. It is common to see several dark debris

SmartFigure 4.16 Glacial erratic This large glacially transported boulder, called Doane Rock, is a prominent feature near Nauset Bay on Cape Cod. Such boulders are called *glacial erratics.* (Photo by Michael Collier) (http://goo.gl/MfwH34)

Mobile Field Trip

Figure 4.17 Formation of a medial moraine Kennicott Glacier is a 43-kilometer- (27-mile-) long valley glacier that is sculpting the mountains in Alaska's Wrangell–St. Elias National Park. The dark stripes of sediment are medial moraines. The geologist's sketch shows how lateral moraines of merging glaciers create a medial moraine. (Photo by Michael Collier)

Geologist's Sketch

stripes within a large alpine glacier because one will form each time a tributary glacier joins the main valley (**Figure 4.17**).

End Moraines and Ground Moraines

An **end moraine** is a ridge of till that accumulates at the terminus of a glacier. These relatively common landforms are deposited when a state of equilibrium is attained between wastage and ice accumulation. That is, the end moraine forms when the ice is melting near the end of the glacier at a rate equal to the forward advance of the glacier from its region of nourishment. Although the terminus of the glacier is stationary, the ice continues to flow forward, delivering a continuous supply of sediment in the same manner a conveyor belt delivers goods to the end of a production line. As the ice melts, the till is dropped and the end moraine grows. The longer the ice front remains stable, the larger the ridge of till will become.

Now, suppose that wastage exceeds nourishment. At this point, the front of the glacier begins to recede in the direction from which it originally advanced. As the ice front retreats, however, the conveyor-belt action of the glacier continues to provide fresh supplies of sediment to the terminus. In this manner, a large quantity of till is deposited as the ice wastes away, creating a rock-strewn, undulating surface. This gently rolling layer of till deposited as the ice front recedes is termed **ground moraine**. Ground moraine has a leveling effect, filling in low spots and clogging old stream channels, often leading to a derangement of the existing drainage system. In areas where this layer of till is still relatively fresh, such as the northern Great Lakes region, poorly drained swampy lands are quite common. Periodically, a glacier will retreat to a point where wastage and nourishment once again balance. When this happens, the ice front stabilizes, and a new end moraine forms.

The pattern of end moraine formation and ground moraine deposition may be repeated many times before the glacier has completely vanished. Such a pattern is illustrated in **Figure 4.18**. The very first end moraine to form marks the farthest advance of the glacier and is called the *terminal end moraine*. End moraines that form as the ice front occasionally stabilizes during retreat are termed *recessional end moraines*. Terminal and recessional moraines are essentially alike; the only difference between them is their relative positions.

End moraines deposited by the most recent major stage of Ice Age glaciation

SmartFigure 4.18 End moraines of the Great Lakes region End moraines deposited during the most recent stage of glaciation are prominent features in many parts of this region. (https://goo.gl/QNwbma)

Animation

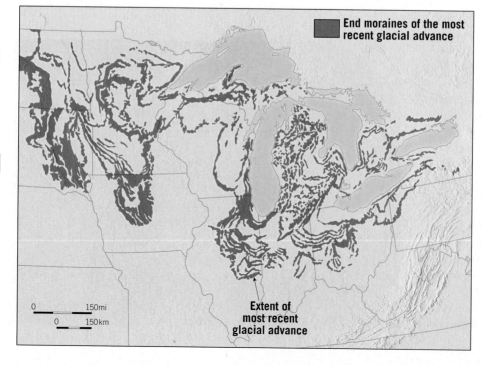

End moraines of the most recent glacial advance

0 ___ 150mi
0 ___ 150km

Extent of most recent glacial advance

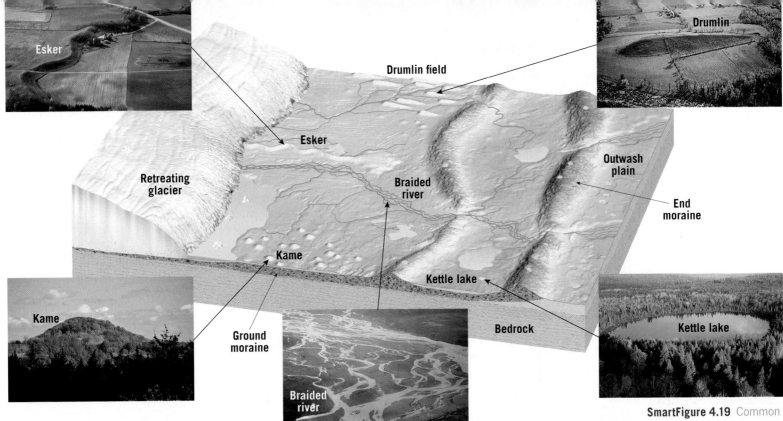

SmartFigure 4.19 Common depositional landforms This diagram depicts a hypothetical area affected by ice sheets in the recent geologic past. (Drumlin photo courtesy of Ward's Science; esker photo by Richard P. Jacobs/JLM Visuals Stock Photos, LLC; kame photo by John Dankwardt; kettle lake photo by Carlyn Iverson/ Science Source; braided river photo by Michael Collier) (https:// goo.gl/7TDPkt)

Tutorial

are prominent features in many parts of the Midwest and Northeast. In Wisconsin, the wooded, hilly terrain of the Kettle Moraine near Milwaukee is a particularly picturesque example. A well-known example in the Northeast is Long Island. This linear strip of glacial sediment that extends northeastward from New York City is part of an end moraine complex that stretches from eastern Pennsylvania to Cape Cod, Massachusetts.

Figure 4.19 represents a hypothetical area during and following glaciation. It shows the end moraines that were just described as well as the depositional features that are discussed in the sections that follow. This figure depicts landscape features similar to those that might be encountered if you were traveling in the upper Midwest or New England. As you read about other glacial deposits, you will be referred to this figure again.

Outwash Plains and Valley Trains At the same time that an end moraine is forming, meltwater emerges from the ice in rapidly moving streams that are often choked with suspended material and carrying a substantial bed load. As the water leaves the glacier and loses velocity, much of its bed load is dropped, often causing the stream channel to become braided. In this way, a broad, ramplike accumulation of stratified drift is built adjacent to the downstream edge of most end moraines. When the feature is formed in association with an ice sheet, it is termed an **outwash plain**, and when it is confined to a mountain valley, it is commonly referred to as a **valley train** (see Figure 4.19).

Kettles Often, end moraines, outwash plains, and valley trains are pockmarked with basins or depressions known as **kettles** (see Figure 4.19). Kettles form when blocks of stagnant ice become buried in drift and eventually melt,

leaving pits in the glacial sediment. Most kettles do not exceed 2 kilometers (1.25 miles) in diameter, and the typical depth of most kettles is less than 10 meters (33 feet). Water often fills a kettle depression and forms a pond or lake.

Drumlins, Eskers, and Kames

Moraines are not the only landforms deposited by glaciers. Some landscapes are characterized by numerous elongate parallel hills made of till. Other areas exhibit conical hills and relatively narrow winding ridges composed mainly of stratified drift.

Drumlins **Drumlins** are streamlined asymmetrical hills composed of till (see Figure 4.19). They range in height from 15 to 60 meters (50 to 200 feet) and average 0.4 to 0.8 kilometer (0.25 to 0.50 mile) in length. The steep side of the hill faces the direction from which the ice advanced, while the gentler slope points in the direction the ice moved. Drumlins are not found singly but rather occur in clusters, called *drumlin fields*. One such cluster, east of Rochester, New York, is estimated to contain about 10,000 drumlins. Their streamlined shape indicates that they were molded in the zone of flow within an active glacier. It is thought that drumlins originate when glaciers advance over previously deposited drift and reshape the material.

Eskers and Kames In some areas that were once occupied by glaciers, sinuous ridges composed largely of sand and gravel may be found. These ridges, called **eskers**, are deposits made by streams flowing in tunnels within or beneath the ice, near the terminus of a glacier (see Figure 4.19). They may be several meters high and extend for many kilometers. In some areas, they are mined for sand and gravel, and for this reason eskers are disappearing in some localities.

Kames are steep-sided hills that, like eskers, are composed largely of stratified drift (see Figure 4.19). Kames originate when glacial meltwater washes sediment into openings and depressions in the stagnant wasting terminus of a glacier. When the ice eventually melts away, the stratified drift is left behind as mounds or hills.

 4.4 CONCEPT CHECKS

1. What is the difference between till and stratified drift?
2. Distinguish among terminal end moraine, recessional end moraine, and ground moraine. Relate these moraines to the budget of a glacier.
3. Describe the formation of a medial moraine.
4. List four depositional features other than moraines.

4.5 Other Effects of Ice Age Glaciers

Describe and explain several important effects of Ice Age glaciers other than the formation of erosional and depositional landforms.

In addition to the massive erosional and depositional work carried out by Ice Age glaciers, ice sheets had other, sometimes profound, effects on the landscape. For example, as the ice advanced and retreated, animals and plants were forced to migrate. This led to stresses that some organisms could not tolerate. Furthermore, many present-day stream courses bear little resemblance to their preglacial routes. The Missouri River once flowed northward, toward Hudson Bay in Canada. The Mississippi River followed a path through central Illinois, and the head of the Ohio River reached only as far as Indiana (**Figure 4.20**). A comparison of the two parts of Figure 4.20 shows that the Great Lakes were created by glacial erosion during the Ice Age. Prior to the Ice Age, the basins occupied by these huge lakes were lowlands with rivers that ran eastward to the Gulf of St. Lawrence.

In areas that were centers of ice accumulation, such as Scandinavia and northern Canada, the land has been slowly rising for the past several thousand years. The land had downwarped under the tremendous weight of 3-kilometer-thick (almost 2-mile-thick) masses of ice. Following the removal of this immense load, the crust has been adjusting by gradually rebounding upward ever since.

Ice sheets and alpine glaciers can act as dams to create lakes by trapping glacial meltwater and blocking the flow of rivers. Some of these lakes are relatively small and short-lived. Others can be large and exist for hundreds or thousands of years.

Figure 4.21 is a map of Lake Agassiz—the largest lake to form during the Ice Age in North America. With the retreat of the ice sheet came enormous volumes of meltwater. The Great Plains generally slope upward to the west. As the terminus of the ice sheet receded northeastward, meltwater was trapped between the ice on one side and the sloping land on the other, causing Lake Agassiz

to deepen and spread across the landscape. It came into existence about 12,000 years ago and lasted for about 4500 years. Such water bodies are termed **proglacial lakes**, referring to their position just beyond the outer limits of a glacier or an ice sheet.

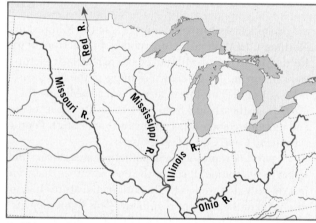

A. This map shows the Great Lakes and the familiar present-day pattern of rivers. Quaternary ice sheets played a major role in creating this pattern.

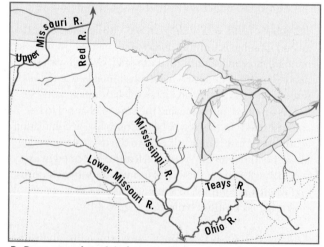

B. Reconstruction of drainage systems prior to the Ice Age. The pattern was very different from today, and the Great Lakes did not exist.

Figure 4.20 Changing rivers The advance and retreat of ice sheets caused major changes in the routes followed by rivers in the central United States.

Figure 4.21 Glacial Lake Agassiz This lake was an immense feature—bigger than all of the present-day Great Lakes combined. Modern-day remnants of this proglacial water body are still major landscape features.

A far-reaching effect of the Ice Age was the world-wide change in sea level that accompanied each advance and retreat of the ice sheets. The snow that nourishes glaciers ultimately comes from moisture evaporated from the oceans. Therefore, when the ice sheets increased

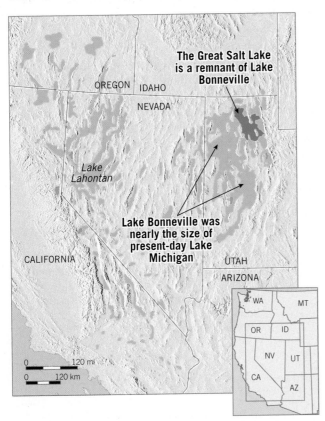

Figure 4.23 Pluvial lakes During the Ice Age, the Basin and Range region experienced a wetter climate than it has today. Many basins turned into large lakes.

During the Last Glacial Maximum, about 18,000 years ago, sea level was nearly 100 meters (330 feet) lower than it is today.

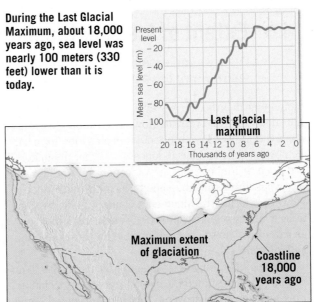

During the Last Glacial Maximum, the shoreline extended out onto the present-day continental shelf.

in size, sea level fell, and the shoreline moved seaward (**Figure 4.22**). Estimates suggest that at the last glacial maximum sea level was as much as 100 meters (330 feet) lower than it is today. Consequently, the Atlantic coast of the United States was located more than 100 kilometers (60 miles) to the east of New York City. Moreover, France and Britain were joined where the English Channel is today. Alaska and Siberia were connected across the Bering Strait, and Southeast Asia was tied by dry land to the islands of Indonesia.

The formation and growth of ice sheets was an obvious response to significant changes in climate. But the existence of the glaciers themselves triggered climatic changes in the regions beyond their margins. In arid and semiarid areas on all continents, temperatures were lower, which meant evaporation rates were also lower. At the same time, precipitation was moderate. This cooler, wetter climate resulted in the formation of many lakes called **pluvial lakes** (from the Latin term *pluvia*, meaning "rain"). In North America, pluvial lakes were concentrated in the vast Basin and Range region of Nevada and Utah (**Figure 4.23**). Although most are now gone, a few remnants remain, the largest of which is Utah's Great Salt Lake.

SmartFigure 4.22
Changing sea level As ice sheets form and then melt away, sea level falls and rises, causing the shoreline to shift. (https://goo.gl/fKN1kJ)

Animation

Did You Know?
Antarctica's ice sheet weighs so much that it depresses Earth's crust by an estimated 900 m (3000 ft) or more.

4.5 CONCEPT CHECKS

1. List four effects of Ice Age glaciers, aside from the formation of major erosional and depositional features.
2. Compare the two parts of Figure 4.20 and identify three major changes to the flow of rivers in the central United States during the Ice Age.
3. How has sea level changed since the Last Glacial Maximum?
4. Contrast proglacial lakes and pluvial lakes.

4.6 Extent of Ice Age Glaciation

Discuss the extent of glaciation and climate variability during the Quaternary Ice Age.

Remember that the Ice Age experienced many cycles of glacial advance and retreat. How do we know this? There was a time when the most popular explanation for what we now know to be glacial deposits was that the material had been drifted in by means of icebergs or perhaps simply swept across the landscape by a catastrophic flood. However, during the nineteenth century, field investigations by many scientists provided convincing proof that an extensive Ice Age was responsible for these deposits and for many other features.

By the beginning of the twentieth century, geologists had largely determined the extent of Ice Age glaciation. Further, they had discovered that many glaciated regions did not have just one layer of drift but several. Close examination of these older deposits showed well-developed zones of chemical weathering and soil formation, as well as the remains of plants that require warm temperatures. The evidence was clear: There had not been just one glacial advance but several, each separated by extended periods when climates were as warm as or warmer than at present. The Ice Age was not simply a time when the ice advanced over the land, lingered for a while, and then receded. Rather, it was a complex period characterized by a number of advances and withdrawals of glacial ice.

The glacial record on land is punctuated by many erosional gaps. This makes it difficult to reconstruct the episodes of the Ice Age. But sediment on the ocean floor provides an uninterrupted record of climate cycles for this period. Studies of these seafloor sediments show that glacial/interglacial cycles have occurred about every 100,000 years. About 20 such cycles of cooling and warming were identified for the span we call the Ice Age.

During the Ice Age, ice left its imprint on almost 30 percent of Earth's land area, including about 10 million square kilometers (nearly 4 million square miles) of North America, 5 million square kilometers (2 million square miles) of Europe, and 4 million square kilometers (1.6 million square miles) of Siberia (**Figure 4.24**). The amount of glacial ice in the Northern Hemisphere was roughly twice that in the Southern Hemisphere. The primary reason is that the Southern Hemisphere has little land in the middle latitudes, and, therefore, the southern polar ice could not spread far beyond the margins of

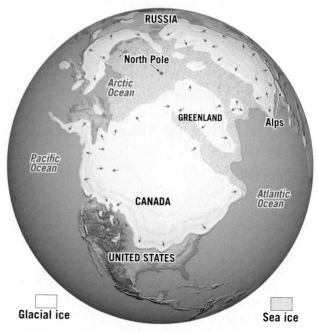

Glacial ice   Sea ice

Figure 4.24 Where was the ice? This map shows the maximum extent of ice sheets in the Northern Hemisphere during the Quaternary Ice Age.

Antarctica. By contrast, North America and Eurasia provided great expanses of land for the spread of ice sheets.

We now know that the Ice Age began between 2 million and 3 million years ago. This means that most of the major glacial episodes occurred during a division of the geologic time scale called the **Quaternary period**. Although the Quaternary is commonly used as a synonym for the Ice Age, this period does not encompass it all. The Antarctic Ice Sheet, for example, formed at least 30 million years ago.

4.6 CONCEPT CHECKS

1. About what percentage of Earth's land surface was affected by glaciers during the Quaternary period?
2. Where were ice sheets more extensive during the Ice Age: the Northern Hemisphere or the Southern Hemisphere? Why?

4.7 Deserts

Describe the general distribution and causes of Earth's dry lands and the role that water plays in modifying desert landscapes.

The dry regions of the world encompass about 42 million square kilometers (more than 16 million square miles), a surprising 30 percent of Earth's land surface. No other climate group covers so large a land area. The word *desert* literally means "deserted," or "unoccupied." For many dry regions, this is a very appropriate description. Yet where water is available in deserts, plants and animals thrive. Nevertheless, the world's dry regions are among the least familiar land areas on Earth outside of the polar realm.

Desert landscapes frequently appear stark. Their profiles are not softened by a carpet of soil and abundant plant life. Instead, barren rocky outcrops with steep, angular slopes are common. Some rocks are tinted orange and red; others in different locations are gray and brown and streaked with black. For many visitors, desert scenery exhibits a striking beauty; to others, the terrain seems bleak. No matter which feeling is elicited, it is clear that deserts are very different from the more humid places where most people live.

As you will see, arid regions are not dominated by a single geologic process. Rather, the effects of tectonic (mountain-building) forces, running water, and wind are all apparent. Because these processes combine in different ways from place to place, the appearance of desert landscapes varies a great deal as well (**Figure 4.25**).

Distribution and Causes of Dry Lands

We all recognize that deserts are dry places, but just what is meant by the word *dry*? That is, how much rain defines the boundary between humid and dry regions?

Sometimes, *dry* is arbitrarily defined by a single rainfall figure, such as 25 centimeters (10 inches) per year of precipitation. However, the concept of dryness is relative; it refers to *any situation in which a water deficiency exists.* Climatologists define **dry climate** as a climate in which yearly precipitation is less than the potential loss of water by evaporation.

Within these water-deficient regions, two climatic types are commonly recognized: **desert**, or arid, and **steppe**, or semiarid. The two categories have many

Figure 4.25 Nevada's Great Basin Desert Mountains separate this area from Pacific moisture and thus contribute to its aridity. When rare storms occur, the sparse vegetation does little to protect the surface from erosion. The appearance of desert landscapes varies a great deal from place to place. (Photo by Dennis Tasa)

features in common; their differences are primarily a matter of degree. The steppe is a marginal and more humid variant of the desert and represents a transition zone that surrounds the desert and separates it from bordering humid climates. Maps showing the distribution of desert and steppe regions reveal that dry lands are concentrated in the subtropics and in the middle latitudes (**Figure 4.26**).

Deserts in places such as Africa, Arabia, and Australia primarily result from the prevailing global

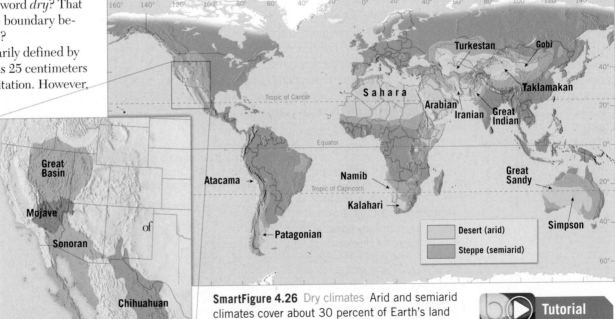

SmartFigure 4.26 Dry climates Arid and semiarid climates cover about 30 percent of Earth's land surface. The dry region of the American West is commonly divided into four deserts, two of which extend into Mexico. (https://goo.gl/HxXqnO)

Tutorial

SmartFigure 4.27
Subtropical deserts
Subtropical deserts and steppes are centered between 20 degrees and 30 degrees north and south latitude in association with belts of high pressure. Dry subsiding air inhibits cloud formation and precipitation. (NASA) (https://goo.gl/ONZDSx)

Did You Know?

The Sahara of North Africa is the world's largest desert. Extending from the Atlantic Ocean to the Red Sea, it covers about 9 million sq km (3.5 million sq mi), an area about the size of the United States. By comparison, the largest desert in the United States, Nevada's Great Basin Desert, has an area that is less than 5 percent as large as the Sahara.

In this view from space, the Sahara Desert, the adjacent Arabian Desert and the Kalahari and Namib deserts are clearly visible as tan-colored, cloud-free zones. The band of clouds across central Africa and the adjacent oceans coincides with the equatorial low-pressure belt.

distribution of air pressure and winds (**Figure 4.27**). Coinciding with dry regions in the lower latitudes are zones of high air pressure known as the *subtropical highs*. These pressure systems are characterized by subsiding air currents (see Figure 13.17). When air sinks, it is compressed and warmed. Such conditions are just the opposite of what is needed to produce clouds and precipitation. Consequently, these regions are known for their clear skies, sunshine, and ongoing dryness.

Middle-latitude deserts and steppes exist principally because they are sheltered in the deep interiors of large landmasses. They are far removed from the ocean, which is the ultimate source of moisture for cloud formation and precipitation. In addition, the presence of high mountains across the paths of prevailing winds further acts to separate these areas from water-bearing,

maritime air masses. In North America, the Coast Range, Sierra Nevada, and Cascades are the foremost mountain barriers to moisture from the Pacific (**Figure 4.28**). In the rainshadow of these mountains lies the dry and expansive Basin and Range region of the American West. Figure 4.25 is a scene of a rainshadow desert in this region. Middle-latitude deserts provide an example of how mountain-building processes affect climate. Without mountains, wetter climates would prevail where dry regions exist today.

The Role of Water in Arid Climates

Permanent streams are normal in humid regions, but practically all desert streambeds are dry most of the time (**Figure 4.29A**). Deserts have **ephemeral streams**, which means that they carry water only in response to specific episodes of rainfall. A typical ephemeral stream may flow only a few days or perhaps just a few hours during the year. In some years, the channel might carry no water at all.

This fact is obvious even to a casual traveler who notices the numerous bridges with no streams beneath them, or numerous dips in the road where dry channels cross. However, when the rare heavy showers occur, so much rain falls in such a short time that it cannot all soak in. Because desert vegetative cover is sparse, runoff is largely unhindered and, therefore, rapid, often creating flash floods along valley floors (**Figure 4.29B**). These floods are quite unlike floods in humid regions. A flood on a river like the Mississippi may take days or weeks to reach its crest and then subside. But desert floods arrive suddenly and subside quickly. Because much surface material in a desert is not anchored by vegetation, the amount of erosional work that occurs during a single short-lived rain event is impressive.

A basic characteristic of desert streams is that they are small and die out before reaching the sea. Because the water table is usually far below the surface, desert streams are not usually fed by groundwater, as occurs in

SmartFigure 4.28 Rainshadow deserts Mountains frequently contribute to the aridity of middle-latitude deserts and steppes by creating a rainshadow. The Great Basin Desert is a rainshadow desert that covers nearly all of Nevada and portions of adjacent states. Figure 4.25 is a scene in the Great Basin Desert. (https://goo.gl/xfTCht)

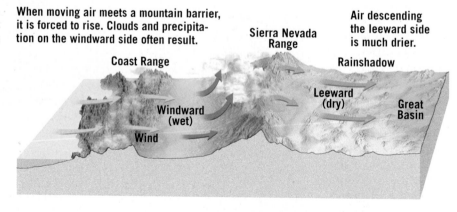

When moving air meets a mountain barrier, it is forced to rise. Clouds and precipitation on the windward side often result.

Air descending the leeward side is much drier.

Sierra Nevada Range

Rainshadow

Coast Range

Windward (wet)

Leeward (dry)

Great Basin

Wind

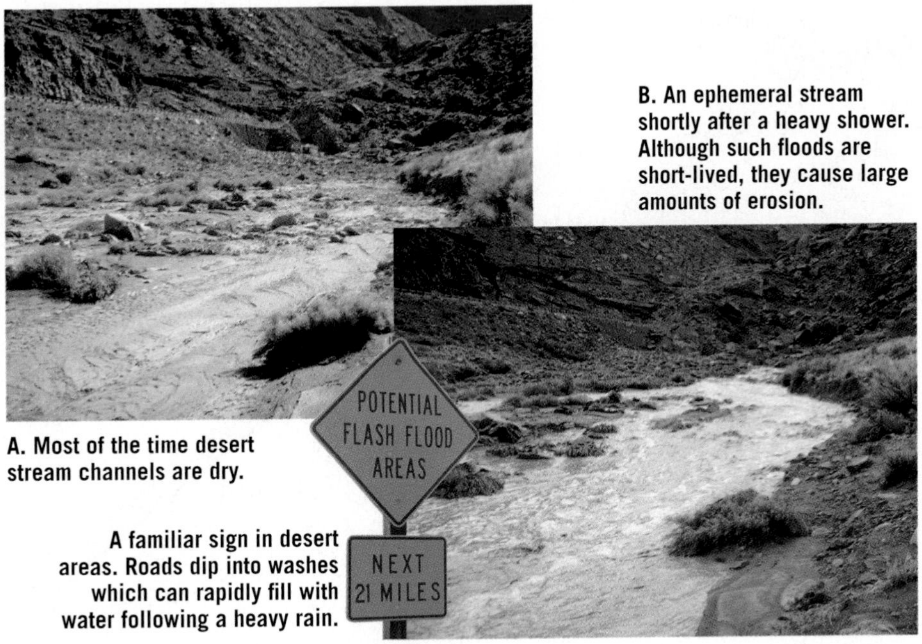
A. Most of the time desert stream channels are dry.
B. An ephemeral stream shortly after a heavy shower. Although such floods are short-lived, they cause large amounts of erosion.
A familiar sign in desert areas. Roads dip into washes which can rapidly fill with water following a heavy rain.
POTENTIAL FLASH FLOOD AREAS
NEXT 21 MILES

Figure 4.29 Ephemeral stream This example is near Arches National Park in southern Utah. In the dry western United States, an ephemeral stream channel is frequently called a *wash* or an *arroyo*. (Photos by E. J. Tarbuck)

streams in humid regions. Without a steady supply of water, the combination of evaporation and soaking in soon depletes the stream.

The few permanent streams that do cross arid regions, such as the Colorado River in the American Southwest and the Nile River in North Africa, originate *outside* the desert, often in mountains where precipitation is plentiful. The supply of water must be great to compensate for the losses that occur as the stream crosses the desert. For example, after the Nile leaves its headwaters in the lakes and mountains of central Africa, it crosses almost 3000 kilometers (nearly 1900 miles) of the Sahara Desert *without a single tributary*.

It should be emphasized that although rainfall is infrequent, *running water is responsible for most of the erosional work in deserts*. This is contrary to a common belief that wind is the most important erosional agent sculpting desert landscapes. Although wind erosion is indeed more significant in dry areas than elsewhere, most desert landforms are carved by running water. The main role of wind, as you will see later in this chapter, is in the transportation and deposition of sediment, which creates and shapes the ridges and mounds we call dunes.

4.7 CONCEPT CHECKS

1. Define *dry climate*. How extensive are the desert and steppe regions of Earth?
2. Discuss the causes of deserts.
3. What is an ephemeral stream?
4. What is the most important erosional agent in deserts?

Did You Know?

Some of the highest temperatures ever recorded have been in deserts. The highest authentically recorded temperature in the United States is also the extreme for the world. On July 10, 1913, the temperature in California's Death Valley reached 134°F.

4.8 Basin and Range: The Evolution of a Mountainous Desert Landscape

Discuss the stages of landscape evolution in the Basin and Range region of the western United States.

Dry regions typically lack permanent streams and often have **interior drainage**. This means that they exhibit a discontinuous pattern of ephemeral streams that do not flow out of the desert to the ocean. In the United States, the dry Basin and Range region provides an excellent example. The region includes southern Oregon, all of Nevada, western Utah, southeastern California, southern Arizona, and southern New Mexico. The chapter-opening photo shows an area in southern California that is part of the Basin and Range. The name Basin and Range is an apt description for this almost 800,000-square-kilometer (312,000-square-mile) region, as it is characterized by more than 200 relatively small mountain ranges that rise 900 to 1500 meters (3000 to 5000 feet) above the basins that separate them. The origin of these *fault-block mountains* is examined in Chapter 6. In this discussion we look at how surface processes change the landscape.

In the Basin and Range region, as in other desert regions with interior drainage, most erosion occurs without reference to the ocean (ultimate base level) because interior-draining streams never reach the sea. The

Did You Know?

Not all deserts are hot. Cold temperatures are experienced in middle-latitude deserts. For example, at Ulaanbaatar in Mongolia's Gobi Desert, the average ***high*** temperature on January days is only −19°C (−2°F)!

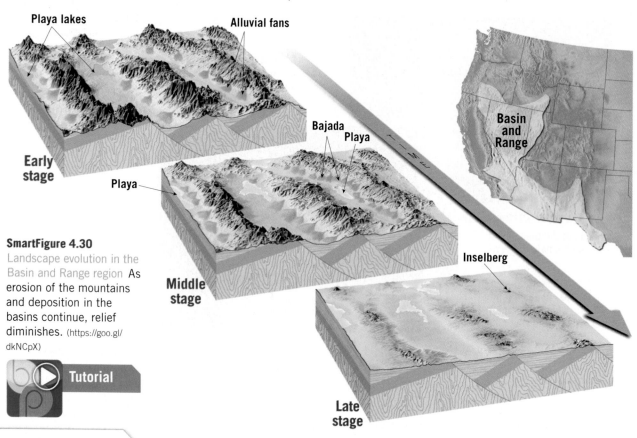

SmartFigure 4.30
Landscape evolution in the Basin and Range region As erosion of the mountains and deposition in the basins continue, relief diminishes. (https://goo.gl/dkNCpX)

Tutorial

block models in **Figure 4.30** depict how the landscape has evolved in the Basin and Range region. During and following uplift of the mountains, running water carves the elevated masses and deposits large quantities of sediment in the basin. In this early stage, relief (difference in elevation between high and low points in an area) is greatest. As erosion lowers the mountains and sediment fills the basins, elevation differences diminish.

When the occasional torrents of water produced by sporadic rains or periods of snowmelt move down the mountain canyons, they are heavily loaded with sediment. Emerging from the confines of the canyon, the runoff spreads over the gentler slopes at the base of the mountains and quickly loses velocity. Consequently, most of its sediment load is dumped within a short distance. The

result is a cone of debris known as an **alluvial fan** at the mouth of a canyon. Over the years, a fan enlarges, eventually coalescing with fans from adjacent canyons to produce an apron of sediment, called a **bajada**, along the mountain front.

On the rare occasions of abundant rainfall, streams may flow across the alluvial fans to the center of the basin, converting the basin floor into a shallow **playa lake**. Playa lakes last only a few days or weeks before evaporation and infiltration remove the water. The dry, flat lake bed that remains is termed a *playa*. Playas occasionally become encrusted with salts (*salt flats*) that are left behind when the water in which the salts were dissolved evaporates. **Figure 4.31** includes a satellite image of a portion of California's Death Valley, a classic Basin and Range landscape. Many of the features just

SmartFigure 4.31 Death Valley: A classic Basin and Range landscape Shortly before the satellite image (left) was taken in February 2005, heavy rains led to the formation of a small playa lake—the pool of greenish water on the basin floor. By May 2005, the lake had reverted to a salt-covered playa. (NASA) The small photo is a closer view of one of Death Valley's many alluvial fans. (Photo by Michael Collier) (https://goo.gl/bqOZ5R)

Condor Video

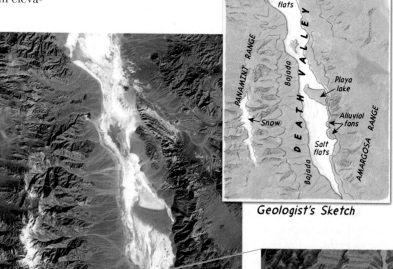
Geologist's Sketch

described are prominent, including a bajada (left side of valley), alluvial fans, a playa lake, and extensive salt flats.

Once a mountain range has stopped rising, ongoing erosion of the mountain mass and the accompanying sedimentation continue to reduce the local relief. Eventually, nearly the entire mountain mass is gone. By the late stages of erosion, the mountain areas are reduced to a few large bedrock knobs (called *inselbergs*) projecting above the sediment-filled basin.

Each of the stages of landscape evolution in an arid climate depicted in Figure 4.30 can be observed in the Basin and Range region. Recently uplifted mountains in an early stage of erosion are found in southern Oregon

and northern Nevada. Death Valley, California, and southern Nevada fit into the more advanced middle stage, whereas the late stage, with its inselbergs, can be seen in southern Arizona.

4.8 CONCEPT CHECKS

1. What is meant by *interior drainage*?
2. Describe the features and characteristics associated with each stage in the evolution of a mountainous desert.
3. Where in the United States can each stage of desert landscape evolution be observed?

4.9 The Work of Wind

Describe the ways in which wind transports sediment and the features created by wind erosion. Distinguish between two basic types of wind deposits.

Like all other erosional agents, wind is a link in the rock cycle in which weathered material is picked up, transported, and deposited. As you will see, the work of wind tends to be most pronounced in arid and semiarid regions.

Wind Erosion

Moving air, like moving water, is turbulent and able to pick up loose debris and transport it to other locations. Just as in a river, the velocity of wind increases with height above the surface. Also like a river, wind transports fine particles in *suspension*, while it carries heavier ones as *bed load*. However, the transport of sediment by wind differs from that by running water in two significant ways. First, wind's lower density compared to water renders it less capable of picking up and transporting coarse materials. Second, because wind is not confined to channels, it can spread sediment over large areas, as well as high into the atmosphere.

Compared to running water and glaciers, wind is a relatively insignificant erosional agent. Recall that even in deserts, most erosion is performed by intermittent

running water, not by wind. Wind erosion is more effective in arid lands than in humid areas because in humid places, moisture binds particles together and vegetation anchors the soil. For wind to be an effective erosional force, dryness and scant vegetation are important prerequisites. When such circumstances exist, wind may pick up, transport, and deposit great quantities of fine sediment. During the 1930s, parts of the Great Plains experienced vast dust storms. The plowing under of the natural vegetative cover for farming, followed by severe drought, exposed the land to wind erosion and led to the area being labeled the Dust Bowl.

Deflation, Blowouts, and Desert Pavement One way that wind erodes is by **deflation**, the lifting and removal of loose material. Wind can suspend only fine sediment such as clay and silt (**Figure 4.32A**). Larger grains of sand roll or bounce along the surface, a process called *saltation*, and comprise the bed load (**Figure 4.32B**). Particles larger than sand are usually not transported by wind. The effects of deflation are sometimes difficult to notice

SmartFigure 4.32
How wind transports sediment **A.** This satellite image shows thick plumes of dust from the Sahara Desert blowing across the Red Sea on June 30, 2009. Such dust storms are common in arid North Africa. In fact, this region is the largest dust source in the world. Satellites are an excellent tool for studying the transport of dust on a global scale. They show that dust storms can cover huge areas and that dust can be transported great distances. (NASA) **B.** The bed load carried by wind consists of sand grains, many of which move by bouncing along the surface. Sand never travels far from the surface, even when winds are very strong. (Photo by Bernd Zoller/Getty Images) (https://goo.gl/ALuj15)

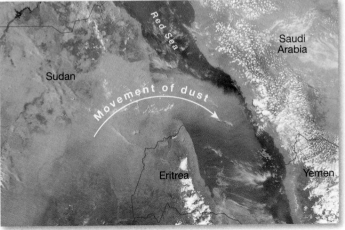

A.

Saltating sand grains.

B.

Video

Figure 4.33 Blowouts Deflation is especially effective in creating these depressions when the land is dry and largely unprotected by anchoring vegetation. (Photo courtesy of USDA/NRCS/Natural Resources Conservation Service)

The man is pointing to where the ground surface was when the grasses began to grow. Wind erosion lowered the land surface to the level of his feet.

Clumps of anchored soil

Unanchored soil

Sand dune

1.2 meters

because the entire surface is being lowered at the same time, but it can be significant.

The most noticeable results of deflation in some places are shallow depressions called **blowouts** (Figure 4.33). In the Great Plains region, from Texas north to Montana, thousands of blowouts can be seen. They range from small dimples less than 1 meter deep and 3 meters (10 feet) wide to depressions that are more than 45 meters (150 feet) deep and several kilometers across. The factor that controls the depths of these basins is the local water table. When blowouts are lowered to the water table, damp ground and vegetation usually prevent further deflation.

In portions of many deserts, the surface is characterized by a layer of coarse pebbles and cobbles that are too large to be moved by the wind. This stony veneer, called **desert pavement**, may form as deflation lowers the surface by removing sand and silt from poorly sorted materials. As **Figure 4.34A** illustrates, the concentration of larger particles at the surface gradually increases as the finer particles are blown away. Eventually, a continuous cover of coarse particles remains.

Studies have shown that the process depicted in Figure 4.34A is not an adequate explanation for all environments in which desert pavement exists. As a result, an alternate explanation was formulated and is illustrated in **Figure 4.34B**. This hypothesis suggests that pavement develops on a surface that initially consists of coarse pebbles. Over time, protruding cobbles trap fine windblown grains that settle and sift downward through the spaces between the larger surface stones. The process is aided by infiltrating rainwater.

Once desert pavement becomes established, a process that might take hundreds of years, the surface is effectively protected from further deflation if left undisturbed. However, because the layer is only one or two stones thick, the passage of vehicles or animals can dislodge the pavement and expose the fine-grained material below. If this happens, the surface is no longer protected from deflation.

Wind Abrasion Like glaciers and streams, wind erodes in part by *abrasion*. In dry regions as well as along some beaches, windblown sand cuts and polishes exposed rock surfaces. Abrasion is often given credit for accomplishments beyond its actual capabilities. Such features as balanced rocks that stand high atop narrow pedestals and intricate detailing on tall pinnacles are not the results of abrasion by windblown sand. Sand is seldom lifted more than a meter above the surface, so the wind's sand-blasting effect is obviously limited in vertical extent. But in areas prone to such activity, telephone poles have actually been cut through near their bases. For this reason, collars may be fitted on poles to protect them from being "sawed" down.

SmartFigure 4.34 Formation of desert pavement **A.** This model shows an area with poorly sorted surface deposits. Over time, deflation lowers the surface, and coarse particles become concentrated. **B.** In this model, the surface is initially covered with cobbles and pebbles. Over time, windblown dust accumulates at the surface and gradually sifts downward. Infiltrating rainwater aids the process. (https://goo.gl/6diZff)

Tutorial

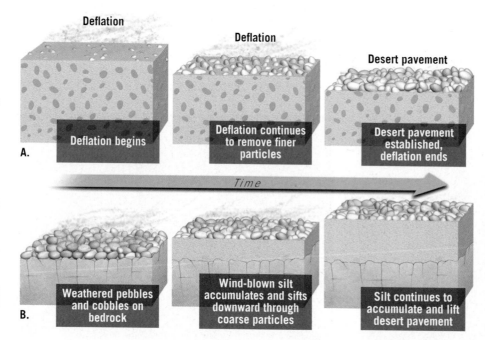

Deflation

Deflation

Desert pavement

A.

Deflation begins

Deflation continues to remove finer particles

Desert pavement established, deflation ends

Time

B.

Weathered pebbles and cobbles on bedrock

Wind-blown silt accumulates and sifts downward through coarse particles

Silt continues to accumulate and lift desert pavement

Figure 4.35 Loess In some regions, the surface is mantled with deposits of windblown silt. (Photo by James E. Patterson, courtesy of F. Lutgens)

This vertical bluff near the Mississippi River in southern Illinois is about 3 meters (10 feet) high.

Wind Deposits

Although wind is relatively unimportant in carving erosional features, significant depositional landforms are created by wind in some regions. Accumulations of windblown sediment are particularly conspicuous in the world's dry lands and along many sandy coasts. Wind deposits are of two distinctive types: (1) extensive blankets of silt, called *loess*, that once were carried in suspension, and (2) mounds and ridges of sand from the wind's bed load, which we call *dunes*.

Loess

In some parts of the world, the surface topography is mantled with deposits of windblown silt, called **loess**. Dust storms deposited this material over thousands of years. When loess is breached by streams or road cuts, it tends to maintain vertical cliffs and lacks any visible layers, as you can see in **Figure 4.35**.

The distribution of loess worldwide indicates two primary sources for this sediment: deserts and glacial deposits of stratified drift. The thickest and most extensive loess deposits occur in western and northern China. They were blown there from the extensive desert basins of central Asia. Accumulations of 30 meters (100 feet) are not uncommon, and thicknesses of more than 100 meters (330 feet) have been measured. It is this fine, buff-colored sediment that gives the Yellow River (Huang He) its name.

In the United States, deposits of loess are significant in many areas, including South Dakota, Nebraska, Iowa, Missouri, and Illinois, as well as portions of the Columbia Plateau in the Pacific Northwest. Unlike the deposits in China, which originated in deserts, the loess in the United States and Europe is an indirect product of glaciation. Its source is deposits of stratified drift.

During the retreat of the ice sheets, many river valleys were choked with glacial sediment. Strong winds sweeping across the barren floodplains picked up the finer sediment and dropped it as a blanket on areas adjacent to the valleys.

Sand Dunes

Like running water, wind deposits its load of sediment when its velocity falls and the energy available for transport diminishes. Thus, sand begins to accumulate wherever an obstruction across the path of the wind slows its movement. Unlike deposits of loess, which form blanketlike layers over broad areas, winds commonly deposit sand in mounds or ridges called **dunes** (**Figure 4.36**).

As moving air encounters an object, such as a clump of vegetation or a rock, the wind sweeps around and over it, leaving a shadow of more slowly moving air behind the obstacle as well as a smaller zone of quieter air just in front of the obstacle. Some of the sand grains moving with the wind come to rest in these *wind shadows*. As the accumulation of sand continues, it forms an increasingly efficient wind barrier that traps even more sand. If there is a sufficient supply of sand and the wind blows steadily long enough, the mound of sand grows into a dune.

Wind

Strong winds move sand up the relatively gentle windward slope.

As sand accumulates at the dune crest, the slope steepens and some of the sand slides down the steep *slip face*.

SmartFigure 4.36 White Sands National Monument The dunes at this landmark in southeastern New Mexico are composed of gypsum. The dunes slowly migrate with the wind. (Photos by Michael Collier) (http://goo.gl/m8eYBH)

Mobile Field Trip

SmartFigure 4.37 Cross-bedding As sand is deposited on the slip face, layers form that are inclined in the direction the wind is blowing. With time, complex patterns develop in response to changes in wind direction. (Photo by Dennis Tasa) (https://goo.gl/Ewb76M)

▶ **Tutorial**

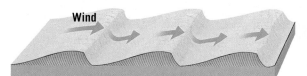

Dunes commonly have an asymmetrical shape and migrate with the wind.

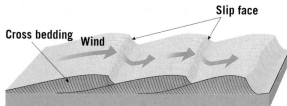

Cross bedding   Wind   Slip face

Sand grains deposited on the slip face at the angle of repose create the cross bedding of dunes.

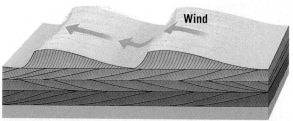

When dunes are buried and become part of the sedimentary rock record, the cross bedding is preserved.

Cross bedding is an obvious characteristic of the Navajo Sandstone in Zion National Park, Utah.

Did You Know?

The highest dunes in the world are located along the southwestern coast of Africa in the Namib Desert. In places, these huge dunes reach heights of 300 to 350 m (1000 to 1167 ft). The dunes at Great Sand Dunes National Park in southern Colorado are the highest in North America, rising more than 210 m (700 ft) above the surrounding terrain.

Many dunes have an asymmetrical profile, with the leeward (sheltered) slope being steep and the windward slope more gently inclined. Sand is rolled up the gentle slope on the windward side by the force of the wind. Just beyond the crest of the dune, the wind velocity is reduced, and the sand accumulates. As more sand collects, the slope steepens, and eventually some of it slides under the pull of gravity. In this way, the leeward slope of the dune, called the **slip face**, maintains a relatively steep angle. Continued sand accumulation, coupled with periodic slides down the slip face, results in the slow migration of the dune in the direction of air movement (see Figure 4.36).

As sand is deposited on the slip face, it forms layers inclined in the direction the wind is blowing. These sloping layers are referred to as **cross bedding**. When the dunes are eventually buried under layers of sediment and become part of the sedimentary rock record, their asymmetrical shape is destroyed, but the cross bedding remains as a testimony to their origin. Nowhere is cross bedding more prominent than in the sandstone walls of Zion Canyon in southern Utah (**Figure 4.37**).

4.9 CONCEPT CHECKS

1. Why is wind erosion relatively more effective in arid regions than in humid areas?
2. What are blowouts? What term describes the process that creates these features?
3. Briefly describe two hypotheses used to explain the formation of desert pavement.
4. Contrast loess and sand dunes.
5. Describe how sand dunes migrate.
6. What is cross bedding?

CONCEPTS IN REVIEW
Glacial and Arid Landscapes

4.1 Glaciers and the Earth System

Explain the role of glaciers in the hydrologic and rock cycles and describe the different types of glaciers and their present-day distribution.

KEY TERMS: glacier, valley (alpine) glacier, ice sheet, sea ice, ice shelf, ice cap, piedmont glacier, outlet glacier

- A glacier is a thick mass of ice originating on land from the compaction and recrystallization of snow, and it shows evidence of past or present flow. Glaciers are part of both the hydrologic cycle and the rock cycle. They store and release freshwater, and they transport and deposit large quantities of sediment.
- Valley glaciers flow down mountain valleys, whereas ice sheets are very large masses, such as those that cover Greenland and Antarctica. During the Last Glacial Maximum, about 18,000 years ago, large areas of Earth were covered by glacial ice.

• North pole

Greenland

NASA

- When valley glaciers exit confining mountains, they may spread out into broad lobes called piedmont glaciers. Similarly, ice shelves form when glaciers flow into the ocean, producing a layer of floating ice.
- Ice caps are like small ice sheets. Both ice sheets and ice caps may be drained by outlet glaciers, which often resemble valley glaciers.

(?) **This satellite image shows ice in the high latitudes of the Northern Hemisphere. What term is applied to the ice at the North Pole? What term best describes Greenland's ice? Are both considered glaciers? Explain.**

4.2 How Glaciers Move

Describe how glaciers move, the rates at which they move, and the significance of the glacial budget.

KEY TERMS: crevasse, zone of accumulation, zone of wastage, iceberg, glacial budget

- Glaciers move in part by flowing under pressure. On the surface of a glacier, ice is brittle. Below about 50 meters (165 feet), pressure is great, and ice behaves like a plastic material and flows. Another type of movement involves the bottom of the glacier sliding along the ground.
- Fast glaciers may move 800 meters (2600 feet) per year, while slow glaciers may move only 2 meters (6.5 feet) per year. Some glaciers experience periodic surges of rapid movement.

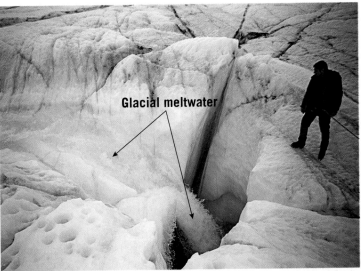

Glacial meltwater

Robbie Shone/Photo Researchers, Inc

- When a glacier has a positive budget, the terminus advances. This occurs when the glacier gains more snow in its upper zone of accumulation that it loses at its downstream zone of wastage. If wastage exceeds the input of new ice, the glacier's terminus retreats.

(?) **The image shows that melting is one mechanism by which glacial ice wastes away. What is another way that ice is lost from a glacier?**

4.3 Glacial Erosion

Discuss the processes of glacial erosion and the major features created by these processes.

KEY TERMS: plucking, abrasion, rock flour, glacial striation, glacial trough, hanging valley, cirque, arête, horn, fiord

- Glaciers acquire sediment through plucking from the bedrock beneath the glacier, by abrasion of the bedrock using

sediment already in the ice, and when mass-wasting processes drop debris on top of the glacier. Grinding of the bedrock produces grooves and scratches called glacial striations.
- Erosional features produced by valley glaciers include glacial troughs, hanging valleys, cirques, arêtes, horns, and fiords.

(?) **Examine the illustration of a mountainous landscape after glaciation. Identify the landforms that resulted from glacial erosion.**

4.4 Glacial Deposits

Distinguish between the two basic types of glacial deposits and briefly describe the features associated with each type.

KEY TERMS: glacial drift, till, stratified drift, glacial erratic, lateral moraine, medial moraine, end moraine, ground moraine, outwash plain, valley train, kettle, drumlin, esker, kame

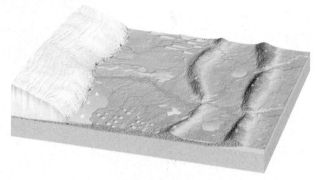

- Any sediment of glacial origin is called drift. The two distinct types of glacial drift are (a) till, which is unsorted material deposited directly by the ice, and (b) stratified drift, which is sediment sorted and deposited by meltwater from a glacier.
- The most widespread features created by glacial deposition are layers or ridges of till, called moraines. Associated with valley glaciers are lateral moraines, formed along the sides of the valley, and medial moraines, formed between two valley glaciers that have merged. End moraines, which mark the former position of the front of a glacier, and ground moraines, undulating layers of till deposited as the ice front retreats, are common to both valley glaciers and ice sheets.

(?) **Examine the illustration of depositional features left behind by a retreating ice sheet. Identify the features and indicate which landforms are composed of till and which are composed of stratified drift.**

4.5 Other Effects of Ice Age Glaciers

Describe and explain several important effects of Ice Age glaciers other than the formation of erosional and depositional landforms.

KEY TERMS: proglacial lake, pluvial lake

- In addition to erosional and depositional features, other effects of Ice Age glaciers include the forced migration of organisms and adjustments of the crust by rebounding upward after removal or the immense load of ice.
- Advance and retreat of ice sheets caused significant changes to the paths followed by rivers. Proglacial lakes formed when glaciers acted as dams to create lakes by trapping glacial meltwater or blocking rivers. In response to the cooler and wetter glacial climate, pluvial lakes formed in areas such as present-day Nevada.
- Ice sheets are nourished by water that ultimately comes from the ocean, so when ice sheets grow, sea level falls, and when they melt, sea level rises.

4.6 Extent of Ice Age Glaciation

Discuss the extent of glaciation and climate variability during the Quaternary Ice Age.

KEY TERMS: Quaternary period

- The Ice Age, which began between 2 and 3 million years ago, was a complex period characterized by numerous advances and withdrawals of glacial ice. Most of the major glacial episodes occurred during a span on the geologic time scale called the Quaternary period. Evidence for the

occurrence of several glacial advances during the Ice Age comes from the existence of multiple layers of drift on land and an uninterrupted record of climate cycles preserved in seafloor sediments.

4.7 Deserts

Describe the general distribution and causes of Earth's dry lands and the role that water plays in modifying desert landscapes.

KEY TERMS: dry climate, desert, steppe, ephemeral stream

- Dry climates cover about 30 percent of Earth's land area. These regions have yearly precipitation totals that are less than the potential loss of water through evaporation. Deserts are drier than steppes, but both climate types are considered water deficient.
- Dry regions in the lower latitudes coincide with zones of subsiding air and high air pressure known as subtropical highs. Middle-latitude deserts exist because of their positions in the deep interiors of large continents far removed from oceans. Mountains also act to shield these regions from humid marine air masses.
- Practically all desert streams are dry most of the time and are said to be ephemeral. Nevertheless, running water is responsible for most of the erosional work in a desert. Although wind erosion is more significant in dry areas than elsewhere, the main role of wind in a desert is transportation and deposition of sediment.

4.8 Basin and Range: The Evolution of a Mountainous Desert Landscape

Discuss the stages of landscape evolution in the Basin and Range region of the western United States.

KEY TERMS: interior drainage, alluvial fan, bajada, playa lake

- The Basin and Range region of the western United States is characterized by interior drainage with streams eroding uplifted mountain blocks and depositing sediment in interior basins. Alluvial fans, bajadas, playas, playa lakes, salt flats, and inselbergs are features often associated with these landscapes.

(?) **Identify the lettered features in this photo. How did they form?**

Michael Collier

4.9 The Work of Wind

Describe the ways in which wind transports sediment and the features created by wind erosion. Distinguish between two basic types of wind deposits.

KEY TERMS: deflation, blowout, desert pavement, loess, dune, slip face, cross bedding

- For wind erosion to be effective, dryness and scant vegetation are essential. Deflation, the lifting and removal of loose material, often produces shallow depressions called blowouts and can also lower the surface by removing sand and silt.
- Abrasion, the "sandblasting" effect of wind, is often given too much credit for producing desert features. However, abrasion does cut and polish rock near the surface.
- Wind deposits are of two distinct types: (1) extensive blankets of silt, called loess, that form when wind deposits its suspended load, and (2) mounds and ridges of sand, called dunes, that form when sediment that was carried as part of the wind's bed load accumulates.

GIVE IT SOME THOUGHT

1. The accompanying diagram shows the results of a classic experiment used to determine how glacial ice moves in a mountain valley. The experiment occurred over an 8-year span. Refer to the diagram and answer the following:
   a. What was the average yearly rate at which ice in the center of the glacier advanced?
   b. About how fast was the center of the glacier advancing *per day*?
   c. Calculate the average rate at which ice along the sides of the glacier moved forward.
   d. Why was the rate at the center different than the rate along the sides?

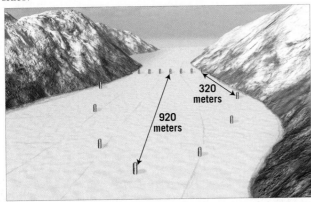

2. Studies have shown that during the Ice Age, the margins of some ice sheets advanced southward from the Hudson Bay region at rates ranging from about 50 to 320 meters per year.
   a. Determine the maximum amount of time required for an ice sheet to move from the southern end of Hudson Bay to the south shore of present-day Lake Erie, a distance of 1600 kilometers.
   b. Calculate the minimum number of years required for an ice sheet to move this distance.

3. If the budget of a valley glacier were balanced for an extended time span, what feature would you expect to find at the terminus of the glacier? Is it composed of till or stratified drift? Now assume that the glacier's budget changes so that wastage exceeds accumulation. How would the terminus of the glacier change? Describe the deposit you would expect to form under these conditions.

4. Is the glacial deposit shown here an example of till or stratified drift? Is it more likely part of an end moraine or an esker?

5. Assume that you and a nongeologist friend are visiting the glacier shown in Figure 4.17. After studying the glacier for quite a long time, your friend asks, "Do these things really move?" How would you convince your friend that this glacier does indeed move, using evidence that is clearly visible in this image?

6. Albuquerque, New Mexico, receives an average of 20.7 centimeters (8.07 inches) of rainfall annually. Albuquerque is considered a desert when the commonly used Köppen climate classification is applied. Yearly precipitation at the Russian city of Verkhoyansk averages 15.5 centimeters (6.05 inches), about 5 centimeters (2 inches) less than Albuquerque, yet is classified as a humid climate. Explain how this can be the case.

7. Bryce Canyon National Park, shown in the accompanying photo, is in dry southern Utah. It is carved into the eastern edge of the Paunsaugunt Plateau. Erosion has sculpted the colorful limestone into bizarre shapes, including spires called "hoodoos." As you and a companion (who has not studied geology) are taking a hike in the national park, your friend says, "It's amazing how wind has created this incredible scenery!" Now that you have studied arid landscapes, how would you respond to your companion's statement?

8. Is either of the following statements true? Are they both true? Explain your answer.
   a. Wind is more effective as an agent of erosion in dry places than in humid places.
   b. Wind is the most important agent of erosion in deserts.

9. The accompanying photo is an aerial view of Preston Mesa dunes in northern Arizona.
   a. Sketch a simple profile (side view) of one of these dunes. Add an arrow to show the prevailing wind direction and label the dune's slip face.
   b. These dunes gradually migrate across the surface. Describe this process.

MasteringGeology™

5

FOCUS ON CONCEPTS

Each statement represents the primary learning objective for the corresponding major heading within the chapter. After you complete the chapter, you should be able to:

5.1 Summarize the view that most geologists held prior to the 1960s regarding the geographic positions of the ocean basins and continents.

5.2 List and explain the evidence Wegener presented to support his continental drift hypothesis.

5.3 List the major differences between Earth's lithosphere and asthenosphere and explain the importance of each in the plate tectonics theory.

5.4 Sketch and describe the movement along a divergent plate boundary that results in the formation of new oceanic lithosphere.

5.5 Compare and contrast the three types of convergent plate boundaries and name a location where each type can be found.

5.6 Describe the relative motion along a transform fault boundary and be able to locate several examples on a plate boundary map.

5.7 Explain why plates such as the African and Antarctic plates are increasing in size, while the Pacific plate is decreasing in size.

5.8 List and explain the evidence used to support the plate tectonics theory.

5.9 Describe plate–mantle convection and explain two of the primary driving forces of plate motion.

Hikers crossing a crevasse in Khumbu Glacier, Mount Everest, Nepal. (Photo by Christian Kober/Robert Harding World Imagery)

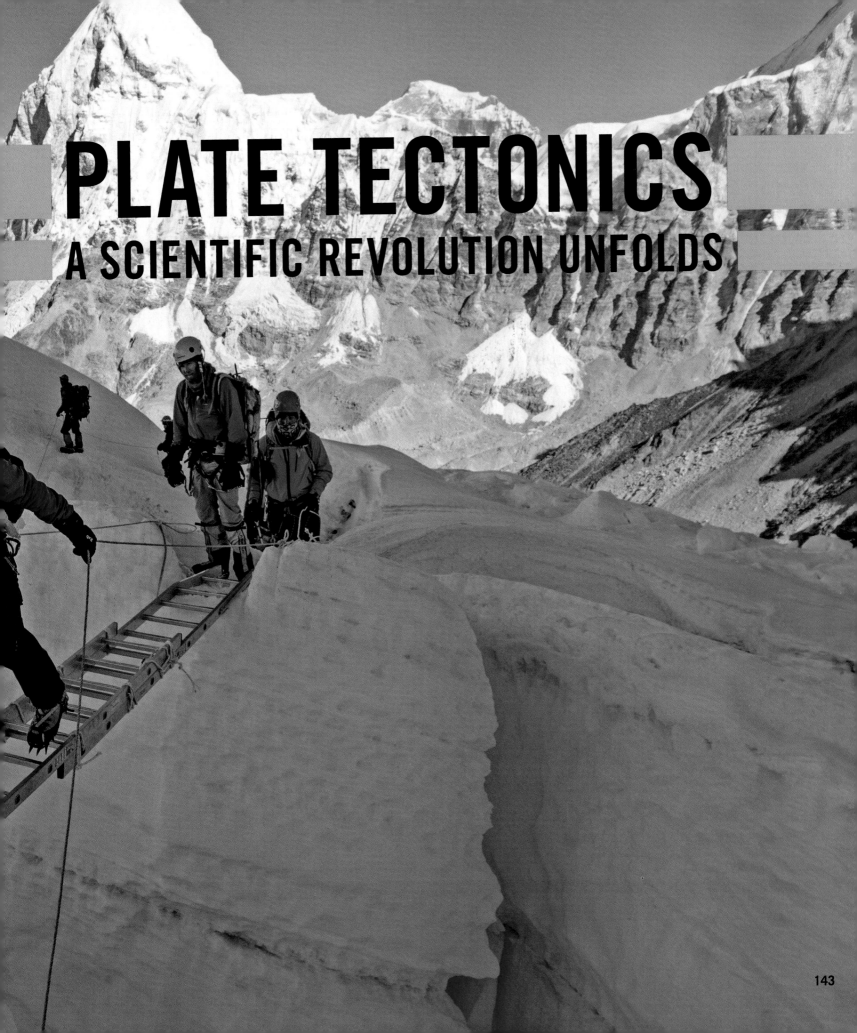

PLATE TECTONICS
A SCIENTIFIC REVOLUTION UNFOLDS

Plate tectonics is the first theory to provide a comprehensive view of the processes that produced Earth's major surface features, including the continents and ocean basins. Within the framework of this model, geologists have found explanations for the basic causes and distribution of earthquakes, volcanoes, and mountain belts. Further, the plate tectonics theory helps explain the formation and distribution of igneous and metamorphic rocks and their relationship with the rock cycle.

5.1 From Continental Drift to Plate Tectonics

Summarize the view that most geologists held prior to the 1960s regarding the geographic positions of the ocean basins and continents.

Figure 5.1 Himalayan mountain range as seen from northern India The tallest mountains on Earth, the Himalayas, were created when the subcontinent of India collided with southeastern Asia. (Photo by Hartmut Postges/Robert Harding World Imagery)

Until the late 1960s most geologists held the view that the ocean basins and continents had fixed geographic positions and were of great antiquity. Over the following decade, scientists came to realize that Earth's continents are not static; instead, they gradually migrate across the globe. These movements cause blocks of continental material to collide, deforming the intervening crust and thereby creating Earth's great mountain chains (**Figure 5.1**). Furthermore, landmasses occasionally split apart. As continental blocks separate, a new ocean basin emerges between them. Meanwhile, other portions of the seafloor plunge into the mantle. In short, a dramatically different model of Earth's tectonic processes emerged. *Tectonic processes* (*tekto* = to build) are those that deform Earth's

crust to create major structural features, such as mountains, continents, and ocean basins.

This profound reversal in scientific thought has been appropriately called a *scientific revolution*. The revolution began early in the twentieth century as a relatively straightforward proposal termed *continental drift*. For more than 50 years, the scientific community categorically rejected the idea that continents are capable of movement. North American geologists in particular had difficulty accepting continental drift, perhaps because much of the supporting evidence had been gathered from Africa, South America, and Australia, continents with which most North American geologists were unfamiliar.

After World War II, modern instruments replaced rock hammers as the tools of choice for many Earth scientists. Armed with more advanced tools, geologists and a new breed of researchers, including *geophysicists* and *geochemists*, made several surprising discoveries that rekindled interest in the drift hypothesis. By 1968 these developments had led to the unfolding of a far more encompassing explanation known as the *theory of plate tectonics*.

In this chapter, we will examine the events that led to this dramatic reversal of scientific opinion. We will also briefly trace the development of the *continental drift hypothesis*, examine why it was initially rejected, and consider the evidence that finally led to the acceptance of its direct descendant—the theory of plate tectonics.

5.1 CONCEPT CHECKS

1. Briefly describe the view held by most geologists regarding the ocean basins and continents prior to the 1960s.
2. What group of geologists were the least receptive to the continental drift hypothesis? Why?

Did You Know?

Alfred Wegener, best known for his continental drift hypothesis, also wrote numerous scientific papers on weather and climate. Following his interest in meteorology, Wegener made four extended trips to the Greenland Ice Sheet to study its harsh winter weather. In November 1930, while making a month-long trek across the ice sheet, Wegener and a companion perished.

5.2 Continental Drift: An Idea Before Its Time

List and explain the evidence Wegener presented to support his continental drift hypothesis.

The idea that continents, particularly South America and Africa, fit together like pieces of a jigsaw puzzle came about during the 1600s, as better world maps became available. However, little significance was given to this notion until 1915, when Alfred Wegener (1880–1930), a German meteorologist and geophysicist, wrote *The Origin of Continents and Oceans*. This book outlined Wegener's hypothesis called **continental drift**, which dared to challenge the long-held assumption that the continents and ocean basins had fixed geographic positions.

Wegener suggested that a single **supercontinent** consisting of all Earth's landmasses once existed.° He named this giant landmass **Pangaea** (pronounced "Pan-jee-ah," meaning "all lands") (**Figure 5.2**). Wegener further hypothesized that about 200 million years ago, during a time period called the *Mesozoic era*, this supercontinent began to fragment into smaller landmasses. These continental blocks then "drifted" to their present positions over a span of millions of years.

Wegener and others who advocated the continental drift hypothesis collected substantial evidence to support their point of view. The fit of South America and Africa and the geographic distribution of fossils and ancient climates all seemed to buttress the idea that these now separate landmasses had once been joined. Let us examine some of this evidence.

Evidence: The Continental Jigsaw Puzzle

Like a few others before him, Wegener suspected that the continents might once have been joined when he noticed the remarkable similarity between the coastlines on opposite sides of the Atlantic Ocean. However, other Earth scientists challenged Wegener's use of present-day shorelines to "fit" these continents together. These opponents correctly argued that wave erosion and depositional processes continually modify shorelines. Even if continental displacement had taken place, a good fit today would be unlikely. Because Wegener's original jigsaw fit of the continents was crude, it is assumed that he was aware of this problem (see Figure 5.2).

Modern reconstruction of Pangaea

Asia

S.E. Asia

Tethys Sea

North America

PANGAEA

Africa

South America

India

Australia

Antarctica

Wegener's Pangaea, redrawn from his book published in 1915.

North America

Europe

Asia

Africa

South America

Australia

Antarctica

SmartFigure 5.2
Reconstructions of Pangaea
The supercontinent of Pangaea, as it is thought to have formed in the late Paleozoic and early Mesozoic eras more than 200 million years ago.
(https://goo.gl/eOttu9)

Tutorial

° Wegener was not the first to conceive of a long-vanished supercontinent. Eduard Suess (1831–1914), a distinguished nineteenth-century geologist, pieced together evidence for a giant landmass comprising South America, Africa, India, and Australia.

Figure 5.3 Two of the puzzle pieces The best fit of South America and Africa occurs along the continental slope at a depth of 500 fathoms (about 900 meters [3000 feet]).

Scientists later determined that a much better approximation of the outer boundary of a continent is the seaward edge of its continental shelf, which lies submerged a few hundred meters below sea level. In the early 1960s, Sir Edward Bullard and two associates constructed a map that pieced together the edges of the continental shelves of South America and Africa at a depth of about 900 meters (3000 feet) (**Figure 5.3**). The remarkable fit obtained was more precise than even these researchers had expected.

Evidence: Fossils Matching Across the Seas

Although the seed for Wegener's hypothesis came from the remarkable similarities of the continental margins on opposite sides of the Atlantic, it was when he learned that identical fossil organisms had been discovered in rocks from both South America and Africa that his pursuit of continental drift became more focused. Wegener learned that most paleontologists (scientists who study the fossilized remains of ancient organisms) agreed that some type of land connection was needed to explain the existence of similar Mesozoic-age life-forms on widely separated landmasses. Just as modern life-forms native to North America are not the same as those of Africa and Australia, during the Mesozoic era, organisms on widely separated continents should have been distinctly different.

Mesosaurus To add credibility to his argument, Wegener documented cases of several fossil organisms found on different landmasses, even though their living forms were unlikely to have crossed the vast ocean presently separating them (**Figure 5.4**). A classic example is *Mesosaurus*, a small aquatic freshwater reptile whose fossil remains are limited to black shales of the Permian period (about 260 million years ago) in eastern South America and southwestern Africa. If *Mesosaurus* had been able to make the long journey across the South Atlantic, its remains should be more widely distributed. As this is not the case, Wegener asserted that South America and Africa must have been joined during that period of Earth history.

How did opponents of continental drift explain the existence of identical fossil organisms in places separated by thousands of kilometers of open ocean? Rafting, transoceanic land bridges (isthmian links), and island stepping stones were the most widely invoked explanations for these migrations (**Figure 5.5**). We know, for example, that during the Ice Age that ended about 8000 years ago, the lowering of sea level allowed mammals (including humans) to cross the narrow Bering Strait that separates Russia and Alaska. Was it possible that land bridges once connected Africa and South America but later subsided below sea level? Modern maps of the seafloor substantiate Wegner's views and show no such sunken land bridges.

Glossopteris Wegener also cited the distribution of the fossil "seed fern" *Glossopteris* [*] as evidence for Pangaea's existence (see Figure 5.4). With tongue-shaped leaves

Figure 5.4 Fossil evidence supporting continental drift Fossils of identical organisms have been discovered in rocks of similar age in Australia, Africa, South America, Antarctica, and India—continents that are currently widely separated by ocean barriers. Wegener accounted for these occurrences by placing these continents in their pre-drift locations.

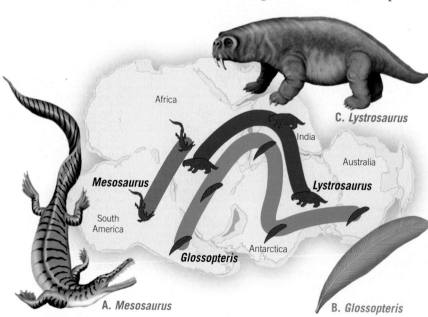

A. *Mesosaurus*
B. *Glossopteris*
C. *Lystrosaurus*

_____

[*] In 1912 Captain Robert Scott and two companions froze to death lying beside 35 pounds (16 kilograms) of rock on their return from a failed attempt to be the first to reach the South Pole. These samples, collected on Beardmore Glacier, contained fossil remains of *Glossopteris*.

and seeds too large to be carried by the wind, this plant was known to be widely dispersed among Africa, Australia, India, and South America. Later, fossil remains of *Glossopteris* were also discovered in Antarctica. Wegener also learned that these seed ferns and associated flora grew only in cool climates—similar to central Alaska. Therefore, he concluded that when these landmasses were joined, they were located much closer to the South Pole.

Figure 5.5 How do land animals cross vast oceans? These sketches illustrate various early proposals to explain the occurrence of similar species on landmasses now separated by vast oceans.
(Used by permission of John C. Holden)

Evidence: Rock Types and Geologic Features

You know that successfully completing a jigsaw puzzle requires maintaining the continuity of the picture while fitting the pieces together. In the case of continental drift, this means that the rocks on either side of the Atlantic that predate the proposed Mesozoic split should match up to form a continuous "picture" when the continents are fitted together as Wegener proposed.

Indeed, Wegener found such "matches" across the Atlantic. For instance, highly deformed igneous rocks in Brazil closely resemble similar rocks of the same age in Africa. Also, the mountain belt that includes the Appalachians trends northeastward through the eastern United States and disappears off the coast of Newfoundland (**Figure 5.6A**). Mountains of comparable age and structure are found in the British Isles and Scandinavia. When these landmasses are positioned as Wegener proposed, (**Figure 5.6B**), the mountain chains form a nearly continuous belt. As Wegener wrote, "It is just as if we were to refit the torn pieces of a newspaper by matching their edges and then check whether the lines of print run smoothly across. If they do, there is nothing left but conclude that the pieces were in fact joined in this way."[†]

[†]Alfred Wegener, *The Origin of Continents and Oceans*, translated from the 4th revised German ed. of 1929 by J. Birman (London: Methuen, 1966).

Evidence: Ancient Climates

Because Alfred Wegener was a student of world climates, he suspected that paleoclimatic (*paleo* = ancient, *climatic* = climate) data might also support the idea of mobile continents. His assertion was bolstered by the discovery of evidence for a glacial period dating to the late Paleozoic era in southern Africa, South America, Australia, and India. This meant that about 300 million years ago, vast ice sheets covered extensive portions of the Southern Hemisphere as well as India

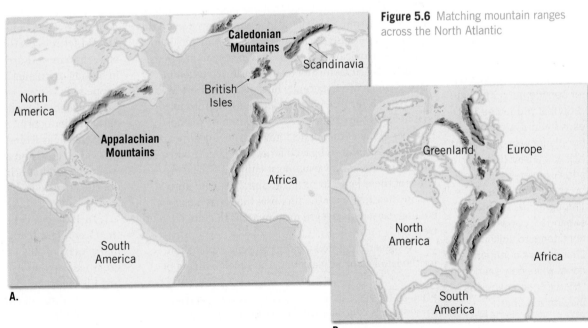

Figure 5.6 Matching mountain ranges across the North Atlantic

A.

B.

Figure 5.7 Paleoclimatic evidence for continental drift **A.** About 300 million years ago, ice sheets covered extensive areas of the Southern Hemisphere and India. Arrows show the direction of ice movement that can be inferred from the pattern of glacial scratches and grooves found in the bedrock. **B.** The continents restored to their pre-drift positions account for tropical coal swamps that existed in areas presently located in temperate climates.

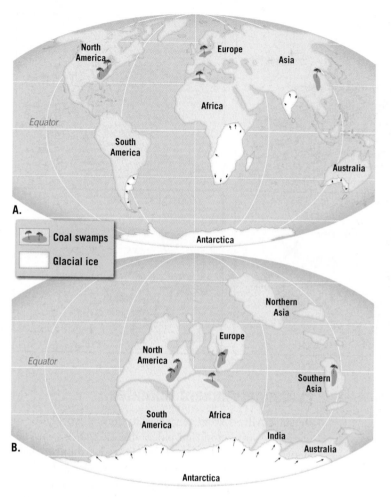

Coal swamps

Glacial ice

(**Figure 5.7A**). Much of the land area that contains evidence of this Paleozoic glaciation presently lies within 30° of the equator in subtropical or tropical climates.

How could extensive ice sheets form near the equator? One proposal suggested that our planet experienced a period of extreme global cooling. Wegener rejected this explanation because during the same span of geologic time, large tropical swamps existed in several locations in the Northern Hemisphere. The lush vegetation in these swamps was eventually buried and converted to coal (**Figure 5.7B**). Today these deposits comprise major coal fields in the eastern United States and Northern Europe. Many of the fossils found in these coal-bearing rocks were produced by tree ferns with large fronds—a fact consistent with warm, moist climates.[°°] The existence of these large tropical swamps, Wegener argued, was inconsistent with the proposal that extreme global cooling caused glaciers to form in areas that are currently tropical.

Wegener suggested a more plausible explanation for the late Paleozoic glaciation: the supercontinent of

_____
[°°] It is important to note that coal can form in a variety of climates, provided that large quantities of plant life are buried.

Pangaea, in which the southern continents were joined together and located near the South Pole (see Figure 5.7B). This would account for the polar conditions required to generate extensive expanses of glacial ice over much of these landmasses. At the same time, this geography places today's northern continents nearer the equator and accounts for the tropical swamps that generated the vast coal deposits.

How does a glacier develop in hot, arid central Australia? How do land animals migrate across wide expanses of the ocean? As compelling as this evidence may have been, 50 years passed before most of the scientific community accepted the concept of continental drift.

The Great Debate

From 1924 when Wegener's book was translated into English, French, Spanish, and Russian until his death in 1930, his proposed drift hypothesis encountered a great deal of hostile criticism. The respected American geologist R. T. Chamberlain stated, "Wegener's hypothesis in general is of the foot-loose type, in that it takes considerable liberty with our globe, and is less bound by restrictions or tied down by awkward, ugly facts than most of its rival theories."

One of the main objections to Wegener's hypothesis stemmed from his inability to identify a credible mechanism for continental drift. Wegener proposed that gravitational forces of the Moon and Sun that produce Earth's tides were also capable of gradually moving the continents across the globe. However, the prominent physicist Harold Jeffreys correctly argued that tidal forces strong enough to move Earth's continents would have resulted in halting our planet's rotation, which, of course, has not happened.

Wegener also incorrectly suggested that the larger and sturdier continents broke through thinner oceanic crust, much as icebreakers cut through ice. However, no evidence existed to suggest that the ocean floor was weak enough to permit passage of the continents without the continents being appreciably deformed in the process.

In 1930, Wegener made his fourth and final trip to the Greenland Ice Sheet (**Figure 5.8**). Although the primary focus of this expedition was to study this great ice cap and its climate, Wegener continued to test his continental drift hypothesis. While returning from Eismitte, an experimental station located in the center of Greenland, Wegener perished along with his

Figure 5.8 Alfred Wegener during an expedition to Greenland (Photo courtesy of Archive of Alfred Wegener Institute)

Greenland companion. His intriguing idea, however, did not die.

Why was Wegener unable to overturn the established scientific views of his day? Foremost was the fact that, although the central theme of Wegener's drift hypothesis was correct, some details were incorrect. For example, continents do not break through the ocean floor, and tidal energy is much too weak to move continents. Moreover, for any comprehensive scientific theory to gain wide acceptance, it must withstand critical testing from all areas of science. Despite Wegener's great contribution to our understanding of Earth, not *all* of the evidence supported the continental drift hypothesis as he had proposed it.

As a result, most of the scientific community (particularly in North America) rejected continental drift or at least treated it with considerable skepticism. However, some scientists recognized the strength of the evidence Wegner had accumulated and continued to pursue the idea.

5.2 CONCEPT CHECKS

1. What was the first line of evidence that led early investigators to suspect that the continents were once connected?
2. Explain why the discovery of the fossil remains of *Mesosaurus* in both South America and Africa, but nowhere else, supports the continental drift hypothesis.
3. Early in the twentieth century, what was the prevailing view of how land animals migrated across vast expanses of open ocean?
4. How did Wegener account for evidence of glaciers in portions of South America, Africa, and India when areas in North America, Europe, and Asia supported lush tropical swamps?
5. Describe two aspects of Wegener's continental drift hypothesis that were objectionable to most Earth scientists.

5.3 The Theory of Plate Tectonics

List the major differences between Earth's lithosphere and asthenosphere and explain the importance of each in the plate tectonics theory.

Following World War II, oceanographers equipped with new marine tools and ample funding from the U.S. Office of Naval Research embarked on an unprecedented period of oceanographic exploration. Over the next two decades, a much better picture of large expanses of the seafloor slowly and painstakingly began to emerge. From this work came the discovery of a global oceanic ridge system that winds through all the major oceans.

In other parts of the ocean, more new discoveries were being made. Studies conducted in the western Pacific demonstrated that earthquakes were occurring at great depths beneath deep-ocean trenches. Of equal importance was the fact that dredging of the seafloor did not bring up any oceanic crust that was older than 180 million years. Further, sediment accumulations in the deep-ocean basins were found to be thin, not the thousands of meters that had been predicted. By 1968 these developments, among others, had led to the unfolding of a far more encompassing theory than continental drift, known as the **theory of plate tectonics**.

Rigid Lithosphere Overlies Weak Asthenosphere

According to the plate tectonics model, the crust and the uppermost, and therefore coolest, part of the mantle constitute Earth's strong outer layer, the **lithosphere** (*lithos* = stone). The lithosphere varies in both thickness and density, depending on whether it is oceanic or

SmartFigure 5.9 Rigid lithosphere overlies the weak asthenosphere (https://goo.gl/KH1iAR)

Tutorial

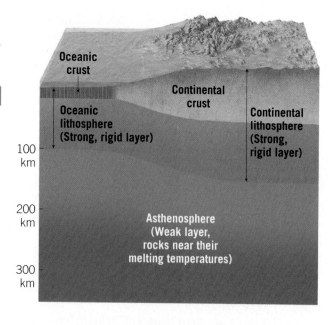

Oceanic crust

Continental crust

Oceanic lithosphere (Strong, rigid layer)

Continental lithosphere (Strong, rigid layer)

100 km

200 km

Asthenosphere (Weak layer, rocks near their melting temperatures)

300 km

continental (**Figure 5.9**). Oceanic lithosphere is about 100 kilometers (60 miles) thick in the deep-ocean basins but is considerably thinner along the crest of the oceanic ridge system—a topic we will consider later. In contrast, continental lithosphere averages about 150 kilometers (90 miles) thick but may extend to depths of 200 kilometers (125 miles) or more beneath the stable interiors of the continents. Further, oceanic and continental crust differ in density. Oceanic crust is composed of basalt and rich in dense iron and magnesium, whereas continental crust is composed largely of less dense granitic rocks. Because of these differences, the overall density of oceanic lithosphere (crust and upper mantle) is greater than the overall density of continental lithosphere. This important difference will be considered in greater detail later in this chapter.

The **asthenosphere** (*asthenos* = weak) is a hotter, weaker region in the mantle that lies below the lithosphere (see Figure 5.9). In the upper asthenosphere (extending in depth from about 100 to 200 kilometers [60 to 125 miles]), the temperatures and pressures are such that rocks are very near their melting temperatures. Consequently, although these rocks are solid, they are quite ductile and respond to forces acting on them by exhibiting a *slow fluid-like flow*. By contrast, the comparatively cool and strong lithosphere tends to respond to forces acting on it by *bending or breaking*. These differences result in Earth's rigid outer shell being effectively detached from the underlying asthenosphere, which allows it to move independently of the layers below.

Earth's Major Plates

The lithosphere is broken into about two dozen segments of irregular size and shape called **lithospheric plates**, or simply **plates**, that are in constant motion with respect to one another (**Figure 5.10**). Seven major lithospheric plates are recognized and account for 94 percent of Earth's surface area: the *North American, South American, Pacific, African, Eurasian, Australian-Indian,* and *Antarctic plates*. The largest is the Pacific plate, which encompasses a significant portion of the Pacific basin. Each of the six other large plates consists of an entire continent, as well as a significant amount of oceanic crust. Notice in Figure 5.10 that the South American plate encompasses almost all of South America and about one-half of the floor of the South Atlantic. Note also that none of the plates are defined entirely by the margins of a single continent. This is a major departure from Wegener's continental drift hypothesis, which proposed that the continents move through the ocean floor, not with it.

Intermediate-sized plates include the *Caribbean, Nazca, Philippine, Arabian, Cocos, Scotia,* and *Juan de Fuca plates*. These plates, with the exception of the Arabian plate, are composed mostly of oceanic lithosphere. In addition, several smaller plates (*microplates*) have been identified but are not shown in Figure 5.10.

Plate Movement

One of the main tenets of the plate tectonics theory is that plates move as somewhat rigid units relative to all other plates. As plates move, the distance between two locations on different plates, such as New York and London, gradually changes, whereas the distance between sites on the same plate—New York and Denver, for example—remains relatively constant. However, parts of some plates are comparatively "weak," such as southern China, which is literally being squeezed as the Indian subcontinent rams into Asia proper.

Because plates are in constant motion relative to each other, most major interactions among them (and, therefore, most deformation) occur along their *boundaries*. In fact, plate boundaries were first established by plotting the locations of earthquakes and volcanoes. Plates are delimited by three distinct types of boundaries, which are differentiated by the type of movement they exhibit. These boundaries are depicted at the bottom of Figure 5.10 and are briefly described here:

- Divergent plate boundaries—where two plates move apart, resulting in upwelling and partial melting of hot material from the mantle to create new seafloor (see Figure 5.10A).
- Convergent plate boundaries—where two plates move together, resulting either in oceanic lithosphere descending beneath an overriding plate, eventually to be reabsorbed into the mantle, or possibly in the collision of two continental blocks to create a mountain belt (see Figure 5.10B).

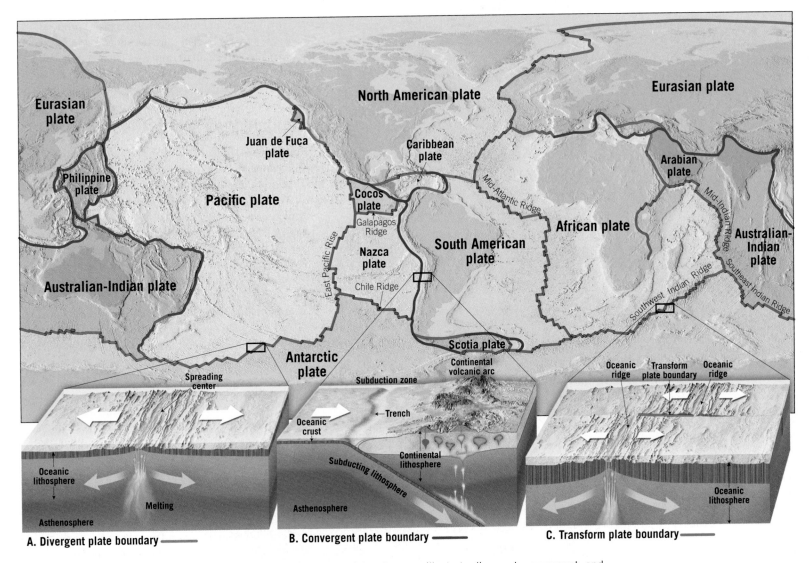

Figure 5.10 Earth's major lithospheric plates The block diagrams below the map illustrate divergent, convergent, and transform plate boundaries.

- Transform plate boundaries—where two plates grind past each other without the production or destruction of lithosphere (see Figure 5.10C).

Divergent and convergent plate boundaries each account for about 40 percent of all plate boundaries. Transform boundaries account for the remaining 20 percent. In the following sections we will discuss the three types of plate boundaries.

5.4 Divergent Plate Boundaries and Seafloor Spreading

Sketch and describe the movement along a divergent plate boundary that results in the formation of new oceanic lithosphere.

Most **divergent plate boundaries** (*di* = apart, *vergere* = to move) are located along the crests of oceanic ridges and can be thought of as *constructive plate margins* because this is where new ocean floor is generated (**Figure 5.11**).

Here, two adjacent plates move away from each other, producing long, narrow fractures in the ocean crust. As a result, hot molten rock from the mantle below migrates upward to fill the voids left as the crust is being ripped

Figure 5.11 Seafloor spreading Most divergent plate boundaries are situated along the crests of oceanic ridges—the sites of seafloor spreading.

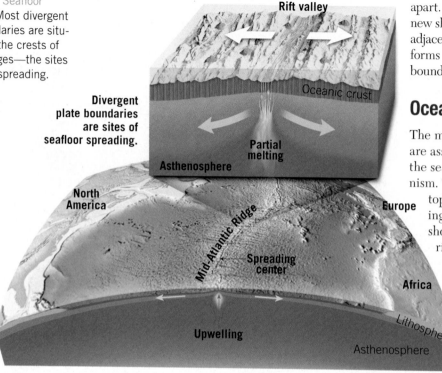

apart. This molten material gradually cools to produce new slivers of seafloor. In a slow yet unending manner, adjacent plates spread apart, and new oceanic lithosphere forms between them. For this reason, divergent plate boundaries are also called **spreading centers**.

Oceanic Ridges and Seafloor Spreading

The majority of, but not all, divergent plate boundaries are associated with *oceanic ridges*: elevated areas of the seafloor characterized by high heat flow and volcanism. The global **oceanic ridge system** is the longest topographic feature on Earth's surface, exceeding 70,000 kilometers (43,000 miles) in length. As shown in Figure 5.10, various segments of the global ridge system have been named, including the Mid-Atlantic Ridge, East Pacific Rise, and Mid-Indian Ridge.

Representing 20 percent of Earth's surface, the oceanic ridge system winds through all major ocean basins, like the seams on a baseball. Although the crest of the oceanic ridge is commonly 2 to 3 kilometers (1 to 2 miles) higher than the adjacent ocean basins, the term *ridge* may be misleading because it implies "narrow" when, in fact, ridges vary in width from 1000 kilometers (600 miles) to more than 4000 (2500 miles) kilometers. Further, along the crest of some ridge segments is a deep canyonlike structure called a **rift valley** (**Figure 5.12**). This structure is evidence that tensional (pulling apart) forces are actively pulling the ocean crust apart at the ridge crest.

The mechanism that operates along the oceanic ridge system to create new seafloor is appropriately called **seafloor spreading**. Spreading typically averages around 5 centimeters (2 inches) per year, roughly the same rate at which human fingernails grow. Comparatively slow spreading rates of 2 centimeters per year are found along the Mid-Atlantic Ridge, whereas spreading rates exceeding 15 centimeters (6 inches) per year have

SmartFigure 5.12 Rift valley in Iceland Thingvellir National Park, Iceland, is located on the western margin of a rift valley roughly 30 kilometers (20 miles) wide. This rift valley is connected to a similar feature that extends along the crest of the Mid-Atlantic Ridge. The cliff in the left half of the image approximates the eastern edge of the North American plate. (Photo by Ragnar Sigurdsson/Arctic Images/ Alamy) (http://goo.gl/RsbHWM)

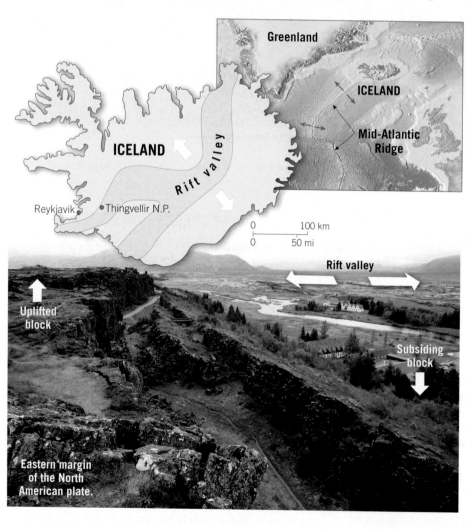

been measured along sections of the East Pacific Rise. Although these rates of seafloor production are slow on a human time scale, they are nevertheless rapid enough to have generated all of Earth's current oceanic lithosphere within the past 200 million years.

The primary reason for the elevated position of the oceanic ridge is that newly created oceanic lithosphere is hot and, therefore, less dense than cooler rocks located away from the ridge axis. (Geologists use the term *axis* to refer to a line that follows the general trend of the ridge crest.) As soon as new lithosphere forms, it is slowly yet continually displaced away from the zone of upwelling. Thus, it begins to cool and contract, thereby increasing in density. This thermal contraction accounts for the increase in ocean depth away from the ridge crest. It takes about 80 million years for the temperature of oceanic lithosphere to stabilize and contraction to cease. By this time, rock that was once part of the elevated oceanic ridge system is located in the deep-ocean basin, where it may be buried by substantial accumulations of sediment.

In addition, as the plate moves away from the ridge, cooling of the underlying asthenosphere causes it to become increasingly more rigid. Thus, oceanic lithosphere is generated by cooling of the asthenosphere from the top down. Stated another way, the thickness

of oceanic lithosphere is age dependent. The older (cooler) it is, the greater its thickness. Oceanic lithosphere that exceeds 80 million years in age is about 100 kilometers (60 miles) thick—approximately its maximum thickness.

Continental Rifting

Divergent boundaries can develop within a continent, in which case the landmass may split into two or more smaller segments separated by an ocean basin. Continental rifting begins when plate motions produce tensional forces that pull and stretch the lithosphere. This stretching, in turn, promotes mantle upwelling and broad upwarping of the overlying lithosphere (**Figure 5.13A**). During this process, the lithosphere is thinned,

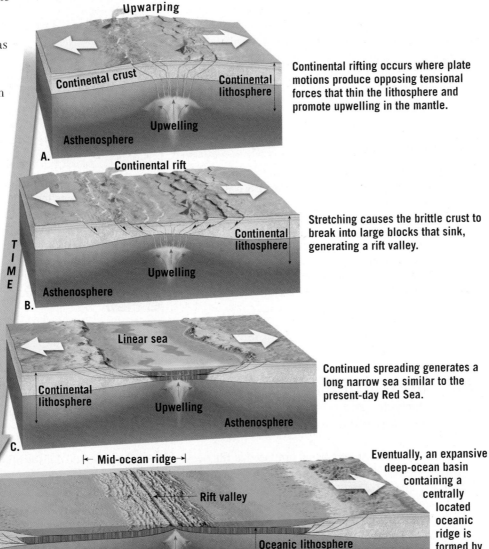

A. Continental rifting occurs where plate motions produce opposing tensional forces that thin the lithosphere and promote upwelling in the mantle.

B. Stretching causes the brittle crust to break into large blocks that sink, generating a rift valley.

C. Continued spreading generates a long narrow sea similar to the present-day Red Sea.

D. Eventually, an expansive deep-ocean basin containing a centrally located oceanic ridge is formed by continued seafloor spreading.

SmartFigure 5.13
Continental rifting: Formation of new ocean basins (https://goo.gl/s4RWua)

Tutorial

while the brittle crustal rocks break into large blocks. As the tectonic forces continue to pull apart the crust, the broken crustal fragments sink, generating an elongated depression called a **continental rift**, which can widen to form a narrow sea (**Figure 5.13B,C**) and eventually a new ocean basin (**Figure 5.13D**).

An example of an active continental rift is the East African Rift (**Figure 5.14**). Whether this rift will eventually result in the breakup of Africa is a topic of ongoing research. Nevertheless, the East African Rift is an excellent model of the initial stage in the breakup of a continent. Here, tensional forces have stretched and thinned the lithosphere, allowing molten rock to ascend from the mantle. Evidence for this upwelling includes several large volcanic mountains, including Mount Kilimanjaro and Mount Kenya, the tallest peaks in Africa. Research suggests that if rifting continues, the rift valley will lengthen and deepen (see Figure 5.13C). At some point, the rift valley will become a narrow sea with an outlet to the ocean. The Red Sea, formed when the Arabian Peninsula split from Africa, is a modern example of such a feature and provides us with a view of how the Atlantic Ocean may have looked in its infancy (see Figure 5.13D).

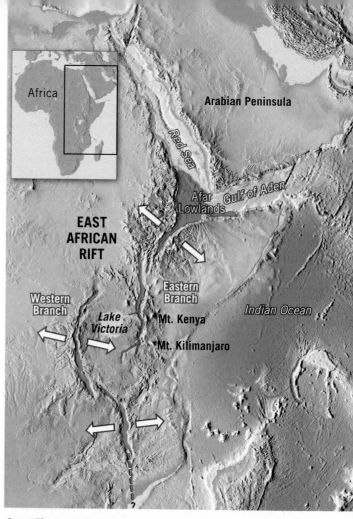

SmartFigure 5.14 East African Rift Valley The East African Rift valley represents the early stage in the breakup of a continent. Areas shown in red consist of lithosphere that has been stretched and thinned, allowing magma to well up from the mantle. (https://goo.gl/Gp4pje)

Condor Video

5.4 CONCEPT CHECKS

1. Sketch or describe how two plates move in relation to each other along divergent plate boundaries.
2. What is the average rate of seafloor spreading in modern oceans?
3. List four features that characterize the oceanic ridge system.
4. Briefly describe the process of continental rifting. Name a location where is it occurring today.

5.5 Convergent Plate Boundaries and Subduction

Compare and contrast the three types of convergent plate boundaries and name a location where each type can be found.

New lithosphere is constantly being produced at the oceanic ridges. However, our planet is not growing larger; its total surface area remains constant. A balance is maintained because older, denser portions of oceanic lithosphere descend into the mantle at a rate equal to seafloor production. This activity occurs along **convergent plate boundaries**, where two plates move toward each other and the leading edge of one is bent downward as it slides beneath the other.

Convergent boundaries are also called **subduction zones** because they are sites where lithosphere is descending (being subducted) into the mantle. Subduction occurs because the density of the descending lithospheric plate is greater than the density of the underlying asthenosphere. Recall that oceanic crust has a greater density than continental crust. In general, old oceanic lithosphere is about 2 percent more dense than the underlying asthenosphere, which causes it to subduct. Continental lithosphere, in contrast, is less dense than the underlying asthenosphere and resists subduction. As a consequence, only oceanic lithosphere will subduct to great depths.

Deep-ocean trenches are long, linear depressions in the seafloor that are generally located only a few hundred kilometers offshore of either a continent or a chain of volcanic islands such as the Aleutian chain. These underwater surface features are produced where oceanic lithosphere bends as it descends into the mantle along subduction zones (see Figure 9.16, page 284). An example is the Peru–Chile trench located along the west coast of South America. It is more than 4500 kilometers (2800 miles) long, and its floor is as much as 8 kilometers (5 miles) below sea level. Western Pacific trenches, including the Mariana and Tonga trenches, are even deeper than those of the eastern Pacific.

Slabs of oceanic lithosphere descend into the mantle at angles that vary from a few degrees to nearly vertical (90 degrees). The angle at which oceanic lithosphere subducts depends largely on its age and, therefore, its density. For example, when seafloor spreading occurs near a subduction zone, as is the case along the coast of Chile, the subducting lithosphere is young and buoyant, which results in a low angle of descent. As the two plates converge, the overriding plate scrapes over the top of the subducting plate below—a type of forced subduction. Consequently, the region around the Peru–Chile trench experiences great earthquakes, including the 2010 Chilean earthquake—one of the 10 largest on record.

As oceanic lithosphere ages (moves farther from the spreading center), it gradually cools, which causes it to thicken and increase in density. In parts of the western Pacific, some oceanic lithosphere is 180 million years old—the thickest and densest in today's oceans. The very dense slabs in this region typically plunge into the mantle at angles approaching 90 degrees. This largely explains why most trenches in the western Pacific are deeper than trenches in the eastern Pacific.

Although all convergent zones have the same basic characteristics, they may vary considerably depending on the type of crustal material involved and the tectonic setting. Convergent boundaries can form *between one oceanic plate and one continental plate, between two oceanic plates,* or *between two continental plates* (**Figure 5.15**).

Oceanic–Continental Convergence

When the leading edge of a plate capped with continental crust converges with a slab of oceanic lithosphere, the buoyant continental block remains "floating," while the denser oceanic slab sinks into the mantle (see Figure

SmartFigure 5.15 Three types of convergent plate boundaries (https://goo.gl/Zlylbf)

Tutorial

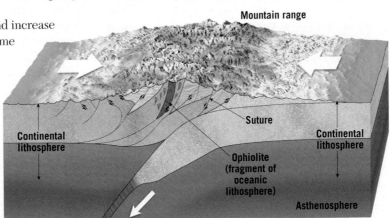

A. Convergent plate boundary where oceanic lithosphere is subducting beneath continental lithosphere.

B. Convergent plate boundary involving two slabs of oceanic lithosphere.

C. Continental collisions occur along convergent plate boundaries when both plates are capped with continental crust.

5.15A). When a descending oceanic slab reaches a depth of about 100 kilometers (60 miles), melting is triggered within the wedge of hot asthenosphere that lies above it. But how does the subduction of a cool slab of oceanic lithosphere cause mantle rock to melt? The answer lies in the fact that water contained in the descending plates acts the way salt does to melt ice. That is, "wet" rock in a high-pressure environment melts at substantially lower temperatures than does "dry" rock of the same composition.

Figure 5.16
Example of an oceanic–continental convergent plate boundary The Cascade Range is a continental volcanic arc formed by the subduction of the Juan de Fuca plate below the North American plate. Mount Hood, Oregon, is one of more than a dozen large composite volcanoes in the Cascade Range.

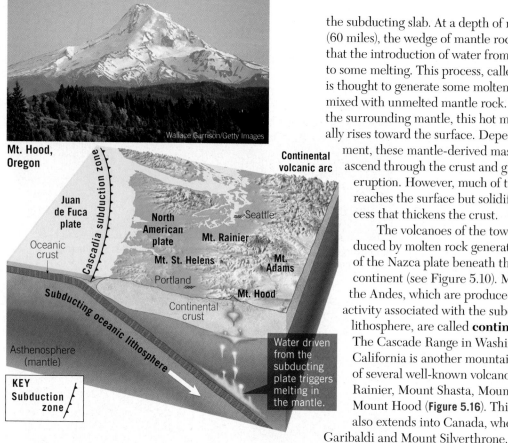

the subducting slab. At a depth of roughly 100 kilometers (60 miles), the wedge of mantle rock is sufficiently hot that the introduction of water from the slab below leads to some melting. This process, called **partial melting**, is thought to generate some molten material, which is mixed with unmelted mantle rock. Being less dense than the surrounding mantle, this hot mobile material gradually rises toward the surface. Depending on the environment, these mantle-derived masses of molten rock may ascend through the crust and give rise to a volcanic eruption. However, much of this material never reaches the surface but solidifies at depth—a process that thickens the crust.

The volcanoes of the towering Andes were produced by molten rock generated by the subduction of the Nazca plate beneath the South American continent (see Figure 5.10). Mountain systems like the Andes, which are produced in part by volcanic activity associated with the subduction of oceanic lithosphere, are called **continental volcanic arcs**. The Cascade Range in Washington, Oregon, and California is another mountain system consisting of several well-known volcanoes, including Mount Rainier, Mount Shasta, Mount St. Helens, and Mount Hood (**Figure 5.16**). This active volcanic arc also extends into Canada, where it includes Mount Garibaldi and Mount Silverthrone.

Sediments and oceanic crust contain large amounts of water, which is carried to great depths by a subducting plate. As the plate plunges downward, heat and pressure drive water from the hydrated (water-rich) minerals in

Oceanic–Oceanic Convergence

An *oceanic–oceanic convergent boundary* has many features in common with oceanic–continental plate margins (see Figure 5.15A,B). Where two oceanic slabs converge, one descends beneath the other, initiating volcanic activity by the same mechanism that operates at all subduction zones (see Figure 5.10). Water released from the subducting slab of oceanic lithosphere triggers melting in the hot wedge of mantle rock above. In this setting, volcanoes grow up from the ocean floor rather than upon a continental platform. When subduction is sustained, it will eventually build a chain of volcanic structures large enough to emerge as islands. The newly formed land, consisting of an arc-shaped chain of volcanic islands, is called a **volcanic island arc** or simply an **island arc** (**Figure 5.17**).

The Aleutian, Mariana, and Tonga Islands are examples of relatively young volcanic island arcs. Island arcs are generally located 120 to 360 kilometers (75 to 225 miles) from a deep-ocean trench. Located adjacent to the island arcs just mentioned are the Aleutian trench, the Mariana trench, and the Tonga trench.

Figure 5.17 Volcanoes in the Aleutian chain The Aleutian Islands are a volcanic island arc produced by the subduction of the Pacific plate beneath the North American plate. Notice that the volcanoes of the Aleutian chain extend into Alaska proper.

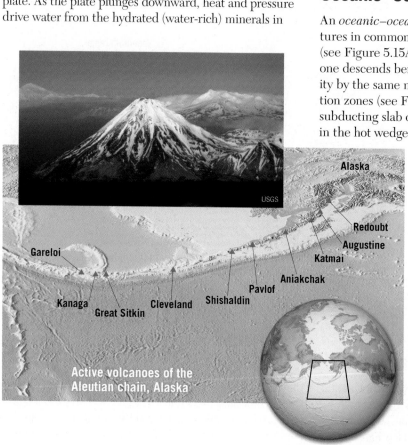

Most volcanic island arcs are located in the western Pacific. Only two are located in the Atlantic—the Lesser Antilles arc, on the eastern margin of the Caribbean Sea, and the Sandwich Islands, located off the tip of South America. The Lesser Antilles are a product of the subduction of the Atlantic seafloor beneath the Caribbean plate. Located within this volcanic arc are the Virgin Islands of the United States and Britain as well as the island of Martinique, where Mount Pelée erupted in 1902, destroying the town of St. Pierre and killing an estimated 28,000 people. This chain of islands also includes Montserrat, where volcanic activity has occurred as recently as 2010.

Island arcs are typically simple structures made of numerous volcanic cones underlain by oceanic crust that is generally less than 20 kilometers (12 miles) thick. Some island arcs, however, are more complex and are underlain by highly deformed crust that may reach 35 kilometers (22 miles) in thickness. Examples include Japan, Indonesia, and the Alaskan Peninsula. These island arcs are built on material generated by earlier episodes of subduction or on small slivers of continental crust that have rafted away from the mainland.

Continental–Continental Convergence

The third type of convergent boundary results when one landmass moves toward the margin of another because of subduction of the intervening seafloor (**Figure 5.18A**). Whereas oceanic lithosphere tends to be dense and sinks into the mantle, the buoyancy of continental material inhibits it from being subducted. Consequently, a collision between two converging continental fragments ensues (**Figure 5.18B**). This process folds and deforms the accumulation of sediments and sedimentary rocks along the continental margins as if they had been placed in a gigantic vise. The result is the formation of a new mountain belt composed of deformed sedimentary and metamorphic rocks that often contain slivers of oceanic lithosphere.

Such a collision began about 50 million years ago, when the subcontinent of India "rammed" into Asia, producing the Himalayas—the most spectacular mountain range on Earth (**Figure 5.18C**). During this collision, the

continental crust buckled and fractured and was generally shortened horizontally and thickened vertically. In addition to the Himalayas, several other major mountain systems, including the Alps, Appalachians, and Urals, formed as continental fragments collided. This topic will be considered further in Chapter 6.

SmartFigure 5.18 The collision of India and Eurasia formed the Himalayas The ongoing collision of the subcontinent of India with Eurasia began about 50 million years ago and produced the majestic Himalayas. It should be noted that both India and Eurasia were moving as these landmasses collided. The map in part C illustrates only the movement of India. (https://goo.gl/9lDLvo)

Animation

5.5 CONCEPT CHECKS

1. Explain why the rate of lithosphere production is roughly equal to the rate of lithosphere destruction.
2. Why does oceanic lithosphere subduct, while continental lithosphere does not?
3. What characteristic of a slab of oceanic lithosphere leads to the formation of deep-ocean trenches instead of shallow trenches?
4. What distinguishes a continental volcanic arc from a volcanic island arc?
5. Briefly describe how mountain belts such as the Himalayas form.

5.6 Transform Plate Boundaries

Describe the relative motion along a transform fault boundary and be able to locate several examples on a plate boundary map.

Along a **transform plate boundary**, also called a **transform fault**, plates slide horizontally past one another without the production or destruction of lithosphere. The nature of transform faults was discovered in 1965 by Canadian geologist J. Tuzo Wilson, who proposed that these large faults connected two spreading centers (divergent boundaries) or, less commonly, two trenches (convergent boundaries). Most transform faults are found on the ocean floor, where they offset segments of the oceanic ridge system, producing a steplike plate margin (**Figure 5.19A**). Notice that the zigzag shape of the Mid-Atlantic Ridge in Figure 5.10 roughly reflects the shape of the original rifting that caused the breakup of the supercontinent of Pangaea. (Compare the shapes of the continental margins of the landmasses on both sides of the Atlantic with the shape of the Mid-Atlantic Ridge.)

Typically, transform faults are part of prominent linear breaks in the seafloor known as **fracture zones**, which include both active transform faults and their inactive extensions into the plate interior (**Figure 5.19B**). Active transform faults lie *only between* the two offset ridge segments and are generally defined by weak, shallow earthquakes. Here seafloor produced at one ridge axis moves in the opposite direction of seafloor produced at an opposing ridge segment. Thus, between the ridge segments, these adjacent slabs of oceanic crust are grinding past each other along a transform fault. Beyond the

ridge crests are inactive zones, where the fractures are preserved as linear topographic depressions. The trend of these fracture zones roughly parallels the direction of plate motion at the time of their formation. Thus, these structures help geologists map the direction of plate motion in the geologic past.

Transform faults also provide the means by which the oceanic crust created at ridge crests can be transported to a site of destruction—the deep-ocean trenches. **Figure 5.20** illustrates this situation. Notice that the Juan de Fuca plate moves in a southeasterly direction, eventually being subducted under the west coast of the United States and Canada. The southern end of this plate is bounded by a transform fault called the Mendocino Fault. This transform boundary connects the Juan de Fuca Ridge to the Cascadia subduction zone. Therefore, it facilitates the movement of the crustal material created at the Juan de Fuca Ridge to its destination beneath the North American continent.

Like the Mendocino Fault, most other transform fault boundaries are located within the ocean basins; however, a few cut through continental crust. Two examples are the earthquake-prone San Andreas Fault of California and New Zealand's Alpine Fault. Notice in Figure 5.20 that the San Andreas Fault connects a spreading center located in the Gulf of California to the Cascadia subduction zone and the Mendocino Fault.

SmartFigure 5.19

Transform plate boundaries

Most transform faults offset segments of a spreading center, producing a plate margin that exhibits a zigzag pattern. (https://goo.gl/SaoJ2o)

Tutorial

A. The Mid-Atlantic Ridge, with its zigzag pattern, roughly reflects the shape of the rifting zone that resulted in the breakup of Pangaea.

B. Fracture zones are long, narrow scar-like features in the seafloor that are roughly perpendicular to the offset ridge segments. They include both the active transform fault and its "fossilized" trace.

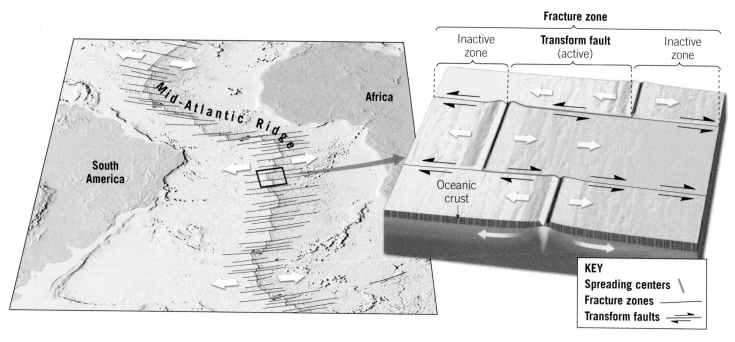

KEY
Spreading centers \
Fracture zones ———
Transform faults ⇄

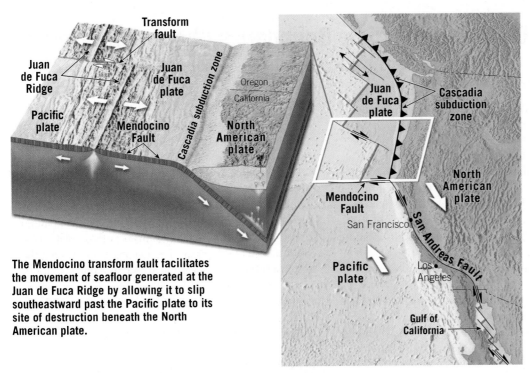

Figure 5.20 Transform faults facilitate plate motion
Seafloor generated along the Juan de Fuca Ridge moves southeastward, past the Pacific plate. Eventually it subducts beneath the North American plate. Thus, this transform fault connects a spreading center (divergent boundary) to a subduction zone (convergent boundary). Also shown is the San Andreas Fault, a transform fault connecting a spreading center located in the Gulf of California with the Mendocino Fault.

The Mendocino transform fault facilitates the movement of seafloor generated at the Juan de Fuca Ridge by allowing it to slip southeastward past the Pacific plate to its site of destruction beneath the North American plate.

Geologist's Sketch

SmartFigure 5.21 Movement along the San Andreas Fault This aerial view shows the offset in the dry channel of Wallace Creek near Taft, California. (http://goo.gl/tKTXky)

Mobile Field Trip

Along the San Andreas Fault, the Pacific plate is moving toward the northwest, past the North American plate (**Figure 5.21**). If this movement continues, the part of California west of the fault zone, including Mexico's Baja Peninsula, will become an island off the West Coast of the United States and Canada. However, a more immediate concern is the earthquake activity triggered by movements along this fault system.

5.6 CONCEPT CHECKS

1. Sketch or describe how two plates move in relation to each other along a transform plate boundary.
2. List two characteristics that differentiate transform faults from the two other types of plate boundaries.

Did You Know?
The Great Alpine Fault of New Zealand is a transform fault that runs through the South Island, marking the boundary line where two plates meet. The northwestern part of the South Island sits on the Australian plate, whereas the rest of the island lies on the Pacific plate. As with its sister fault, California's San Andreas, displacement has been measured in hundreds of kilometers.

5.7 How Do Plates and Plate Boundaries Change?

Explain why plates such as the African and Antarctic plates are increasing in size, while the Pacific plate is decreasing in size.

Although Earth's total surface area does not change, the size and shape of individual plates are constantly changing. For example, the African and Antarctic plates, which are mainly bounded by divergent boundaries—sites of seafloor production—are continually growing in size as new lithosphere is added to their margins. By contrast, the Pacific plate is being consumed into the mantle along much of its flanks faster that it is being generated along the East Pacific Rise and thus is diminishing in size.

Another result of plate motion is that boundaries also migrate. For example, the position of the Peru–Chile trench, which is the result of the Nazca plate being bent downward as it descends beneath the South American plate, has changed over time (see Figure 5.10). Because of the westward drift of the South American plate

relative to the Nazca plate, the Peru–Chile trench has migrated in a westerly direction as well.

Plate boundaries can also be created or destroyed in response to changes in the forces acting on the lithosphere. For example, some plates carrying continental crust are presently moving toward one another. In the South Pacific, Australia is moving northward toward southern Asia. If Australia continues its northward migration, the boundary separating it from Asia will disappear as these plates become one. Other plates are moving apart. Recall that the Red Sea is the site of a relatively new spreading center that came into existence less than 20 million years ago, when the Arabian Peninsula began to break apart from Africa. The breakup of Pangaea is a classic example of how plate boundaries change through geologic time.

The Breakup of Pangaea

Wegener used evidence from fossils, rock types, and ancient climates to create a jigsaw-puzzle fit of the continents, thereby creating his supercontinent of Pangaea. By employing modern tools not available to Wegener, geologists have re-created the steps in the breakup of this supercontinent, an event that began about 180 million

Figure 5.22 The breakup of Pangaea

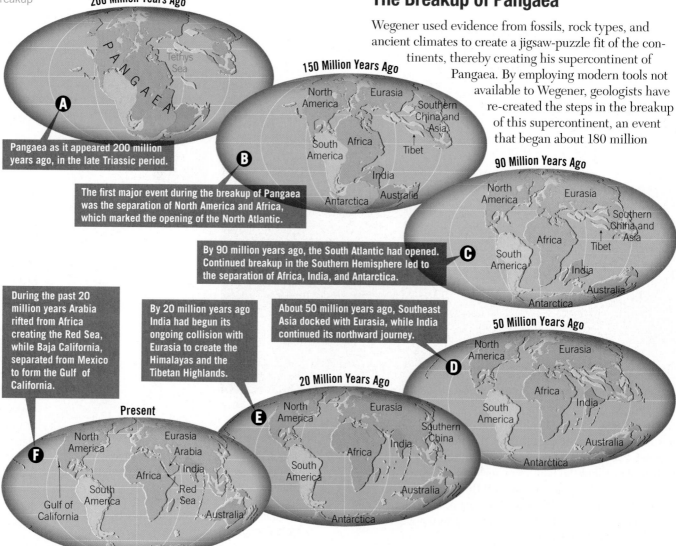

200 Million Years Ago

A Pangaea as it appeared 200 million years ago, in the late Triassic period.

150 Million Years Ago

B The first major event during the breakup of Pangaea was the separation of North America and Africa, which marked the opening of the North Atlantic.

90 Million Years Ago

C By 90 million years ago, the South Atlantic had opened. Continued breakup in the Southern Hemisphere led to the separation of Africa, India, and Antarctica.

50 Million Years Ago

D About 50 million years ago, Southeast Asia docked with Eurasia, while India continued its northward journey.

20 Million Years Ago

E By 20 million years ago India had begun its ongoing collision with Eurasia to create the Himalayas and the Tibetan Highlands.

Present

F During the past 20 million years Arabia rifted from Africa creating the Red Sea, while Baja California, separated from Mexico to form the Gulf of California.

years ago. From this work, the dates when individual crustal fragments separated from one another and their relative motions have been well established (**Figure 5.22**).

An important consequence of Pangaea's breakup was the creation of a "new" ocean basin: the Atlantic. As you can see in Figure 5.22, splitting of the supercontinent did not occur simultaneously along the margins of the Atlantic. The first split developed between North America and Africa. Here, the continental crust was highly fractured, providing pathways for huge quantities of fluid lavas to reach the surface. Today, these lavas are represented by weathered igneous rocks found along the eastern seaboard of the United States—primarily buried beneath the sedimentary rocks that form the continental shelf. Radiometric dating of these solidified lavas indicates that rifting began between 200 million and 190 million years ago. This time span represents the "birth date" for this section of the North Atlantic.

By 130 million years ago, the South Atlantic began to open near the tip of what is now South Africa. As this zone of rifting migrated northward, it gradually opened the South Atlantic (**Figures 5.22B,C**). Continued breakup of the southern landmass led to the separation of Africa and Antarctica and sent India on a northward journey. By the early Cenozoic, about 50 million years ago, Australia had separated from Antarctica, and the South Atlantic had emerged as a full-fledged ocean (**Figure 5.22D**).

India eventually collided with Asia (**Figure 5.22E**), an event that began about 50 million years ago and created the Himalayas and the Tibetan Highlands. About the same time, the separation of Greenland from Eurasia completed the breakup of the northern landmass. During the past 20 million years or so of Earth's history, Arabia has rifted from Africa to form the Red Sea, and Baja California has separated from Mexico to form the Gulf of California (**Figure 5.22F**). Meanwhile, the Panama Arc joined North America and South America to produce our globe's familiar modern appearance.

Plate Tectonics in the Future

Geologists have extrapolated present-day plate movements into the future. **Figure 5.23** illustrates where Earth's landmasses may be 50 million years from now if present plate movements persist during this time span.

In North America we see that the Baja Peninsula and the portion of southern

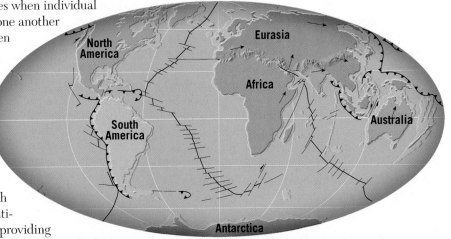

California that lies west of the San Andreas Fault will have slid past the North American plate. If this northward migration takes place, Los Angeles and San Francisco will pass each other in about 10 million years, and in about 60 million years the Baja Peninsula will begin to collide with the Aleutian Islands.

If Africa continues on a northward path, it will continue to collide with Eurasia. The result will be the closing of the Mediterranean, the last remnant of a once-vast ocean called the Tethys Ocean, and the initiation of another major mountain-building episode (see Figure 5.23). Australia will be astride the equator and, along with New Guinea, will be on a collision course with Asia. Meanwhile, North and South America will begin to separate, while the Atlantic and Indian Oceans will continue to grow, at the expense of the Pacific Ocean.

A few geologists have even speculated on the nature of the globe 250 million years in the future. In this scenario the Atlantic seafloor will eventually become old and dense enough to form subduction zones around much of its margins, not unlike the present-day Pacific basin. Continued subduction of the floor of the Atlantic Ocean will result in the closing of the Atlantic basin and the collision of the Americas with the Eurasian–African landmass to form the next supercontinent, shown in **Figure 5.24**. Support for the possible

Figure 5.23 The world as it may look 50 million years from now This reconstruction is highly idealized and based on the assumption that the processes that caused the breakup of Pangaea will continue to operate. (Based on Robert S. Dietz, John C. Holden, C. Scotese, and others)

Figure 5.24 Earth as it may appear 250 million years from now

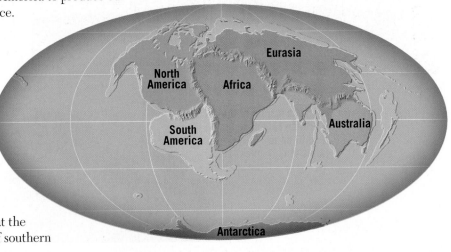

closing of the Atlantic comes from a similar event, when an ocean predating the Atlantic closed during Pangaea's formation. Australia is also projected to collide with Southeast Asia by that time. If this scenario is accurate, the dispersal of Pangaea will end when the continents reorganize into the next supercontinent.

Such projections, although interesting, must be viewed with considerable skepticism because many assumptions must be correct for these events to unfold as just described. Nevertheless, changes in the shapes and positions of continents that are equally profound will undoubtedly occur for many hundreds of millions of years to come. Only after much more of Earth's internal

heat has been lost will the engine that drives plate motions cease.

 5.7 CONCEPT CHECKS

1. Name two plates that are growing in size. Name a plate that is shrinking in size.
2. What new ocean basin was created by the breakup of Pangaea?
3. Briefly describe changes in the positions of the continents if we assume that the plate motions we see today continue 50 million years into the future.

5.8 Testing the Plate Tectonics Model

List and explain the evidence used to support the plate tectonics theory.

Some of the evidence supporting continental drift was presented earlier in this chapter. With the development of plate tectonics theory, researchers began testing this new model of how Earth works. In addition to new supporting data, new interpretations of already existing data had often swayed the tide of opinion.

Evidence: Ocean Drilling

Some of the most convincing evidence for seafloor spreading came from the Deep Sea Drilling Project, which operated from 1966 until 1983. One of the early

goals of the project was to gather samples of the ocean floor in order to establish its age. To accomplish this, the *Glomar Challenger*, a drilling ship capable of working in water thousands of meters deep, was built. Hundreds of holes were drilled through the layers of sediments that blanket the ocean crust, as well as into the basaltic rocks below. Rather than use radiometric dating, which can be unreliable on oceanic rocks because of the alteration of basalt by seawater, researchers dated the seafloor by examining the fossil remains of microorganisms found in the sediments resting directly on the crust at each site.

When the oldest sediment from each drill site was plotted against its distance from the ridge crest, the plot showed that the sediments increased in age with increasing distance from the ridge (**Figure 5.25A**). This finding supported the seafloor-spreading hypothesis, which predicted that the youngest oceanic crust would be found at the ridge crest, the site of seafloor production, and the oldest oceanic crust would be located adjacent to the continents.

The distribution and thickness of ocean-floor sediments provided additional verification of seafloor spreading. Drill cores from the *Glomar Challenger* revealed that sediments are almost entirely absent on the ridge crest and that sediment thickness increases with increasing distance from the ridge (see Figure 5.25A). This pattern of sediment distribution should be expected if the seafloor-spreading hypothesis is correct.

The data collected by the Deep Sea Drilling Project also reinforced the idea

Figure 5.25 Deep-sea drilling **A.** Data collected through deep-sea drilling have shown that the ocean floor is indeed youngest at the ridge axis. **B.** The Japanese deep-sea drilling ship *Chikyu* became operational in 2007. (Photo by Itsuo Inouye/AP Images)

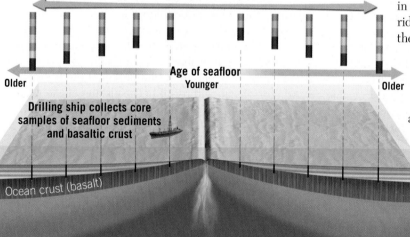

Core samples show that the thickness of sediments increases with increasing distance from the ridge crest.

Older

Age of seafloor
Younger

Older

Drilling ship collects core samples of seafloor sediments and basaltic crust

Ocean crust (basalt)

A.

Chikyu is a state-of-the-art drilling ship designed to drill up to 7000 meters (more than 4 miles) below the seafloor.

B.

that the ocean basins are geologically young because no seafloor older than 180 million years was found. By comparison, most continental crust exceeds several hundred million years in age, and some samples are more than 4 billion years old.

In 1983, a new ocean-drilling program was launched by the Joint Oceanographic Institutions for Deep Earth Sampling (JOIDES). Now the International Ocean Discovery Program (IODP), this ongoing international effort uses multiple vessels for exploration, including the massive 210-meter-long (nearly 690-foot-long) *Chikyu* ("planet Earth" in Japanese), which began operations in 2007 (**Figure 5.25B**). One of the goals of the IODP is to recover a complete section of the ocean crust, from top to bottom.

Evidence: Mantle Plumes and Hot Spots

Mapping volcanic islands and *seamounts* (submarine volcanoes) in the Pacific Ocean revealed several linear chains of volcanic structures. One of the most-studied chains consists of at least 129 volcanoes that extend from the Hawaiian Islands to Midway Island and continue northwestward toward the Aleutian trench (**Figure 5.26**). Radiometric dating of this linear structure, called the Hawaiian Island–Emperor Seamount chain, showed that the volcanoes increase in age with increasing distance from the Big Island of Hawaii. The youngest volcanic island in the chain (Hawaii) rose from the ocean floor less than 1 million

years ago, whereas Midway Island is 27 million years old, and Detroit Seamount, near the Aleutian trench, is about 80 million years old (see Figure 5.26).

Most researchers agree that a cylindrically shaped upwelling of hot rock, called a **mantle plume**, is located beneath the island of Hawaii. As the hot, rocky plume ascends through the mantle, the confining pressure drops, which triggers partial melting. (This process, called *decompression melting*, is discussed in Chapter 7.) The surface manifestation of this activity is a **hot spot**, an area of volcanism, high heat flow, and crustal uplifting that is a few hundred kilometers across. As the Pacific plate moved over a hot spot, a chain of volcanic structures known as a **hot-spot track** was built. As shown in Figure 5.26, the age of each volcano indicates how much time has elapsed since it was situated over the mantle plume. Of approximately 40 hot spots that are thought to have formed because of upwelling of hot mantle plumes, most, but not all, have hot-spot tracks.

A closer look at the five largest Hawaiian Islands reveals a similar pattern of ages, from the volcanically active island of Hawaii to the inactive volcanoes that make up the oldest island, Kauai (see Figure 5.26). Five million

Did You Know?
Olympus Mons is a huge volcano on Mars that strongly resembles the Hawaiian volcanoes. Rising 25 km (16 mi) above the surrounding plains, Olympus Mons owes its massive size to the fact that plate tectonics *does not* operate on Mars. Consequently, instead of being carried away from the hot spot by plate motion, as occurred with the Hawaiian volcanoes, Olympus Mons remained fixed and grew to a gigantic size.

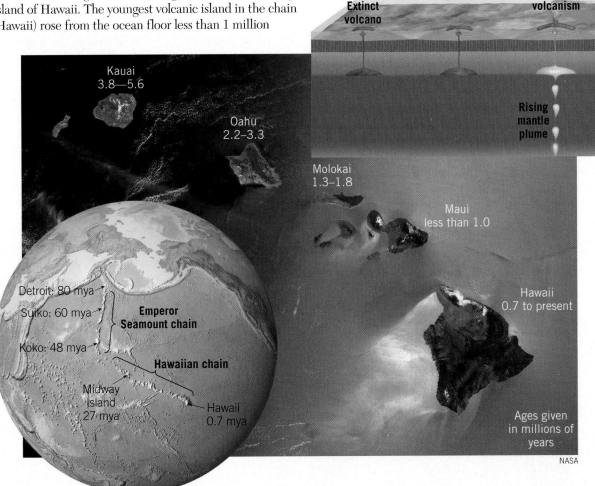

Figure 5.26 Hot-spot volcanism and the formation of the Hawaiian chain Radiometric dating of the Hawaiian Islands shows that volcanic activity increases in age moving away from the Big Island of Hawaii.

Figure 5.27 Earth's magnetic field Earth's magnetic field consists of lines of force much like those a giant bar magnet would produce if placed at the center of Earth.

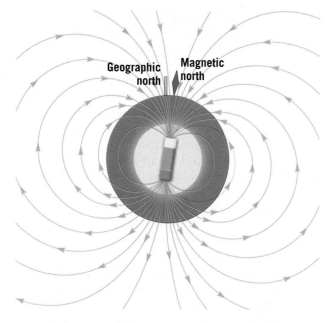

years ago, when Kauai was positioned over the hot spot, it was the *only* modern Hawaiian island in existence. We can see evidence of Kauai's age by examining its extinct volcanoes, which have been eroded into jagged peaks and vast canyons. By contrast, the relatively young island of Hawaii exhibits many fresh lava flows, and one of its five major volcanoes, Kilauea, remains active today.

Evidence: Paleomagnetism

You are probably familiar with how a compass operates and know that Earth's magnetic field has north and south magnetic poles. Today these magnetic poles roughly align with the geographic poles that are located where Earth's rotational axis intersects the surface. Earth's magnetic field is similar to that produced by a simple bar magnet. Invisible lines of force pass through the planet and extend from one magnetic pole to the other (**Figure 5.27**). A compass needle, itself a small magnet free to rotate on

an axis, becomes aligned with the magnetic lines of force and points to the magnetic poles.

Earth's magnetic field is less obvious than the pull of gravity because humans cannot feel it. Movement of a compass needle, however, confirms its presence. In addition, some naturally occurring minerals are magnetic and are influenced by Earth's magnetic field. One of the most common is the iron-rich mineral *magnetite*, which is abundant in lava flows of basaltic composition.* Basaltic lavas erupt at the surface at temperatures greater than 1000°C (1800°F), exceeding a threshold temperature for magnetism known as the **Curie point** (about 585°C [1085°F]). Consequently, the magnetite grains in molten lava are nonmagnetic. However, as the lava cools, these iron-rich grains become magnetized and align themselves in the direction of the existing magnetic lines of force. Once the minerals solidify, the magnetism they possess usually remains "frozen" in this position. Thus, they act like a compass needle because they "point" toward the position of the magnetic poles at the time of their formation. Rocks that formed thousands or millions of years ago and contain a "record" of the direction of the magnetic poles at the time of their formation are said to possess **paleomagnetism**, or **fossil magnetism**.

Apparent Polar Wandering A study of paleomagnetism in ancient lava flows throughout Europe led to an interesting discovery. Taken at face value, the magnetic alignment of iron-rich minerals in lava flows of different ages indicated that the position of the paleomagnetic poles had changed through time. A plot of the location of the magnetic north pole, as measured from Europe, indicated that during the past 500 million years, the pole had gradually "wandered" from a location near Hawaii northeastward to its present location over the Arctic Ocean (**Figure 5.28**). This was strong evidence that either the magnetic north pole had migrated, an idea known as *polar wandering*, or that the poles had remained in place and the continents had drifted beneath them—in other words, Europe had drifted relative to the magnetic north pole.

Although the magnetic poles are known to move in a somewhat erratic path, studies of paleomagnetism from numerous locations show that the positions of the magnetic poles, averaged over thousands of years, correspond closely to the positions of the geographic

Figure 5.28 Apparent polar-wandering path **A.** Scientists believe that the more westerly path determined from North American data was caused by the westward drift of North America by about 24 degrees from Eurasia. **B.** The positions of the wandering paths when the landmasses are reassembled in their pre-drift locations.

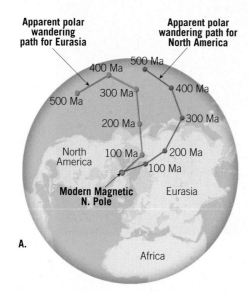

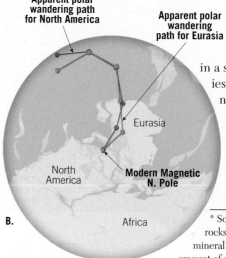

* Some sediments and sedimentary rocks also contain enough iron-bearing mineral grains to acquire a measurable amount of magnetization.

poles. Therefore, a more acceptable explanation for the apparent polar wander was provided by Wegener's hypothesis: If the magnetic poles remain stationary, their *apparent movement* is produced by the drift of the seemingly fixed continents.

Further evidence for continental drift came a few years later, when a polar-wandering path was constructed for North America (see Figure 5.28A). For the first 300 million years or so, the paths for North America and Europe were found to be similar in direction—but separated by about 5000 kilometers (3000 miles). Then, during the middle of the Mesozoic era (180 million years ago), they began to converge on the present North Pole. The explanation for these curves is that North America and Europe were joined until the Mesozoic, when the Atlantic began to open. From this time forward, these continents continuously moved apart. When North America and Europe are moved back to their pre-drift positions, as shown in Figure 5.28B, these paths of apparent polar wandering coincide. This is evidence that North America and Europe were once joined and moved relative to the poles as part of the same continent.

Magnetic Reversals and Seafloor Spreading More evidence emerged when geophysicists learned that over periods of hundreds of thousands of years, Earth's magnetic field periodically reverses polarity. During a **magnetic reversal**, the magnetic north pole becomes the magnetic south pole and vice versa. Lava solidifying during a period of reverse polarity will be magnetized with the polarity opposite that of volcanic rocks being formed

today. When rocks exhibit the same magnetism as the present magnetic field, they are said to possess **normal polarity**, whereas rocks exhibiting the opposite magnetism are said to have **reverse polarity**.

Once the concept of magnetic reversals was confirmed, researchers set out to establish a time scale for these occurrences. The task was to measure the magnetic polarity of hundreds of lava flows and use radiometric dating techniques to establish the age of each flow. **Figure 5.29** shows the **magnetic time scale** established using this technique for the past few million years. The major divisions of the magnetic time scale are called *chrons* and last for roughly 1 million years. As more measurements became available, researchers realized that several short-lived reversals (less than 200,000 years long) often occurred during a single chron.

Meanwhile, oceanographers had begun magnetic surveys of the ocean floor in conjunction with their efforts to construct detailed maps of seafloor topography. These magnetic surveys were accomplished by towing very sensitive instruments, called **magnetometers**, behind research vessels (**Figure 5.30A**). The goal of these geophysical surveys was to map variations in the strength of Earth's magnetic field that arise from differences in the magnetic properties of the underlying crustal rocks.

The first comprehensive study of this type was performed off the Pacific coast of North America and had an unexpected outcome. Researchers discovered alternating stripes of high- and low-intensity magnetism, as shown in **Figure 5.30B**. This relatively simple pattern of magnetic variation defied explanation until 1963, when Fred Vine and D. H. Matthews demonstrated that the high- and low-intensity stripes supported the concept of seafloor spreading. Vine and Matthews suggested

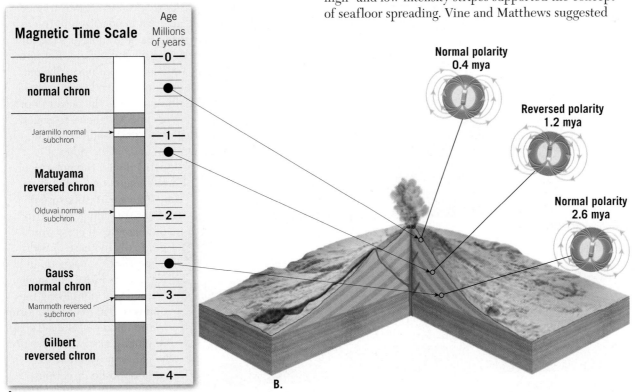

A.

B.

SmartFigure 5.29 Time scale of magnetic reversals **A.** Time scale of Earth's magnetic reversals for the past 4 million years. **B.** This time scale was developed by establishing the magnetic polarity for lava flows of known age. (Data from Allen Cox and G. B. Dalrymple) (https://goo.gl/e9Qwe2)

Tutorial

Figure 5.30 Ocean floor as a magnetic recorder **A.** Magnetic intensities are recorded when a magnetometer is towed across a segment of the oceanic floor. **B.** Notice the symmetrical stripes of low- and high-intensity magnetism that parallel the axis of the Juan de Fuca Ridge. The stripes of high-intensity magnetism occur where normally magnetized oceanic rocks enhanced the existing magnetic field. Conversely, the low-intensity stripes are regions where the crust is polarized in the reverse direction, which weakens the existing magnetic field.

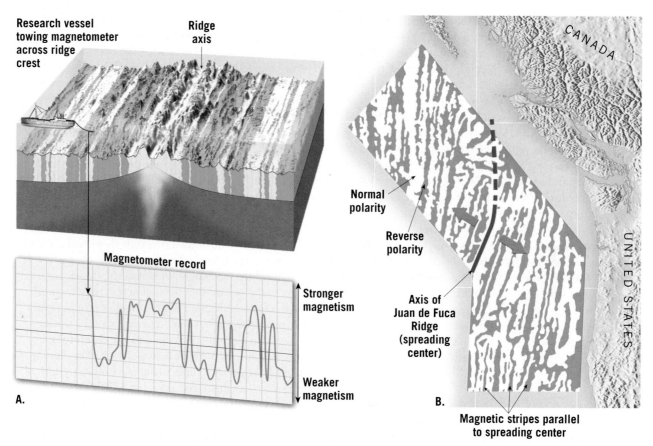

Research vessel towing magnetometer across ridge crest

Ridge axis

Magnetometer record

Stronger magnetism

Weaker magnetism

A.

Normal polarity

Reverse polarity

Axis of Juan de Fuca Ridge (spreading center)

B.

Magnetic stripes parallel to spreading center

SmartFigure 5.31 Magnetic reversals and seafloor spreading When new basaltic rocks form at mid-ocean ridges, they magnetize according to Earth's existing magnetic field. Hence, oceanic crust provides a permanent record of each reversal of our planet's magnetic field over the past 200 million years. (https://goo.gl/5gKsdz)

 **Animation**

A. Normal polarity

Magma

B. Reverse polarity

Magma

C. Normal polarity

Magma

TIME

that the stripes of high-intensity magnetism are regions where the paleomagnetism of the ocean crust exhibits normal polarity (see Figure 5.29A). Consequently, these rocks *enhance* (reinforce) Earth's magnetic field. Conversely, the low-intensity stripes are regions where the ocean crust is polarized in the reverse direction and therefore *weaken* the existing magnetic field. But how do parallel stripes of normally and reversely magnetized rock become distributed across the ocean floor?

Vine and Matthews reasoned that as magma solidifies at the crest of an oceanic ridge, it is magnetized with the polarity of Earth's magnetic field at that time (**Figure 5.31**). Because of seafloor spreading, this strip of magnetized crust would gradually increase in width. When Earth's magnetic field reverses polarity, any newly formed seafloor having the opposite polarity would form in the middle of the old strip. Gradually, the two halves of the old strip would be carried in opposite directions, away from the ridge crest. Subsequent reversals would build a pattern of normal and reverse magnetic stripes, as shown in Figure 5.31. Because new rock is added in equal amounts to both trailing edges of the spreading ocean floor, we should expect the pattern of stripes (width and polarity) found on one side of an oceanic ridge to be a mirror image of those on the other side. In fact, a survey across the Mid-Atlantic Ridge just south of Iceland reveals a pattern of magnetic stripes exhibiting a remarkable degree of symmetry in relation to the ridge axis.

5.8 CONCEPT CHECKS

1. What is the age of the oldest sediments recovered using deep-ocean drilling? How do the ages of these sediments compare to the ages of the oldest continental rocks?
2. How do sedimentary cores from the ocean floor support the concept of seafloor spreading?
3. Assuming that hot spots remain fixed, in what direction was the Pacific plate moving while the Hawaiian Islands were forming?
4. Describe how Fred Vine and D. H. Matthews related the seafloor-spreading hypothesis to magnetic reversals.

5.9 What Drives Plate Motions?

Describe plate–mantle convection and explain two of the primary driving forces of plate motion.

Researchers are in general agreement that some type of *convection*—where hot mantle rocks rise and cold, dense oceanic lithosphere sinks—is the ultimate driver of plate tectonics. Many of the details of this convective flow, however, remain topics of debate in the scientific community.

Forces That Drive Plate Motion

Geophysical evidence confirms that although the mantle consists almost entirely of solid rock, it is hot and weak enough to exhibit a slow, fluid-like convective flow. The simplest type of **convection** is analogous to heating a pot of water on a stove (**Figure 5.32**). Heating the base of a pot warms the water, making it less dense (more buoyant) and causing it to rise in relatively thin sheets or blobs that spread out at the surface. As the surface layer cools, its density increases, and the cooler water sinks back to the bottom of the pot, where it is reheated until it achieves enough buoyancy to rise again. Mantle convection is similar to, but considerably more complex than, the model just described.

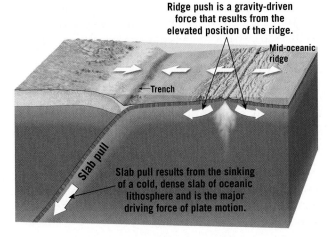

Ridge push is a gravity-driven force that results from the elevated position of the ridge.

Mid-oceanic ridge

Trench

Slab pull

Slab pull results from the sinking of a cold, dense slab of oceanic lithosphere and is the major driving force of plate motion.

Figure 5.33 Forces that act on lithospheric plates

Geologists generally agree that subduction of cold, dense slabs of oceanic lithosphere is a major driving force of plate motion (**Figure 5.33**). This phenomenon, called **slab pull**, occurs because cold slabs of oceanic lithosphere are more dense than the underlying warm asthenosphere and hence "sink like a rock"—meaning that they are pulled down into the mantle by gravity.

Another important driving force is **ridge push** (see Figure 5.33). This gravity-driven mechanism results from the elevated position of the oceanic ridge, which causes slabs of lithosphere to "slide" down the flanks of the ridge. Despite its importance, ridge push contributes far less to plate motions than slab pull. The primary evidence for this is that the fastest-moving plates—Pacific, Nazca, and Cocos plates—have extensive subduction zones along their margins. By contrast, the spreading rate in the North Atlantic basin, which is nearly devoid of subduction zones, is one of the lowest, at about 2.5 centimeters (1 inch) per year.

Models of Plate–Mantle Convection

Although *convection* in the mantle has yet to be fully understood, researchers generally agree on the following:

Cooler water sinks

Warm water rises

Convection is a type of heat transfer that involves the movement of a substance.

Figure 5.32 Convection in a cooking pot As a stove warms the water in the bottom of a cooking pot, the heated water expands, becomes less dense (more buoyant), and rises. Simultaneously, the cooler, denser water near the top sinks.

Did You Know?
Because tectonic processes are powered by heat from Earth's interior, the forces that drive plate motion will cease sometime in the distant future. The work of external forces (such as wind, water, and ice), however, will continue to erode Earth's surface. Eventually, landmasses will be nearly flat. What a different world it will be—an Earth with no earthquakes, no volcanoes, and no mountains.

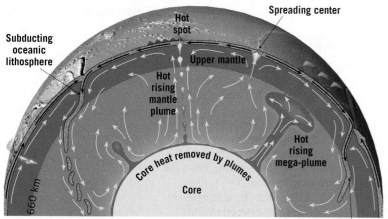

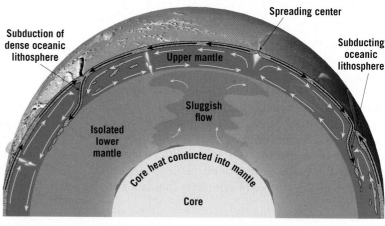

A. In the "whole-mantle model," sinking slabs of cold oceanic lithosphere are the downward limbs of convection cells, while rising mantle plumes carry hot material from the core–mantle boundary toward the surface.

B. The "layer cake model" has two largely disconnected convective layers; a dynamic upper layer driven by descending slabs of cold oceanic lithosphere and a sluggish lower layer that carries heat upward without appreciably mixing with the layer above.

Figure 5.34 Models of mantle convection

- Convective flow—in which warm, buoyant mantle rocks rise while cool, dense lithospheric plates sink—is the underlying driving force for plate movement.
- Mantle convection and plate tectonics are part of the same system. Subducting oceanic plates drive the cold downward-moving portion of convective flow, while shallow upwelling of hot rock along the oceanic ridge and buoyant mantle plumes are the upward-flowing arms of the convective mechanism.
- Convective flow in the mantle is a major mechanism for transporting heat away from Earth's interior to the surface, where it is eventually radiated into space.

What is not known with certainty is the exact structure of this convective flow. Several models have been proposed for plate–mantle convection, and we will look at two of them.

Whole-Mantle Convection One group of researchers favor some type of *whole-mantle convection* model, also called the *plume model*, in which cold oceanic lithosphere sinks to great depths and stirs the entire mantle (**Figure 5.34A**). The whole-mantle model suggests that the ultimate burial ground for these subducting lithospheric slabs is the core–mantle boundary. The downward flow of these subducting slabs is balanced by buoyantly rising mantle plumes that transport hot mantle rock toward the surface.

Two kinds of plumes have been proposed—narrow tube-like plumes and giant upwellings, often referred to as *mega-plumes*. The long, narrow plumes are thought to originate from the core–mantle boundary and produce hot-spot volcanism of the type associated with the Hawaiian Islands, Iceland, and Yellowstone. Scientists believe that areas of large mega-plumes, as shown in Figure 5.34A, occur beneath the Pacific basin and southern Africa. The latter structure is thought to explain why southern Africa has an elevation much higher than would be predicted for a stable continental landmass. In the whole-mantle convection model, heat for both the narrow plumes and mega-plumes is thought to arise mainly from Earth's core, while the deep mantle provides a source for chemically distinct magmas. However, some researchers

have questioned that idea and instead propose that the source of magma for most hot-spot volcanism is found in the upper mantle (asthenosphere).

Layer Cake Model Some researchers argue that the mantle resembles a "layer cake" divided at a depth of perhaps 660 kilometers (410 miles) but no deeper than 1000 kilometers (620 miles). As shown in **Figure 5.34B**, this layered model has two zones of convection—a thin, dynamic layer in the upper mantle and a thick, larger, sluggish one located below. As with the whole-mantle model, the downward convective flow is driven by the subduction of cold, dense oceanic lithosphere. However, rather than reach the lower mantle, these subducting slabs penetrate to depths of no more than 1000 kilometers (620 miles). Notice in Figure 5.34B that the upper layer in the layer cake model is littered with recycled oceanic lithosphere of various ages. Melting of these fragments is thought to be the source of magma for some of the volcanism that occurs away from plate boundaries, such as the hot-spot volcanism of Hawaii.

In contrast to the active upper mantle, the lower mantle is sluggish and does not provide material to support volcanism at the surface. Very slow convection within this layer likely carries heat upward, but very little mixing between these two layers occurs.

Geologists continue to debate the nature of the convective flow in the mantle. As they investigate the possibilities, perhaps a hypothesis that combines features from the layer cake model and the whole-mantle convection model will emerge.

5.9 CONCEPT CHECKS

1. Define *slab pull* and *ridge push*. Which of these forces appears to contribute more to plate motion?
2. Briefly describe the two models of plate–mantle convection.
3. What geological processes are associated with the upward and downward circulation in the mantle?

CONCEPTS IN REVIEW
Plate Tectonics: A Scientific Revolution Unfolds

5.1 From Continental Drift to Plate Tectonics

Summarize the view that most geologists held prior to the 1960s regarding the geographic positions of the ocean basins and continents.

- Fifty years ago, most geologists thought that ocean basins were very old and that continents were fixed in place. Those ideas were discarded with a scientific revolution that revitalized geology: the theory of plate tectonics. Supported by multiple kinds of evidence, plate tectonics is the foundation of modern Earth science.

5.2 Continental Drift: An Idea Before Its Time

List and explain the evidence Wegener presented to support his continental drift hypothesis.

KEY TERMS: continental drift, supercontinent, Pangaea

- German meteorologist Alfred Wegener formulated the continental drift hypothesis in 1915. He suggested that Earth's continents are not fixed in place but moved slowly over geologic time.
- Wegener proposed a supercontinent called Pangaea that existed about 200 million years ago, during the late Paleozoic and early Mesozoic eras.
- Wegener's evidence that Pangaea existed but later broke into pieces that drifted apart included (1) the shape of the continents, (2) continental fossil organisms that matched across oceans, (3) matching rock types and modern mountain belts, and (4) sedimentary rocks that recorded ancient climates, including glaciers on the southern portion of Pangaea.
- Wegener's hypothesis suffered from two flaws: It proposed tidal forces as the mechanism for the motion of continents, and it implied that the continents would have plowed their way through weaker oceanic crust, like boats cutting through a thin layer of sea ice. Most geologists rejected the idea of continental drift when Wegener proposed it, and it wasn't resurrected for another 50 years.

0   10 cm
0   5 inches

John Cancalosi/AGE Fotostock

(?) Why did Wegener choose organisms such as *Glossopteris* and *Mesosaurus* as evidence for continental drift, as opposed to other fossil organisms such as sharks or jellyfish?

5.3 The Theory of Plate Tectonics

List the major differences between Earth's lithosphere and asthenosphere and explain the importance of each in the plate tectonics theory.

KEY TERMS: theory of plate tectonics, lithosphere, asthenosphere, lithospheric plate (plate)

- Research conducted after World War II led to new insights that helped revive Wegener's hypothesis of continental drift. Exploration of the seafloor revealed previously unknown features, including an extremely long mid-ocean ridge system. Sampling of the oceanic crust revealed that it was quite young relative to the continents.
- The lithosphere, Earth's outermost rocky layer, is relatively stiff and deforms by bending or breaking. The lithosphere consists both of crust (either oceanic or continental) and underlying upper mantle. Beneath the lithosphere is the asthenosphere, a relatively weak layer that deforms by flowing.
- The lithosphere consists of about two dozen segments of irregular size and shape. There are seven large lithospheric plates, another seven intermediate-size plates, and numerous relatively small microplates. Plates meet along boundaries that may be divergent (moving apart from each other), convergent (moving toward each other), or transform (moving laterally past each other).

5.4 Divergent Plate Boundaries and Seafloor Spreading

Sketch and describe the movement along a divergent plate boundary that results in the formation of new oceanic lithosphere.

KEY TERMS: divergent plate boundary (spreading center), oceanic ridge system, rift valley, seafloor spreading, continental rift

- Seafloor spreading leads to the formation of new oceanic lithosphere at mid-ocean ridge systems. As two plates move apart from one another, tensional forces open cracks in the plates, allowing magma to well up and generate new slivers of seafloor. This process generates new oceanic lithosphere at a rate of 2 to 15 centimeters (1 to 6 inches) each year.
- As it ages, oceanic lithosphere cools and becomes denser. It therefore subsides as it is transported away from the mid-ocean ridge. At the same time, new material is added to its underside, causing the plate to grow thicker.
- Divergent boundaries are not limited to the seafloor. Continents can break apart, too, starting with a continental rift (like modern-day east Africa) and may eventually lead to a new ocean basin opening between the two sides of the rift.

5.5 Convergent Plate Boundaries and Subduction

Compare and contrast the three types of convergent plate boundaries and name a location where each type can be found.

KEY TERMS: convergent plate boundary (subduction zones), deep-ocean trenches, partial melting, continental volcanic arcs, volcanic island arc (island arc)

- When plates move toward one another, oceanic lithosphere is subducted into the mantle, where it is recycled. Subduction manifests itself on the ocean floor as a deep linear trench. The subducting slab of oceanic lithosphere can descend at a variety of angles, from nearly horizontal to nearly vertical.
- Aided by the presence of water, the subducted oceanic lithosphere triggers melting in the mantle, which produces magma. The magma is less dense than the surrounding rock and will rise. It may cool at depth, thickening the crust, or it may make it all the way to Earth's surface, where it erupts as a volcano.

- A line of volcanoes that emerge through continental crust is termed a continental volcanic arc, while a line of volcanoes that emerge through an overriding plate of oceanic lithosphere is a volcanic island arc.
- Continental crust resists subduction due to its relatively low density, and so when an intervening ocean basin is completely destroyed through subduction, the continents on either side collide, generating a new mountain range.

(?) Sketch a typical continental volcanic arc and label the key parts. Then repeat the drawing with an overriding plate made of oceanic lithosphere.

5.6 Transform Plate Boundaries

Describe the relative motion along a transform fault boundary and be able to locate several examples on a plate boundary map.

KEY TERMS: transform plate boundary (transform fault), fracture zone

- At a transform boundary, lithospheric plates slide horizontally past one another. No new lithosphere is generated, and no old lithosphere is consumed. Shallow earthquakes signal the movement of these slabs of rock as they grind past their neighbors.
- The San Andreas Fault in California is an example of a transform boundary in continental crust, while the fracture zones between segments of the Mid-Atlantic Ridge are transform faults in oceanic crust.

(?) On the accompanying tectonic map of the Caribbean, find the Enriquillo Fault. (The location of the 2010 Haiti earthquake is shown as a yellow star.) What kind of plate boundary is shown here? Are there any other faults in the area that show the same type of motion?

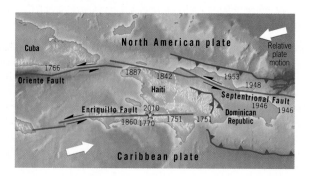

5.7 How Do Plates and Plate Boundaries Change?

Explain why plates such as the African and Antarctic plates are increasing in size, while the Pacific plate is decreasing in size.

- Although the total surface area of Earth does not change, the shape and size of individual plates are constantly changing as a result of subduction

and seafloor spreading. Plate boundaries can also be created or destroyed in response to changes in the forces acting on the lithosphere.
- The breakup of Pangaea and the collision of India with Eurasia are two examples of how plates change through geologic time.

5.8 Testing the Plate Tectonics Model

List and explain the evidence used to support the plate tectonics theory.

KEY TERMS: mantle plume, hot spot, hot-spot track, Curie point, paleomagnetism (fossil magnetism), magnetic reversal, normal polarity, reverse polarity, magnetic time scale, magnetometer

- Multiple lines of evidence have verified the plate tectonics model. For instance, the Deep Sea Drilling Project found that the age of the seafloor increases with distance from a mid-ocean ridge. The thickness of sediment atop this seafloor is also proportional to distance from the ridge: Older lithosphere has had more time to accumulate sediment.
- A hot spot is an area of volcanic activity where a mantle plume reaches Earth's surface. Volcanic rocks generated by hot-spot volcanism provide evidence of both the direction and rate of plate movement over time.
- Magnetic minerals such as magnetite align themselves with Earth's magnetic field as rock forms. These fossil magnets are records of the ancient orientation of Earth's magnetic field. This is useful to geologists in two ways: (1) It allows a given stack of rock layers to be interpreted in terms of their orientation relative to the magnetic poles through time, and (2) reversals in the orientation of the magnetic field are preserved as "stripes" of normal and reversed polarity in the oceanic crust. Magnetometers reveal this signature of seafloor spreading as a symmetrical pattern of magnetic stripes parallel to the axis of the mid-ocean ridge.

5.9 What Drives Plate Motions?

Describe plate–mantle convection and explain two of the primary driving forces of plate motion.

KEY TERMS: convection, slab pull, ridge push

- Some kind of convection (upward movement of less dense material and downward movement of more dense material) appears to drive the motion of plates.
- Slabs of oceanic lithosphere sink at subduction zones because the subducted slab is denser than the underlying asthenosphere. In this process, called slab pull, Earth's gravity tugs at the slab, drawing the rest of the plate toward the subduction zone. As oceanic lithosphere slides down the mid-ocean ridge, it exerts a small additional force, called ridge push.
- Convection may occur throughout the entire mantle, as suggested by the whole-mantle model. Or it may occur in two layers within the mantle—the active upper mantle and the sluggish lower mantle—as proposed in the layer cake model.

(?) Compare and contrast mantle convection with the operation of a lava lamp.

GIVE IT SOME THOUGHT

1. Refer to Section I.5 in the Introduction, titled "The Nature of Scientific Inquiry," to answer the following:
   a. What observation led Alfred Wegener to develop his continental drift hypothesis?
   b. Why did the majority of the scientific community reject the continental drift hypothesis?
   c. Do you think Wegener followed the basic principles of scientific inquiry? Support your answer.

2. Volcanic islands that form over mantle plumes, such as the Hawaiian chain, are home to some of Earth's largest volcanoes. However, several volcanoes on Mars are gigantic compared to any on Earth. What does this difference tell us about the role of plate motion in shaping the Martian surface?

3. Some people predict that California will sink into the ocean. Is this idea consistent with the theory of plate tectonics? Explain.

4. Refer to the accompanying hypothetical plate map to answer the following questions:

   a. How many portions of plates are shown?

   b. Explain why active volcanoes are more likely to be found on continents A and B than on continent C.

   c. Provide one scenario in which volcanic activity might be triggered on continent C.

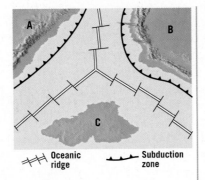

⫫⫫ Oceanic ridge   ⌒⌒⌒ Subduction zone

5. Referring to the accompanying diagrams that illustrate the three types of convergent plate boundaries, complete the following:

   a. Identify each type of convergent boundary.

   b. On what type of crust do volcanic island arcs develop?

   c. Why are volcanoes largely absent where two continental blocks collide?

   d. Describe two ways that oceanic–oceanic convergent boundaries are different from oceanic–continental boundaries. How are they similar?

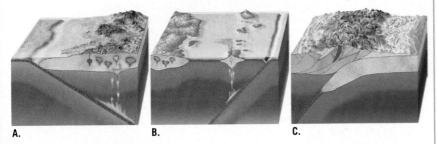

A.   B.   C.

6. Imagine that you are studying seafloor spreading along two different oceanic ridges. Using data from a magnetometer, you produced the two accompanying maps. From these maps, what can you determine about the relative rates of seafloor spreading along these two ridges? Explain.

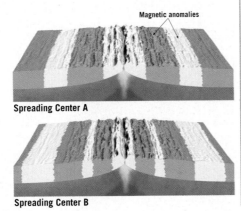

Magnetic anomalies

Spreading Center A

Spreading Center B

7. Australian marsupials (kangaroos, koalas, etc.) have direct fossil links to marsupial opossums found in the Americas. Yet the modern marsupials in Australia are markedly different from their American relatives. How does the breakup of Pangaea help to explain these differences? (*Hint:* See Figure 5.22.)

8. Density is a key component in the behavior of Earth materials and is especially important in understanding important aspects of the plate tectonics model. Describe three different ways that density and/or density differences play a role in plate tectonics.

9. Explain how the processes that create hot-spot volcanic chains differ from the processes that generate volcanic island arcs.

10. Refer to the accompanying map and the pairs of cities below to complete the following:
    (Boston, Denver), (London, Boston), (Honolulu, Beijing)

    a. Which pair of cities is moving apart as a result of plate motion?

    b. Which pair of cities is moving closer as a result of plate motion?

    c. Which pair of cities is not presently moving relative to each other?

Plate motion measured in centimeters per year

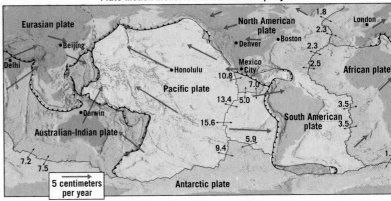

MasteringGeology™

6

FOCUS ON CONCEPTS

Each statement represents the primary learning objective for the corresponding major heading within the chapter. After you complete the chapter, you should be able to:

6.1 Sketch and describe the mechanism that generates most earthquakes.

6.2 Compare and contrast the types of seismic waves and describe the principle of the seismograph.

6.3 Locate Earth's major earthquake belts on a world map and label the regions associated with the largest earthquakes.

6.4 Distinguish between intensity scales and magnitude scales.

6.5 List and describe the major destructive forces that earthquake vibrations can trigger.

6.6 Explain how Earth acquired its layered structure and list and describe each of its major layers.

6.7 Compare and contrast brittle and ductile deformation.

6.8 List and describe the major types of folds.

6.9 Sketch and describe the relative motion of rock bodies located on opposite sides of normal, reverse, and strike-slip faults.

6.10 Locate and identify Earth's major mountain belts on a world map.

6.11 Sketch an Andean-type mountain belt and describe how each of its major features is generated.

6.12 Summarize the stages in the development of a collisional mountain belt such as the Himalayas.

Tsunami striking the coast of Japan on March 11, 2011.
(Photo by Sadatasgu Tomizawa/AFP/Getty Images)

RESTLESS EARTH
EARTHQUAKES AND MOUNTAIN BUILDING

On April 25, 2015, Nepal experienced its worst natural disaster in over 80 years, when an earthquake measuring M 7.8* struck the mountainous region of Southern Asia. Geologists had anticipated a significant earthquake for decades, because of Nepal's location high in the Himalayas, above a collisional boundary where India is being pushed into Asia. The quake's death toll exceeded 9000 and resulted in injuries to nearly 23,000 people. The shallow nature of the 50-second quake resulted in widespread destruction and triggered avalanches on Mt. Everest, claiming the lives of 19 people, including international hikers and Sherpa guides. Nepal's capital, Kathmandu, reportedly shifted 3 meters (10 feet) to the south during the course of the event.

Earthquakes are striking examples of movement within our planet, but it is Earth's spectacular mountain belts, such as the still-developing Himalayan range, that illustrate the dynamic processes that result from tectonic activity. This chapter explains the mechanisms by which Earth's interior processes lead to earthquakes and mountain building.

6.1 What Is an Earthquake?

Sketch and describe the mechanism that generates most earthquakes.

An **earthquake** is ground shaking caused by the sudden and rapid movement of one block of rock slipping past another along fractures in Earth's crust, called **faults**. Most faults are *locked*, except for brief, abrupt movements when sudden slippage produces an earthquake. Faults are locked because the confining pressure exerted by the overlying crust is enormous, causing these fractures in the crust to be "squeezed shut."

Earthquakes tend to occur along preexisting faults where internal stresses cause the crustal rocks to rupture or break into two or more units. The location where slippage begins is called the **hypocenter**, or **focus**. Earthquake waves initially radiate out from this spot into the surrounding rock. The point on Earth's surface directly above the hypocenter is called the **epicenter** (**Figure 6.1**).

Large earthquakes release huge amounts of stored-up energy as **seismic waves**—a form of energy that travels through the lithosphere and Earth's interior. The energy carried by these waves causes the material that transmits them to shake. Seismic waves are analogous to waves produced when a stone is dropped into a calm pond. Just as the impact of the stone creates a pattern of circular waves, an earthquake generates waves that radiate outward in all directions from the hypocenter. Although seismic energy dissipates rapidly as it moves away from the quake's hypocenter, sensitive instruments can detect earthquakes even when they occur on the opposite side of Earth.

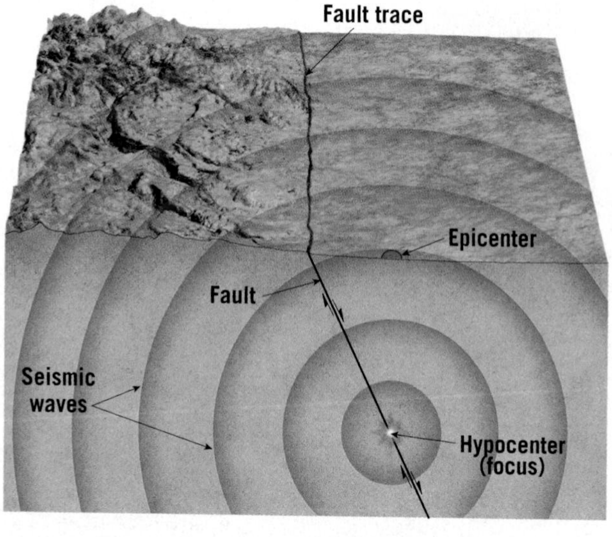

Thousands of earthquakes occur around the world every day. Fortunately, most are small enough that people cannot detect them. Only about 15 strong earthquakes (magnitude 7 or greater) are recorded each year, many of them occurring in remote regions. The occasional large earthquakes that are triggered near major population centers are among the most destructive natural events on Earth. The shaking of the ground, coupled with the liquefaction of soils, wreaks havoc on buildings, roadways, and other structures. In addition, a quake occurring in a populated area can rupture power and gas lines, causing numerous fires. In the famous 1906 San Francisco earthquake, much of the damage was caused by fires that became uncontrollable when broken water mains left firefighters with only trickles of water (**Figure 6.2**).

Discovering the Causes of Earthquakes

The energy released by volcanic eruptions, massive landslides, and meteorite impacts can generate earthquake-like waves, but these events are usually weak. What

Figure 6.1 Earthquake's hypocenter and epicenter The *hypocenter* is the zone at depth where the initial displacement occurs. The *epicenter* is the surface location directly above the hypocenter.

Fault trace

Epicenter

Fault

Seismic waves

Hypocenter (focus)

*An earthquake's *magnitude* (abbreviation M) is a measure of earthquake strength that will be discussed later in this chapter.

San Francisco in flames following the 1906 quake. Broken water mains left firefighters without water.

Figure 6.2 Earthquakes can trigger fires (Reproduced from the collection of the Library of Congress Prints and Photographs Division; inset photo by Hal Garb/ Getty Images)

Fire triggered when a gas line ruptured during the Northridge earthquake in southern California in 1994.

mechanism produces a destructive earthquake? As you have learned, Earth is not a static planet. Because fossils of marine organisms have been discovered thousands of meters above sea level, we know that large sections of Earth's crust have been thrust upward. Other regions, such as California's Death Valley, exhibit evidence of extensive subsidence. In addition to these vertical displacements, offsets in fences, roads, and other structures indicate that horizontal movements between blocks of Earth's crust are also common (**Figure 6.3**).

The actual mechanism of earthquake generation eluded geologists until H. F. Reid conducted a landmark study following the 1906 San Francisco earthquake. This earthquake was accompanied by horizontal surface displacements of several meters along the northern portion of the San Andreas Fault. Field studies determined that during this single earthquake, the Pacific plate lurched as much as 9.7 meters (32 feet) northward past the adjacent North American plate. To better visualize this, imagine standing on one side of the fault and watching a person on the other side suddenly slide horizontally 32 feet to your right.

What Reid concluded from his investigations is illustrated in **Figure 6.4**. Over tens to hundreds of years, differential stress slowly bends the crustal rocks on both sides of a fault. This is much like a person bending a limber wooden stick, as shown in Figure 6.4A,B. Frictional resistance keeps the fault from rupturing and

slipping. (Friction inhibits slippage and is enhanced by irregularities that occur along the fault surface.) At some point, the stress along the fault overcomes the frictional resistance, and slip initiates. Slippage allows the deformed (bent) rock to "snap back" to its original, stress-free, shape; a series of earthquake waves radiate as it slides (see Figure 6.4C,D). Reid termed this "springing back" **elastic rebound** because the rock behaves elastically, much as a stretched rubber band does when it is released.

Figure 6.3 Displacement of structures along a fault (Color photo by University of Washington Libraries, Special Collections, KC6874; inset photo by G. K. Gilbert/USGS)

Slippage along a fault produced this offset in an orange grove east of Calexico, California.

This fence was offset 2.5 meters (8.5 feet) during the 1906 San Francisco earthquake.

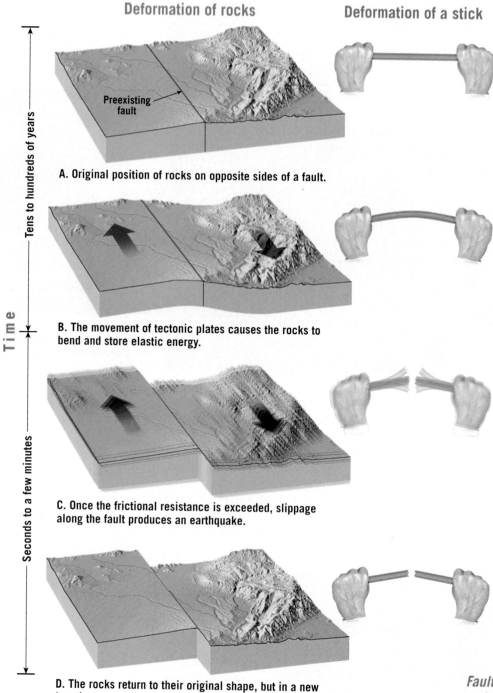

Deformation of rocks   **Deformation of a stick**

Tens to hundreds of years

Time

Seconds to a few minutes

A. Original position of rocks on opposite sides of a fault.

Preexisting fault

B. The movement of tectonic plates causes the rocks to bend and store elastic energy.

C. Once the frictional resistance is exceeded, slippage along the fault produces an earthquake.

D. The rocks return to their original shape, but in a new location.

SmartFigure 6.4 Elastic rebound (https://goo.gl/ooefwf)

Tutorial

Faults and Large Earthquakes

The slippage that occurs along faults can be explained by the plate tectonics theory, which states that large slabs of Earth's lithosphere are continually grinding past one another. These mobile plates interact with neighboring plates, straining and deforming the rocks at their edges. Faults associated with convergent and transform plate boundaries are the source of most large earthquakes.

Convergent Plate Boundaries Earth's strongest earthquakes most often occur along large faults associated with convergent plate boundaries. Compressional forces associated with continental collisions that result in mountain building generate numerous *reverse faults* and *thrust faults*, discussed in more detail in Section 6.9. The 2015 Nepal earthquake is an example of an earthquake that occurred along a thrust fault. The epicenter of the quake was located about 80 kilometers (50 miles) north of Kathmandu, where the Indian plate is converging with the Eurasian plate at a rate of 4.5 centimeters (about 2 inches) per year, driving the uplift of the Himalayas.

In addition, the plate boundary between a subducting slab of oceanic lithosphere and an overlying continental plate form a fault termed a **megathrust fault** (**Figure 6.5**). Megathrust faults have produced the majority of Earth's most powerful and destructive earthquakes, including the 2011 Japan quake (M 9.0), the 2004 Indian Ocean (Sumatra) quake (M 9.1), the 1964 Alaska quake (M 9.2), and the largest earthquake yet recorded, the 1960 Chile quake (M 9.5).

Transform Fault Boundaries Faults in which the dominant displacement is horizontal and parallel to the direction of the fault trace are called *strike-slip faults* (discussed in more detail in Section 6.9). Recall from Chapter 5 that *transform plate boundaries* or simply *transform faults* accommodate motion between two tectonic plates. For example, the San Andreas Fault is a large transform fault that separates the North American plate and the Pacific plate (**Figure 6.6**). Most large transform faults are not perfectly straight or continuous; instead, they consist of numerous branches and smaller fractures that display kinks and offsets (see Figure 6.6). Earthquakes can occur along any of these branches.

Fault Rupture and Propagation By studying earthquakes around the globe, geologists have learned that displacement along large faults occurs along discrete fault segments that often behave differently from one another. Some sections of the San Andreas, for example, exhibit slow, gradual displacement known as **fault creep** and produce little seismic shaking. Other segments slip at relatively closely spaced intervals, producing numerous small to moderate earthquakes. Still other segments remain locked and store elastic energy for a few hundred years before they break loose. Ruptures on segments that have been locked for a few hundred years or longer usually result in major earthquakes.

Geologists have also discovered that slippage along large faults, such as the San Andreas, does not occur

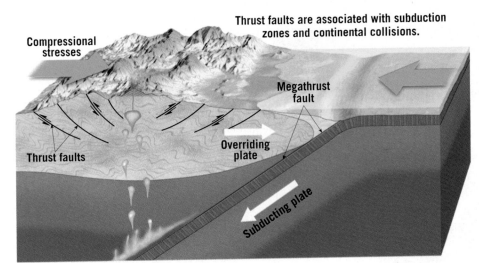

Thrust faults are associated with subduction zones and continental collisions.

Compressional stresses

Megathrust fault

Overriding plate

Subducting plate

Thrust faults

Figure 6.5 Megathrust faults are the sites of Earth's largest earthquakes A convergent plate boundary is a site where one plate is subducting beneath another, and this subduction produces some of Earth's largest earthquakes.

instantaneously. The initial slip occurs at the hypocenter and propagates (travels) along the fault surface; as each section slips, it puts strain on the next section, causing it to slip, too. Slippage propagates at 2 to 4 kilometers per second—faster than a rifle shot. Rupture of a 100-kilometer (60-mile) fault segment takes about 30 seconds, and rupture of a 300-kilometer (200-mile) segment takes about 90 seconds. As rupturing progresses, it can slow down, speed up, or even jump to a nearby fault segment. Earthquake waves are generated at every point along the fault as that portion of the fault begins to slip.

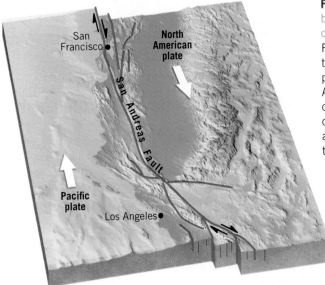

Figure 6.6 Transform plate boundaries and large earthquakes The San Andreas Fault is a large fault system separating the Pacific plate from the North American plate. This type of large strike-slip fault is called a transform fault and can generate destructive earthquakes.

San Francisco

North American plate

San Andreas Fault

Pacific plate

Los Angeles

6.1 CONCEPT CHECKS

1. What is an earthquake? Under what circumstances do most large earthquakes occur?
2. How are faults, hypocenters, and epicenters related?
3. Who was first to explain the mechanism by which most earthquakes are generated?
4. Explain what is meant by *elastic rebound*.
5. What type of fault tends to produce the most destructive earthquakes?

6.2 Seismology: The Study of Earthquake Waves

Compare and contrast the types of seismic waves and describe the principle of the seismograph.

The study of earthquake waves, **seismology**, dates back to attempts made in China almost 2000 years ago to determine the direction from which these waves originated. The earliest-known instrument, invented by Zhang Heng, was a large hollow jar containing a weight suspended from the top (**Figure 6.7**). The suspended weight (similar to a clock pendulum) was connected to the jaws of several large dragon figurines that encircled the container. The jaws of each dragon held a metal ball. When earthquake waves reached the instrument, the relative motion

Figure 6.7 Ancient Chinese seismograph During an Earth tremor, the dragons located in the direction of the main vibrations would drop a ball into the mouth of a frog below. (Photo by James E. Patterson Collection)

Ball dropping

SmartFigure 6.8 Principle of the seismograph The inertia of the suspended weight tends to keep it motionless while the recording drum, which is anchored to bedrock, vibrates in response to seismic waves. The stationary weight provides a reference point from which to measure the amount of displacement occurring as a seismic wave passes through the ground. (Photo courtesy of Zephyr/Science Source) (https://goo.gl/IuclEa)

Animation

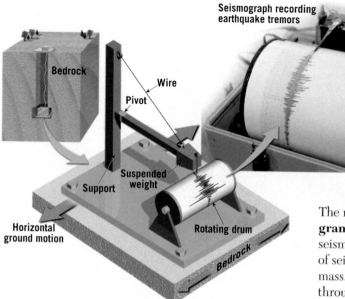

Did You Know?
Literally thousands of earthquakes occur daily! Fortunately, the majority of them are too small for people to feel, and the majority of larger ones occur in remote regions. Their existence is known only because of sensitive seismographs.

between the suspended mass and the jar would dislodge some of the metal balls into the waiting mouths of frogs directly below.

Instruments That Record Earthquakes

In principle, modern **seismographs**, or **seismometers**, are similar to the instruments used in ancient China. A seismograph has a weight freely suspended from a support that is securely attached to bedrock (**Figure 6.8**). When vibrations from an earthquake reach the instrument, the **inertia** of the weight keeps it relatively stationary, while Earth and the support move. Inertia can be simply described by this statement: *Objects at rest tend to stay at rest, and objects in motion tend to remain in motion, unless acted upon by an outside force.* You have experienced

inertia when you have tried to stop your automobile quickly and your body continued to move forward.

Most seismographs are designed to amplify ground motion in order to detect very weak earthquakes or a great earthquake that has occurred in another part of the world. Instruments used in earthquake-prone areas are designed to withstand the violent shaking that can occur near a quake's epicenter.

Seismic Waves

The records obtained from seismographs, called **seismograms**, provide useful information about the nature of seismic waves. Seismograms reveal that two main types of seismic waves are generated by the slippage of a rock mass. One of these wave types, called **body waves**, travel through Earth's interior. The other type, called **surface waves**, travel in the rock layers just below Earth's surface (**Figure 6.9**).

Body Waves Body waves are further divided into two types—called **primary waves**, or **P waves**, and **secondary waves**, or **S waves**—and are identified by their mode of travel through intervening materials. P waves are "push/pull" waves; they momentarily push (compress) and pull (stretch) rocks in the direction the wave is traveling (**Figure 6.10A**). This wave motion is similar to that generated by striking a drum, which moves air back and forth to create sound. Solids, liquids, and gases resist stresses that change their volume when compressed and, therefore, elastically spring back once the stress is removed. Therefore, P waves can travel through all these materials.

By contrast, S waves "shake" the particles at right angles to their direction of travel. This can be illustrated by fastening one end of a rope and shaking the other end, as shown in **Figure 6.10B**. Unlike P waves, which temporarily change the *volume* of intervening material by alternately squeezing and stretching it, S waves change the *shape* of the material that transmits them. Because fluids (gases and liquids) do not resist stresses that cause changes in shape—meaning fluids do not return to their original shape once the stress is removed—liquids and gases do not transmit S waves.

Surface Waves There are two types of surface waves. One type causes Earth's surface and anything resting on it to move up and down, much as ocean swells toss a ship (**Figure 6.11A**). The second type of surface wave causes Earth's surface to move

SmartFigure 6.9 Body waves (P and S waves) versus surface waves P and S waves travel through Earth's interior, while surface waves travel in the layer directly below the surface. P waves are the first to arrive at a seismic station, followed by S waves, and then surface waves. (https://goo.gl/y6owpc)

Tutorial

Hypocenter

P waves / S waves / Surface waves

P waves / S waves / Surface waves
Seismograph #1

P waves / S waves / Surface waves
Seismograph #2

P waves / S waves / surface waves have not yet arrived
Seismograph #3

Core   Mantle

A. As illustrated by a toy Slinky, P waves alternately compress and expand the material through which they pass.

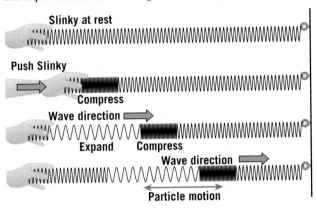

B. S waves cause material to oscillate at right angles to the direction of wave motion.

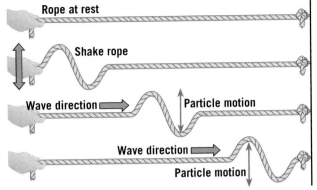

Figure 6.10 The characteristic motion of P and S waves During a strong earthquake, ground shaking consists of a combination of various kinds of seismic waves.

side to side. This motion is particularly damaging to the foundations of structures (**Figure 6.11B**).

Comparing the Speed and Size of Seismic Waves By examining the seismogram shown in **Figure 6.12**, you can see that the major difference among seismic waves is their speed of travel. P waves are the

A. One type of surface wave travels along Earth's surface similar to rolling ocean waves. The red arrows show the movement of rock as the wave passes.

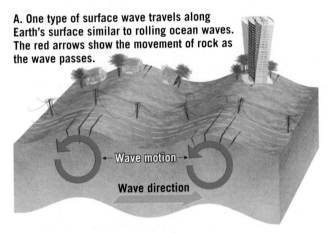

B. A second type of surface wave moves the ground from side to side and can be particularly damaging to building foundations.

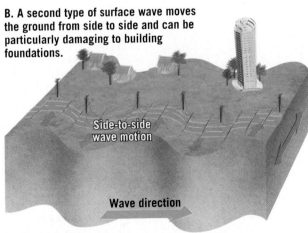

SmartFigure 6.11 Two types of surface waves
(https://goo.gl/5nRPa8)

Animation

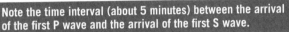
Note the time interval (about 5 minutes) between the arrival of the first P wave and the arrival of the first S wave.

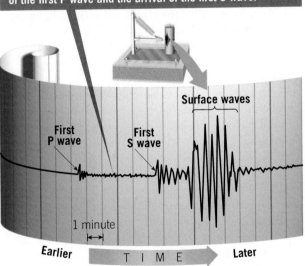

Figure 6.12 Typical seismogram

first to arrive at a recording station, then S waves, and finally surface waves. Generally, in any solid Earth material, P waves travel about 70 percent faster than S waves, and S waves are roughly 10 percent faster than surface waves.

In addition to the velocity differences in the waves, notice in Figure 6.12 that the height, or *amplitude*, of these wave types also varies. S waves have slightly greater amplitudes than P waves, and surface waves exhibit even greater amplitudes. Surface waves also retain their maximum amplitude longer than P and S waves. As a result, surface waves tend to cause greater ground shaking and, hence, greater property damage, than either P or S waves.

6.2 **CONCEPT CHECKS**

1. Describe how a seismograph works.
2. List the major differences between P, S, and surface waves.
3. Which type of seismic wave tends to cause the greatest destruction to buildings?

Did You Know?
Some of the most interesting uses of seismographs involve the reconstruction of human catastrophes, such as airline crashes, pipeline explosions, and mining disasters. For example, a seismologist helped with the investigation of Pan Am Flight 103, which was brought down by a terrorist's bomb in 1988 over Lockerbie, Scotland. Nearby seismographs recorded six separate impacts, indicating that the plane had broken into that many large pieces before it crashed.

6.3 Where Do Most Earthquakes Occur?

Locate Earth's major earthquake belts on a world map and label the regions associated with the largest earthquakes.

About 95 percent of the energy released by earthquakes originates in a few relatively narrow zones, shown in **Figure 6.13**. These zones of earthquake activity are located along fault surfaces where tectonic plates interact along one of the three types of plate boundaries—convergent, divergent, and transform plate boundaries.

Earthquakes Associated with Plate Boundaries

The previous section described the relationship between plate tectonics and seismic activity. The zone of greatest seismic activity, called the **circum-Pacific belt**, encompasses the coastal regions of Chile, Central America, Indonesia, Japan, and Alaska, including the Aleutian Islands (see Figure 6.13). Most earthquakes in the circum-Pacific belt occur along convergent plate boundaries, where one plate subducts at a low angle beneath another. Recall that the contacts between the subducting and overlying plates are called *megathrust faults* (see Figure 6.5). There are more than 40,000 kilometers (25,000 miles) of subduction boundaries in the circum-Pacific belt where displacement is dominated by thrust faulting. Ruptures occasionally occur along segments that are nearly 1000 kilometers (600 miles) long, generating catastrophic earthquakes with magnitudes of 8 or greater.

Another major concentration of strong seismic activity, referred to as the *Alpine–Himalayan belt*, runs through the mountainous regions that flank the Mediterranean Sea and extends past the Himalaya Mountains (see Figure 6.13). Tectonic activity in this region is mainly attributed to collisions of the African plate and the Indian subcontinent with the vast Eurasian plate (see Figure 5.10, page 151). These plate interactions created many thrust and strike-slip faults that remain active.

Transform faults that run through continental crust are also a source of some large earthquakes. Examples include California's San Andreas Fault, New Zealand's Alpine Fault, and Turkey's North Anatolian Fault, which produced a deadly earthquake in 1999.

Figure 6.13 shows another continuous earthquake belt that extends thousands of kilometers through the world's oceans. This zone coincides with the oceanic ridge system—a divergent plate boundary—an area of frequent but weak seismic activity.

Damaging Earthquakes East of the Rockies

When you think "earthquake," you probably think of the western United States or Japan. However, six major earthquakes and several others that inflicted considerable damage have occurred in the central and eastern United States since colonial times (**Figure 6.14**).

Three of these quakes, which occurred as a cluster, had estimated magnitudes of 7.5, 7.3, and 7.8 and destroyed what was then the frontier town of New Madrid, Missouri, located in the Mississippi River Valley. It has been estimated that if an earthquake the size of the 1811–1812 New Madrid event were to strike in the same location in the next decade, it would result in casualties in the thousands and damages in the tens of billions of dollars.

The greatest historical earthquake in the eastern states occurred on August 31, 1886, in Charleston, South Carolina. This earthquake resulted in 60 deaths, numerous injuries, and great economic loss over an area up to 200 kilometers (120 miles) from Charleston. Within 8 minutes, effects were felt as far away as Chicago and St. Louis, where strong vibrations shook the upper floors of

Figure 6.13 Global earthquake belts Distribution of nearly 15,000 earthquakes with magnitudes equal to or greater than 5 for a 10-year period. (Data from USGS)

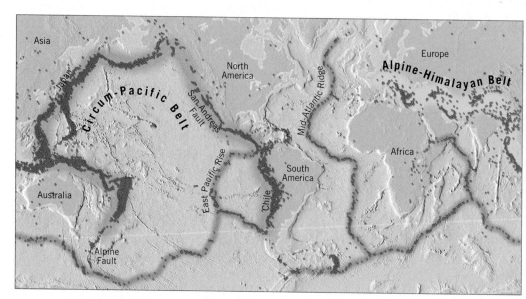

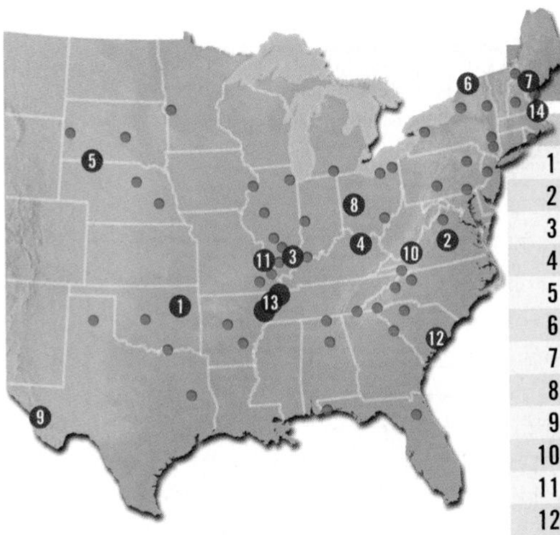

Historic Earthquakes East of the Rockies 1755–2011

	LOCATION	DATE	INTENSITY	MAGNITUDE*	COMMENTS
1	East of Oklahoma City	2011	VII	5.6	Fourteen homes destroyed
2	Mineral, Virginia	2011	VII	5.8	Felt by many
3	Southeastern Illinois	2008	VII	5.4	Occurred along the Wabash Valley Seismic Zone
4	Northeast Kentucky	1980	VII	5.2	Largest earthquake ever recorded in Kentucky
5	Merriman, Nebraska	1964	VII	5.1	Largest earthquake ever recorded in Nebraska
6	Northern New York	1944	VIII	5.8	Left several structures unsafe for occupancy
7	Ossipee Lake, New Hampshire	1947	VII	5.5	Two earthquakes occurred four days apart
8	Western Ohio	1937	VIII	5.4	Extensive damage to plaster walls
9	Valentine, Texas	1931	VIII	5.8	Brick buildings were severely damaged
10	Giles County, Virginia	1897	VIII	5.9	Changed the flow of natural springs
11	Charleston, Missouri	1895	VIII	6.6	Structural damage and liquefaction reported
12	Charleston, South Carolina	1886	X	7.3	Caused 60 deaths, destroyed many buildings
13	New Madrid, Missouri	1811–1812	X	7.0–7.7	Three strong earthquakes occurred
14	Cape Ann, Massachusetts	1755	VIII	?	Buildings damaged in Boston

Source: U.S. Geological Survey
*Intensity and magnitudes have been estimated for many of these events.

Figure 6.14 Historical earthquakes east of the Rockies Large earthquakes are uncommon in the middle of continents, far from the places where plates collide or grind past one another, or where one plate slides beneath another. Nevertheless, several damaging earthquakes have occurred in the central and eastern United States since colonial times.

buildings, causing people to rush outdoors. In Charleston alone, more than 100 buildings were destroyed, and 90 percent of the remaining structures were damaged (**Figure 6.15**).

Earthquakes in the central and eastern United States occur far less frequently than in California, yet history indicates that the East is vulnerable. Further, these shocks east of the Rockies have generally produced structural damage over a larger area than earthquakes of similar magnitude in California. This is because the underlying bedrock in the central and eastern United States is older and more rigid. As a result, seismic waves can travel greater distances with less attenuation (loss of strength) than in the western United States.

Earthquakes that occur away from plate boundaries are called *intraplate earthquakes*. Intraplate earthquakes can be caused by a variety of factors and tend to be weaker than those associated with plate boundaries. For example, the process called fracking, in which a solution is injected into the ground under high pressure to enhance oil and gas production, appears to have contributed to the recent increase in earthquake activity east of the Rockies.

Locating the Source of an Earthquake

When seismologists analyze an earthquake, they first determine its *epicenter*, the point on Earth's surface directly above the focus (see Figure 6.1). One method used for locating an earthquake's epicenter relies on the fact that P waves travel faster than S waves.

The traveling waves are analogous to two racing automobiles, one faster than the other. The first P wave, like the faster automobile, always wins the race, arriving ahead

Figure 6.15 Damage to Charleston, South Carolina, caused by the August 31, 1886, earthquake (Photo courtesy of USGS)

Did You Know?
During the 1811–1812 New Madrid earthquake, some areas subsided as much as 4.5 m (15 ft). This subsidence created Lake St. Francis west of the Mississippi and enlarged Reelfoot Lake to the east of the river. Other regions rose, creating temporary waterfalls in the bed of the Mississippi River.

Figure 6.16 Seismograms of the same earthquake recorded at three different locations

THREE SEISMOGRAMS

Seismogram A – New York, NY

|← 1 minute →|

FIRST P WAVE   FIRST S WAVE

Seismogram B – Nome, Alaska

FIRST P WAVE   FIRST S WAVE

Seismogram C – Mexico City, Mexico

FIRST P WAVE   FIRST S WAVE

TRAVEL–TIME GRAPH

Distance from epicenter in miles

S-WAVE CURVE

5 min. time interval

P-WAVE CURVE

Time in minutes since earthquake

Distance from epicenter in kilometers

Figure 6.17 Travel–time graph A travel–time graph is used to determine the distance to an earthquake's epicenter. The difference in arrival times of the first P and S waves in the example shown is 5 minutes.

of the first S wave. The greater the length of the race, the greater the difference in their arrival times at the finish line (the seismic station). Therefore, the longer the interval between the arrival of the first P wave and the arrival of the first S wave, the greater the distance to the epicenter. **Figure 6.16** shows three simplified seismograms for the same earthquake. Based on the P–S interval, which city—New York, Nome, or Mexico City—is farthest from the epicenter?

The system for locating earthquake epicenters was developed by using seismograms from earthquakes whose epicenters could be easily pinpointed based on physical evidence. From these seismograms, travel–time graphs were constructed (**Figure 6.17**). Using the sample seismogram for New York in Figure 6.16 and the travel–time curve in Figure 6.17, we can determine the distance separating the recording station from the earthquake in three steps:

1. Using the seismogram for New York, we determine that the time interval between the arrival of the first P wave and the arrival of the first S wave is 5 minutes.

2. Using the travel–time graph, we find the location where the vertical separation between P and S curves is equal to the P–S time interval (5 minutes in this example).

Figure 6.18 Triangulation to locate an earthquake This method involves using the distance obtained from three or more seismic stations to establish the location of an earthquake.

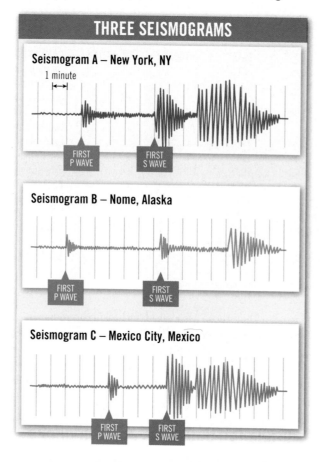

Nome

4400 km
2750 miles

Epicenter

New York

3700 km
2300 miles

2900 km
1800 mi

Mexico City

0   1000   2000   3000 MILES
0   1000   2000   3000   4000   5000 KILOMETERS

3. From the position in step 2, we draw a vertical line to the horizontal axes and read the distance to the epicenter.

Using these steps, we determine that the earthquake occurred 3700 kilometers (2300 miles) from the recording instrument in New York City.

Now we know the *distance*, but what about *direction*? The epicenter could be in any direction from the seismic station. Using a method called *triangulation*, we can determine the location of an epicenter if we know the distance to it from two or more additional seismic stations (**Figure 6.18**). On a map or globe, we draw a circle around each seismic station with a radius equal to the distance from that station to the epicenter. The point where the three circles intersect is the approximate epicenter of the quake.

6.3 CONCEPT CHECKS

1. What zone on Earth has the greatest amount of seismic activity?
2. What type of plate boundary is associated with Earth's largest earthquakes?
3. Explain why an earthquake east of the Rockies may produce damage over a larger area than one of similar magnitude in California.
4. Briefly describe how *triangulation* is used to locate the epicenter of an earthquake.

6.4 Determining the Size of an Earthquake

Distinguish between intensity scales and magnitude scales.

Seismologists use a variety of methods to determine two fundamentally different measures that describe the size of an earthquake: *intensity* and *magnitude*. An **intensity** scale uses observed property damage to estimate the amount of ground shaking at a particular location. **Magnitude** scales, which were developed more recently, use data from seismographs to estimate the amount of energy released at an earthquake's source.

Intensity Scales

Until the mid-1800s, historical records provided the only accounts of the severity of earthquake shaking and destruction. Perhaps the first attempt to scientifically describe the aftermath of an earthquake came following the great Italian earthquake of 1857. By systematically mapping effects of the earthquake, a measure of the intensity of ground shaking was established. The map generated by this study used lines to connect places of equal damage and hence equal ground shaking. Using this technique, zones of intensity were identified, with the zone of highest intensity located near the center of maximum ground shaking and often (but not always) the earthquake epicenter.

In 1902, Giuseppe Mercalli developed a more reliable intensity scale, which is still used today in a modified form. The **Modified Mercalli Intensity scale**, shown in **Table 6.1**, was developed using California buildings as its standard. Based on the 12-point Mercalli Intensity scale, an area in which some well-built wood structures and most masonry buildings are destroyed by an earthquake would be assigned a Roman numeral X (10).

More recently, the U.S. Geological Survey has developed a website called "Did You Feel It," where Internet users enter their zip code and answer questions

Table 6.1 Modified Mercalli Intensity Scale	
I	Not felt except by a very few under especially favorable circumstances.
II	Felt only by a few persons at rest, especially on upper floors of buildings.
III	Felt quite noticeably indoors, especially on upper floors of buildings, but many people do not recognize it as an earthquake.
IV	During the day felt indoors by many, outdoors by few. Sensation like heavy truck striking building.
V	Felt by nearly everyone, many awakened. Disturbances of trees, poles, and other tall objects sometimes noticed.
VI	Felt by all; many frightened and run outdoors. Some heavy furniture moved; few instances of fallen plaster or damaged chimneys. Damage slight.
VII	Everybody runs outdoors. Damage negligible in buildings of good design and construction; slight to moderate in well-built ordinary structures; considerable in poorly built or badly designed structures.
VIII	Damage slight in specially designed structures; considerable, with partial collapse, in ordinary substantial buildings; great in poorly built structures. (Falling chimneys, factory stacks, columns, monuments, walls.)
IX	Damage considerable in specially designed structures. Buildings shifted off foundations. Ground cracked conspicuously.
X	Some well-built wooden structures destroyed. Most masonry and frame structures destroyed. Ground badly cracked.
XI	Few, if any, (masonry) structures remain standing. Bridges destroyed. Broad fissures in ground.
XII	Damage total. Waves seen on ground surfaces. Objects thrown upward into air.

The green dots on the national map show locations of people who reported feeling earthquakes of similar magnitude, one that occurred in California versus one that occurred in Virginia. The difference is attributable to the rigidity of the bedrock.

USGS Community Internet Intensity Map

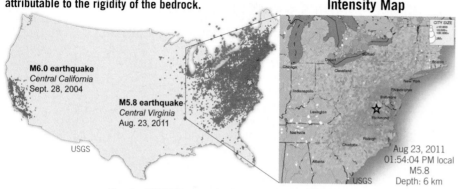

Key for USGS Community Internet Intensity Map

INTENSITY	I	II-III	IV	V	VI	VII	VIII	IX	X
SHAKING	Not felt	Weak	Light	Moderate	Strong	Very strong	Severe	Violent	Extreme
DAMAGE	none	none	none	Very light	Light	Moderate	Moderate/Heavy	Heavy	Very Heavy

SmartFigure 6.19 USGS Community Internet Intensity Map Maps like this one are prepared using data collected over the Internet from people responding to questions such as, "Did objects fall off shelves?" (https://goo.gl/Pdseso)

Tutorial

such as "Did objects fall off shelves?" Within a few hours, a Community Internet Intensity Map, like the one in **Figure 6.19** for the 2011 central Virginia earthquake (M 5.8), is generated. As shown in Figure 6.19, shaking strong enough to be felt was reported from Maine to Florida, an area occupied by one-third of the U.S. population. Several national landmarks were damaged, including the Washington Monument and the National Cathedral, located about 130 kilometers (80 miles) away from the epicenter.

Magnitude Scales

To more accurately compare earthquakes around the globe, scientists searched for a way to describe the energy released by earthquakes that did not rely on factors such as building practices, which vary considerably from one part of the world to another. As a result, several magnitude scales were developed.

Richter Magnitude   In 1935 Charles Richter of the California Institute of Technology developed the first magnitude scale to use seismic records. As shown in **Figure 6.20** (top), the **Richter scale** is calculated by measuring the amplitude of the largest seismic wave (usually an S wave or a surface wave) recorded on a seismogram. Because seismic waves weaken as the distance between the hypocenter and the seismograph increases, Richter developed a method that accounts for the decrease in wave amplitude with increasing distance. Theoretically, as long as equivalent instruments are used, monitoring stations at different locations will obtain the same Richter magnitude for each recorded earthquake. In practice, however, different recording stations often obtain slightly different magnitudes for the same earthquake—a result of the variations in the rock types through which the waves travel.

Earthquakes vary enormously in strength, and great earthquakes produce wave amplitudes thousands of times larger than those generated by weak tremors. To accommodate this wide variation, Richter used a *logarithmic scale* to express magnitude, in which a *10-fold* increase in wave amplitude corresponds to an increase of 1 on the magnitude scale. Thus, the intensity of ground shaking for a magnitude 5 earthquake is 10 times greater than that produced by an earthquake having a Richter magnitude (M_L) of 4 (**Figure 6.21**).

In addition, each unit of increase in Richter magnitude equates to roughly a *32-fold increase in the energy released*. Thus, an earthquake with a magnitude of 6.5 releases 32 times more energy than one with a magnitude of 5.5 and roughly 1000 times (32×32) more energy than a magnitude 4.5 quake. A major earthquake with a magnitude of 8.5 releases millions of times more energy than the smallest earthquakes felt by humans (**Figure 6.22**).

Figure 6.20 Determining the Richter magnitude of an earthquake

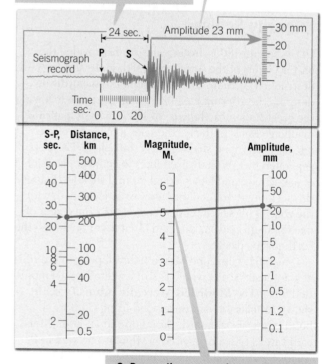

1. Measure the height (amplitude) of the largest wave on the seismogram (23 mm) and plot it on the amplitude scale (right).

2. Determine the distance to the earthquake using the time interval separating the arrival of the first P wave and the arrival of the first S wave (24 seconds) and plot it on the distance scale (left).

3. Draw a line connecting the two plots and read the Richter magnitude (M_L 5) from the magnitude scale (center).

Magnitude vs. Ground Motion and Energy

Magnitude Change	Ground Motion Change (amplitude)	Energy Change (approximate)
4.0	10,000 times	1,000,000 times
3.0	1000 times	32,000 times
2.0	100 times	1000 times
1.0	10 times	32 times
0.5	3.2 times	5.5 times
0.1	1.3 times	1.4 times

Figure 6.21 Magnitude versus ground motion and energy released An earthquake that is one magnitude stronger than another (M 6 versus M 5) produces seismic waves that have a maximum amplitude 10 times greater and releases about 32 times more energy than the weaker quake.

The convenience of describing the size of an earthquake by a single number that can be calculated quickly from seismograms makes the Richter scale a powerful tool. Seismologists have since modified Richter's work and developed other Richter-like magnitude scales.

Despite its usefulness, the Richter scale is not adequate for describing very large earthquakes. For example, the 1906 San Francisco earthquake and the 1964 Alaska earthquake have roughly the same Richter magnitudes.

However, based on the relative size of the affected areas and the associated tectonic changes, the Alaska earthquake released considerably more energy than the San Francisco quake. Thus, the Richter scale is considered *saturated* for major earthquakes because it cannot distinguish among them. Despite this shortcoming, Richter-like scales are still used because they allow for quick calculations.

Moment Magnitude For measuring medium and large earthquakes, seismologists now favor a newer scale called **moment magnitude** (M_W), which estimates the total energy released during an earthquake. Moment magnitude is calculated by determining the average amount of slip on the fault, the area of the fault surface that slipped, and the strength of the faulted rock.

Moment magnitude can also be calculated by modeling data obtained from seismograms. The results are converted to a magnitude number, similar to other magnitude scales. Also, as with the Richter scale, each unit increase on the moment magnitude scale equates to roughly a 32-fold increase in the energy released.

Because the moment magnitude is better at estimating the relative size of very large earthquakes, seismologists have used the moment magnitude scale to

Did You Know?
It is a commonly held belief that moderate earthquakes decrease the chances of a major earthquake in the same region, but this is not the case. When you compare the amount of energy released by earthquakes of different magnitudes, it turns out that thousands of moderate tremors would be needed to release the huge amount of energy released during one "great" earthquake.

Figure 6.22 Annual occurrence of earthquakes with various magnitudes

Frequency and Energy Released by Earthquakes of Different Magnitudes

Magnitude (Mw)	Average Per Year	Description	Examples	Energy Release (equivalent kilograms of explosive)
9	<1	**Largest recorded earthquakes—** destruction over vast area massive loss of life possible	Chile, 1960 (M 9.5); Alaska, 1964 (M 9.2); Sumatra, 2004 (M 9.1); Japan, 2011 (M 9.0)	56,000,000,000,000
8	1	**Great earthquakes—** severe economic impact large loss of life	Chile, 2010 (M 8.8); Mexico City, 1980 (M 8.1)	1,800,000,000,000
7	15	**Major earthquakes—** damage ($ billions) loss of life	New Madrid, Missouri 1812 (M 7.7); Haiti, 2012 (M 7.0); Charleston, South Carolina, 1886 (M 7.3)	56,000,000,000
6	134	**Strong earthquakes—** can be destructive in populated areas	Kobe, Japan, 1995 (M 6.9); Loma Prieta, California, 1989 (M 6.9); Northridge, California, 1994 (M 6.7)	1,800,000,000
5	1319	**Moderate earthquakes—** property damage to poorly constructed buildings	Mineral, Virginia, 2011 (M 5.8); Northern New York, 1994 (M 5.8); East of Oklahoma City, Oklahoma, 2011 (M 5.6)	56,000,000
4	13,000	**Light earthquakes—** noticable shaking of items indoors, some property damage	Western Minnesota, 1975 (M 4.6); Arkansas, 2011 (M 4.7)	1,800,000
3	130,000	**Minor earthquakes—** felt by humans, very light property damage, if any	New Jersey, 2009 (M 3.0); Maine, 2006 (M 3.8); Texas, 2015 (M 3.6)	56,000
2	1,300,000	**Very minor earthquakes—** felt by humans, no property damage		1,800
	Unknown	**Very minor earthquakes—** generally not felt by humans, but may be recorded		56

Data from USGS

recalculate the magnitudes of older strong earthquakes. For example, the 1964 Alaska earthquake, given a Richter magnitude of 8.3, has since been recalculated using the moment magnitude scale, resulting in an upgrade to M_W 9.2. Conversely, the 1906 San Francisco earthquake, having a Richter magnitude of 8.3, was downgraded to M_W 7.9. The strongest earthquake on record is the 1960 Chilean subduction zone earthquake, which has a moment magnitude of 9.5.

6.4 CONCEPT CHECKS

1. What does the Modified Mercalli Intensity scale tell us about an earthquake?
2. What information is used to establish the lower numbers on the Mercalli scale?
3. How much more energy does a magnitude 7.0 earthquake release than a magnitude 6.0 earthquake?
4. Why is the moment magnitude scale favored over the Richter scale for large earthquakes?

6.5 Earthquake Destruction

List and describe the major destructive forces that earthquake vibrations can trigger.

The most violent earthquake ever recorded in North America—the Alaska earthquake—occurred at 5:36 P.M. on March 27, 1964. Felt over most of the state, the earthquake had a moment magnitude (M_W) of 9.2 and lasted 3 to 4 minutes. This event left 128 people dead and thousands homeless, and it badly disrupted the state's economy. Within 24 hours of the initial shock, 28 aftershocks were recorded, 10 of them exceeding magnitude 6. The epicenter and towns hardest hit by the quake are shown in **Figure 6.23**.

Many factors determine the degree of destruction that accompanies an earthquake. The most obvious is the *magnitude of the earthquake and its proximity to a populated area.* During an earthquake, the region within 20 to 50 kilometers (12 to 30 miles) of the epicenter tends to experience roughly the same degree of ground shaking, and beyond that limit, vibrations usually diminish rapidly. As described earlier, earthquakes that occur in the stable continental interior, such as the New Madrid, Missouri, earthquakes of 1811–1812, are generally felt more over a much larger area than those in earthquake-prone areas such as California.

Destruction from Seismic Vibrations

The 1964 Alaska earthquake provided geologists with insights into the role of ground shaking as a destructive force. As the energy released by an earthquake travels along Earth's surface, it causes the ground to vibrate in a complex manner involving up-and-down as well as side-to-side motion. The amount of damage to human-made structures attributable to the vibrations depends on several factors, including (1) the *intensity* and (2) *duration of the vibrations,* (3) *the nature of the material on which structures rest,* and (4) the *building materials and construction practices of the region.*

All the multistory structures in Anchorage were damaged by the vibrations. The more flexible wood-frame residential buildings fared best. A striking example of how construction variations affect earthquake damage is shown in **Figure 6.24**. You can see that the steel-frame building on the left withstood the vibrations, whereas the poorly designed JCPenney building was badly damaged. Engineers have learned that buildings constructed

Figure 6.24 Comparing damage to structures The poorly designed five-story JCPenney building in Anchorage, Alaska, sustained extensive damage. The steel-frame adjacent building incurred very little structural damage. (Courtesy of Seattle/NOAA)

 Figure 6.23 Region most affected by the Alaska earthquake, 1964

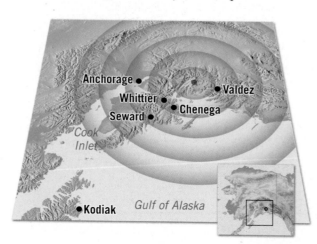

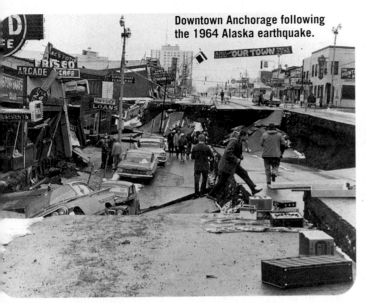

Downtown Anchorage following the 1964 Alaska earthquake.

Figure 6.25 Ground failure caused this street in Anchorage, Alaska, to collapse (Photo by USGS)

These tilted buildings rested on unconsolidated sediment that behaved like quicksand during the 1964 Niigata, Japan, earthquake.

Figure 6.26 Effects of liquefaction on buildings (Photo courtesy of USGS)

of blocks and bricks that are not reinforced with steel rods are the most serious safety threats in earthquakes. Unfortunately, most of the structures in the developing world are constructed of unreinforced concrete slabs and bricks made of dried mud—a primary reason the death toll in poor countries such as Haiti and Nepal is usually higher than for earthquakes of similar size in Japan and the United States.

The 1964 Alaska earthquake damaged most large structures in Anchorage, even though they were built according to the earthquake provisions of the Uniform Building Code. Perhaps some of that destruction can be attributed to the unusually long duration of the earthquake. Most quakes involve tremors that last less than a minute. For example, the 1994 Northridge earthquake was felt for about 40 seconds, and the strong vibrations of the 1989 Loma Prieta earthquake lasted less than 15 seconds. But the Alaska quake reverberated for 3 to 4 minutes.

Amplification of Seismic Waves Although the whole region near the epicenter experiences about the same intensity of ground shaking, destruction may vary considerably in this area due to the nature of the ground on which the structures are built. Soft sediments, for example, generally amplify the vibrations more than solid bedrock. Thus, the buildings in Anchorage that were situated on unconsolidated sediments experienced heavy structural damage (**Figure 6.25**). In contrast, most of the town of Whittier, though much nearer the epicenter, rested on a firm foundation of solid bedrock and, therefore, suffered much less damage from seismic vibrations.

Liquefaction The intense shaking of an earthquake can cause loosely packed water-logged materials, such as sandy stream deposits or fill, to be transformed into a substance that acts like a fluid. The phenomenon of transforming a somewhat stable soil into mobile material capable of rising toward Earth's surface is known as **liquefaction**. When

liquefaction occurs, the ground may not be capable of supporting buildings, and underground storage tanks and sewer lines may literally float toward the surface (**Figure 6.26**).

During the 1989 Loma Prieta earthquake, in San Francisco's Marina District, foundations failed, and geysers of sand and water shot from the ground, evidence that liquefaction had occurred (**Figure 6.27**). Liquefaction

Figure 6.27 Liquefaction These "sand volcanoes," produced by the Christchurch, New Zealand, earthquake of 2011, formed when "geysers" of sand and water shot from the ground, an indication that liquefaction occurred. (Photo by Diarmuid/Alamy)

Sand volcanoes

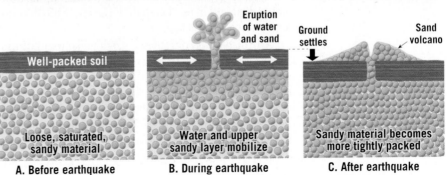

Eruption of water and sand

Ground settles

Sand volcano

Well-packed soil

Loose, saturated, sandy material

A. Before earthquake

Water and upper sandy layer mobilize

B. During earthquake

Sandy material becomes more tightly packed

C. After earthquake

A. Vibrations from the Alaska earthquake caused cracks to appear near the edge of the Turnagain Heights bluff.

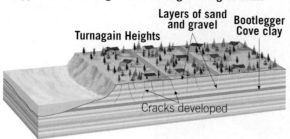

B. Blocks of land began to slide toward the sea on a weak layer called the Bootlegger Cove clay and in less than 5 minutes, as much as 200 meters of the Turnagain Heights bluff area had been destroyed.

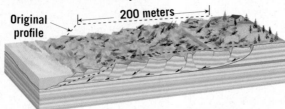

C. Photo of a small area of destruction caused by the Turnagain Heights slide.

SmartFigure 6.28 Turnagain Heights slide caused by the 1964 Alaska earthquake (Photo courtesy of USGS) (https://goo.gl/35qQXQ)

also contributed to the damage inflicted on San Francisco's water system during the 1906 earthquake. During the 2011 Japan earthquake, liquefaction caused entire buildings to sink several feet.

Landslides and Ground Subsidence

The greatest earthquake-related damage to structures is often caused by landslides and ground subsidence triggered by earthquake vibrations. This was the case during the 1964 Alaska earthquake in Valdez and Seward, where the violent shaking caused coastal sediments to slump, carrying away both waterfronts. In Valdez, 31 people died when a dock slid into the sea. Because of the threat of recurrence, the entire town of Valdez was relocated to more stable ground several kilometers away.

Much of the damage in Anchorage was attributed to landslides. Homes in Turnagain Heights were destroyed when a layer of clay lost its strength and over 200 acres of land slid toward the ocean (**Figure 6.28**). A portion of this spectacular landslide was left in its natural condition, as a reminder of this destructive event. The site was appropriately named "Earthquake Park." Downtown Anchorage was also disrupted as sections of the main business district dropped by as much as 3 meters (10 feet).

Fire

More than a century ago, San Francisco was the economic center of the western United States, largely because of gold and silver mining. Then, at dawn on April 18, 1906, a violent earthquake struck, triggering an enormous firestorm (see Figure 6.2). Much of the city was reduced to ashes and ruins. It is estimated that 3000 people died and more than half of the city's 400,000 residents were left homeless.

The historic San Francisco earthquake reminds us of the formidable threat of fire, which started during that quake when gas and electrical lines were severed. The initial ground shaking broke the city's water lines into hundreds of disconnected pieces, which made controlling the fires virtually impossible. The fires, which raged out of control for 3 days, were finally contained when expensive houses along Van Ness Avenue were dynamited to

provide a fire break, similar to the strategy used in fighting forest fires.

While few deaths were attributed to the San Francisco fires, other earthquake-initiated fires have been more destructive, claiming many more lives. For example, the 1923 earthquake in Japan triggered an estimated 250 fires, devastating the city of Yokohama and destroying more than half the homes in Tokyo. More than 100,000 deaths were attributed to the fires, which were driven by unusually high winds.

Tsunamis

Major undersea earthquakes often set in motion a series of large ocean waves that are known by the Japanese name **tsunami** ("harbor wave"). Most tsunamis are generated by displacement along a megathrust fault that suddenly lifts a large slab of seafloor (**Figure 6.29**). Once generated, a tsunami resembles a series of ripples formed when a pebble is dropped into a pond. In contrast to ripples, however, tsunamis advance across the ocean at amazing speeds, about 800 kilometers (500 miles) per hour—equivalent to the speed of a commercial airliner. Despite this striking characteristic, a tsunami in the open ocean can pass undetected because its height (amplitude) is usually less than 1 meter (3 feet), and the distance separating wave crests ranges from 100 to 700 kilometers (60 to 425 miles). However, upon entering shallow coastal waters, these destructive waves "feel bottom" and slow, causing the water to pile up (see Figure 6.29). A few exceptional tsunamis have exceeded 40 meters (130 feet) in height. As the crest of a tsunami approaches the shore, it appears as a rapid rise in sea level with a turbulent and chaotic surface; it does not resemble a breaking wave (**Figure 6.30**).

The first warning of an approaching tsunami is often the rapid withdrawal of water from beaches, the result of the trough of the first large wave preceding the crest. Some inhabitants of the Pacific basin have learned to heed this warning and quickly move to higher ground. Approximately 5 to 30 minutes after the retreat of water, a surge capable of extending several kilometers inland occurs. In a successive fashion, each surge is followed by a rapid oceanward retreat of the sea. Therefore, people experiencing a tsunami should not return to the shore when the first surge of water retreats.

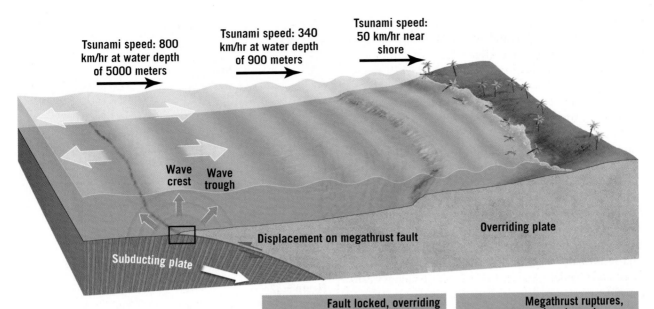

Tsunami speed: 800 km/hr at water depth of 5000 meters

Tsunami speed: 340 km/hr at water depth of 900 meters

Tsunami speed: 50 km/hr near shore

Wave crest

Wave trough

Displacement on megathrust fault

Overriding plate

Subducting plate

Fault locked, overriding plate deforms

Bulge

Subducting plate

Overriding plate

Megathrust ruptures, strain released

Subducting plate

Overriding plate

SmartFigure 6.29 Tsunami generated by displacement of the ocean floor The speed of a tsunami wave correlates with ocean depth. In deep water, these waves can advance at speeds in excess of 800 kilometers (500 miles) per hour. When they enter coastal waters and begin to "feel bottom," they slow down and grow in height. They are still very fast moving, with a speed of 50 kilometers (30 miles) per hour at a depth of 20 meters (65 feet). The size and spacing of the swells in this figure are not to scale. (https://goo.gl/3lphb3)

Tsunami Damage from the 2004 Indonesia Earthquake A massive undersea earthquake of M_W 9.1 occurred near the island of Sumatra on December 26, 2004, sending waves of water racing across the Indian Ocean and Bay of Bengal. It was one of the deadliest natural disasters of any kind in modern times, claiming more than 230,000 lives. As water surged several kilometers inland, cars and trucks were flung around like toys in a bathtub, and fishing boats were rammed into homes. In some locations, the backwash of water dragged bodies and huge amounts of debris out to sea.

The destruction was indiscriminate, destroying luxury resorts as well as poor fishing hamlets along the Indian Ocean. Damage was reported as far away as the coast of Somalia in Africa, 4100 kilometers (2500 miles) west of the earthquake epicenter.

Japan Tsunami Because of Japan's location along the circum-Pacific belt and its extensive coastline, it is especially vulnerable to tsunami destruction. The most powerful earthquake to strike Japan in the age of modern seismology was the 2011 Tohoku earthquake (M_W 9.0). This historic earthquake and devastating tsunami resulted in at least 15,890 deaths, more than 3000 people missing, and 6107 injured. Nearly 400,000 buildings, 56 bridges, and 26 railways were destroyed or damaged.

The majority of human casualties and damage were caused by a Pacific-wide tsunami that reached a

SmartFigure 6.30 Tsunami generated off the coast of Sumatra, 2004 (AFP/Stringer/Getty Images) (https://goo.gl/b8OPkN)

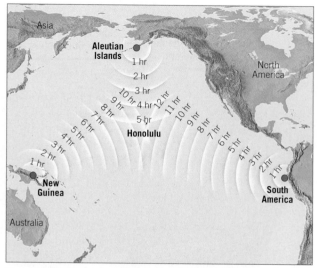

Figure 6.32 Tsunami travel times Travel times to Honolulu, Hawaii, from selected locations throughout the Pacific. (Data from NOAA)

Did You Know?

Although there are no historical records of tsunamis produced by meteorite impacts in the oceans, such events have occurred. Geologic evidence indicates that the most recent occurrence was a megatsunami that devastated portions of the Australian coast around the year 1500. A meteorite impact 65 million years ago near Mexico's Yucatan Peninsula created one of the largest impact-induced tsunamis ever. The giant wave swept hundreds of kilometers inland around the shore of the Gulf of Mexico.

maximum height of about 40 meters (130 feet) above sea-level and traveled inland 10 kilometers (6 miles) in the region of Sendai, Japan (**Figure 6.31**). In addition, meltdowns occurred at three inundated nuclear reactors in Japan's Fukushima Daiichi Nuclear Complex. Across the Pacific in California, Oregon, Peru, and Chile, some loss of life occurred, and several houses, boats, and docks were destroyed. The tsunami was generated when a slab of seafloor located 60 kilometers (37 miles) off the east coast of Japan was suddenly "thrust up" an estimated 5 to 8 meters (16 to 26 feet).

Tsunami Warning System In 1946, a large tsunami struck the Hawaiian Islands without warning. A wave more than 15 meters (50 feet) high left several coastal villages in shambles. This destruction motivated the U.S. Coast and Geodetic Survey to establish a tsunami warning system for coastal areas of the Pacific that today includes 26 countries. Seismic observatories throughout the region report large earthquakes to the Tsunami Warning Center in Honolulu. Scientists at the center use deep-sea buoys equipped with pressure sensors to detect energy released by an earthquake. In addition, tidal gauges measure the rise and fall in sea level that accompany tsunamis, and warnings are issued within an hour. Although tsunamis travel very rapidly, there is sufficient time to warn all except those in the areas nearest the epicenter. For example, a tsunami generated near the Aleutian Islands would take 5 hours to reach Hawaii, and one generated near the coast of Chile would travel 15 hours before reaching the shores of Hawaii (**Figure 6.32**).

6.5 CONCEPT CHECKS

1. List four factors that influence the amount of destruction that seismic vibrations cause to human-made structures.
2. In addition to the destruction created directly by seismic vibrations, list three other types of destruction associated with earthquakes.
3. What is a tsunami? How are tsunamis generated?
4. List at least three reasons an earthquake with a magnitude of 7.0 might result in more death and destruction than a quake with a magnitude of 8.0.

Figure 6.31 Japan tsunami, March 2011 This tsunami breached an embankment and devastated the city of Miyako, Japan shortly after a 9.0 magnitude earthquake hit northern Japan in 2011. (JIJI PRESS/AFP/Getty Images)

6.6 Earth's Interior

Explain how Earth acquired its layered structure and list and describe each of its major layers.

The detailed studies of seismic waves described above help scientists locate and measure earthquakes. Knowing how seismic waves behave has also given us a better understanding of the nature of Earth's interior.

If we could slice Earth in half, the first thing we would notice is that it has distinct layers. The heaviest materials (metals) are in the center. Less dense solids (rocks) are in the middle, and lighter liquids (mainly water) and gases are on top. Within Earth we know these layers as the iron-rich core, the rocky mantle and crust, the liquid ocean, and the gaseous atmosphere.

Probing Earth's Interior: "Seeing" Seismic Waves

How do we know about the structure and properties of Earth's deep interior? Light does not travel through rock, so we must find other ways to "see" into our planet. The best way to learn about Earth's interior is to dig or drill a hole and examine it directly. Unfortunately, this is possible only at shallow depths. The deepest a drilling rig has ever penetrated is only 12.3 kilometers (7.6 miles), which is about 1/500 of the way to Earth's center! Even this was an extraordinary accomplishment because temperature and pressure increase rapidly with depth.

About 3000 earthquakes occur each year that are large enough (about M_w 5.5) to travel all the way through Earth and be recorded by seismographs at the other side of the globe (**Figure 6.33**). The P and S waves from these large earthquakes can be used to "see" into our planet in much the same way that doctors use ultrasound waves to generate sonograms.

Using the waves recorded on seismograms to visualize Earth's interior structure is challenging. Seismic waves do not travel along straight paths; instead, they are *reflected*, *refracted*, and *diffracted* as they pass through our planet. They reflect off boundaries between different layers; they refract (change direction) when passing from one layer to another layer; and they diffract (follow a curved path) around obstacles they encounter (see Figure 6.33). These different wave behaviors have been used to identify the boundaries that exist within Earth.

One of the most noticeable behaviors of seismic waves is that they follow strongly curved paths (see Figure 6.33) because the velocity of seismic waves generally increases with depth. Seismic waves also travel faster when rock is stiffer or less compressible. These properties of stiffness and compressibility can be used to interpret the composition and temperature of the rock. For instance, when rock is hotter, it becomes less stiff (imagine a chocolate bar left out in the Sun), and waves

travel more slowly. Waves also travel at different speeds through rocks of different compositions. Thus, the speed at which seismic waves travel can help determine both the kind of rock that is inside Earth and how hot it is.

Earth's Layered Structure

According to the *nebular theory*, Earth and the solar system began to form nearly 5 billion years ago due to the gravitational collapse of a huge cloud of dust and gases called a *nebula*. As material accumulated to form Earth (and for a short period afterward), the high-velocity impact of nebular debris and the decay of radioactive elements caused the temperature of our planet to increase steadily. During this time of intense heating, Earth became hot enough that iron and nickel began to melt. Melting produced liquid blobs of heavy metal that sank toward the center of the planet. This process occurred rapidly on the scale of geologic time and produced Earth's dense iron-rich core.

The early period of heating resulted in another process, called *chemical differentiation*, whereby melting formed buoyant masses of molten rock that rose toward Earth's surface and solidified to produce a primitive crust. These rocky materials were rich in oxygen and "oxygen-seeking" elements, particularly silicon and aluminum, along with lesser amounts of calcium, sodium, potassium, iron, and magnesium. In addition, some heavy metals such as gold, lead, and uranium, which have low melting points or were highly soluble in the ascending molten masses, were scavenged from Earth's interior and

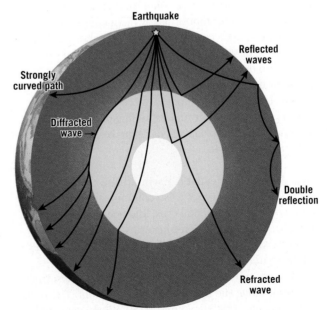

Figure 6.33 Possible paths that earthquake waves take

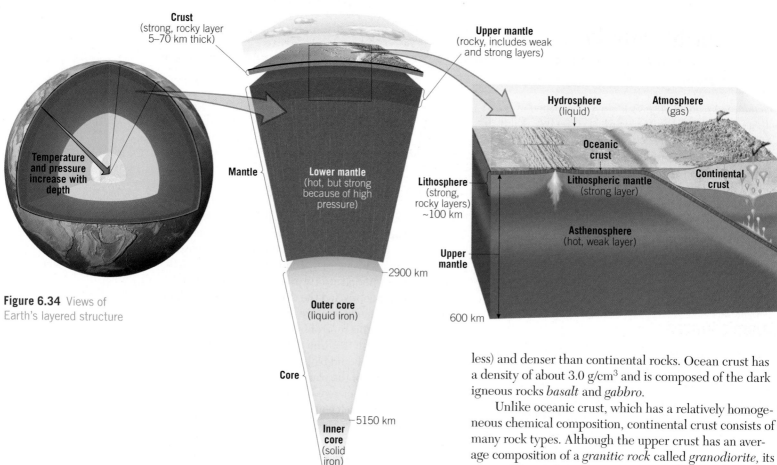

Figure 6.34 Views of Earth's layered structure

concentrated in the developing crust. This early period of chemical segregation established the three basic divisions of Earth's interior: (1) the iron-rich *core*, (2) the thin *primitive crust*, and (3) Earth's largest layer, called the *mantle*, which is located between the core and crust.

Earth's three compositionally distinct layers—the crust, mantle, and core—can be further subdivided into zones based on *physical* properties. The physical properties used to define such regions include whether the layer is solid or liquid and how weak or strong it is. Knowledge of both the chemical and physical properties of Earth's layers is essential to our understanding of basic geologic processes, such as volcanism, earthquakes, and mountain building (**Figure 6.34**).

Earth's Crust The **crust** is Earth's relatively thin, rocky outer skin, and two types exist: continental crust and oceanic crust. Continental crust and oceanic crust have very different compositions, histories, and ages. In fact, oceanic crust is compositionally more similar to Earth's mantle than to continental crust.

The ocean crust is about 7 kilometers (4 miles) thick and forms along the mid-oceanic ridge system. The rocks of the oceanic crust are younger (180 million years old or

less) and denser than continental rocks. Ocean crust has a density of about 3.0 g/cm^3 and is composed of the dark igneous rocks *basalt* and *gabbro*.

Unlike oceanic crust, which has a relatively homogeneous chemical composition, continental crust consists of many rock types. Although the upper crust has an average composition of a *granitic rock* called *granodiorite*, its composition and structure varies considerably from place to place.

Continental crust averages about 40 kilometers (25 miles) thick but can be more than 70 kilometers (40 miles) thick in mountainous regions such as the Himalayas and the Andes. Continental crust has an average density of about 2.7 g/cm^3, which is much lower than the density of mantle rock. The low density of the continents relative to the mantle explains why continents are buoyant—acting like giant rafts, floating atop tectonic plates—and why they cannot be readily subducted into the mantle. Because continental rocks cannot be easily recycled into the mantle, continental rocks that exceed 4 billion years in age have been found.

Earth's Mantle More than 82 percent of Earth's volume is contained in the **mantle**, a solid, rocky shell that extends to a depth of about 2900 kilometers (1800 miles) beneath Earth's crust. The boundary between the crust and mantle represents a marked change in chemical composition. The dominant rock type in the uppermost mantle is *peridotite*, which is richer in the metals iron and magnesium than are the rocks found in either the continental or oceanic crust.

The upper mantle extends from the crust–mantle boundary down to a depth of about 660 kilometers (410 miles). Recall that the upper mantle is divided into two different parts. The top portion of the upper mantle is part of the stiff *lithosphere*, and beneath that is the weaker *asthenosphere*. The **lithosphere** ("sphere of rock") consists of the crust and uppermost mantle and forms Earth's relatively cool, rigid outer shell. Averaging about 100 kilometers (62 miles) thick, the lithosphere is more than 250 kilometers (155 miles) thick below the oldest portions of the continents (see Figure 6.34).

Beneath Earth's stiff outer layer lies a solid but comparatively weak layer termed the **asthenosphere** ("weak sphere"). Because the asthenosphere and lithosphere are mechanically detached from each other, the lithosphere is able to move independently of the asthenosphere.

From 660 kilometers (410 miles) deep to the top of the core, at a depth of 2900 kilometers (1800 miles), is the lower mantle. Because the pressure in the mantle increases with depth (due to the weight of the overlying rock), the mantle becomes stronger with depth. Despite their strength, however, the rocks within the lower mantle are hot and capable of very gradual flow.

Inner and Outer Core The **core** is thought to consist mainly of iron combined with an unknown quantity of nickel, as well as minor amounts of oxygen, silicon, and sulfur—elements that readily form compounds with iron. Because of the extreme pressure found in the core, this iron-rich material has an average density of more than 10 times the density of water, or 10 grams per cubic

centimeter. The density at Earth's center is about 13 times that of water, or 13 grams per cubic centimeter.

The **outer core** is a liquid iron-rich layer 2270 kilometers (1410 miles) thick. The liquid nature of the outer core was discovered when researchers found that S waves do not penetrate the core. Because S waves do not pass through liquids, their inability to pass though the outer core indicates that it must be a liquid.

At Earth's center lies the **inner core**, a solid dense metallic sphere with a radius of 1216 kilometers (754 miles). Despite its higher temperature, the inner core is solid, not molten like the outer core, due to the immense pressures that exist at the center of our planet. Because the inner core is a sphere, whereas Earth's other layers are shells, drawings make the inner core appear much larger than it really is (see Figure 6.34).

6.6 CONCEPT CHECKS

1. How did Earth acquire its layered structure?
2. How do continental crust and oceanic crust differ?
3. Contrast the physical characteristics of the asthenosphere and the lithosphere.
4. How are Earth's inner and outer cores different? How are they similar?

6.7 Rock Deformation

Compare and contrast brittle and ductile deformation.

Earth is a dynamic planet. Shifting lithospheric plates continually change the face of our planet by moving continents across the globe. The results of this tectonic activity are perhaps most strikingly apparent in Earth's major mountain belts. Rocks containing fossils of marine organisms are found thousands of meters above sea level, and massive rock units are bent, contorted, overturned, and sometimes rife with fractures.

Why Rocks Deform

Every body of rock, no matter how strong, has a point at which it will fracture or flow. **Deformation** is a general

term that refers to all changes in the shape, position, or orientation of a rock mass. Significant crustal deformation occurs along plate margins. Plate motions and the interactions along plate boundaries generate the tectonic forces that cause rock to deform.

When rocks are subjected to forces (stresses) greater than their own strength, they begin to deform, usually by bending or breaking (**Figure 6.35**). It is easy to visualize how rocks break because we normally think of them as being brittle. But how can rock masses be bent into intricate folds without being broken during the process? To answer this question, geologists performed laboratory experiments in which they subjected rocks to forces under

SmartFigure 6.35
Deformed sedimentary strata These deformed strata are exposed in a road cut near Palmdale, California. In addition to the obvious folding, light-colored beds are offset along a fault located on the right side of the photograph. (Photo by E. J. Tarbuck) (https://goo.gl/vwWFKx)

Tutorial

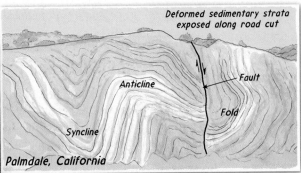

Deformed sedimentary strata exposed along road cut

Anticline
Fault
Fold
Syncline
Palmdale, California

Geologist's Sketch

conditions that simulated those existing at various depths within the crust. Although each rock type deforms somewhat differently, the general characteristics of rock deformation were determined from these experiments.

Types of Rock Deformation

Geologists discovered that when stress is gradually applied to a rock, the rock first responds by deforming elastically. Changes that result from elastic deformation are recoverable; that is, like a rubber band, the rock will return to nearly its original size and shape when the force is removed. During **elastic deformation**, the chemical bonds of the minerals within a rock are stretched but do not break. Once the elastic limit (strength) of a rock is surpassed, it either flows (**ductile deformation**) or fractures (**brittle deformation**).

The factors that influence the strength of a rock and how it will deform include temperature, confining pressure, rock type, and time. Rocks near the surface, where temperatures and confining pressures are low, tend to behave like a brittle solid and fracture once their strength is exceeded. This type of deformation is called brittle deformation. From our everyday experience, we know that glass objects, wooden pencils, china plates, and even our bones exhibit brittle failure once their strength is surpassed.

By contrast, at depth, where temperatures and confining pressures are high, rocks exhibit *ductile* behavior. Ductile deformation is a type of solid-state flow that produces a change in the size and shape of an object without fracturing. Ordinary objects that display ductile behavior include modeling clay, bee's wax, caramel candy, and most metals. For example, an automobile's fender may be dented but not broken when hit by another vehicle. Ductile deformation of a rock—strongly aided by high temperature and high confining pressure—is somewhat similar to the deformation of a penny flattened by a train. During ductile deformation, the chemical bonds within mineral grains break and re-form, allowing the grains to gradually change shape. Rocks that display evidence of ductile flow were usually deformed at great depth and may exhibit contorted folds that give the impression that the strength of the rock was akin to the strength of soft putty. The following sections describe how rock responds to both ductile and brittle deformation.

 6.7 CONCEPT CHECKS

1. Define *deformation*.
2. Describe elastic deformation.
3. How is brittle deformation different from ductile deformation?

6.8 Folds: Structures Formed by Ductile Deformation

List and describe the major types of folds.

Along convergent plate boundaries, flat-lying sedimentary and volcanic rocks are often bent into a series of wavelike undulations called **folds**. Folds in sedimentary strata are much like those that would form if you were to hold the ends of a sheet of paper and then push them together. In nature, folds come in a wide variety of sizes and configurations. Some folds are broad flexures in which rock units hundreds of meters thick have been slightly warped. Others are very tight microscopic structures found in metamorphic rocks. Size differences notwithstanding, most folds are caused by *compressional forces that result in the shortening and thickening of the crust.*

Anticlines and Synclines

The two most common types of folds are anticlines and synclines (**Figure 6.36**). **Anticlines** usually arise by upfolding, or arching, of sedimentary layers and are sometimes spectacularly displayed along highways that have been cut through deformed strata. Almost always found in association with anticlines are downfolds, or

troughs, called **synclines**.[†] Notice in Figure 6.36 that the *limb* (or flank) of an anticline is also a limb of the adjacent syncline.

Depending on their orientation, these basic folds are described as *symmetrical* when the limbs are mirror images of each other and *asymmetrical* when they are not. An asymmetrical fold is said to be *overturned* if one or both limbs are tilted beyond the vertical (see Figure 6.36). An overturned fold can also "lie on its side" so that a plane extending through the axis of the fold would be horizontal. Such *recumbent* folds are common in highly deformed mountainous regions such as the Alps.

Domes and Basins

Broad upwarps in basement rock may deform the overlying cover of sedimentary strata and generate large

---

[†] By strict definition, an *anticline* is a structure in which the oldest strata are found in the center. This most typically occurs when strata are upfolded. A *syncline* is strictly defined as a structure in which the youngest strata are found in the center. This occurs most commonly when strata are downfolded.

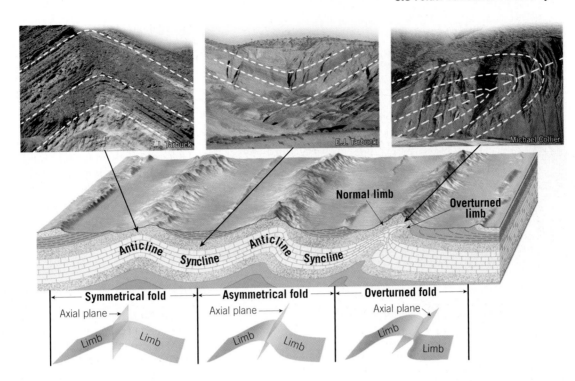

folds. When this upwarping produces a circular or elongated structure, the feature is called a **dome**. Downwarped structures having a similar shape are termed **basins**.

The Black Hills of western South Dakota constitute a large domed structure thought to have been generated by upwarping. Erosion has stripped away the highest portions of the upwarped sedimentary beds, exposing older igneous and metamorphic rocks in the center (**Figure 6.37**). Remnants of these once-continuous sedimentary layers flank the crystalline core of these mountains.

Several large basins exist in the United States. The basins of Michigan and Illinois have very gently sloping beds similar to saucers. These basins are thought to have resulted from large, heavy accumulations of sediment that caused the crust to subside.

Because large basins usually contain sedimentary beds that slope at very low angles, they are usually identified by the age of the rocks comprising them. The youngest rocks are found near the center, and the oldest rocks are at the flanks. This is just the opposite of a domed structure, such as the Black Hills, where the oldest rocks form the core.

Monoclines

Although we discuss folds and faults separately, folds can be uniquely coupled with faults. Examples include prominent features of the Colorado Plateau, called **monoclines**, which are large, step-like folds in

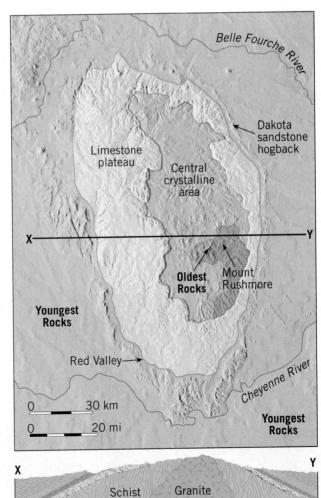

SmartFigure 6.38 The East Kaibab Monocline, Arizona This monocline consists of bent sedimentary beds that were deformed by faulting in the bedrock below. The thrust fault shown is called a *blind thrust* because it does not reach the surface. (Photo by Michael Collier) (https://goo.gl/4FVx8Y)

East Kaibab Monocline, Arizona

the plateau. As large blocks of basement rock were displaced upward, the comparatively ductile sedimentary strata above responded by draping over the fault like clothes hanging over a bench. Displacement along these reactivated faults often exceeds 1 kilometer (0.6 miles).

Examples of monoclines found on the Colorado Plateau include the East Kaibab Monocline, the Raplee Anticline, the Waterpocket Fold, and the San Rafael Swell. The inclined sedimentary beds shown in Figure 6.38 once formed a continuous flat-lying layer that blanketed much of the Colorado Plateau.

otherwise horizontal sedimentary strata (**Figure 6.38**). These folds result from the reactivation of ancient, steep-dipping faults located in basement rocks beneath

6.8 CONCEPT CHECKS

1. Distinguish between anticlines and synclines, domes and basins, anticlines and domes.
2. The Black Hills of South Dakota is a good example of what type of geologic structure?
3. Where do we find the youngest rocks in an eroded basin: near the center or near the flanks?

6.9 Faults and Joints: Structures Formed by Brittle Deformation

Sketch and describe the relative motion of rock bodies located on opposite sides of normal, reverse, and strike-slip faults.

Brittle deformation results in the fracturing of rock material. Fractures in the crust along which appreciable displacement has taken place are called **faults**. Occasionally, small faults can be recognized in road cuts where sedimentary beds have been offset a few meters, as shown in **Figure 6.39A**. Faults of this scale usually occur as single discrete breaks. By contrast, large faults, such as the San Andreas Fault in California, have displacements

of hundreds of kilometers and consist of many interconnecting fault surfaces. These *fault zones* can be several kilometers wide and are often easier to identify from high-altitude photographs than at ground level.

Dip-Slip Faults

Faults in which movement is primarily parallel to the *dip* (or inclination) of the fault surface are called **dip-slip faults**. It has become common practice to call the rock surface that is immediately above the fault the **hanging wall block** and to call the rock surface below the **footwall block** (**Figure 6.39B**). This nomenclature arose from prospectors and miners excavating shafts and tunnels along fault zones that were sites of ore deposits. In these tunnels, the miners would walk on the rocks below the mineralized fault zone (the footwall block) and hang their lanterns on the rocks above (the hanging wall block).

Vertical displacements along dip-slip faults may produce long, low cliffs

SmartFigure 6.39 Faults: Fractures where appreciable displacement has occurred This photo and diagram illustrate the relative displacement between the blocks on either side of a normal dip-slip fault: The hanging wall block drops relative to the footwall block. Notice the low cliff called a fault scarp in the block diagram. Fault scarps result from rapid vertical displacement along a fault. (Photo by Marli Miller) (https://goo.gl/vFFfSV)

A.

B.

called **fault scarps**. Fault scarps, such as the one illustrated in Figure 6.39B, are produced by displacements that generate earthquakes.

Normal Faults Dip-slip faults are classified as **normal faults** when the hanging wall block moves down relative to the foot-wall block (see Figure 6.39B). Because of the downward motion of the hanging wall block, normal faults accommodate stretching, or extension, of the crust.

Normal faults are found in a variety of sizes. Some are small, having displacements of only 1 meter (3 feet) or so, like the one shown in the road cut in Figure 6.39A. Others extend for tens of kilometers and may sinuously trace the boundary of a mountain front. Most large normal faults have relatively steep dips that tend to flatten out with depth.

Fault-Block Mountains In the western United States, large normal faults are associated with structures called **fault-block mountains**. Excellent examples of fault-block mountains are found in the Basin and Range Province, a region that encompasses Nevada and portions of the surrounding states (**Figure 6.40**). Here the crust has been elongated and broken to create more than 200 relatively small mountain ranges. Averaging about 80 kilometers (50 miles) in length, these ranges rise 900 to 1500 meters (3000 to 5000 feet) above the adjacent down-faulted basins.

The topography of the Basin and Range Province evolved in association with a system of normal faults that trend roughly north–south. Movements along these faults produced alternating uplifted fault blocks called **horsts** and down-dropped blocks called **grabens**. Horsts generate elevated topography, whereas grabens form basins. As Figure 6.40 illustrates, tilted fault blocks also contribute to the alternating topographic highs and lows in the Basin and Range Province. In this setting the down-dropped side of a block, called a *half-graben*, creates a valley. The horsts and the higher ends of the tilted fault blocks are the source of sediments that have accumulated in the adjacent basins.

Notice in Figure 6.40 that the slopes of the normal faults decrease with depth and eventually join to form a nearly horizontal fault called a *detachment fault*. This type of fault forms a major boundary between the rocks below, which exhibit ductile deformation, and the rocks above, which exhibit brittle deformation.

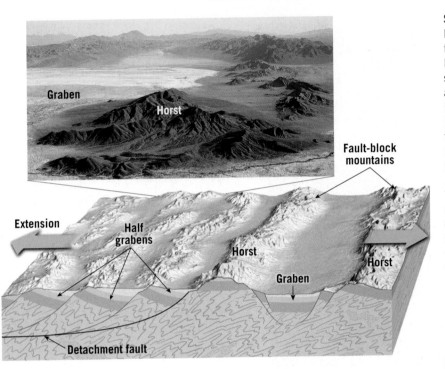

Mobile Field Trip

SmartFigure 6.40 Fault-block mountains of the Basin and Range Province Here, tensional stresses have elongated and fractured the crust into numerous blocks. Movement along these faults has tilted the blocks, producing parallel mountain ranges called *fault-block mountains*. The down-faulted blocks (*grabens*) form basins, whereas the up-faulted blocks (*horsts*) erode to form rugged mountainous topography. In addition, numerous tilted blocks (*half-grabens*) form both basins and mountains. (Photo by Michael Collier) (http://goo.gl/FbBhca)

Fault motion provides geologists with a method of determining the nature of the tectonic forces at work within Earth. Normal faults are associated with tensional forces that pull the crust apart. This "pulling apart" can be caused either by uplifting, which causes the surface to stretch and break, or by opposing horizontal forces.

Reverse and Thrust Faults Dip-slip faults in which the hanging wall block moves up relative to the footwall block are called **reverse faults** (**Figure 6.41**). Whereas normal faults occur in tensional environments, reverse faults result from strong compressional stresses. Because

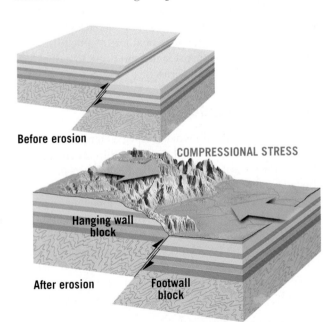

Reverse faults are dip-slip faults in which the hanging wall block moves up relative to the footwall block.

Figure 6.41 Reverse faults Reverse faults are a type of dip-slip fault generated by compressional stresses that displace the hanging wall block upward relative to the footwall block.

Figure 6.42 Strike-slip faulting **A.** The block diagram illustrates the features associated with large strike-slip faults. Notice how the stream channels have been offset by fault movement. **B.** Aerial view of the San Andreas Fault. (Photo by D. Parker/ Science Source)

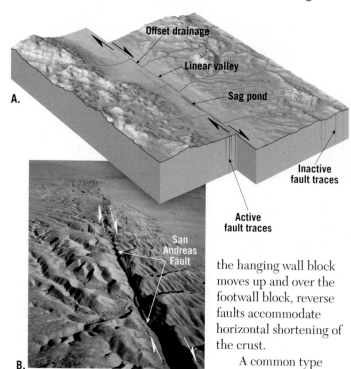

crustal block on the opposite side of the fault to move to the right as you face the fault, it is called a *right-lateral* strike-slip fault (**Figure 6.42B**).

The Great Glen Fault in Scotland is a well-known example of a *left-lateral* strike-slip fault, which exhibits displacement in the opposite direction. The total displacement along the Great Glen Fault is estimated to exceed 100 kilometers (60 miles). Also associated with this fault trace are numerous lakes, including Loch Ness, the home of the legendary monster.

Recall that some strike-slip faults, called **transform faults**, cut through the lithosphere and accommodate motion between two large tectonic plates. Numerous transform faults cut the oceanic lithosphere and link segments of oceanic ridges. Others accommodate displacement between continental plates that move horizontally with respect to each other. One of the best-known transform faults is the San Andreas Fault (see Figure 6.42).

Joints

Unlike faults, **joints** are fractures along which no appreciable displacement has occurred. Most joints develop in response to barely perceptible regional upwarping and downwarping of the crust, which causes rock near the surface to fail by brittle fracture.

Joints often consist of two or even three sets of intersecting fractures that slice the rock into numerous regularly shaped blocks. These joint sets often exert a strong influence on other geologic processes. For example, chemical weathering tends to be concentrated along joints, and joint patterns influence how groundwater moves through the crust.

We have already considered one type of joint. Recall from Chapter 2 that sheeting produces a pattern of gently curved joints that develop more or less parallel to the surface of large exposed igneous bodies such as batholiths. Here the jointing results from the gradual expansion that occurs when erosion removes the overlying load (see Figure 2.15, page 57). In Chapter 7 we will examine another type, called *columnar joints*, that form when lava cools and develops shrinkage fractures that produce elongated, pillarlike columns (see Figure 7.30, page 235).

the hanging wall block moves up and over the footwall block, reverse faults accommodate horizontal shortening of the crust.

A common type of reverse fault, called a **thrust fault**, has a low angle of dip (less than 45 degrees), which allows the overlying block to slide over the top of the underlying block. While most high-angle reverse faults are small, thrust faults exist at all scales, with some large thrust faults having displacements ranging from of tens to hundreds of kilometers. In mountainous regions such as the Alps, Northern Rockies, Himalayas, and Appalachians, thrust faults have displaced strata as far as 100 kilometers (60 miles) over adjacent rock units. The result of this large-scale movement is that older strata end up overlying younger rocks.

Thrust faulting is most pronounced in subduction zones where oceanic lithosphere is descending into the mantle, as well as along convergent boundaries where plates are colliding. Compressional forces generally produce folds as well as thrust faults and result in a thickening and shortening of the crust.

Strike-Slip Faults

A fault in which the dominant displacement is horizontal and parallel to the *strike* (the compass direction) of the fault surface is called a **strike-slip fault** (**Figure 6.42A**). Recall from Section 6.1 that the earliest scientific records of strike-slip faulting were made following surface ruptures that produced large earthquakes, including the study of the great San Francisco earthquake of 1906. During this strong earthquake, structures such as fences that were built across the San Andreas Fault were displaced as much as 4.7 meters (15 feet) (see Figure 6.3). Because movement along the San Andreas causes the

6.9 CONCEPT CHECKS

1. Contrast the movements that occur along normal and reverse faults.
2. What type of faults are associated with fault-block mountains?
3. How are reverse faults different from thrust faults? In what way are they the same?
4. Describe the relative movement along a strike-slip fault.
5. How are joints different from faults?

6.10 Mountain Building

Locate and identify Earth's major mountain belts on a world map.

Mountain building has occurred in the recent geologic past in several locations around the world (**Figure 6.43**). One example is the American Cordillera, which runs along the western margin of the Americas from Cape Horn at the tip of South America to Alaska and includes the Andes and Rocky Mountains. Other examples include the Alpine–Himalaya chain, which extends along the margin of the Mediterranean, through Iran to northern India, and into Indochina; and the mountainous terrains of the western Pacific, which include the volcanic island arcs that comprise Japan, the Philippines, and Sumatra. Most of these young mountain belts have come into existence within the past 100 million years.

In addition to these young mountain belts, there are several chains of Paleozoic-age mountains on Earth. Although these older structures are deeply eroded and topographically less prominent, they exhibit the same structural features found in younger mountains. The Appalachians in the eastern United States and the Urals in Russia are classic examples of this group of older and well-worn mountain belts.

The term for the processes that collectively produce a mountain belt is **orogenesis** (*oros* = mountain, *genesis* = to come into being). Most major mountain belts display striking visual evidence of great horizontal forces that have shortened and thickened the crust. These **compressional mountains** contain large quantities of preexisting sedimentary and crystalline rocks that have been faulted and contorted into a series of folds. Although folding and thrust faulting are often the most conspicuous signs of orogenesis, varying degrees of metamorphism and igneous activity are always present.

How do mountain belts form? This question has intrigued some of the greatest philosophers and scientists since the time of the ancient Greeks. One early proposal suggested that mountains are simply wrinkles in Earth's crust, produced as the planet cooled from its original semimolten state. According to this idea, Earth contracted and shrank as it lost heat, which caused the crust to deform in a manner similar to how an orange peel wrinkles as the fruit dries out. However, neither this nor any other early hypothesis withstood scientific scrutiny.

The development of the theory of plate tectonics produced a model for orogenesis with excellent explanatory power. According to this model, the tectonic processes that generate Earth's major mountainous terrains begin at convergent plate boundaries, where oceanic lithosphere descends into the mantle.

6.10 CONCEPT CHECKS

1. Define *orogenesis*.
2. In the plate tectonics model, which type of plate boundary is most directly associated with Earth's major mountain belts?

> **Did You Know?**
> Edmund Hillary of New Zealand and Tenzing Norgay of Nepal were the first to reach the summit of Mount Everest, on May 29, 1953. Not one to rest on his laurels, Hillary later went on to lead the first crossing of Antarctica.

Figure 6.43 Earth's major mountain belts Notice the east–west trend of major mountain belts in Eurasia in contrast to the north–south trend of the North and South American Cordillera. The shields and stable platforms shown are composed of old crustal rocks that have been highly deformed during ancient mountain-building events.

Key
- Young mountain belts (less than 100 million years old)
- Old mountain belts
- Shields
- Stable platforms (shields covered by sedimentary rock)

6.11 Subduction and Mountain Building

Sketch an Andean-type mountain belt and describe how each of its major features is generated.

The subduction of oceanic lithosphere is the main driving force of orogenesis. Where oceanic lithosphere subducts beneath an oceanic plate, a *volcanic island arc* and related tectonic features develop. Subduction beneath a continental block, on the other hand, forms a *continental volcanic arc* and mountainous topography along the margin of a continent. In addition, volcanic island arcs and other crustal fragments "drift" across the ocean basin until they reach a subduction zone, where they collide and become welded to another crustal fragment or larger continental block. If subduction continues long enough, it can ultimately lead to the collision of two or more continents.

Figure 6.44 Andean-type mountain building

A. Passive continental margin with an extensive platform of sediments and sedimentary rocks.

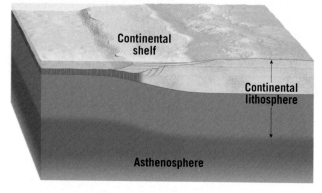

B. Plate convergence generates a subduction zone, and partial melting produces a continental volcanic arc. Compressional forces and igneous activity further deform and thicken the crust, elevating the mountain belt.

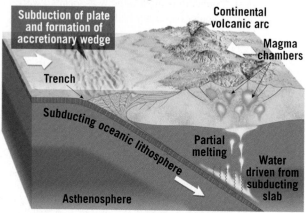

C. Subduction ends and is followed by a period of uplift.

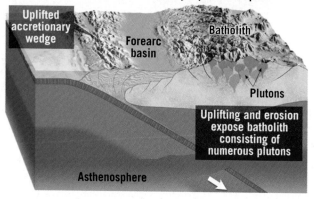

Island Arc–Type Mountain Building

Island arcs result from the steady subduction of oceanic lithosphere, which may last for 200 million years or more (see Figure 5.15B, page 155). Periodic volcanic activity, the emplacement of igneous plutons at depth, and the accumulation of sediment that is scraped from the subducting plate gradually increase the volume of crustal material capping the upper plate. Some large volcanic island arcs, such as Japan, owe their size to having been built upon a preexisting fragment of continental crust or to the joining of two or more island arcs.

The continued growth of a volcanic island arc can generate mountainous topography consisting of belts of igneous and metamorphic rocks. This activity, however, is viewed as just one phase in the development of a major mountain belt.

Andean-Type Mountain Building

Mountain building along continental margins involves the convergence of an oceanic plate and a plate whose leading edge contains continental crust. Exemplified by the Andes Mountains, an *Andean-type convergent zone* results in the formation of a continental volcanic arc and related tectonic features inland of the continental margin.

The first stage in the development of an idealized Andean-type mountain belt occurs prior to the formation of the subduction zone. During this period, the continental margin is a **passive continental margin**; that is, it is not a plate boundary but a part of the same plate as the adjoining oceanic crust. The east coast of North America provides a present-day example of a passive continental margin. Here, as at other passive continental margins surrounding the Atlantic, deposition of sediment on the continental shelf is producing a thick wedge of shallow-water sandstones, limestones, and shales (**Figure 6.44A**). Beyond the continental shelf, turbidity currents are depositing sediments on the continental slope and rise (see Chapter 9).

At some point, the continental margin becomes active. A subduction zone forms, and the deformation process begins (**Figure 6.44B**). A good place to examine an **active**

continental margin is the west coast of South America, where the Nazca plate is being subducted beneath the South American plate along the Peru–Chile trench.

In an idealized Andean-type subduction, convergence of the continental block and the subducting oceanic plate leads to deformation and metamorphism of the continental margin. Once the oceanic plate descends to about 100 kilometers (62 miles), partial melting of mantle rock above the subducting slab generates magma that migrates upward (see Figure 6.44B). Thick continental crust greatly impedes the ascent of magma. Consequently, a high percentage of the magma that intrudes the crust never reaches the surface; instead, it crystallizes at depth to form intrusive igneous rocks (plutons). Eventually, uplifting and erosion exhume these igneous bodies and associated metamorphic rocks. Once they are exposed at the surface, these massive structures are called *batholiths* (**Figure 6.44C**). Composed of numerous plutons, batholiths form the core of California's Sierra Nevada and are prevalent in the Peruvian Andes.

During the development of this continental volcanic arc, sediment derived from the land and scraped from the subducting plate is plastered against the landward side of the trench like piles of dirt in front of a bulldozer. This chaotic accumulation of sedimentary and metamorphic rocks with occasional scraps of ocean crust is called an **accretionary wedge** (see Figure 6.44B). Prolonged subduction can build an accretionary wedge that is large enough to stand above sea level (see Figure 6.44C).

Andean-type mountain belts are composed of two roughly parallel zones. The volcanic arc develops on the continental block. It consists of volcanoes and large intrusive bodies intermixed with high-temperature metamorphic rocks. The seaward segment is the accretionary wedge. It consists of folded and faulted sedimentary and metamorphic rocks (see Figure 6.44C).

Sierra Nevada and Coast Ranges One of the best examples of an inactive Andean-type orogenic belt is found in the western United States. It includes the Sierra Nevada and the Coast Ranges in California. These parallel mountain belts were produced by the subduction of a portion of the Pacific basin under the western edge of the North American plate. The Sierra Nevada batholith is a remnant of a portion of the continental volcanic arc that was produced by several surges of magma over tens of millions of years. Subsequent uplifting and erosion have removed most of the evidence of past volcanic activity and exposed a core of crystalline, igneous, and associated metamorphic rocks.

In the trench region, sediments scraped from the subducting plate, plus those provided by the eroding continental volcanic arc, were intensely folded and faulted into an accretionary wedge. This chaotic mixture of rocks presently constitutes the Franciscan Formation of California's Coast Ranges. Uplifting of the Coast Ranges took place only recently, as evidenced by the young unconsolidated sediments that still mantle portions of these highlands.

6.11 CONCEPT CHECKS

1. Describe and give an example of a passive continental margin.
2. What is an accretionary wedge? Briefly describe its formation.
3. In what ways are the Sierra Nevada and the Andes similar?

6.12 Collisional Mountain Belts

Summarize the stages in the development of a collisional mountain belt such as the Himalayas.

Most major mountain belts are generated when one or more buoyant crustal fragments collide with a continental margin as a result of subduction. Oceanic lithosphere, which is relatively dense, readily subducts, whereas continental lithosphere, which contains significant amounts of low-density crustal rocks, is too buoyant to undergo subduction. Consequently, a crustal fragment arriving at a trench will collide with the margin of the overlying continental block, which usually ends further subduction at that location.

Cordilleran-Type Mountain Building

Cordilleran-type mountain building, named after the North American Cordillera, occurs in Pacific-like ocean basins, where a rapid rate of seafloor spreading is roughly balanced by an equally high rate of subduction. In this setting, it is highly likely that island arcs or small crustal fragments will be carried along until they collide with an active continental margin. The process of collision and accretion (joining together) of comparatively small crustal fragments to a continental margin has generated many of the mountainous regions that rim the Pacific.

Geologists call these accreted crustal blocks *terranes*. The term **terrane** is used to describe a crustal fragment consisting of a distinct and recognizable series of rock formations that has been transported by plate tectonic processes. By contrast, the term *terrain* is used when describing the shape of the surface topography, or the "lay of the land."

The Nature of Terranes What is the nature of these crustal fragments, and where did terranes originate?

SmartFigure 6.45 Collision and accretion of small crustal fragments to a continental margin (https://goo.gl/RtF14w)

A. A microcontinent and a volcanic island arc are being carried toward a subduction zone.

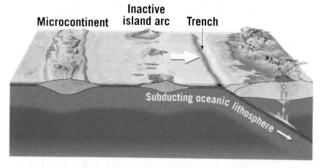

B. The volcanic island arc is sliced off the subducting plate and thrust onto the continent.

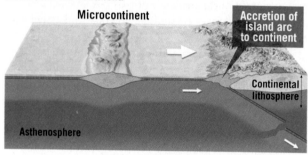

C. A new subduction zone forms seaward of the old subduction zone.

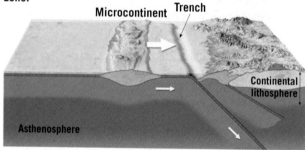

D. The accretion of the microcontinent to the continental margin shoves the remnant island arc further inland and grows the continental margin seaward.

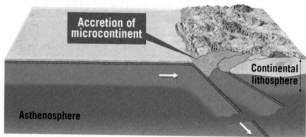

Research suggests that prior to their accretion to a continental block, some of these fragments may have been **microcontinents** similar to the modern-day island of Madagascar, located east of Africa in the Indian Ocean. Many others were island arcs similar to Japan, the Philippines, and the Aleutian Islands. Still others may have been submerged oceanic plateaus created by massive outpourings of basaltic lavas associated with mantle plumes. As shown in Figure 9.15, on page 284, quite a few small-to medium-size crustal

fragments are known to exist—many of them submerged below the Pacific Ocean.

Accretion and Orogenesis As oceanic plates move, they carry embedded oceanic plateaus, volcanic island arcs, and microcontinents to an Andean-type subduction zone. When an oceanic plate contains small seamounts, these structures are generally subducted along with the descending oceanic slab. However, large, thick units of oceanic crust, such as the Ontong Java Plateau, which is the size of Alaska, or an island arc composed of abundant "light" igneous rocks, render the oceanic lithosphere too buoyant to subduct. In these situations, a collision between the crustal fragment and the continental margin occurs.

The sequence of events that happen when small crustal fragments reach a continental margin is shown in **Figure 6.45**. Rather than subduct, the upper crustal layers of these thickened zones are peeled from the descending plate and thrust in relatively thin sheets upon the adjacent continental block. Convergence does not generally end with the accretion of a crustal fragment. Rather, new subduction zones typically form, and they can carry other island arcs or microcontinents toward a collision with the continental margin. Each collision displaces earlier accreted terranes further inland, adding to the zone of deformation as well as to the thickness (height) and lateral extent (width) of the continental margin.

The North American Cordillera The correlation between mountain building and the accretion of crustal fragments arose primarily from studies conducted in the North American Cordillera (**Figure 6.46**). Researchers determined that some of the rocks in the orogenic belts of Alaska and British Columbia contained fossil and paleomagnetic evidence indicating that these strata previously lay much closer to the equator.

It is now known that many of the terranes that make up the North American Cordillera were scattered throughout the eastern Pacific, like the island arcs and oceanic plateaus currently distributed in the western Pacific. During the breakup of Pangaea, the eastern portion of the Pacific basin (Farallon plate) began to subduct under the western margin of North America. This activity resulted in the piecemeal addition of crustal fragments to the entire Pacific margin of the continent—from Mexico's Baja Peninsula to northern Alaska (see Figure 6.46). Geologists expect that many modern microcontinents will likewise be accreted to active continental margins surrounding the Pacific, producing new orogenic belts.

Alpine-Type Mountain Building: Continental Collisions

Alpine-type orogenies, named after the European Alps that geologists have intensively studied for more than 200 years, occur where two continental masses collide. This type of orogeny may also involve the accretion of continental fragments or island arcs that occupied the ocean basin

Did You Know?
During the assembly of Pangaea, the landmasses of Europe and Siberia collided to produce the Ural Mountains. Long before the discovery of plate tectonics, this extensively eroded mountain chain was regarded as the boundary between Europe and Asia.

Figure 6.46 Terranes that have been added to western North America during the past 200 million years

ALASKA

Wrangellia

Yukon-Tanana

Cache Creek

Eastern limit of Cordilleran deformation

NORTH AMERICAN CORDILLERA

Stikinia

Wrangellia

Franciscan

ARCTIC OCEAN

PACIFIC OCEAN

CANADA

UNITED STATES

Oceanic terranes

Area deformed by the accretion of terranes

Other colored areas are accreted terranes

that once separated the two continental blocks. Mountain belts formed by the closure of major ocean basins include the Himalayas, Appalachians, Urals, and Alps. Continental collisions result in the development of mountains characterized by shortened and thickened crust, achieved through folding and large-scale thrust faulting.

We will take a closer look at two examples of collisional mountains—the Himalayas and the Appalachians. The Himalayas, one of Earth's youngest collisional mountains, are still rising. By contrast, the Appalachians are a much older mountain belt, in which active mountain building ceased about 250 million years ago.

The Himalayas The mountain-building episode that created the Himalayas began sometime between 50 and 30 million years ago, when the northern margin of India began to collide with Asia. Prior to the breakup of Pangaea, India was located between Africa and Antarctica in the Southern Hemisphere. As Pangaea fragmented, India moved rapidly, geologically speaking, a few thousand kilometers in a northward direction.

The subduction zone that facilitated India's northward migration was near the southern margin of Asia (**Figure 6.47A**). Continued subduction along Asia's margin created an Andean-type plate margin containing a well-developed continental volcanic arc and an accretionary wedge. India's northern margin, on the other hand, was a passive continental margin consisting of a thick platform of shallow-water sediments and sedimentary rocks.

Geologists have determined that one or perhaps more small continental fragments were positioned on the subducting plate somewhere between India and Asia. During the closing of the intervening ocean basin, a small crustal fragment, which now forms southern Tibet, reached the trench. This event was followed by the collision of India with Asia. The tectonic forces involved in this collision were immense, causing the more deformable materials located on the seaward edges of these landmasses to become highly folded and faulted (**Figure 6.47B**). The shortening and thickening of the crust elevated great quantities of crustal material, thereby generating the spectacular Himalaya Mountains.

In addition to uplift, crustal shortening caused rocks at the "bottom of the pile" to become deeply buried—an environment where they experienced elevated temperatures and pressures (see Figure 6.47B). Partial melting within the deepest and most deformed region of the developing mountain belt produced magmas that intruded the overlying rocks. These environments generate the metamorphic and igneous cores of collisional mountains.

The formation of the Himalayas was followed by a period of uplift that raised the Tibetan Plateau. Seismic evidence suggests that a portion of the Indian subcontinent was thrust beneath Tibet—a distance of perhaps 400 kilometers (250 miles). If this occurred, the added crustal thickness would account for the lofty landscape of southern Tibet, which has an average elevation higher than Mount Whitney, the highest point in the contiguous United States.

The collision with Asia slowed but did not stop the northward migration of India, which has since penetrated at least 2000 kilometers (1200 miles) into the mainland of Asia. Crustal shortening accommodated some of this motion. Much of the remaining penetration into Asia caused lateral displacement of large blocks of the Asian crust by a mechanism described as *continental escape*. These displaced crustal blocks include much of Southeast Asia (the region between India and China) and sections of China.

The Appalachians The Appalachian Mountains provide great scenic beauty near the eastern margin of North America, from Alabama to Newfoundland. The orogenies that generated this extensive mountain system lasted a few hundred million years and were one of the stages in assembling the supercontinent of Pangaea. Detailed studies of the Appalachians indicate that the formation of this mountain belt was complex and resulted from three distinct episodes of mountain building.

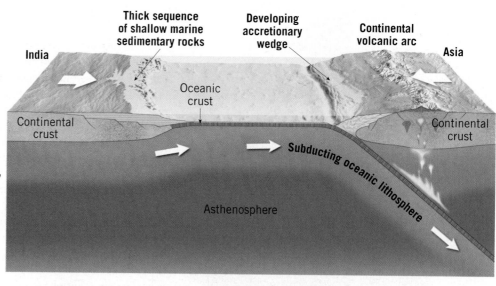

A. Prior to the collision of India and Asia, India's northern margin consisted of a thick platform of continental shelf sediments, whereas Asia's was an active continental margin with a well developed accretionary wedge and volcanic arc.

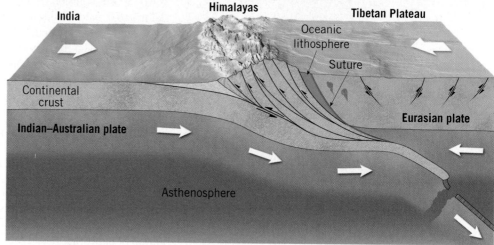

B. The continental collision folded and faulted the crustal rocks that lay along the margins of these continents to form the Himalayas. This event was followed by the gradual uplift of the Tibetan Plateau as the subcontinent of India was shoved under Asia.

Our simplified overview begins roughly 750 million years ago, with the breakup of a supercontinent called Rodinia that predates Pangaea. Similar to the breakup of Pangaea, this episode of continental rifting and seafloor spreading generated a new ocean between the rifted continental blocks. Located within this developing ocean basin was an active volcanic arc that lay off the coast of North America and a microcontinent situated closer to Africa (**Figure 6.48A**).

About 600 million years ago, for reasons geologists do not completely understand, plate motion changed dramatically, and this ancient ocean basin began to close. This led to three main orogenic events that culminated in the collision of North America and Africa. Around 450 million years ago, the marginal sea between the volcanic island arc and ancestral North America began to close. The collision that ensued, called the *Taconic orogeny*, caused the volcanic arc along with ocean sediments located on the upper plate to be thrust over the larger continental block (**Figure 6.48B**).

A second episode of mountain building, called the *Acadian orogeny*, occurred about 350 million years ago as the continued closing of the ancient ocean basin led to the collision of the microcontinent with North America (**Figure 6.48C**). This orogeny involved thrust faulting,

metamorphism, and the intrusion of several large granite bodies. In addition, this event added substantially to the width of North America.

The final orogeny, called the *Alleghanian orogeny*, occurred between 250 and 300 million years ago, when Africa collided with North America. The material that had been accreted earlier was displaced by as much as 250 kilometers (155 miles) toward the interior of North America. This event also displaced and further deformed the shelf sediments and sedimentary rocks that had once flanked the eastern margin of North America (**Figure 6.48D**). Today these folded and thrust-faulted sandstones, limestones, and shales make up the largely unmetamorphosed rocks of the Valley and Ridge Province.

With the collision of Africa and North America, the Appalachians, then perhaps as majestic as the Himalayas today, lay in the interior of Pangaea. Then, about 180 million years ago, this supercontinent began to break into smaller fragments, a process that ultimately created the modern Atlantic Ocean. Because this new zone of rifting occurred east of the suture that formed when Africa and North America collided, remnants of Africa remain "welded" to the North American plate (**Figure 6.48E**). The crust underlying Florida is an example.

Other mountain ranges that exhibit evidence of continental collisions include the Alps and the Urals. The Alps formed as Africa and at least two smaller crustal fragments collided with Europe during the closing of the Tethys Sea. Similarly, the Urals were uplifted during the assembly of Pangaea, when northern Europe and northern Asia collided, forming a major portion of Eurasia.

6.12 **CONCEPT CHECKS**

1. Differentiate between terrane and terrain.
2. Where might magma be generated in a newly formed collisional mountain belt?
3. How does the plate tectonics theory help explain the existence of fossil marine life in rocks atop compressional mountains?
4. List four mountain belts that formed as a result of continental collisions.

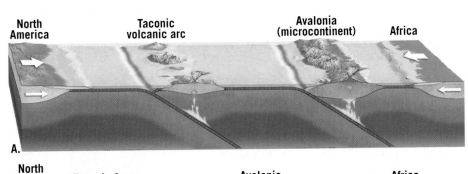

A.

Closing of an Ocean Basin About 600 million years ago, the precursor to the North Atlantic began to close. Located within this ocean basin was an active volcanic arc that lay off the coast of North America and a microcontinent situated closer to Africa.

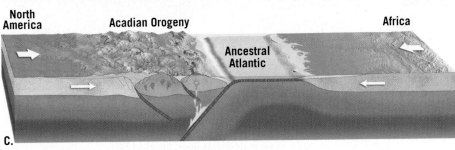

B.

Taconic Orogeny Around 450 million years ago, the marginal sea between the volcanic island arc and North America closed. The collision, called the Taconic Orogeny, thrust the island arc over the eastern margin of North America.

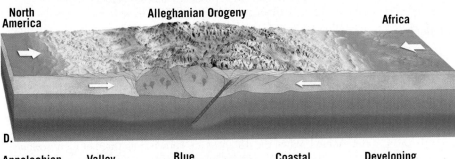

C.

Acadian Orogeny A second episode of mountain building, called the Acadian Orogeny, occurred about 350 million years ago and involved the collision of a microcontinent with North America.

D.

Alleghanian Orogeny The final event, the Alleghanian Orogeny, occurred between 250 and 300 million years ago, when Africa collided with North America. The result was the formation of the Appalachian Mountains, perhaps once as majestic as the Himalayas. The Appalachians lay in the interior of the newly assembled supercontinent of Pangaea.

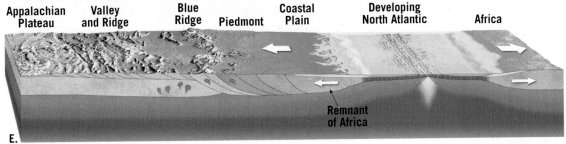

E.

Rifting of Pangaea About 180 million years ago, Pangaea began to break into smaller fragments, a process that ultimately created the modern Atlantic Ocean. Because this new zone of rifting occurred east of the suture that formed when Africa and North America collided, remnants of African crust remain "welded" to the North American plate.

SmartFigure 6.48
Formation of the Appalachian Mountains
The Appalachians formed during the closing of a precursor to the Atlantic Ocean. This event involved three separate stages of mountain building that spanned more than 300 million years. (Based on Zve Ben Avraham, Jack Oliver, Larry Brown, and Frederick Cook) (https://goo.gl/b2Deug)

Tutorial

CONCEPTS IN REVIEW
Restless Earth: Earthquakes and Mountain Building

6.1 What Is an Earthquake?

Sketch and describe the mechanism that generates most earthquakes.

KEY TERMS: earthquake, fault, hypocenter (focus), epicenter, seismic wave, elastic rebound, megathrust fault, fault creep

- The sudden movements of large blocks of rock on opposite sides of faults cause most earthquakes. The location where the rock begins to slip is called the hypocenter, or focus. During an earthquake, seismic waves radiate outward from the hypocenter into the surrounding rock. The point on Earth's surface directly above the hypocenter is the epicenter.
- Over tens to hundreds of years, differential stresses gradually bend Earth's crust. Frictional resistance keeps the rock from rupturing and slipping. At some point, the stress overcomes the frictional resistance, and slippage allows the deformed (bent) rock to "spring back" to its original shape, generating an earthquake. The springing back is called elastic rebound.
- Convergent plate boundaries and associated subduction zones are marked by megathrust faults. These large faults are responsible for most of the largest earthquakes in recorded history. Megathrust faults are also capable of generating tsunamis.
- The San Andreas Fault in California is an example of a large strike-slip fault that forms a transform plate boundary capable of generating destructive earthquakes.

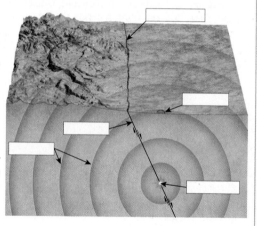

(?) **Label the blanks on the diagram to show the relationship between earthquakes and faults using the following terms: epicenter, seismic waves, fault, fault trace, and hypocenter.**

6.2 Seismology: The Study of Earthquake Waves

Compare and contrast the types of seismic waves and describe the principle of the seismograph.

KEY TERMS: seismology, seismograph (seismometer), inertia, seismogram, body waves, surface waves, P waves (primary waves), S waves (secondary waves)

- Seismology is the study of seismic waves. A seismograph measures these waves, using the principle of inertia. While the body of the instrument moves with the waves, the inertia of a suspended weight keeps a sensor stationary to record the displacement between the two.
- A seismogram, a record of seismic waves, reveals two main categories of earthquake waves: body waves (P waves and S waves), which are capable of moving through Earth's interior, and surface waves, which travel only along the upper layers of the crust. P waves are the fastest, S waves are intermediate in speed, and surface waves are the slowest. However, surface waves tend to have the greatest amplitude, S waves are intermediate, and P waves have the lowest amplitude. Large-amplitude waves produce the most shaking, so surface waves usually account for most damage during earthquakes.

- P waves and S waves exhibit different kinds of motion. P waves momentarily push (compress) and pull (stretch) rocks as they travel through a rock body, thereby changing the volume of the rock. S waves impart a shaking motion as they pass through rock, changing the rock's shape but not its volume. Because fluids do not resist forces that change their shape, S waves cannot travel through fluids, whereas P waves can.

(?) **How could you physically demonstrate the difference between P waves and S waves to a friend who hasn't taken a geology course? (Caution: Don't hurt your friend!)**

6.3 Where Do Most Earthquakes Occur?

Locate Earth's major earthquake belts on a world map and label the regions associated with the largest earthquakes.

KEY TERM: circum-Pacific belt

- Most earthquake energy is released in the circum-Pacific belt, the ring of megathrust faults rimming the Pacific Ocean. Another earthquake belt is the Alpine–Himalayan belt, which runs along the zone where the Eurasian plate collides with the Indian–Australian and African plates.

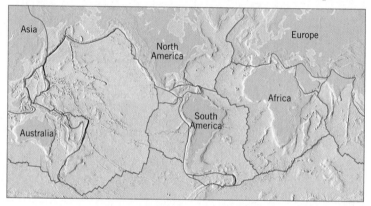

- Earth's oceanic ridge system produces another belt of earthquake activity. Here seafloor spreading generates many frequent, small-magnitude quakes. Transform faults in the continental crust, including the San Andreas Fault, can produce large earthquakes.
- Although most destructive earthquakes are produced along plate boundaries, some occur at considerable distances from plate boundaries. Examples include the 1811–1812 New Madrid, Missouri, earthquakes and the 1886 Charleston, South Carolina, earthquake.
- Using the difference in arrival times between P and S waves, the distance separating a recording station from an earthquake's epicenter can be determined. When the distances are known from three or more seismic stations, the epicenter can be located using a method called triangulation.

(?) **Outline Earth's major earthquake belts on the accompanying map. Name the type of plate boundary responsible for producing each belt.**

6.4 Determining the Size of an Earthquake

Distinguish between intensity scales and magnitude scales.

KEY TERMS: intensity, magnitude, Modified Mercalli Intensity scale, Richter scale, moment magnitude

- Intensity and magnitude are different measures of earthquake strength. Intensity measures the amount of ground shaking at a location due to an earthquake, and magnitude is an estimate of the actual amount of energy released during an earthquake.

- The Modified Mercalli Intensity scale is a tool for measuring an earthquake's intensity at different locations. The scale is based on verifiable physical evidence that is used to quantify intensity on a 12-point scale.
- The Richter scale takes into account both the maximum amplitude of the seismic waves measured at a given seismograph and that seismograph's distance from the earthquake. The Richter scale is logarithmic, meaning that the next higher number on the scale represents seismic amplitudes that are 10 times greater than those represented by the number below. Furthermore, each larger number on the Richter scale represents the release of about 32 times more energy than the number below it.
- Because the Richter scale does not effectively differentiate between very large earthquakes, the moment magnitude scale was devised. This scale measures the total energy released from an earthquake by considering the strength of the faulted rock, the amount of slippage, and the area of the fault that slipped. Moment magnitude is the modern standard for measuring the size of earthquakes.

6.5 Earthquake Destruction

List and describe the major destructive forces that earthquake vibrations can trigger.

KEY TERMS: liquefaction, tsunami

- Factors influencing how much destruction an earthquake might inflict on a human-made structure include (1) intensity of the shaking, (2) how long shaking persists, (3) the nature of the ground that underlies the structure, and (4) building construction. Buildings constructed of unreinforced bricks and blocks are more likely than other types of structures to be severely damaged in a quake.
- In general, bedrock-supported buildings fare best in an earthquake, as loose sediments amplify seismic shaking.
- Liquefaction may occur when water-logged sediment or soil is severely shaken during an earthquake. Liquefaction can reduce the strength of the ground to the point that it may not support buildings.
- Earthquakes may also trigger landslides or ground subsidence, and they may break gas lines, which can initiate devastating fires.
- Tsunamis are large ocean waves that form when water is displaced, usually by a megathrust fault rupturing on the seafloor. Traveling at the speed of a jet aircraft, a tsunami is hardly noticeable in deep water. However, upon arrival in shallower coastal waters, the tsunami slows down and piles up, producing a wall of water sometimes more than 30 meters (100 feet) in height. Tsunamis cause major destruction in coastal areas if they strike the shoreline. Tsunami warning systems have been established in most of the large ocean basins.

❓ Of the secondary earthquake hazards discussed in this section, which is (are) the greatest concern in the region where you live? Why?

6.6 Earth's Interior

Explain how Earth acquired its layered structure and list and describe each of its major layers.

KEY TERMS: crust, mantle, lithosphere, asthenosphere, core, outer core, inner core

- The layered internal structure of Earth developed due to gravitational sorting of Earth materials early in the history of the planet. The densest material settled to form Earth's core, while the least dense material rose to form Earth's crust, oceans, and atmosphere.
- Seismic waves allow geoscientists to "look into" Earth's interior. Like sonograms used to image human organs, seismic waves generated by large earthquakes reveal details about Earth's layered structure.

- Earth has two distinct kinds of crust: oceanic and continental. Oceanic crust is thinner, denser, and younger than continental crust. Oceanic crust also readily subducts, whereas the less dense continental crust does not.
- The uppermost mantle and crust make up Earth's rigid outer shell, called the lithosphere, which overlies the asthenosphere—a solid but relatively weak layer. The lower mantle is a strong solid layer but capable of very gradual flow.
- Earth's core is very dense and composed of a mixture of iron and nickel, with minor amounts of lighter elements. The outer core is liquid, whereas the inner core is solid.

6.7 Rock Deformation

Compare and contrast brittle and ductile deformation.

KEY TERMS: deformation, elastic deformation, ductile deformation, brittle deformation

- Elastic deformation is a temporary bending of rock that does not go past the breaking point. When the stress is released, the rock snaps back to its original shape.
- When stress is greater than the strength of a rock, the rock will deform in either a brittle or ductile fashion. Brittle deformation is the breaking of rocks into smaller pieces, whereas ductile deformation is flow in the manner of modeling clay or warm wax.

Anthony Pleva/Alamy

❓ Which type of deformation (elastic, brittle, or ductile) is best displayed in the accompanying photograph of a flattened quarter?

6.8 Folds: Structures Formed by Ductile Deformation

List and describe the major types of folds.

KEY TERMS: fold, anticline, syncline, dome, basin, monocline

- Folds are wavelike undulations in layered rocks that develop through ductile deformation in rocks undergoing compressional stress.
- Folds that have an arch-like structure are called anticlines, while folds that have a trough-like structure are called synclines. Monoclines are large step-like folds in otherwise horizontal strata; they result from subsurface faulting.
- Domes and basins are large folds that produce "bull's-eye"–shaped outcrop patterns. The overall shape of a dome or basin is like a saucer or a bowl, either right side up (basin) or inverted (dome).

E. J. Tarbuck

❓ What name is given to the rock structure shown in the accompanying photo?

6.9 Faults and Joints: Structures Formed by Brittle Deformation

Sketch and describe the relative motion of rock bodies located on opposite sides of normal, reverse, and strike-slip faults.

KEY TERMS: fault, dip-slip fault, hanging wall block, footwall block, fault scarp, normal fault, fault-block mountain, horst, graben, reverse fault, thrust fault, strike-slip fault, transform fault, joint

- Faults are fractures in rock along which there has been displacement (movement) of the rocks on both sides of the fracture. Joints are fractures along which no appreciable displacement has occurred.
- If the movement is in the direction of the fault's dip (or inclination), then the rock above the fault plane is called the hanging wall block, and the rock below the fault is the footwall block. If the hanging wall moves down relative to the footwall, the fault is a normal fault. If the hanging wall moves up relative to the footwall, the fault is a reverse fault. Reverse faults with low dip angles are thrust faults.
- Areas of tectonic extension, such as the Basin and Range Province, produce fault-block mountains: horsts separated by neighboring grabens or half-grabens.
- Strike-slip faults have most of their movement in the direction of the trend of the fault trace. Transform faults are large strike-slip faults that serve as tectonic boundaries between lithospheric plates.

(?) **What name is given to the cliff-like geologic feature shown by the arrows?**

6.10 Mountain Building

Locate and identify Earth's major mountain belts on a world map.

KEY TERMS: orogenesis, compressional mountains

- Most mountain building (termed orogenesis) occurs along convergent plate boundaries, where compressional forces cause folding and faulting of the rock, thickening the crust vertically while shortening it horizontally. Some mountain belts are very old, while others are actively forming today.

6.11 Subduction and Mountain Building

Sketch an Andean-type mountain belt and describe how each of its major features is generated.

KEY TERMS: passive continental margin, active continental margin, accretionary wedge

- Subduction leads to orogenesis. If a subducted plate is overridden by oceanic lithosphere, an island arc–type results.
- Andean-type mountain building occurs where subduction takes place beneath continental lithosphere and results in a continental volcanic arc.
- Sediment scraped off the subducting plate builds up in an accretionary wedge. The geography of central California preserves an accretionary wedge (Coast Ranges) and roots of a continental volcanic arc (Sierra Nevada).

6.12 Collisional Mountain Belts

Summarize the stages in the development of a collisional mountain belt such as the Himalayas.

KEY TERMS: terrane, microcontinent

- Terranes are relatively small crustal fragments (microcontinents, volcanic island arcs, or oceanic plateaus). Terranes may be accreted to continents when subduction brings them to a trench, but they usually do not subduct due to their relatively low density. The terranes are "peeled off" the subducting slab and thrust onto the leading edge of the continent.
- The Himalayas and Appalachians have similar origins; both were formed by the collision of landmasses formerly separated by now-subducted ocean basins. The Himalayas are younger, having been formed by the collision of India and Eurasia starting between 50 and 30 million years ago. In contrast, the Appalachians are older, caused by the collision of ancestral North America with ancestral Africa more than 250 million years ago.

GIVE IT SOME THOUGHT

1. Describe the concept of elastic rebound. Develop an analogy other than a rubber band to illustrate this concept.

2. The accompanying map shows the locations of many of the largest earthquakes in the world since 1900. Refer to the map of Earth's plate boundaries in Figure 5.10, page 151, and determine which type of plate boundary is most often associated with these destructive events.

3. You go for a jog on a beach and choose to run near the water, where the sand is well packed and solid under your feet. With each step, you notice that your footprint quickly fills with water, but not water coming in from the ocean. What is this water's source? For what earthquake-related hazard is this phenomenon a good analogy?

4. Use the accompanying seismogram to answer the following questions:
   a. Which of the three types of seismic waves reached the seismograph first?
   b. What is the time interval between the arrival of the first P wave and the arrival of the first S wave?
   c. Use your answer from Question b and the travel–time graph in Figure 6.17 to determine the distance from the seismic station to the earthquake.
   d. Which of the three types of seismic waves had the highest amplitude when they reached the seismic station?

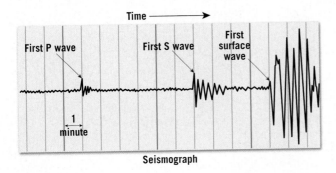

Seismograph

5. Refer to the accompanying diagrams to answer the following:
   a. What type of dip-slip fault is shown in Diagram 1?
   b. What type of dip-slip fault is shown in Diagram 2?
   c. Match the correct pair of arrows in Diagram 3 to the faults in Diagrams 1 and 2.

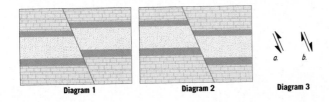

Diagram 1   Diagram 2   Diagram 3

6. Match each of the accompanying sketches to the type of orogeny it illustrates: Andean, a Cordilleran, and an Alpine.

7. The Ural Mountains exhibit a north–south orientation through Eurasia (see Figure 6.43). How does the theory of plate tectonics explain the existence of this mountain belt in the interior of an expansive landmass?

8. Briefly describe the major differences between the formation of the Appalachian Mountains and the North American Cordillera.

9. Refer to the accompanying diagram. Which of these features—the *Galapagos Rise* or the *Rio Grande Rise*—will likely end up accreted to a continent? Explain your choice. In the future, how might a geologist determine that this accreted terrane is distinct from the continental crust to which it accreted? (*Hint:* See the plate map in Figure 5.10, page 151.)

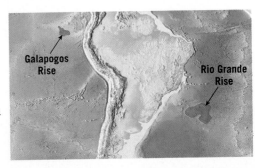

Galapogos Rise

Rio Grande Rise

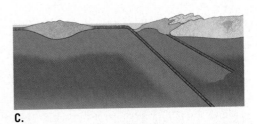

10. Describe the fault shown by the dashed line using selected terms from the following list: dip-slip, strike-slip, normal, reverse, right-lateral, left-lateral.

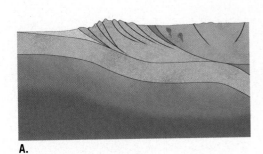

A.

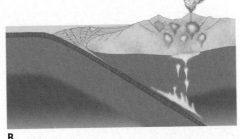

B.

C.

MasteringGeology™

www.masteringgeology.com Looking for additional review and test prep materials? With individualized coaching on the toughest topics of the course, MasteringGeology offers a wide variety of ways for you to move beyond memorization to begin thinking like a geologist. Visit the Study Area in www.masteringgeology.com to find practice quizzes, study tools, and multimedia that will improve your understanding of this chapter's content. Sign in today to enjoy the following features: **Self Study Quizzes, SmartFigure: Tutorials/Animations/Condor Videos/Mobile Field Trips, Geoscience Animation Library, GEODe, RSS Feeds, Digital Study Modules,** and an optional **Pearson eText.**

UNIT
III

FORCES WITHIN

7

FOCUS ON CONCEPTS

Each statement represents the primary learning objective for the corresponding major heading within the chapter. After you complete the chapter, you should be able to:

7.1 Compare and contrast the 1980 eruption of Mount St. Helens with the most recent eruption of Kilauea, which began in 1983.

7.2 Explain why some volcanic eruptions are explosive and others are quiescent.

7.3 List and describe the three categories of materials extruded during volcanic eruptions.

7.4 Draw and label a diagram that illustrates the basic features of a typical volcanic cone.

7.5 Summarize the characteristics of shield volcanoes and provide one example of this type of volcano.

7.6 Describe the formation, size, and composition of cinder cones.

7.7 List the characteristics of composite volcanoes and describe how these volcanoes form.

7.8 Describe the major geologic hazards associated with volcanoes.

7.9 List volcanic landforms other than shield, cinder, and composite volcanoes and describe their formation.

7.10 Compare and contrast these intrusive igneous structures: dikes, sills, batholiths, stocks, and laccoliths.

7.11 Summarize the major processes that generate magma from solid rock.

7.12 Explain how the global distribution of volcanic activity is related to plate tectonics.

Anak Krakatau ejecting volcanic bombs and incandescent lava in 2012. This volcano, situated in the Sunda Strait, Indonesia, occupies the site of a former volcano that erupted in 1883, killing 30,000 people. (Photo by Claudia Bucarey/Reuters)

VOLCANOES AND OTHER IGNEOUS ACTIVITY

The significance of igneous activity may not be obvious at first glance. However, because volcanoes extrude molten rock that formed at great depth, they provide our only means of directly observing processes that occur many kilometers below Earth's surface. Furthermore, Earth's atmosphere and oceans have evolved from gases emitted during volcanic eruptions. Either of these facts is reason enough for igneous activity to warrant our attention.

7.1 Mount St. Helens Versus Kilauea

Compare and contrast the 1980 eruption of Mount St. Helens with the most recent eruption of Kilauea, which began in 1983.

On Sunday, May 18, 1980, the largest volcanic eruption to occur in North America in historic times transformed a picturesque volcano into a decapitated remnant (**Figure 7.1**). On that date in southwestern Washington State, Mount St. Helens erupted with tremendous force. The blast blew out the entire north flank of the volcano, leaving a gaping hole. In one brief moment, a prominent volcano whose summit had been more than 2900 meters (9500 feet) above sea level was lowered by more than 400 meters (1350 feet).

The event devastated a wide swath of timber-rich land on the north side of the mountain (**Figure 7.2**). Trees within a 400-square-kilometer (160-square-mile) area lay intertwined and flattened, stripped of their branches and appearing from the air like toothpicks strewn about. The accompanying mudflows carried ash, trees, and water-saturated rock debris 29 kilometers (18 miles) down the Toutle River. The eruption claimed 59 lives; some died from the intense heat and the suffocating cloud of ash and gases, others from the impact of the blast, and still others from being trapped in mudflows.

The eruption ejected nearly a cubic kilometer of ash and rock debris. Following the devastating explosion, Mount St. Helens continued to emit great quantities of hot gases and ash. The force of the blast was so strong that some ash was propelled more than 18 kilometers (over 11 miles), entering the stratosphere. During the next few days, this very fine-grained material was carried around Earth by strong upper-air winds. Measurable deposits were reported in Oklahoma and Minnesota, and crop damage occurred as far

Figure 7.1 Before-and-after photographs show the transformation of Mount St. Helens The May 18, 1980, eruption of Mount St. Helens occurred in southwestern Washington.

Spirit Lake

USGS

The blast blew out the entire north flank of Mount St. Helens, leaving a gaping hole. In a brief moment, a prominent volcano was lowered by 1350 feet.

1350 feet

Spirit Lake

USGS

Figure 7.2 Douglas fir trees snapped off or uprooted by the lateral blast of Mount St. Helens (Large photo by Lyn Topinka/USGS Cascades Volcano Observatory; inset photo by John M. Burnley/Science Source)

Not all volcanic eruptions are as violent as the 1980 Mount St. Helens event. Some volcanoes, such as Hawaii's Kilauea Volcano, generate relatively quiet outpourings of fluid lavas. These quiescent eruptions are not without some fiery displays; occasionally fountains of incandescent lava spray hundreds of meters into the air (see Figure 7.4), but most lava pours from the vent and flows downslope. During Kilauea's most recent active phase, which began in 1983, more than 180 homes and a national park visitor center have been destroyed by flowing lava igniting material in its path.

Testimony to the quiescent nature of Kilauea's eruptions is the fact that the Hawaiian Volcanoes Observatory has operated on its summit since 1912, despite the fact that Kilauea has had more than 50 eruptive phases since record keeping began in 1823.

away as central Montana. Meanwhile, ash fallout in the immediate vicinity exceeded 2 meters (6 feet) in depth. The air over Yakima, Washington (130 kilometers [80 miles] to the east), was so filled with ash that residents experienced midnight-like darkness at noon.

> **Did You Know?**
> The eruption of Tambora, Indonesia, in 1815 is the largest known volcanic event in modern history. About 20 times more ash and rock were explosively ejected during this eruption than were emitted during the 1980 Mount St. Helens event. The sound of the explosion was heard an incredible 4800 km (3000 mi) away, about the distance across the conterminous United States.

7.1 CONCEPT CHECKS

1. Briefly compare the May 18, 1980, eruption of Mount St. Helens to a typical eruption of Hawaii's Kilauea Volcano.

7.2 The Nature of Volcanic Eruptions

Explain why some volcanic eruptions are explosive and others are quiescent.

Volcanic activity is commonly perceived as a process that produces a picturesque, cone-shaped structure that periodically erupts in a violent manner. However, many eruptions are not explosive, as indicated by Kilauea's activity. What determines the manner in which volcanoes erupt?

Magma: Source Material for Volcanic Eruptions

Recall that **magma**, molten rock that contains some solid crystalline material and varying amounts of dissolved gas

(mainly water vapor), is the parent material of igneous rocks. Erupted magma is called **lava**.

Composition of Magma As we covered in Chapter 2, basaltic igneous rocks contain a high percentage of dark silicate minerals and calcium-rich plagioclase feldspar (but no quartz) and tend to be dark in color. By contrast, the rock rhyolite and its intrusive equivalent, granite, contains mainly light-colored silicate minerals—quartz and potassium feldspar. Andesitic rocks have a composition between basaltic and rhyolitic rocks. Correspondingly, basaltic (mafic) magmas contain a much lower percentage of silica (SiO_2) than do rhyolitic (felsic) magmas.

Figure 7.3 Compositional differences of magma bodies cause their properties to vary

Properties of Magma Bodies with Differing Compositions

Composition	Silica Content (SiO₂)	Gas Content (% by weight)	Eruptive Temperature	Viscosity	Tendency to Form Pyroclastics	Volcanic Landform
Basaltic (mafic) High in Fe, Mg, Ca, low in K, Na	**Least** (~50%)	**Least** (0.5–2%)	**Highest** 1000–1250°C	**Least**	**Least**	Shield volcanoes, basalt plateaus, cinder cones
Andesitic Intermediate amounts of Fe, Mg, Ca, K, Na	**Intermediate** (~60%)	**Intermediate** (3–4%)	**Intermediate** 800–1050°C	**Intermediate**	**Intermediate**	Composite cones
Rhyolitic (felsic) High in K, Na, low in Fe, Mg, Ca	**Most** (~70%)	**Most** (5–8%)	**Lowest** 650–900°C	**Greatest**	**Greatest**	Pyroclastic flow deposits, lava domes

The compositional differences between magmas also affect other properties, as shown in **Figure 7.3**.

Where Is Magma Generated? Most magma is generated in Earth's upper mantle (asthenosphere) by *partial melting of solid rock*—a process that will be considered in more detail later in this chapter. The magmas generated by melting mantle rocks tend to have a *basaltic composition*. Once formed, the basaltic magma, which is less dense than the surrounding rock, slowly rises toward Earth's surface. In some settings, this hot molten rock reaches the surface, where it usually produces fluid basaltic lavas. The largest quantity of basaltic magma erupts on the ocean floor along divergent plate boundaries in association with seafloor spreading. Extensive basaltic flows are also the product of hot-spot volcanism generated by rising hot *mantle plumes*.

In continental settings, however, overlying crustal rocks are usually less dense than the ascending basaltic magma, and as a result, the rising molten rock ponds at the crust–mantle boundary. Because the newly formed magma is much hotter than the melting temperature of the overlying crustal rocks, they begin to melt. This process generates a less dense, more *silica-rich magma of andesitic or rhyolitic composition*, which continues the journey toward Earth's surface.

Quiescent Versus Explosive Eruptions

Volcanic eruptions exhibit a range of behavior from *quiescent* (nonexplosive) eruptions that produce outpourings of fluid lava (geologists often refer to these as *effusive* eruptions) to *explosive eruptions*. The two primary factors that determine how magma erupts are its *viscosity* and *gas content*. **Viscosity** (*viscos* = sticky) is a measure of a fluid's mobility. The more viscous a material, the greater is its resistance to flow. For example, syrup is more viscous, and thus more resistant to flow, than water.

Factors Affecting Viscosity Magma's viscosity depends primarily on its temperature and silica content: *The more silica in magma, the greater its viscosity*. Silicate structures begin to link together into long chains early in the crystallization process, which makes the magma more rigid and impedes its flow. Consequently, silica-rich rhyolitic lavas are the most viscous and tend to form comparatively short, thick flows. By contrast, basaltic lavas, which contain much less silica, are relatively fluid and have been known to travel 150 kilometers (90 miles) or more before solidifying.

The effect of *temperature* on magma viscosity can be compared to the effect of temperature on pancake syrup. Just as heating syrup makes it more fluid (less viscous), temperature also strongly influences the viscosity and mobility of lava. As lava cools and begins to congeal, its viscosity increases, and the flow eventually halts.

Role of Gases The nature of volcanic eruptions also depends on the amount of dissolved gases held in the magma body by the pressure exerted by the overlying rock. This is analogous to how carbon dioxide is retained in cans and bottles of soft drinks. When the pressure is reduced on a soft drink by opening the cap, the carbon dioxide begins to bubble out of solution. Likewise, the gases dissolved in magma tend to come out of solution when the confining pressure is reduced. The most abundant gas found in most magma is water vapor.

The viscosity and gas content of magma is directly related to its composition, as shown in Figure 7.3. At one end of the spectrum are basaltic magmas, which are very fluid and have a low gas content, sometimes as little as 0.5 percent by weight. At the other extreme are rhyolitic magmas, which are highly viscous (sticky) and contain a lot of gas, as much as 8 percent by weight.

Quiescent Hawaiian-Type Eruptions

All magmas contain some water vapor and other gases that are kept in solution by the immense pressure of the overlying rock. As magma rises (or the rocks confining the magma fail), the confining pressure drops, causing the dissolved gases to separate from the melt and form large numbers of tiny bubbles. This is similar to opening a warm can of soda: The dissolved carbon dioxide in the soda quickly forms bubbles that rise and escape.

When fluid basaltic magmas erupt, the pressurized gases readily escape. At temperatures that often exceed 1100°C (2000°F), these gases can quickly expand to occupy hundreds of times their original volumes. Occasionally, these expanding gases propel incandescent lava hundreds of meters into the air, producing lava fountains (**Figure 7.4**). Although spectacular, these fountains are usually harmless and generally not associated with major explosive events that cause great loss of life and property.

Eruptions that involve very fluid basaltic lavas, such as the recent eruptions of Kilauea on Hawaii's Big Island, are often triggered by the arrival of a new batch of molten rock, which accumulates in a near-surface magma chamber. Geologists can usually detect such an event because the summit of the volcano begins to inflate and rise months or even years before an eruption. The injection of a fresh supply of hot molten rock heats and remobilizes the semiliquid magma in the chamber. In addition, swelling of the magma chamber fractures the rock above, allowing the fluid magma to move upward along the newly formed fissures, often generating effusions of fluid lava for weeks, months, or possibly years. The eruption of Kilauea that began in 1983 is ongoing.

How Explosive Eruptions Are Triggered

Recall that silica-rich rhyolitic magmas have a high gas content and are quite viscous (sticky) compared to basaltic magmas. As rhyolitic magma rises, the gases remain dissolved until the confining pressure drops sufficiently, at which time tiny bubbles begin to form and increase in size. Because of the high viscosity of rhyolitic magma, gas bubbles tend to remain trapped in the magma, forming a sticky froth.

When the pressure exerted by the expanding gases exceeds the strength of the overlying rock, fracturing

Gases readily escape hot fluid basaltic flows, producing lava fountains. Although often spectacular, these features do not cause great loss of life or property.

Figure 7.4 Lava fountain produced by gases escaping fluid basaltic lava Kilauea, on Hawaii's Big Island, is one of the most active volcanoes on Earth. (Photo by David Reggie/ Getty Images)

occurs. As the frothy magma moves up through the fractures, a further drop in confining pressure creates additional gas bubbles. This chain reaction often generates an explosive event in which magma is literally blown into fragments (ash and pumice) that are carried to great heights by the escaping hot gases. (The collapse of a volcano's flank can also greatly reduce the pressure on the magma below, causing an explosive eruption, as exemplified by the 1980 eruption of Mount St. Helens.)

When molten rock in the uppermost portion of the magma chamber is forcefully ejected by the escaping gases, the confining pressure on the magma directly below also drops suddenly. Thus, rather than being a single "bang," an explosive eruption is really a series of violent explosions that can last for days.

Because highly gaseous magmas expel fragmented lava at nearly supersonic speeds, they are associated with hot buoyant **eruption columns** consisting mainly

Eruptions of highly viscous lavas may produce explosive clouds of hot ash and gases called eruption columns.

SmartFigure 7.5 Eruption column generated by viscous, silica-rich magma Steam and ash eruption column from Mount Sinabung, Indonesia, 2014. A deadly cloud of fiery ash can be seen racing down the volcano's slope in the foreground. (Photo by Beawiharta/Reuters) (https://goo.gl/Gd11FL)

Video

of volcanic ash and gases (**Figure 7.5**). Eruption columns can rise perhaps 40 kilometers (25 miles) into the atmosphere. Frequently, these eruption columns collapse, sending hot ash rushing down the volcanic slope at speeds exceeding 100 kilometers (60 miles) per hour. As a result, volcanoes that erupt highly viscous magmas having a high gas content are the most destructive to property and human life.

Following explosive eruptions, partially degassed lava may slowly ooze out of the vent to form thick lava flows or dome-shaped lava bodies that grow over the vent.

7.2 CONCEPT CHECKS

1. List these magmas in order, from the most silica-rich to the least silica-rich composition: basaltic magma, rhyolitic magma, andesitic magma.
2. List the two primary factors that determine the manner in which magma erupts.
3. Define *viscosity*.
4. Are volcanoes fed by highly viscous magma *more likely* or *less likely* to be a greater threat to life and property than volcanoes supplied with very fluid magma?

7.3 Materials Extruded During an Eruption

List and describe the three categories of materials extruded during volcanic eruptions.

Volcanoes erupt lava, large volumes of gas, and pyroclastic materials (broken rock, lava "bombs," and ash). In this section we will examine each of these materials.

Lava Flows

The vast majority of Earth's lava, more than 90 percent of the total volume, is estimated to be basaltic (mafic) in composition. Most of these basaltic lavas erupt on the seafloor, via a process known as *submarine volcanism*. Andesitic lavas account for most of the rest, while rhyolitic (felsic) flows make up as little as 1 percent of the total. (Rhyolitic magmas tend to extrude mostly volcanic ash rather than lava.)

On land, hot basaltic lavas, which are usually very fluid, generally flow in thin, broad sheets or stream-like ribbons. Fluid basaltic lavas have been clocked at speeds exceeding 30 kilometers (19 miles) per hour down steep slopes. However, flow rates of 10 to 300 meters (30 to 1000 feet) per hour are more common. Silica-rich, rhyolitic lava, by contrast, often moves too slowly to be perceived. Furthermore, most rhyolitic lavas seldom travel more than a few kilometers from their vents. As you might expect, andesitic lavas, which are intermediate in composition, exhibit flow characteristics between these extremes.

Aa and Pahoehoe Flows Fluid basaltic magmas tend to generate two types of lava flows, which are known by their Hawaiian names. The first, called **aa** (pronounced "ah-ah") **flows**, have surfaces of rough jagged blocks with dangerously sharp edges and spiny projections (**Figure 7.6A**). Crossing a hardened aa flow can be a trying and miserable experience. The second type, **pahoehoe** (pronounced "pah-hoy-hoy") **flows**, exhibit smooth surfaces that sometimes resemble twisted braids of ropes (**Figure 7.6B**). Pahoehoe means "on which one can walk."

Although both lava types can erupt from the same volcano, pahoehoe lavas are hotter and more fluid than aa flows. In addition, pahoehoe lavas can change into aa lava flows, although the reverse (aa to pahoehoe) does not occur.

Cooling that occurs as the flow moves away from the vent is one factor that facilitates the change from pahoehoe to aa. The lower temperature increases viscosity and promotes bubble formation. Escaping gas bubbles produce numerous voids (vesicles) and sharp spines in the surface of the congealing lava. As the molten interior advances, the outer crust is broken, transforming the relatively smooth surface of a pahoehoe flow into an aa flow made up of an advancing mass of rough, sharp, broken lava blocks.

Figure 7.6 Lava flows
A. A slow-moving, basaltic aa flow advancing over hardened pahoehoe lava. **B.** A typical fluid pahoehoe (ropy) lava. Both of these lava flows erupted from a rift on the flank of Hawaii's Kilauea Volcano. (Photos courtesy of U.S. Geological Survey)

A. Active aa flow overriding an older pahoehoe flow.

Aa flow

Pahoehoe flow

TASA Piece

B. Pahoehoe flow displaying the characteristic ropy appearance.

TASA Piece

A. Lava tubes are cave-like tunnels that once served as conduits carrying lava from an active vent to the flow's leading edge.

Valentine Cave, a lava tube at Lava Beds National Monument, California.

B. Skylights develop where the roofs of lava tubes collapse and reveal the hot lava flowing through the tube.

Figure 7.7 Lava tubes **A.** A lava flow may develop a solid upper crust, while the molten lava below continues to advance in a conduit called a lava tube. Some lava tubes exhibit extraordinary dimensions. Kazumura Cave, located on the southeastern slope of Hawaii's Mauna Loa Volcano, is a lava tube extending more than 60 kilometers (40 miles). (Photo by Dave Bunell) **B.** The collapsed section of the roof of a lava tunnel results in a skylight. (Photo courtesy of U.S. Geological Survey)

Pahoehoe flows often develop cavelike tunnels called **lava tubes** that start as conduits for carrying lava from an active vent to the flow's leading edge (**Figure 7.7**). Lava tubes form in the interior of a lava flow, where the temperature remains high long after the exposed surface cools and hardens. Because they serve as insulated pathways that allow lava to flow great distances from its source, lava tubes are important features of fluid lava flows.

Pillow Lavas Recall that most of Earth's volcanic output occurs along oceanic ridges (divergent plate boundaries),

Pillow lavas form on the ocean floor and have elongated shapes, resembling toothpaste coming out of a tube.

generating new oceanic crust. When outpourings of lava occur on the ocean floor, the flow's outer skin quickly freezes (solidifies) to form volcanic glass. However, the interior lava is able to move forward by breaking through the hardened surface. This process occurs over and over, as molten basalt is extruded—like toothpaste from a tightly squeezed tube. The result is a lava flow composed of numerous tube-like structures called **pillow lavas**, stacked one atop the other (**Figure 7.8**). Pillow lavas are useful when reconstructing geologic history because their presence indicates that the lava flow formed below the surface of a water body.

Gases

Recall that magmas contain varying amounts of dissolved gases, called **volatiles**. These gases are held in the molten rock by confining pressure, just as carbon dioxide is held in cans of soft drinks. As with soft drinks, as soon as the pressure is reduced, the gases begin to escape. Obtaining gas samples from an erupting volcano is difficult and dangerous, so geologists usually must estimate the amount of gas originally contained in the magma.

The gaseous portion of most magma bodies ranges from less than 1 percent to about 8 percent of the total volume, with most of this in the form of water vapor. Although the percentage may be small, the actual quantity of emitted gas can exceed thousands of tons per day. Occasionally, eruptions emit colossal amounts of volcanic gases that rise high into the atmosphere, where they may reside for several years.

Figure 7.8 Pillow lava A diagram showing how pillow lava forms and a photo of undersea pillow lava off the coast of Hawaii. (Photo courtesy of U.S. Geological Survey)

The composition of volcanic gases is important because these gases contribute significantly to our planet's atmosphere. The most abundant gas typically released into the atmosphere from volcanoes is water vapor (H_2O), followed by carbon dioxide (CO_2) and sulfur dioxide (SO_2), with lesser amounts of hydrogen sulfide (H_2S), carbon monoxide (CO), and helium (H_2). (The relative proportion of each gas varies significantly from one volcanic region to another.) Sulfur compounds are easily recognized by their pungent odor. Volcanoes are also natural sources of air pollution; some emit large quantities of sulfur dioxide (SO_2), which readily combines with atmospheric gases to form toxic sulfuric acid and other sulfate compounds.

Pyroclastic Materials

When volcanoes erupt energetically, they eject pulverized rock and fragments of lava and glass from the vent. The particles produced are called **pyroclastic materials** (*pyro* = fire, *clast* = fragment) and are also called **tephra**. These fragments range in size from very fine dust and sand-sized volcanic ash (less than 2 millimeters) to pieces that weigh several tons (**Figure 7.9**).

Ash and *dust* particles are produced when gas-rich viscous magma erupts explosively. As magma moves up in the vent, the gases rapidly expand, generating a melt that resembles the froth that flows from a bottle of champagne. As the hot gases expand explosively, the froth is blown into very fine glassy fragments. When the hot ash falls, the glassy shards often fuse to form a rock called *welded tuff*. Sheets of this material, as well as ash deposits that later consolidate, cover vast portions of the western United States.

Somewhat larger pyroclasts that range in size from small beads to walnuts (2–64 millimeters [0.08–2.5

Pyroclastic Materials (Tephra)		
Particle name	**Particle size**	**Image**
Volcanic ash*	Less than 2 mm (0.08 inch)	
Lapilli (Cinders)	Between 2 mm and 64 mm (0.08–2.5 inches)	
Volcanic bombs	More than 64 mm (2.5 inches)	
Volcanic blocks		

Figure 7.9 Types of pyroclastic materials Pyroclastic materials are also commonly referred to as tephra.

*The term volcanic dust is used for fine volcanic ash less than 0.063 mm (0.0025 inch).

inches] in diameter) are known as *lapilli* ("little stones") or *cinders*. Particles larger than 64 millimeters (2.5 inches) in diameter are called *blocks* when they are made of hardened lava and *bombs* when they are ejected as incandescent lava (see Figure 7.7). Because bombs are semimolten upon ejection, they often take on a streamlined shape as they hurl through the air. Because of their size, bombs and blocks usually fall near the vent; however, they are occasionally propelled great distances. For instance, bombs 6 meters (20 feet) long and weighing about 200 tons were blown 600 meters (2000 feet) from the vent during an eruption of the Japanese volcano Asama.

Pyroclastic materials can be classified by texture and composition as well as by size. For instance, **scoria** is the term for vesicular ejecta produced most often during the eruption of basaltic magmas (**Figure 7.10A**). These black to reddish-brown fragments are generally found in the size range of lapilli and resemble cinders and clinkers produced by furnaces used to smelt iron.

By contrast, when magmas with an andesitic (intermediate) or rhyolitic (felsic) compositions erupt explosively, they emit ash and the vesicular rock **pumice** (**Figure 7.10B**). Pumice is usually lighter in color and less dense than scoria, and many pumice fragments have so many vesicles that they are light enough to float (see Figure 2.10, page 54).

A. Scoria is a vesicular rock commonly having a basaltic composition. Pea-to-basketball size scoria fragments make up a large portion of most cinder cones (also called *scoria cones*).

(7.3) CONCEPT CHECKS

1. Contrast pahoehoe and aa lava flows.
2. How do lava tubes form?
3. List the main gases released during a volcanic eruption.
4. How do volcanic bombs differ from blocks of pyroclastic debris?
5. What is scoria? How is scoria different from pumice?

B. Pumice is a low density vesicular rock that forms during explosive eruptions of viscous magma having an andesitic to rhyolitic composition.

Figure 7.10 Common vesicular rocks Scoria and pumice are volcanic rocks that exhibit a vesicular texture. Vesicles are small holes left by escaping gas bubbles. (Photos by E.J. Tarbuck)

(7.4) Anatomy of a Volcano
Draw and label a diagram that illustrates the basic features of a typical volcanic cone.

A popular image of a volcano is a solitary, graceful, snow-capped cone, such as Mount Hood in Oregon or Japan's Fujiyama. These picturesque, conical mountains are produced by volcanic activity that occurred intermittently over thousands, or even hundreds of thousands, of years. However, many volcanoes do not fit this image. Cinder cones are quite small and form during a single eruptive phase that lasts a few days to a few years. Alaska's Valley of Ten Thousand Smokes is a flat-topped ash deposit that blanketed a river valley to a depth of 200 meters (600 feet). The eruption that produced it lasted less than 60 hours yet emitted more than 20 times more volcanic material than the 1980 Mount St. Helens eruption.

Volcanic landforms come in a wide variety of shapes and sizes, and each volcano has a unique eruptive history. Nevertheless, volcanologists have been able to classify volcanic landforms and determine their eruptive patterns. In this section we will consider the general anatomy of an idealized volcanic cone. We will follow this discussion by exploring the three major types of volcanic cones—shield volcanoes, cinder cones, and composite volcanoes—as well as their associated hazards.

Volcanic activity frequently begins when a **fissure** (crack) develops in Earth's crust as magma moves

forcefully toward the surface. As the gas-rich magma moves up through a fissure, its path is usually localized into a somewhat pipe-shaped **conduit** that terminates at a surface opening called a **vent** (**Figure 7.11**). The cone-shaped structure we call a **volcanic cone** is often created by successive eruptions of lava, pyroclastic material, or frequently a combination of both, often separated by long periods of inactivity.

Located at the summit of most volcanic cones is a somewhat funnel-shaped depression called a **crater** (*crater* = bowl). Volcanoes built primarily of pyroclastic materials typically have craters that form by gradual accumulation of volcanic debris on the surrounding rim. Other craters form during explosive eruptions, as the rapidly ejected particles erode the crater walls. Craters also form when the summit area of a volcano collapses following an eruption. Some volcanoes have very large circular depressions, called **calderas**, which have diameters that are greater than 1 kilometer (0.6 mile) and that in rare cases exceed 50 kilometers (30 miles). The formation of various types of calderas will be considered later in this chapter.

During early stages of growth, most volcanic discharges come from a central summit vent. As a volcano

SmartFigure 7.11 Anatomy of a volcano Compare the structure of the "typical" composite cone shown here to that of a shield volcano (Figure 7.12) and a cinder cone (Figure 7.15). (https://goo.gl/NI9iNq)

Tutorial

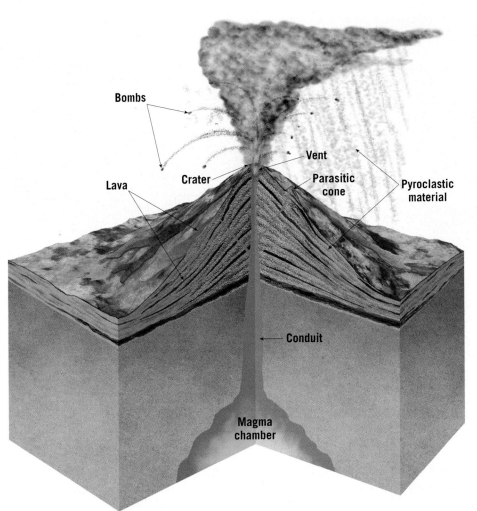

Bombs

Lava

Crater

Vent

Parasitic cone

Pyroclastic material

Conduit

Magma chamber

matures, material also tends to be emitted from fissures that develop along the flanks (sides) or at the base of the volcano. Continued activity from a flank eruption may produce one or more small **parasitic cones**. Italy's Mount Etna, for example, has more than 200 secondary vents, some of which have built parasitic cones. Many of these vents, however, emit only gases and are appropriately called **fumaroles** (*fumus* = smoke).

7.4 CONCEPT CHECKS

1. Distinguish among a conduit, a vent, and a crater.
2. How is a crater different from a caldera?
3. What is a parasitic cone, and where does it form?
4. What is emitted from a fumarole?

7.5 Shield Volcanoes

Summarize the characteristics of shield volcanoes and provide one example of this type of volcano.

Shield volcanoes are produced by the accumulation of fluid basaltic lavas. This type of volcano is a broad, slightly domed structure that resembles a warrior's shield (**Figure 7.12**). Most shield volcanoes begin on the ocean floor as **seamounts** (submarine volcanoes), a few of which grow large enough to form volcanic islands. In fact, many oceanic islands are either a single shield volcano or, more often, the coalescence of two or more shields built upon massive amounts of pillow lavas. Examples include the Hawaiian Islands, the Canary Islands, Iceland, the

Galápagos Islands, and Easter Island. Although less common, some shield volcanoes form on continental crust. Included in this group are Nyamuragira, Africa's most active volcano, and Newberry Volcano, Oregon.

Mauna Loa: Earth's Largest Shield Volcano

Extensive study of the Hawaiian Islands has revealed that they are constructed of a myriad of thin basaltic

Figure 7.12 Volcanoes of Hawaii Mauna Loa, Earth's largest volcano, is one of five shield volcanoes that collectively make up the Big Island of Hawaii. Shield volcanoes are built primarily of fluid basaltic lava flows and contain only a small percentage of pyroclastic materials.

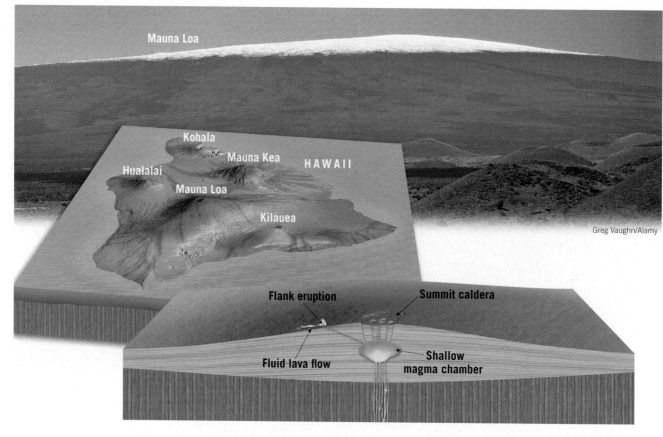

Greg Vaughn/Alamy

SmartFigure 7.13
Comparing scales of different volcanoes
A. Profile of Mauna Loa, Hawaii, the largest shield volcano in the Hawaiian chain. Note the size comparison with Mount Rainier, Washington, a large composite cone.
B. Profile of Mount Rainier, Washington. Note how it dwarfs a typical cinder cone. **C.** Profile of Sunset Crater, Arizona, a typical steep-sided cinder cone.
(http://goo.gl/VXEa4X)

lava flows averaging a few meters thick intermixed with relatively minor amounts of ejected pyroclastic material. Mauna Loa is the largest of five overlapping shield volcanoes that comprise the Big Island of Hawaii (see Figure 7.12). From its base on the floor of the Pacific Ocean to its summit, Mauna Loa is over 9 kilometers (6 miles) high, exceeding the height of Mount Everest above sea level. The volume of material composing Mauna Loa is roughly 200 times greater than that of the large composite cone Mount Rainier, located in Washington State (**Figure 7.13**).

Like Hawaii's other shield volcanoes, the flanks of Mauna Loa have gentle slopes of only a few degrees. This low angle is due to the very hot, fluid lava that traveled "fast and far" from the vent. In addition, most of the lava (perhaps 80 percent) flowed through a well-developed system of lava tubes. Another feature common to active shield volcanoes is one or more large, steep-walled calderas that occupy the summit (see Figure 7.22, page 230). Calderas on shield volcanoes usually form when the roof above the magma chamber collapses. This occurs after the magma reservoir empties, either following a large eruption or as magma migrates to the flank of a volcano to feed a fissure eruption.

In their final stage of growth, shield volcanoes erupt more sporadically, and pyroclastic ejections are more common. The lava emitted later tends to be more viscous, resulting in thicker, shorter flows. These eruptions steepen the slope of the summit area, which often becomes capped with clusters of cinder cones. This explains why Mauna Kea, a more mature volcano that has not erupted in historic times, has a steeper summit than Mauna Loa, which erupted as recently as 1984. Astronomers are so certain that Mauna Kea is "over the hill" that they built an elaborate astronomical observatory on its summit to house some of the world's most advanced and expensive telescopes.

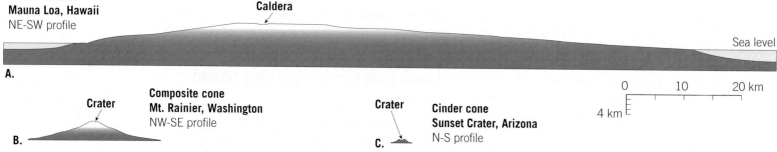

SmartFigure 7.14 Lava "curtain" extruded along the East Rift Zone, Kilauea, Hawaii (Photo by Greg Vaughn/Alamy) (http://goo.gl/UHGrvC)

Mobile Field Trip

Kilauea: Hawaii's Most Active Volcano

Volcanic activity on the Big Island of Hawaii began on what is now the northwestern flank of the island, and it has gradually migrated southeastward. It is currently centered on Kilauea Volcano, one of the most active and intensely studied shield volcanoes in the world. Kilauea, located in the shadow of Mauna Loa, has experienced more than 50 eruptions since record keeping began in 1823.

Several months before each eruptive phase, Kilauea inflates as magma gradually migrates upward and accumulates in a central reservoir located a few kilometers below the summit. For up to 24 hours before an eruption, swarms of small earthquakes warn of the impending activity. Most of the recent activity on Kilauea has occurred along the flanks of the volcano, in a region called the *East Rift Zone* (**Figure 7.14**). The longest and largest rift eruption ever recorded on Kilauea began in 1983 and continues to this day, with no signs of abating.

 7.5 CONCEPT CHECKS

1. Describe the composition and viscosity of the lava associated with shield volcanoes.
2. Are pyroclastic materials a significant component of shield volcanoes?
3. Where do most shield volcanoes form—on the ocean floor or on the continents?
4. Where are the best-known shield volcanoes in the United States? Name some examples in other parts of the world.

 7.6 Cinder Cones

Describe the formation, size, and composition of cinder cones.

As the name suggests, **cinder cones** (also called **scoria cones**) are built from ejected lava fragments that begin to harden in flight to produce the vesicular rock *scoria* (**Figure 7.15**). These pyroclastic fragments range in size from fine ash to bombs that may exceed 1 meter (3 feet) in diameter. However, most of the volume of a cinder cone consists of pea- to walnut-sized fragments that are markedly vesicular and have a black to reddish-brown color (see Figure 7.10A). In addition, this pyroclastic material tends to have basaltic composition.

Although cinder cones are composed mostly of loose scoria fragments, some produce extensive lava fields. These lava flows generally form in the final stages of the volcano's life span, when the magma body has lost most of its gas content. Because cinder cones are composed of loose fragments rather than solid rock, the lava usually flows out from the unconsolidated base of the cone rather than from the crater.

Cinder cones have very simple, distinct shapes (see Figure 7.15). Because cinders have a high angle of repose (the steepest angle at which material remains stable), cinder cones are steep-sided, having slopes between 30 and 40 degrees. In addition, cinder cone have large, deep craters in relation to the overall size of the structure. Although relatively symmetrical, some cinder cones are

Mobile Field Trip

Lava flow

Crater

Pyroclastic material

SP Crater is a classic cinder cone located north of Flagstaff, Arizona.

Central vent filled with rock fragments

Michael Collier

elongated and higher on the side that was downwind during the final eruptive phase.

Most cinder cones are produced by a single, short-lived eruptive event. One study found that half of all cinder cones examined were constructed in less than 1 month, and 95 percent of them formed in less than 1 year. Once the event ceases, the magma in the "plumbing" connecting the vent to the magma source solidifies, and the volcano usually does not erupt again. (One exception is Cerro Negro, a cinder cone in Nicaragua, which has erupted more than 20 times since it formed in 1850.)

As a result of this typically short life span, cinder cones are small, usually between 30 and 300 meters (100 and 1000 feet) tall. A few rare examples exceed 700 meters (2300 feet) in height.

Cinder cones number in the thousands around the globe. Some occur in groups, such as the volcanic field near Flagstaff, Arizona, which consists of about 600 cones. Others are parasitic cones that are found on the flanks or within the calderas of larger volcanic structures.

Condor Video

Parícutin, a cinder cone located in Mexico, erupted for nine years.

An aa flow emanating from the base of the cone buried much of the village of San Juan Parangaricutiro, leaving only remnants of the village's church.

Parícutin: Life of a Garden-Variety Cinder Cone

One of the very few volcanoes studied by geologists from its very beginning is the cinder cone called Parícutin, located about 320 kilometers (200 miles) west of Mexico City. In 1943, its eruptive phase began in a cornfield owned by Dionisio Pulido, who witnessed the event.

For 2 weeks prior to the first eruption, numerous tremors caused apprehension in the nearby village of Parícutin. Then, on February 20, sulfurous gases began billowing from a small depression that had been in the cornfield for as long as local residents could remember. During the night, hot, glowing rock fragments were ejected from the vent, producing a spectacular fireworks display. Explosive discharges continued, throwing hot fragments and ash occasionally as high as 6000 meters (20,000 feet) into the air. Larger fragments fell near the crater, some remaining incandescent as they rolled down the slope. These built an aesthetically pleasing cone, while finer ash fell over a much larger area, burning and eventually covering the village of Parícutin. In the first day, the cone grew to 40 meters (130 feet), and by the fifth day it was more than 100 meters (330 feet) high.

The first lava flow came from a fissure that opened just north of the cone, but after a few months, flows began to emerge from the base of the cone. In June 1944, a clinkery aa flow 10 meters (30 feet) thick moved over much of the village of San Juan Parangaricutiro, leaving only remnants of the church exposed (**Figure 7.16**). After 9 years of intermittent pyroclastic explosions and nearly continuous discharge of lava from vents at its base, the activity ceased almost as quickly as it had begun. Today, Parícutin is just another one of the scores of cinder cones dotting the landscape in this region of Mexico. Like the others, it will not erupt again.

7.6 CONCEPT CHECKS

1. Describe the composition of a cinder cone.
2. How do cinder cones compare in size and steepness of their flanks with shield volcanoes?
3. Over what time span does a typical cinder cone form?

7.7 Composite Volcanoes

List the characteristics of composite volcanoes and describe how these volcanoes form.

Earth's most picturesque yet potentially dangerous volcanoes are **composite volcanoes**, also known as **stratovolcanoes**. Most are located in a relatively narrow zone that rims the Pacific Ocean, appropriately called the *Ring of Fire* (see Figure 7.35, page 239). This active zone includes a chain of continental volcanoes distributed along the west coast of the Americas, including the large cones of the Andes in South America and the Cascade Range of the western United States and Canada.

Classic composite cones are large, nearly symmetrical structures consisting of alternating layers of explosively erupted cinders and ash interbedded with lava flows. A few composite cones, notably Italy's Etna and Stromboli, display very persistent eruption activity, and molten lava has been observed in their summit craters for decades. Stromboli is so well known for eruptions that eject incandescent blobs of lava that it has been called the "Lighthouse of the Mediterranean." Mount Etna has erupted, on average, once every 2 years since 1979.

Just as shield volcanoes owe their shape to fluid basaltic lavas, composite cones reflect the viscous nature of the material from which they are made. In general, composite cones are the product of silica-rich magma having an andesitic composition. However, many composite cones also emit various amounts of fluid basaltic lava and, occasionally, pyroclastic material having a felsic (rhyolitic) composition. The silica-rich magmas typical of composite cones generate thick, viscous lavas that travel less than a few kilometers. Composite cones are also noted for generating explosive eruptions that eject huge quantities of pyroclastic material.

A conical shape, with a steep summit area and gradually sloping flanks, is typical of most large composite cones. This classic profile, which adorns calendars and postcards, is partially a result of the way viscous lavas and pyroclastic ejected materials contribute to the cone's growth. Coarse fragments ejected from the summit crater tend to accumulate near their source and contribute to the steep slopes around the summit. Finer ejected materials, on the other hand, are deposited as a thin layer over a large area and hence tend to flatten the flank of the cone. In addition, during the early stages of growth, lavas tend to be more abundant and flow greater distances from the vent than they do later in the volcano's history, which contributes to the cone's broad base. As a composite volcano matures, the shorter flows that come from the central vent serve to armor and strengthen the summit area. Consequently, steep slopes exceeding 40 degrees are possible. Two of the most perfect cones—Mount Mayon in the Philippines and Fujiyama in Japan—exhibit the classic form we expect of composite

Figure 7.17 Fujiyama, a classic composite volcano Japan's Fujiyama exhibits the classic form of a composite cone—a steep summit and gently sloping flank. (Photo by Koji Nakano/Sebun Photo/Getty Images)

cones, with steep summits and gently sloping flanks (**Figure 7.17**).

Despite the symmetrical forms of many composite cones, most have complex histories. Many composite volcanoes have secondary vents on their flanks that have produced cinder cones or even much larger volcanic structures. Huge mounds of volcanic debris surrounding these structures provide evidence that large sections of these volcanoes slid downslope as massive landslides. Some develop amphitheater-shaped depressions at their summits as a result of explosive lateral eruptions—as occurred during the 1980 eruption of Mount St. Helens. Often, so much rebuilding has occurred since these eruptions that no trace

of these amphitheater-shaped scars remain. Others, such as Crater Lake, have been truncated by the collapse of their summit (see Figure 7.22).

 7.7 CONCEPT CHECKS

1. What name is given to the region having the greatest concentration of composite volcanoes?
2. Describe the materials that compose composite volcanoes.
3. How does the composition and viscosity of lava flows differ between composite volcanoes and shield volcanoes?

 7.8 Volcanic Hazards

Describe the major geologic hazards associated with volcanoes.

Roughly 1500 of Earth's known volcanoes have erupted at least once, and some several times, in the past 10,000 years. Based on historical records and studies of active volcanoes, 70 volcanic eruptions can be expected each year and 1 large-volume eruption every decade. The large eruptions account for the vast majority of volcano-related human fatalities.

Today, an estimated 500 million people from Japan to Indonesia, and from Italy to Oregon, live near active volcanoes. They face a number of volcanic hazards, such as destructive pyroclastic flows, molten lava flows, mudflows called lahars, and falling ash and volcanic bombs.

Pyroclastic Flow: A Deadly Force of Nature

Some of the most destructive forces of nature are **pyroclastic flows**, which consist of hot gases infused with incandescent ash and larger lava fragments. Also referred to as **nuée ardentes** (*glowing avalanches*), these fiery flows can race down steep volcanic slopes at speeds exceeding 100 kilometers (60 miles) per hour (**Figure 7.18**). Pyroclastic flows have two components—a low-density cloud of hot expanding gases containing fine ash particles and a ground-hugging portion composed of pumice and other vesicular pyroclastic material.

Jeep

Figure 7.18 Pyroclastic flows, one of the most destructive volcanic forces This pyroclastic flow occurred on Mount Pinatubo, Philippines, during the 1991 eruption. Pyroclastic flows are composed of hot ash and pumice and/or blocky lava fragments that race down the slopes of volcanoes. (Photo by Alberto Garcia/Corbis)

Driven by Gravity Pyroclastic flows are propelled by the force of gravity and tend to move in a manner similar to snow avalanches. They are mobilized by expanding volcanic gases released from the lava fragments and by the expansion of heated air that is overtaken and trapped in the moving front. These gases reduce friction between ash and pumice fragments, which gravity propels downslope in a nearly frictionless environment. This is why some pyroclastic flow deposits are found many miles from their source.

Occasionally, powerful hot blasts that carry small amounts of ash separate from the main body of a pyroclastic flow. These low-density clouds, called *surges*, can be deadly but seldom have sufficient force to destroy buildings in their paths. Nevertheless, in 2014, a hot ash cloud from Japan's Mount Ontake killed 47 hikers and injured 69 more.

Pyroclastic flows may originate in a variety of volcanic settings. Some occur when a powerful eruption blasts pyroclastic material out of the side of a volcano. More frequently, however, pyroclastic flows are generated by the collapse of tall eruption columns during an explosive event. When gravity eventually overcomes the initial upward thrust provided by the escaping gases, the ejected materials begin to fall, sending massive amounts of incandescent blocks, ash, and pumice cascading down slope.

The Destruction of St. Pierre In 1902, an infamous pyroclastic flow and associated surge from Mount Pelée, a small volcano on the Caribbean island of Martinique, destroyed the port town of St. Pierre.

Although the main pyroclastic flow was largely confined to the valley of Riviere Blanche, a low-density fiery surge spread south of the river and quickly engulfed the entire city. The destruction happened in moments and was so devastating that nearly all of St. Pierre's 28,000 inhabitants were killed. Only 1 person on the outskirts of town—a prisoner protected in a dungeon—and a few people on ships in the harbor were spared (**Figure 7.19**).

Scientists arrived on the scene within days and found that although St. Pierre was mantled by only a thin layer of volcanic debris, masonry walls nearly 1 meter (3 feet) thick had been knocked over like dominoes, large trees had been uprooted, and cannons had been torn from their mounts.

Figure 7.19 Destruction of St. Pierre **A.** St. Pierre as it appeared shortly after the eruption of Mount Pelée in 1902. (Reproduced from the collection of the Library of Congress Prints and Photographs Division) **B.** St. Pierre before the eruption. Many vessels were anchored offshore when this photo was taken, as was the case on the day of the eruption. (Photo by UPPA/Photoshot)

B. St. Pierre before the 1902 eruption.

A. St. Pierre following the eruption of Mount Pelée.

Lahars: Mudflows on Active and Inactive Cones

In addition to violent eruptions, large composite cones may generate a type of fluid mudflow, known by its Indonesian name, **lahar**. These destructive flows occur when volcanic debris becomes saturated with water and rapidly moves down steep volcanic slopes, generally following stream valleys. Some lahars are triggered when magma nears the surface of a glacially clad volcano, causing large volumes of ice and snow to melt. Others are generated when heavy rains saturate weathered volcanic deposits. Thus, lahars may occur even when a volcano is *not* erupting.

When Mount St. Helens erupted in 1980, several lahars were generated. These flows and accompanying floodwaters raced down nearby river valleys at speeds exceeding 30 kilometers (20 miles) per hour. These raging rivers of mud destroyed or severely damaged nearly all the homes and bridges along their paths (**Figure 7.20**). Fortunately, the area was not densely populated.

In 1985, deadly lahars were produced during a small eruption of Nevado del Ruiz, a 5300-meter (17,400-foot) volcano in the Andes Mountains of Colombia. Hot pyroclastic material melted ice and snow that capped the mountain (*nevado* means *snow* in Spanish) and sent torrents of ash and debris down three major river valleys that flank the volcano. Reaching speeds of 100 kilometers (60 miles) per hour, these mudflows tragically claimed 25,000 lives.

Many consider Mount Rainier, Washington, to be America's most dangerous volcano because, like Nevado del Ruiz, it has a thick, year-round mantle of snow and glacial ice. Adding to the risk is the fact that more than 100,000 people live in the valleys around Rainier, and many homes are built on deposits left by lahars that flowed down the volcano hundreds or thousands of years ago. A future eruption, or perhaps just a period of heavier-than-average rainfall, may produce lahars that could be similarly destructive.

Other Volcanic Hazards

Volcanoes can be hazardous to human health and property in other ways. Ash and other pyroclastic material can collapse the roofs of buildings or can be drawn into the lungs of humans and other animals or into aircraft engines (**Figure 7.21**). Volcanic gases, most notably sulfur dioxide, pollute the air—and when mixed with rainwater, can destroy vegetation and reduce the quality of groundwater. Despite the known risks, millions of people live in close proximity to active volcanoes.

Volcano-Related Tsunamis Although **tsunamis** are most often associated with displacement along a fault located on the seafloor (see Chapter 6), some result from the collapse of a volcanic cone. This was dramatically demonstrated during the 1883 eruption on Krakatau, when the northern half of a volcano plunged into the Sunda Strait, creating a tsunami that exceeded 30 meters (100 ft) in height. Although the island of Krakatau was uninhabited, an estimated 36,000 people were killed along the coastline of the Indonesian islands of Java and Sumatra.

Volcanic Ash and Aviation During the past 15 years, at least 80 commercial jets have been damaged by inadvertently flying into clouds of volcanic ash. For example, in 1989, a Boeing 747 carrying more than 300 passengers encountered an ash cloud from Alaska's Redoubt Volcano; all four engines clogged with ash and stalled mid-air. Fortunately, the pilots were able to restart the engines and safely landed the aircraft in Anchorage.

More recently, the 2010 eruption of Iceland's Eyjafjallajökull Volcano sent ash high into the atmosphere. This thick plume of ash drifted over Europe, causing airlines to cancel thousands of flights and leaving hundreds of thousands of travelers stranded. Several weeks passed before air travel resumed its normal schedule.

Volcanic Gases and Respiratory Health One of the most destructive volcanic events, called the Laki eruptions, began along a

Figure 7.20 Lahars, mudflows that originate on volcanic slopes This lahar raced down the Muddy River, located southeast of Mount St. Helens, following the May 18, 1980, eruption. Notice the former height of this fluid mudflow, as recorded by the mudflow line on the tree trunks. Note the person for scale. (Photo by Lyn Topinka/U.S. Geological Survey)

Ash and other pyroclastic materials can collapse roofs, or completely cover buildings.

Lava flows can destroy homes, roads, and other structures in their paths.

Labels in Figure 7.21: Prevailing wind, Eruption cloud, Ash fall, Acid rain, Eruption column, Bombs, Pyroclastic flow, Collapse of flank, Lava dome collapse, Emission of sulfur dioxide gases, Lava flow, Lahar (mudflow)

Figure 7.21 Volcanic hazards In addition to generating destructive pyroclastic flows and lahars, volcanoes can be hazardous to human health and property in many other ways.

large fissure in southern Iceland in 1783. An estimated 14 cubic kilometers of fluid basaltic lavas were released along with 130 million tons (3.4 cubic miles) of sulfur dioxide and other poisonous gases. When sulfur dioxide is inhaled, it reacts with moisture in the lungs to produce sulfuric acid, a deadly toxin. More than half of Iceland's livestock died, and the ensuing famine killed 25 percent of the island's human population.

This huge eruption also endangered people and property all across Europe. Crop failure occurred in parts of Western Europe, and thousands of residents perished from lung-related diseases. A recent report estimated that a similar eruption today would cause more than 140,000 cardiopulmonary fatalities in Europe alone.

Effects of Volcanic Ash and Gases on Weather and Climate Volcanic eruptions can eject dust-sized particles of volcanic ash and sulfur dioxide gas high into the atmosphere. The ash particles reflect sunlight back to space, producing temporary atmospheric cooling. The 1783 Laki eruptions in Iceland appear to have affected atmospheric circulation around the globe. Drought conditions prevailed in the Nile River valley and India, and the winter

of 1784 saw the longest period of below-zero temperatures in New England's history.

Other eruptions that have produced significant effects on climate worldwide include the eruption of Indonesia's Mount Tambora in 1815, which produced the "year without a summer" (1816), and the eruption of El Chichón in Mexico in 1982. El Chichón's eruption, although small, emitted an unusually large quantity of sulfur dioxide that reacted with water vapor in the atmosphere to produce a dense cloud of tiny sulfuric-acid droplets. These particles, called aerosols, take several years to settle out of the atmosphere. Like fine ash, these aerosols lower the mean temperature of the atmosphere by reflecting solar radiation back to space.

7.8 CONCEPT CHECKS

1. Describe pyroclastic flows and explain why they are capable of traveling great distances.
2. What is a lahar?
3. List at least three volcanic hazards besides pyroclastic flows and lahars.

7.9 Other Volcanic Landforms

List volcanic landforms other than shield, cinder, and composite volcanoes and describe their formation.

The most widely recognized volcanic structures are the cone-shaped edifices of composite volcanoes that dot Earth's surface. However, volcanic activity produces other distinctive and important landforms.

Calderas

Recall that *calderas* are large steep-sided depressions that have diameters exceeding 1 kilometer (0.6 miles) and have a somewhat circular form. Those less than 1 kilometer across are called *collapse pits*, or *craters*. Most calderas are formed by one of the following processes: (1) the collapse of the summit of a large composite volcano following an explosive eruption of silica-rich pumice and ash fragments (*Crater Lake–type calderas*); (2) the collapse of the top of a shield volcano caused by subterranean drainage from a central magma chamber (*Hawaiian-type calderas*); and (3) the collapse of a large area, caused by the discharge of colossal volumes of silica-rich pumice and ash along ring fractures (*Yellowstone-type calderas*).

Crater Lake–Type Calderas Crater Lake, Oregon, is situated in a caldera approximately 10 kilometers (6

miles) wide and 1175 meters (more than 3800 feet) deep. This caldera formed about 7000 years ago, when a composite cone named Mount Mazama violently extruded 50 to 70 cubic kilometers of pyroclastic material (**Figure 7.22**). With the loss of support, 1500 meters (nearly 1 mile) of the summit of this once-prominent cone collapsed, producing a caldera that eventually filled with rainwater and snowmelt. Later, volcanic activity built a small cinder cone in the caldera. Today this cone, called Wizard Island, provides a mute reminder of past activity.

Hawaiian-Type Calderas Unlike Crater Lake–type calderas, many calderas form gradually because of the loss of lava from a shallow magma chamber underlying a volcano's summit. For example, Hawaii's active shield volcanoes, Mauna Loa and Kilauea, both have large calderas at their summits. Kilauea's measures 3.3 by 4.4 kilometers (about 2 by 3 miles) and is 150 meters (500 feet) deep. The walls are almost vertical, and as a result, the caldera looks like a vast, nearly flat-bottomed pit. Kilauea's caldera formed by gradual subsidence as magma slowly drained laterally from the underlying magma chamber, leaving the summit unsupported.

SmartFigure 7.22
Formation of Crater Lake–type calderas About 7000 years ago, a violent eruption partly emptied the magma chamber of former Mount Mazama, causing its summit to collapse. Precipitation and groundwater contributed to forming Crater Lake, the deepest lake in the United States—594 meters (1949 feet) deep—and the ninth-deepest lake in the world. (Inset photo courtesy of USGS) (https://goo.gl/FGvNQJ)

Animation

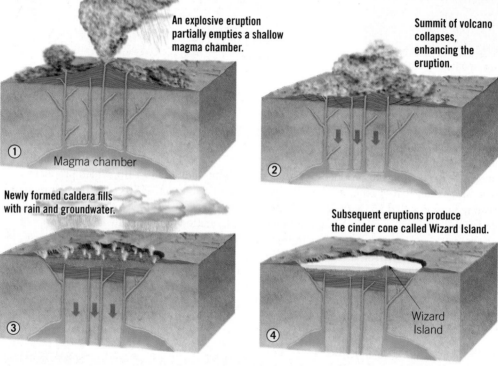

An explosive eruption partially empties a shallow magma chamber.

① Magma chamber

Summit of volcano collapses, enhancing the eruption.

②

Newly formed caldera fills with rain and groundwater.

③

Subsequent eruptions produce the cinder cone called Wizard Island.

Wizard Island

④

Yellowstone-Type Calderas Historic and destructive eruptions such as Mount St. Helens pale in comparison to what happened 630,000 years ago in the region now occupied by Yellowstone National Park, when approximately 1000 cubic kilometers of pyroclastic material erupted. This catastrophic eruption sent showers of ash as far as the Gulf of Mexico and formed a caldera 70 kilometers (43 miles) across (**Figure 7.23A**). Vestiges

Crater Lake

Wizard Island

Michael Collier

Close-up view of Wizard Island.

of this event are the many hot springs and geysers in the Yellowstone region.

Yellowstone-type eruptions eject huge volumes of pyroclastic materials, mainly in the form of ash and pumice fragments. Typically, these materials are ejected as *pyroclastic flows* that sweep across the landscape, destroying most living things in their paths. Upon coming to rest, the hot fragments of ash and pumice fuse together, forming a welded tuff that closely resembles a solidified lava flow. Despite the immense size of these calderas, the eruptions that produce them are brief, lasting hours to perhaps a few days.

Large calderas tend to exhibit a complex eruptive history. In the Yellowstone region, for example, three caldera-forming episodes are known to have occurred over the past 2.1 million years (**Figure 7.23B**). The most recent eruption (630,000 years ago) was followed by episodic outpourings of degassed rhyolitic and basaltic lavas. In the intervening years, a slow upheaval of the floor of the caldera has produced two elevated regions called *resurgent domes* (see Figure 7.23A). Geologic evidence suggests that a magma reservoir still exists beneath Yellowstone; thus, another caldera-forming eruption is likely—but not necessarily imminent.

Unlike calderas associated with shield volcanoes or composite cones, Yellowstone-type calderas are so large and poorly defined that many were undetected until high-quality aerial and satellite images became available. Other examples of Yellowstone-type calderas are California's Long Valley Caldera; LaGarita Caldera, located in the San Juan Mountains of southern Colorado; and the Valles Caldera, west of Los Alamos, New Mexico. These and similar calderas found around the globe are among the largest volcanic structures on Earth, hence the name "supervolcanoes." Volcanologists compare their destructive force to that of the impact of a small asteroid. Fortunately, no Yellowstone-type eruption has occurred in historic times.

Fissure Eruptions and Basalt Plateaus

The greatest volume of volcanic material is extruded from fractures in Earth's crust, called *fissures*. Rather than building cones, **fissure eruptions** usually emit fluid basaltic lavas that blanket wide areas (**Figure 7.24**). In some locations, extraordinary amounts of lava have been extruded along fissures

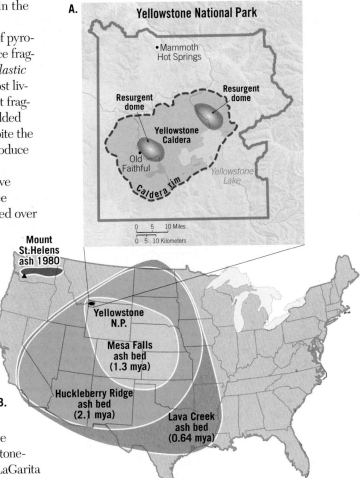

SmartFigure 7.23 Supereruptions at Yellowstone
A. This map shows Yellowstone National Park and the location and size of the Yellowstone caldera. **B.** Three huge eruptions, separated by relatively regular intervals of about 700,000 years, were responsible for the ash layers shown. The largest of these eruptions was 10,000 times greater than the 1980 eruption of Mount St. Helens. (https://goo.gl/6al3WM)

Tutorial

Figure 7.24 Basaltic fissure eruptions Lava fountaining from a fissure and formation of fluid lava flows called *flood basalts*. The lower photo shows flood basalt flows near Idaho Falls.

Figure 7.25 Columbia River basalts **A.** The Columbia River basalts cover an area of nearly 164,000 square kilometers (63,000 square miles) that is commonly called the Columbia Plateau. Activity here began about 17 million years ago, as lava began to pour out of large fissures, eventually producing a basalt plateau with an average thickness of more than 1 kilometer (0.6 mile). **B.** Columbia River basalt flows exposed in the Palouse River Canyon in southwestern Washington State. (Photo by Williamborg)

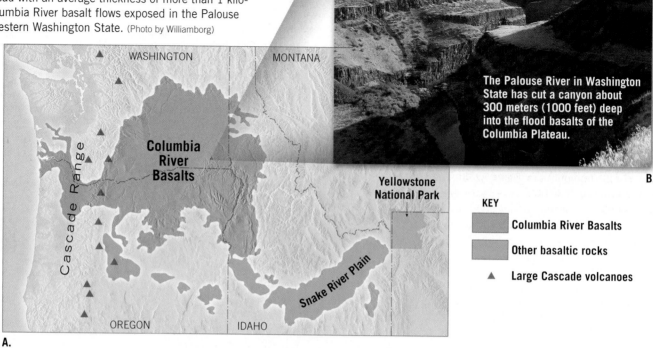

The Palouse River in Washington State has cut a canyon about 300 meters (1000 feet) deep into the flood basalts of the Columbia Plateau.

B.

A.

KEY
- Columbia River Basalts
- Other basaltic rocks
- ▲ Large Cascade volcanoes

in a relatively short time, geologically speaking. These voluminous accumulations are commonly referred to as **basalt plateaus** because most have a basaltic composition and tend to be rather flat and broad. The Columbia Plateau in the northwestern United States, which consists of the Columbia River basalts, is a product of this type of activity (**Figure 7.25**). Numerous fissure eruptions have buried the landscape, creating a lava plateau nearly 1500 meters (1 mile) thick. Some of the lava remained molten long enough to flow 150 kilometers (90 miles) from its source. The term **flood basalts** appropriately describe these extrusions.

Massive accumulations of basaltic lava, similar to those of the Columbia Plateau, occur elsewhere in the world. One of the largest is the Deccan Traps (*traps* = stairs), a thick sequence of flat-lying basalt flows covering nearly 500,000 square kilometers (195,000 square miles) of west-central India. When the Deccan Traps formed about 66 million years ago, nearly

SmartFigure 7.26 Volcanic neck Shiprock, New Mexico, is a volcanic neck that stands about 520 meters (1700 feet) high. It consists of igneous rock that crystallized in the vent of a volcano that has long since been eroded. (https://goo.gl/XSRxzc)

Tutorial

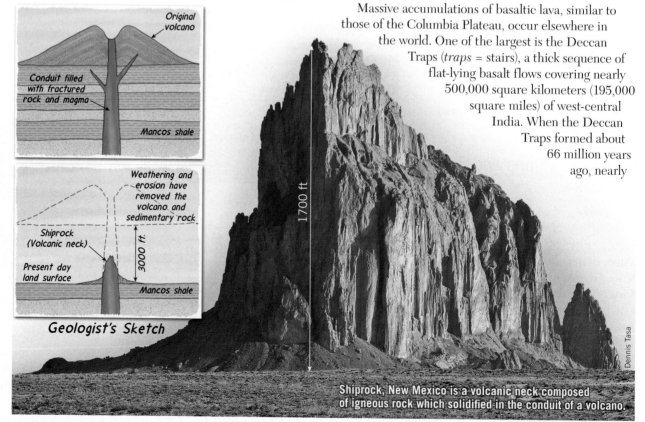

Original volcano

Conduit filled with fractured rock and magma

Mancos shale

Weathering and erosion have removed the volcano and sedimentary rock

Shiprock (Volcanic neck)

Present day land surface

Mancos shale

3000 ft.

Geologist's Sketch

1700 ft

Shiprock, New Mexico is a volcanic neck composed of igneous rock which solidified in the conduit of a volcano.

2 million cubic kilometers of lava were extruded over a period of approximately 1 million years. Several other massive accumulations of flood basalts, including the Ontong Java Plateau, have been discovered in the deep ocean basins (see Figure 9.15, page 284).

Volcanic Necks

Most volcanic eruptions are fed lava through short conduits that connect shallow magma chambers to vents located at the surface. When a volcano becomes inactive, congealed magma is often preserved in the feeding conduit of the volcano as a crudely cylindrical mass. As the volcano succumbs to forces of weathering and erosion, the rock occupying the volcanic conduit, which is highly resistant to weathering, may remain standing above the surrounding terrain long after the cone has been worn away. Shiprock, New Mexico, is a widely recognized and

spectacular example of these structures, which geologists call **volcanic necks** (or **plugs**) (**Figure 7.26**). More than 510 meters (1700 feet) high, Shiprock is taller than most skyscrapers and is one of many such landforms that protrude conspicuously from the red desert landscapes of the American Southwest.

7.9 CONCEPT CHECKS

1. Describe the formation of Crater Lake. Compare it to the calderas found on shield volcanoes such as Kilauea.
2. Other than composite volcanoes, what volcanic landform can generate a pyroclastic flow?
3. How do the eruptions that created the Columbia Plateau differ from the eruptions that create large composite volcanoes?
4. What type of volcanic structure is Shiprock, New Mexico, and how did it form?

7.10 Intrusive Igneous Activity

Compare and contrast these intrusive igneous structures: dikes, sills, batholiths, stocks, and laccoliths.

Although volcanic eruptions can be violent and spectacular events, most magma does not erupt at the surface as lava but is emplaced and crystallized at depth, without fanfare. Therefore, understanding the *intrusive igneous processes* that occur deep underground is as important to geologists as studying volcanic events.

Nature of Intrusive Bodies

When magma rises through the crust, it forcefully displaces preexisting crustal rocks, termed **host rock** (or **country rock**). The structures that result from the emplacement of magma into preexisting rocks are called **intrusions**, or **plutons**. Because all intrusions form far below Earth's surface, they are studied primarily after uplifting and erosion have exposed them. The challenge lies in reconstructing the events that generated these structures in vastly different conditions deep underground, millions of years ago.

Intrusions occur in a great variety of sizes and shapes; some of the most common are illustrated in **Figure 7.27**. Notice that some of these bodies cut across existing structures, such as sedimentary strata, whereas others form when magma is injected between sedimentary layers. Because of these differences, intrusive igneous bodies are generally classified according to their shape as either **tabular** (*tabula* = table) or **massive** (large and blob shaped) and by their orientation with respect to the host rock. Igneous bodies are said to be **discordant** (*discordare* = to disagree) if they cut *across* existing structures and

concordant (*concordare* = to agree) if they intrude *parallel* to features such as sedimentary strata.

Tabular Intrusive Bodies: Dikes and Sills

Dikes and Sills Tabular intrusive bodies are produced when magma is forcibly injected into a fracture or zone of weakness, such as a bedding surface (see Figure 7.27A). **Dikes** are discordant bodies that form when magma is forcibly injected into fractures and that cut across bedding surfaces and other structures in the host rock. By contrast, **sills** are nearly horizontal, concordant bodies that form when magma exploits weaknesses between sedimentary beds or other rock structures (**Figure 7.28**). In general, dikes serve as tabular conduits that *transport magma* toward Earth's surface, whereas sills tend to *accumulate magma* and gradually increase in thickness. Consequently, sills only form near Earth's surface where the force of the injected magma is great enough to lift the overlying rocks.

Dikes and sills are typically shallow features, occurring where the host rocks are sufficiently brittle to fracture. They can range in thickness from less than 1 millimeter to more than 1 kilometer.

While dikes and sills can occur as solitary bodies, dikes tend to form in roughly parallel groups called *dike swarms*. These multiple structures reflect the tendency for fractures to form in sets when tensional forces pull apart brittle country rock. Dikes can also radiate from an eroded volcanic neck, like spokes on a wheel. Where this occurs, the active ascent of magma

A. Relationship between volcanism and intrusive igneous activity.

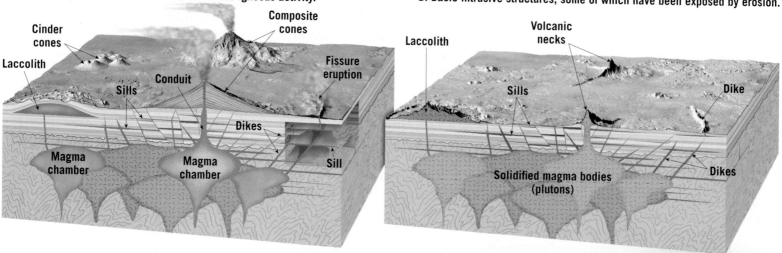

B. Basic intrusive structures, some of which have been exposed by erosion.

C. Extensive uplift and erosion exposed a batholith composed of several smaller intrusive bodies (plutons).

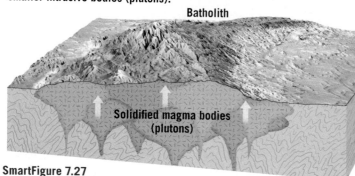

Exposed portion of the Sierra Nevada Batholith

Belinda Images/SuperStock

SmartFigure 7.27
Intrusive igneous structures
(https://goo.gl/3sS9U5)

SmartFigure 7.28 Sill exposed in Sinbad County, Utah The dark, essentially horizontal band is a sill of basaltic composition that intruded horizontal layers of sedimentary rock. (Photo by Michael Collier) (http://goo.gl/4MZelh)

generated fissures in the volcanic cone out of which lava flowed (see Figure 7.27A). Dikes frequently are more resistant and thus weather more slowly than the surrounding rock. Consequently, when exposed by erosion, dikes tend to have a wall-like appearance, as shown in **Figure 7.29**.

Because dikes and sills are relatively uniform in thickness and can extend for many kilometers, they are assumed to be the product of very fluid, and therefore mobile, magmas. One of the largest and most studied of all sills in the United States is the Palisades Sill. Exposed for 80 kilometers (50 miles) along the west bank of the Hudson River in southeastern New York and northeastern New Jersey, this sill is about 300 meters (1000 feet) thick.

Columnar Jointing In many respects, sills closely resemble buried lava flows. Both are tabular and can extend over a wide area, and both may exhibit columnar jointing. **Columnar jointing** occurs when igneous rocks cool and develop shrinkage fractures that produce elongated, pillar-like columns that most often have six sides (**Figure 7.30**). Further, because sills and dikes generally form in near-surface environments and may be only a few meters thick, the emplaced magma often cools quickly enough to generate a fine-grained texture. (Recall that most intrusive igneous bodies have a coarse-grained texture.)

Massive Plutons: Batholiths, Stocks, and Laccoliths

Batholiths and Stocks By far the largest intrusive igneous bodies are **batholiths** (see Figure 7.27C). Batholiths occur as mammoth linear structures several hundred kilometers long and up to 100 kilometers (60 miles) wide. The Sierra Nevada batholith, for example, is a continuous granitic structure that forms much of the "backbone" of the Sierra Nevada in California. An even larger batholith extends for over 1800 kilometers (1100 miles) along the Coast Mountains of western Canada and into southern Alaska. Although batholiths can cover a large area, recent gravitational studies indicate that most are less than 10 kilometers (6 miles) thick. Some are even thinner; the coastal batholith of Peru, for example, is essentially a flat slab with an average thickness of only 2 to 3 kilometers (1 to nearly 2 miles). Batholiths are typically composed of felsic (granitic) and intermediate rock types and are often called "granite batholiths."

Early investigators thought the Sierra Nevada batholith was a huge single body of intrusive igneous rock. Today we know that large batholiths are produced by hundreds of discrete injections of magma that form smaller intrusive bodies (plutons) that intimately crowd against or penetrate one another. These bulbous masses are emplaced over spans of millions of years. The intrusive activity that created the Sierra Nevada batholith, for example, occurred nearly continuously over a 130-million-year period that ended about 80 million years ago.

By definition, to be considered a batholith, a plutonic body must have a surface exposure greater than 100 square kilometers (40 square miles). Smaller plutons are termed **stocks**. However, many stocks appear to be portions of much larger intrusive bodies that would be classified as batholiths if they were fully exposed.

Dike

SmartFigure 7.29 Dike exposed in the Spanish Peaks, Colorado This wall-like dike is composed of igneous rock that is more resistant to weathering than the surrounding material. (Photo by Michael Collier) (https://goo.gl/MJHkem)

Condor Video

Geologist's Sketch

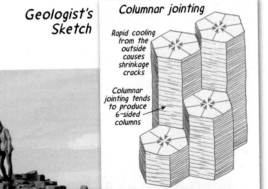

Columnar jointing

Rapid cooling from the outside causes shrinkage cracks

Columnar jointing tends to produce 6-sided columns

Figure 7.30 Columnar jointing Giant's Causeway in Northern Ireland is an excellent example of columnar jointing. (Photo by E. J. Tarbuck)

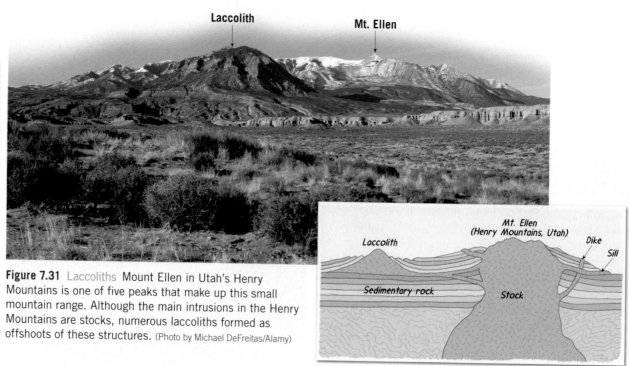

Figure 7.31 Laccoliths Mount Ellen in Utah's Henry Mountains is one of five peaks that make up this small mountain range. Although the main intrusions in the Henry Mountains are stocks, numerous laccoliths formed as offshoots of these structures. (Photo by Michael DeFreitas/Alamy)

Geologist's Sketch

Laccoliths A nineteenth-century study by G. K. Gilbert of the U.S. Geological Survey in the Henry Mountains of Utah produced the first clear evidence that igneous intrusions can lift the sedimentary strata they penetrate. Gilbert named the igneous intrusions he observed **laccoliths**, which he envisioned as igneous rock forcibly injected between sedimentary strata, so as to arch the beds above while leaving those below relatively flat. It is now known that the five major peaks of the Henry Mountains are not laccoliths but stocks. However, these central magma bodies are the source material for branching offshoots that are true laccoliths, as Gilbert defined them (**Figure 7.31**).

Numerous other granitic laccoliths have since been identified in Utah. The largest is a part of the Pine Valley Mountains located north of St. George, Utah. Others are found in the La Sal Mountains near Arches National Park and in the Abajo Mountains directly to the south.

7.10 CONCEPT CHECKS

1. What is meant by the term *host rock*?
2. Describe *dikes* and *sills*, using the appropriate terms from the following list: massive, discordant, tabular, and concordant.
3. Distinguish among batholiths, stocks, and laccoliths in terms of size and shape.

7.11 Origin of Magma

Summarize the major processes that generate magma from solid rock.

Recall that most magma originates from partial melting of solid rock in Earth's uppermost mantle. The focus of this section is the process of partial melting and the mechanisms that trigger it.

Partial Melting

Because igneous rocks are composed of various minerals with different melting points, they tend to melt within a temperature range of about 200°C (360°F). As rock begins to melt, the minerals with the lowest melting temperatures melt first. If melting continues, minerals with higher melting points begin to melt, and the composition of the melt steadily approaches the overall composition of the rock from which it was derived.

Most often, however, melting is not complete. The incomplete melting of rocks is known as **partial melting**, a process that produces most magma. Partial melting can be likened to a chocolate chip cookie set out in the Sun. The chocolate chips represent the minerals with the lowest melting points because they begin to melt before the other ingredients. However, when rock partially melts, the molten material melts and separates from the surrounding solid components. Further, because molten rock is less dense than the surrounding solids, it rises toward Earth's surface.

Generating Magma from Solid Rock

Workers in underground mines know that temperatures increase as they descend deeper below Earth's surface. Although the rate of temperature change varies considerably from place to place, it *averages* about 25°C per kilometer (1°F per 70 feet) in the *upper* crust.

This increase in temperature with depth is known as the **geothermal gradient**. As shown in **Figure 7.32**, when a typical geothermal gradient is compared to the melting point curve for the mantle rock peridotite, the temperature at which peridotite melts is higher than the geothermal gradient. Thus, under normal conditions, the mantle is solid. However, tectonic processes trigger melting though various means, including reducing the melting point (temperature) of mantle rock.

Decrease in Pressure: De-compression Melting If temperature were the only factor determining whether rock melts, our planet would be a molten ball covered with a thin, solid outer shell. This is not the case because pressure, which also increases with depth, influences the melting temperatures of rocks.

Melting, which is accompanied by an increase in volume, *occurs at progressively higher temperatures with increased depth*. This is the result of the steady increase in confining pressure exerted by the weight of overlying rocks (see Chapter 2). Conversely, reducing confining pressure lowers a rock's melting temperature. When confining pressure drops sufficiently, **decompression melting** is triggered.

Decompression melting occurs wherever hot, solid mantle rock ascends, thereby moving into regions of lower pressure. The process of decompression melting is responsible for generating magma along divergent plate boundaries (oceanic ridges) where plates are rifting apart (**Figure 7.33**). Below the ridge crest, hot mantle rock rises and melts, replacing the material that shifted horizontally away from the ridge axis. Decompression melting also occurs when ascending mantle plumes reach the uppermost mantle.

Addition of Water Another important factor affecting the melting temperature of rock is its water content. Water and other volatiles act as salt does to melt ice. Water causes rock to melt at lower temperatures, just as putting rock salt on an icy sidewalk induces melting.

The introduction of water to generate magma occurs mainly at convergent

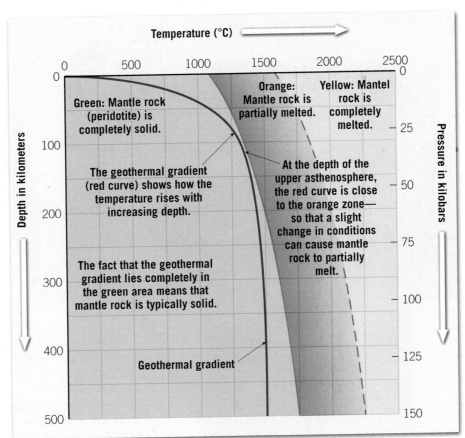

Figure 7.32 Why Earth's mantle is mainly solid This diagram shows the geothermal gradient (increase in temperature with depth) for the crust and upper mantle. Also illustrated is the melting point curve for the mantle rock peridotite.

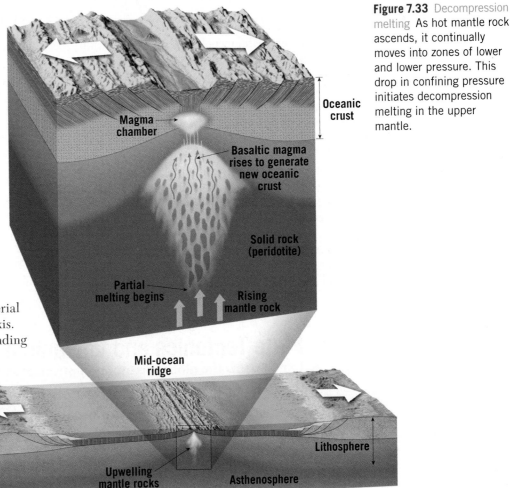

Figure 7.33 Decompression melting As hot mantle rock ascends, it continually moves into zones of lower and lower pressure. This drop in confining pressure initiates decompression melting in the upper mantle.

Figure 7.34 Water lowers the melting temperature of hot mantle rock to trigger partial melting As an oceanic plate descends into the mantle, water and other volatiles are driven from the subducting crustal rocks into the mantle above.

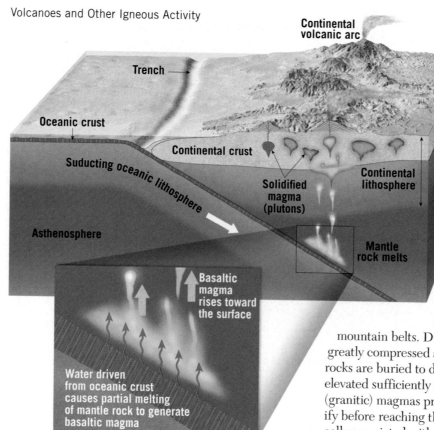

plate boundaries where cool slabs of oceanic lithosphere descend into the mantle (**Figure 7.34**). As an oceanic plate sinks, heat and pressure drive water from hydrated (water-rich) minerals found in the subducting oceanic crust and overlying sediments. These fluids migrate upward into the wedge of hot mantle that lies directly above. At a depth of about 100 kilometers (60 miles), the addition of water lowers the melting temperature of mantle rock sufficiently to trigger partial melting. Partial melting of the mantle rock peridotite generates hot basaltic magmas with temperatures that may exceed 1250°C (2300°F).

Temperature Increase: Melting Crustal Rocks As mantle-derived basaltic magma forms, it buoyantly rises toward the surface. In continental settings, basaltic

magma often "ponds" beneath crustal rocks, which have a lower density and are already near their melting temperature. The hot basaltic magma is thought to heat the overlying crustal rocks sufficiently to generate a secondary, silica-rich magma. If these low-density, silica-rich magmas reach the surface, they tend to produce the explosive eruptions that we associate with convergent plate boundaries.

Crustal rocks can also melt during continental collisions that result in the formation of large mountain belts. During such events, the crust is greatly compressed and thickened, and some crustal rocks are buried to depths where the temperatures are elevated sufficiently to cause partial melting. The felsic (granitic) magmas produced in this manner usually solidify before reaching the surface, so volcanism is not typically associated with these collision-type mountain belts.

In summary, magma can be generated three ways: (1) *A decrease in pressure* (without an increase in temperature) can result in *decompression melting*; (2) the *introduction of water* can lower the melting temperature of hot mantle rock sufficiently to generate magma; and (3) *heating* crustal rocks above their melting temperature produces magma.

7.11 CONCEPT CHECKS

1. What is the geothermal gradient? Describe how the geothermal gradient compares with the melting temperatures of the mantle rock peridotite at various depths.
2. Describe the process of decompression melting.
3. What role do water and other volatiles play in the formation of magma?
4. Name two plate tectonic settings in which you would expect magma to be generated.

7.12 Plate Tectonics and Volcanism

Explain how the global distribution of volcanic activity is related to plate tectonics.

Geologists have known for decades that the global distribution of most of Earth's volcanoes is not random. Most active volcanoes on land are located along the margins of the ocean basins—notably within the circum-Pacific belt known as the **Ring of Fire** (**Figure 7.35**). Another group of volcanoes includes the innumerable seamounts

that form along the crest of the mid-ocean ridges. There are some volcanoes, however, that appear to be randomly distributed around the globe. These volcanic structures comprise most of the islands of the deep-ocean basins, including the Hawaiian Islands, the Galapagos Islands, and Easter Island.

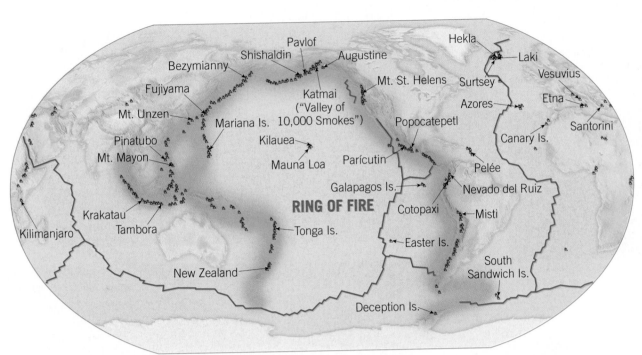

Figure 7.35 Ring of Fire Most of Earth's major volcanoes are located in a zone around the Pacific called the Ring of Fire. Another large group of active volcanoes lie undiscovered along the mid-ocean ridge system.

The development of the theory of plate tectonics provided geologists with a plausible explanation for the distribution of Earth's volcanoes and established the basic connection between plate tectonics and volcanism: *Plate motions provide the mechanisms by which mantle rocks undergo partial melting to generate magma.*

Volcanism at Divergent Plate Boundaries

The greatest volume of magma erupts along divergent plate boundaries associated with seafloor spreading—out of human sight (**Figure 7.36B**). Below the ridge axis where lithospheric plates are continually being pulled apart, the solid, yet mobile, mantle rises to fill the rift. Recall that as hot rock rises, it experiences a decrease in confining pressure and may undergo *decompression melting*. This activity continuously adds new basaltic rock to plate margins, temporarily welding them together, only to have them break again as spreading continues. Along some ridge segments, extrusions of pillow lavas build numerous volcanic structures, the largest of which is Iceland.

Although most spreading centers are located along the axis of an oceanic ridge, some are not. In particular, the East African Rift is a site where continental lithosphere is being pulled apart (see **Figure 7.36F** and Figure 5.14, page 154). Vast outpourings of fluid basaltic lavas as well as several active volcanoes are found in this region of the globe.

Volcanism at Convergent Plate Boundaries

Recall that along convergent plate boundaries, two plates move toward each other, and a slab of dense oceanic lithosphere descends into the mantle. In these settings, water driven from hydrated (water-rich) minerals found in the subducting oceanic crust and overlying sediments triggers partial melting in the hot mantle above (**Figure 7.36A**).

Volcanism at a convergent plate margin results in the development of a slightly curved chain of volcanoes called a *volcanic arc*. These volcanic chains develop roughly parallel to the associated trench—at distances of 200 to 300 kilometers (100 to 200 miles). Volcanic arcs that develop within the ocean and grow large enough for their tops to rise above the surface are labeled *archipelagos* in most atlases. Geologists prefer the more descriptive term **volcanic island arcs**, or simply **island arcs** (see Figure 7.36A). Several young volcanic island arcs border the western Pacific basin, including the Aleutians, the Tongas, and the Marianas.

Volcanism associated with convergent plate boundaries may also develop where slabs of oceanic lithosphere are subducted under continental lithosphere to produce a **continental volcanic arc** (**Figure 7.36E**). The mechanisms that generate these mantle-derived magmas are essentially the same as those that create volcanic island arcs. The most significant difference is that continental crust is much thicker and composed of rocks having higher silica content than oceanic crust.

Did You Know?
At 4392 m (14,411 ft) in altitude, Washington's Mount Rainier is the tallest of the 15 great volcanoes that make up the backbone of the Cascade Range. Although Mount Rainier is considered an active volcano, its summit is covered by more than 25 alpine glaciers.

A. Convergent Plate Volcanism When an oceanic plate subducts, melting in the mantle produces magma that gives rise to a volcanic island arc on the overlying oceanic crust.

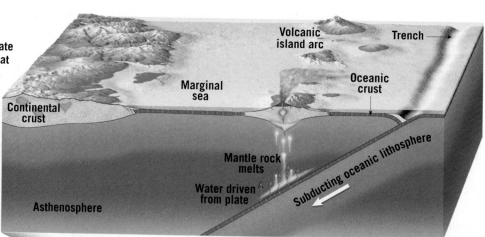

Cleveland Volcano, Aleutian Islands (USGS)

C. Intraplate Volcanism When an oceanic plate moves over a hot spot, a chain of volcanic structures such as the Hawaiian Islands is created.

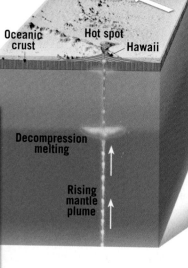

Kilauea, Hawaii (USGS)

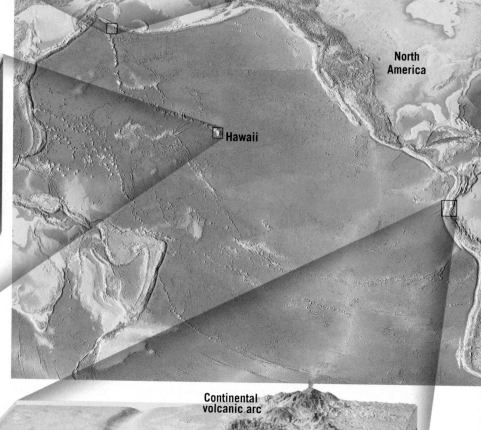

E. Convergent Plate Volcanism When oceanic lithosphere descends beneath a continent, magma generated in the mantle rises to form a continental volcanic arc.

SmartFigure 7.36 Earth's zones of volcanism (https://goo.gl/MoqLrr)

Tutorial

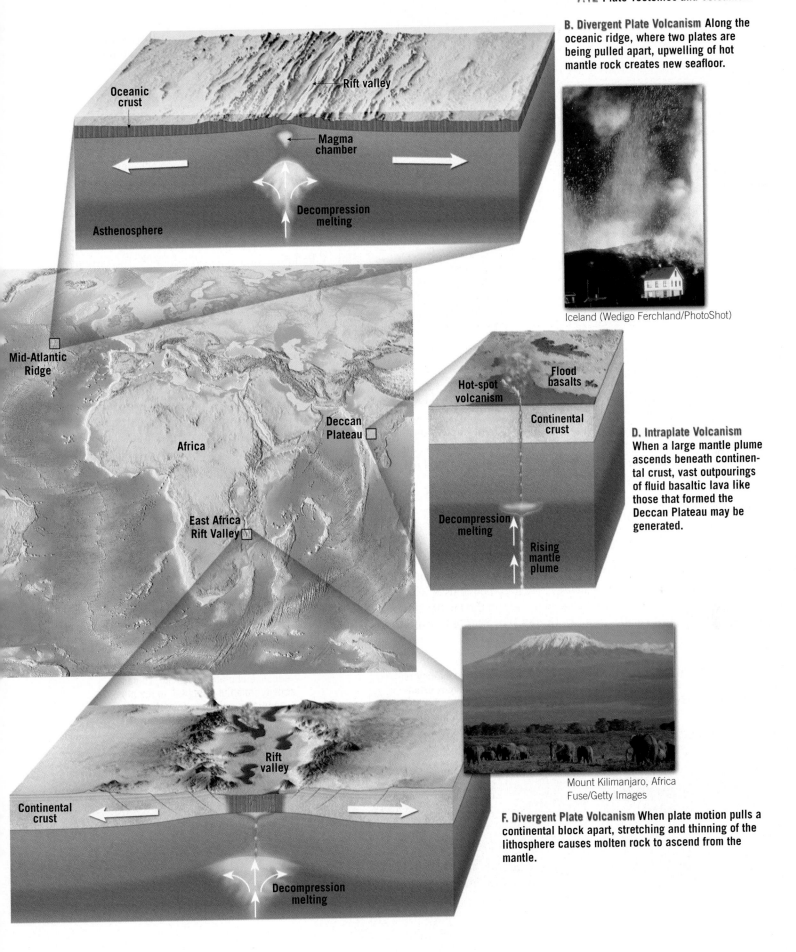

B. Divergent Plate Volcanism Along the oceanic ridge, where two plates are being pulled apart, upwelling of hot mantle rock creates new seafloor.

Oceanic crust

Rift valley

Magma chamber

Decompression melting

Asthenosphere

Iceland (Wedigo Ferchland/PhotoShot)

Mid-Atlantic Ridge

Africa

Deccan Plateau

East Africa Rift Valley

Hot-spot volcanism

Flood basalts

Continental crust

D. Intraplate Volcanism When a large mantle plume ascends beneath continental crust, vast outpourings of fluid basaltic lava like those that formed the Deccan Plateau may be generated.

Decompression melting

Rising mantle plume

Rift valley

Continental crust

Decompression melting

Mount Kilimanjaro, Africa
Fuse/Getty Images

F. Divergent Plate Volcanism When plate motion pulls a continental block apart, stretching and thinning of the lithosphere causes molten rock to ascend from the mantle.

SmartFigure 7.37
Subduction-produced
Cascade Range
volcanoes Subduction
of the Juan de Fuca
plate along the Cascadia
subduction zone produced
the Cascade volcanoes.
(https://goo.gl/4To5ak)

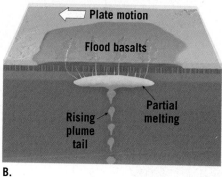

Hence, by melting the surrounding silica-rich crustal rocks, mantle-derived magma changes composition as it rises through the crust. The volcanoes of the Cascade Range in the northwestern United States, including Mount Hood, Mount Rainier, Mount Shasta, and Mount St. Helens, are examples of volcanoes generated at a convergent plate boundary along a continental margin (**Figure 7.37**).

Intraplate Volcanism

We know why igneous activity is initiated along plate boundaries, but why do eruptions occur in the interiors of plates? Hawaii's Kilauea, considered one of the world's most active volcanoes, is situated thousands of kilometers from the nearest plate boundary, in the middle of the vast Pacific plate (see **Figure 7.36C**). Sites of **intraplate**

(meaning "within the plate") **volcanism** include the large outpourings of fluid basaltic lavas such as those that compose the Columbia Plateau, the Siberian Traps in Russia, India's Deccan Traps, and several submerged oceanic plateaus, including the Ontong Java Plateau in the western Pacific.

Most intraplate volcanism occurs when a relatively narrow mass of hot **mantle plume** ascends toward the surface (**Figure 7.38**).° Although the depth at which mantle plumes originate is a topic of debate, some are thought to form deep within Earth, at the core–mantle boundary. These plumes of solid yet mobile rock rise toward the surface in a manner similar to the blobs that form within a lava lamp, which contain two immiscible liquids in a glass container. As the base of the lamp is heated, the denser liquid at the bottom becomes buoyant and forms blobs that rise to the top. Like the blobs in a lava lamp, a mantle plume has a bulbous head that draws out a narrow stalk beneath it as it rises. The surface manifestation of this activity is called a **hot spot**, an area of volcanism, high heat flow, and crustal uplifting a few hundred kilometers wide.

Large mantle plumes, dubbed **superplumes**, are thought to be responsible for the vast outpourings of basaltic lava that created the large basalt plateaus. When the head of the plume reaches the base of the lithosphere, decompression melting progresses rapidly. This causes the burst of volcanism that emits voluminous flows of lava over a period of 1 million or so years (see Figure 7.38B). Extreme eruptions of this type would have affected Earth's climate, causing (or at least contributing to) the extinction events recorded in the fossil record.

°Some geologists question the role of mantle plumes in the formation of Earth's volcanic landforms.

SmartFigure 7.38 Mantle plumes and large basalt provinces Model of hot-spot volcanism thought to explain the formation of large basalt plateaus and the chains of volcanic islands associated with these features. (https://goo.gl/NFTuK7)

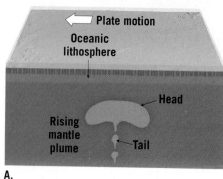

A rising mantle plume with a large bulbous head is thought to generate Earth's large basalt plateaus.

A.

Rapid decompression melting of the plume head produces extensive outpourings of flood basalts over a relatively short time span.

B.

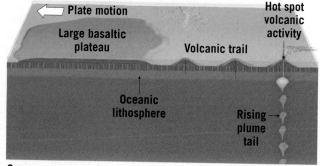

Because of plate movement, volcanic activity from the rising tail of the plume generates a linear chain of smaller volcanic structures.

C.

The comparatively short initial eruptive phase is often followed by millions of years of less voluminous activity, as the plume tail slowly rises to the surface. Extending away from some large flood basalt plateaus is a chain of volcanic structures, similar to the Hawaiian chain (see Figure 7.38C).

Intraplate volcanism associated with mantle plumes is also thought to be responsible for the massive eruptions of silica-rich pyroclastic material that occurred in continental settings. Perhaps the best known of these hot-spot eruptions are the three caldera-forming eruptions that occurred in the Yellowstone region over the past 2.1 million years (see Figure 7.23).

7.12 CONCEPT CHECKS

1. Are volcanoes in the Ring of Fire generally described as quiescent or explosive? Name an example that supports your answer.
2. How is magma generated along convergent plate boundaries?
3. Volcanism at divergent plate boundaries is most often associated with which magma type? What causes rocks to melt in these settings?
4. What is thought to be the source of magma for most intraplate volcanism?
5. Which type of plate boundary generates the greatest quantity of magma?

CONCEPTS IN REVIEW
Volcanoes and Other Igneous Activity

7.1 Mount St. Helens Versus Kilauea

Compare and contrast the 1980 eruption of Mount St. Helens with the most recent eruption of Kilauea, which began in 1983.

- Volcanic eruptions cover a broad spectrum from explosive eruptions, like that of Mount St. Helens in 1980, to the quiescent eruptions of Kilauea.

7.2 The Nature of Volcanic Eruptions

Explain why some volcanic eruptions are explosive and others are quiescent.

KEY TERMS: magma, lava, viscosity, eruption column

- The two primary factors determining the nature of a volcanic eruption are the lava's viscosity (resistance to flow) and its gas content. In general, the more silica in the lava, the more viscous is the lava; lavas with lower silica content are more fluid. Temperature also influences viscosity. Hot lavas are more fluid, while cool lavas are more viscous.
- Basaltic magmas, which are fluid and have low gas content, tend to generate quiescent (nonexplosive) eruptions. In contrast, silica-rich magmas (andesitic and rhyolitic), which are the most viscous and contain the greatest quantity of gases, are the most explosive.

(?) Although Kilauea mostly erupts in a gentle manner, what risks might you encounter if you chose to live nearby?

USGS

7.3 Materials Extruded During an Eruption

List and describe the three categories of materials extruded during volcanic eruptions.

KEY TERMS: aa flow, pahoehoe flow, lava tube, pillow lava, volatiles, pyroclastic materials, tephra, scoria, pumice

- Volcanoes bring molten lava, gases, and solid pyroclastic materials to Earth's surface.
- Low-viscosity basaltic lava flows can extend great distances from a volcano. On the surface, they travel as pahoehoe or aa flows. Sometimes the surface of the flow congeals, but lava continues to flow below in tunnels called lava tubes. When lava erupts underwater, the outer surface is chilled instantly to obsidian, while the inside continues to flow, producing pillow lavas.
- The gases most commonly emitted by volcanoes are water vapor and carbon dioxide. Upon reaching the surface, these gases rapidly expand, leading to explosive eruptions that can generate a mass of lava fragments called pyroclastic materials.
- Pyroclastic materials come in several sizes. From smallest to largest, they are ash, lapilli, and blocks or bombs. Blocks exit the volcano as solid fragments, whereas bombs exit as liquid blobs.
- If bubbles of gas in lava don't pop before the lava solidifies, they are preserved as voids called vesicles. Especially frothy, silica-rich lava can cool to make lightweight pumice, while basaltic lava with lots of bubbles cools to make scoria.

(?) This photo shows layers of volcanic material ejected by a violent eruption and deposited roughly horizontally. What term is used to describe this type of volcanic material?

Erik Klemetti

7.4 Anatomy of a Volcano

Draw and label a diagram that illustrates the basic features of a typical volcanic cone.

KEY TERMS: fissure, conduit, vent, volcanic cone, crater, caldera, parasitic cone, fumarole

- Volcanoes vary in size and form but share a few common features. Most are roughly conical piles of extruded material that collect around a central vent. The vent is usually within a summit crater or caldera. On the flanks of the volcano, there may be smaller vents marked by small parasitic cones, or there may be fumaroles, spots where gas is expelled.

(?) **Label the diagram using the following terms: conduit, vent, lava, parasitic cone, bombs, pyroclastic material.**

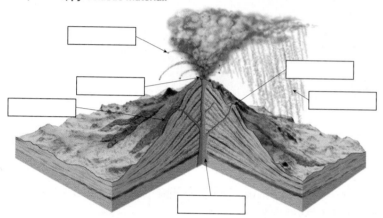

7.5 Shield Volcanoes

Summarize the characteristics of shield volcanoes and provide one example of this type of volcano.

KEY TERMS: shield volcano, seamount

- Shield volcanoes consist of many successive lava flows of low-viscosity basaltic lava but lack significant amounts of pyroclastic debris. Lava tubes help transport lava far from the main vent, resulting in very gentle, shield-like profiles.
- Most shield volcanoes begin as seamounts that grow from Earth's seafloor. Mauna Loa, Mauna Kea, and Kilauea in Hawaii are classic examples of the low, wide form characteristic of shield volcanoes.

7.6 Cinder Cones

Describe the formation, size, and composition of cinder cones.

KEY TERM: cinder cone (scoria cone)

- Cinder cones are steep-sided structures composed mainly of pyroclastic debris, typically having a basaltic composition. Lava flows sometimes emerge from the base of a cinder cone but typically do not flow out of the crater.
- Cinder cones are small relative to the other major kinds of volcanoes, reflecting the fact that they form quickly, as single eruptive events. Because they are unconsolidated, cinder cones easily succumb to weathering and erosion.

7.7 Composite Volcanoes

List the characteristics of composite volcanoes and describe how these volcanoes form.

KEY TERM: composite volcano (stratovolcano)

- Composite volcanoes are called "composite" because they consist of both pyroclastic material and lava flows. They typically erupt silica-rich lavas that cool to produce andesite or rhyolite. They are much larger than cinder cones and form from multiple eruptions over millions of years.
- Because andesitic and rhyolitic lavas are more viscous than basaltic lava, they accumulate at a steeper angle than does the lava from shield volcanoes. Over time, a composite volcano's combination of lava and cinders produces a towering volcano with a classic symmetrical shape.
- Mount Rainier and the other volcanoes of the Cascade Range in the northwest United States are good examples of composite volcanoes.

(?) **If your family had to live next to a volcano, would you rather it be a shield volcano, cinder cone, or composite volcano? Explain.**

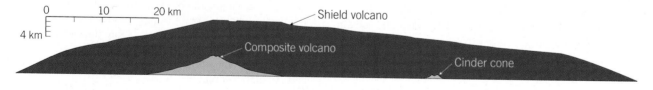

7.8 Volcanic Hazards

Describe the major geologic hazards associated with volcanoes.

KEY TERMS: pyroclastic flow (nuée ardente), lahar, tsunami

- The greatest volcanic hazard to human life is the pyroclastic flow, or nuée ardente. This dense mix of hot gas and pyroclastic fragments races downhill at great speed and incinerates everything in its path. A pyroclastic flow can travel many miles from its source volcano. Because pyroclastic flows are hot, their deposits frequently "weld" together into a solid rock called welded tuff.
- Lahars are mudflows that form on volcanoes. These rapidly moving slurries of ash and debris suspended in water tend to follow stream valleys and can result in loss of life and/or significant damage to structures.
- Volcanic ash in the atmosphere can be a risk to air travel when it is sucked into airplane engines. Volcanoes at sea level can generate tsunamis when they erupt or when their flanks collapse into the ocean. Those that spew large amounts of gas such as sulfur dioxide can cause respiratory problems. If volcanic gases reach the stratosphere, they screen out a portion of incoming solar radiation and can trigger short-term cooling at Earth's surface.

(?) **What do lahars and pyroclastic flows have in common?**

7.9 Other Volcanic Landforms

List volcanic landforms other than shield, cinder, and composite volcanoes and describe their formation.

KEY TERMS: fissure eruption, basalt plateau, flood basalt, volcanic neck (plug)

- Calderas, which can be among the largest volcanic structures, form when the rigid, cold rock above a magma chamber cannot be supported and collapses, creating a broad, roughly circular depression. On shield volcanoes, calderas form slowly as lava drains from the magma chamber beneath the volcano. On a composite volcano, caldera collapse often follows an explosive eruption that can result in significant loss of life and destruction of property.
- Fissure eruptions occasionally produce massive floods of fluid basaltic lava from large cracks, called fissures, in the crust. Layer upon layer of these flood basalts may accumulate to significant thicknesses and blanket a wide area. The Columbia Plateau located in northwestern United States is an example.
- Shiprock, New Mexico, is an example of a volcanic neck where the lava in the "throat" of an ancient volcano crystallized to form a "plug" of solid rock that weathered more slowly than the surrounding volcanic rocks. After the pyroclastic debris has eroded, the resistant neck has produced a distinctive landform.

7.10 Intrusive Igneous Activity

Compare and contrast these intrusive igneous structures: dikes, sills, batholiths, stocks, and laccoliths.

KEY TERMS: host (country) rock, intrusion (pluton), tabular, massive, discordant, concordant, dike, sill, columnar jointing, batholith, stock, laccolith

- When magma intrudes other rocks, it may cool and crystallize before reaching the surface to produce intrusions called plutons. Plutons come in many shapes. They may cut across the host rocks without regard for preexisting structures, or the magma may flow along weak zones in the host rock, such as between the horizontal layers of sedimentary bedding.
- Tabular intrusions may be concordant (sills) or discordant (dikes). Massive plutons may be small (stocks) or very large (batholiths). Blister-like intrusions also exist (laccoliths). As solid igneous rock cools, its volume decreases. Contraction can produce a distinctive fracture pattern called columnar jointing.

(?) Label the intrusive igneous structures in the accompanying diagram, using the following terms: sill, batholith, volcanic neck, laccolith.

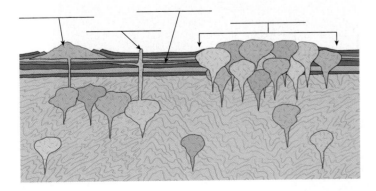

7.11 Origin of Magma

Summarize the major processes that generate magma from solid rock.

KEY TERMS: partial melting, geothermal gradient, decompression melting

- Solid rock may melt under three geologic circumstances: when heat is added to the rock, raising its temperature; when already hot rock experiences lower pressures (decompression, as seen at mid-ocean ridges); and when water is added to hot rock that is near its melting point (as occurs at subduction zones).

7.12 Plate Tectonics and Volcanism

Explain how the global distribution of volcanic activity is related to plate tectonics.

KEY TERMS: Ring of Fire, volcanic island arc (island arc), continental volcanic arc, intraplate volcanism, mantle plume, hot spot, superplume

- Volcanoes occur at both convergent and divergent plate boundaries, as well as in intraplate settings.
- Convergent plate boundaries that involve the subduction of oceanic crust are the most common site for explosive volcanoes—most prominently in the Pacific Ring of Fire. The release of water from the subducting plate triggers melting in the overlying mantle. The ascending magma interacts with the lower crust of the overlying plate and can form a volcanic arc at the surface.
- At divergent boundaries, where lithosphere is being rifted apart, decompression melting is the dominant generator of magma. As warm rock rises, it can begin to melt without the addition of heat.
- In intraplate settings, the source of magma is a mantle plume—a column of warmer-than-normal, rising mantle rock.

(?) The accompanying diagram shows one of the tectonic settings where volcanism is a dominant process. Name the tectonic setting and briefly explain how magma is generated in this setting.

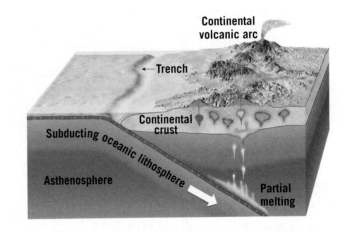

Continental volcanic arc

← Trench

Continental crust

Subducting oceanic lithosphere

Asthenosphere

Partial melting

GIVE IT SOME THOUGHT

1. Examine the accompanying photo and complete the following:
   a. What type of volcano is it? What features helped you classify it as such?
   b. What is the eruptive style of such volcanoes? Describe the likely composition and viscosity of its magma.
   c. Which type of plate boundary is the likely setting for this volcano?
   d. Name a city that is vulnerable to the effects of a volcano of this type.

2. Answer the following questions about divergent boundaries, such as the Mid-Atlantic Ridge, and their associated lavas:
   a. Divergent boundaries are characterized by eruptions of what type of lava: andesitic, basaltic, or rhyolitic?
   b. What is the source of the lavas at divergent boundaries?
   c. What process causes the source rocks to melt?

3. For each of the accompanying four sketches, identify the geologic setting (zone of volcanism). Which of these settings will most likely generate explosive eruptions? Which will produce outpouring of fluid basaltic lavas?

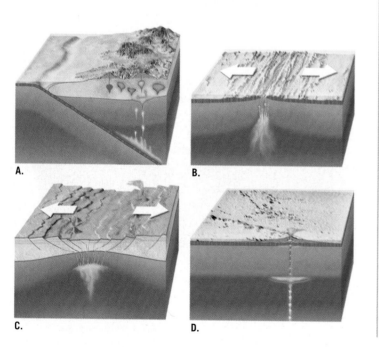

A.

B.

C.

D.

4. Match each of these volcanoes or volcanic regions with one of the three zones of volcanism (convergent plate boundaries, divergent plate boundaries, or intraplate volcanism):
   a. Crater Lake
   b. Hawaii's Kilauea
   c. Mount St. Helens
   d. East African Rift
   e. Yellowstone
   f. Mount Pelée
   g. Deccan Traps
   h. Fujiyama

5. Explain why an eruption of Mount Rainier that resembled the 1980 eruption of Mount St. Helens would be considerably more destructive.

6. This image shows the Buddhist monastery Taung Kalat, located in central Myanmar (Burma). The monastery sits high on a sheer-sided rock made mainly of magmas that solidified in the conduit of an ancient volcano. The volcano has since been worn away.
   a. Based on this information, what igneous structure do you think is shown in this photo?
   b. Would this volcanic structure most likely have been associated with a composite volcano or a cinder cone? Explain how you arrived at your answer.

Rene Drouyer/Dreamstime

7. The accompanying image shows an igneous structure (dark color) located in southeastern Utah that was intruded into sedimentary strata.
   a. What name is given to this intrusive igneous feature?
   b. The light bands above and below the dark igneous body are metamorphic rock. Identify this type of metamorphism and briefly explain how it alters rock. (Hint: Refer to Chapter 2, if necessary.)

8. During a field trip with your geology class, you visit an exposure of rock layers similar to the one sketched here. A fellow student suggests that the layer of basalt is a sill. You disagree. Why do you think the other student is incorrect? What is a more likely explanation for the basalt layer?

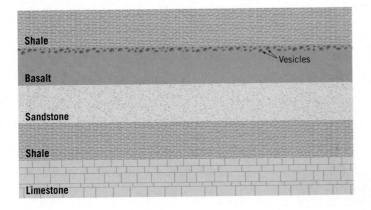

9. The formula for the volume of a cone is $V = 1/3\pi r^2 h$ (where V = volume, $\pi = 3.14$, r = radius, and h = height). If Mauna Loa is 9 kilometers high and has a radius of roughly 85 kilometers, what is its approximate total volume?

10. Different processes produce magma in different tectonic settings. Consider magma bodies found at locations A, B, and C in the accompanying diagram and describe the processes that would be most likely to trigger the melting that produced each.

MasteringGeology™

www.masteringgeology.com   Looking for additional review and test prep materials? With individualized coaching on the toughest topics of the course, MasteringGeology offers a wide variety of ways for you to move beyond memorization to begin thinking like a geologist. Visit the Study Area in www.masteringgeology. com to find practice quizzes, study tools, and multimedia that will

improve your understanding of this chapter's content. Sign in today to enjoy the following features: **Self Study Quizzes, SmartFigure: Tutorials/Animations/Condor Videos/Mobile Field Trips, Geoscience Animation Library, GEODe, RSS Feeds, Digital Study Modules,** and an optional **Pearson eText.**

8

FOCUS ON CONCEPTS

Each statement represents the primary learning objective for the corresponding major heading within the chapter. After you complete the chapter, you should be able to:

8.1 Explain the principle of uniformitarianism and discuss how it differs from catastrophism.

8.2 Distinguish between numerical and relative dating and apply relative dating principles to determine a time sequence of geologic events.

8.3 Define *fossil* and discuss the conditions that favor the preservation of organisms as fossils. List and describe various fossil types.

8.4 Explain how rocks of similar age that are in different places can be matched up.

8.5 Discuss three types of radioactive decay and explain how radioactive isotopes are used to determine numerical dates.

8.6 Distinguish among the four basic time units that make up the geologic time scale and explain why the time scale is considered to be a dynamic tool.

8.7 Explain how reliable numerical dates are determined for layers of sedimentary rock.

This hiker is on the Kaibab Trail in Arizona's Grand Canyon National Park. Millions of years of Earth history are exposed in the canyon's rock walls. (Photo by Michael Collier)

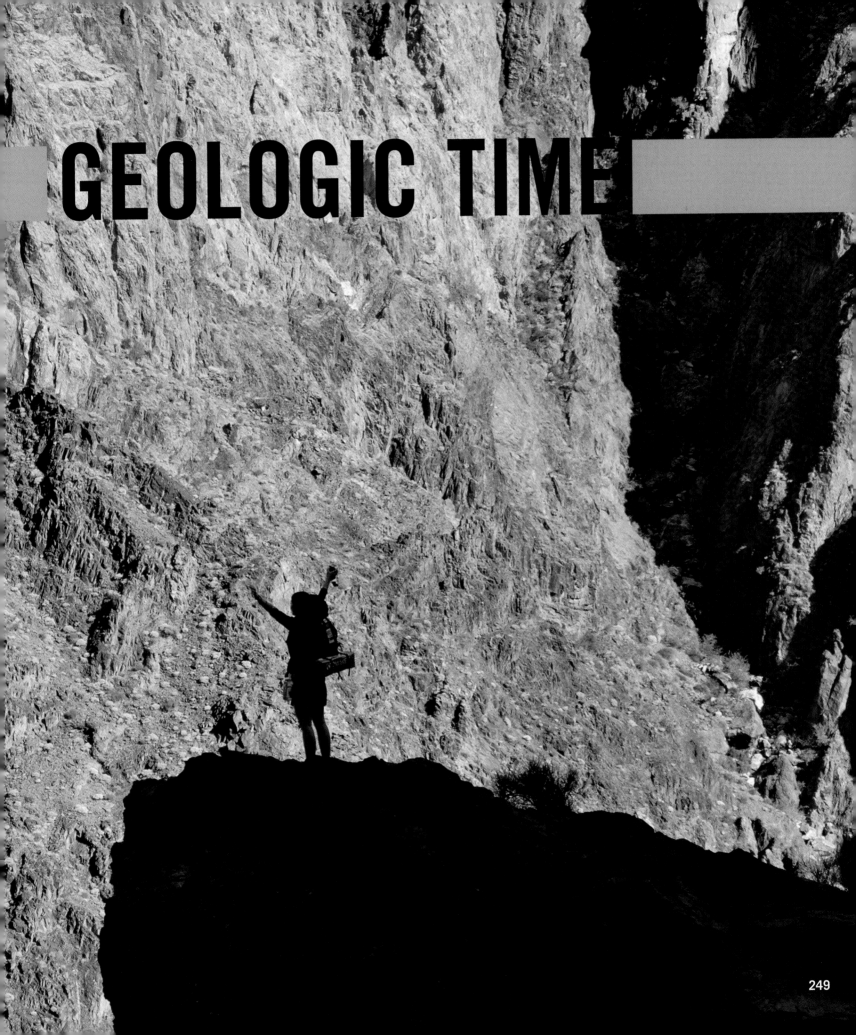

GEOLOGIC TIME

In the eighteenth century, James Hutton recognized the immensity of Earth history and the importance of time as a component in all geologic processes. In the nineteenth century, others effectively demonstrated that Earth had experienced many episodes of mountain building and erosion, which must have required great spans of geologic time. Although these pioneering scientists understood that Earth was very old, they had no way of knowing its true age. Was it tens of millions, hundreds of millions, or even billions of years old? Because they didn't know exactly how old Earth was, they developed a geologic time scale that showed the sequence of events based on relative dating principles. What were these principles? What part do fossils play? With the discovery of radioactivity and the development of radiometric dating techniques, geologists can now assign quite accurate dates to many of the events in Earth history. What is radioactivity? Why is it a good "clock" for dating the geologic past? This chapter will answer these questions.

8.1 A Brief History of Geology

Explain the principle of uniformitarianism and discuss how it differs from catastrophism.

This chapter begins with two images from the Grand Canyon. The chapter-opening photo shows a hiker along the Kaibab Trail, deep in the canyon. By contrast, the hiker in **Figure 8.1** is perched on the rim of the canyon. The layers of sedimentary rock separating the two hikers represent millions of years of Earth history. Geologists have developed the knowledge and skills needed to decipher the clues contained in these strata that tell the long and complex story of this region.

The nature of our Earth—its materials and processes— has been a focus of study for centuries. However, the late 1700s is generally regarded as the beginning of modern geology. It was during this time that James Hutton published his important work *Theory of the Earth*. Prior to that time, a great many explanations about Earth history relied on supernatural events.

Catastrophism

In the mid-1600s, James Ussher, Anglican Archbishop of Armagh, Primate of All Ireland, published a work that had immediate and profound influence. A respected scholar of the Bible, Ussher constructed a chronology of human and Earth history in which he determined that Earth was only a few thousand years old, having been created in 4004 B.C. Ussher's treatise earned widespread acceptance among Europe's scientific and religious leaders, and his chronology was soon printed in the margins of the Bible itself.

During the 1600s and 1700s, the doctrine of **catastrophism** strongly influenced people's thinking about Earth. Briefly stated, catastrophists believed that Earth's varied landscapes had been fashioned primarily by great catastrophes. Features such as mountains and canyons, which today we know take great periods of time to form, were explained as having been produced by sudden and often worldwide disasters of unknowable causes that no longer operate. This philosophy was an attempt to fit the rate of Earth's processes to the prevailing ideas about Earth's age.

The Birth of Modern Geology

Modern geology began in the late 1700s, when James Hutton, a Scottish physician and gentleman farmer, published his *Theory of the Earth*. In this work, Hutton put forth a fundamental principle that is a pillar of geology today: **uniformitarianism**. It simply states that *the physical, chemical, and biological laws that operate today have also operated in the geologic past.* This means that the forces and processes that we observe presently shaping our planet have been at work for a very long time. Thus, to understand ancient rocks, we must first understand present-day processes and their results. This idea is commonly expressed by saying "The present is the key to the past."

Prior to Hutton's *Theory of the Earth*, no one had effectively demonstrated that geologic processes occur over extremely long periods of time. However, Hutton persuasively argued that processes that appear to be weak and slow acting can, over long spans of time, produce effects that are just as great as those resulting from sudden catastrophic events. Unlike his predecessors, Hutton cited verifiable observations to support his ideas.

For example, when he argued that mountains are sculpted and ultimately destroyed by weathering and the erosional work of running water and that their wastes are carried to the oceans by processes that can be observed, Hutton said, "We have a chain of facts which clearly demonstrates that the materials of the wasted mountains

Figure 8.1 Contemplating geologic time This hiker is resting atop the Kaibab Formation, the uppermost layer in the Grand Canyon. Beneath him are thousands of meters of sedimentary strata that go back more than 540 million years. These strata rest atop even older sedimentary, metamorphic, and igneous rocks, some as old as 2 billion years. Although the canyon's rock record has numerous interruptions, the rocks beneath the hiker contain clues to great spans of Earth history. (Photo by Michael Collier)

have traveled through the rivers"; and, further, "There is not one step in all this progress that is not to be actually perceived." He went on to summarize these thoughts by asking a question and immediately providing the answer: "What more can we require? Nothing but time."

Geology Today

The basic tenets of uniformitarianism are just as viable today as in Hutton's day. We realize more strongly than ever before that the present gives us insight into the past and that the physical, chemical, and biological laws that govern geologic processes remain unchanging through time. However, we also understand that the doctrine should not be taken too literally. To say that geologic processes in the past were the same as those occurring today is not to suggest that they always had the same relative importance or that they operated at precisely the same rate. Moreover, some important geologic processes are not currently observable, but evidence that they occur is well established. For example, we know that Earth has experienced impacts from large meteorites even though we have no human witnesses. Such events altered Earth's crust, modified its climate, and strongly influenced life on the planet.

The acceptance of the concept of uniformitarianism, however, meant the acceptance of a very long history for Earth. Although Earth's processes vary in their intensity, they still take a long time to create or destroy major landscape features. For example, geologists have established that mountains once existed in portions of present-day Minnesota, Wisconsin, Michigan, and Manitoba. Today, the region consists of low hills and plains. Erosion (processes that wear away land) gradually destroyed those peaks. The rock record contains evidence that shows Earth has experienced *many cycles* of mountain building and erosion. Concerning the ever-changing nature of Earth through great expanses of geologic time, Hutton made a statement that was to become his most famous. In concluding his classic 1788 paper published in the *Transactions of the Royal Society of Edinburgh*, he stated, "The results, therefore, of our present enquiry is, that we find no vestige of a beginning—no prospect of an end."

It is important to remember that although many features of our physical landscape may seem to be unchanging over the decades we observe them, they are nevertheless changing—but on time scales of hundreds, thousands, or even many millions of years.

Did You Know?

Early attempts at determining Earth's age proved to be unreliable. One method reasoned that if the rate at which sediment accumulates could be determined, as well as the total thickness of sedimentary rock that had been deposited during Earth history, an estimate of Earth's age could be made. All that was necessary was to divide the rate of sediment accumulation into the total thickness of sedimentary rock. This method was riddled with difficulties. Can you think of some?

8.1 CONCEPT CHECKS

1. Contrast catastrophism and uniformitarianism.
2. How did each philosophy view the age of Earth?

8.2 Creating a Time Scale—Relative Dating Principles

Distinguish between numerical and relative dating and apply relative dating principles to determine a time sequence of geologic events.

Like the pages in a long and complicated history book, rocks record the geologic events and changing life-forms of the past. The book, however, is not complete. Many pages, especially in the early chapters, are missing. Others are tattered, torn, or smudged. Yet enough of the book remains to allow much of the story to be deciphered.

Interpreting Earth history is a prime goal of the science of geology. Like a modern-day sleuth, a geologist must interpret the clues found preserved in the rocks. By studying rocks and the features they contain, geologists can unravel the complexities of the past.

Geologic events by themselves, however, have little meaning until they are put into a time perspective. Studying history, whether it is the Civil War or the age of dinosaurs, requires a calendar. Among geology's major contributions to human knowledge are the *geologic time scale* and the discovery that Earth history is exceedingly long.

Numerical and Relative Dates

The geologists who developed the geologic time scale revolutionized the way people think about time and perceive our planet. They learned that Earth is much older than anyone had previously imagined and that its surface and interior have been changed over and over again by the same geologic processes that operate today.

Numerical Dates During the late 1800s and early 1900s, attempts were made to determine Earth's age. Although some of the methods appeared promising at the time, none of these early efforts proved to be reliable. What these scientists were seeking was a **numerical date**. Such dates specify the actual number of years that have passed since an event occurred. Today, our understanding of radiometric dating techniques allows us to accurately determine numerical dates for rocks that represent important events in Earth's distant past. We will study these techniques later in this chapter. Prior to

the development of radiometric dating, geologists had no reliable method of numerical dating and had to rely solely on relative dating.

Relative Dates When we place rocks in their proper *sequence of formation*—which formed first, second, third, and so on—we are establishing **relative dates**. Such dates do not indicate how long ago something took place, only that it followed one event and preceded another. The relative dating techniques that were developed are valuable and still widely used. Numerical dating methods did not replace these techniques; rather, they supplemented them. To establish a relative time scale, a few basic principles or rules had to be discovered and applied. Although they may seem obvious to us today, they were major breakthroughs in thinking at the time, and their discovery was an important scientific achievement.

Principle of Superposition

Nicolas Steno, a Danish anatomist, geologist, and priest (1638–1686), is credited with being the first to recognize a sequence of historical events in an outcrop of sedimentary rock layers. Working in the mountains of western Italy, Steno applied a very simple rule that has come to be the most basic principle of relative dating—the **principle of superposition**. The principle simply states that in an undeformed sequence of sedimentary rocks, each bed is older than the one above and younger than the one below. Although it may seem obvious that a rock layer could not be deposited with nothing beneath it for support, it was not until 1669 that Steno clearly stated this principle.

This rule also applies to other surface-deposited materials, such as lava flows and beds of ash from volcanic eruptions. Applying superposition to the beds exposed in the upper portion of the Grand Canyon, we can easily place the layers in their proper order. Among those that are pictured in **Figure 8.2**, the sedimentary rocks in the Supai Group are the oldest, followed in order by the Hermit Shale, Coconino Sandstone, Toroweap Formation, and Kaibab Limestone.

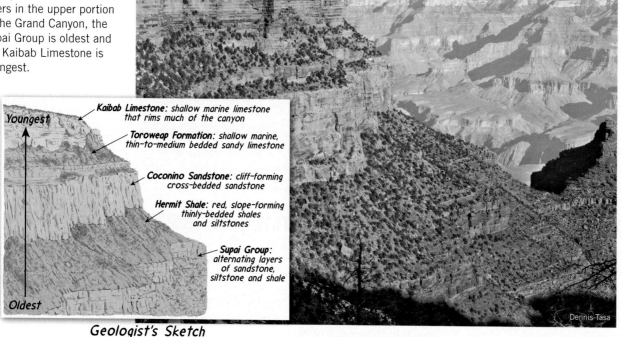

Figure 8.2 Superposition Applying the principle of superposition to these layers in the upper portion of the Grand Canyon, the Supai Group is oldest and the Kaibab Limestone is youngest.

Youngest

Kaibab Limestone: shallow marine limestone that rims much of the canyon

Toroweap Formation: shallow marine, thin-to-medium bedded sandy limestone

Coconino Sandstone: cliff-forming cross-bedded sandstone

Hermit Shale: red, slope-forming thinly-bedded shales and siltstones

Supai Group: alternating layers of sandstone, siltstone and shale

Oldest

Geologist's Sketch

Dennis Tasa

Principle of Original Horizontality

Steno is also credited with recognizing the importance of another basic principle, called the **principle of original horizontality**. It simply states that layers of sediment are generally deposited in a horizontal position. Thus, if we observe rock layers that are flat, we know that they have not been disturbed and still have their *original* horizontality. The layers in the Grand Canyon shown in Figures 8.1 and 8.2 illustrate this. But if layers are folded or inclined at a steep angle, they must have been moved into that position by crustal disturbances sometime *after* their deposition (**Figure 8.3**).

Figure 8.3 Original horizontality Most layers of sediment are deposited in a nearly horizontal position. When we see strata that are folded or tilted, we can assume that they were moved into that position by crustal disturbances *after* their deposition. (Photo by Marco Simoni/ Robert Harding World Imagery)

Principle of Lateral Continuity

The **principle of lateral continuity** refers to the fact that sedimentary beds originate as continuous layers that extend in all directions until they eventually grade into a different type of sediment or until they thin out at the edge of the basin of deposition (**Figure 8.4**). For example,

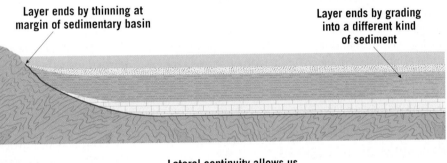

Layer ends by thinning at margin of sedimentary basin

Layer ends by grading into a different kind of sediment

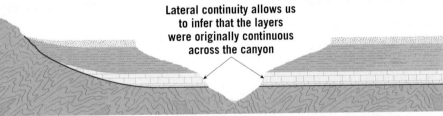

Lateral continuity allows us to infer that the layers were originally continuous across the canyon

Figure 8.4 Lateral continuity Sediments are deposited over a large area in a continuous sheet. Sedimentary strata extend continuously in all directions until they thin out at the edge of a depositional basin or grade into a different type of sediment.

when we look at the two walls of a canyon and see strata that are identical or similar, we can assume that those strata were continuous before the canyon was carved. Although rock outcrops may be separated by a considerable distance, the principle of lateral continuity tells us that they once formed a continuous layer. This principle allows geologists to relate rocks in isolated outcrops to one another. Combining the principles of lateral continuity and superposition lets us extend relative age relationships over broad areas. This process, called *correlation*, is examined later in the chapter.

Principle of Cross-Cutting Relationships

Figure 8.5 shows a mass of rock that is offset by a fault, a fracture in rock along which displacement occurs. It is clear that the rocks must be older than the fault that

Fault

SmartFigure 8.5 Cross-cutting fault The rocks are older than the fault that displaced them. (Morley Read/Alamy) (https://goo.gl/ BiFVHa)

Condor Video

Figure 8.6 Cross-cutting dikes An igneous intrusion is younger than the rocks that are intruded. (Photo by Jonathan. S Kt)

Dikes

broke them. The **principle of cross-cutting relationships** states that geologic features that cut across rocks must form *after* the rocks they cut through. Igneous intrusions provide another example. The dikes shown in **Figure 8.6** are tabular masses of igneous rock that cut through the surrounding rocks. The magmatic heat from igneous intrusions often creates a narrow "baked" zone of contact metamorphism on the adjacent rock, also indicating that the intrusion occurred after the surrounding rocks were in place.

Principle of Inclusions

Sometimes inclusions can aid in the relative dating process. **Inclusions** are fragments of one rock unit that have been enclosed within another. The principle of inclusions

is logical and straightforward. The rock mass adjacent to the one containing the inclusions must have been there first in order to provide the rock fragments. Therefore, the rock mass that contains the inclusions is the younger of the two. For example, when magma intrudes into surrounding rock, blocks of the surrounding rock may become dislodged and incorporated into the magma. If these pieces do not melt, they remain as inclusions, known as *xenoliths*. In another example, when sediment is deposited atop a weathered mass of bedrock, pieces of the weathered rock become incorporated into the younger sedimentary layer (**Figure 8.7**).

Unconformities

When we observe layers of rock that have been deposited essentially without interruption, we call them **conformable**. Particular sites exhibit conformable beds representing certain spans of geologic time. However, no place on Earth has a complete set of conformable strata.

Throughout Earth history, the deposition of sediment has been interrupted over and over again. All such breaks in the rock record are termed unconformities. An **unconformity** represents a long period during which deposition ceased, erosion removed previously formed rocks, and then deposition resumed. In each case, uplift and erosion are followed by subsidence and renewed sedimentation. Unconformities are important features because they represent significant geologic events in Earth history. Moreover, their recognition helps us identify what intervals of time are not represented by strata and thus are missing from the geologic record.

There are three basic types of unconformities.

Angular Unconformity Perhaps the most easily recognized unconformity is an **angular unconformity**. It consists of tilted or folded sedimentary rocks that are overlain by younger, more flat-lying strata. An angular unconformity indicates that during the pause in deposition, a period of deformation (folding or tilting) and erosion occurred (**Figure 8.8**).

When James Hutton studied an angular unconformity in Scotland more than 225 years ago, it was clear to him that it represented a major episode of geologic activity (**Figure 8.9**). He and his colleagues also appreciated the

SmartFigure 8.7 Inclusions The rock containing inclusions is younger than the inclusions. (https://goo.gl/Okfrm6)

Tutorial

These inclusions of igneous rock contained in the adjacent sedimentary layer indicate that the sediments were deposited atop the weathered igneous mass and thus are younger.

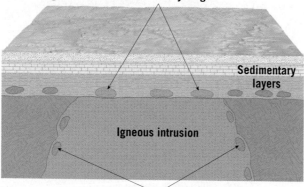

Sedimentary layers

Igneous intrusion

Xenoliths are inclusions in an igneous intrusion that form when pieces of surrounding rock are incorporated into magma.

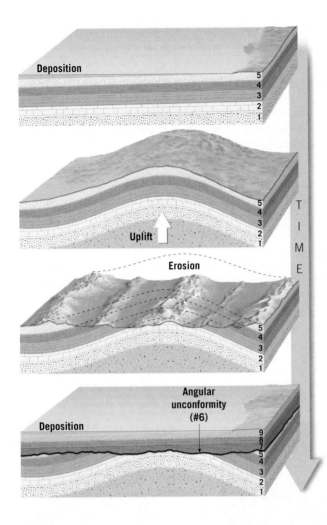

SmartFigure 8.8 Formation of an angular unconformity An angular unconformity represents an extended period during which deformation and erosion occurred. (https://goo.gl/arrwhC)

Figure 8.9 Siccar Point, Scotland James Hutton studied this famous unconformity in the late 1700s. (Photo by Marli Miller)

Above the unconformity lie gently dipping beds of reddish sandstone and conglomerate

Angular unconformity

Rock hammer

Below the unconformity lie nearly vertical beds of sandstone and shale

immense time span implied by such relationships. When a companion later wrote of their visit to this site, he stated that "the mind seemed to grow giddy by looking so far into the abyss of time."

Disconformity A **disconformity** is a gap in the rock record that represents a period during which erosion rather than deposition occurred. Imagine that a series of sedimentary layers are deposited in a shallow marine setting. Following this period of deposition, sea level falls or the land rises, exposing some of the sedimentary layers. During this span, when the sedimentary beds are above sea level, no new sediment accumulates, and some of the existing layers are eroded away. Later sea level rises or the land subsides, submerging the landscape. Now the surface is again below sea level, and a new series of sedimentary beds is deposited. The boundary separating the two sets of beds is a disconformity— a span for which there is no rock record (**Figure 8.10**). Because the layers above and below a disconformity are parallel, these features are sometimes difficult to identify unless you notice evidence of erosion such as a buried stream channel.

Disconformity
Gap in the rock record represents a period of nondeposition and erosion

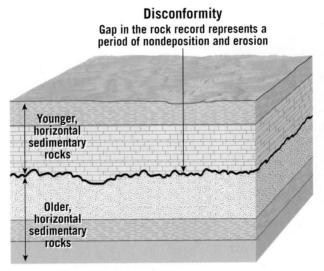

Younger, horizontal sedimentary rocks

Older, horizontal sedimentary rocks

Figure 8.10 Disconformity The layers on both sides of this gap in the rock record are essentially parallel.

Figure 8.11 Nonconformity Younger sedimentary rocks rest atop older metamorphic or igneous rocks.

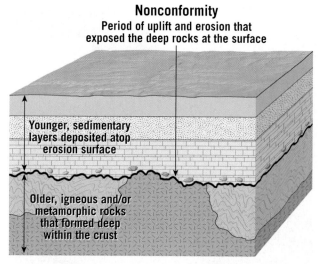

Nonconformity
Period of uplift and erosion that exposed the deep rocks at the surface

Younger, sedimentary layers deposited atop erosion surface

Older, igneous and/or metamorphic rocks that formed deep within the crust

Nonconformity The third basic type of unconformity is a **nonconformity**, in which younger sedimentary strata overlie older metamorphic or intrusive igneous rocks (**Figure 8.11**). Just as angular unconformities and some

disconformities imply crustal movements, so too do nonconformities. Intrusive igneous masses and metamorphic rocks originate far below the surface. Thus, for a nonconformity to develop, there must be a period of uplift and erosion of overlying rocks. Once the igneous or metamorphic rocks are exposed at the surface, they are subjected to weathering and erosion prior to subsidence and a period of sedimentation.

Unconformities in the Grand Canyon The rocks exposed in the Grand Canyon of the Colorado River represent a tremendous span of geologic history. It is a wonderful place to take a trip through time. The canyon's colorful strata record a long history of sedimentation in a variety of environments—advancing seas, rivers and deltas, tidal flats, and sand dunes. But the record is not continuous. Unconformities represent vast amounts of time that have not been recorded in the canyon's layers. **Figure 8.12** is a geologic cross section of the Grand Canyon. All three types of unconformities can be seen in the canyon walls.

Figure 8.12 Cross section of the Grand Canyon **All three types of unconformities are present.** (Center photo by Marli Miller; other photos by E. J. Tarbuck)

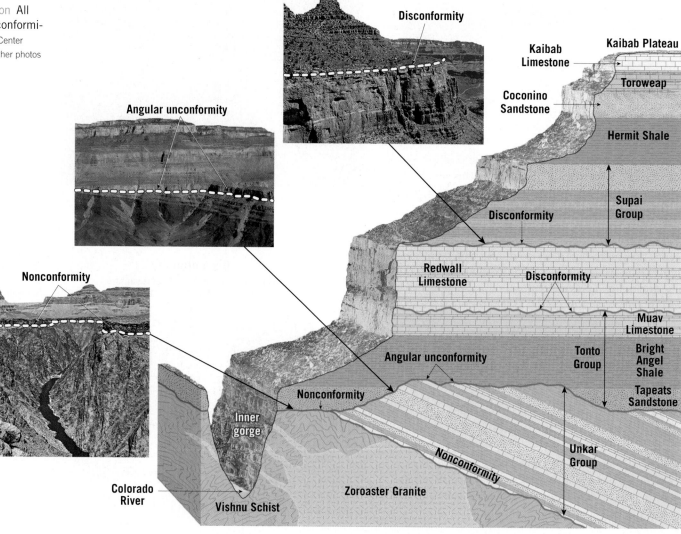

Disconformity

Angular unconformity

Nonconformity

Kaibab Plateau

Kaibab Limestone

Toroweap

Coconino Sandstone

Hermit Shale

Disconformity

Supai Group

Redwall Limestone

Disconformity

Muav Limestone

Angular unconformity

Tonto Group

Bright Angel Shale

Nonconformity

Tapeats Sandstone

Inner gorge

Unkar Group

Nonconformity

Colorado River

Vishnu Schist

Zoroaster Granite

SmartFigure 8.13 Applying principles of relative dating (https://goo.gl/w4HtAw)

▶ Tutorial

Working out the geologic history of a hypothetical region

Interpretation:

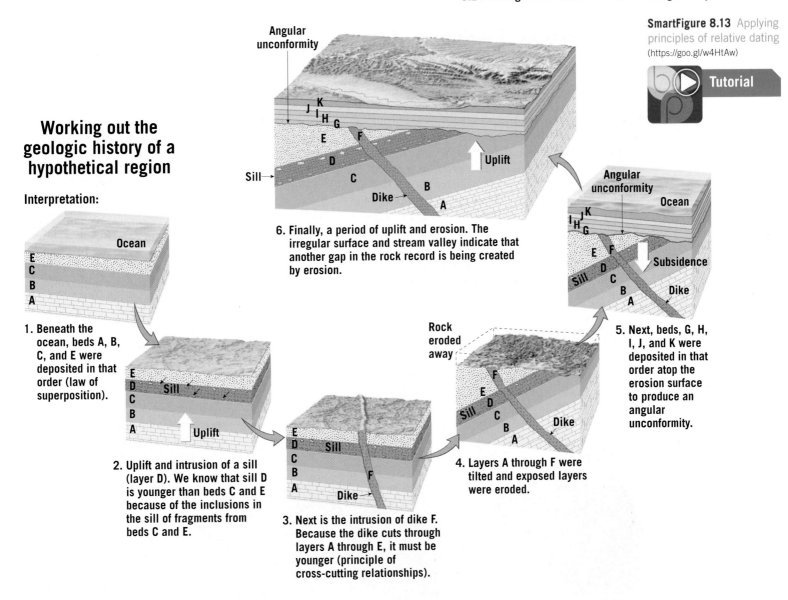

1. Beneath the ocean, beds A, B, C, and E were deposited in that order (law of superposition).

2. Uplift and intrusion of a sill (layer D). We know that sill D is younger than beds C and E because of the inclusions in the sill of fragments from beds C and E.

3. Next is the intrusion of dike F. Because the dike cuts through layers A through E, it must be younger (principle of cross-cutting relationships).

4. Layers A through F were tilted and exposed layers were eroded.

5. Next, beds, G, H, I, J, and K were deposited in that order atop the erosion surface to produce an angular unconformity.

6. Finally, a period of uplift and erosion. The irregular surface and stream valley indicate that another gap in the rock record is being created by erosion.

Applying Relative Dating Principles

By applying the principles of relative dating to the hypothetical geologic cross section shown in **Figure 8.13**, the rocks and the events in Earth history they represent can be placed into their proper sequence. The statements in the figure summarize the logic used to interpret the cross section. In this example, we establish a relative time scale for the rocks and events in the area of the cross section. Remember, we do not know how many years of Earth history are represented, nor do we know how this area compares to any other.

8.2 CONCEPT CHECKS

1. Distinguish between numerical dates and relative dates.
2. Sketch and label five simple diagrams that illustrate each of the following: superposition, original horizontality, lateral continuity, cross-cutting relationships, and inclusions.
3. What is the significance of an unconformity?
4. Distinguish among angular unconformity, disconformity, and nonconformity.

8.3 Fossils: Evidence of Past Life

Define *fossil* and discuss the conditions that favor the preservation of organisms as fossils. List and describe various fossil types.

Fossils, the remains or traces of prehistoric life, are important inclusions in sediment and sedimentary rocks. They are basic and important tools for interpreting the geologic past. The scientific study of fossils is called **paleontology**. It is an interdisciplinary science that blends geology and biology in an attempt to understand all aspects of the succession of life over the vast expanse of geologic time. Knowing the nature of the life-forms that existed at a particular time helps researchers understand past environmental conditions. Further, fossils are important time indicators and play a key role in correlating rocks of similar ages that are from different places.

Types of Fossils

Fossils are of many types. The remains of relatively recent organisms may not have been altered at all. Such objects as teeth, bones, and shells are common examples. Far less common are entire animals, flesh included, that have been preserved because of rather unusual circumstances. Remains of prehistoric elephants called mammoths that were frozen in the Arctic tundra of Siberia and Alaska are examples, as are the mummified remains of sloths preserved in a dry cave in Nevada.

Permineralization When mineral-rich groundwater permeates porous tissue such as bone or wood, minerals precipitate out of solution and fill pores and empty spaces in a process called *permineralization*. The formation of *petrified wood* involves permineralization with silica, often from a volcanic source such as a surrounding layer of volcanic ash. The wood is gradually transformed into chert, sometimes with colorful bands from impurities such as iron or carbon (**Figure 8.14A**). The word *petrified* literally means "turned into stone." Sometimes the microscopic details of the petrified structure are faithfully retained.

Molds and Casts Another common class of fossils is *molds* and *casts*. When a shell or another structure is buried in sediment and then dissolved by underground water, a *mold* is created. The mold faithfully reflects only the shape and surface marking of the organism; it does not reveal any information concerning its internal structure. If these hollow spaces are subsequently filled with mineral matter, *casts* are created (**Figure 8.14B**).

Carbonization and Impressions A type of fossilization called *carbonization* is particularly effective in preserving leaves and delicate animal forms. It occurs when fine sediment encases the remains of an organism. As time passes, pressure squeezes out the liquid and gaseous components and leaves behind a thin residue of carbon (**Figure 8.14C**). Black shale deposited as organic-rich mud in oxygen-poor environments often contains abundant carbonized remains. If the film of carbon is lost from a fossil preserved in fine-grained sediment, a replica of the surface, called an *impression*, may still show considerable detail (**Figure 8.14D**).

Figure 8.14 Types of fossils (Photo A by Bernhard Edmaier/Science Source; photo B by E. J. Tarbuck; photo C by Florissant Fossil Beds National Monument; photo D by E. J. Tarbuck; photo E by Colin Keates/Dorling Kindersley Ltd; photo F by E. J. Tarbuck)

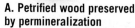

A. Petrified wood preserved by permineralization

B. A trilobite preserved as a mold and cast

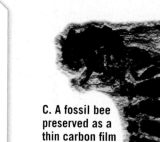

C. A fossil bee preserved as a thin carbon film

D. Fishes preserved as detailed impressions

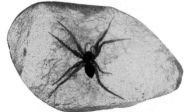

E. Spider preserved in amber

F. A coprolite (fossil dung)— an example of a trace fossil

Amber Delicate organisms, such as insects, are difficult to preserve, and consequently they are relatively rare in the fossil record. However, *amber*—the hardened resin of ancient trees—can preserve them in exquisite three-dimensional detail. The spider in **Figure 8.14E** was preserved after being trapped in a drop of sticky resin. Resin sealed off the spider from the atmosphere and protected the remains from damage by water and air. As the resin hardened, a protective pressure-resistant case was formed.

Trace Fossils In addition to the fossils already mentioned, there are numerous other types, many of them only traces of prehistoric life. Examples of such indirect evidence include:

- Tracks—animal footprints made in soft sediment that later turned into sedimentary rock.
- Burrows—tubes in sediment, wood, or rock made by an animal. These holes may later become filled with mineral matter and preserved.
- Coprolites—fossil dung and stomach contents that can provide useful information pertaining to the size and food habits of organisms (**Figure 8.14F**).
- Gastroliths—highly polished stomach stones that some organisms use in the grinding of food.

Conditions Favoring Preservation

Only a tiny fraction of the organisms that lived during the geologic past have been preserved as fossils. The remains of an animal or a plant are normally destroyed. Under what circumstances are they preserved? Two special conditions favor preservation: rapid burial and the possession of hard parts.

When an organism perishes, its soft parts are usually quickly eaten by scavengers or decomposed by bacteria. Occasionally, however, the remains are buried by sediment. When this occurs, the remains are protected from the environment, where destructive processes operate. Rapid burial, therefore, is an important condition favoring preservation.

In addition, animals and plants have a much better chance of being preserved as part of the fossil record if they have hard parts. Although traces and imprints of soft-bodied animals such as jellyfish, worms, and insects exist, they are not common. Flesh usually decays so rapidly that preservation is exceedingly unlikely. Hard parts such as shells, bones, and teeth predominate in the record of life in the past.

Because preservation is contingent on special conditions, the record of life in the geologic past is biased. The fossil record of those organisms with hard parts that lived in areas of sedimentation is quite abundant. However, we get only an occasional glimpse of the vast array of other life-forms that did not meet the special conditions favoring preservation.

8.3 CONCEPT CHECKS

1. Describe several ways that an animal or a plant can be preserved as a fossil.
2. List three examples of trace fossils.
3. What conditions favor the preservation of an organism as a fossil?

8.4 Correlation of Rock Layers

Explain how rocks of similar age that are in different places can be matched up.

To develop a geologic time scale that is applicable to the entire Earth, rocks of similar age in different regions must be matched up. Such a task is referred to as **correlation**. Correlating the rocks from one place to another makes possible a more comprehensive view of the geologic history of a region. **Figure 8.15**, for example, shows the correlation of strata at three sites on the Colorado Plateau in southern Utah and northern Arizona. No single locale exhibits the entire sequence, but correlation reveals a more complete picture of the sedimentary rock record.

Correlation Within Limited Areas

Within a limited area, correlating rocks of one locality with those of another may be done simply by walking along the outcropping edges. However, this may not be possible when the rocks are mostly concealed by soil and vegetation. Correlation over short distances is often achieved by noting the position of a bed in a sequence of strata. Or a layer may be identified in another location if it is composed of distinctive or uncommon minerals.

Many geologic studies involve relatively small areas. Although they are important in their own right, their full value is realized only when they are correlated with other regions. Although the methods just described are sufficient to trace a rock formation over relatively short distances, they are not adequate for matching up rocks that are separated by great distances. When correlation between widely separated areas or between continents is the objective, geologists must rely on fossils.

Fossils and Correlation

The existence of fossils had been known for centuries, but it was not until the late 1700s and early 1800s that their significance as geologic tools was made evident. During this period, an English engineer and canal builder, William Smith, discovered that each rock formation in the canals he worked on contained fossils unlike those in the

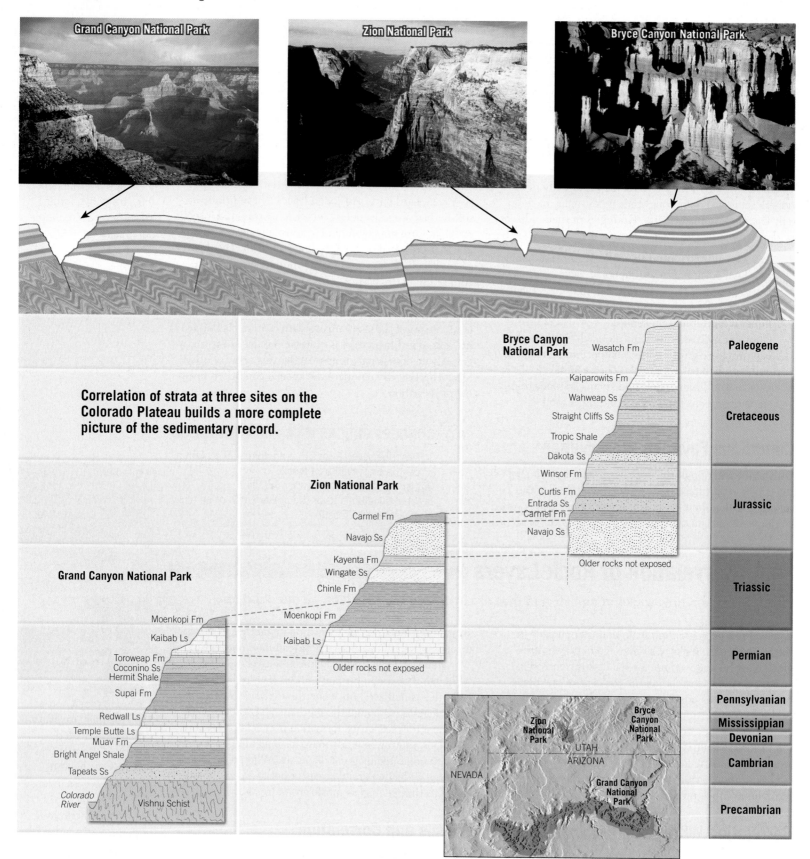

Figure 8.15 Correlation Matching strata at three locations on the Colorado Plateau. (Photos by E. J. Tarbuck)

beds either above or below. Further, he noted that sedimentary strata in widely separated areas could be identified—and correlated—by their distinctive fossil content.

Principle of Fossil Succession Based on Smith's classic observations and the findings of many geologists who followed, one of the most important and basic principles in historical geology was formulated: *Fossil organisms succeed one another in a definite and determinable order, and therefore any time period can be recognized by its fossil content.* This has come to be known as the **principle of fossil succession**. In other words, when fossils are arranged according to their age, they do not present a random or haphazard picture. To the contrary, fossils document the evolution of life through time.

For example, an Age of Trilobites is recognized quite early in the fossil record. Then, in succession, paleontologists recognize an Age of Fishes, an Age of Coal Swamps, an Age of Reptiles, and an Age of Mammals. These "ages" pertain to groups that were especially plentiful and characteristic during particular time periods. Within each of the "ages" are many subdivisions based, for example, on certain species of trilobites and certain types of fish, reptiles, and so on. This same succession of dominant organisms, never out of order, is found on every continent.

Index Fossils and Fossil Assemblages When fossils were found to be time indicators, they became the most useful means of correlating rocks of similar age in different regions. Geologists pay particular attention to certain fossils called **index fossils** (**Figure 8.16**). These fossils are widespread geographically and are limited to a short span of geologic time, so their presence provides an important method of matching rocks of the same age. Rock formations, however, do not always contain an index fossil. In such situations a group of fossils, called a **fossil assemblage**, is used to establish the age of the bed. **Figure 8.17** illustrates how an assemblage of fossils may be used to date rocks more precisely than could be accomplished by the use of any one of the fossils.

Environmental Indicators In addition to being important, and often essential, tools for correlation, fossils are important environmental indicators. Although much can be deduced about past environments by studying the nature and characteristics of sedimentary rocks, a close examination of the fossils present can usually provide a great deal more information. For example, when the remains of certain clam shells are found in limestone, a geologist quite reasonably assumes that the region was once covered by a shallow

Figure 8.16 Index fossils Since microfossils are often very abundant, widespread, and quick to appear and become extinct, they constitute ideal index fossils. This scanning electron micrograph shows marine microfossils from the Miocene epoch (see the geologic time scale in Figure 8.23). (Photo by Biophoto Associates/Science Source)

sea. Also, by using what we know of living organisms, we can conclude that fossil animals with thick shells, capable of withstanding pounding and surging waves, inhabited shorelines. On the other hand, animals with thin, delicate shells probably indicate deep, calm offshore waters. Hence, by looking closely at the types of fossils, the approximate position of an ancient shoreline may be identified.

Further, fossils can be used to indicate the former temperature of the water. Certain kinds of present-day corals must live in warm and shallow tropical seas like those around Florida and the Bahamas. When similar types of coral are found in ancient limestones, they indicate the marine environment that must have existed when they were alive. These examples illustrate how fossils can help unravel the complex story of Earth history.

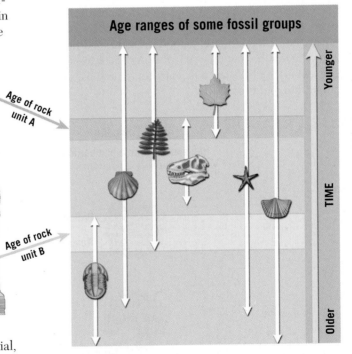

Age ranges of some fossil groups

SmartFigure 8.17 Fossil assemblage Overlapping ranges of fossils help date rocks more exactly than using a single fossil. (https://goo.gl/dUqgP3)

Tutorial

8.4 CONCEPT CHECKS

1. What is the goal of correlation?
2. State the principle of fossil succession in your own words.
3. Contrast *index fossil* and *fossil assemblage*.
4. In addition to being important time indicators, how else are fossils useful to geologists?

8.5 Determining Numerical Dates with Radioactivity

Discuss three types of radioactive decay and explain how radioactive isotopes are used to determine numerical dates.

In addition to establishing relative dates by using the principles described in the preceding sections, it is also possible to obtain reliable numerical dates for events in the geologic past. We know that Earth is about 4.6 billion years old and that dinosaurs became extinct about 66 million years ago. Dates that are expressed in millions and billions of years truly stretch our imagination because our personal calendars involve time measured in hours, weeks, and years. Nevertheless, the vast expanse of geologic time is a reality, and radiometric dating allows us to measure it accurately. In this section, you will learn about radioactivity and its application in radiometric dating.

Reviewing Basic Atomic Structure

Recall from Chapter 1 that each atom has a *nucleus* that contains protons and neutrons and that the nucleus is orbited by electrons. An *electron* has a negative electrical charge, and a *proton* has a positive charge. A *neutron* has no charge (it is electrically neutral), but it can be converted to a positively charged proton plus a negatively charged electron.

An element's *atomic number* (an element's identifying number) is the number of protons in the nucleus. Every element has a different number of protons in the nucleus and thus a different atomic number (hydrogen = 1, oxygen = 8, uranium = 92, etc.). Atoms of the same element always have the same number of protons, so the atomic number is constant.

Practically all (99.9 percent) of an atom's mass is found in the nucleus, indicating that electrons have practically no mass at all. By adding together the number of protons and neutrons in the nucleus, the *mass number* of the atom is determined. The number of neutrons in the nucleus can vary. These variants, called *isotopes*, have different mass numbers.

We can summarize with an example. Uranium's nucleus always has 92 protons, so its atomic number always is 92. But its neutron population varies, and uranium has three isotopes: uranium-234 (number of protons + neutrons = 234), uranium-235, and uranium-238. All three isotopes are found together in nature. They look the same and behave the same in chemical reactions.

Radioactivity

Some isotopes have a composition that makes them inherently unstable, so that they can change spontaneously to a different isotope or to a different element, frequently by absorbing or emitting one or more particles from their nuclei. We say these isotopes are **radioactive** and undergo **radioactive decay**.

Common Examples of Radioactive Decay What happens when unstable nuclei undergo decay? Three common types of radioactive decay are illustrated in **Figure 8.18** and are summarized as follows:

Figure 8.18 Common types of radioactive decay Notice that in each example, the number of protons (atomic number) in the nucleus changes, thus producing a different element.

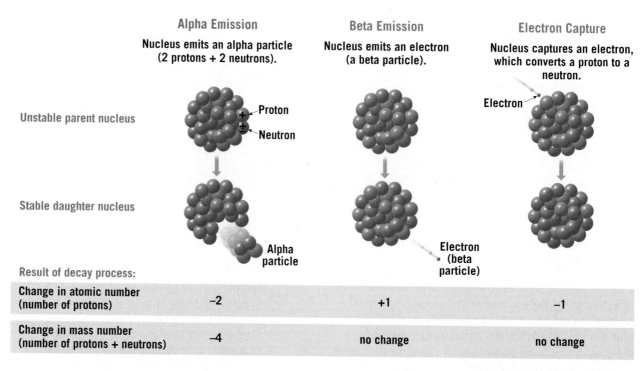

	Alpha Emission	Beta Emission	Electron Capture
	Nucleus emits an alpha particle (2 protons + 2 neutrons).	Nucleus emits an electron (a beta particle).	Nucleus captures an electron, which converts a proton to a neutron.
Change in atomic number (number of protons)	−2	+1	−1
Change in mass number (number of protons + neutrons)	−4	no change	no change

- *Alpha particles* (α particles) may be emitted from the nucleus. An alpha particle consists of 2 protons and 2 neutrons. Consequently, the emission of an alpha particle means that (a) the mass number of the isotope is reduced by 4 and (b) the atomic number is decreased by 2.
- When a *beta particle* (β particle), or an electron, is given off from a nucleus, the mass number remains unchanged because electrons have practically no mass. The electron that is emitted comes from a neutron, which causes this neutral particle to become a positively charged proton. Therefore, the atomic number increases by 1.
- Sometimes an electron is captured by the nucleus. The electron combines with a proton and forms an additional neutron. As in the last example, the mass number remains unchanged. However, because the nucleus now contains one fewer proton, the atomic number decreases by 1.

An unstable (radioactive) isotope is referred to as the *parent*. The isotopes resulting from the decay of the parent are the *daughter products*. **Figure 8.19** provides an example of radioactive decay. When the radioactive parent, uranium-238 (atomic number 92, mass number 238), decays, it follows a number of steps, emitting eight alpha particles and 6 beta particles before finally becoming the stable daughter product lead-206 (atomic number 82, mass number 206).

Radiometric Dating

Certainly among the most important properties of radioactivity is that it provides a reliable method of calculating the ages of rocks and minerals that contain particular radioactive isotopes. The procedure is called **radiometric dating**. Why is radiometric dating reliable? The rates of decay for many isotopes have been precisely measured and do not vary under the physical conditions that exist in Earth's outer layers. Therefore, each radioactive isotope used for dating has been decaying at a fixed rate ever since the formation of the rocks in which it occurs, and the products of decay have been accumulating at a corresponding rate. For example, when uranium is incorporated into a mineral that crystallizes from magma, there is no lead (the stable daughter product) from previous decay. The radiometric "clock" starts at this point. As the uranium in this newly formed mineral disintegrates, atoms of the daughter product are trapped, and measurable amounts of lead eventually accumulate.

Half-Life

The time required for one-half of the nuclei in a sample to decay is called the **half-life** of the isotope. Half-life is a common way of expressing the rate of radioactive disintegration. **Figure 8.20** illustrates what occurs when a

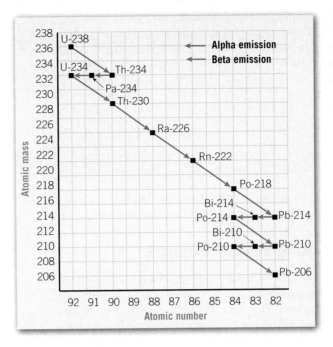

Figure 8.19 Decay of U-238 Uranium-238 is an example of a radioactive decay series. Before the stable end product (Pb-206) is reached, many different isotopes are produced as intermediate steps.

radioactive parent decays directly into its stable daughter product. When the quantities of parent and daughter are equal (ratio 1:1), we know that one half-life has transpired. When one-quarter of the original parent atoms remain and three-quarters have decayed to the daughter product, the parent–daughter ratio is 1:3, and we know that two half-lives have passed. After three half-lives, the ratio of parent atoms to daughter atoms is 1:7 (one parent atom for every seven daughter atoms).

If the half-life of a radioactive isotope is known and the parent–daughter ratio can be determined, the age of the sample can be calculated. For example, assume that the half-life of a hypothetical unstable isotope is 1 million years and the parent–daughter ratio in a sample is 1:15.

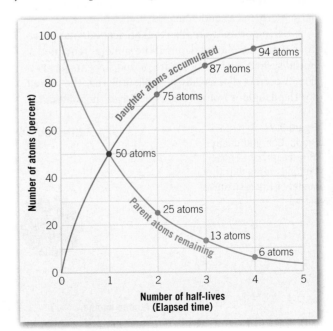

SmartFigure 8.20
Radioactive decay curve Change is exponential. Half of the radioactive parent remains after one half-life. After a second half-life, one-quarter of the parent remains, and so forth. (https://goo.gl/DFHxZg)

Tutorial

Table 8.1 Radioactive Isotopes Frequently Used in Radiometric Dating		
Radioactive Parent	**Stable Daughter Product**	**Currently Accepted Half-Life Values**
Uranium-238	Lead-206	4.5 billion years
Uranium-235	Lead-207	704 million years
Thorium-232	Lead-208	14.1 billion years
Rubidium-87	Strontium-87	47.0 billion years
Potassium-40	Argon-40	1.3 billion years

Such a ratio indicates that four half-lives have passed and that the sample must be 4 million years old.

Using Various Isotopes

Notice that the *percentage* of radioactive atoms that decay during one half-life is always the same: 50 percent. However, the *actual number* of atoms that decay with the passing of each half-life continually decreases. As the percentage of radioactive parent atoms declines, the proportion of stable daughter atoms rises, with the increase in daughter atoms just matching the drop in parent atoms. This fact is the key to radiometric dating.

Of the many radioactive isotopes that exist in nature, five have proved particularly important in providing radiometric ages for ancient rocks (**Table 8.1**). Rubidium-87, uranium-238, and uranium-235 are used for dating rocks that are millions of years old, but potassium-40 is more versatile. Although the half-life of potassium-40 is 1.3 billion years, analytical techniques make possible the detection of tiny amounts of its stable daughter product, argon-40, in some rocks that are younger than 100,000 years. Another important reason for its frequent use is that potassium is abundant in many common minerals, particularly micas and feldspars.

It is important to realize that an accurate radiometric date can be obtained only if the mineral remained a closed system during the entire period since its formation. A correct date is only possible if neither the parent nor the daughter isotope enters or leaves the mineral.

This is not always the case. In fact, an important limitation of the potassium–argon method arises from the fact that argon is a gas and may leak from minerals, throwing off measurements.

A Complex Process Remember that although the basic principle of radiometric dating is simple, the actual procedure is quite complex. The analysis that determines the quantities of parent and daughter must be painstakingly precise. In addition, some radioactive materials do not decay directly into the stable daughter product. As you saw in Figure 8.19, uranium-238 produces 13 intermediate unstable daughter products before the fourteenth and final daughter product, the stable isotope lead-206, is produced.

Earth's Oldest Rocks Radiometric dating methods have produced literally thousands of dates for events in Earth history. Rocks exceeding 3.5 billion years in age are found on all of the continents. Earth's oldest rocks (so far) may be as old as 4.28 billion years (b.y.). Discovered in northern Quebec, Canada, on the shores of Hudson Bay, these rocks may be remnants of Earth's earliest crust. Rocks from western Greenland have been dated at 3.7 to 3.8 b.y., and rocks nearly as old are found in the Minnesota River valley and northern Michigan (3.5 to 3.7 b.y.), in southern Africa (3.4 to 3.5 b.y.), and in western Australia (3.4 to 3.6 b.y.). Tiny crystals of the mineral zircon with radiometric ages as old as 4.4 b.y. have been found in younger sedimentary rocks in western Australia. The source rocks for these tiny durable grains either no longer exist or have not yet been found.

Dating with Carbon-14

To date very recent events, carbon-14 is used. Carbon-14 is the radioactive isotope of carbon. The process is often called **radiocarbon dating**. Because the half-life of carbon-14 is only 5,730 years, it can be used for dating events from the historic past as well as those from very recent geologic history. In some cases, carbon-14 can be used to date events as far back as 70,000 years.

Figure 8.21 Carbon-14 Production and decay of radiocarbon. These sketches represent the nuclei of the respective atoms.

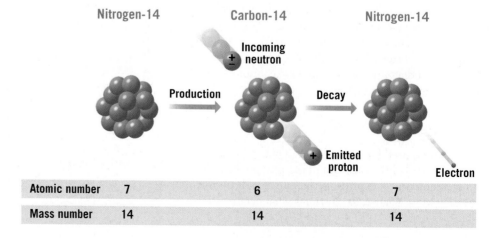

	Nitrogen-14	Carbon-14	Nitrogen-14
Atomic number	7	6	7
Mass number	14	14	14

The production and decay of carbon-14 is illustrated in **Figure 8.21**. Carbon-14 is continuously produced in the upper atmosphere as a result of cosmic-ray bombardment. Cosmic rays, which are high-energy particles, shatter the nuclei of gas atoms, releasing neutrons. Some of the neutrons are absorbed by nitrogen atoms (atomic number 7), causing their nuclei to emit a proton. As a result, the atomic number decreases by 1 (to 6), and a different element, carbon-14, is created. This isotope of carbon quickly becomes incorporated into carbon dioxide, which circulates in the atmosphere and is absorbed by living matter. As a result, all organisms—including you—contain a small amount of carbon-14.

While an organism is alive, the decaying radiocarbon is continually replaced, and the proportions of carbon-14 and carbon-12 remain constant. Carbon-12 is the stable and most common isotope of carbon. However, when any plant or animal dies, the amount of carbon-14 gradually decreases as it decays to nitrogen-14 by beta emission. By comparing the proportions of carbon-14 and carbon-12 in a sample, radiocarbon dates can be determined.

Although carbon-14 is useful in dating only the last small fraction of geologic time, it has become a valuable tool for anthropologists, archaeologists, and historians, as well as for geologists who study very recent Earth history (**Figure 8.22**). In fact, the development of radiocarbon dating was considered so important that the chemist who discovered this application, Willard F. Libby, received a Nobel Prize for it.

Figure 8.22 Cave art Chauvet Cave in southern France, discovered in 1994, contains some of the earliest-known cave paintings. Radiocarbon dating indicates that most of the images were drawn between 30,000 and 32,000 years ago. (Photo by Javier Trueba/MSF/Science Source)

 8.5 CONCEPT CHECKS

1. List three types of radioactive decay. For each type, describe how the atomic number and the atomic mass change.
2. Sketch a simple diagram that explains the idea of half-life.
3. Why is radiometric dating a reliable method for determining numerical dates?
4. For what time span does radiocarbon dating apply?

8.6 The Geologic Time Scale

Distinguish among the four basic time units that make up the geologic time scale and explain why the time scale is considered to be a dynamic tool.

Geologists have divided the whole of geologic history into units of varying magnitude. Together, they comprise the **geologic time scale** of Earth history (**Figure 8.23**). The major units of the time scale were determined during the nineteenth century, principally by scientists in Western Europe and Great Britain. Because radiometric dating was unavailable at that time, the entire time scale was created using methods of relative dating. It was only in the twentieth century that radiometric dating permitted numerical dates to be added.

Structure of the Time Scale

The geologic time scale divides the 4.6-billion-year history of Earth into many different units and provides a meaningful time frame within which the events of the geologic past are arranged. As shown in Figure 8.23, **eons** represent the greatest expanses of time. The eon that began about 541 million years ago is the **Phanerozoic**, a term derived from Greek words meaning "visible life." It is an appropriate description because the rocks and deposits of

Figure 8.23 Geologic time scale: A basic reference The time scale divides the vast 4.6-billion-year history of Earth into eons, eras, periods, and epochs. Numbers on the time scale represent time in millions of years before the present. The Precambrian accounts for more than 88 percent of geologic time. Numerical dates were added long after the time scale was established using relative dating techniques. The dates appearing on this time scale are those currently accepted by the International Committee on Stratigraphy (ICS) in 2015. The color scheme used on this chart was selected because it is similar to that used by the ICS.

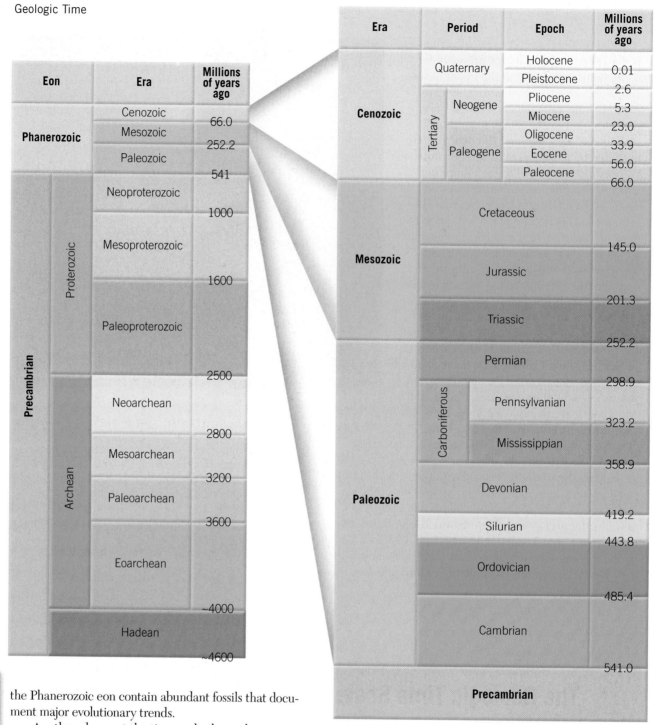

the Phanerozoic eon contain abundant fossils that document major evolutionary trends.

Another glance at the time scale shows that eons are divided into **eras**. The three eras within the Phanerozoic are the **Paleozoic** ("ancient life"), the **Mesozoic** ("middle life"), and the **Cenozoic** ("recent life"). As the names imply, these eras are bounded by profound worldwide changes in life-forms.

Each era of the Phanerozoic eon is divided into units known as **periods**. The Paleozoic has seven, and the Mesozoic and Cenozoic each have three. Each of these periods is characterized by a somewhat less profound change in life-forms as compared with the eras.

Each of the periods is divided into still smaller units called **epochs**. As you can see in Figure 8.23, seven epochs have been named for the periods of the Cenozoic era. The epochs of other periods, however, are not usually referred to by specific names. Instead, the terms *early*, *middle*, and *late* are generally applied to the epochs of these earlier periods.

Precambrian Time

Notice that the geologic time scale is considerably less detailed prior to the beginning of the Cambrian period, 541 million years ago. The 4 billion years that precede the Cambrian are divided into two eons, the *Archean* and the *Proterozoic*, which together are divided into seven eras. It is also common for this vast expanse of time to simply be referred to as the **Precambrian**.

Why is the huge expanse of Precambrian time, which represents about 88 percent of Earth history, not divided into numerous eras, periods, and epochs? The reason is that Precambrian history is not known in enough detail. In geology, as in human history, the farther back we go, the less we know. We know much more about the past decade than about the first century A.D. Equally, in Earth history, the more recent past has the freshest, least disturbed, and most observable record. The farther back in time a geologist goes, the more fragmented the record and clues become.

Terminology and the Geologic Time Scale

Some terms are associated with the geologic time scale but are not officially recognized as being a part of it. The best-known and the most common example is *Precambrian*—the informal name for the eons that came before the current Phanerozoic eon. Although the term *Precambrian* has no formal status on the geologic time scale, it has been traditionally used as though it does.

Hadean is another informal term that is found on some versions of the geologic time scale and is used by some geologists. It refers to the earliest interval (eon) of Earth history—before the oldest-known rocks. When the term was coined in 1972, the age of Earth's oldest-known rocks was about 3.8 billion years. Today that number stands at slightly greater than 4 billion, and, of course, is subject to revision. The name *Hadean* derives from *Hades*, Greek for "underworld"—a reference to the "hellish" conditions that prevailed on Earth early in its history.

Effective communication in the geosciences requires that the geologic time scale consist of standardized divisions and dates. So, who determines which names and dates on the geologic time scale are "official"? The organization that is largely responsible for maintaining and updating this important document is the International Committee on Stratigraphy (ICS), a committee of the International Union of Geological Sciences.° Advances in the geosciences require that the scale be periodically updated to include changes in unit names and boundary age estimates.

If you were to examine a geologic time scale from just a few years ago, it is quite possible that you would see the Cenozoic era divided into the Tertiary and Quaternary periods. However, on more recent versions, the space formerly designated as Tertiary is divided into the Paleogene and Neogene periods. As our understanding of this time span has changed, so too has its designation on the geologic time scale. Today, the Tertiary period is considered a "historic" name and is given no official status on the ICS version of the time scale. Many time scales still contain references to the Tertiary period, though, including Figure 8.23. One reason for this is that a great deal of past (and some current) geologic literature uses this name.

For those who study historical geology, it is important to realize that the geologic time scale is a dynamic tool that continues to be refined as our knowledge and understanding of Earth history evolves.

8.6 CONCEPT CHECKS

1. What are the four basic units that make up the geologic time scale? List the specific ones that apply to the present day.
2. Why is *zoic* part of so many names on the geologic time scale?
3. What term applies to *all* geologic time prior to the Phanerozoic eon? Why is this span not divided into as many smaller units as the Phanerozoic eon?
4. To what does *Hadean* apply? It is an "official" part of the time scale?

° To view the current version of the ICS time scale, go to www.stratigraphy.org. *Stratigraphy* is the branch of geology that studies rock layers (strata) and layering (stratification), and thus its primary focus is sedimentary and layered volcanic rocks.

8.7 Determining Numerical Dates for Sedimentary Strata

Explain how reliable numerical dates are determined for layers of sedimentary rock.

Although reasonably accurate numerical dates have been worked out for the periods of the geologic time scale, the task is not without difficulty. The primary problem in assigning numerical dates is the fact that not all rocks can be dated using radiometric methods. For a radiometric date to tell us the age of a rock, all minerals in the rock must have formed at about the same time. For this reason, radioactive isotopes can be used to determine when minerals in an igneous rock crystallized and when pressure and heat created new minerals in a metamorphic rock.

However, samples of sedimentary rock can only rarely be dated directly by radiometric means. A sedimentary rock may include particles that contain

Figure 8.24 Dating sedimentary strata Numerical dates for sedimentary layers are usually determined by examining their relationship to igneous rocks. Based on this diagram, what can you determine about the numerical age of the Dakota Sandstone?

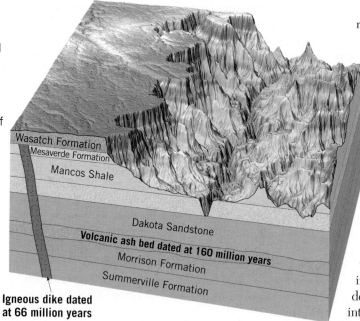

Wasatch Formation
Mesaverde Formation
Mancos Shale
Dakota Sandstone
Volcanic ash bed dated at 160 million years
Morrison Formation
Summerville Formation

Igneous dike dated at 66 million years

radioactive isotopes, but the rock's age cannot be accurately determined because the grains that make up the rock are not the same age as the rock in which they occur. Rather, the sediments have been weathered from rocks of diverse ages.

Radiometric dates obtained from metamorphic rocks may also be difficult to interpret because the age of a particular mineral in a metamorphic rock does not necessarily represent the time when the rock initially formed. Instead, the date may indicate any one of a number of subsequent metamorphic phases.

If samples of sedimentary rocks rarely yield reliable radiometric ages, how can numerical dates be assigned to sedimentary layers? Usually a geologist must relate them to datable igneous masses, as in **Figure 8.24**. In this example, radiometric dating has determined the ages of the volcanic ash bed and the dike. The sedimentary beds below the ash are older than the ash, whereas the layers above the ash are younger (based on the principle of superposition). The dike is younger than all of the layers it cuts through but is older than the Wasatch Formation because the dike does not intrude this topmost layer (based on the principle of cross-cutting relationships).

From this kind of evidence, geologists know that the Summerville and Morrison Formations were deposited more than 160 million years ago, as indicated by the ash bed. Further, they conclude that deposition of the Wasatch Formation began after the intrusion of the dike, 66 million years ago. This is one example of literally thousands that illustrate how datable materials are used to "bracket" the various episodes in Earth history within specific time periods. It shows the necessity of combining laboratory dating methods with field observations of rocks.

8.7 **CONCEPT CHECKS**

1. Briefly explain why it is often difficult to assign a reliable numerical date to a sample of sedimentary rock.
2. How might a numerical date for a layer of sedimentary rock be determined?

CONCEPTS IN REVIEW
Geologic Time

8.1 A Brief History of Geology

Explain the principle of uniformitarianism and discuss how it differs from catastrophism.

KEY TERMS: catastrophism, uniformitarianism

- Early ideas about the nature of Earth were based on religious traditions and notions of great catastrophes.
- In the late 1700s, James Hutton emphasized that the same slow processes have acted over great spans of time and are responsible for Earth's rocks, mountains, and landforms. This similarity of processes over vast spans of time led to this principle being called uniformitarianism.

8.2 Creating a Time Scale—Relative Dating Principles

Distinguish between numerical and relative dating and apply relative dating principles to determine a time sequence of geologic events.

KEY TERMS: numerical date, relative date, principle of superposition, principle of original horizontality, principle of lateral continuity, principle of cross-cutting relationships, inclusion, conformable, unconformity, angular unconformity, disconformity, nonconformity

- The two types of dates that geologists use to interpret Earth history are (1) relative dates, which put events in their proper sequence of formation, and (2) numerical dates, which pinpoint the time in years when an event took place.
- Relative dates can be established using the principles of superposition, original horizontality, lateral continuity, cross-cutting relationships, and inclusions. Unconformities, gaps in the geologic record, may be identified during the relative dating process.

(?) **The accompanying photo shows four features. Place the features in the proper sequence, from oldest to youngest. Explain your reasoning.**

Mike Beauregard, Split Thrice

8.3 Fossils: Evidence of Past Life

Define *fossil* and discuss the conditions that favor the preservation of organisms as fossils. List and describe various fossil types.

KEY TERMS: fossil, paleontology

- Fossils are remains or traces of ancient life. Paleontology is the branch of science that studies fossils.

(8.3 continued)

- Fossils can form through many processes. For an organism to be preserved as a fossil, it usually needs to be buried rapidly. Also, an organism's hard parts are most likely to be preserved because soft tissue decomposes rapidly in most circumstances.

E.J. Tarbuck

(?) **What term is used to describe the type of fossil that is shown here? Briefly describe how it formed.**

8.4 Correlation of Rock Layers

Explain how rocks of similar age that are in different places can be matched up.

KEY TERMS: correlation, principle of fossil succession, index fossil, fossil assemblage

- Matching up exposures of rock that are the same age but are in different places is called correlation. By correlating rocks from around the world, geologists developed the geologic time scale and obtained a fuller perspective on Earth history.
- Fossils can be used to correlate sedimentary rocks in widely separated places by using the rocks' distinctive fossil content and applying the principle of fossil succession. The principle states that fossil organisms succeed one another in a definite and determinable order, and, therefore, a time period can be recognized by examining its fossil content.
- Index fossils are particularly useful in correlation because they are widespread and associated with a relatively narrow time span. The overlapping ranges of fossils in an assemblage may be used to establish an age for a rock layer that contains multiple fossils.
- Fossils may be used to establish ancient environmental conditions that existed when sediment was deposited.

8.5 Determining Numerical Dates with Radioactivity

Discuss three types of radioactive decay and explain how radioactive isotopes are used to determine numerical dates.

KEY TERMS: radioactivity, radioactive decay, radiometric dating, half-life, radiocarbon dating

- Radioactivity is the spontaneous decay of certain unstable atomic nuclei. Three common forms of radioactive decay are (1) emission of an alpha particle from the nucleus, (2) emission of a beta particle (electron) from the nucleus, and (3) capture of an electron by the nucleus.
- An unstable radioactive isotope, called a parent, will decay and form daughter products. The length of time for one-half of the nuclei of a radioactive isotope to decay is called the half-life of the isotope. If the half-life of an isotope is known, and the parent–daughter ratio can be measured, the age of the sample can be calculated.

8.6 The Geologic Time Scale

Distinguish among the four basic time units that make up the geologic time scale and explain why the time scale is considered to be a dynamic tool.

KEY TERMS: geologic time scale, eon, Phanerozoic eon, era, Paleozoic era, Mesozoic era, Cenozoic era, period, epoch, Precambrian

- Earth history is divided into units of time on the geologic time scale. Eons are divided into eras, which each contain multiple periods. Periods are divided into epochs.
- Precambrian time includes the Archean and Proterozoic eons. It is followed by the Phanerozoic eon, which is well documented by abundant fossil evidence, resulting in many subdivisions.
- The geologic time scale is a work in progress, continually being refined as new information becomes available.

(?) **Is the Mesozoic an example of an eon, an era, a period, or an epoch? What about the Jurassic?**

8.7 Determining Numerical Dates for Sedimentary Strata

Explain how reliable numerical dates are determined for layers of sedimentary rock.

- Sedimentary strata are usually not directly datable using radiometric techniques because they consist of the material produced by the weathering of other rocks. A particle in a sedimentary rock comes from some older source rock. If you were to date the particle using isotopes, you would get the age of the source rock, not the age of the sedimentary deposit.
- One way geologists assign numerical dates to sedimentary rocks is to use relative dating principles to relate them to datable igneous masses, such as dikes, lava flows, and volcanic ash beds. A layer may be older than one igneous feature and younger than another.

(?) **Determine the age of the sandstone layer in the figure as accurately as possible.**

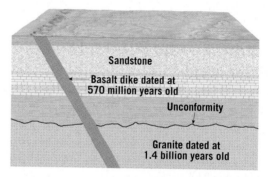

Sandstone

Basalt dike dated at 570 million years old

Unconformity

Granite dated at 1.4 billion years old

GIVE IT SOME THOUGHT

1. The accompanying image shows the metamorphic rock gneiss, a basaltic dike, and a fault. Place these three features in their proper sequence (which came first, second, and third) and explain your logic.

Gneiss

Dike

Fault

Dike

Gneiss

Marli Miller

2. A mass of granite is in contact with a layer of sandstone. Using a principle described in this chapter, explain how you might determine whether the sandstone was deposited on top of the granite or whether the magma that formed the granite was intruded after the sandstone was deposited.

3. This scenic image is from Monument Valley in the northeastern corner of Arizona. The bedrock in this region consists of layers of sedimentary rocks. Although the prominent rock exposures ("monuments") in this photo are widely separated, we can infer that they represent a once-continuous layer. Discuss the principle that allows us to make this inference.

Michael Collier

4. The accompanying photo shows two layers of sedimentary rock. The lower layer is shale from the late Mesozoic era. Note the old river channel that was carved into the shale after it was deposited. Above is a younger layer of boulder-rich breccia. Are these layers conformable? Explain why or why not. What term from relative dating applies to the line separating the two layers?

Callan Bentley

5. Refer to Figure 8.9, which shows the historic angular unconformity at Scotland's Siccar Point that James Hutton studied in the late 1700s. Refer to this photo for the following exercises.
   a. Describe in general what occurred to produce this feature.
   b. Suggest ways in which at least three of Earth's four spheres could have been involved.
   c. The Earth system is powered by energy from two sources. How are both sources represented in the Siccar Point unconformity?

6. These polished stones are called *gastroliths*. Explain how such objects can be considered fossils. What category of fossil are they? Name another example of a fossil in this category.

Francois Gohier/Science Source

7. If a radioactive isotope of thorium (atomic number 90, mass number 232) emits 6 alpha particles and 4 beta particles during the course of radioactive decay, what are the atomic number and mass number of the stable daughter product?

8. A hypothetical radioactive isotope has a half-life of 10,000 years. If the ratio of radioactive parent to stable daughter product is 1:3, how old is the rock that contains the radioactive material?

9. Solve the problems below that relate to the magnitude of Earth history. To make calculations easier, round Earth's age to 5 billion years.
   a. What percentage of geologic time is represented by recorded history? (Assume 5000 years for the length of recorded history.)
   b. Humans and their close relatives (hominins) have been around for roughly 5 million years. What percentage of geologic time is represented by the history of this group?
   c. The first abundant fossil evidence does not appear until the beginning of the Cambrian period, about 540 million years ago. What percentage of geologic time is represented by abundant fossil evidence?

10. A portion of a popular college text in historical geology includes 10 chapters (281 pages) in a unit titled "The Story of Earth." Two chapters (49 pages) are devoted to Precambrian time. By contrast, the last two chapters (67 pages) focus on the most recent 23 million years, with 25 of those pages devoted to the Holocene Epoch, which began 10,000 years ago.
   a. Compare the percentage of pages devoted to the Precambrian to the actual percentage of geologic time that this span represents.
   b. How does the number of pages about the Holocene compare to its actual percentage of geologic time?
   c. Suggest some reasons why the text seems to have such an unequal treatment of Earth history.

11. The accompanying diagram is a cross section of a hypothetical area. Place the lettered features in the proper sequence, from oldest to youngest. Where in the sequence can you identify an unconformity?

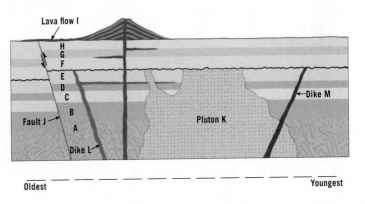

MasteringGeology™

9

FOCUS ON CONCEPTS

Each statement represents the primary learning objective for the corresponding major heading within the chapter. After you complete the chapter, you should be able to:

9.1 Discuss the extent and distribution of oceans and continents on Earth. Identify Earth's four main ocean basins.

9.2 Define *salinity* and list the main elements that contribute to the ocean's salinity. Describe the sources of dissolved substances in seawater and causes of variations in salinity.

9.3 Discuss temperature, salinity, and density changes with depth in the open ocean.

9.4 Define *bathymetry* and summarize the various techniques used to map the ocean floor.

9.5 Compare a passive continental margin with an active continental margin and list the major features of each.

9.6 List and describe the major features associated with deep-ocean basins.

9.7 Summarize the basic characteristics of oceanic ridges.

9.8 Distinguish among three categories of seafloor sediment and explain why some of these sediments can be used to study climate change.

Sunset at Fuerteventura in the Canary Islands in the eastern Atlantic Ocean. (Photo by Michael Weber/imageBroker/Alamy)

OCEANS: THE LAST FRONTIER

How much of Earth is covered by the ocean? What is the difference between freshwater and seawater? How deep is the ocean? What does the seafloor look like? Answers to these and other questions about the oceans and the basins they occupy are sometimes elusive and often difficult to determine. Suppose that all of the water were drained from the ocean. What would we see? Plains? Mountains? Canyons? Plateaus? Indeed, the ocean conceals all these features—and more. And what about the carpet of sediment that covers most of the ocean floor? Where did it come from, and what can be learned by studying it? This chapter provides answers to these questions.

9.1 The Vast World Ocean

Discuss the extent and distribution of oceans and continents on Earth. Identify Earth's four main ocean basins.

Oceans are a major part of our planet. In fact, Earth is frequently called the *water planet* or the *blue planet*. There is a good reason for these names: Nearly 71 percent of Earth's surface is covered by the global ocean (**Figure 9.1**). The focus of this chapter and Chapter 10 is oceanography. **Oceanography** is an interdisciplinary science that draws on the methods and knowledge of geology, chemistry, physics, and biology to study all aspects of the world ocean.

Geography of the Oceans

The area of Earth is about 510 million square kilometers (197 million square miles). Of this total, approximately 360 million square kilometers (140 million square miles),

or 71 percent, is represented by oceans and marginal seas (seas around the ocean's margin, such as the Mediterranean Sea and the Caribbean Sea). Continents and islands comprise the remaining 29 percent, or 150 million square kilometers (58 million square miles).

By studying a globe or world map, we can easily see that the continents and oceans are not evenly divided between the Northern and Southern Hemispheres (see Figure 9.1). When we compute the percentages of land and water in the Northern Hemisphere, we find that nearly 61 percent of the surface is water, and about 39 percent is land. In the Southern Hemisphere, on the other hand, almost 81 percent of the surface is water, and only 19 percent is land. It is no wonder then that the Northern Hemisphere is called the *land hemisphere* and the Southern Hemisphere the *water hemisphere*.

Figure 9.2A shows the distribution of land and water in the Northern and Southern Hemispheres by way of a graph. Between latitudes 45° north and 70° north, there is actually more land than water, whereas between 40° south and 65° south, there is almost no land to interrupt the oceanic and atmospheric circulation.

The world ocean can be divided into four main ocean basins (**Figure 9.2B**):

1. The *Pacific Ocean*, which is the largest ocean (and the largest single geographic feature on the planet), covers more than half of the ocean surface area on Earth. In fact, the Pacific Ocean is so large that all of the continents could fit into the space occupied by it—with room left over! It is also the world's deepest ocean, with an average depth of 3940 meters (12,927 feet).

2. The *Atlantic Ocean* is about half the size of the Pacific Ocean and is not quite as deep. It is a relatively narrow ocean compared to the Pacific and is bounded by almost parallel continental margins.

3. The *Indian Ocean* is slightly smaller than the Atlantic Ocean but has about the same average depth. Unlike the Pacific and Atlantic Oceans, it is largely a Southern Hemisphere water body.

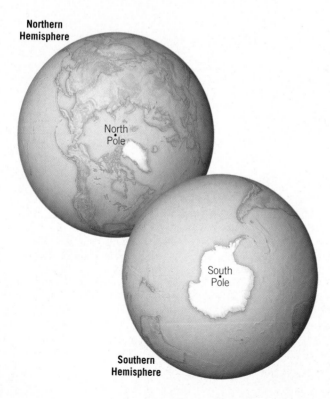

Figure 9.1 North versus south These views of Earth show the uneven distribution of land and water between the Northern and Southern Hemispheres. Almost 81 percent of the Southern Hemisphere is covered by the oceans—20 percent more than the Northern Hemisphere.

Northern Hemisphere

North Pole

South Pole

Southern Hemisphere

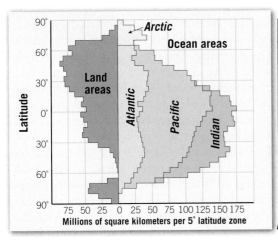

SmartFigure 9.2 Distribution of land and water **A.** The graph shows the amount of land and water in each 5° latitude belt. **B.** The world map provides a more familiar view. (http://goo.gl/xvv2Rv)

4. The *Arctic Ocean* is about 7 percent of the size of the Pacific Ocean and is only a little more than one-quarter as deep as the rest of the oceans.

Oceanographers also recognize an additional ocean near the continent of Antarctica in the Southern Hemisphere. Defined by the meeting of currents near Antarctica called the Antarctic Convergence, the *Southern Ocean*, or *Antarctic Ocean*, is actually those portions of the Pacific, Atlantic, and Indian Oceans south of about 50° south latitude.

Comparing the Oceans to the Continents

A major difference between continents and the ocean basins is their relative levels. The average elevation of the continents above sea level is about 840 meters (2756 feet), whereas the average depth of the oceans is nearly four and a half times this amount—3729 meters (12,234 feet). The volume of ocean water is so large that if Earth's solid mass were perfectly smooth (level) and spherical, the oceans would cover Earth's entire surface to a uniform depth of more than 2000 meters (1.2 miles)!

9.1 CONCEPT CHECKS

1. How does the area of Earth's surface covered by oceans compare with the area covered by continents?
2. Contrast the distribution of land and water in the Northern Hemisphere and the Southern Hemisphere.
3. Excluding the Southern Ocean, name the four main ocean basins. Contrast them in terms of area and depth.
4. How does the average depth of the oceans compare to the average elevation of the continents?

Did You Know?
The Bering Sea is the most northerly marginal sea of the Pacific Ocean and connects to the Arctic Ocean through the Bering Strait. This water body is effectively cut off from the Pacific basin by the Aleutian Islands, which were created by volcanic activity associated with northward subduction of the Pacific plate.

9.2 Composition of Seawater

Define *salinity* and list the main elements that contribute to the ocean's salinity. Describe the sources of dissolved substances in seawater and causes of variations in salinity.

What is the difference between freshwater and seawater? One of the most obvious differences is that seawater contains dissolved substances that give it a distinctly salty taste. These dissolved substances are not simply sodium chloride (table salt); they include various other salts, metals, and even dissolved gases. In fact, every known naturally occurring element is found dissolved in at least trace amounts in seawater. Unfortunately, the salt content of seawater makes it unsuitable for drinking or for irrigating most crops and causes it to be highly corrosive to many materials. Yet, many parts of the ocean are teeming with life that is superbly adapted to the marine environment.

Salinity

Seawater consists of about 3.5 percent (by weight) dissolved mineral substances that are collectively termed *salts*. Although the percentage of dissolved components may seem small, the actual quantity is huge because the ocean is so vast.

Salinity is the total amount of solid material dissolved in water. More specifically, it is the ratio of the mass of dissolved substances to the mass of the water sample. Many common quantities are expressed in percent (%), which is really *parts per hundred*. Because the proportion of dissolved substances in seawater is such a

Figure 9.3 Composition of seawater The diagram shows the relative proportions of water and dissolved components in a typical sample of seawater.

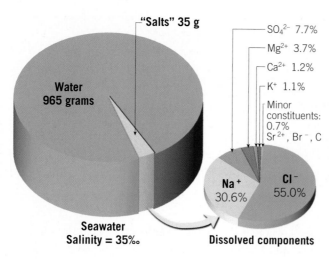

Seawater Salinity = 35‰

Dissolved components

small number, oceanographers typically express salinity *in parts per thousand* (‰). Thus, the average salinity of seawater is 3.5%, or 35‰.

Figure 9.3 shows the principal elements that contribute to the ocean's salinity. Artificial seawater could be approximated by following the recipe in **Table 9.1**. This table shows that most of the salt in seawater is sodium chloride—common table salt. Sodium chloride together with the next four most abundant salts comprise more than 99 percent of all dissolved substances in the sea. Although only seven elements make up these five most abundant salts, seawater contains all of Earth's other naturally occurring elements. Despite their presence in minute quantities, many of these elements are very important in maintaining the necessary chemical environment for life in the sea.

Sources of Sea Salts

What are the primary sources for the vast quantities of dissolved substances in the ocean? Chemical weathering of rocks on the continents is one source. These dissolved materials are delivered to the oceans by streams at an estimated rate of more than 2.5 billion tons annually. The second major source of elements found in ocean water is

Earth's interior. Through volcanic eruptions, large quantities of water and dissolved gases have been emitted during much of geologic time. This process, called *outgassing*, is the principal source of water in the oceans and in the atmosphere. Certain elements—notably chlorine, bromine, sulfur, and boron—were outgassed along with water and exist in the ocean in much greater abundance than could be explained by weathering of rocks alone.

Although rivers and volcanic activity continually contribute material to the oceans, the salinity of seawater is not increasing. In fact, evidence suggests that the composition of seawater has been relatively stable for millions of years. Why doesn't the sea get saltier? The answer is that material is being removed just as rapidly as it is added. For example, some dissolved components are withdrawn from seawater by plants and animals as they build hard parts. Other components are removed when they chemically precipitate from the water as sediment. The net effect is that the overall makeup of seawater remains relatively constant through time.

Processes Affecting Seawater Salinity

Because the ocean is well mixed, the relative abundances of the major components in seawater are essentially constant, no matter where the ocean is sampled. Variations in salinity, therefore, are primarily a result of changes in the water content of the solution.

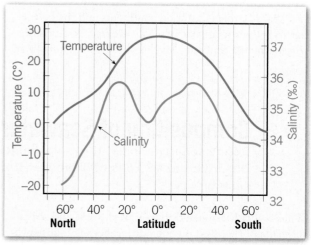

SmartFigure 9.4 Variations in surface temperature and salinity with latitude Average temperatures are highest near the equator and get colder toward the poles. Important factors influencing salinity are variations in rainfall and rates of evaporation. For example, in the dry subtropics in the vicinity of the Tropics of Cancer and Capricorn, high evaporation rates remove more water than is replaced by the meager rainfall, resulting in high surface salinities. In the wet equatorial region, abundant rainfall reduces surface salinities. (goo.gl/0adhQG)

Tutorial

Did You Know?

Some of the saltiest water in the world is found in arid regions that have inland lakes, which are often called "seas." The Great Salt Lake in Utah, for example, has a salinity of 280‰, and the Dead Sea on the border of Israel and Jordan has a salinity of 330‰. The water in the Dead Sea, therefore, contains 33 percent dissolved solids and is almost 10 times saltier than seawater. As a result, these waters have such high density and are so buoyant that you can easily float while lying down in the water—with your arms and legs sticking up above water level (see Figure 9.8).

Table 9.1 Recipe for Artificial Seawater	
Ingredient	**Amount (grams)**
Sodium chloride (NaCl)	23.48
Magnesium chloride ($MgCl_2$)	4.98
Sodium sulfate (Na_2SO_4)	3.92
Calcium chloride ($CaCl_2$)	1.10
Potassium chloride (KCl)	0.66
Sodium bicarbonate ($NaHCO_3$)	0.192
Potassium bromide (KBr)	0.096
Hydrogen borate (H_3BO_3)	0.026
Strontium chloride ($SrCl_2$)	0.024
Sodium fluoride (NaF)	0.003
Mix together these ingredients and then add:	
Pure water (H_2O) to form 1000 grams of solution.	

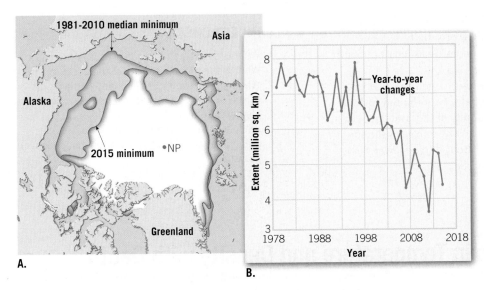

A.

B.

Video

SmartFigure 9.5 Tracking sea ice changes Sea ice is frozen seawater. The formation and melting of sea ice influence the surface salinity of the ocean. In winter the Arctic Ocean is almost completely ice covered. In summer, a portion of the ice melts. **A.** This map shows the extent of sea ice in early September 2015 compared to the average extent of the period 1981 to 2010. The sea ice that does not melt in summer is getting thinner. **B.** The graph clearly depicts the trend in the area covered by sea ice at the end of the summer melt period. The downward trend is related to global warming. (Data from National Snow and Ice Data Center) (https://goo.gl/Cxmt5k)

Various surface processes alter the amount of water in seawater, thereby affecting salinity. Processes that add large amounts of freshwater to seawater—and thereby *decrease* salinity—include precipitation, runoff from land, iceberg melting, and sea ice melting. Processes that remove large amounts of freshwater from seawater—and thereby *increase* seawater salinity—include evaporation and the formation of sea ice. High salinities, for example, are found where evaporation rates are high, such as dry subtropical regions (roughly between 25° and 35° north or south latitude). Conversely, where large amounts of precipitation dilute ocean waters, as in the midlatitudes (between 35° and 60° north or south latitude) and near the equator, lower salinities prevail (**Figure 9.4**).

Surface salinity in polar regions varies seasonally due to the formation and melting of sea ice. When seawater freezes in winter, sea salts do not become part of the ice. Therefore, the salinity of the remaining seawater increases. In summer, when sea ice melts, the addition of the relatively freshwater dilutes the solution, and salinity decreases (**Figure 9.5**).

Surface salinity variation in the open ocean normally ranges from 33‰ to 38‰. Some marginal seas, however, demonstrate extraordinary extremes. For example, in the restricted waters of the Middle East's Persian Gulf and Red Sea—where evaporation far exceeds precipitation— salinity may exceed 42‰. Conversely, very low salinities occur where large quantities of freshwater are supplied by rivers and precipitation. Such is the case for northern Europe's Baltic Sea, where salinity is often below 10‰.

The map in **Figure 9.6** was produced with data from the *Aquarius* satellite. It shows how ocean surface salinity varies worldwide. Notice how the overall pattern on the map matches the graph in Figure 9.4. Also notice the following:

- The Atlantic Ocean is saltiest, greater than 37‰ in some places. This is because compared to Earth's other oceans, the Atlantic has a greater inequality between the loss of freshwater via evaporation and the addition of freshwater from rainfall and river runoff.

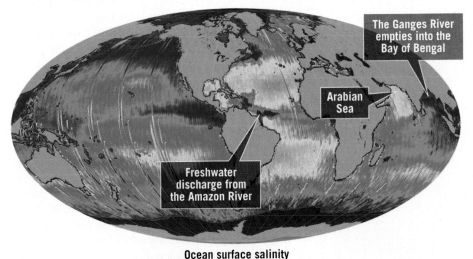

Figure 9.6 Surface salinity of the oceans This map shows average surface salinity for September 2012. It was created using data from the *Aquarius* satellite. (NASA)

Ocean surface salinity

30 32 33   34 34.5 35 35.5 36   37 38 40

parts per thousand (‰)

- The huge discharge of freshwater from the Amazon River produces a zone of lower-salinity seawater in the western Atlantic off the northeast coast of South America.
- The impact of freshwater discharge from rivers is also illustrated when the salinity of the Bay of Bengal is compared to the salinity of the Arabian Sea. Freshwater from the Ganges River causes the surface salinity of the Bay of Bengal to be lower than the salinity of the Arabian Sea.

9.2) CONCEPT CHECKS

1. What is salinity, and how is it usually expressed? What is the average salinity of the ocean?
2. What are the two most abundant elements dissolved in seawater? What is produced when these elements combine?
3. What are the two primary sources for the elements that comprise the dissolved components in seawater?
4. List several factors that cause salinity to vary from place to place and from time to time.

9.3) Variations in Temperature and Density with Depth

Discuss temperature, salinity, and density changes with depth in the open ocean.

Temperature and density are basic ocean-water properties that influence such things as deep-ocean circulation and the distribution and types of life-forms. A sampling of the open ocean from the surface to the seafloor shows that these basic properties change with depth and that the changes are not the same everywhere. How and why temperature and density change with depth is the focus of this section.

Temperature Variations

If a thermometer were lowered from the surface of the ocean into deeper water, what temperature pattern would be found? Surface waters are warmed by the Sun, so they generally have higher temperatures than deeper waters. Surface-water temperatures are also higher in the tropics and get colder toward the poles.

Figure 9.7 shows two graphs of temperature versus depth: one for high-latitude regions and one for low-latitude regions. The low-latitude curve begins at the surface with high temperature, but the temperature decreases rapidly with depth because of the inability of the Sun's rays to penetrate very far into the ocean. At a depth of about 1000 meters (3300 feet), the temperature

remains just a few degrees above freezing and is relatively constant from this level down to the ocean floor. The layer of ocean water between about 300 meters (980 feet) and 1000 meters (3300 feet), where there is a rapid change of temperature with depth, is called the **thermocline** (*thermo* = heat, *cline* = slope). The thermocline is a very important zone in the ocean because it creates a vertical barrier to many types of marine life.

The high-latitude curve in Figure 9.7 displays a pattern quite different from the low-latitude curve. Surface-water temperatures in high latitudes are much colder than in low latitudes, so the curve begins at the surface with low temperature. Deeper in the ocean, the temperature of the water is similar to that at the surface (just a few degrees above freezing), so the curve remains vertical, and there is no rapid change of temperature with depth. A thermocline is not present in high latitudes; instead, the water column is *isothermal* (*iso* = same, *thermal* = heat).

Some high-latitude waters can experience minor warming during the summer months. Thus, certain high-latitude regions experience an extremely weak seasonal thermocline. Midlatitude waters, on the other hand, experience a more dramatic seasonal thermocline and exhibit characteristics intermediate between high- and low-latitude regions.

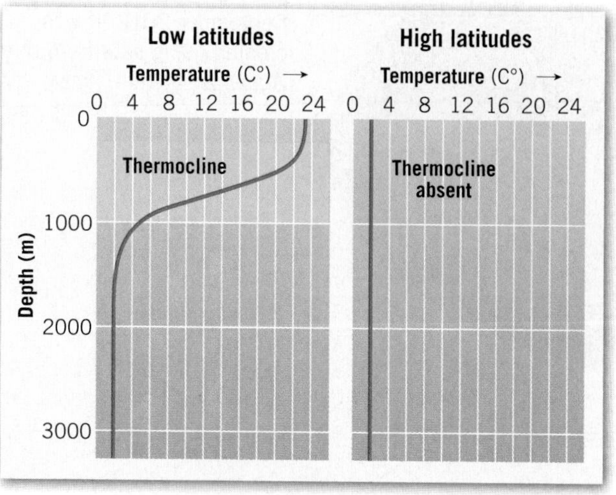

Density Variations

Density is defined as mass per unit volume but can be thought of as a measure of *how heavy something is for its size*. For instance, an object that has low density is lightweight for its size—for example, a dry sponge, foam packing, or a surfboard. Conversely, an object that has high density is heavy for its size—for example, cement and many metals.

Density is an important property of ocean water because it determines the water's vertical position in the ocean. Furthermore, density differences cause large areas of ocean water to sink or float. For example, when high-density seawater is added to low-density freshwater, the denser seawater sinks below the freshwater.

Figure 9.7 Variations in ocean-water temperature with depth for low- and high-latitude regions The layer of rapidly changing temperature called the *thermocline* is not present in the high latitudes.

Low latitudes
Temperature (C°) →
0 4 8 12 16 20 24

High latitudes
Temperature (C°) →
0 4 8 12 16 20 24

Thermocline

Thermocline absent

Depth (m)
0
1000
2000
3000

Figure 9.8 The Dead Sea This water body, which has a salinity of 330‰ (almost 10 times the average salinity of seawater), has high density. As a result, it also has high buoyancy that allows swimmers to float easily. (Photo by Peter Guttman/Corbis)

Factors Affecting Seawater Density Seawater density is influenced by two main factors: *salinity* and *temperature*. An increase in salinity adds dissolved substances and results in an increase in seawater density (**Figure 9.8**). An increase in temperature, on the other hand, causes water to expand and results in a decrease in seawater density. Such a relationship, where one variable decreases as a result of another variable's increase, is known as an *inverse relationship*, where one variable is *inversely proportional* with the other.

Temperature has the greatest influence on surface seawater density because variations in surface seawater temperature are greater than salinity variations. In fact, only in the extreme polar areas of the ocean, where temperatures are low and remain relatively constant, does salinity significantly affect density. Cold water that also has high salinity is some of the highest-density water in the world.

Density Variation with Depth By extensively sampling ocean waters, oceanographers have learned that temperature and salinity—and the water's resulting density—vary with depth. **Figure 9.9** shows two graphs of density versus depth: one for high-latitude regions and one for low-latitude regions.

The low-latitude curve in Figure 9.9 begins at the surface with low density (related to high surface-water temperatures). However, density increases rapidly with depth because the water temperature is getting colder. At

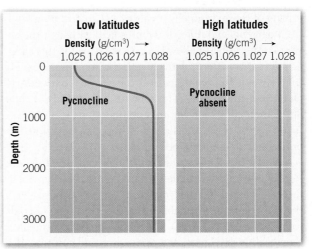

SmartFigure 9.9 Variations in ocean-water density with depth for low- and high-latitude regions The layer of rapidly changing density called the *pycnocline* is present in the low latitudes but absent in the high latitudes. (goo.gl/bsoS7C)

a depth of about 1000 meters (3300 feet), seawater density reaches a maximum value related to the water's low temperature. From this depth to the ocean floor, density remains constant and high. The layer of ocean water between about 300 meters (980 feet) and 1000 meters (3300 feet), where there is a rapid change of density with depth, is called the **pycnocline** (*pycno* = density, *cline* = slope). A pycnocline presents a significant barrier to mixing between low-density ("lighter weight") water above and high-density ("heavier") water below.

The high-latitude curve in Figure 9.9 is also related to the temperature curve for high latitudes shown in Figure 9.7. Figure 9.9 shows that in high latitudes, there is high-density (cold) water at the surface and high-density (cold) water below. Thus, the high-latitude density curve remains vertical, and there is no rapid change of density with depth. A pycnocline is not present in high latitudes; instead, the water column is *isopycnal* (*iso* = same, *pycno* = density).

Ocean Layering

The ocean, like Earth's interior, is layered according to density. Low-density water exists near the surface, and higher-density water occurs below. Except for some shallow inland seas with a high rate of evaporation, the highest-density water is found at the greatest ocean depths. Oceanographers generally recognize a three-layered structure in most parts of the open ocean: a shallow surface mixed zone, a transition zone, and a deep zone (**Figure 9.10**).

Because solar energy is received at the ocean surface, it is here that water temperatures are warmest. The

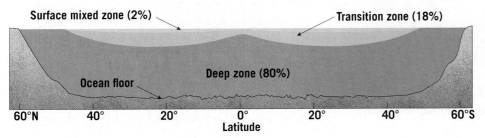

Figure 9.10 The ocean's layers Oceanographers recognize three main layers in the ocean, based on water density, which varies with temperature and salinity. The warm *surface mixed layer* accounts for only 2 percent of ocean water; the *transition zone* includes the thermocline and the pycnocline and accounts for 18 percent of ocean water; and the *deep zone* contains cold, high-density water that accounts for 80 percent of ocean water.

mixing of these waters by waves as well as turbulence from currents and tides distributes heat gained at the surface through a shallow layer. Consequently, this *surface mixed zone* has relatively uniform temperatures. The thickness and temperature of this layer vary, depending on latitude and season. The zone usually extends to about 300 meters (980 feet) but may attain a thickness of 450 meters (1500 feet). The surface mixed zone accounts for only about 2 percent of ocean water.

Below the Sun-warmed zone of mixing, the temperature falls abruptly with depth (see Figure 9.7). Here, a distinct layer called the *transition zone* exists between the warm surface layer above and the deep zone of cold water below. The transition zone includes a prominent thermocline and associated pycnocline and accounts for about 18 percent of ocean water.

Below the transition zone is the *deep zone*, where sunlight never reaches and water temperatures are just a few degrees above freezing. As a result, water density remains constant and high. The deep zone includes about 80 percent of ocean water, indicating the immense depth of the ocean. (The average depth of the ocean is more than 3700 meters [12,200 feet].)

In high latitudes, the three-layer structure does not exist because the water column is isothermal and isopycnal, which means that there is no rapid change in temperature or density with depth. Consequently, good vertical mixing between surface and deep waters can occur in high-latitude regions. Here, cold high-density water forms at the surface, sinks, and initiates deep-ocean currents, which are discussed in Chapter 10.

9.3 CONCEPT CHECKS

1. Contrast temperature variations with depth in the high and low latitudes. Why do high-latitude waters generally lack a thermocline?
2. What two factors influence seawater density? Which one has the greater influence on surface seawater density?
3. Contrast density variations with depth in the high and low latitudes. Why do high-latitude waters generally lack a pycnocline?
4. Describe the ocean's layered structure. Why does the three-layer structure not exist in high latitudes?

9.4 An Emerging Picture of the Ocean Floor

Define *bathymetry* and summarize the various techniques used to map the ocean floor.

If all the water were removed from the ocean basins, a great variety of features would be visible, including broad volcanic peaks, deep trenches, extensive plains, mountain chains, and large plateaus. In fact, the scenery would be nearly as diverse as that on the continents.

Mapping the Seafloor

The complex nature of ocean-floor topography did not begin to become known until the historic 3½-year

voyage of the HMS *Challenger* (**Figure 9.11**). From December 1872 to May 1876, the *Challenger* expedition made the first comprehensive study of the global ocean ever attempted. During the 127,500-kilometer (79,200-mile) voyage, the ship and its crew of scientists traveled to every ocean except the Arctic. Throughout the voyage, they sampled many ocean properties, including water depth, which was accomplished by lowering a long weighted line overboard and then retrieving it. Using this laborious process, the *Challenger* made the first recording of the deepest-known point on the ocean floor in 1875. This spot, on the floor of the western Pacific, was later named the *Challenger Deep.*

Modern Bathymetric Techniques The measurement of ocean depths and the charting of the shape (topography) of the ocean floor is known as **bathymetry** (*bathos* = depth, *metry* = measurement). Today, sound energy is used to measure water depths. The basic approach employs **sonar**, an acronym for *so*und *na*vigation and *rang*ing. The first devices that used sound to measure water depth, called **echo sounders**, were developed early in

Figure 9.11 HMS *Challenger* The first systematic bathymetric measurements of the ocean were made aboard the HMS *Challenger*, which departed England in December 1872 and returned in May 1876. (Image courtesy of the Library of Congress Prints and Photographs Division)

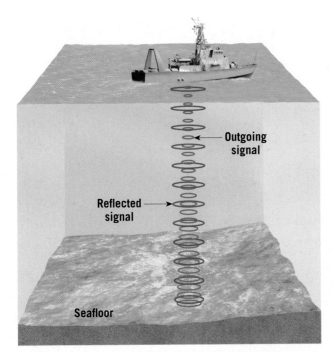

Figure 9.12 Echo sounder An echo sounder determines water depth by measuring the time interval required for an acoustic wave to travel from a ship to the seafloor and back. The speed of sound in water is 1500 meters per second. Therefore, Depth = ½ (1500 m/sec × Echo travel time).

the twentieth century. Echo sounders work by transmitting a sound wave (called a *ping*) into the water in order to produce an echo when it bounces off any object, such as a large marine organism or the ocean floor (**Figure 9.12**).

A sensitive receiver intercepts the echo reflected from the bottom, and a clock precisely measures the travel time to fractions of a second. By knowing the speed of sound waves in water—about 1500 meters (4900 feet) per second—and the time required for the energy pulse to reach the ocean floor and return, depth can be calculated. Depths determined from continuous monitoring of these echoes are plotted to create a profile of the ocean floor. By laboriously combining profiles from many adjacent traverses, a chart of the seafloor was produced.

Following World War II, the U.S. Navy developed *sidescan sonar* to look for explosive devices that had been deployed in shipping lanes (**Figure 9.13A**). These torpedo-shaped instruments can be towed behind a ship, where they send out a fan of sound extending to either side of the ship's track. By combining swaths of sidescan sonar data, researchers produced the first photograph-like images of the seafloor. Although sidescan sonar provides valuable views of the seafloor, it does not provide bathymetric (water depth) data.

This drawback was resolved with the development of *high-resolution multibeam* instruments (see Figure 9.13A). These systems use hull-mounted sound sources that send out a fan of sound and then record reflections from the seafloor through a set of narrowly focused receivers aimed

at different angles. Rather than obtain the depth of a single point every few seconds, this technique makes it possible for a survey ship to map the features of the ocean floor along a strip tens of kilometers wide. These systems can collect bathymetric data of such high resolution that they can distinguish depths that differ by less than a meter (**Figure 9.13B**). When multibeam sonar is used to make a map of a section of seafloor, the ship travels through the area in a regularly spaced back-and-forth pattern known as "mowing the lawn."

Despite their greater efficiency and enhanced detail, research vessels equipped with multibeam sonar travel at a mere 10 to 20 kilometers (6 to 12 miles) per hour. It would take at least 100 vessels outfitted with this equipment hundreds of years to map the entire seafloor. This explains why only about 5 percent of the seafloor has been mapped in detail—and why large areas of the seafloor have not been mapped with sonar at all.

Figure 9.13 Sidescan and multibeam sonar

A. Sidescan sonar and multibeam sonar operating from the same research vessel.

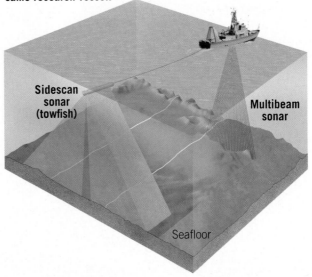

B. Color-enhanced perspective map of the seafloor and coastal landforms in the Los Angeles area of California.

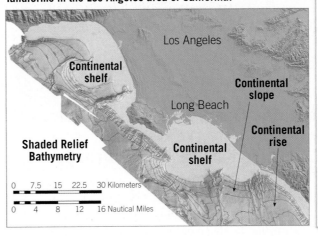

Did You Know?
Ocean depths are often expressed in *fathoms*. One fathom equals 1.8 m (6 ft), which is about the distance of a person's outstretched arms. The term is derived from how depth-sounding lines were brought back on board a vessel by hand. As the line was hauled in, a worker counted the number of arm lengths collected. By knowing the length of the person's outstretched arms, the amount of line taken in could be calculated. The length of 1 fathom was later standardized to 1.8 m (6 ft).

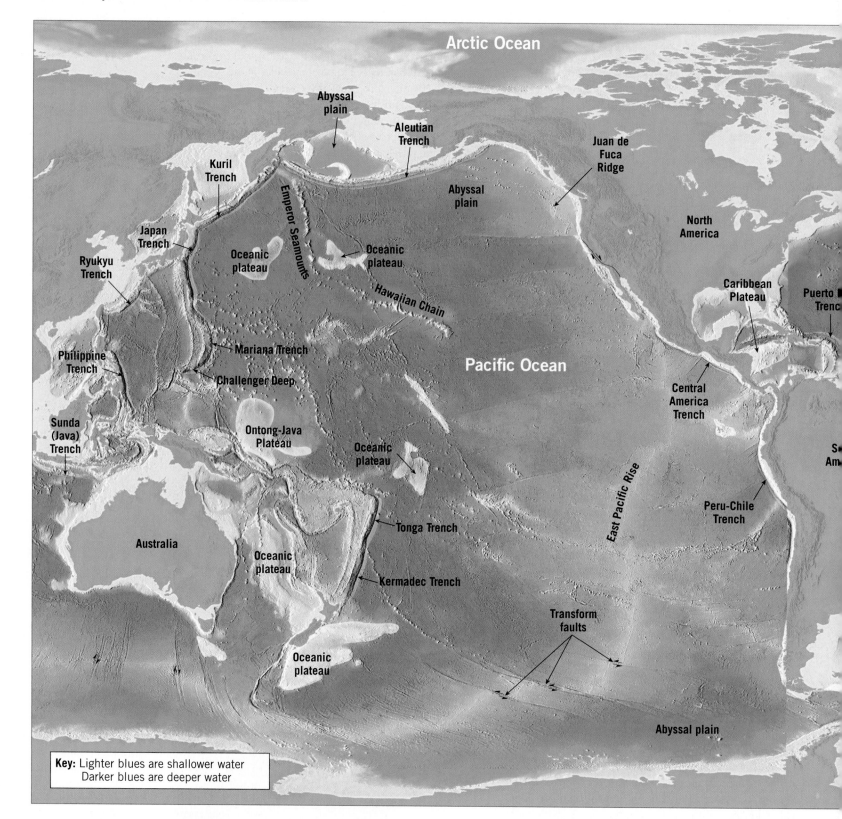

Arctic Ocean

Abyssal plain

Aleutian Trench

Juan de Fuca Ridge

Kuril Trench

Emperor Seamounts

Abyssal plain

North America

Japan Trench

Oceanic plateau

Oceanic plateau

Ryukyu Trench

Hawaiian Chain

Caribbean Plateau

Puerto Rico Trench

Mariana Trench

Pacific Ocean

Philippine Trench

Challenger Deep

Central America Trench

Ontong-Java Plateau

Oceanic plateau

Sunda (Java) Trench

Oceanic plateau

East Pacific Rise

Peru-Chile Trench

South America

Tonga Trench

Australia

Oceanic plateau

Kermadec Trench

Transform faults

Oceanic plateau

Abyssal plain

Key: Lighter blues are shallower water
Darker blues are deeper water

Mapping the Ocean Floor from Space

Breakthroughs in technology have led to more detailed ocean floor maps (**Figure 9.14**). One breakthrough that enhanced our understanding of the seafloor involves measuring the shape of the ocean surface from space.

After compensating for waves, tides, currents, and atmospheric effects, it was discovered that the ocean surface is not perfectly "flat." Because massive seafloor features exert stronger-than-average gravitational attraction, they produce elevated areas on the ocean surface. Conversely, canyons and trenches create slight depressions.

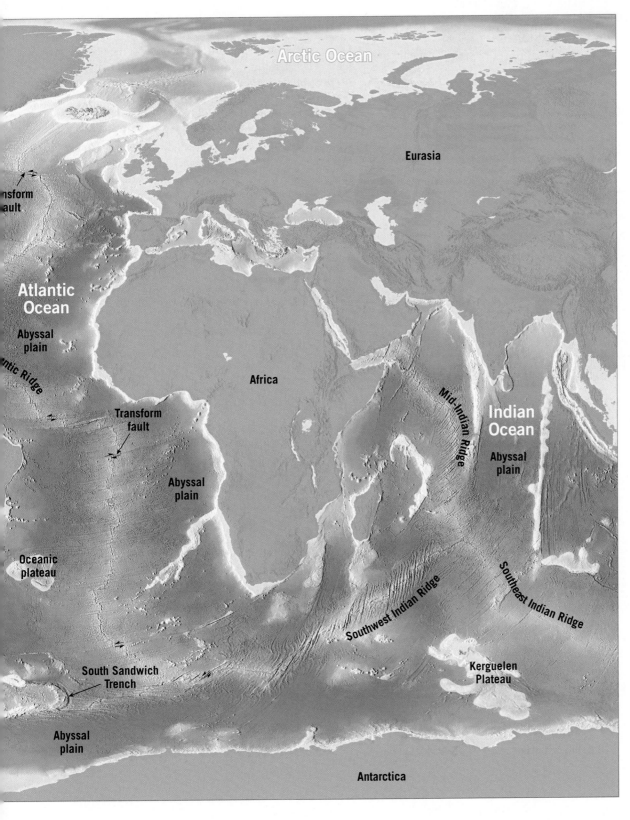

Figure 9.14 Major features of the seafloor

Satellites equipped with *radar altimeters* are able to measure these subtle differences by bouncing microwaves off the sea surface (**Figure 9.15**). These devices can measure variations as small as a few centimeters. Such data have added greatly to the knowledge of ocean-floor topography. Combined with traditional sonar depth measurements, the data are used to produce detailed ocean-floor maps, such as the one shown in Figure 9.14.

Tutorial

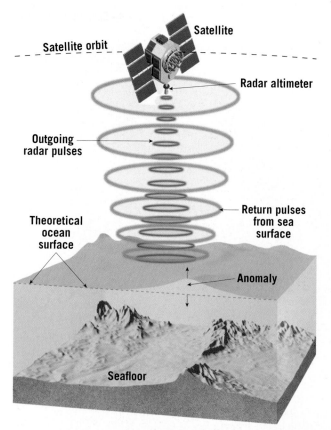

Provinces of the Ocean Floor

Oceanographers studying the topography of the ocean floor have delineated three major areas: *continental margins*, the *deep-ocean basins*, and the *oceanic (mid-ocean) ridges*. The map in **Figure 9.16** outlines these provinces for the North Atlantic Ocean, and the profile at the bottom of the illustration shows the varied topography. Such profiles usually have their vertical dimension exaggerated many times to make topographic features more conspicuous. Vertical exaggeration, however, makes slopes shown in seafloor profiles appear to be *much* steeper than they actually are.

9.4 CONCEPT CHECKS

1. Define *bathymetry*.
2. Describe how satellites orbiting Earth can determine features on the seafloor without being able to directly observe them beneath several kilometers of seawater.
3. List the three major provinces of the ocean floor.

Figure 9.16 Major topographic divisions of the North Atlantic Map view (*top*) and corresponding profile view (*bottom*). On the profile, the vertical scale has been exaggerated to make topographic features more conspicuous.

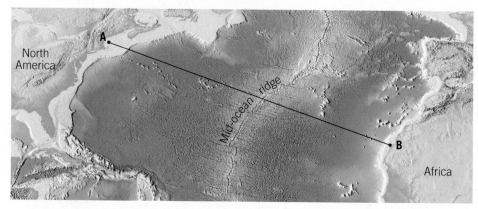

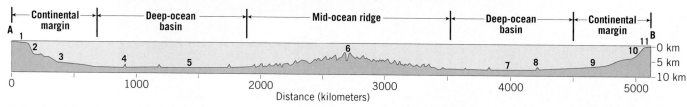

1. Continental shelf
2. Continental slope
3. Continental rise
4. Seamount
5. Abyssal plain
6. Rift valley
7. Abyssal plain
8. Seamount
9. Continental rise
10. Continental slope
11. Continental shelf

9.5 Continental Margins

Compare a passive continental margin with an active continental margin and list the major features of each.

As the name implies, **continental margins** are the outer margins of the continents where continental crust transitions to oceanic crust. Two types of continental margin have been identified: *passive* and *active*. Nearly the entire Atlantic Ocean and a large portion of the Indian Ocean are surrounded by passive continental margins (see Figure 9.15). By contrast, most of the Pacific Ocean is bordered by active continental margins, which are represented by subduction zones in **Figure 9.17**. Notice that some of the active subduction zones lie far beyond the margins of the continents.

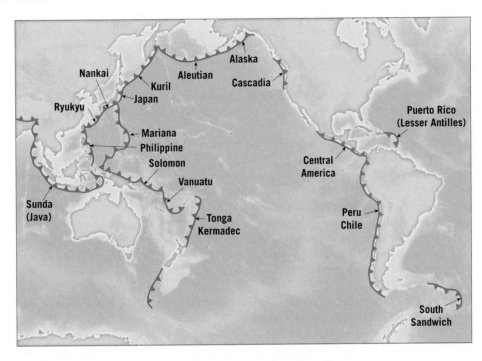

Figure 9.17 Distribution of Earth's subduction zones Most of the active subduction zones surround the Pacific basin.

Passive Continental Margins

Passive continental margins are geologically inactive regions located some distance from plate boundaries. As a result, they are not associated with strong earthquakes or volcanic activity. Passive continental margins develop when continental blocks rift apart and are separated by continued seafloor spreading. As a result, the continental blocks are firmly attached to the adjacent oceanic crust.

Most passive margins are relatively wide and are sites where large quantities of sediments are deposited. The features comprising passive continental margins include the continental shelf, the continental slope, and the continental rise (**Figure 9.18**).

Continental Shelf The **continental shelf** is a gently sloping, submerged surface extending from the shoreline toward the deep-ocean basin. It consists mainly of continental crust capped with sedimentary rocks and sediments eroded from adjacent landmasses.

The continental shelf varies greatly in width. Although almost

nonexistent along portions of some continents, the shelf extends seaward more than 1500 kilometers (930 miles) along others. The average inclination of the continental shelf is only about one-tenth of 1 degree, a slope so slight that it would appear to an observer to be a horizontal surface.

The continental shelf tends to be relatively featureless; however, some areas are mantled by extensive glacial deposits and thus are quite rugged. In addition, some continental shelves are dissected by large valleys running from the coastline into deeper waters. Many of these *shelf valleys* are the seaward extensions of river valleys on the adjacent landmass. They were eroded during the last Ice Age (Quaternary period), when enormous

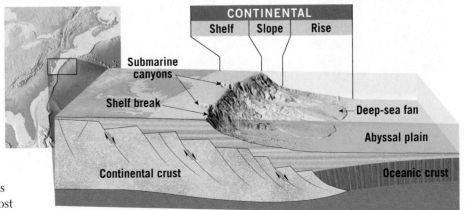

Figure 9.18 Passive continental margin The slopes shown for the continental shelf and continental slope are greatly exaggerated. The continental shelf has an average slope of one-tenth of 1 degree, whereas the continental slope has an average slope of about 5 degrees.

quantities of water were stored in vast ice sheets on the continents, causing sea levels to drop at least 100 meters (330 feet). Because of this sea-level drop, rivers extended their courses, and land-dwelling plants and animals migrated to the newly exposed portions of the continents. Dredging off the coast of North America has retrieved the ancient remains of numerous land dwellers, including mammoths, mastodons, and horses, providing further evidence that portions of the continental shelves were once above sea level.

Although continental shelves represent only 7.5 percent of the total ocean area, they have economic and political significance because they contain important reservoirs of oil and natural gas, and they support important fishing grounds.

Continental Slope   Marking the seaward edge of the continental shelf is the **continental slope**, a relatively steep feature that marks the boundary between continental crust and oceanic crust. Although the inclination of the continental slope varies greatly from place to place, it averages about 5 degrees and in places exceeds 25 degrees.

Continental Rise   The continental slope merges into a more gradual incline known as the **continental rise** that may extend seaward for hundreds of kilometers. The continental rise consists of a thick accumulation of sediment that has moved down the continental slope and onto deep-ocean floor. Most of the sediments are delivered to the seafloor by *turbidity currents* that

periodically flow down *submarine canyons*. (We will discuss these shortly.) When these muddy slurries emerge from the mouth of a canyon onto the relatively flat ocean floor, they deposit sediment that forms a **deep-sea fan**. As fans from adjacent submarine canyons grow, they merge to produce a continuous wedge of sediment at the base of the continental slope, forming the continental rise.

Submarine Canyons and Turbidity Currents   Deep, steep-sided valleys known as **submarine canyons** are cut into the continental slope and may extend across the entire continental rise to the deep-ocean basin (**Figure 9.19**). Although some of these canyons appear to be the seaward extensions of river valleys, many others do not line up in this manner. Furthermore, submarine canyons extend to depths far below the maximum lowering of sea level during the Ice Age, so we cannot attribute their formation to stream erosion.

Most submarine canyons have likely been excavated by turbidity currents. **Turbidity currents** are downslope movements of dense, sediment-laden water. They are created when sand and mud on the continental shelf and slope are dislodged and thrown into suspension. Because the mud-choked water is denser than normal seawater, it moves downslope as a mass, eroding and accumulating more sediment as it goes. The erosional work repeatedly carried on by these muddy torrents is thought to be the major force in the excavation of most submarine canyons.

Figure 9.19 Turbidity currents and submarine canyons Turbidity currents are an important factor in the formation of submarine canyons. (Photo by Marli Miller)

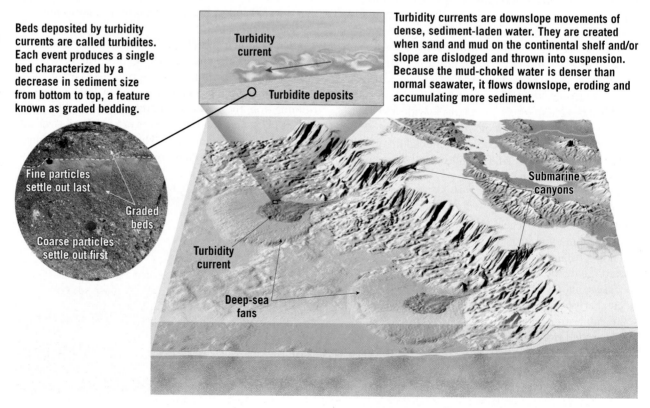

Beds deposited by turbidity currents are called turbidites. Each event produces a single bed characterized by a decrease in sediment size from bottom to top, a feature known as graded bedding.

Fine particles settle out last

Graded beds

Coarse particles settle out first

Turbidity current

Turbidite deposits

Turbidity currents are downslope movements of dense, sediment-laden water. They are created when sand and mud on the continental shelf and/or slope are dislodged and thrown into suspension. Because the mud-choked water is denser than normal seawater, it flows downslope, eroding and accumulating more sediment.

Turbidity current

Deep-sea fans

Submarine canyons

Turbidity currents usually originate along the continental slope and continue across the continental rise, still cutting channels. Eventually, they lose momentum and come to rest along the floor of the deep-ocean basin. As these currents slow, suspended sediments begin to settle out. First, the coarser sand is dropped, followed by successively finer accumulations of silt and then clay. These deposits, called *turbidites*, display a decrease in sediment grain size from bottom to top, a feature known as *graded bedding*. Turbidity currents are an important mechanism of sediment transport in the ocean. By the action of turbidity currents, submarine canyons are excavated, and sediments are carried to the deep-ocean floor.

Active Continental Margins

Active continental margins are located along convergent plate boundaries where oceanic lithosphere is being subducted beneath the leading edge of a continent (**Figure 9.20**). Deep-ocean trenches are the major topographic expression at convergent plate boundaries. These deep, narrow furrows surround most of the Pacific Rim.

Along some subduction zones, sediments from the ocean floor and pieces of oceanic crust are scraped from the descending oceanic plate and plastered against the edge of the overriding plate. This chaotic accumulation of deformed sediment and scraps of oceanic crust is called an **accretionary wedge** (*ad* = toward, *crescere* = to grow). Prolonged plate subduction can produce massive accumulations of sediment along active continental margins.

The opposite process, known as **subduction erosion**, characterizes many other active continental margins. Rather than sediment accumulating along the front of the overriding plate, sediment and rock are scraped off the bottom of the overriding plate and transported into the mantle by the subducting plate. Subduction erosion is

A. Active continental margins are located along convergent plate boundaries, where oceanic lithosphere is being subducted beneath the leading edge of a continent.

B. Accretionary wedges develop along subduction zones, where sediments from the ocean floor are scraped from the descending oceanic plate and pressed against the edge of the overriding plate.

C. Subduction erosion occurs where sediment and rock scraped off the bottom of the overriding plate are carried into the mantle by the subducting plate.

SmartFigure 9.20 Active continental margins (goo.gl/WJn92)

Tutorial

particularly effective when the angle of descent is steep. Sharp bending of the subducting plate causes faulting in the ocean crust and a rough surface, as shown in Figure 9.20.

9.5 CONCEPT CHECKS

1. List the three major features of a passive continental margin. Which of these features is considered a flooded extension of the continent? Which one has the steepest slope?
2. Describe the differences between active and passive continental margins. Where is each type found?
3. Discuss the process that is responsible for creating most submarine canyons.
4. How are active continental margins related to plate tectonics?
5. Briefly explain how an accretionary wedge forms. What is meant by *subduction erosion*?

9.6 Features of Deep-Ocean Basins

List and describe the major features associated with deep-ocean basins.

Between the continental margin and the oceanic ridge lies the **deep-ocean basin** (see Figure 9.16). The size of this region—almost 30 percent of Earth's surface—is roughly comparable to the percentage of the surface that presently projects above sea level as land. This region includes *deep-ocean trenches*, which are extremely deep linear depressions in the ocean floor; remarkably flat regions, known as *abyssal plains*; tall volcanic peaks, called *seamounts* and *guyots*; and extensive areas of lava flows piled one atop the other, called *oceanic plateaus*.

Deep-Ocean Trenches

Deep-ocean trenches are long, relatively narrow troughs that are the deepest parts of the ocean. Most trenches are located along the margins of the Pacific Ocean, where some exceed 10,000 meters (33,000 feet) in depth (see Figure 9.14). A portion of one—the Challenger Deep in the Mariana trench—has been measured at 10,994 meters (36,070 feet) below sea level, making it the deepest known part of the world ocean (**Figure 9.21**). Only two trenches are located in the Atlantic—the Puerto Rico trench and the South Sandwich trench.

Although deep-ocean trenches represent only a very small portion of the area of the ocean floor, they are nevertheless significant geologic features. Trenches are sites of plate convergence where slabs of oceanic lithosphere subduct and plunge back into the mantle. In addition to earthquakes being created as one plate "scrapes" beneath another, volcanic activity is also associated with these regions. Thus, trenches are often paralleled by an arc-shaped row of active volcanoes called a **volcanic island arc**. Furthermore, **continental volcanic arcs**, such as those making up portions of the Andes and Cascades, are located parallel to trenches that lie adjacent to active continental margins (see Figure 9.20). The volcanic activity associated with the trenches that surround the Pacific Ocean explains why the region is called the *Ring of Fire*.

Abyssal Plains

Abyssal plains are deep and flat features; in fact, these regions are likely among the most level places on Earth. The abyssal plain found off the coast of Argentina, for example, has less than 3 meters (10 feet) of relief over a distance exceeding 1300 kilometers (800 miles). The monotonous topography of abyssal plains is occasionally interrupted by the protruding summit of a buried volcanic peak (seamount).

Using *seismic reflection profilers*, instruments whose signals penetrate far below the ocean floor, researchers have determined that abyssal plains consist of thick accumulations of sediment that have buried an otherwise rugged ocean floor (**Figure 9.22**). The nature of the sediment indicates that these plains consist primarily of three materials: (1) fine sediments transported far out to sea by turbidity currents, (2) mineral matter that has precipitated out of seawater, and (3) shells and skeletons of microscopic marine organisms.

Abyssal plains are found in all oceans. However, the floor of the Atlantic has the most extensive abyssal plains because it has few trenches to act as traps for sediment carried down the continental slope.

Volcanic Structures on the Ocean Floor

Dotting the seafloor are numerous volcanic structures of various sizes. Many occur as isolated features that resemble volcanoes on land. Others occur as long, narrow chains that stretch for thousands of kilometers, while still others are massive structures that cover areas the size of Texas.

Seamounts and Volcanic Islands Submarine volcanoes, called **seamounts**, may rise hundreds of meters above the surrounding topography. It is estimated that more than 1 million seamounts exist. Some grow large enough to become oceanic islands, but most do not have

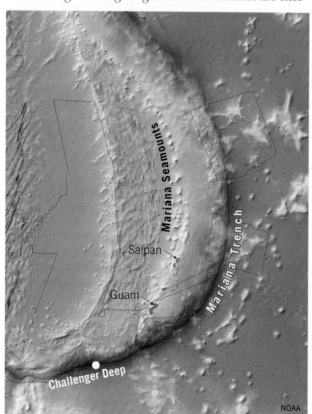

Figure 9.21 The Challenger Deep Located near the southern end of the Mariana trench, the Challenger Deep is the deepest place in the global ocean, about 10,994 meters (36,070 feet) deep. Film director James Cameron (*Titanic* and *Avatar*) made news in March 2012 as the first person to take a deep-diving submersible to the bottom of the Challenger Deep in more than 50 years.

Mariana Seamounts

Saipan

Guam

Mariana Trench

Challenger Deep

NOAA

a sufficiently long eruptive history to build a structure above sea level. Although seamounts are found on the floors of all the oceans, they are most common in the Pacific.

Some, like the Hawaiian Island–Emperor seamount chain, which stretches from the Hawaiian Islands to the Aleutian trench, form over volcanic hot spots (see Figure 9.15 and Figure 5.26). Others are born near oceanic ridges.

If a volcano grows large enough before it is carried from its magma source by plate motion, the structure may emerge as a *volcanic island*. Examples of volcanic islands include Easter Island, Tahiti, Bora Bora, the Galapagos Islands, and the Canary Islands.

Guyots During their existence, inactive volcanic islands are gradually but inevitably lowered to near sea level by the forces of weathering and erosion. As a moving plate slowly carries inactive volcanic islands away from the elevated oceanic ridge or hot spot over which they formed, they gradually sink and disappear below the water surface. Submerged, flat-topped seamounts that formed in this manner are called **guyots**.*

Oceanic Plateaus The ocean floor contains several massive **oceanic plateaus**, which resemble lava plateaus composed of flood basalts found on the continents. Oceanic plateaus, which can be more than 30 kilometers (20 miles) thick, are generated from vast outpourings of fluid basaltic lavas.

Some oceanic plateaus appear to have formed quickly in geologic terms. Examples include the Ontong Java Plateau, which formed in less than 3 million years, and the Kerguelen Plateau, which formed in 4.5 million

---

* The term *guyot* is named after Arnold Guyot, Princeton University's first geology professor. It is pronounced "GEE-oh" with a hard *g*, as in "give."

Seismic reflection profile

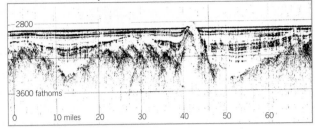

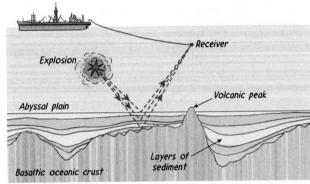

Geologist's Sketch

Figure 9.22 Seismic profile This seismic cross section and matching sketch across a portion of the Madeira abyssal plain in the eastern Atlantic show the irregular oceanic crust buried by sediments. (Image courtesy Woods Hole Oceanographic Institution, Charles Hollister, translator. Used with permission of the Woods Hole Oceanographic Institution.)

years (see Figure 9.15). Like basalt plateaus on land, oceanic plateaus are thought to form when the bulbous head of a rising mantle plume melts and produces a vast outpouring of basaltic lava.

9.6 CONCEPT CHECKS

1. Explain how deep-ocean trenches are related to plate boundaries.
2. Why are abyssal plains more extensive on the floor of the Atlantic than on the floor of the Pacific?
3. How does a flat-topped seamount, called a *guyot*, form?
4. What features on the ocean floor most resemble basalt plateaus on the continents?

9.7 The Oceanic Ridge System

Summarize the basic characteristics of oceanic ridges.

Along well-developed oceanic divergent plate boundaries, the seafloor is elevated, forming a broad linear swell called the **oceanic ridge** or **rise**, or **mid-ocean ridge**. Our knowledge of the oceanic ridge system comes from soundings of the ocean floor, core samples from deep-sea drilling, visual inspection using deep-diving submersibles, and firsthand study of slices of ocean floor that have been thrust onto dry land during continental collisions. At oceanic ridges we find extensive faulting, earthquakes, high heat flow, and volcanism.

Anatomy of the Oceanic Ridge System

The oceanic ridge system winds through all major oceans in a manner similar to the seam on a baseball and is the longest topographic feature on Earth, exceeding 70,000 kilometers (43,000 miles) in length (**Figure 9.23**). The crest of the ridge typically stands 2 to 3 kilometers above the adjacent deep-ocean basins and marks the divergent plate boundary where new oceanic crust is created.

Notice in Figure 9.23 that large sections of the oceanic ridge system have been named based on their

Figure 9.23 Distribution of the oceanic ridge system The map shows ridge segments that exhibit slow, intermediate, and fast spreading rates.

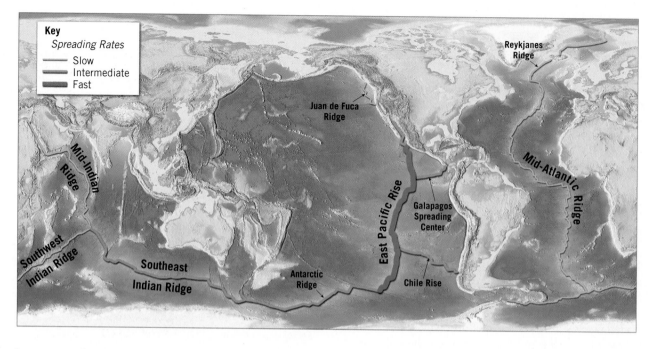

Key
Spreading Rates
— Slow
═ Intermediate
━ Fast

Reykjanes Ridge

Mid-Indian Ridge

Juan de Fuca Ridge

East Pacific Rise

Galapagos Spreading Center

Mid-Atlantic Ridge

Southwest Indian Ridge

Southeast Indian Ridge

Antarctic Ridge

Chile Rise

locations in the various ocean basins. Some ridges run through the middle of ocean basins, where they are appropriately called *mid-ocean* ridges. The Mid-Atlantic Ridge and the Mid-Indian Ridge are examples. By contrast, the East Pacific Rise is *not* a "mid-ocean" feature. Rather, as its name implies, it is located in the eastern Pacific, far from the center of the ocean.

The term *ridge* is somewhat misleading because these features are not narrow and steep, as the term implies, but have widths of 1000 to 4000 kilometers (about 620 to 2500 miles) and the appearance of broad, elongated swells that exhibit varying degrees of ruggedness.

Furthermore, the ridge system is broken into segments that range from a few tens to hundreds of kilometers in length. Each segment is offset from an adjacent segment by a transform fault.

Oceanic ridges are as high as some mountains on the continents, but the similarities end there. Whereas most mountain ranges on land form when the compressional forces associated with continental collisions fold and metamorphose thick sequences of sedimentary rocks, oceanic ridges form where upwelling from the mantle generates new oceanic crust. Oceanic ridges consist of layers and piles of newly formed basaltic rocks that are buoyantly uplifted by the hot mantle rocks from which they formed.

Along the axis of some segments of the oceanic ridge system are deep, down-faulted structures that are called **rift valleys** because of their striking similarity to the continental rift valleys found in East Africa (**Figure 9.24**). Some rift valleys, including those along the rugged Mid-Atlantic Ridge, are typically 30 to 50 kilometers (20 to 30 miles) wide and have walls that tower 500 to 2500 meters (1640 to 8200 feet) above the valley floor. This makes them comparable to the deepest and widest part of Arizona's Grand Canyon.

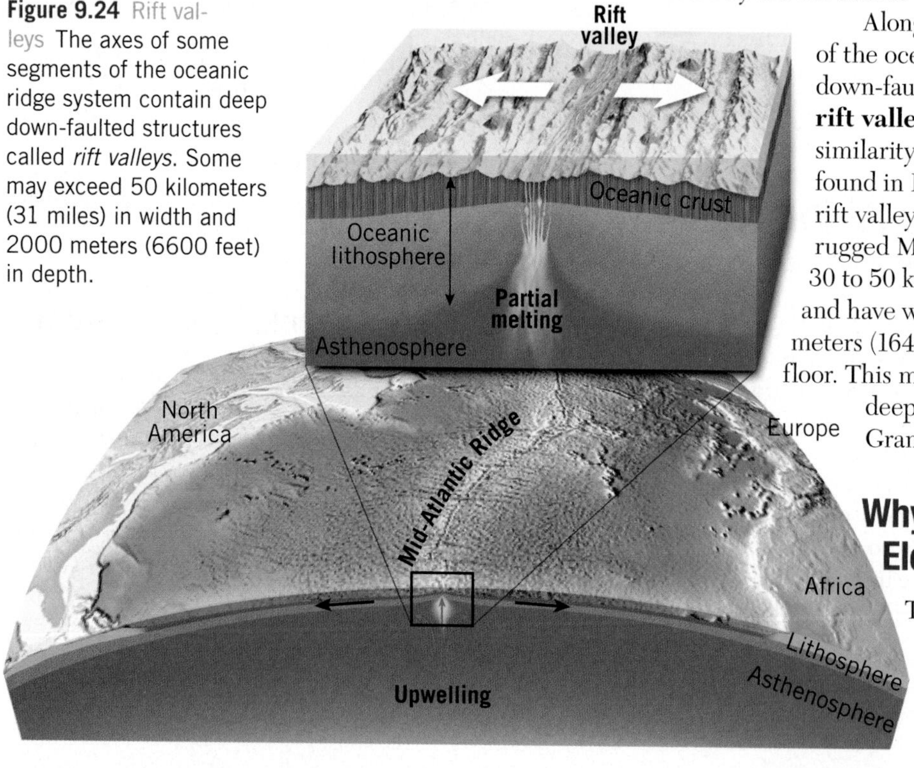

Figure 9.24 Rift valleys The axes of some segments of the oceanic ridge system contain deep down-faulted structures called *rift valleys*. Some may exceed 50 kilometers (31 miles) in width and 2000 meters (6600 feet) in depth.

Rift valley

Oceanic crust

Oceanic lithosphere

Partial melting

Asthenosphere

North America

Mid-Atlantic Ridge

Europe

Africa

Lithosphere

Asthenosphere

Upwelling

Why Is the Oceanic Ridge Elevated?

The primary reason for the elevated position of the ridge system is that newly created oceanic lithosphere is hot and therefore less dense than cooler rocks of the deep-ocean basin. As the newly formed basaltic crust travels away from the ridge crest, it is cooled from above as seawater circulates through the pore spaces and fractures in the rock. In addition, it cools because it gets farther and farther from the zone of hot mantle upwelling. As a result, the lithosphere gradually cools, contracts, and becomes denser. This thermal contraction accounts for the greater ocean depths that occur away from the ridge. After about 80 million years of cooling and contraction, rock that was once part of an elevated oceanic-ridge system becomes part of the deep-ocean basin.

As lithosphere is displaced away from the ridge crest, cooling also causes a gradual increase in lithospheric thickness. This happens because the boundary between the lithosphere and asthenosphere is a thermal (temperature) boundary. Recall that the lithosphere is Earth's cool, stiff outer layer, whereas the asthenosphere is a comparatively hot and weak layer. As material in the uppermost asthenosphere ages (cools), it becomes stiff and rigid. Thus, the upper portion of the asthenosphere is gradually converted to lithosphere simply by cooling. Oceanic lithosphere continues to thicken until it is about 80 to 100 kilometers (50 to 60 miles) thick. Thereafter, its thickness remains relatively unchanged until it is subducted.

 9.7 CONCEPT CHECKS

1. Write a brief summary describing oceanic ridges.
2. Although oceanic ridges can be as tall as some mountains on the continents, list some ways that oceanic ridges are different.
3. Where do rift valleys form along the oceanic ridge system?
4. What is the primary reason for the elevated position of the oceanic ridge system?

9.8 Seafloor Sediments

Distinguish among three categories of seafloor sediment and explain why some of these sediments can be used to study climate change.

Except for steep areas of the continental slope and areas near the crest of the mid-ocean ridge, the ocean floor is covered with sediment. Part of this material has been deposited by turbidity currents, and the rest has slowly settled to the seafloor from above. The thickness of this carpet of debris varies greatly. In some trenches, which act as traps for sediments originating on the continental margin, accumulations may approach 10 kilometers (6 miles). In general, however, sediment accumulations are considerably less. In the Pacific Ocean, for example, uncompacted sediment measures about 600 meters (2000 feet) or less, whereas on the floor of the Atlantic, the thickness varies from 500 to 1000 meters (1600 to 3300 feet).

Types of Seafloor Sediments

Seafloor sediments can be classified according to their origin into three broad categories: (1) **terrigenous** ("derived from land"), (2) **biogenous** ("derived from organisms"), and (3) **hydrogenous** ("derived from water"). Although each category is discussed separately, remember that all seafloor sediments are mixtures. No body of sediment comes entirely from a single source.

Terrigenous Sediment Terrigenous sediment consists primarily of mineral grains that were weathered from continental rocks and transported to the ocean. Larger particles (sand and gravel) usually settle rapidly near shore, whereas the very smallest particles take years to settle to the ocean floor and may be carried thousands of kilometers by ocean currents. As a result, virtually every area of the ocean receives some terrigenous sediment. The rate at which this sediment accumulates on the deep-ocean floor, though, is very slow. Forming a 1-centimeter (0.4-inch) abyssal clay layer, for example, requires as much as 50,000 years. Conversely, on the continental margins near the mouths of large rivers, terrigenous sediment accumulates rapidly and forms thick deposits.

Biogenous Sediment Biogenous sediment consists of shells and skeletons of marine animals and algae (**Figure 9.25**). This debris is produced mostly by microscopic organisms living in the sunlit waters near the ocean surface. Once these organisms die, their hard *tests* (*testa* = shell) continually "rain" down and accumulate on the seafloor.

The most common biogenous sediment is *calcareous* ($CaCO_3$) *ooze*, which, as the name implies, has the consistency of thick mud. This sediment is produced from

Figure 9.25 Marine microfossils: An example of biogenous sediment These tiny, single-celled organisms are sensitive to even small fluctuations in temperature. Seafloor sediments containing these fossils are useful sources of data on climate change. This is a false-color image. (Photo by Mary Martin/Biophoto Associates/Science Source)

the tests of organisms that inhabit warm surface waters. When calcareous hard parts slowly sink through a cool layer of water, they begin to dissolve. This results because the deeper cold seawater is richer in carbon dioxide and is thus more acidic than warm water. In seawater deeper than about 4500 meters (15,000 feet), calcareous tests completely dissolve before they reach bottom. Consequently, calcareous ooze does not accumulate at these greater depths.

Other biogenous sediments include *siliceous* (SiO_2) *ooze* and phosphate-rich material. The former is composed primarily of tests of diatoms (single-celled algae) and radiolaria (single-celled protozoa), whereas the latter is derived from a variety of sources, including the bones, teeth, and scales of fish and other marine organisms.

Hydrogenous Sediment Hydrogenous sediment consists of minerals that crystallize directly from seawater through various chemical reactions. For example, some limestones are formed when calcium carbonate precipitates directly from the water; however, most limestone is composed of biogenous sediment.

Some of the most common types of hydrogenous sediment include the following:

- *Manganese nodules* are rounded, hard lumps of manganese, iron, and other metals that precipitate in concentric layers around a central object (such as a volcanic pebble or a grain of sand). The nodules can be up to 20 centimeters (8 inches) in diameter and are often littered across large areas of the deep seafloor (**Figure 9.26A**).
- *Calcium carbonates* form by precipitating directly from seawater in warm climates. If this material is buried and hardened, it forms limestone. Most limestone, however, is composed of biogenous sediment.

- *Metal sulfides* are often precipitated near ocean floor vents that spew mineral-rich hot water. These hydrothermal (hot water) vents are usually associated with the crest of a mid-ocean ridge. When the hot fluid comes in contact with the much colder seawater, mineral matter precipitates to form smoke-like clouds called *black smokers* (**Figure 9.26B**). The particles that compose black smokers eventually settle out. The deposits may contain economically significant amounts of iron, copper, zinc, lead, and occasionally silver and gold.
- *Evaporites* form where evaporation rates are high and there is restricted open-ocean circulation. As water evaporates from such areas, the remaining seawater becomes saturated with dissolved minerals, which then begin to precipitate. Heavier than seawater, they sink to the bottom or form a characteristic white crust of evaporite minerals around the edges of these areas. Collectively termed *salts*, some evaporite minerals taste salty, such as *halite* (common table salt, NaCl), and some do not, such as the calcium sulfate minerals *anhydrite* ($CaSO_4$) and *gypsum* ($CaSO_4 \cdot 2H_2O$).

Seafloor Sediment—A Storehouse of Climate Data

Reliable climate records go back only a couple hundred years, at best. How do scientists learn about climates and climate change prior to that time? They must reconstruct past climates from *indirect evidence*; that is, they must analyze phenomena that respond to and reflect changing atmospheric conditions. An interesting and important technique for analyzing Earth's climate history is the study of biogenous seafloor sediments.

We know that the parts of the Earth system are linked so that a change in one part can produce changes in any or all of the other parts. In this example, you will see how changes in atmospheric and oceanic temperatures are reflected in the nature of life in the sea.

Most seafloor sediments contain the remains of organisms that once lived near the sea surface (the ocean–atmosphere interface). When such near-surface organisms die, their shells slowly settle to the

SmartFigure 9.26 Examples of hydrogenous sediment
A. Manganese nodules. (Photo by Charles A. Winter/Science Source)
B. This black smoker is spewing hot, mineral-rich water. When the heated solutions meet cold seawater, metal sulfides precipitate and form mounds of minerals around these hydrothermal vents. (Photo by Verena Tunnicliffe/Uvic/Fisheries and Oceans Canada/Newscom) (goo.gl/QH1xDk)

 Tutorial

A.

B.

floor of the ocean, where they become part of the sedimentary record. These seafloor sediments are useful recorders of worldwide climate change because the numbers and types of organisms living near the sea surface change with the climate.

Thus, in seeking to understand climate change as well as other environmental transformations, scientists are tapping the huge reservoir of data in seafloor sediments (**Figure 9.27**). The sediment cores gathered by research vessels have provided invaluable data that have significantly expanded our knowledge and understanding of past climates.

One notable example of the importance of seafloor sediments to our understanding of climate change relates to unraveling the fluctuating atmospheric conditions that caused the alternating glacial and interglacial periods of the Quaternary Ice Age. The records of temperature changes contained in cores of sediment from the ocean floor have proven critical to our present understanding of this recent span of Earth history.

Figure 9.27 Data from the seafloor This scientist is holding a sediment core obtained by an oceanographic research vessel. Cores of sediment provide data that allow for a more complete understanding of past climates. Notice the "library" of sediment cores in the background. (Photo by Ingo Wagner/EPA/Newscom)

(9.8) CONCEPT CHECKS

1. List and describe the three basic types of seafloor sediments. Give at least one example of each.
2. Why are seafloor sediments useful in studying past climates?

CONCEPTS IN REVIEW
Oceans: The Last Frontier

9.1 The Vast World Ocean

Discuss the extent and distribution of oceans and continents on Earth. Identify Earth's four main ocean basins.

KEY TERM: oceanography

- Oceanography is an interdisciplinary science that draws on the methods and knowledge of biology, chemistry, physics, and geology to study all aspects of the world ocean.
- Earth's surface is dominated by oceans. Nearly 71 percent of the planet's surface area is oceans and marginal seas. In the Southern Hemisphere, about 81 percent of the surface is water.
- Of the three major oceans—the Pacific, Atlantic, and Indian—the Pacific Ocean is the largest, contains slightly more than half of the water in the world ocean, and has the greatest average depth—3940 meters (12,927 feet).

(?) **Name the four principal oceans identified with letters on this world map.**

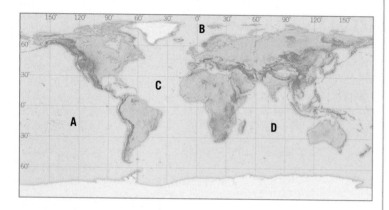

9.2 Composition of Seawater

Define *salinity* and list the main elements that contribute to the ocean's salinity. Describe the sources of dissolved substances in seawater and causes of variations in salinity.

KEY TERM: salinity

- Salinity is the proportion of dissolved salts to pure water, usually expressed in parts per thousand (‰). The average salinity in the open ocean ranges from 35‰ to 37‰. The principal elements that contribute to the ocean's salinity are chlorine (55%) and sodium (31%). The primary sources for the elements in sea salt are chemical weathering of rocks on the continents and volcanic outgassing on the ocean floor.
- Variations in seawater salinity are primarily caused by changing the water content. Natural processes that add large amounts of freshwater to seawater and decrease salinity include precipitation, runoff from land, iceberg melting, and sea ice melting. Processes that remove large amounts of freshwater from seawater and increase salinity include the formation of sea ice and evaporation.

9.3 Variations in Temperature and Density with Depth

Discuss temperature, salinity, and density changes with depth in the open ocean.

KEY TERMS: thermocline, density, pycnocline

- The ocean's surface temperature is related to the amount of solar energy received and varies as a function of latitude. Low-latitude regions have relatively warm surface water and distinctly colder water at depth, creating a thermocline, which is a layer of rapid temperature change. No thermocline exists in high-latitude regions because there is little temperature difference between the top and bottom of the water column.
- Seawater density is mostly affected by water temperature but also by salinity. Cold, high-salinity water is densest. Low-latitude regions have distinctly denser (colder) water at depth than at the surface, creating a pycnocline, which is a layer of rapidly changing density. No pycnocline exists in high-latitude regions because there is little density difference between the top and bottom of the water column.
- Most open-ocean regions exhibit a three-layered structure based on water density. The shallow surface mixed zone has warm and nearly uniform temperatures. The transition zone includes a prominent thermocline and associated pycnocline. The deep zone is continually dark and cold and accounts for 80 percent of the water in the ocean. In high latitudes, the three-layered structure does not exist.

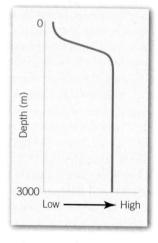

(?) **Does this graph represent changes in ocean-water temperature or ocean-water density with depth? Does it more likely represent a location in the tropics or near the poles? Explain.**

9.4 An Emerging Picture of the Ocean Floor

Define *bathymetry* and summarize the various techniques used to map the ocean floor.

KEY TERMS: bathymetry, sonar, echo sounder

- Seafloor mapping is done with sonar—shipboard instruments that emit pulses of sound that "echo" off the bottom. Satellites are also used to map the ocean floor. Their instruments measure slight variations in sea level that result from differences in the gravitational pull of features on the seafloor. Accurate maps of seafloor topography can be made using these data.
- Mapping efforts have revealed three major areas of the ocean floor: continental margins, deep-ocean basins, and oceanic ridges.

(?) **Do the math: How many seconds would it take an echo sounder's ping to make the trip from a ship to the Challenger Deep (10,994 meters [36,070 feet]) and back? Recall that depth = ½ (1500 m/sec × Echo travel time).**

9.5 Continental Margins

Compare a passive continental margin with an active continental margin and list the major features of each.

KEY TERMS: continental margin, passive continental margin, continental shelf, continental slope, continental rise, deep-sea fan, submarine canyon, turbidity current, active continental margin, accretionary wedge, subduction erosion

- Continental margins are transition zones between continental and oceanic crust. Active continental margins occur where a plate boundary and the edge of a continent coincide, usually on the leading edge of a plate. Passive continental margins are on the trailing edges of continents, far from plate boundaries.

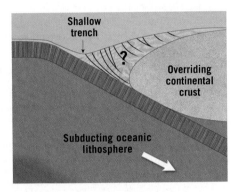

- Heading offshore from the shoreline of a passive margin, a submarine traveler would first encounter the gently sloping continental shelf and then the steeper continental slope, which marks the end of the continental crust and the beginning of the oceanic crust. Beyond the continental slope is another gently sloping section, the continental rise: It is made of sediment transported by turbidity currents through submarine canyons and piled up in deep-sea fans atop the oceanic crust.
- Submarine canyons are deep, steep-sided valleys that originate on the continental slope and may extend to the deep-ocean basin. Many submarine canyons have been excavated by turbidity currents (downslope movements of dense, sediment-laden water).
- At an active continental margin, material may be added to the leading edge of a continent in the form of an accretionary wedge (common at shallow-angle subduction zones), or material may be scraped off the edge of a continent by subduction erosion (common at steeply dipping subduction zones).

(?) **What type of continental margin is depicted by this diagram? Name the feature indicated by the question mark.**

9.6 Features of Deep-Ocean Basins

List and describe the major features associated with deep-ocean basins.

KEY TERMS: deep-ocean basin, deep-ocean trench, volcanic island arc, continental volcanic arc, abyssal plain, seamount, guyot, oceanic plateau

- The deep-ocean basin makes up about half of the ocean floor's area. Much of it is abyssal plain (deep, featureless sediment-draped crust). Subduction zones and deep-ocean trenches also occur in deep-ocean basins. Paralleling trenches are volcanic island arcs (if the subduction goes underneath oceanic lithosphere) or continental

volcanic arcs (if the overriding plate has continental lithosphere on its leading edge).
- There are a variety of volcanic structures on the deep-ocean floor. Seamounts are submarine volcanoes; if they pierce the surface of the ocean, we call them volcanic islands. Guyots are old volcanic islands that have had their tops eroded off before sinking below sea level. Oceanic plateaus are unusually thick sections of oceanic crust formed by massive underwater eruptions of lava.

(?) **On the diagrammatic cross-sectional view of a deep-ocean basin, match each of the five lettered features to the correct name: seamount, guyot, volcanic island, oceanic plateau, abyssal plain.**

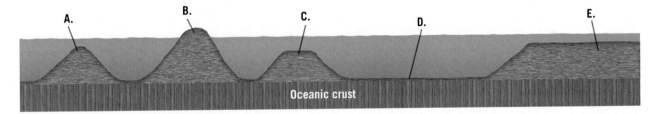

9.7 The Oceanic Ridge System

Summarize the basic characteristics of oceanic ridges.

KEY TERMS: oceanic ridge or rise (mid-ocean ridge), rift valley

- The oceanic ridge system is the longest topographic feature on Earth, wrapping around the world through all major ocean basins. Oceanic ridges are a few kilometers tall, a few thousand kilometers wide, and up to tens of thousands of kilometers long. The summit of a ridge is the place where new oceanic crust is generated, often marked by a rift valley.
- Oceanic ridges are elevated features because they are warm and therefore less dense than older, colder oceanic lithosphere. As oceanic crust moves away from the ridge crest, heat loss causes the oceanic crust to become denser and subside. After 80 million years, crust that was once part of an oceanic ridge is in the deep ocean basin, far from the ridge.

9.8 Seafloor Sediments

Distinguish among three categories of seafloor sediment and explain why some of these sediments can be used to study climate change.

KEY TERMS: terrigenous sediment, biogenous sediment, hydrogenous sediment

- There are three broad categories of seafloor sediments. Terrigenous sediment consists primarily of mineral grains that were weathered from continental rocks and transported to the ocean; biogenous sediment consists of shells and skeletons of marine organisms; and hydrogenous sediment includes minerals that crystallize directly from seawater through various chemical reactions.
- Seafloor sediments are helpful in studying worldwide climate change because they often contain the remains of organisms that once lived near the sea surface. The numbers and types of these organisms change as the climate changes, and their remains in seafloor sediments record these changes.

GIVE IT SOME THOUGHT

1. Refer to Figure 9.2 to answer the following questions:
   a. Water dominates Earth's surface, but not everywhere. In what Northern Hemisphere latitude belt is there more land than water?
   b. In what latitude belt is there no land at all?

2. Assuming that the average speed of sound waves in water is 1500 meters per second, determine the water depth if a signal sent out by an echo sounder on a research vessel requires 6 seconds to strike bottom and return to the recorder aboard the ship.

3. The accompanying photo shows sea ice in the Beaufort Sea near Barrow, Alaska. How do seasonal changes in the amount of sea ice influence the salinity of the remaining surface water? Is water density greater before or after sea ice forms? Explain.

Michael Collier

4. Assume that someone brings several water samples to your laboratory. His problem is that the labels are incomplete. He knows that samples A and B are from the Atlantic Ocean; he also knows that one came from near the equator, and the other came from near the Tropic of Cancer. But he does not know which one is which. He has a similar problem with samples C and D. One is from the Red Sea, and the other is from the Baltic Sea. Applying your knowledge of ocean salinity, how would you identify the location of each sample? How were you able to figure this out?

5. Refer to the accompanying map and diagram showing the eastern seaboard of the United States to complete the following:
   a. Which letter is associated with each of the following: continental shelf, continental slope, shelf break?
   b. How does the size of the continental shelf surrounding Florida compare to the size of the Florida peninsula?
   c. Why are there no deep-ocean trenches on this map?

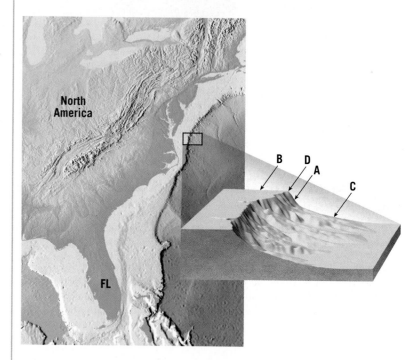

6. Are the continental margins surrounding the Atlantic Ocean primarily active or passive? How about the margins surrounding the Pacific Ocean? Based on your response to the previous questions, indicate whether each ocean basin is getting larger, shrinking, or staying the same size. Explain your answer.

7. Examine the accompanying sketch, which shows three sediment layers on the ocean floor. What term is applied to such layers? What was your clue? What process was responsible for creating these layers? Are these layers more likely part of a deep-sea fan or an accretionary wedge?

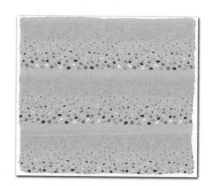

8. Imagine that while you and a passenger are in a deep-diving submersible in the North Pacific near Alaska's Aleutian Islands, you encounter a long, narrow depression on the ocean floor. Your passenger asks whether you think it is a submarine canyon, a rift valley, or a deep-ocean trench. How would you respond? Explain your response.

9. This satellite image from February 4, 2013, shows a plume of dust moving from the southern Arabian Peninsula over the Arabian Sea (an arm of the Indian Ocean). When the wind subsided, the dust settled onto the water surface and slowly sank, eventually reaching the ocean floor. To which category of seafloor sediment does this material belong?

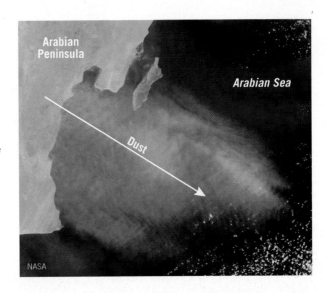

MasteringGeology™

www.masteringgeology.com   Looking for additional review and test prep materials? With individualized coaching on the toughest topics of the course, MasteringGeology offers a wide variety of ways for you to move beyond memorization to begin thinking like a geologist. Visit the Study Area in www.masteringgeology.com to find practice quizzes, study tools, and multimedia that will improve your understanding of this chapter's content. Sign in today to enjoy the following features: **Self Study Quizzes, SmartFigure: Tutorials/ Animations/Condor Videos/Mobile Field Trips, Geoscience Animation Library, GEODe, RSS Feeds, Digital Study Modules,** and an optional **Pearson eText.**

10

FOCUS ON CONCEPTS

Each statement represents the primary learning objective for the corresponding major heading within the chapter. After you complete the chapter, you should be able to:

10.1 Discuss the factors that create and influence ocean currents and describe the influence ocean currents have on climate.

10.2 Explain the processes that produce coastal upwelling and the ocean's deep circulation.

10.3 Explain why the shoreline is considered a dynamic interface. List the factors that influence the height, length, and period of a wave and describe the motion of water within a wave.

10.4 Describe how waves erode and move sediment along the shore.

10.5 Describe the features typically created by wave erosion and those resulting from sediment deposited by longshore transport processes.

10.6 Distinguish between emergent and submergent coasts. Contrast the erosion problems faced along different parts of America's coasts.

10.7 Summarize the ways in which people deal with shoreline erosion problems.

10.8 Explain the cause of tides as well as their monthly cycles and other patterns. Describe the horizontal flow of water that accompanies the rise and fall of tides.

In contrast to the gently sloping coastal plains of the Atlantic and Gulf coasts, the Pacific coast is characterized by relatively narrow beaches that are often backed by steep cliffs and mountain ranges. These crashing waves and sea stacks are at Soberanes Point along the California coast. (Photo by Jamie Pham/Zoonar/ AGE Fotostock)

THE RESTLESS OCEAN

The restless waters of the ocean are constantly in motion, powered by many different forces. Winds, for example, generate surface currents, which influence coastal climate and provide nutrients that affect the abundance of algae and other marine life in surface waters. Winds also produce waves that carry energy from storms to distant shores, where their impact erodes the land. In some regions, density differences create deep-ocean circulation, which is important for ocean mixing and nutrient recycling. In addition, the Moon and the Sun produce tides, which periodically raise and lower sea level. This chapter will examine these movements of ocean waters and their effects on coastal regions.

10.1 The Ocean's Surface Circulation

Discuss the factors that create and influence ocean currents and describe the influence ocean currents have on climate.

You may have heard of the Gulf Stream, an important surface current in the Atlantic Ocean that flows northward along the east coast of the United States (**Figure 10.1**). Surface currents like this one are set in motion by the wind. At the water surface, where the atmosphere and ocean meet, energy is passed from moving air to the water through friction. The drag exerted by winds blowing steadily across the ocean causes the surface layer of water to move. Thus, major horizontal movements of surface waters are closely related to the global pattern of prevailing winds.* As an example, the small map in **Figure 10.2** shows how the wind belts known as the *trade winds* and the *westerlies* create large, circular-moving loops of water in the Atlantic Ocean. The same wind belts influence the other oceans as well, so that a similar pattern of currents can also be seen in the Pacific and Indian Oceans. Essentially, the pattern of surface-ocean circulation closely matches the pattern of global winds but is also strongly influenced by the distribution of major landmasses and by Earth's spinning on its axis.

The Pattern of Ocean Currents

Huge circular-moving current systems dominate the surfaces of the oceans. These large loops of water within an ocean basin are called **gyres** (*gyros* = circle). The large map in Figure 10.2 shows the world's five main gyres: the *North Pacific Gyre*, the *South Pacific Gyre*, the *North Atlantic Gyre*, the *South Atlantic Gyre*, and the *Indian Ocean Gyre* (which exists mostly within the Southern Hemisphere). The center of each gyre coincides with the subtropics at about 30° north or south latitude, and these gyres are therefore often called *subtropical gyres*.

Coriolis Effect As shown in Figure 10.2, subtropical gyres rotate clockwise in the Northern Hemisphere and counterclockwise in the Southern Hemisphere. Why do the gyres flow in different directions in the two hemispheres? Although wind is the force that generates surface currents, other factors also influence the movement of ocean waters. The most significant of these is the **Coriolis effect**. Because of Earth's rotation, currents are deflected to the *right* in the Northern Hemisphere and to the *left* in the Southern Hemisphere. You can see the influence of the Coriolis effect by comparing the wind arrows and the arrows representing the ocean currents on the small map in Figure 10.2. (The Coriolis effect is more fully explained in Chapter 13.) As a result, gyres flow in opposite directions in the two hemispheres.

North Pacific Currents Four main currents generally exist within each gyre (see Figure 10.2). The North Pacific Gyre, for example, consists of the North Equatorial Current, the Kuroshio Current, the North Pacific Current, and the California Current. The tracking of floating objects that are released into the ocean intentionally or accidentally reveals that it takes about 6 years for the objects to go all the way around the loop.

North Atlantic Currents Like the North Pacific, the North Atlantic Ocean has four main currents (see

Gulf Stream

SmartFigure 10.1 The Gulf Stream In this satellite image off the east coast of the United States, orange and yellow represent higher water temperatures, and blue indicates cooler water temperatures. The current transports heat from the subtropics far into the North Atlantic. (NOAA) (https://goo.gl/LwqnV4)

Animation

---

* Details about the global pattern of winds appear in Chapter 13.

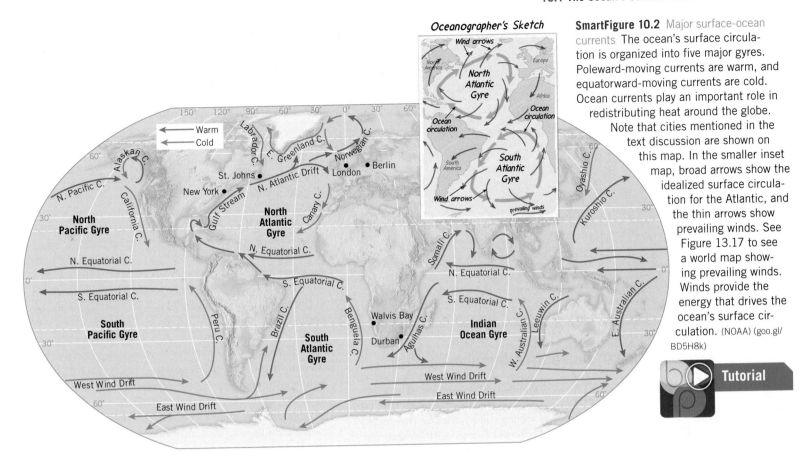

Oceanographer's Sketch

The ocean's surface circulation is organized into five major gyres. Poleward-moving currents are warm, and equatorward-moving currents are cold. Ocean currents play an important role in redistributing heat around the globe. Note that cities mentioned in the text discussion are shown on this map. In the smaller inset map, broad arrows show the idealized surface circulation for the Atlantic, and the thin arrows show prevailing winds. See Figure 13.17 to see a world map showing prevailing winds. Winds provide the energy that drives the ocean's surface circulation. (NOAA) (goo.gl/BD5H8k)

Tutorial

Figure 10.2). Beginning near the equator, the North Equatorial Current is deflected northward through the Caribbean, where it becomes the Gulf Stream. As the Gulf Stream moves along the east coast of the United States, it is strengthened by the prevailing westerly winds and is deflected to the east (to the right) offshore of North Carolina into the North Atlantic. As it continues northeastward, it gradually widens and slows until it becomes a vast, slowly moving current known as the North Atlantic Current, which because of its sluggish nature is also known as the North Atlantic Drift.

As the North Atlantic Current approaches Western Europe, it splits, and part of it moves northward past Great Britain, Norway, and Iceland, carrying heat to these otherwise chilly areas. The other part is deflected southward as the cool Canary Current. As the Canary Current moves southward, it eventually merges into the North Equatorial Current, completing the gyre. Because the North Atlantic Ocean basin is about half the size of the North Pacific basin, it takes floating objects about 3 years to go completely around this gyre.

The circular motion of gyres leaves a large central area that has no well-defined currents. In the North Atlantic, this zone of calmer waters is known as the Sargasso Sea, named for the large quantities of *Sargassum*, a type of floating seaweed encountered there.

Southern Hemisphere Currents The ocean basins in the Southern Hemisphere exhibit a pattern of flow similar to that in the Northern Hemisphere basins, with surface currents influenced by wind belts, the position of continents, and the Coriolis effect. In the South Atlantic and South Pacific, for example, surface-ocean circulation is very much the same as in their Northern Hemisphere counterparts, except that the direction of flow is counterclockwise (see Figure 10.2).

The West Wind Drift is the only current that completely encircles Earth (see Figure 10.2). It flows around the ice-covered continent of Antarctica, where no large landmasses are in the way, so its cold surface waters circulate in a continuous loop. It moves in response to the Southern Hemisphere's prevailing westerly winds, and portions of it split off into the adjoining southern ocean basins.

Indian Ocean Currents The Indian Ocean exists mostly in the Southern Hemisphere, and it follows a surface circulation pattern similar to those in other Southern Hemisphere ocean basins (see Figure 10.2). The small portion of the Indian Ocean in the Northern Hemisphere, however, is influenced by the seasonal wind shifts known as the summer and winter *monsoons* (*mausim* = season). During the summer, winds blow from the Indian Ocean toward the Asian landmass. In the winter, the winds reverse and blow out from Asia over the Indian Ocean. You can see this reversal when you compare the January and July winds in Figure 13.18 on page 406. When the winds change direction, the surface currents also reverse direction.

Did You Know?
In 1768, as deputy postmaster of the colonies, Benjamin Franklin, together with a Nantucket ship captain, produced the first map of the Gulf Stream. Franklin's interest began when he realized that ships carrying the mail took 2 weeks longer going from England to America than in the other direction.

Figure 10.3 The chilling effect of a cold current Monthly mean temperatures for Rio de Janeiro, Brazil, and Arica, Chile, both of which are coastal cities near sea level. Even though Arica is closer to the equator, its temperatures are cooler than Rio de Janeiro's. Arica is influenced by the cold Peru Current, whereas Rio de Janeiro is adjacent to the warm Brazil Current.

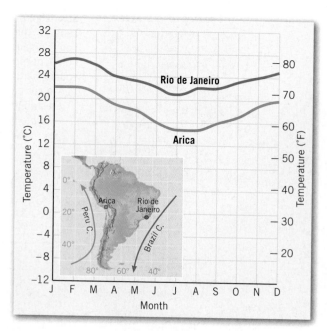

movements account for about one-quarter of this total heat transport, and winds account for the remaining three-quarters.

The Effect of Warm Currents The moderating effect of poleward-moving warm ocean currents is well known. The North Atlantic Drift, an extension of the warm Gulf Stream, keeps wintertime temperatures in Great Britain and much of Western Europe warmer than would be expected for their latitudes. London is farther north than St. John's, Newfoundland, yet is not nearly so frigid in winter. (Cities mentioned in this section are shown in Figure 10.2.) Because of the prevailing westerly winds, the moderating effects are carried far inland. For example, Berlin (52° north latitude) has a mean January temperature similar to that experienced at New York City, which lies 12° latitude farther south. The January mean at London (51° north latitude) is 4.5°C (8.1°F) higher than at New York City.

Ocean Currents Influence Climate

Surface-ocean currents have an important effect on climate. When Earth is considered as a whole, the energy gained from incoming solar radiation is equal to the energy radiated back out to space. However, that is not true for most individual latitudes. Low latitudes gain more solar energy than they radiate to space; the reverse is true for high latitudes. Because the tropics are not heating up, nor the polar regions getting colder, there must be a large-scale transfer of heat from areas of excess to areas of deficit. This is indeed the case. *The transfer of heat by winds and ocean currents equalizes these latitudinal energy imbalances.* Ocean water

Cold Currents Chill the Air In contrast to warm ocean currents like the Gulf Stream, the effects of which are felt most during the winter, cold currents exert their greatest influence in the subtropics or during the summer months in the middle latitudes. For example, the cool Benguela Current off the western coast of southern Africa moderates the tropical heat along this coast. Walvis Bay (23° south latitude), a town adjacent to the Benguela Current, is 5°C (9°F) cooler in summer than Durban, which is 6° latitude farther poleward but on the eastern side of South Africa, away from the influence of the cold current. The east and west coasts of South America provide another example. **Figure 10.3** shows monthly mean temperatures for Rio de Janeiro, Brazil, which is influenced by the warm Brazil Current, and Arica, Chile, which is adjacent to the cold Peru Current. Closer to home, because of the cold California Current, summer temperatures in subtropical coastal southern California are lower by 6°C (10.8°F) or more compared to east coast stations.

Cold Currents Increase Aridity In addition to influencing temperatures of adjacent land areas, cold currents have other climatic influences. For example, where subtropical deserts exist along the west coasts of continents, cold ocean currents have a dramatic impact. The principal west coast deserts are the Atacama in Peru and Chile and the Namib in southwestern Africa (**Figure 10.4**). The aridity along these coasts is intensified because the lower atmosphere is chilled by cold offshore waters. When this occurs, the air becomes very stable and resists the upward movement necessary to create precipitation-producing clouds. In addition,

Figure 10.4 Chile's Atacama Desert This is the driest desert on Earth. Average rainfall at the wettest locations is not more than 3 millimeters (0.12 inch) per year. Stretching nearly 1000 kilometers (600 miles), the Atacama is situated between the Pacific Ocean and the towering Andes Mountains. The cold Peru Current makes this slender zone cooler and drier than it would otherwise be. (Photo by Jacques Jangoux/Science Source)

the presence of cold currents causes temperatures to approach and often reach the dew point, the temperature at which water vapor condenses. As a result, these areas are characterized by high relative humidities and frequent fogs. Thus, not all subtropical deserts are hot with low humidities and clear skies. Rather, the presence of cold currents transforms some subtropical deserts into relatively cool, damp places that are often shrouded in fog.

10.1 CONCEPT CHECKS

1. What is the primary driving force of surface-ocean currents?
2. How does the Coriolis effect influence ocean currents?
3. Name the five subtropical gyres and identify the main surface currents in each.
4. How do ocean currents influence climate? Provide at least three examples.

10.2 Upwelling and Deep-Ocean Circulation

Explain the processes that produce coastal upwelling and the ocean's deep circulation.

The preceding discussion focused mainly on the horizontal movements of the ocean's surface waters. In this section, you will learn that the ocean also exhibits significant vertical movements and a slow-moving, multilayered deep-ocean circulation. Some vertical movements are related to wind-driven surface currents, whereas deep-ocean circulation is strongly influenced by density differences.

Coastal Upwelling

In addition to producing surface currents, winds can also cause *vertical* water movements. *Upwelling*, the rising of cold water from deeper layers to replace warmer surface water, is a common wind-induced vertical movement. One type of upwelling, called **coastal upwelling**, is most characteristic along the west coasts of continents, most notably along California, western South America, and West Africa.

Coastal upwelling occurs in these areas when winds blow toward the equator and parallel to the coast (**Figure 10.5**). Coastal winds combined with the Coriolis effect cause surface water to move away from shore. As the surface layer moves away from the coast, it is replaced by water that wells up from below the surface. This slow upward movement of water from depths of 50 to 300 meters (165 to 1000 feet) brings water that is cooler than the original surface water and results in lower surface-water temperatures near the shore.

For swimmers who are accustomed to the warm waters along the mid-Atlantic shore of the United States, a swim in the Pacific off the coast of central California can be a chilling surprise. In August, when temperatures in the Atlantic are 21°C (70°F) or higher, central California's surf is only about 15°C (60°F).

Upwelling brings greater concentrations of dissolved nutrients, such as nitrates and phosphates, to the ocean surface. These nutrient-enriched waters from below promote the growth of microscopic plankton, which in turn support extensive populations of fish and other marine organisms. Figure 10.5 includes a satellite image that

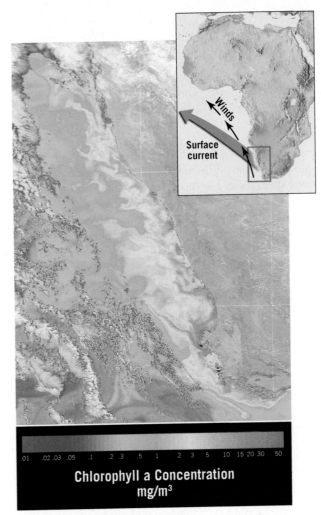

Chlorophyll a Concentration
mg/m³

shows high productivity due to coastal upwelling off the southwest coast of Africa.

SmartFigure 10.5 Coastal upwelling Coastal upwelling occurs along the west coasts of continents, where winds blow toward the equator and parallel to the coast. The Coriolis effect (deflection to the left in the Southern Hemisphere) causes surface water to move away from the shore, which brings cold, nutrient-rich water to the surface. This satellite image shows chlorophyll concentration along the southwest coast of Africa (February 21, 2001). An instrument aboard the satellite detects changes in seawater color caused by changing concentrations of chlorophyll. High chlorophyll concentrations indicate high amounts of photosynthesis, which is linked to the upwelling nutrients. Red indicates high concentrations, and blue indicates low concentrations. (Provided by the SeaStar Satellite. Chlorophyll Concentration/NASA) (goo.gl/y5TnCZ)

Tutorial

Deep-Ocean Circulation

Deep-ocean circulation has a significant vertical component and accounts for the thorough mixing of deep-water

Figure 10.6 Sea ice near Antarctica When seawater freezes, sea salts do not become part of the ice. Consequently, the surface salinity of the remaining seawater increases, which makes it denser and prone to sink. (Photo by John Higdon/AGE Fotostock)

masses. This component of ocean circulation is a response to density differences among water masses that cause denser water to sink and slowly spread out beneath the surface. Because the density variations that cause deep-ocean circulation are caused by differences in temperature and salinity, deep-ocean circulation is also referred to as **thermohaline** (thermo = heat, haline = salt) **circulation**.

An increase in seawater density can be caused by either a decrease in temperature or an increase in salinity. Density changes due to salinity variations are important in very high latitudes, where water temperature remains low and relatively constant throughout the year.

Much of the water involved in deep-ocean currents (thermohaline circulation) begins in high latitudes at the surface. In these regions, where surface waters are cold, salinity increases when sea ice forms (**Figure 10.6**). When seawater freezes to form sea ice, salts do not become part of the ice. As a result, the salinity (and therefore the density) of the remaining seawater increases. When this surface water becomes dense enough, it sinks, initiating deep-ocean currents. Once this water sinks, it is removed from the physical processes that increased its density in the first place, and so its temperature and salinity remain largely unchanged for the duration of the time it spends in the deep ocean.

Near Antarctica, surface conditions create the highest-density water in the world. This cold saline brine slowly sinks to the seafloor, where it moves throughout the ocean basins in sluggish currents. After sinking from the surface of the ocean, deep waters will not reappear at the surface for an average of 500 to 2000 years.

A simplified model of ocean circulation is similar to a conveyor belt that travels from the Atlantic Ocean through the Indian and Pacific Oceans and back again (**Figure 10.7**). In

SmartFigure 10.7 The ocean conveyor belt Source areas for dense water masses exist in high-latitude regions where cold, high-salinity water sinks and flows into all the oceans. This water eventually ascends and completes the conveyor by returning to the source areas as warm surface currents. (https://goo.gl/ojdtTq)

Animation

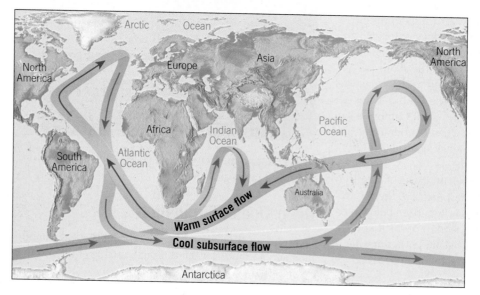

this model, warm water in the ocean's upper layers flows poleward, converts to dense water, and returns toward the equator as cold deep water that eventually upwells to complete the circuit. As this "conveyor belt" moves around the globe, it influences global climate by converting warm water to cold and liberating heat to the atmosphere.

CONCEPT CHECKS

1. Describe the process of coastal upwelling. Why is an abundance of marine life associated with these areas?
2. Why is deep-ocean circulation referred to as *thermohaline circulation*?
3. How does the ocean's surface salinity change when sea ice forms? How does this change influence density?

10.3 The Shoreline

Explain why the shoreline is considered a dynamic interface. List the factors that influence the height, length, and period of a wave and describe the motion of water within a wave.

Shorelines are dynamic environments. Their topography, geologic makeup, and climate vary greatly from place to place. Continental and oceanic processes converge along coasts to create landscapes that frequently undergo rapid change. When it comes to the deposition of sediment, they are transition zones between marine and continental environments.

A Dynamic Interface

Nowhere is the restless nature of the ocean's water more noticeable than along the shore—the dynamic interface among air, land, and sea. An **interface** is a common boundary where different parts of a system interact. This is certainly an appropriate designation for the coastal zone, where we can see the rhythmic rise and fall of tides and observe waves rolling in and breaking. Sometimes, the waves are low and gentle. At other times, they pound the shore with awesome fury.

Although it may not be obvious, the shoreline is constantly being modified by waves. Crashing surf erodes the land. Wave activity also moves sediment toward and away from the shore, as well as along it. Such activity sometimes produces narrow sandbars and fragile offshore islands that frequently change size and shape as storm waves come and go.

The nature of present-day shorelines is not just the result of the relentless attack of

the land by the sea. The shore has a complex character that results from multiple geologic processes. For example, practically all coastal areas were affected by the worldwide rise in sea level that accompanied the melting of glaciers following the Last Glacial Maximum (see Figure 4.22). As the sea encroached landward, the shoreline retreated, becoming superimposed upon existing landscapes that had resulted from such diverse processes as stream erosion, glaciation, volcanic activity, and the forces of mountain building.

Today, the coastal zone is experiencing intensive human activity. Unfortunately, people often treat the shoreline as if it were a stable platform on which structures can safely be built. This attitude inevitably leads to conflicts between people and nature. In October 2012, this fact was tragically reinforced when the storm surge from Hurricane Sandy struck parts of New York City and the narrow barrier islands along the coast of New Jersey (**Figure 10.8**). Many coastal landforms, especially beaches and barrier islands,

Figure 10.8 Hurricane Sandy A portion of the New Jersey shoreline shortly after this huge storm, unofficially called "Superstorm Sandy," made landfall just south of New York City in late October 2012. The extraordinary storm surge caused much of the damage pictured here. Many shoreline areas are intensively developed. Often the shifting shoreline sands and the desire of people to occupy these areas are in conflict. There is more about hurricanes and hurricane damage in Chapter 14. (Photo by Mario Tama/Getty Images)

are relatively fragile, short-lived features that are frequently inappropriate sites for development.

Ocean Waves

Ocean waves travel along the interface between ocean and atmosphere. They can carry energy from a storm far out at sea over distances of several thousand kilometers. That's why even on calm days, the ocean still has waves that travel across its surface. When observing waves, always remember that you are watching *energy* travel through a medium (water). If you make waves by tossing a pebble into a pond, or by splashing in a pool, or by blowing across the surface of a cup of coffee, you are imparting *energy* to the water, and the waves you see are the visible evidence of the energy passing through.

Wind-generated waves provide most of the energy that shapes and modifies shorelines. Where the land and sea meet, waves that may have traveled unimpeded for hundreds or thousands of kilometers suddenly encounter a barrier that will not allow them to advance farther and must absorb their energy. Stated another way, the shore is the location where a practically irresistible force confronts an almost immovable object. The conflict that results is never ending and sometimes dramatic.

Wave Characteristics

Most ocean waves derive their energy and motion from the wind. When a breeze has a speed less than 3 kilometers (2 miles) per hour, only small wavelets appear. At greater wind speeds, more stable waves gradually form and advance in the direction of the wind.

Characteristics of ocean waves are illustrated in **Figure 10.9**, which shows a simple, nonbreaking waveform. The tops of the waves are the *crests*, which are separated by *troughs*. Halfway between the crests and troughs is the *still water level*, which is the level that the water would occupy if there were no waves. The vertical distance between trough and crest is called the **wave height**, and the horizontal distance between successive

crests (or troughs) is the **wavelength**. The time it takes one full wave—one wavelength—to pass a fixed position is the **wave period**.

The height, length, and period that are eventually achieved by a wave depend on three factors: (1) wind speed, (2) length of time the wind has blown, and (3) **fetch**, the distance that wind has traveled across open water. As the quantity of energy transferred from the wind to the water increases, both the height and steepness of the waves increase. Eventually, a critical point is reached where waves grow so tall that they topple over, forming ocean breakers called *whitecaps*.

For a particular wind speed, there is a maximum fetch and duration of wind beyond which waves will no longer increase in size. When the maximum fetch and duration are reached for a given wind velocity, the waves are said to be "fully developed." The reason that waves can grow no further is that they are losing as much energy through the breaking of whitecaps as they are receiving from the wind.

When the wind stops or changes direction, or the waves leave the storm area where they were created, the waves continue on without relation to local winds. The waves also undergo a gradual change to *swells*, which are lower in height and longer in length and may carry a storm's energy to distant shores. Because many independent wave systems exist at the same time, the sea surface acquires a complex and irregular pattern, sometimes producing very large waves. The sea waves that are seen from shore are usually a mixture of swells from faraway storms and waves created by local winds.

Circular Orbital Motion

Waves can travel great distances across ocean basins. In one study, waves generated near Antarctica were tracked as they traveled through the Pacific Ocean basin. After more than 10,000 kilometers (more than 6000 miles), the waves finally expended their energy a week later, along the shoreline of the Aleutian Islands of Alaska. It is important to realize that what crossed the Pacific was the waveform, not the water. As a wave passes, each bit of water moves in a near-circle and returns to about where it started. This movement, which transfers wave energy, is called *circular orbital motion*.

Observation of an object floating in waves reveals that it moves not only up and down but also slightly forward and backward with each successive wave. **Figure 10.10** shows that a floating object moves up and backward as the crest approaches, up and forward

SmartFigure 10.9 Wave basics An idealized nonbreaking wave, showing its basic parts and the movement of water with increasing depth. (https://goo.gl/57GfHl)

Animation

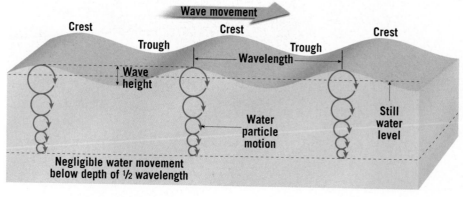

Wave movement

Crest · Crest · Crest

Trough · Trough

Wavelength

Wave height

Water particle motion

Still water level

Negligible water movement below depth of ½ wavelength

as the crest passes, down and forward after the crest, and down and backward as the trough approaches, and then it rises and moves backward again as the next crest advances. When the movement of the toy boat shown in Figure 10.10 is traced as a wave passes, it can be seen that the boat moves in a circle, and it returns to nearly the same place. Circular orbital motion allows a wave-form (the wave's shape) to move forward *through the water*, while the individual water particles that transmit the wave move around in a circle. Wind moving across a field of wheat causes a similar phenomenon: The wheat itself does not travel across the field, but the waves do.

The energy contributed by the wind to the water is transmitted not only along the surface of the sea but also downward. However, beneath the surface, the circular motion rapidly diminishes until, at a depth equal to one-half the wavelength measured from the still water level, the movement of water particles becomes negligible. This depth is known as the *wave base*. The dramatic decrease of wave energy with depth is shown by the rapidly diminishing diameters of water-particle orbits in Figure 10.9.

Waves in the Surf Zone

As long as a wave is in deep water, it is unaffected by water depth (**Figure 10.11**, *left*). However, when a wave approaches the shore, the water becomes shallower and influences wave behavior. The wave begins to "feel bottom" at a water depth equal to its wave base. Such depths interfere with water movement at the base of the wave and slow its advance (see Figure 10.11, *center*).

As a wave advances toward the shore, the slightly faster waves farther out to sea catch up, decreasing the wavelength. As the speed and length of the wave diminish, the wave steadily grows higher. Finally, a critical point is reached when the wave is too steep to support itself and the wave front collapses, or *breaks* (see Figure 10.11, *right*), causing water to advance up the shore.

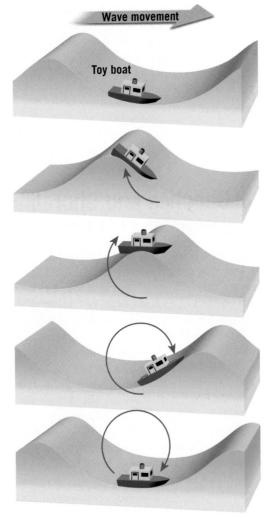

SmartFigure 10.10 Passage of a wave The movements of the toy boat show that the wave form advances, but the water does not advance appreciably from the original position. In this sequence, the wave moves from left to right as the boat (and the water in which it is floating) rotates in an imaginary circle. (http://goo.gl/yT2d1q)

SmartFigure 10.11 Waves approaching the shore Waves touch bottom as they encounter water depths that are less than half a wavelength. As a result, the wave speed decreases, and the faster-moving waves farther from shore begin to catch up, which causes the distance between waves (the wavelength) to decrease. This causes an increase in wave height, to the point where the waves finally pitch forward and break in the surf zone. The first portion of the animation deals with the ideas presented in this figure. (https://goo.gl/57GfHI)

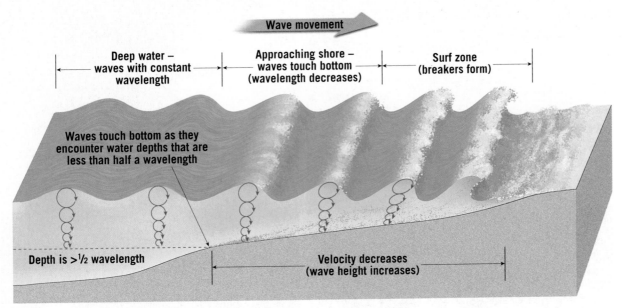

The turbulent water created by breaking waves is called **surf**. On the landward margin of the surf zone, the turbulent sheet of water from collapsing breakers, called *swash*, moves up the slope of the beach. When the energy of the swash has been expended, the water flows back down the beach toward the surf zone as *backwash*.

10.3 CONCEPT CHECKS

1. Define *interface*.
2. Aside from ocean waves, what other factors influence the nature of present-day shorelines?
3. List three factors that determine the height, length, and period of a wave.
4. Describe the motion of a floating object as a wave passes.
5. How do a wave's speed, wavelength, and height change as it moves into shallow water and breaks?

10.4 Beaches and Shoreline Processes

Describe how waves erode and move sediment along the shore.

For many, a beach is a sandy area where people lie in the sun and walk along the water's edge. Technically, a **beach** is an accumulation of sediment found along the landward margins of an ocean or a lake. Along straight coasts, beaches may extend for tens or hundreds of kilometers. Where coasts are irregular, beach formation may be confined to the relatively quiet waters of bays.

Beaches are composed of whatever material is locally abundant. The sediment for some beaches is derived from the erosion of adjacent cliffs or nearby coastal mountains. Other beaches are built from sediment delivered to the coast by rivers. Although the mineral makeup of many beaches is dominated by durable quartz grains, other minerals may be dominant. For example, in areas such as southern Florida, where there are no mountains or other sources of rock-forming minerals nearby, most beaches are composed largely of shell fragments and the remains of organisms that live in coastal waters (**Figure 10.12A**). Some beaches on volcanic islands in the open ocean are composed of weathered grains of the basaltic lava that comprise the islands or of coarse debris eroded from the coral reefs that develop around many tropical islands (**Figure 10.12B**).

Regardless of its composition, the material that comprises a beach does not stay in one place. Instead, the waves that crash along the shoreline are constantly moving it. Thus, beaches can be thought of as material in transit along the shoreline.

Wave Erosion

During calm weather, wave action is minimal. During storms, however, waves can cause much erosion. The impact of large, high-energy waves against the shore

Figure 10.12 Beaches A beach is an accumulation of sediment on the landward margin of an ocean or a lake and can be thought of as material in transit along the shore. Beaches are composed of whatever material is locally available. (Photo A by David R. Frazier/Photo Library/Alamy Images; photo B by E. J. Tarbuck)

This beach on Florida's Sanibel Island consists of shells and shell fragments.

A.

The black sands on this beach in Hawaii were derived from the weathering of nearby basaltic lava flows.

B.

can be awesome in its violence. Each breaking wave may hurl thousands of tons of water against the land, sometimes causing the ground literally to tremble. The pressures exerted by Atlantic waves in winter, for example, average nearly 10,000 kilograms per square meter (more than 2000 pounds per square foot). The force during storms is even greater.

It is no wonder that cracks and crevices are quickly opened in cliffs, coastal structures, and anything else that is subjected to these enormous shocks (**Figure 10.13**). Water is forced into every opening, causing air in the cracks to become highly compressed by the thrust of crashing waves. When a wave subsides, the air expands rapidly, dislodging rock fragments and enlarging and extending fractures.

In addition to the erosion caused by wave impact and pressure, **abrasion**, the sawing and grinding action of the water armed with rock fragments, is also important. In fact, abrasion is probably more intense in the surf zone than in any other environment. Smooth, rounded stones and pebbles along the shore are obvious reminders of the relentless grinding action of rock against rock in the

Figure 10.13 Storm waves When large waves break against the shore, the force of the water can be powerful and the erosional work that is accomplished can be great. These storm waves are breaking along the coast of Wales. (The Photo Library Wales/Alamy)

surf zone (**Figure 10.14A**). Rock fragments are also used as "tools" by the waves as they cut horizontally into the land (**Figure 10.14B**).

Sand Movement on the Beach

Beaches are sometimes called "rivers of sand." This is an appropriate description because the energy from breaking waves often causes large quantities of sand to move roughly parallel to the shoreline along the beach face and in the offshore surf zone. Wave energy also causes sand to move perpendicular to (toward and away from) the shoreline.

A.

Smooth, rounded rocks along the shore are an obvious reminder that abrasion can be intense in the surf zone.

B.

This sandstone cliff at Gabriola Island, British Columbia, was undercut by wave action.

Figure 10.14 Abrasion: Sawing and grinding Breaking waves armed with rock debris can do a great deal of erosional work. (Photo A by Michael Collier; photo B by Fletcher and Baylis/Science Source)

Movement Perpendicular to the Shoreline If you stand ankle deep in water at the beach, you will see that swash and backwash move sand toward and away from the shoreline. Whether there is a net loss or addition of sand depends on the level of wave activity. When wave activity is relatively light (less energetic waves), much of the swash soaks into the beach, which reduces the backwash. Consequently, the swash dominates and causes a net movement of sand up the beach face.

When high-energy waves prevail, the beach is saturated from previous waves, so much less of the swash soaks in. As a result, the beach erodes because backwash is strong and causes a net movement of sand toward open water.

Along many beaches, light wave activity is the rule during the summer. Therefore, a wide sandy beach gradually develops. During winter, when storms are frequent and more powerful, strong wave activity erodes and narrows the beach. A wide beach that may have taken months to build can be dramatically narrowed in just a few hours by the high-energy waves created by a strong winter storm.

Wave Refraction The bending of waves, called **wave refraction**, plays an important part in shoreline processes (**Figure 10.15**). It affects the distribution of energy along the shore and thus strongly influences where and to what degree erosion, sediment transport, and deposition will take place.

Waves seldom approach the shore straight on. Rather, most waves move toward the shore at a slight angle. However, when they reach the shallow water of a smoothly sloping bottom, the wave crests are refracted (bent) and tend to become parallel to the shore. Such

bending occurs because the part of the wave nearest the shore touches bottom and slows first, whereas the part of the wave that is still in deeper water continues forward at its full speed. The net result is a wave that approaches nearly parallel to the shore, regardless of its original orientation.

Because of refraction, wave energy is concentrated against the sides and ends of headlands that project into the water, whereas wave attack is weakened in bays. This differential wave attack along irregular coastlines is illustrated in Figure 10.15. Because the waves reach the shallow water in front of the headland sooner than they do in adjacent bays, they are bent more nearly parallel to the protruding land and strike it from all three sides. By contrast, refraction in the bays causes waves to diverge and expend less energy. In these zones of weakened wave activity, sediments can accumulate and form sandy beaches. Over a long period, erosion of the headlands and deposition in the bays will straighten an irregular shoreline.

Longshore Transport Although waves are refracted, most still reach the shore at a slight angle. Consequently, the uprush of water from each breaking wave (the swash) is at an oblique angle to the shoreline. However, the backwash is straight down the slope of the beach. The effect of this pattern of water movement is to transport sediment in a zigzag pattern along the beach face (**Figure 10.16**). This movement is called **beach drift**, and it can transport sand and pebbles hundreds or even thousands of meters daily. However, a more typical rate is 5 to 10 meters (16 to 33 feet) per day.

Waves that approach the shore at an angle also produce currents within the surf zone that flow parallel to the shore. These currents move substantially more

SmartFigure 10.15 Wave refraction As waves first touch bottom in the shallows along an irregular coast, they are slowed; they then bend (refract) and align nearly parallel to the shoreline. (Photo by Rich Reid/National Geographic/Getty Images) (https://goo.gl/3jIK4V)

Tutorial

As these waves approach nearly straight on, refraction causes the wave energy to be concentrated at headlands (resulting in erosion) and dispersed in bays (resulting in deposition).

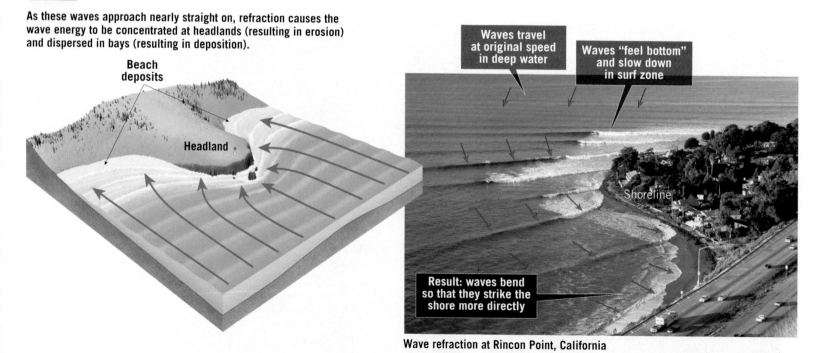

Beach deposits

Headland

Waves travel at original speed in deep water

Waves "feel bottom" and slow down in surf zone

Shoreline

Result: waves bend so that they strike the shore more directly

Wave refraction at Rincon Point, California

sediment than does beach drift (see Figure 10.16). Because water in the surf zone is turbulent, these **longshore currents** easily move the fine suspended sand and larger sand and gravel along the bottom. When the sediment transported by longshore currents is added to the quantity moved by beach drift, the total amount can be very large. At Sandy Hook, New Jersey, example, the quantity of sand transported along the shore over a 48-year period averaged almost 680,000 metric tons (750,000 short tons) annually. For a 10-year period at Oxnard, California, more than 1.4 million metric tons (1.5 million short tons) of sediment moved along the shore each year.

Both rivers and coastal zones move water and sediment from one area (*upstream*) to another (*downstream*). That is why the beach is often characterized as a "river of sand." Beach drift and longshore currents, however, move in a zigzag pattern, whereas rivers flow mostly in a turbulent, swirling fashion. In addition, the direction of flow of longshore currents along a shoreline can change, whereas rivers always flow in the same direction (downhill). Longshore currents change direction when waves approach the beach in different directions at different seasons. Nevertheless, longshore currents generally flow southward along both the Atlantic and Pacific shores of the United States.

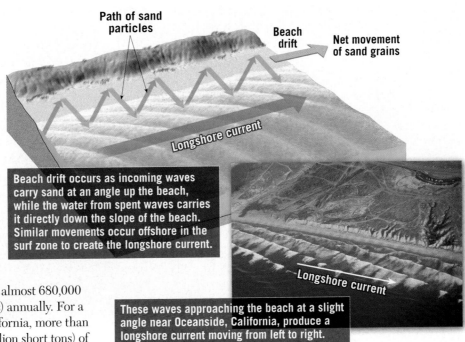

Path of sand particles

Beach drift

Net movement of sand grains

Longshore current

Beach drift occurs as incoming waves carry sand at an angle up the beach, while the water from spent waves carries it directly down the slope of the beach. Similar movements occur offshore in the surf zone to create the longshore current.

Longshore current

These waves approaching the beach at a slight angle near Oceanside, California, produce a longshore current moving from left to right.

SmartFigure 10.16 The longshore transport system The two components of the transport system, beach drift and longshore currents, are created by breaking waves that approach the beach at an angle. These processes transport large quantities of material along the beach and in the surf zone. (Photo by University of Washington Libraries, Special Collections, KC14461) (https://goo.gl/Z0EFUs)

Tutorial

10.4 CONCEPT CHECKS

1. What is a beach?
2. Why do waves approaching the shoreline often bend?
3. What is the effect of wave refraction along an irregular coastline?
4. Describe the two processes that contribute to longshore transport.

10.5 Shoreline Features

Describe the features typically created by wave erosion and those resulting from sediment deposited by longshore transport processes.

A fascinating assortment of shoreline features can be observed along the world's coastal regions. Although the same processes cause change along every coast, not all coasts respond in the same way. Interactions among different processes and the relative importance of each process depend on local factors. The factors include (1) the proximity of a coast to sediment-laden rivers, (2) the degree of tectonic activity, (3) the topography and composition of the land, (4) prevailing winds and weather patterns, and (5) the configuration of the coastline and near-shore areas. Features that form primarily due to erosion are called *erosional features*, whereas deposits of sediment produce *depositional features*.

Erosional Features

Many coastal landforms owe their origin to erosional processes. Such erosional features are common along the rugged and irregular New England coast and along the steep shorelines of the west coast of the United States.

Wave-Cut Cliffs, Wave-Cut Platforms, and Marine Terraces As the name implies, **wave-cut cliffs** originate in the cutting action of the surf against the base of coastal land. As erosion progresses, rocks overhanging the notch at the base of the cliff crumble into the surf, and the cliff retreats. A relatively flat, benchlike surface, called a **wave-cut platform**, is left behind by the receding cliff (**Figure 10.17**, *left*). The platform broadens as wave attack continues. Some debris produced by the breaking waves remains along the water's edge as sediment on the beach, and the remainder is transported farther seaward. If a wave-cut platform is uplifted above sea level by tectonic forces, it becomes a **marine terrace** (see Figure 10.17, *right*). Marine terraces are easily recognized by their

Did You Know?
Along shorelines composed of unconsolidated material rather than hard rock, the rate of erosion by breaking waves can be extraordinary. In parts of Britain, where waves have the easy task of eroding glacial deposits of sand, gravel, and clay, the coast has been worn back 3 to 5 km (2 to 3 mi) since Roman times (2000 years ago), sweeping away many villages and ancient landmarks.

Figure 10.17 Wave-cut platform and marine terrace This wave-cut platform is exposed at low tide along the California coast at Bolinas Point near San Francisco. A wave-cut platform was uplifted to create the marine terrace. (Photo by University of Washington Libraries, Special Collections, KC5902)

gentle seaward-sloping shape and are often perceived as desirable sites for coastal development.

Sea Arches and Sea Stacks Headlands that extend into the sea are vigorously attacked by waves because of refraction. The surf erodes the rock selectively, wearing away the softer or more highly fractured rock at the fastest rate. At first, sea caves may form. When two caves on opposite sides of a headland unite, a **sea arch** results (**Figure**

10.18). Eventually, the arch falls in, leaving an isolated remnant, or **sea stack**, on the wave-cut platform. In time, it, too, will be consumed by the action of the waves.

Depositional Features

Sediment from a beach is transported along the shore and deposited in areas where wave energy is low. Such processes produce a variety of depositional features.

Figure 10.18 Sea stack and sea arch These features at the tip of Mexico's Baja Peninsula resulted from vigorous wave attack on a headland. Sea stacks are also shown in the chapter-opening photo. (Photo by Lew Robertson/Getty Images)

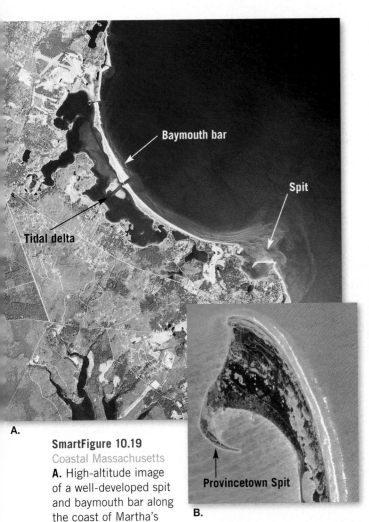

sand parallel the coast at distances from 3 to 30 kilometers (2 to 20 miles) offshore. From Cape Cod, Massachusetts, to Padre Island, Texas, nearly 300 barrier islands rim the coast (**Figure 10.20**).

Most barrier islands are from 1 to 5 kilometers (0.6 to 3 miles) wide and 15 to 30 kilometers (9 to 19 miles) long. The tallest features are sand dunes, which usually reach heights of 5 to 10 meters (16 to 33 feet). The lagoons that separate these narrow islands from the shore are zones of relatively quiet water that allow small craft traveling between New York and northern Florida to avoid the rough waters of the North Atlantic.

Barrier islands probably form in several ways. Some originate as spits that were subsequently severed from the mainland by wave erosion or by the general rise in sea level following the last episode of glaciation. Others were created when turbulent waters in the line of breakers heaped up sand that had been scoured from the bottom. Finally, some barrier islands may be former sand-dune ridges that originated along the shore during the last glacial period, when sea level was lower. As the ice

Figure 10.20 Barrier islands Nearly 300 barrier islands rim the Gulf and Atlantic coasts. The islands along the coast of North Carolina are excellent examples. This view is looking south. (Photo by Michael Collier)

SmartFigure 10.19
Coastal Massachusetts
A. High-altitude image of a well-developed spit and baymouth bar along the coast of Martha's Vineyard. (Image courtesy of USDA-ASCS) **B.** This photograph, taken from the International Space Station, shows Provincetown Spit at the tip of Cape Cod. (NASA image) (http://goo.gl/MfwH34)

Mobile Field Trip

Spits, Bars, and Tombolos Where beach drift and longshore currents are active, several features related to the movement of sediment along the shore may develop. A **spit** is an elongated ridge of sand that projects from the land into the mouth of an adjacent bay. Often the end in the water hooks landward in response to the dominant direction of the longshore current. Both images in **Figure 10.19** show spits. The term **baymouth bar** is applied to a sandbar that completely crosses a bay, sealing it off from the open ocean. Such a feature tends to form across bays where currents are weak enough to allow a spit to extend to the other side (see Figure 10.19A). A **tombolo**, a ridge of sand that connects an island to the mainland or to another island, forms in much the same manner as a spit.

Barrier Islands The Atlantic and Gulf coastal plains are relatively flat and slope gently seaward. The shore zone is characterized by **barrier islands**. These low ridges of

Figure 10.21 The evolving shore These diagrams illustrate changes that can take place through time along an initially irregular coastline that remains relatively stable. The diagrams also illustrate many of the features described in the section on shoreline features. (Top and bottom photos by E. J. Tarbuck; middle photo by Michael Collier)

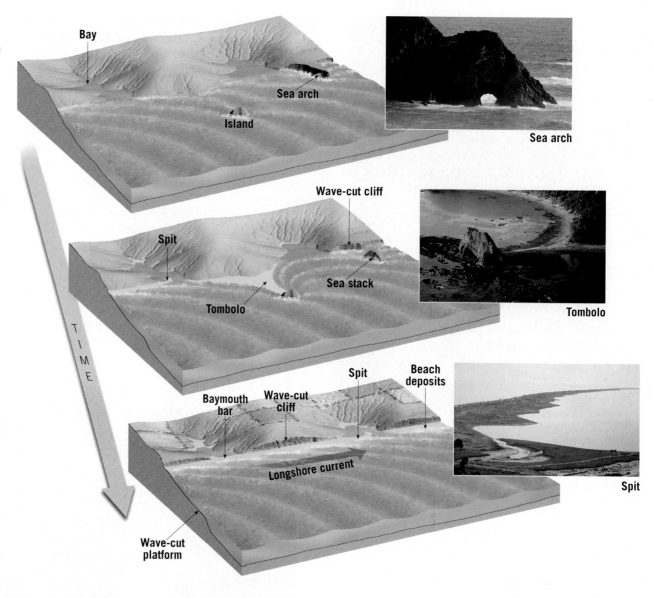

sheets melted, sea level rose and flooded the area behind the beach–dune complex.

The Evolving Shore

A shoreline continually undergoes modification, regardless of its initial configuration. At first, most coastlines are irregular, although the degree of and reason for the irregularity may differ considerably from place to place. Along a coastline that is characterized by varied geology, the pounding surf may initially increase its irregularity as the waves erode the weaker rocks more easily than the stronger ones. However, if a shoreline remains stable, marine erosion and deposition will eventually produce a straighter, more regular coast.

Figure 10.21 illustrates the evolution of an initially irregular coast that remains relatively stable and shows many of the coastal features discussed in the previous

section. As headlands are eroded and erosional features such as wave-cut cliffs and wave-cut platforms are created, sediment is produced that is carried along the shore by beach drift and longshore currents. Some material is deposited in the bays, whereas other debris is formed into depositional features such as spits and baymouth bars. At the same time, rivers fill the bays with sediment. Eventually, a generally straight, smooth coast results.

10.5 CONCEPT CHECKS

1. Explain how the marine terrace in Figure 10.17 formed.
2. Describe the formation of each labeled feature in Figure 10.21.
3. Along which of America's coastal areas are barrier islands common?

10.6 Contrasting America's Coasts

Distinguish between emergent and submergent coasts. Contrast the erosion problems faced along different parts of America's coasts.

The shoreline along the Pacific coast of the United States is strikingly different from that of the Atlantic and Gulf coast regions. Some of the differences are related to plate tectonics. The west coast is the leading edge of the North American plate, and because of this, it experiences active uplift and deformation. By contrast, the east coast is far from any active plate boundary and is relatively quiet tectonically. Because of this basic geologic difference, the nature of shoreline erosion problems along America's opposite coasts is different.

Coastal Classification

The great variety of shorelines demonstrates their complexity. Indeed, to understand any particular coastal area, many factors must be considered, including rock types, size and direction of waves, frequency of storms, tidal range, and offshore topography. In addition, practically all coastal areas were affected by the worldwide rise in sea level that accompanied the melting of Ice Age glaciers at the close of the Pleistocene epoch. Finally, tectonic events that elevate or drop the land or change the volume of ocean basins must be taken into account. The large number of factors that influence coastal areas make shoreline classification difficult.

Many geologists classify coasts based on the changes that have occurred with respect to sea level. This commonly used classification divides coasts into two general categories: emergent and submergent. **Emergent coasts** develop either because an area experiences uplift or as a result of a drop in sea level. Conversely, **submergent coasts** are created when sea level rises or the land adjacent to the sea subsides.

Emergent Coasts In some areas, the coast is clearly emergent because rising land or a falling water level exposes wave-cut cliffs and platforms above sea level. Excellent examples include portions of coastal California, where uplift has occurred in the recent geologic past. The marine terrace shown in Figure 10.17 illustrates this situation. In the case of the Palos Verdes Hills, south of Los Angeles, seven different terrace levels exist, indicating seven episodes of uplift. The ever-persistent sea is now cutting a new platform at the base of the cliff. If uplift follows, this platform, too, will become an elevated marine terrace.

Other examples of emergent coasts include regions that were once buried beneath great ice sheets. When glaciers were present, their weight depressed the crust, and when the ice melted, the crust began gradually to spring back. Consequently, prehistoric shoreline features may now be found high above sea level. The Hudson Bay region of Canada is such an area; portions of it are still rising at a rate of more than 1 centimeter per year.

Submergent Coasts In contrast to the preceding examples, other coastal areas show definite signs of submergence. Shorelines that have been submerged in the relatively recent past are often highly irregular because the sea typically floods the lower reaches of river valleys flowing into the ocean. The ridges separating the valleys, however, remain above sea level and project into the sea as headlands. These drowned river mouths, which are called **estuaries**, characterize many coasts today. Along the Atlantic coastline, the Chesapeake and Delaware Bays are examples of large estuaries created by submergence (**Figure 10.22**). The picturesque coast of Maine, particularly in the vicinity of Acadia National Park, is

SmartFigure 10.22 East coast estuaries The lower portions of many river valleys were flooded by the rise in sea level that followed the end of the Quaternary Ice Age, creating large estuaries such as Chesapeake and Delaware Bays. (https://goo.gl/iYlc7z)

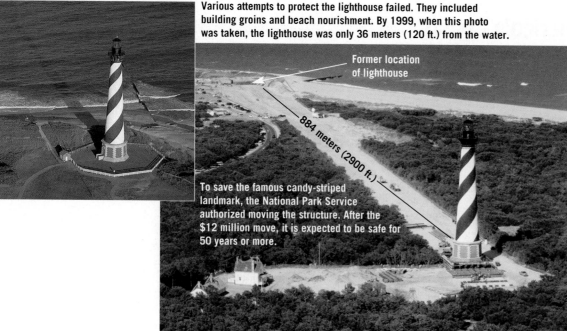

Various attempts to protect the lighthouse failed. They included building groins and beach nourishment. By 1999, when this photo was taken, the lighthouse was only 36 meters (120 ft.) from the water.

Former location of lighthouse

884 meters (2900 ft.)

To save the famous candy-striped landmark, the National Park Service authorized moving the structure. After the $12 million move, it is expected to be safe for 50 years or more.

Figure 10.23 Relocating the Cape Hatteras lighthouse After the failure of a number of efforts to protect this 21-story lighthouse, the nation's tallest lighthouse, from being destroyed due to a receding shoreline, the structure finally had to be moved. (Photo showing original location by Don Smetzer/PhotoEdit Inc.; photo showing new location by Drew C. Wilson/Virginian-Pilot/AP Images)

another excellent example of an area that was flooded by the postglacial rise in sea level and transformed into a highly irregular coastline.

Keep in mind that most coasts have complicated geologic histories. With respect to sea level, many have, at various times, emerged and then submerged. Each time, they may retain some of the features created during the previous situation.

Atlantic and Gulf Coasts

Much of the coastal development along the Atlantic and Gulf coasts has occurred on barrier islands. Typically, a barrier island, also termed a *barrier beach* or *coastal barrier*, consists of a wide beach that is backed by dunes and separated from the mainland by marshy lagoons. The broad expanses of sand and exposure to the ocean have made barrier islands exceedingly attractive sites for development. Unfortunately, development has taken place more rapidly than increases in our understanding of barrier island dynamics.

Because barrier islands face the open ocean, they receive the full force of major storms that strike the coast. When a storm occurs, the barriers absorb the energy of the waves primarily through the movement of sand. **Figure 10.23**, which shows changes at Cape Hatteras National Seashore, reinforces this point. The process and the dilemma that results were recognized years ago and accurately described as follows:

> Waves may move sand from the beach to offshore areas or, conversely, into the dunes; they may erode the dunes, depositing sand onto the beach or carrying it out to sea; or they may carry sand from the

beach and the dunes into the marshes behind the barrier, a process known as overwash. The common factor is movement. Just as a flexible reed may survive a wind that destroys an oak tree, so the barriers survive hurricanes and nor'easters not through unyielding strength but by giving before the storm.

This picture changes when a barrier is developed for homes or as a resort. Storm waves that previously rushed harmlessly through gaps between the dunes now encounter buildings and roadways. Moreover, since the dynamic nature of the barriers is readily perceived only during storms, homeowners tend to attribute damage to a particular storm, rather than to the basic mobility of coastal barriers. With their homes or investments at stake, local residents are more likely to seek to hold the sand in place and the waves at bay than to admit that development was improperly placed to begin with.[*]

Pacific Coast

In contrast to the broad, gently sloping coastal plains of the Atlantic and Gulf coasts, much of the Pacific coast is characterized by relatively narrow beaches that are backed by steep cliffs and mountain ranges (**Figure 10.24**). Recall that America's western margin is a more rugged and tectonically active region than the eastern margin. Because uplift continues, a rise in sea level in the West is not so readily apparent. Nevertheless, like the shoreline erosion problems facing the East's barrier islands, west coast difficulties also stem largely from the alteration of a natural system by people.

A major problem facing the Pacific shoreline, and especially portions of southern California, is a significant

---

[*]Frank Lowenstein, "Beaches or Bedrooms—The Choice as Sea Level Rises," *Oceanus* 28 (No. 3, Fall 1985): p. 22 © Woods Hole Oceanographic Institute.

Figure 10.24 Pacific coast Wave refraction along the coast of California south of Shelter Cove. These steep cliffs are a sharp contrast to typical scenes along the Atlantic and Gulf coasts. (Photo by Michael Collier)

narrowing of many beaches. The bulk of the sand on many of these beaches is supplied by rivers that transport it from the mountains to the coast. Over the years, this natural flow of material to the coast has been interrupted

by dams built for irrigation and flood control. The reservoirs effectively trap the sand that would otherwise nourish the beach environment. When the beaches were wider, they protected the cliffs behind them from the force of storm waves. Now, however, the waves move across the narrowed beaches without losing much energy and cause more rapid erosion of the sea cliffs.

Although the retreat of the cliffs provides material to replace some of the sand impounded behind dams, it also endangers homes and roads built on the bluffs. In addition, development atop the cliffs aggravates the problem. Urbanization increases runoff, which, if not carefully controlled, can result in serious bluff erosion. Watering lawns and gardens adds significant quantities of water to the slope. This water percolates downward toward the base of the cliff, where it may emerge in small seeps. This action reduces the slope's stability and facilitates mass wasting.

Shoreline erosion along the Pacific coast varies considerably from one year to the next, largely because of the sporadic occurrence of storms. As a consequence, when the infrequent but serious episodes of erosion occur, the damage is often blamed on the unusual storms and not on coastal development or the sediment-trapping dams that may be great distances away. If, as predicted, sea level rises at an increasing rate in the years to come, increased shoreline erosion and sea-cliff retreat should be expected along many parts of the Pacific coast.

10.6 CONCEPT CHECKS

1. Are estuaries associated with submergent or emergent coasts? Explain.
2. What observable features would lead you to classify a coastal area as emergent?
3. Briefly describe what happens when storm waves strike an undeveloped barrier island.
4. How might building a dam on a river that flows to the sea affect a coastal beach?

10.7 Stabilizing the Shore

Summarize the ways in which people deal with shoreline erosion problems.

The coastal zone teems with human activity. Unfortunately, people often treat the shoreline as if it were a stable platform on which structures can be built safely. This approach jeopardizes both people and the shoreline because many coastal landforms are relatively fragile, short-lived features that are easily damaged by development. As anyone who has endured a strong coastal storm knows, the shoreline is not always a safe place to live. A glance back at Figure 10.8 reminds us of that fact.

Compared with other natural hazards, such as earthquakes, volcanic eruptions, and landslides, shoreline erosion appears to be a more continuous and predictable

process that causes relatively modest damage to limited areas. In reality, the shoreline is one of Earth's most dynamic places, changing rapidly in response to natural forces. Exceptional storms are capable of eroding beaches and cliffs at rates that far exceed the long-term average. Such bursts of accelerated erosion not only have a significant impact on the natural evolution of a coast but can also have a profound impact on people who reside in the coastal zone. Erosion along the coast causes significant property damage. Huge sums are spent annually not only to repair damage but also to prevent or control erosion. Already a problem at many sites, shoreline

erosion is certain to become increasingly serious as extensive coastal development continues.

During the past 100 years, growing affluence and increasing demands for recreation have brought unprecedented development to many coastal areas. As both the number and the value of buildings have increased, so, too, have efforts to protect property from storm waves by stabilizing the shore. Also, controlling the natural migration of sand is an ongoing struggle in many coastal areas. Such interference can result in unwanted changes that are difficult and expensive to correct.

Hard Stabilization

Structures built to protect a coast from erosion or to prevent the movement of sand along a beach are known as **hard stabilization**. Hard stabilization can take many forms and often results in predictable yet unwanted outcomes. Hard stabilization includes groins, breakwaters, and seawalls.

Groins To maintain or widen beaches that are losing sand, groins are sometimes constructed. A **groin** is a barrier built at a right angle to the beach to trap sand that is moving parallel to the shore. Groins are usually constructed of large rocks but may also be composed of wood. These structures often do their job so effectively that the longshore current beyond the groin becomes sand-starved. As a result, the current erodes sand from the beach on the downstream side of the groin.

To offset this effect, property owners downstream from the structure may erect a groin on their property. In this manner, the number of groins multiplies, resulting in a *groin field* (**Figure 10.25**). An example of such proliferation is the shoreline of New Jersey, where hundreds of these structures have been built. Because it has been shown that groins often do not provide a satisfactory solution, using them is no longer the preferred method of keeping beach erosion in check.

Breakwaters and Seawalls Hard stabilization can be built parallel to the shoreline. One such structure is a **breakwater**, which is designed to protect boats from the force of large breaking waves by creating a quiet-water zone near the shore. However, the reduced wave activity along the shore behind the breakwater may allow sand to accumulate. If this happens, the boat anchorage will eventually fill with sand, while the downstream beach erodes and retreats. At Santa Monica, California, where a breakwater has created such a problem, the city uses a dredge to remove sand from the protected quiet-water zone and deposit it farther downstream, where longshore currents continue to move the sand down the coast (**Figure 10.26**).

Another type of hard stabilization built parallel to the shore is a **seawall**, which is designed to armor the coast and defend property from the force of breaking waves. Waves expend much of their energy as they move across an open beach. Seawalls cut this process short by reflecting the force of unspent waves seaward. As a consequence, the beach to the seaward side of the seawall experiences significant erosion and may, in some instances, be eliminated entirely (**Figure 10.27**). Once the width of the beach is reduced, the seawall is subjected to even greater pounding by the waves. Eventually, this battering will take its toll; the seawall will fail and will need to be replaced with a larger, more expensive structure.

The wisdom of building temporary protective structures along shorelines is increasingly questioned. The opinions of many coastal scientists and engineers is that halting an eroding shoreline with protective structures

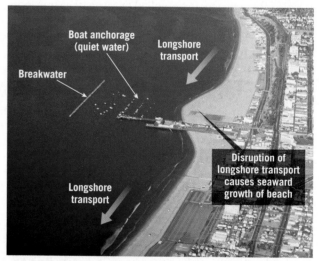

Figure 10.26 Breakwater Aerial view of a breakwater at Santa Monica, California. The structure appears as a line in the water behind which many boats are anchored. The construction of the breakwater disrupted longshore transport and caused the seaward growth of the beach. (Photo by University of Washington Libraries, Special Collections, KC8275)

Figure 10.27 Seawall Seabright in northern New Jersey once had a broad, sandy beach. A seawall 5 to 6 meters (16 to 18 feet) high and 8 kilometers (5 miles) long was built to protect the town and the railroad that brought tourists to the beach. After the wall was built, the beach narrowed dramatically. (Photo by Rafael Macia/Science Source)

benefits only a few and seriously degrades or destroys the natural beach and the value it holds for the majority. Protective structures divert the ocean's energy temporarily from private properties but usually refocus that energy on the adjacent beaches. Many structures interrupt the natural sand flow in coastal currents, robbing affected beaches of vital sand replacement.

Alternatives to Hard Stabilization

Armoring the coast with hard stabilization has several potential drawbacks, including the cost of the structure and the loss of sand on the beach. Alternatives to hard stabilization include beach nourishment and changing land use.

Beach Nourishment One approach to stabilizing shoreline sands without hard stabilization is **beach nourishment**. As the term implies, this practice involves adding large quantities of sand to the beach system (**Figure 10.28**). Extending beaches seaward makes buildings along the shoreline less vulnerable to destruction by storm waves and enhances recreational uses. Without sandy beaches, tourism suffers.

The process of beach nourishment is straightforward. Sand is pumped by dredges from offshore or trucked from inland locations. The "new" beach, however, will not be the same as the former beach. Because replenishment sand is from somewhere else, typically not another beach, it is new to the beach environment. The new sand is often different in size, shape, sorting, and composition. Such differences pose problems in terms of erodibility and the kinds of life the new sand will support.

Beach nourishment is not a permanent solution to the problem of shrinking beaches. The same processes that removed the sand in the first place will eventually remove the replacement sand as well. Nevertheless, the number of nourishment projects has increased in recent years, and many beaches, especially along the Atlantic coast, have had their sand replenished many times. Virginia Beach, Virginia, has been nourished more than 50 times.

Beach nourishment is costly. For example, a modest project might involve 38,000 cubic meters (50,000

Figure 10.28 Beach nourishment If you visit a beach along the Atlantic coast, it is more and more likely that you will walk into the surf zone atop an artificial beach. (Photo by Michael Weber/Image Broker/Alamy Images)

cubic yards) of sand distributed across about 1 kilometer (0.6 mile) of shoreline. A good-sized dump truck holds about 7.6 cubic meters (10 cubic yards) of sand. So this small project would require about 5000 dump-truck loads. Many projects extend for many miles. Nourishing beaches typically costs millions of dollars per mile.

Changing Land Use Instead of building structures such as groins and seawalls to hold the beach in place or adding sand to replenish eroding beaches, another option is available. Many coastal scientists and planners are calling for a policy shift from defending and rebuilding beaches and coastal property in high-hazard areas to relocating or abandoning storm-damaged buildings and letting nature reclaim the beach. This approach is similar to an approach the federal government adopted for river floodplains following the devastating 1993 Mississippi River floods, in which vulnerable structures are abandoned and/or relocated on higher, safer ground.

A recent example of changing land use occurred on New York's Staten Island following Hurricane Sandy in 2012. The state turned some vulnerable shoreline areas of the island into waterfront parks. The parks act as buffers to protect inland homes and businesses from strong storms while providing the community with needed open space and access to recreational opportunities.

Land use changes are sometimes controversial. People with significant near-shore investments want to rebuild and defend coastal developments from the erosional wrath of the sea. Others, however, argue that with sea level rising, the impact of coastal storms will get worse in the decades to come, and vulnerable or oft-damaged structures should be abandoned or relocated to improve personal safety and reduce costs. Such ideas will no doubt be the focus of much study and debate as states and communities evaluate and revise coastal land use policies.

10.7 CONCEPT CHECKS

1. List two examples of hard stabilization and describe what each is intended to do. How does each affect the distribution of sand on a beach?
2. What are two alternatives to hard stabilization, and what potential problems are associated with each?

10.8 Tides

Explain the cause of tides as well as their monthly cycles and other patterns. Describe the horizontal flow of water that accompanies the rise and fall of tides.

Tides are daily changes in the elevation of the ocean surface. Their rhythmic rise and fall along coastlines have been known since antiquity. Other than waves, they are the easiest ocean movements to observe (**Figure 10.29**).

Although known throughout human history, tides were not explained satisfactorily until Sir Isaac Newton applied the law of gravitation to them. Newton showed that there is a mutual attractive force between two objects, in this case Earth and the Moon. Because both the atmosphere and the ocean are fluids and are free to move, both are deformed by this force. Hence, ocean tides result from the gravitational attraction exerted upon Earth by the Moon and, to a lesser extent, by the Sun.

In addition, however, an equally large tidal bulge is produced on the side of Earth directly opposite the Moon (**Figure 10.30**).

Both tidal bulges are caused, as Newton discovered, by the pull of gravity. The strength of gravity is inversely

Causes of Tides

It is easy to see how the Moon's gravitational force can cause the water to bulge on the side of Earth nearest the Moon.

Figure 10.29 Bay of Fundy tides High tide and low tide at Hopewell Rocks on the Bay of Fundy. Tidal flats are exposed during low tide. (High tide photo courtesy of Ray Coleman/Science Source; low tide photo courtesy of Jeffrey Greenberg/Science Source)

proportional to the square of the distance between two objects, meaning simply that it weakens quickly with distance. For this reason, the Moon's gravitational pull is slightly greater on the near side of Earth than on the far side. This differential pulling tends to stretch (elongate) Earth. The solid Earth responds only very slightly, but the world ocean, which is mobile, is deformed quite dramatically, producing the two opposing tidal bulges.

Because the position of the Moon relative to Earth changes only moderately in a single day, the tidal bulges remain in place while Earth rotates "through" them. If you stand on the seashore for 24 hours, Earth will rotate you through alternating areas of higher and lower water. As you are carried into each tidal bulge, the tide rises, and as you are carried into the intervening troughs between the tidal bulges, the tide falls. Therefore, most places on Earth experience two high tides and two low tides each day.

In addition, the tidal bulges migrate as the Moon revolves around Earth about every 29 days. As a result, the tides, like the time of moonrise, shift about 50 minutes later each day. After 29 days the cycle is complete, and a new one begins.

Many locations may show an inequality between the high tides during a given day. Depending on the Moon's position, the tidal bulges may be inclined to the equator as in Figure 10.30. This figure illustrates that one high

SmartFigure 10.30 Idealized tidal bulges caused by the Moon If Earth were covered to a uniform depth with water, there would be two equal tidal bulges: one on the side of Earth facing the Moon (right) and the other on the opposite side of Earth (left). Depending on the Moon's position, tidal bulges may be inclined relative to Earth's equator. In this situation, Earth's rotation causes an observer to experience two unequal high tides during a day. (https://goo.gl/ElkpUH)

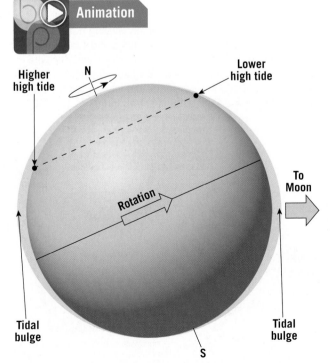

SmartFigure 10.31 Spring and neap tides Earth–Moon–Sun positions influence the tides. (https://goo.gl/YLHO8T)

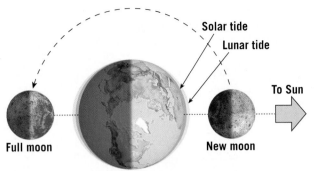

A. Spring Tide When the Moon is in the full or new position, the tidal bulges created by the Sun and Moon are aligned, and there is a large tidal range.

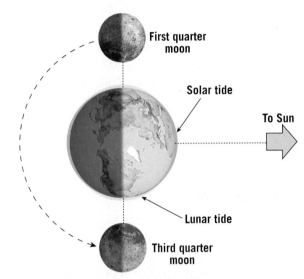

B. Neap Tide When the Moon is in the first- or third-quarter position, the tidal bulges produced by the Moon are at right angles to the bulges created by the Sun, and the tidal range is smaller.

tide experienced by an observer in the Northern Hemisphere is considerably higher than the high tide half a day later. In contrast, a Southern Hemisphere observer would experience the opposite effect.

Monthly Tidal Cycle

The primary body that influences the tides is the Moon, which makes one complete revolution around Earth every 29 days. The Sun, however, also influences the tides. It is far larger than the Moon, but because it is much farther away, it has considerably less effect. In fact, the Sun's tide-generating effect is only about 46 percent that of the Moon.

Near the times of new and full moons, the Sun and Moon are aligned, and their forces on tides add together (**Figure 10.31A**). The combined gravity of these two tide-producing bodies causes larger tidal bulges (higher high tides) and larger tidal troughs (lower low tides), producing a larger tidal range. These are called the **spring tides**, which have no connection with the spring season but occur twice a month, during the times when the Earth–Moon–Sun system is aligned. Conversely,

SmartFigure 10.32 Tidal patterns A diurnal tidal pattern (lower right) features one high tide and one low tide each tidal day. A semidiurnal pattern (upper right) features two high tides and two low tides of approximately equal heights during each tidal day. A mixed tidal pattern (left) features two high tides and two low tides of unequal heights during each tidal day.

(https://goo.gl/NNoCGJ)

Tutorial

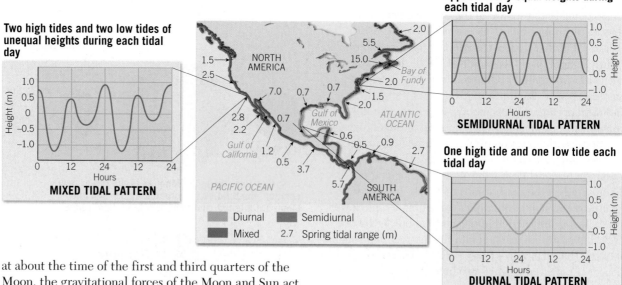

at about the time of the first and third quarters of the Moon, the gravitational forces of the Moon and Sun act on Earth at right angles, and each partially offsets the influence of the other (**Figure 10.31B**). As a result, the daily tidal range is less. These are called **neap tides**, and they also occur twice each month. Each month, then, there are two spring tides and two neap tides, each about 1 week apart.

Tidal Patterns

The basic causes and types of tides have been explained. Keep in mind, however, that these theoretical considerations cannot be used to predict either the height or the time of actual tides at a particular place. Many factors—including the shape of the coastline, the configuration of ocean basins, and water depth—greatly influence the tides. Consequently, tides at various locations respond differently to tide-producing forces. Thus, the nature of the tide at any coastal location can be determined most accurately by actual observation. The predictions in tidal tables and tidal data on nautical charts are based on such observations.

Three main tidal patterns exist worldwide (**Figure 10.32**). A **diurnal** (diurnal = daily) **tidal pattern** is characterized by a single high tide and a single low tide each tidal day. Tides of this type occur along the northern shore of the Gulf of Mexico, among other locations. A **semidiurnal** (semi = twice, diurnal = daily) **tidal pattern** exhibits two high tides and two low tides each tidal day, with the two highs about the same height and the two lows about the same height. This type of tidal pattern is common along the Atlantic coast of the United States. A **mixed tidal pattern** is similar to a semidiurnal pattern except that it is characterized by a large inequality in high-water heights, low-water heights, or both. In this case, there are usually two high tides and two low tides each day, with high tides of different heights and low tides of different heights. Such tides are prevalent along the Pacific coast of the United States and in many other parts of the world.

Tidal Currents

Tidal current is the term used to describe the *horizontal* flow of water accompanying the rise and fall of the tides. These water movements induced by tidal forces can be important in some coastal areas. Tidal currents that advance into the coastal zone as the tide rises are called *flood currents*. As the tide falls, seaward-moving water generates *ebb currents*. Periods of little or no current, called *slack water*, separate flood and ebb. Flat areas that are alternately covered and uncovered by the tides are called **tidal flats** (see Figure 10.29). Depending on the nature of the coastal zone, tidal flats vary from narrow strips seaward of the beach to zones that may extend for several kilometers.

Although tidal currents are not important in the open sea, they can be rapid in bays, river estuaries,

Figure 10.33 Tidal deltas As a rapidly moving tidal current (flood current) moves through a barrier island's inlet into the quiet waters of the lagoon, the current slows and deposits sediment, creating a tidal delta. Because this tidal delta has developed on the landward side of the inlet, it is called a *flood delta*. Such a tidal delta is shown in Figure 10.19A.

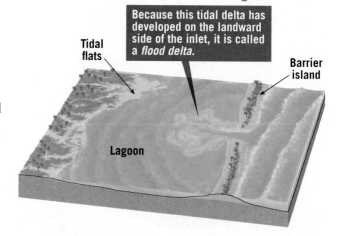

straits, and other narrow places. Off the coast of Brittany in France, for example, tidal currents that accompany a high tide of 12 meters (40 feet) may attain a speed of 20 kilometers (12 miles) per hour. Tidal currents are not generally considered to be major agents of erosion and sediment transport. Notable exceptions occur where tides move through narrow inlets, where they scour the narrow entrances to many harbors that would otherwise be blocked.

Sometimes deposits called **tidal deltas** are created by tidal currents (**Figure 10.33**). They may develop either as *flood deltas* landward of an inlet or as *ebb deltas* on the seaward side of an inlet. Because wave activity and longshore currents are reduced on the sheltered

landward side, flood deltas are more common and actually more prominent (see Figure 10.19A). They form after the tidal current moves rapidly through an inlet. As the current emerges into more open waters from the narrow passage, it slows and deposits its load of sediment.

(10.8) CONCEPT CHECKS

1. Explain why an observer can experience two *unequal* high tides during one day.
2. Distinguish between neap tides and ebb tides.
3. How is a mixed tidal pattern different from a semidiurnal tidal pattern?
4. Contrast flood current and ebb current.

> **Did You Know?**
> The world's largest tidal range (that is, the largest difference between successive high and low tides) is found in the northern end of Nova Scotia's Bay of Fundy. Here, the maximum spring tidal range is about 17 m (56 ft).

CONCEPTS IN REVIEW
The Restless Ocean

10.1 The Ocean's Surface Circulation

Discuss the factors that create and influence ocean currents and describe the influence ocean currents have on climate.

KEY TERMS: gyre, Coriolis effect

- The ocean's surface currents follow the general pattern of the world's major wind belts. Surface currents are parts of huge, slowly moving loops of water called *gyres* that are centered in the subtropics of each ocean basin. The positions of the continents and the Coriolis effect also influence the movement of ocean water within gyres. Because of the Coriolis effect, subtropical gyres move clockwise in the Northern Hemisphere and counterclockwise in the Southern Hemisphere. Generally, four main currents comprise each subtropical gyre.
- Ocean currents can have a significant effect on climate. Poleward-moving *warm* ocean currents moderate winter temperatures in the middle latitudes. Cold currents exert their greatest influence during summer in middle latitudes and year-round in the tropics. In addition to cooler temperatures, cold currents are associated with greater fog frequency and drought.

(?) **Assume that arrow A represents prevailing winds in a Northern Hemisphere ocean. Which arrow, A, B, or C, best represents the surface-ocean current in this region? Explain.**

10.2 Upwelling and Deep-Ocean Circulation

Explain the processes that produce coastal upwelling and the ocean's deep circulation.

KEY TERMS: coastal upwelling, thermohaline circulation

- Upwelling, the rising of colder water from deeper layers, is a wind-induced movement that brings cold, nutrient-rich water to the surface.

Coastal upwelling is most characteristic along the west coasts of continents.

- In contrast to surface currents, deep-ocean circulation is governed by gravity and driven by density differences. The two factors that are most significant in creating a dense mass of water are temperature and salinity, so the movement of deep-ocean water is often termed thermohaline circulation.
- Most water involved in thermohaline circulation begins in high latitudes at the surface, when the salinity of the cold water increases as a result of sea ice formation. This dense water sinks, initiating deep-ocean density currents.

10.3 The Shoreline

Explain why the shoreline is considered a dynamic interface. List the factors that influence the height, length, and period of a wave and describe the motion of water within a wave.

KEY TERMS: interface, wave height, wavelength, wave period, fetch, surf

- The shoreline is a transition zone between marine and continental environments. It is a dynamic interface, a boundary where land, sea, and air meet and interact.
- Energy from waves plays an important role in shaping the shoreline, but many factors contribute to the character of particular shorelines.
- Waves are moving energy, and most ocean waves are initiated by wind. The three factors that influence the height, wavelength, and period of a wave are (1) wind speed, (2) length of time the wind has blown, and (3) fetch, the distance that the wind has traveled across open water. Once waves leave a storm area, they are termed *swells*, which are symmetrical, longer-wavelength waves.
- As waves travel, water particles transmit energy by circular orbital motion, which extends to a depth equal to about one-half the wavelength (the wave base). When a wave enters water that is shallower than the wave base, it slows down, which allows waves farther from shore to catch up. As a result, wavelength decreases and wave height increases. Eventually the wave breaks, creating turbulent surf in which water rushes toward the shore.

10.4 Beaches and Shoreline Processes

Describe how waves erode and move sediment along the shore.

KEY TERMS: beach, abrasion, wave refraction, beach drift, longshore current

- A beach is composed of any locally derived material that is in transit along the shore. Wave erosion is caused by wave impact pressure and abrasion (the sawing and grinding action of water armed with rock fragments). The bending of waves, called wave refraction, causes wave impact to be concentrated against the sides and ends of headlands and dispersed in bays.

- Wind-generated waves provide most of the energy that modifies shorelines. Each time a wave hits, it can impart tremendous force. The impact of waves, coupled with abrasion from the grinding action of rock particles, erodes material exposed along the shoreline.

- Wave refraction is a consequence of a wave encountering shallower water as it approaches shore. The shallowest part of the wave (closest to shore) slows the most, allowing the faster part (still in deeper water) to catch up. This modifies a wave's trajectory so that the wave front becomes almost parallel to the shore by the time it hits.

- The term beach drift describes the movement of sediment in a zigzag pattern along a beach face. The swash of incoming waves pushes the sediment up the beach at an oblique angle, but the backwash transports it directly downhill. Net movement along the beach can be many meters per day. Longshore currents are a similar phenomenon in the surf zone, capable of transporting very large quantities of sediment parallel to a shoreline.

(?) **What process is causing wave energy to be concentrated on the headland? Predict how this area will appear in the future.**

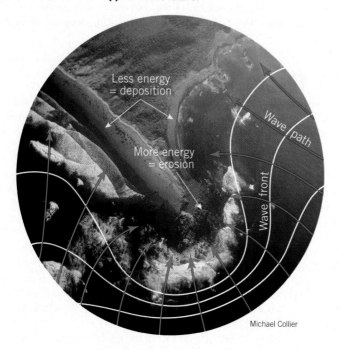

Michael Collier

10.5 Shoreline Features

Describe the features typically created by wave erosion and those resulting from sediment deposited by longshore transport processes.

KEY TERMS: wave-cut cliff, wave-cut platform, marine terrace, sea arch, sea stack, spit, baymouth bar, tombolo, barrier island

- Erosional features include wave-cut cliffs (which originate from the cutting action of the surf against the base of coastal land), wave-cut platforms (relatively flat, bench-like surfaces left behind by receding cliffs), and marine terraces (uplifted wave-cut platforms). Erosional features also include sea arches (formed when a headland is eroded and two sea caves from opposite sides unite) and sea stacks (formed when the roof of a sea arch collapses).

- Some of the depositional features that form when sediment is moved by beach drift and longshore currents are spits (elongated ridges of sand that project from the land into the mouth of an adjacent bay), baymouth bars (sandbars that completely cross a bay), and tombolos (ridges of sand that connect an island to the mainland or to another island). Along the Atlantic and Gulf coastal plains, the coastal region is characterized by offshore barrier islands, which are low ridges of sand that parallel the coast.

(?) **Identify the lettered features in this diagram.**

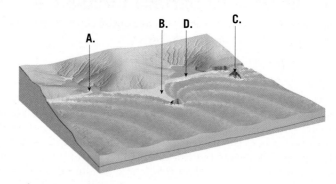

10.6 Contrasting America's Coasts

Distinguish between emergent and submergent coasts. Contrast the erosion problems faced along different parts of America's coasts.

KEY TERMS: emergent coast, submergent coast, estuary

- Coasts may be classified by their changes relative to sea level. Emergent coasts are sites of either land uplift or sea-level fall. Marine terraces are features of emergent coasts. Submergent coasts are sites of land subsidence or sea-level rise. One characteristic of submergent coasts is drowned river valleys called estuaries.

- The Atlantic and Gulf coasts of the United States are markedly different from the Pacific coast. The Atlantic and Gulf coasts are lined in many places by barrier islands—dynamic expanses of sand that see a lot of change during storm events. Many of these low and narrow islands have also been prime sites for real estate development.

(10.6 continued)

- The Pacific coast's big issue is the narrowing of beaches due to sediment starvation. Rivers that drain to the coast (bringing it sand) have been dammed, resulting in reservoirs that trap sand before it can make it to the coast. Narrower beaches offer less resistance to incoming waves, often leading to erosion of bluffs behind the beach.

(?) Is this an emergent coast or a submergent coast? Provide an easily seen line of evidence to support your answer. Is the location more likely along the coast of North Carolina or California? Explain.

Michael Collier

10.7 Stabilizing the Shore

Summarize the ways in which people deal with shoreline erosion problems.

KEY TERMS: hard stabilization, groin, breakwater, seawall, beach nourishment

- Local factors that influence shoreline erosion are (1) the proximity of a coast to sediment-laden rivers, (2) the degree of tectonic activity, (3) the topography and composition of the land, (4) prevailing winds and weather patterns, and (5) the configuration of the coastline and near-shore areas.
- Hard stabilization is a term that pertains to any structures built along the coastline to prevent movement of sand. Groins are oriented perpendicular to the coast, with the goal of slowing beach erosion by longshore currents. Breakwaters are parallel to the coast but located some distance offshore; their goal is to blunt the force of incoming ocean waves, often to protect boats. Like breakwaters, seawalls are parallel to the coast, but they are built on the shoreline itself. Often the installation of hard stabilization actually leads to increased erosion.
- Beach nourishment is an expensive alternative to hard stabilization. Sand is pumped onto the beach from some other area, temporarily replenishing the sediment supply. An alternative to hard stabilization and beach nourishment is relocating buildings away from high-risk areas and leaving the beach to be shaped by natural processes.

(?) Based on their position and orientation, identify the three types of hard stabilization illustrated in this diagram.

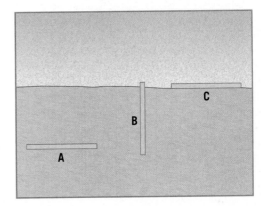

10.8 Tides

Explain the cause of tides as well as their monthly cycles and other patterns. Describe the horizontal flow of water that accompanies the rise and fall of tides.

KEY TERMS: tide, spring tide, neap tide, diurnal tidal pattern, semidiurnal tidal pattern, mixed tidal pattern, tidal current, tidal flat, tidal delta

- Tides are daily changes in ocean surface elevation. They are caused by gravitational pull on ocean water by the Moon and, to a lesser extent, the Sun. When the Sun, Earth, and Moon all line up about every 2 weeks (full moon or new moon), the tides are most exaggerated. When a quarter moon is in the sky, the Moon is pulling on Earth's water at a right angle relative to the Sun, and the daily tidal range is minimized as the two forces partially counteract one another.
- Tides are strongly influenced by local conditions, including the shape of the local coastline and the depth of the ocean basin. Tidal patterns may be diurnal (one high tide per day), semidiurnal (two high tides per day), or mixed (similar to semidiurnal but with significant inequality between high tides).
- A flood current is the landward movement of water during the shift between low tide and high tide. When high tide transitions to low tide again, the movement of water away from the land is an ebb current. Ebb currents may expose tidal flats to the air. If a tide passes through an inlet, the current may carry sediment that gets deposited as a tidal delta.

GIVE IT SOME THOUGHT

1. In this chapter you learned that global winds are the force that drives surface-ocean currents. A glance at the accompanying map, however, shows a surface current that does not exactly coincide with the prevailing wind. Provide an explanation.

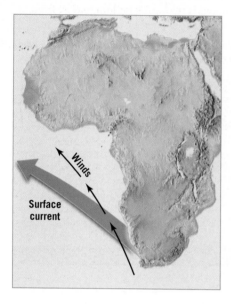

2. If the North Atlantic Drift were to cease, how might the climate of Western Europe change?

3. During a visit to the beach, you and a friend get in a rubber raft and paddle out into deep water *beyond* the surf zone. Tiring, you stop and take a rest. Describe the movement of the raft during your rest. How does this movement differ, if at all, from what you would have experienced if you had stopped paddling while *in* the surf zone?

4. Answer these questions that relate to the photo of a surfer enjoying a ride on a large wave.

Ron Dahlquist/Getty Images

a. What is the likely source of energy that created this wave?
b. How was the wavelength changing just prior to the time this photo was taken?
c. Why was the wavelength changing?
d. Many waves exhibit circular orbital motion. Is that true of the wave in this photo? Explain.

5. Examine the accompanying aerial photo, which shows a portion of the New Jersey coast. What term is applied to the wall-like structures that extend into the water? What is their purpose? In what direction are beach drift and longshore currents moving sand—toward the top or toward the bottom of the photo?

University of Washington Libraries, Special Collections, KCS2-26

6. You and a friend set up an umbrella and chairs at a beach. Your friend then goes into the surf zone to play Frisbee with another person. Several minutes later, your friend looks back toward the beach and is surprised to see that she is no longer near where the umbrella and chairs are set up. Although she is still in the surf zone, she is 30 yards away from where she started. How would you explain to your friend why she moved along the shore?

7. A friend wants to purchase a vacation home on a barrier island. If consulted, what advice would you give your friend?

8. If wastage (melting and calving) of the Greenland Ice Sheet were to dramatically increase, how would the salinity of the adjacent North Atlantic be affected? How might this influence thermohaline circulation?

9. The force of gravity plays a critical role in creating ocean tides. The more massive an object, the stronger its gravitational pull. Explain why the Sun's influence is much less than that of the Moon, even though the Sun is much more massive than the Moon.

10. This photo shows a portion of the Maine coast. The brown muddy area in the foreground is influenced by tidal currents. What term is applied to this muddy area? Name the type of tidal current this area will experience in the hours to come.

Marli Miller

MasteringGeology™

www.masteringgeology.com Looking for additional review and test prep materials? With individualized coaching on the toughest topics of the course, MasteringGeology offers a wide variety of ways for you to move beyond memorization to begin thinking like a geologist. Visit the Study Area in www.masteringgeology.com to find practice quizzes, study tools, and multimedia that will improve your understanding of this chapter's content. Sign in today to enjoy the following features: **Self Study Quizzes, SmartFigure: Tutorials/Animations/Condor Videos/Mobile Field Trips, Geoscience Animation Library, GEODe, RSS Feeds, Digital Study Modules**, and an optional **Pearson eText.**

11

FOCUS ON CONCEPTS

Each statement represents the primary learning objective for the corresponding major heading within the chapter. After you complete the chapter, you should be able to:

11.1 Distinguish between weather and climate and name the basic elements of weather and climate.

11.2 List the major gases composing Earth's atmosphere and identify the components that are most important to understanding weather and climate.

11.3 Interpret a graph that shows changes in air pressure from Earth's surface to the top of the atmosphere. Sketch and label a graph that shows atmospheric layers based on temperature.

11.4 Explain what causes the Sun angle and length of daylight to change during the year and describe how these changes produce the seasons.

11.5 Distinguish between heat and temperature. List and describe the three mechanisms of heat transfer.

11.6 Sketch and label a diagram that shows the paths taken by incoming solar radiation. Summarize the greenhouse effect.

11.7 Summarize the nature and cause of the atmosphere's changing composition since about 1750. Describe the atmosphere's response and some possible future consequences.

11.8 Calculate five commonly used types of temperature data and interpret a map that depicts temperature data using isotherms.

11.9 Discuss the principal controls of temperature and use examples to describe their effects.

11.10 Interpret the patterns depicted on world maps of January and July temperatures.

This snow-capped mountain in Alaska's Denali National Park reminds us that temperatures decrease with an increase in altitude in the lowest layer of the atmosphere. Altitude is one of several factors that cause air temperatures to vary from place to place. (Photo by Clement Philippe/Arterra Picture Library/Alamy)

HEATING THE ATMOSPHERE

E arth's atmosphere is unique. No other planet in our solar system has an atmosphere with the exact mixture of gases or the heat and moisture conditions necessary to sustain life as we know it. The gases that make up Earth's atmosphere and the controls to which they are subject are vital to our existence. In this chapter, we will begin our examination of the ocean of air in which we all must live. What is the composition of the atmosphere? Where does the atmosphere end, and where does outer space begin? What causes the seasons? How is air heated? What factors control temperature variations around the globe?

11.1 Focus on the Atmosphere

Distinguish between weather and climate and name the basic elements of weather and climate.

Weather influences our everyday activities, our jobs, and our health and comfort. Many of us pay little attention to the weather unless we are inconvenienced by it or it adds to our enjoyment outdoors. Nevertheless, few other aspects of our physical environment affect our lives more than the phenomena we collectively call the weather.

Weather in the United States

The United States occupies an area that stretches from the tropics to the Arctic Circle. It has thousands of miles of coastline and extensive regions that are far from the influence of the ocean. Some landscapes are mountainous, and others are dominated by plains. It is a place where Pacific storms strike the west coast, and the East is sometimes influenced by events in the Atlantic and the Gulf of Mexico. For those in the center of the country, it is common to experience weather events triggered when frigid southward-bound Canadian air masses clash with northward-moving ones.

Stories about weather are a routine part of the daily news. Articles and items about the effects of heat, cold, floods, drought, fog, snow, ice, and strong winds are commonplace. Of course, storms of all kinds are frequently front-page news (**Figure 11.1**). Beyond its direct impact on the lives of individuals, the weather has a strong effect on the world economy, influencing agriculture, energy use, water resources, transportation, and industry.

Weather clearly influences our lives a great deal. Yet, it is important to realize that people influence the atmosphere and its behavior as well (**Figure 11.2**). Today and in the future, significant political and scientific decisions must be made involving these impacts. Important examples are air pollution control and the effects of human activities on global climate. So there is a need for increased awareness and understanding of our atmosphere and its behavior.

Weather and Climate

Acted on by the combined effects of Earth's motions and energy from the Sun, our planet's formless and invisible envelope of air reacts by producing an infinite variety of weather, which in turn creates the basic patterns of global climates. Although not identical, weather and climate have much in common.

Weather is constantly changing, sometimes from hour to hour and at other times from day to day. It is a term that refers to the state of the atmosphere at a given time and place. Whereas changes in the weather are continuous and sometimes seemingly erratic, it is nevertheless possible to arrive at a generalization of these variations. Such a description of aggregate weather conditions is termed **climate**. It is based on observations that have been accumulated over many years. Climate is often defined simply as "average weather," but this is an inadequate definition. To more accurately portray the character of an area, variations and extremes must also be included, as well as

Figure 11.1 Memorable weather events Few aspects of our physical environment influence our daily lives more than the weather. During the winter of 2014–2015, Boston experienced record-breaking snows that at times paralyzed the city. A total of 280.9 centimeters (110.6 inches) of snow fell, the most in 120 years of record keeping. This image is from February 15, 2015. (Photo by Brian Snyder/Reuters)

the probabilities that such departures will take place. For example, it is not only necessary for farmers to know the average rainfall during the growing season but also for them to know the frequency of extremely wet and extremely dry years. Thus, climate is the sum of all statistical weather information that helps describe a place or region.

Suppose you were planning a vacation trip to an unfamiliar place. You would probably want to know what kind of weather to expect. Such information would help as you selected clothes to pack and could influence decisions regarding activities you might engage in during your stay. Unfortunately, weather forecasts that go beyond a few days are not very dependable. Therefore, you might ask someone who is familiar with the area about what kind of weather to expect. "Are thunderstorms common?" "Does it get cold at night?" "Are the afternoons sunny?" What you are seeking is information about the climate, the conditions that are typical for that place. Another useful source of information is the great variety of climate tables, maps, and graphs that are available. For example, the graph in **Figure 11.3** shows average daily high and low temperatures for each month, as well as extremes, for New York City. Such information could no doubt help as you planned your trip. But it is important to realize that *climate data cannot predict the weather.* Although a place may usually (climatically) be warm, sunny, and dry during the time of your planned vacation, you may actually experience cool, overcast, and rainy weather. There is a well-known saying that summarizes this idea: "Climate is what you expect, but weather is what you get."

Weather and climate are expressed in terms of the same basic **elements**—quantities or properties that are measured regularly. The most important elements are (1) air temperature, (2) humidity, (3) type and amount of cloudiness, (4) type and amount of precipitation, (5) air pressure, and (6) speed and direction of the wind. These elements are the major variables by which weather patterns and climate types are depicted. Although you will study these elements separately at first, keep in mind that they are very much interrelated. A change in one of the elements often produces changes in the others.

Figure 11.3 Graphs can display climate data This graph shows daily temperature data for New York City. In addition to the average daily maximum and minimum temperatures for each month, extremes are also shown. As this graph shows, there can be significant departures from the average.

Figure 11.2 People influence the atmosphere China is plagued by air quality issues. Coal-burning power plants are major contributors. (Photo by AFP/Stringer/Getty Images)

11.1 **CONCEPT CHECKS**

1. Distinguish between weather and climate.
2. Write two brief statements about your current location: one that relates to weather and one that relates to climate.
3. What is an *element*?
4. List the basic elements of weather and climate.

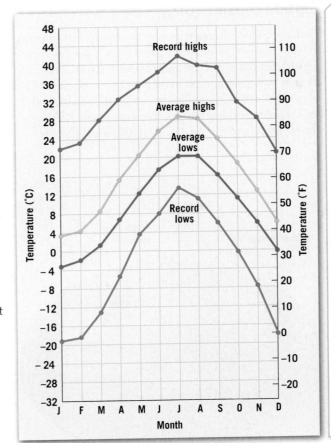

Did You Know?

Because data from every part of the globe are needed to produce accurate weather forecasts, the World Meteorological Organization (WMO) was established by the United Nations to coordinate scientific activity related to weather and climate. It consists of 187 member states and territories. Its World Weather Watch provides up-to-the-minute standardized observations through member-operated observation systems. This global system involves 10 satellites, 10,000 land-observation stations, and 7000 ship stations, as well as hundreds of automated data buoys and thousands of aircraft.

(11.2) Composition of the Atmosphere

List the major gases composing Earth's atmosphere and identify the components that are most important to understanding weather and climate.

Sometimes the term *air* is used as if it were a specific gas, but it is not. Rather, **air** is a *mixture* of many discrete gases, each with its own physical properties, in which varying quantities of tiny solid and liquid particles are suspended.

Major Components

The composition of air is not constant; it varies from time to time and from place to place. However, if the water vapor, dust, and other variable components were removed from the atmosphere, we would find that its makeup is very stable worldwide, up to an altitude of about 80 kilometers (50 miles).

As you can see in **Figure 11.4**, two gases—nitrogen and oxygen—make up 99 percent of the volume of clean, dry air. Although these gases are the most plentiful components of the atmosphere and are of great significance to life on Earth, they are of little or no importance in affecting weather phenomena. The remaining 1 percent of dry air is mostly the inert gas argon (0.93 percent) plus tiny quantities of a number of other gases.

Carbon Dioxide (CO_2)

Carbon dioxide, although present in only minute amounts (0.0400 percent, or 400 parts per million [ppm]), is nevertheless an important constituent of air. Carbon dioxide is of great interest to meteorologists because it is an efficient absorber of energy emitted by Earth and thus influences the heating of the atmosphere. Although the proportion of carbon dioxide in the atmosphere is

relatively uniform, its percentage has been rising steadily for 200 years. The graph in **Figure 11.5** shows the growth in atmospheric CO_2 since 1958. Much of this rise is attributed to the burning of ever-increasing quantities of fossil fuels, such as coal and oil. Some of this additional carbon dioxide is absorbed by the ocean or is used by plants, but about 45 percent remains in the air. Estimates project that by sometime in the second half of the twenty-first century, CO_2 levels will be twice as high as the pre-industrial level.

Most atmospheric scientists agree that increased carbon dioxide concentrations have contributed to a warming of Earth's atmosphere over the past several decades and will continue to do so in the decades to come. The magnitude of such temperature changes is uncertain and depends partly on the quantities of CO_2 contributed by human activities in the years ahead. The role of carbon dioxide in the atmosphere and its possible effects on climate are examined later in the chapter.

Variable Components

Air includes many gases and particles that vary significantly from time to time and from place to place. Important examples include water vapor, dust particles, and ozone. Although usually present in small percentages, they can have significant effects on weather and climate.

Water Vapor You are probably familiar with the term *humidity* from watching or listening to media weather reports. Humidity is a reference to water vapor in the air. As you will learn in Chapter 12, there are several ways to express humidity. The amount of water vapor in the air varies considerably, from practically none at all up to about 4 percent by volume. Why is such a small fraction of the atmosphere so significant? The fact that water vapor is the source of all clouds and precipitation would be enough to explain its importance. However, water vapor has other roles. Like carbon dioxide, water vapor absorbs heat emitted by Earth as well as some solar energy. It is therefore important when we examine the heating of the atmosphere.

When water changes from one state to another (see Figure 12.2), it absorbs or releases heat. This energy is termed *latent heat,* which means "hidden" heat. As we will see in later chapters, water vapor in the atmosphere transports this latent heat from one region to another, and it is the energy source that helps drive many storms.

Aerosols The movements of the atmosphere are sufficient to keep a large quantity of solid and liquid particles suspended within it. Although visible dust sometimes clouds the sky, these relatively large particles are too

Figure 11.4 Composition of the atmosphere
Proportional volume of gases composing dry air. Nitrogen and oxygen obviously dominate.

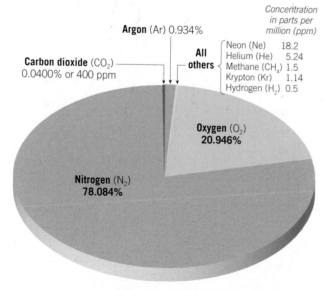

Concentration in parts per million (ppm)

Neon (Ne)	18.2
Helium (He)	5.24
Methane (CH_4)	1.5
Krypton (Kr)	1.14
Hydrogen (H_2)	0.5

Argon (Ar) 0.934%

All others

Carbon dioxide (CO_2)
0.0400% or 400 ppm

Oxygen (O_2)
20.946%

Nitrogen (N_2)
78.084%

heavy to stay in the air very long. Many other particles are microscopic and remain suspended for considerable periods of time. They may originate from many sources, both natural and human made, and include sea salts from breaking waves, fine soil blown into the air, smoke and soot from fires, pollen and microorganisms lifted by the wind, ash and dust from volcanic eruptions, and more (**Figure 11.6**). Collectively, these tiny solid and liquid particles are called **aerosols**.

From a meteorological standpoint, these tiny, often invisible particles can be significant. First, many act as surfaces on which water vapor can condense, an important function in the formation of clouds and fog. Second, aerosols can absorb or reflect incoming solar radiation. Thus, when an air pollution episode is occurring or when ash fills the sky following a volcanic eruption, the amount of sunlight reaching Earth's surface can be measurably reduced. Finally, aerosols contribute to an optical phenomenon we have all observed—the varied hues of red and orange at sunrise and sunset.

Ozone Another important component of the atmosphere is **ozone**. It is a form of oxygen that combines three oxygen atoms into each molecule (O_3). Ozone is not the same as the oxygen we breathe, which has two atoms per molecule (O_2). There is very little ozone in the atmosphere, and its distribution is not uniform. It is concentrated between 10 and 50 kilometers (6 and 31 miles) above the surface, in a layer called the *stratosphere*.

In this altitude range, oxygen molecules (O_2) are split into single atoms of oxygen (O) when they absorb ultraviolet radiation emitted by the Sun. Ozone is then created when a single atom of oxygen (O) and a molecule of oxygen (O_2) collide. This must happen in the presence of a third, neutral molecule that acts as a *catalyst* by allowing the reaction to take place without itself being consumed in the process. Ozone is concentrated in the 10- to 50-kilometer (6- to 31-mile) height range because a crucial balance exists there: The ultraviolet radiation from the Sun is sufficient to produce single atoms of oxygen, and there are enough gas molecules to bring about the required collisions.

The presence of the ozone layer in our atmosphere is crucial to land-dwelling organisms. The reason is that ozone absorbs the potentially harmful ultraviolet (UV) radiation from the Sun. If ozone did not filter a great deal of the ultraviolet radiation, and if the Sun's UV rays reached the surface of Earth undiminished,

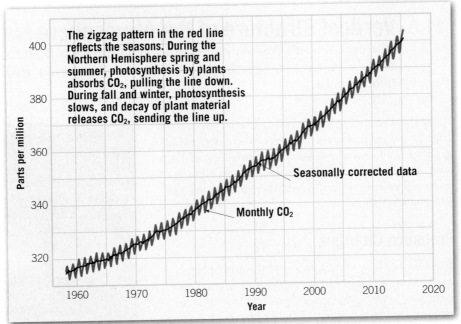

The zigzag pattern in the red line reflects the seasons. During the Northern Hemisphere spring and summer, photosynthesis by plants absorbs CO_2, pulling the line down. During fall and winter, photosynthesis slows, and decay of plant material releases CO_2, sending the line up.

Seasonally corrected data

Monthly CO_2

SmartFigure 11.5 Monthly CO_2 concentrations Atmospheric CO_2 has been measured at Mauna Loa Observatory, Hawaii, since 1958. There has been a consistent increase since monitoring began. This graphic portrayal is known as the *Keeling Curve*, in honor of the scientist who originated the measurements. (NOAA) (https://goo.gl/5HSuIS)

Tutorial

our planet would be uninhabitable for most life on land. Since water effectively filters UV radiation, ocean life would be affected far less.

(11.2) CONCEPT CHECKS

1. Is *air* a specific gas? Explain.
2. What are the two major components of clean, dry air? What proportion does each represent?
3. Why are water vapor and aerosols important constituents of Earth's atmosphere?
4. What is ozone? Why is ozone important to life on Earth?

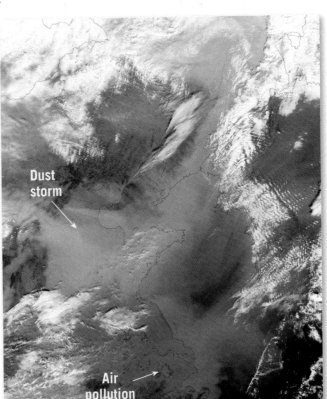

Dust storm

Air pollution

SmartFigure 11.6 Aerosols This satellite image shows two examples of aerosols. First, a dust storm is blowing across northeastern China toward the Korean Peninsula. Second, a dense haze toward the south (bottom center) is human-generated air pollution. (NASA) (http://goo.gl/P4peqp)

Video

Vertical Structure of the Atmosphere

Interpret a graph that shows changes in air pressure from Earth's surface to the top of the atmosphere. Sketch and label a graph that shows atmospheric layers based on temperature.

To say that the atmosphere begins at Earth's surface and extends upward is obvious. However, where does the atmosphere end, and where does outer space begin? There is no sharp boundary; the atmosphere rapidly thins as you travel away from Earth, until there are too few gas molecules to detect.

Pressure Changes

To understand the vertical extent of the atmosphere, let us examine changes in atmospheric pressure with increasing height. Atmospheric pressure is simply the force exerted by the weight of the air above. At sea level, the average pressure is slightly more than 1000 millibars (mb). This corresponds to a weight of slightly more than 1 kilogram per square centimeter (14.7 pounds per square inch). The pressure at higher altitudes is less (**Figure 11.7**).

One-half of the atmosphere lies below an altitude of 5.6 kilometers (3.5 miles). At about 16 kilometers (10 miles), 90 percent of the atmosphere has been traversed, and above 100 kilometers (62 miles), only 0.00003 percent of all the gases making up the atmosphere remain. Even so, traces of our atmosphere

Figure 11.8 Temperatures drop with an increase in altitude in the troposphere Snow-capped mountains and snow-free lowlands are a reminder that temperatures decrease as we go higher in the troposphere. (Photo by David Wall/Alamy)

extend far beyond this altitude, gradually merging with the emptiness of space.

Temperature Changes

By the early twentieth century, much had been learned about the lower atmosphere. The upper atmosphere was partly known from indirect methods. Data from balloons and kites showed that near Earth's surface, air temperature drops with increasing height. This phenomenon is felt by anyone who has climbed a high mountain and is obvious in pictures of snowcapped mountaintops rising above snow-free lowlands, as in the chapter-opening photo and **Figure 11.8**. We divide the atmosphere vertically into four layers, on the basis of temperature (**Figure 11.9**).

Troposphere The lowermost layer in which we live, where temperature decreases with an increase in altitude, is the **troposphere**. The term literally means the region where air "turns over," a reference to the turbulent weather in this lowermost zone. The troposphere is the chief focus of meteorologists because it is in this layer that essentially all important weather phenomena occur.

The temperature decrease in the troposphere is called the **environmental lapse rate**. Although its average value is 6.5°C per kilometer (3.5°F per 1000 feet), a figure known as the *normal lapse rate*, its value is

Figure 11.7 Air pressure changes with altitude The rate at which air pressure decreases with an increase in altitude is not constant. Pressure decreases rapidly near Earth's surface and more gradually at greater heights. Put another way, the graph shows that the vast bulk of the gases making up the atmosphere is very near Earth's surface and that the gases gradually merge with the emptiness of space.

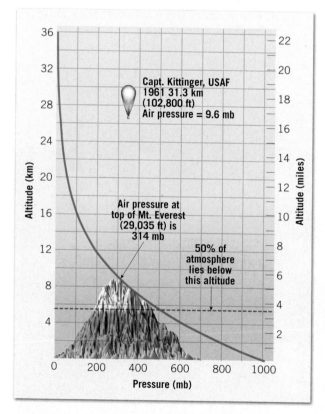

variable. To determine the actual environmental lapse rate for any particular time and place, as well as to gather information about vertical changes in pressure, wind, and humidity, radiosondes are used. A **radiosonde** is an instrument package that is attached to a balloon and transmits data by radio as it ascends through the atmosphere (**Figure 11.10**).

The thickness of the troposphere is not the same everywhere; it varies with latitude and the season. On average, the temperature drop continues to a height of about 12 kilometers (7.4 miles). The outer boundary of the troposphere is the *tropopause*.

Stratosphere Beyond the tropopause is the **stratosphere**. In the stratosphere, the temperature remains constant to a height of about 20 kilometers (12 miles) and then begins an increase that continues until the *stratopause*, at a height of about 50 kilometers (31 miles) above Earth's surface. Below the tropopause, atmospheric properties such as temperature and humidity are readily transferred by large-scale turbulence and mixing. Above the tropopause, in the stratosphere, they are not. The reason for the increased temperatures in the stratosphere is that the atmosphere's ozone is concentrated in this layer. Recall that ozone absorbs ultraviolet radiation from the Sun. As a consequence, the stratosphere is heated.

Mesosphere In the third layer, the **mesosphere**, temperatures again decrease with height until, at the *mesopause*—approximately 80 kilometers (50 miles) above the surface—the temperature approaches –90°C (–130°F). The coldest temperatures anywhere in the atmosphere occur at the mesopause. Because accessibility is difficult, the mesosphere is one of the least-explored regions of the atmosphere. It cannot be reached by the highest research balloons, nor is it accessible to the lowest orbiting satellites. Recent technical developments are just beginning to fill this knowledge gap.

Thermosphere The fourth layer extends outward from the mesopause and has no well-defined upper limit. This is the **thermosphere**, a layer that contains only a tiny fraction of the atmosphere's mass. In the extremely rarefied air of this outermost layer, temperatures again increase, due to the absorption of short-wave high-energy solar radiation by atoms of oxygen and nitrogen.

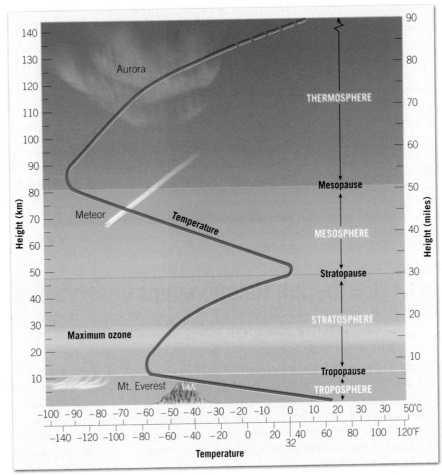

Figure 11.9 Thermal structure of the atmosphere Earth's atmosphere is traditionally divided into four layers, based on temperature.

Figure 11.10 Radiosonde A radiosonde is a lightweight package of instruments that is carried aloft by a small weather balloon. It transmits data on vertical changes in temperature, pressure, and humidity in the troposphere. The troposphere is where practically all weather phenomena occur, so it is very important to have frequent measurements. (Photo by David R. Frazier/Danita Delimont/Newscom)

Temperatures rise to extremely high levels of more than 1000°C (1832°F) in the thermosphere. But such temperatures are not comparable to those experienced near Earth's surface. Temperature is defined in terms of the average speed at which molecules move. Because the gases of the thermosphere are moving at very high speeds, the temperature is very high. But the gases are so sparse that, collectively, they possess only an insignificant quantity of heat. For this reason, the temperature of a satellite orbiting Earth in the thermosphere is determined chiefly by the amount of solar radiation it absorbs and not by the high temperature of the almost nonexistent surrounding air. If an astronaut inside a satellite were to expose his or her hand, the atmosphere would not feel hot.

11.3 CONCEPT CHECKS

1. Does air pressure increase or decrease with an increase in altitude? Is the rate of change constant or variable? Explain.
2. Is the outer edge of the atmosphere clearly defined? Explain.
3. The atmosphere is divided vertically into four layers, on the basis of temperature. List and describe these layers in order, from lowest to highest. In which layer does practically all of our weather occur?
4. What is the environmental lapse rate, and how is it determined?
5. Why are temperatures in the thermosphere not strictly comparable to those experienced near Earth's surface?

11.4 Earth–Sun Relationships

Explain what causes the Sun angle and length of daylight to change during the year and describe how these changes produce the seasons.

Nearly all the energy that drives Earth's variable weather and climate comes from the Sun. Earth intercepts only a minute percentage of the energy given off by the Sun—less than 1 two-billionth. This may seem to be an insignificant amount, until we realize that it is several hundred thousand times the electrical-generating capacity of the United States.

Solar energy is not distributed evenly over Earth's land–sea surface. The amount of energy received varies with latitude, time of day, and season of the year. Contrasting images of polar bears on ice rafts and palm trees along a remote tropical beach serve to illustrate the extremes. It is the unequal heating of Earth that ultimately creates winds and drives ocean currents. These movements, in turn, transport heat from the tropics toward the poles in an unending attempt to balance energy inequalities. The consequences of these processes are the phenomena we call weather.

If the Sun were to be "turned off," global winds would quickly subside. Yet, as long as the Sun shines, the winds will blow and the phenomena we know as weather will persist. So to understand how the atmosphere's dynamic weather machine works, we must first know why different latitudes receive varying quantities of solar energy and why the amount of solar energy changes to produce the seasons. As we will see, the variations in solar heating are caused by the motions of Earth relative to the Sun and by variations in Earth's land–sea surface.

Earth's Motions

Earth has two principal motions—its **rotation** about its axis and its orbital motion around the Sun. The axis is an imaginary line that runs through the poles. Our planet rotates on its axis once every 24 hours, producing the daily cycle of daylight and darkness. At any moment, half of Earth is experiencing daylight and the other half darkness. The line separating the dark half of Earth from the lighted half is called the **circle of illumination**.

Each year, Earth makes one slightly elliptical orbit around the Sun. The distance between Earth and Sun averages about 150 million kilometers (93 million miles).

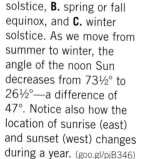

SmartFigure 11.11 The changing Sun angle Daily paths of the Sun for a place located at 40° north latitude for **A.** summer solstice, **B.** spring or fall equinox, and **C.** winter solstice. As we move from summer to winter, the angle of the noon Sun decreases from 73½° to 26½°—a difference of 47°. Notice also how the location of sunrise (east) and sunset (west) changes during a year. (goo.gl/pjB346)

Tutorial

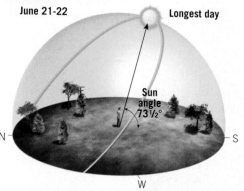

June 21-22   Longest day

Sun angle 73½°

A. Summer solstice at 40°N latitude

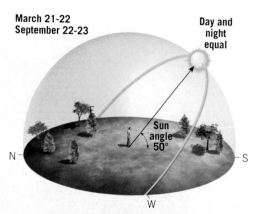

March 21-22
September 22-23   Day and night equal

Sun angle 50°

B. Spring or fall equinox at 40°N latitude

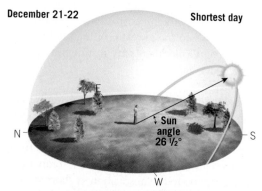

December 21-22   Shortest day

Sun angle 26½°

C. Winter solstice at 40°N latitude

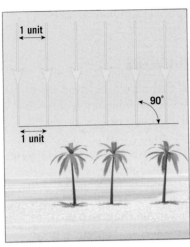
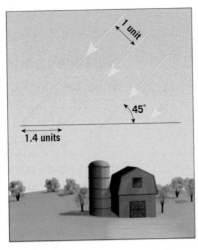
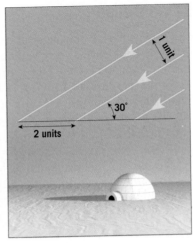
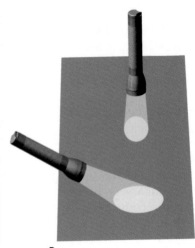

A.

Because Earth's orbit is not perfectly circular, however, the distance varies during the course of a year. Each year, on about January 3, our planet is about 147.3 million kilometers (91.5 million miles) from the Sun, closer than at any other time—a position known as *perihelion*. About 6 months later, on July 4, Earth is about 152 million kilometers (94.5 million miles) from the Sun, farther away than at any other time—a position called *aphelion*. Although Earth is closest to the Sun and receives up to 7 percent more energy in January than in July, this difference plays only a minor role in producing seasonal temperature variations, as evidenced by the fact that Earth is closest to the Sun during the Northern Hemisphere winter.

What Causes the Seasons?

If variations in the distance between the Sun and Earth do not cause seasonal temperature changes, what does? The gradual but significant change in the length of daylight certainly accounts for some of the difference we notice between summer and winter. Furthermore, a gradual change in the angle (altitude) of the Sun above the horizon is also a major contributing factor (**Figure 11.11**). For example, someone living in Chicago, Illinois, experiences the noon Sun highest in the sky in late June. But as summer gives way to autumn, the noon Sun appears lower in the sky, and sunset occurs earlier each evening.

The seasonal variation in the angle of the Sun above the horizon influences the amount of energy received at Earth's surface in two ways. First, when the Sun is directly overhead (at a 90-degree angle), the solar rays are most concentrated and thus most intense. The lower the angle, the more spread out and less intense is the solar radiation that reaches the surface (**Figure 11.12A**). To illustrate this principle, hold a flashlight at a right angle to a surface and then change the angle (**Figure 11.12B**).

Second, but of lesser importance, the angle of the Sun determines the path solar rays take as they pass

through the atmosphere (**Figure 11.13**). When the Sun is directly overhead, the rays strike the atmosphere at a 90-degree angle and travel the shortest possible route to the surface. However, rays entering at a 30-degree angle must travel twice this distance before reaching the surface, while rays at a 5-degree angle travel through a distance roughly equal to the thickness of 11 atmospheres. The longer the path, the greater the chance that sunlight will be dispersed by the atmosphere, which reduces the intensity at the surface. These conditions account for the fact that we cannot look directly at the midday Sun, but we can enjoy gazing at a sunset.

It is important to remember that Earth's shape is spherical. On any given day, the only places that will receive vertical (90-degree) rays from the Sun are located along one particular line of latitude. As we move either north or south of that location, the Sun's rays strike at ever-decreasing angles. The nearer a place is to the latitude receiving vertical rays of the Sun, the higher will be its noon Sun and the more concentrated will be the radiation it receives (see Figure 11.13).

SmartFigure 11.12
Changes in the Sun's angle cause variations in the amount of solar energy that reaches Earth's surface
A. The higher the angle, the more intense the solar radiation reaching the surface. **B.** If a flashlight beam strikes a surface at a 90° angle, a small intense spot is produced; if it strikes at any other angle, the area illuminated is larger—but noticeably dimmer. (https://goo.gl/OkyJFB)

Animation

Figure 11.13 The amount of atmosphere sunlight must traverse before reaching the Earth's surface affects its intensity Rays striking Earth at a low angle (near the poles) must traverse more of the atmosphere than rays striking at a high angle (around the equator) and thus are subject to greater depletion by reflection, scattering, and absorption. The thickness of the atmosphere is greatly exaggerated to illustrate this effect.

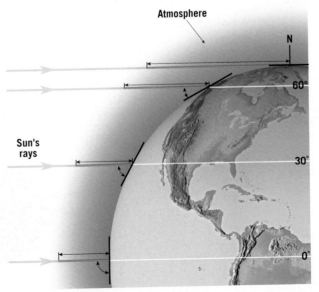

Earth's Orientation

What causes fluctuations in Sun angle and length of daylight during the course of a year? Variations occur because Earth's orientation to the Sun continually changes as the planet travels along its orbit. Earth's axis (the imaginary line through the poles around which Earth rotates) is not perpendicular to the plane of its orbit around the Sun. Instead, it is tilted 23½° from the perpendicular. This is termed the **inclination of the axis**. If the axis were not inclined, Earth would lack seasons. Because the axis remains pointed in the same direction (toward the North Star), the orientation of Earth's axis to the Sun's rays is constantly changing (**Figure 11.14**).

For example, on one day in June each year, Earth's position in orbit is such that the Northern Hemisphere is "leaning" 23½° *toward* the Sun (see Figure 11.14, *left*). Six months later, in December, when Earth has moved to the opposite side of its orbit, the Northern Hemisphere leans 23½° *away* from the Sun (see Figure 11.14, *right*). On days between these extremes, Earth's axis is leaning at amounts less than 23½° to the rays of the Sun. This change in orientation causes the spot where the Sun's rays are vertical to make a yearly migration from 23½° north of the equator to 23½° south of the equator.

In turn, this migration causes the angle of the noon Sun to vary by as much as 47° (23½ + 23½) during the year at places located poleward of latitude 23½°. For example, a midlatitude city such as New York (about 40° north latitude) has a maximum noon Sun angle of 73½° when the Sun's vertical rays reach their farthest

northward location in June and a minimum noon Sun angle of 26½° 6 months later.

Solstices and Equinoxes

Historically, 4 days each year have been given special significance, based on the annual migration of the direct rays of the Sun and its importance to the yearly weather cycle. On June 21 or 22, Earth is in a position such that the north end of its axis is tilted 23½° *toward* the Sun (**Figure 11.15A**). At this time, the vertical rays of the Sun strike 23½° north latitude (23½° north of the equator), a latitude known as the **Tropic of Cancer**. For people in the Northern Hemisphere, June 21 or 22 is known as the **summer solstice**, the first "official" day of summer.

Six months later, on about December 21 or 22, Earth is in the opposite position, with the Sun's vertical rays striking at 23½° south latitude (**Figure 11.15B**). This parallel is known as the **Tropic of Capricorn**. For those in the Northern Hemisphere, December 21 and 22 is the **winter solstice**. However, at the same time in the Southern Hemisphere, people are experiencing just the opposite—the summer solstice.

Midway between the solstices are the equinoxes. September 22 or 23 is the date of the **autumnal equinox** in the Northern Hemisphere, and March 21 or 22 is the date of the **spring equinox**. On these dates, the vertical rays of the Sun strike the equator (0° latitude) because Earth is in such a position in its orbit that the axis is tilted neither toward nor away from the Sun (**Figure 11.15C**).

SmartFigure 11.14
Earth–Sun relationships
(goo.gl/oYidn8)

Animation

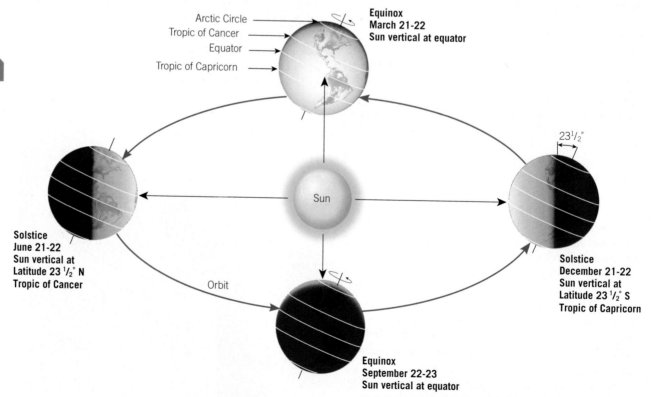

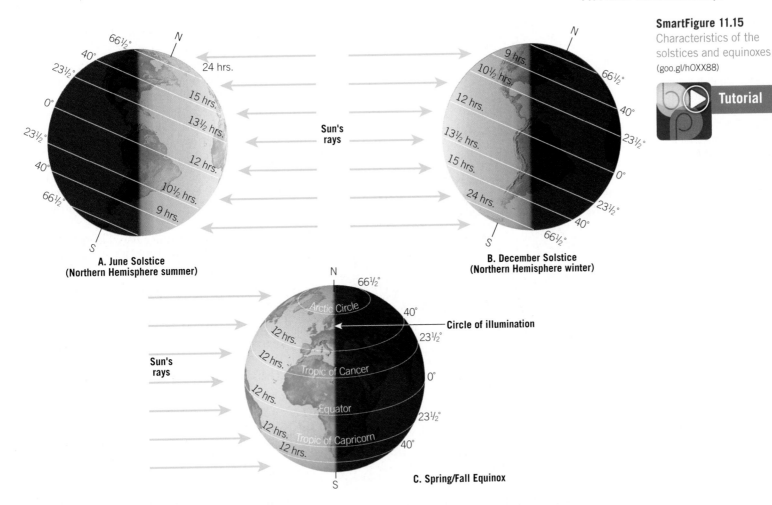

SmartFigure 11.15
Characteristics of the solstices and equinoxes (goo.gl/hOXX88)

Tutorial

A. June Solstice
(Northern Hemisphere summer)

B. December Solstice
(Northern Hemisphere winter)

C. Spring/Fall Equinox

The length of daylight versus darkness is also determined by Earth's position in orbit. The length of daylight on June 21, the summer solstice in the Northern Hemisphere, is greater than the length of night. This fact can be established from Figure 11.15A by comparing the fraction of a given latitude that is on the "day" side of the circle of illumination with the fraction on the "night" side. The opposite is true for the winter solstice, when the nights are longer than the days. Again for comparison, let us consider New York City, which has about 15 hours of daylight on June 21 and only about 9 hours on December 21 (you can see this in Figure 11.15 and **Table 11.1**). Also note from Table 11.1 that the farther north of the equator you are on June 21, the longer the period of daylight. When you reach the Arctic Circle (66½° north latitude), the length of daylight is 24 hours. This is the land of the "midnight Sun," which does not set for about 6 months at the North Pole.

During an equinox (meaning "equal night"), the length of daylight is 12 hours *everywhere* on Earth because the circle of illumination passes directly through the poles, dividing the latitudes in half (see Figure 11.15C).

As a review of the characteristics of the summer solstice for the Northern Hemisphere, examine Figure 11.15A and Table 11.1 and consider the following facts:

- The solstice occurs on June 21 or 22.
- The vertical rays of the Sun are striking the Tropic of Cancer (23½° north latitude).
- Locations in the Northern Hemisphere are experiencing their greatest length of daylight (opposite for the Southern Hemisphere).
- Locations north of the Tropic of Cancer are experiencing their highest noon Sun angles (opposite for places south of the Tropic of Capricorn).
- The farther you are north of the equator, the longer the period of daylight, until the Arctic Circle is

Table 11.1 Length of Daylight

Latitude (degrees)	Summer Solstice	Winter Solstice	Equinoxes
0	12 h	12 h	12 h
10	12 h 35 min	11 h 25 min	12
20	13 h 12 min	10 h 48 min	12
30	13 h 56 min	10 h 04 min	12
40	14 h 52 min	9 h 08 min	12
50	16 h 18 min	7 h 42 min	12
60	18 h 27 min	5 h 33 min	12
70	24 h (for 2 mo)	0 00	12
80	24 h (for 4 mo)	0 00	12
90	24 h (for 6 mo)	0 00	12

Figure 11.16 Monthly temperatures for cities at different latitudes Places located at higher latitudes experience larger temperature differences between summer and winter. Note that Cape Town, South Africa, experiences winter in June, July, and August.

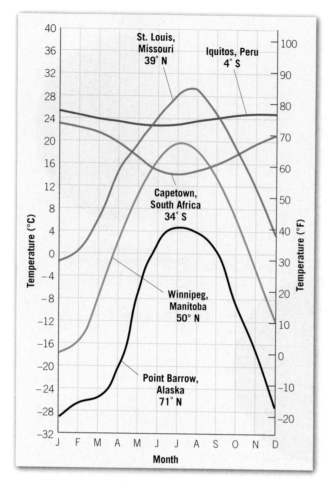

location is warmest in the summer, for it is then that days are longest and Sun's altitude is highest.

These seasonal changes, in turn, cause the month-to-month variations in temperature observed at most locations outside the tropics. **Figure 11.16** shows mean monthly temperatures for selected cities at different latitudes. Notice that the cities located at more poleward latitudes experience larger temperature differences from summer to winter than do cities located nearer the equator. Also notice that temperature minimums for Southern Hemisphere locations occur in July, whereas they occur in January for most places in the Northern Hemisphere.

All places at the same latitude have identical Sun angles and lengths of daylight. If the Earth–Sun relationships just described were the only controls of temperature, we would expect these places to have identical temperatures as well. Obviously, this is not the case. Other factors that influence temperature are discussed later in the chapter.

11.4 CONCEPT CHECKS

1. Do the annual variations in Earth–Sun distance adequately account for seasonal temperature changes? Explain.
2. Use a simple sketch to show why the intensity of solar radiation striking Earth's surface changes when the Sun angle changes.
3. Briefly explain the primary cause of the seasons.
4. What is the significance of the Tropic of Cancer and the Tropic of Capricorn?
5. After examining Table 11.1, write a general statement that relates season, latitude, and length of daylight.

reached, where daylight lasts for 24 hours (opposite for the Southern Hemisphere).

The facts about the winter solstice are just the opposite. It should now be apparent why a midlatitude

11.5 Energy, Heat, and Temperature

Distinguish between heat and temperature. List and describe the three mechanisms of heat transfer.

The universe is made up of a combination of matter and energy. The concept of matter is easy to grasp because it is the "stuff" we can see, smell, and touch. Energy, on the other hand, is abstract and therefore more difficult to describe. For our purposes, we will define energy simply as *the capacity to do work*. We can think of work as being accomplished whenever matter is moved. You are likely familiar with some of the common forms of energy, such as thermal, chemical, nuclear, radiant (light), and gravitational energy. One type of energy is described as *kinetic energy*, which is energy of motion. Recall that matter is composed of atoms or molecules that are constantly in motion and therefore possesses kinetic energy.

Heat is a term that is commonly used synonymously with *thermal energy*. In this usage, heat is energy possessed by a material arising from the internal motions of

its atoms or molecules. Whenever a substance is heated, its atoms move faster and faster, which leads to an increase in its heat content. **Temperature**, on the other hand, is related to the average kinetic energy of a material's atoms or molecules. Stated another way, the term *heat* generally refers to the quantity of thermal energy present, whereas the word *temperature* refers to the intensity—that is, the degree of "hotness."

Heat and temperature are closely related concepts. Heat is the energy that will flow in response to a difference of temperature. In all situations, *heat is transferred from warmer to cooler objects*. Thus, if two objects of different temperatures are in contact, the warmer object will become cooler and the cooler object will become warmer until they both reach the same temperature.

The three mechanisms of heat transfer are conduction, convection, and radiation. Although we present them separately, all three processes go on simultaneously in the atmosphere. In addition, these mechanisms operate to transfer heat between Earth's surface (both land and water) and the atmosphere.

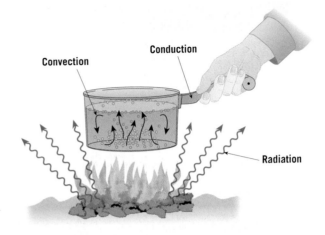

SmartFigure 11.17 Three mechanisms of heat transfer: Conduction, convection, and radiation (goo.gl/Tyjzwl)

Mechanisms of Heat Transfer: Conduction

Conduction is familiar to all of us. Anyone who has touched a metal spoon that was left in a hot pan has discovered that heat was conducted through the spoon. **Conduction** *is the transfer of heat through matter by molecular activity.* The energy of molecules is transferred through collisions between one molecule and another, with the heat flowing from the area of higher temperature to the area of lower temperature.

The ability of substances to conduct heat varies considerably. Metals are good conductors, as those of us who have touched hot metal have quickly learned (**Figure 11.17**). Air, on the other hand, is a very poor conductor of heat. Consequently, conduction is important only between Earth's surface and the air directly in contact with the surface. As a means of heat transfer for the atmosphere as a whole, conduction is the least significant.

Mechanisms of Heat Transfer: Convection

Much of the heat transport in Earth's atmosphere and oceans is carried on by convection. **Convection** *is the transfer of heat by mass movement or circulation within a substance.* It takes place in fluids (for example, liquids such as the ocean and gases such as air) where the material is able to flow.

The pan of water being heated over the campfire in Figure 11.17 illustrates the nature of simple convective circulation. Radiation from the fire warms the bottom of the pan, which conducts heat to the water near the bottom of the container. As the water is heated, it expands and becomes less dense than the water above. Because of this buoyancy, the warmer water rises. At the same time, cooler, denser water near the top of the pan sinks to the bottom, where it becomes heated. As long as the water is heated unequally—that is, from the bottom up—the water will continue to "turn over," producing a *convective circulation.*

In a similar manner, some of the air in the lowest layer of the atmosphere that is heated by radiation and conduction is then transported upward by convection. For example, on a hot, sunny day, the air above a plowed field will be heated more than the air above the surrounding croplands. As warm, less-dense air above the plowed field buoys upward, it is replaced by the cooler air above the croplands (**Figure 11.18**). In this way, a convective flow is established. The warm parcels of rising air are called *thermals* and are what hang-glider pilots use to keep their crafts soaring. Convection of this type not only transfers heat but also transports moisture (water vapor) aloft. The result is the increase in cloudiness that frequently can be observed on warm summer afternoons.

On a much larger scale, the global convective circulation of the atmosphere is driven by the unequal heating of Earth's surface. These complex movements are responsible for the redistribution of heat between hot equatorial regions and frigid polar latitudes and will be discussed in detail in Chapter 13.

Mechanisms of Heat Transfer: Radiation

The third mechanism of heat transfer is **radiation**. As shown in Figure 11.17, radiation travels out in all

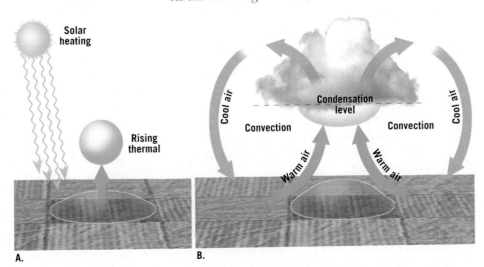

Figure 11.18 Rising warmer air and descending cooler air are examples of convective circulation **A.** Heating of Earth's surface produces thermals of rising air that transport heat and moisture aloft. **B.** The rising air cools, and if it reaches the condensation level, clouds form.

Figure 11.19 The electromagnetic spectrum This diagram illustrates the wavelengths and names of various types of radiation. Visible light consists of an array of colors we commonly call the "colors of the rainbow."

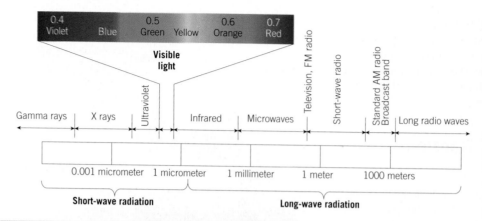

directions from its source. Unlike conduction and convection, which need a medium to travel through, radiant energy readily travels through the vacuum of space. Thus, radiation is the heat-transfer mechanism by which solar energy reaches our planet.

Solar Radiation From our everyday experience, we know that the Sun emits light and heat as well as the ultraviolet rays that cause sunburn. Although these forms of energy comprise a major portion of the total energy that radiates from the Sun, they are only part of a large continuum of energy called **electromagnetic radiation**. This continuum, or spectrum, of electromagnetic energy is shown in **Figure 11.19**. All electromagnetic radiation, whether x-rays, microwaves, or radio waves, transmits energy through the vacuum of space at 300,000 kilometers (186,000 miles) per second and only slightly slower through our atmosphere.

Nineteenth-century physicists were so puzzled by the seemingly impossible phenomenon of energy traveling through the vacuum of space without a medium to transmit it that they assumed that a material, which they named *ether*, existed in outer space. This medium was thought to transmit radiant energy in much the same

way that air transmits sound waves. Of course, this was incorrect. We now know that, like gravity, radiation requires no material for transmission.

In some respects, the transmission of radiant energy parallels the motion of the gentle swells in the open ocean. Like ocean swells, electromagnetic waves come in various sizes. For our purpose, the most important characteristic is their *wavelength*, or the distance from one crest to the next. Radio waves have the longest wavelengths, ranging to tens of kilometers, whereas gamma waves are the shortest, being less than one-billionth of a centimeter long.

Visible light, as the name implies, is the only portion of the spectrum we can see. We often refer to visible light as "white" light since it appears white in color. However, it is easy to show that white light is really a mixture of colors, each corresponding to a particular wavelength. Using a prism, we can divide white light into a rainbow color array. Figure 11.19 shows that violet has the shortest wavelength of visible light—0.4 micrometer—and red has the longest—0.7 micrometer.

Located adjacent to red and having a longer wavelength is **infrared radiation**, which we cannot see but which we can detect as heat. The closest invisible waves to violet are called **ultraviolet (UV) radiation** and are responsible for sunburn after an intense exposure to the Sun. Despite these divisions, all forms of radiation are basically the same. When any form of radiant energy is absorbed by an object, the result is an increase in molecular motion, which causes a corresponding increase in temperature.

Laws of Radiation To obtain a better understanding of how the Sun's radiant energy interacts with Earth's atmosphere and land–sea surface, it is helpful to have a general understanding of the basic laws governing radiation:

1. *All objects, at whatever temperature, emit radiant energy.* Hence, not only hot objects such as the Sun but also Earth, including its polar ice caps, continually emit energy.

2. *Hotter objects radiate more total energy per unit area than do colder objects.* The Sun, which has a surface temperature of nearly 6000°C (10,000°F), emits about 160,000 times more energy per unit area than does Earth, which has an average surface temperature of about 15°C (59°F).

3. *Hotter objects radiate more energy in the form of short-wavelength radiation than do cooler objects.* We can visualize this law by imagining a piece of metal that, when heated sufficiently (as occurs in a blacksmith shop), produces a white glow. As the metal cools, it emits more of its energy in longer wavelengths

and glows a reddish color. Eventually, no light is given off, but if you place your hand near the metal, the still-longer infrared radiation will be detected as heat. The Sun radiates maximum energy at 0.5 micrometer, which is in the visible range. The maximum radiation for Earth occurs at a wavelength of 10 micrometers, well within the infrared (heat) range. Because the maximum Earth radiation is roughly 20 times longer than the maximum solar radiation, Earth radiation is often called *long-wave radiation*, and solar radiation is called *short-wave radiation*.

4. *Objects that are good absorbers of radiation are good emitters as well.* Earth's surface and the Sun approach being perfect radiators because they absorb and radiate with nearly 100 percent efficiency for their respective temperatures. On the other hand, *gases are selective absorbers and radiators.* Thus, the atmosphere, which is nearly transparent (does not absorb) to certain wavelengths of radiation, is nearly opaque (a good absorber) to others. Our experience tells us that the atmosphere is transparent to visible light; that is why visible light readily reaches Earth's surface. The atmosphere is much less transparent to the longer-wavelength radiation emitted by Earth.

Figure 11.17 summarizes the various mechanisms of heat transfer. A portion of the radiant energy generated by the campfire is absorbed by the pan. This energy is readily transferred through the metal container by the process of conduction. Conduction also increases the temperature of the water at the bottom. Once warmed, this layer of water moves upward and is replaced by cool water descending from above. Thus, convection currents that redistribute the newly acquired energy throughout the pan are established. Meanwhile, the camper is warmed by radiation emitted by the fire and the pan. Furthermore, because metals are good conductors, the camper's hand is likely to be burned if he or she does not use a potholder. As in this example, the heating of Earth's atmosphere involves the processes of conduction, convection, and radiation, all of which occur simultaneously.

11.5 CONCEPT CHECKS

1. Distinguish between heat and temperature.
2. Describe the three mechanisms of heat transfer. Which mechanism is *least* important as a means of heat transfer in the atmosphere?
3. In what part of the electromagnetic spectrum does the Sun radiate maximum energy? How does this compare to Earth?
4. Describe the relationship between the temperature of a radiating body and the wavelengths it emits.

11.6 Heating the Atmosphere

Sketch and label a diagram that shows the paths taken by incoming solar radiation. Summarize the greenhouse effect.

The goal of this section is to describe how energy from the Sun heats Earth's surface and atmosphere. It is important to know the paths taken by incoming solar radiation and the factors that cause the amount of solar radiation taking each path to vary.

What Happens to Incoming Solar Radiation?

When radiation strikes an object, three different results usually occur. First, some of the energy is *absorbed* by the object. Recall that when radiant energy is absorbed, it is converted to heat, which causes an increase in temperature. Second, substances such as water and air are transparent to certain wavelengths of radiation. Such materials simply *transmit* this energy. Radiation that is transmitted does not contribute energy to the object. Third, some radiation may "bounce off" the object without being absorbed or transmitted. *Reflection* and *scattering* are responsible for redirecting incoming solar radiation. In summary, *radiation may be absorbed, transmitted, or redirected (reflected or scattered).*

Figure 11.20 shows the fate of incoming solar radiation averaged for the entire globe. Notice that the atmosphere is quite transparent to incoming solar radiation. On average, about 50 percent of the solar energy that reaches the top of the atmosphere is absorbed at Earth's surface. Another 30 percent is reflected back to space by the atmosphere,

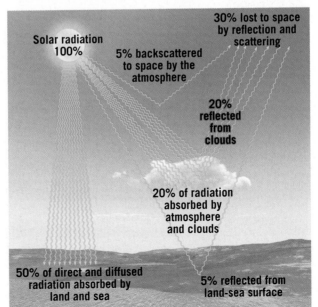

SmartFigure 11.20 Paths taken by solar radiation This diagram shows the average distribution of incoming solar radiation, by percentage. More solar radiation is absorbed by Earth's surface than by the atmosphere. (https://goo.gl/REGnn9)

Tutorial

Figure 11.21 Scattering by atmospheric particles When sunlight is scattered, rays travel in different directions. Usually more energy is scattered in the forward direction than is backscattered.

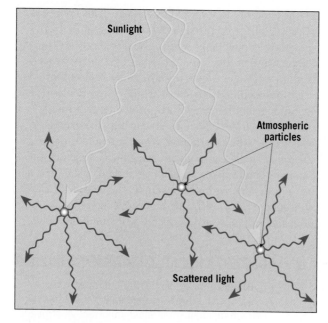

Sunlight

Atmospheric particles

Scattered light

clouds, and reflective surfaces. The remaining 20 percent is absorbed by clouds and the atmosphere's gases. What determines whether solar radiation will be transmitted to the surface, scattered, reflected outward, or absorbed by the atmosphere? As you will see, the answer depends greatly on the wavelength of the energy being transmitted, as well as on the nature of the intervening material.

Reflection and Scattering

Reflection is the process whereby light bounces back from an object at the same angle at which it was received. By contrast, **scattering** is a general process in which radiation is forced to deviate from a straight trajectory. When a beam of light strikes an atom, a molecule, or a

Figure 11.22 Albedos (reflectivity) of various surfaces In general, light-colored surfaces tend to be more reflective than dark-colored surfaces and thus have higher albedos.

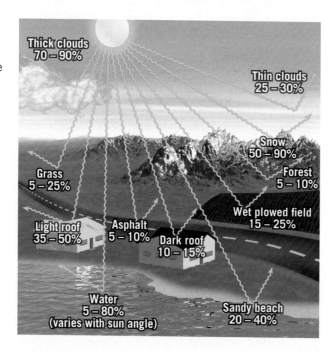

Thick clouds
70–90%

Thin clouds
25–30%

Snow
50–90%

Grass
5–25%

Forest
5–10%

Wet plowed field
15–25%

Light roof
35–50%

Asphalt
5–10%

Dark roof
10–15%

Water
5–80%
(varies with sun angle)

Sandy beach
20–40%

tiny particle in the atmosphere, it can spread out in all directions (**Figure 11.21**). Scattering disperses light both forward and backward. Whether solar radiation is reflected or scattered depends largely on the size of the intervening particles and the wavelength of the light.

Reflection and Earth's Albedo Energy is returned to space from Earth in two ways: reflection and emission of radiant energy. The portion of solar energy that is reflected back to space leaves in the same short wavelengths in which it came to Earth. About 30 percent of the solar energy that reaches the outer atmosphere is reflected back to space. Included in this figure is the amount sent skyward by backscattering. This energy is lost to Earth and does not play a role in heating the atmosphere.

The fraction of the total radiation that is reflected by a surface is called its **albedo**. Thus, the albedo for Earth as a whole (the *planetary albedo*) is 30 percent. However, the albedo varies considerably both from place to place and from one time to another, depending on the amount of cloud cover and particulate matter in the air, on the angle of the Sun's rays, and on the nature of the surface. A lower Sun angle means that more atmosphere must be penetrated, thus making the "obstacle course" longer and the loss of solar radiation greater (see Figure 11.13). **Figure 11.22** shows the albedos for various surfaces. Note that the angle at which the Sun's rays strike a water surface greatly affects its albedo.

Scattering Although incoming solar radiation travels in a straight line, small dust particles and gas molecules in the atmosphere scatter some of this energy in all directions. The result, called **diffused light**, explains how light reaches into the area beneath a shade tree and how a room is lit in the absence of direct sunlight. Further, scattering accounts for the brightness and even the blue color of the daytime sky. In contrast, bodies such as the Moon and Mercury, which are without atmospheres, have dark skies and "pitch-black" shadows, even during daylight hours. Overall, about half of the solar radiation that is absorbed at Earth's surface arrives as diffused (scattered) light.

Absorption

As stated earlier, gases are selective absorbers, meaning that they absorb strongly in some wavelengths, moderately in others, and only slightly in still others. When a gas molecule absorbs electromagnetic radiation, the energy is transformed into internal molecular motion, which is detectable as a rise in temperature.

Nitrogen, the most abundant constituent in the atmosphere, is a poor absorber of all types of incoming solar radiation. Oxygen and ozone are efficient absorbers of ultraviolet radiation. Oxygen removes most of the shorter ultraviolet radiation high in the atmosphere, and ozone absorbs most of the remaining UV rays in the stratosphere. The absorption of UV radiation in the

Airless bodies like the Moon All incoming solar radiation reaches the surface. Some is reflected back to space. The rest is absorbed by the surface and radiated directly back to space. As a result the lunar surface has a much lower average surface temperature than Earth.

Bodies with modest amounts of greenhouse gases like Earth The atmosphere absorbs some of the long-wave radiation emitted by the surface. A portion of this energy is radiated back to the surface and is responsible for keeping Earth's surface 33°C (59°F) warmer than it would otherwise be.

Bodies with abundant greenhouse gases like Venus Venus experiences extraordinary greenhouse warming, which is estimated to raise its surface temperature by 523°C (941°F).

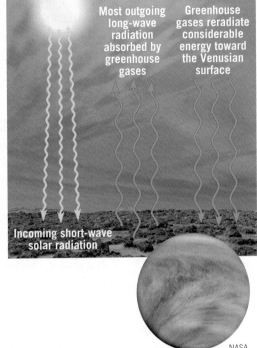

SmartFigure 11.23
The greenhouse effect
Earth's greenhouse effect is compared with that of our close solar system neighbors. (https://goo.gl/e150CY)

stratosphere accounts for the high temperatures experienced there. The only other significant absorber of incoming solar radiation is water vapor, which, along with oxygen and ozone, accounts for most of the solar radiation absorbed directly by the atmosphere.

For the atmosphere as a whole, none of the gases are effective absorbers of visible radiation. This explains why most visible radiation reaches Earth's surface and why we say that the atmosphere is *transparent* to incoming solar radiation. Thus, the atmosphere does not acquire the bulk of its energy directly from the Sun. Rather, it is heated chiefly by energy that is first absorbed by Earth's surface and then reradiated to the sky.

Heating the Atmosphere: The Greenhouse Effect

Approximately 50 percent of the solar energy that strikes the top of the atmosphere reaches Earth's surface and is absorbed. Most of this energy is then reradiated skyward. Because Earth has a much lower surface temperature than the Sun, the radiation that it emits has longer wavelengths than solar radiation.

The atmosphere as a whole is an efficient absorber of the longer wavelengths emitted by Earth (*terrestrial radiation*). Water vapor and carbon dioxide are the principal absorbing gases. Water vapor absorbs roughly five times more terrestrial radiation than do all the other

gases combined and accounts for the warm temperatures found in the lower troposphere, where it is most highly concentrated. Because the atmosphere is quite transparent to shorter-wavelength solar radiation and more readily absorbs longer-wavelength radiation emitted by Earth, the atmosphere is heated from the ground up rather than vice versa. This explains the general drop in temperature with increasing altitude experienced in the troposphere. The farther from the "radiator," the colder it is.

When the gases in the atmosphere absorb radiation emitted by Earth, they warm; but they eventually radiate this energy away. Some energy travels skyward, where it may be reabsorbed by other gas molecules, although that happens progressively less often with increasing height as the atmosphere thins and the concentration of water vapor decreases. The remainder travels Earthward and is again absorbed by Earth. For this reason, Earth's surface is continually being supplied with heat from the atmosphere as well as from the Sun. Without these absorptive gases in our atmosphere, Earth would be a truly frigid place.

This very important phenomenon, illustrated in **Figure 11.23**, received the name **greenhouse effect** because people at the time thought it was the main way greenhouses work. Like Earth's atmosphere, the glass of a greenhouse is largely transparent to visible light but somewhat opaque to the longer-wavelength radiation emitted by materials in the greenhouse. Greenhouse glass allows short-wavelength light to enter and heat plants and soil but prevents much of the longer-wavelength radiation these objects emit from leaving. In the case of greenhouses, we now know that this effect is much less important than the simple fact that the greenhouse prevents the warmer interior air from mixing with cooler outside air. Nevertheless, in the case of the atmosphere, the term *greenhouse effect* is still used.

11.6 CONCEPT CHECKS

1. What are the three paths taken by incoming solar radiation?
2. What factors cause albedo to vary from time to time and from place to place?
3. Explain why the atmosphere is heated chiefly by radiation emitted from Earth's surface rather than by direct solar radiation.
4. Prepare and label a sketch that explains the greenhouse effect.

11.7 Human Impact on Global Climate

Summarize the nature and cause of the atmosphere's changing composition since about 1750. Describe the atmosphere's response and some possible future consequences.

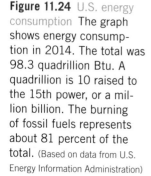

Did You Know?

If Earth's atmosphere had no greenhouse gases, Earth's average surface temperature would be a frigid –18°C (–0.4°F) instead of the relatively comfortable 15°C (59°F) that it is today.

Climate not only varies from place to place but is naturally variable over time. During the great expanse of Earth history, and long before humans were roaming the planet, there were many shifts—from warm to cold and from wet to dry and back again. Today scientists understand that, in addition to the natural forces that cause variations in climate, humans are playing a significant role as well. A key example involves the addition of carbon dioxide and other trace gases to the atmosphere.

In the preceding section, you learned that carbon dioxide (CO_2) absorbs some of the radiation emitted by Earth and thus contributes to the greenhouse effect. Because CO_2 is an important heat absorber, it follows that a change in the atmosphere's CO_2 content could influence air temperature.

Rising CO_2 Levels

Earth's tremendous industrialization of the past two centuries has been fueled—and still is fueled—by burning fossil fuels: coal, natural gas, and petroleum (**Figure 11.24**). Combustion of these fuels has added great quantities of carbon dioxide to the atmosphere. Figure 11.5 shows changes in CO_2 concentrations at Hawaii's Mauna Loa Observatory, where measurements have been made since 1958.

The use of coal and other fuels is the most prominent means by which humans add CO_2 to the atmosphere, but it is not the only way. The clearing of forests also contributes substantially because CO_2 is released as vegetation is burned or decays. Deforestation is particularly pronounced in the tropics, where vast tracts are cleared for ranching and agriculture or subjected to inefficient commercial logging operations (**Figure 11.25**).

Some of the excess CO_2 is taken up by plants or is dissolved in the ocean, but an estimated 45 percent remains in the atmosphere. **Figure 11.26** is a graphic record of changes in atmospheric CO_2 extending back about 800,000 years. Over this long span, natural fluctuations varied from about 160 to about 300 parts per million (ppm). As a result of human activities, the present CO_2 level is more than 30 percent higher than its highest level over at least the past 800,000 years. The rapid increase in CO_2 concentrations since the onset of industrialization is obvious. The annual rate at which atmospheric CO_2 concentrations are growing has been increasing over the past several decades.

The Atmosphere's Response

Given the increase in the atmosphere's carbon dioxide content, have global temperatures actually increased? The answer is "yes." According to a 2013 report by the Intergovernmental Panel on Climate Change (IPCC), "Warming of the climate system is unequivocal, as is now evident from observations of increases in global average air and ocean temperatures, widespread melting of snow and ice, and rising global sea level."[*] Most of the observed increase in global average temperatures since

Figure 11.24 U.S. energy consumption The graph shows energy consumption in 2014. The total was 98.3 quadrillion Btu. A quadrillion is 10 raised to the 15th power, or a million billion. The burning of fossil fuels represents about 81 percent of the total. (Based on data from U.S. Energy Information Administration)

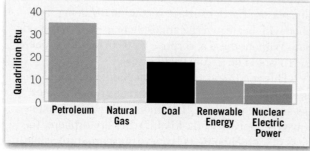

---

[*] IPCC, "Summary for Policymakers," in *Climate Change 2013: The Physical Science Basis. The* IPCC is an authoritative group of scientists that provides advice to the world community through periodic reports that assess the state of knowledge of the causes and effects of climate change.

the mid-twentieth century is *extremely likely* due to the observed increase in human-generated greenhouse gas concentrations. As used by the IPCC, *extremely likely* indicates a probability of 95–100 percent. Global warming since the mid-1970s is now about 0.6°C (1°F), and total warming in the past century is about 0.8°C (1.4°F). The upward trend in surface temperatures is shown in **Figure 11.27A**. With the exception of 1998, the 10 warmest years in the 135-year record have all occurred since 2000, with 2014 ranking as the warmest year on record.

Figure 11.25 Tropical deforestation Clearing the tropical rain forest is a serious environmental issue. In addition to causing a loss of biodiversity, it is a significant source of carbon dioxide. Fires are frequently used to clear the land. This scene is in Brazil's Amazon basin. (Photo by Nigel Dickinson/Alamy)

Weather patterns and other natural cycles cause fluctuations in average temperatures from year to year. This is especially true on regional and local levels. For example, while the globe experienced notably warm temperatures in 2013, the continental United States had its 42nd-warmest year. By contrast, 2013 was the hottest year in Australia's recorded history. Regardless of regional differences in any year, increases in greenhouse gas levels are causing a long-term rise in global temperatures. Although each calendar year will not necessarily be warmer than the one before, scientists expect each decade to be warmer than the previous one. An examination of the decade-by-decade temperature trend in **Figure 11.27B** bears this out.

What about the future? Projections for the years ahead depend in part on the quantities of greenhouse gases that are emitted. **Figure 11.28** shows the best estimates of global warming for several different scenarios. The 2013 IPCC report also states that if there is a doubling of the pre-industrial level of carbon dioxide (280 ppm) to 560 ppm, the *likely* temperature increase will be in the range of 2° to 4.5°C (3.5° to 8.1°F). The increase is *very unlikely* (1 to 10 percent probability) to be less than 1.5°C (2.7°F), and values higher than 4.5°C (8.1°F) are possible.

Some Possible Consequences

What consequences can be expected if the carbon dioxide content of the atmosphere reaches a level that is twice what it was early in the twentieth century? Because the climate system is so complex, predicting the distribution of particular regional

changes is speculative. It is not yet possible to pinpoint specifics, such as where or when it will become drier or wetter. Nevertheless, plausible scenarios can be given for larger scales of space and time.

One important impact of human-induced global warming is a rise in sea level. Potential weather changes include shifts in the paths of large-scale storms, which in turn would affect the distribution of precipitation and the occurrence of severe weather. Other possibilities include stronger tropical storms and increases in the frequency and intensity of heat waves and droughts.

> **Did You Know?**
> Carbon dioxide is not the only gas contributing to global warming. Scientists have come to realize that industrial and agricultural activities are causing a buildup of several trace gases, including methane (CH_4) and nitrous oxide (N_2O), that also play a significant role. These gases absorb wavelengths of outgoing Earth radiation that would otherwise escape into space.

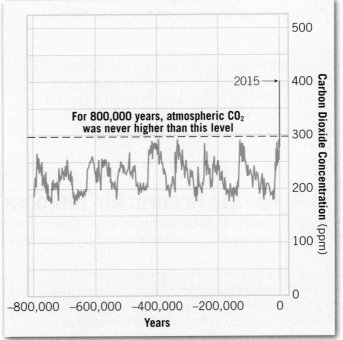

For 800,000 years, atmospheric CO_2 was never higher than this level

2015 →

Figure 11.26 CO_2 concentrations over the past 800,000 years Most of these data come from the analysis of air bubbles trapped in ice cores. The record since 1958 comes from direct measurements at Mauna Loa Observatory, Hawaii. The rapid increase in CO_2 concentrations since the onset of the Industrial Revolution is obvious. (Based on data from NOAA)

SmartFigure 11.27
Changes in global temperatures **A.** This graph shows changes in global temperatures between 1880 and 2014. With the exception of 1998, 10 of the warmest years in this 135-year temperature record have occurred since 2000.
B. Continued increases in the atmosphere's greenhouse gas levels are driving a long-term increase in global temperatures. Each calendar year is not necessarily warmer than the year before, but since 1950, each decade has been warmer than the previous one. The graph clearly shows this. (Based on data provided by NASA) (https://goo.gl/V60gFp)

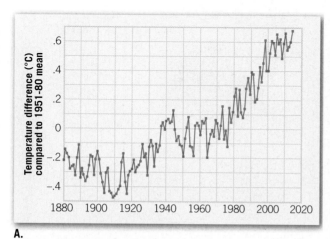
A.

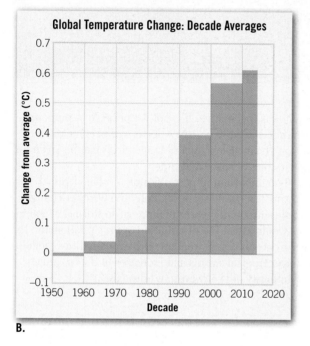
B.

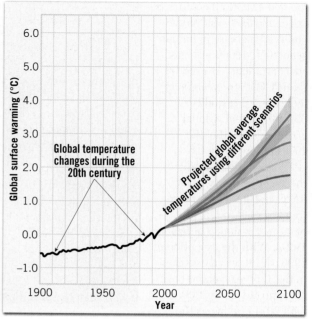

SmartFigure 11.28 Temperature projections to 2100 The right half of the graph shows projected global warming based on different emissions scenarios. The shaded zone adjacent to each colored line shows the uncertainty range for each scenario. The basis for comparison (0.0 on the vertical axis) is the global average for the period 1980 to 1999. The orange line represents the scenario in which CO_2 concentrations were held constant at values for the year 2000. (NOAA) (http://goo.gl/B3UbWC)

Many of the changes will probably take the form of gradual environmental shifts that will be imperceptible to most people from year to year. Although the changes may seem gradual, the effects will clearly have powerful economic, social, and political consequences.

11.7 CONCEPT CHECKS

1. Why has the CO_2 level of the atmosphere been increasing for the past 200 years?
2. How has the atmosphere responded to the growing CO_2 levels?
3. How are temperatures in the lower atmosphere likely to change as CO_2 levels continue to increase?

11.8 For the Record: Air Temperature Data

Calculate five commonly used types of temperature data and interpret a map that depicts temperature data using isotherms.

Changes in air temperature are probably noticed by people more often than changes in any other element of weather. At a weather station, the temperature is monitored on a regular basis from instruments mounted in an instrument shelter (**Figure 11.29**). The shelter protects the instruments from direct sunlight and allows a free flow of air.

Daily maximum and minimum temperatures underlie much of the basic temperature data compiled by meteorologists:

• By adding the maximum and minimum temperatures and then dividing by two, the **daily mean temperature** is calculated.

- The **daily range** of temperature is computed by finding the difference between the maximum and minimum temperatures for a given day.
- The **monthly mean** is calculated by adding together the daily means for each day of the month and dividing by the number of days in the month.
- The **annual mean** is an average of the 12 monthly means.
- The **annual temperature range** is computed by finding the difference between the highest and lowest monthly means.

Mean temperatures are particularly useful for making comparisons, whether on a daily, monthly, or annual basis. It is quite common to hear a weather reporter state, "Last month was the hottest July on record" or "Today, Chicago was 10 degrees warmer than Miami." Temperature ranges are also useful statistics because they give an indication of extremes.

To examine the distribution of air temperatures over large areas, isotherms are commonly used. An **isotherm** is a line that connects points on a map that have the same temperature (*iso* = equal, *therm* = temperature). Therefore, all points through which an isotherm passes have identical temperatures for the time period indicated. Generally, isotherms representing 5° or 10° temperature differences are used, but any interval may be chosen. **Figure 11.30** illustrates how isotherms are drawn on a map. Notice that most isotherms do not pass directly through the observing stations because the station readings may not coincide with the values chosen for the isotherms. Only an occasional station temperature will be exactly the same as the value of the isotherm, so it is usually necessary to draw the lines by estimating the proper position between stations.

Figure 11.29 Measuring temperature electrically This modern shelter contains an electrical thermometer called a *thermistor*. A shelter protects instruments from direct sunlight and allows for the free flow of air. (Photo by GIPhotoStock/Science Source)

Maps with isotherms are valuable tools because they make the temperature distribution visible at a glance. Areas of low and high temperatures are easy to pick out. In addition, the amount of temperature change per unit of distance, called the **temperature gradient**, is easy to visualize. Closely spaced isotherms indicate a rapid rate of temperature change, whereas more widely spaced lines

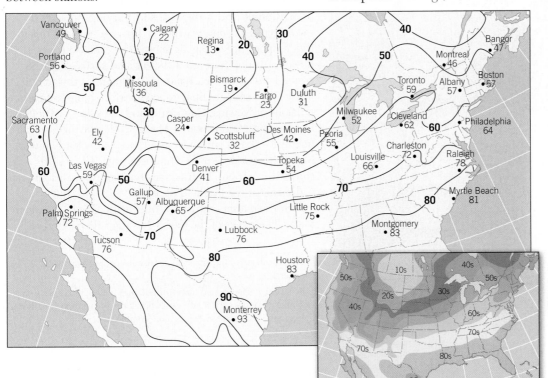

Tutorial

SmartFigure 11.30 Isotherms The large map shows high temperatures for a spring day. Isotherms are lines that connect points of equal temperature. Showing temperature distribution in this way makes patterns easier to see. Notice that most isotherms do not pass directly through the observing stations. It is usually necessary to draw isotherms by estimating their proper position between stations. On television and in many newspapers, temperature maps are in color, as shown in the inset map. Rather than label isotherms, these maps label the area *between* isotherms. For example, the zone between the 60° and 70° isotherms is labeled "60s." (goo.gl/Lx1TM8)

indicate a more gradual rate of change. You can see this in Figure 11.30. The isotherms are closer in Colorado and Utah (steeper temperature gradient), whereas the isotherms are spread farther in Texas (gentler temperature gradient). Without isotherms, a map would be covered with numbers representing temperatures at dozens or hundreds of places, which would make patterns difficult to see.

11.9 Why Temperatures Vary: The Controls of Temperature ▰▰

Discuss the principal controls of temperature and use examples to describe their effects.

A **temperature control** is any factor that causes temperature to vary from place to place and from time to time. Earlier in this chapter, we examined the single greatest cause for temperature variations—differences in the receipt of solar radiation. Because variations in Sun angle and length of daylight are a function of latitude, they are responsible for warm temperatures in the tropics and colder temperatures at more poleward locations. Of course, seasonal temperature changes at a given latitude occur as the Sun's vertical rays migrate toward and away from a place during the year.

But latitude is not the only control of temperature; if it were, we would expect that all places along the same parallel of latitude would have identical temperatures. This is clearly not the case. For example, Eureka,

California, and New York City are both coastal cities at about the same latitude, and both have an average mean temperature of 11°C (52°F). However, New York City is 9°C (16°F) warmer than Eureka in July and 10°C cooler in January. In another example, two cities in Ecuador—Quito and Guayaquil—are relatively close to one another, but the mean annual temperatures at these two cities differ by 12°C (21°F). To explain these situations and countless others, we must understand that factors other than latitude exert a strong influence on temperature. Among the most important are the differential heating of land and water, altitude, geographic position, and ocean currents.[*]

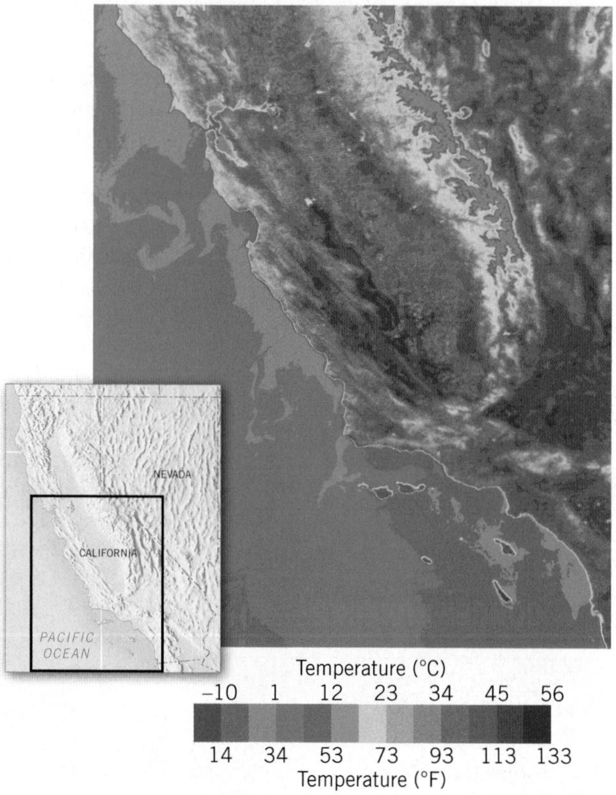

Differential Heating of Land and Water

The heating of Earth's surface directly influences the heating of the air above it. Therefore, to understand variations in air temperature, we must understand the variations in heating properties of the different surfaces that Earth presents to the Sun—soil, water, trees, ice, and so on. Different land surfaces absorb varying amounts of incoming solar energy, which in turn cause variations in the temperature of the air above. The greatest contrast, however, is not between different land surfaces but between land and water. **Figure 11.31** illustrates this idea nicely. This satellite image shows surface temperatures in portions of Nevada, California, and the adjacent Pacific Ocean on the afternoon of May 2, 2004, during a spring heat wave. Land-surface temperatures are clearly much higher than water-surface temperatures. The image shows the extreme high surface temperatures in southern California and Nevada in dark red.[**]

Surface temperatures in the Pacific Ocean are much lower. The peaks of the Sierra Nevada, still capped with

Figure 11.31 Differential heating of land and water This satellite image shows land- and water-surface temperatures (not air temperatures) for the afternoon of May 2, 2004. Water-surface temperatures in the Pacific Ocean are much lower than land-surface temperatures in California and Nevada. The narrow band of cool temperatures in the center of the image is associated with snow-capped mountains (the Sierra Nevada). The cooler water temperatures immediately offshore represent the California Current and its associated upwelling of deep cold water (see Figure 10.2). (NASA)

Temperature (°C)

| -10 | 1 | 12 | 23 | 34 | 45 | 56 |

| 14 | 34 | 53 | 73 | 93 | 113 | 133 |

Temperature (°F)

---

[*]For a discussion of the effects of ocean currents on temperatures, see Chapter 10.

[**]Realize that when a land surface is hot, the air above is cooler. For example, while the surface of a sandy beach can be painfully hot, the air temperature above the surface is more comfortable.

snow, form a cool blue line down the eastern side of California.

In side-by-side areas of land and water, such as those shown in Figure 11.31, *land heats more rapidly and to higher temperatures than water, and it cools more rapidly and to lower temperatures than water.* Variations in air temperatures, therefore, are much greater over land than over water.

Why do land and water heat and cool differently? Several factors are responsible:

- The **specific heat** (the amount of energy needed to raise the temperature of 1 gram of a substance 1°C) is far greater for water than for land. Thus, water requires a great deal more heat to raise its temperature the same amount than does an equal quantity of land.
- Land surfaces are opaque, so heat is absorbed only at the surface. Water, being more transparent, allows heat to penetrate to a depth of many meters.
- The water that is heated often mixes with water below, thus distributing the heat through an even larger mass.
- Evaporation (a cooling process) from water bodies is greater than that from land surfaces.

All these factors collectively cause water to warm more slowly, store greater quantities of heat, and cool more slowly than land.

Monthly temperature data for two cities will demonstrate the moderating influence of a large water body and the extremes associated with land (**Figure 11.32**). Vancouver, British Columbia, is located along the windward Pacific coast, whereas Winnipeg, Manitoba, is in a continental position far from the influence of water. Both cities are at about the same latitude and thus experience similar Sun angles and lengths of daylight. Winnipeg, however, has a mean January temperature that is 20°C lower than Vancouver's. Conversely, Winnipeg's July mean is 2.6°C higher than Vancouver's. Although their latitudes are nearly the same, Winnipeg, which has no water influence, experiences much greater temperature extremes than does Vancouver. The key to Vancouver's moderate year-round climate is the Pacific Ocean.

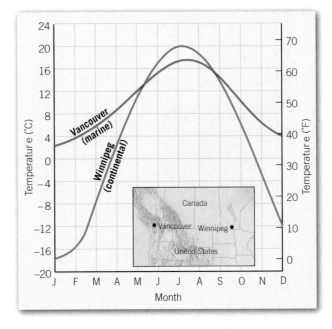

SmartFigure 11.32 Mean monthly temperatures for Vancouver, British Columbia, and Winnipeg, Manitoba Vancouver has a much smaller annual temperature range due to the strong marine influence of the Pacific Ocean. Winnipeg is subject to the greater extremes associated with an interior location. (goo.gl/qYUXGB)

Tutorial

On a different scale, the moderating influence of water may also be demonstrated when temperature variations in the Northern and Southern Hemispheres are compared. In the Northern Hemisphere, 61 percent is covered by water, and land accounts for the remaining 39 percent. However, in the Southern Hemisphere, 81 percent is covered by water and 19 percent by land. The Southern Hemisphere is correctly called the *water hemisphere* (see Figures 9.1 and 9.2). **Table 11.2** portrays the considerably smaller annual temperature variations in the water-dominated Southern Hemisphere as compared with the Northern Hemisphere.

Altitude

The chapter-opening photo and Figure 11.8 remind us that air temperatures drop with an increase in altitude. The two cities in Ecuador mentioned earlier—Quito and Guayaquil—also demonstrate the influence of altitude on mean temperatures. Although both cities are near the equator and not far apart, the annual mean at Guayaquil is 25°C (77°F), as compared to Quito's mean of 13°C (55°F). The difference is explained largely by the difference in the cities' elevations: Guayaquil is only 12 meters (40 feet) above sea level, whereas Quito is high in the Andes Mountains, at 2800 meters (9200 feet). **Figure 11.33** provides another example.

Recall that temperatures drop an average of 6.5°C per kilometer (3.5°F per 1000 feet) in the troposphere; thus, cooler temperatures are to be expected at greater heights. Yet, the magnitude of the difference is not totally explained by the normal lapse rate. If this figure were used, we would expect Quito to be about 18°C (nearly 33°F) cooler than Guayaquil; the difference, however, is only 12°C. The fact that high-altitude places, such as Quito, are warmer than the value calculated using the

Table 11.2 Variation in Mean Annual Temperature Range (°C) with Latitude

Latitude	Northern Hemisphere	Southern Hemisphere
0	0	0
15	3	4
30	13	7
45	23	6
60	30	11
75	32	26
90	40	31

Figure 11.33 Monthly mean temperatures for Concepción and La Paz, Bolivia Both cities have nearly the same latitude (about 16° south). However, because La Paz is high in the Andes, at 4103 meters (13,461 feet), it experiences much cooler temperatures than Concepción, which is at an elevation of 490 meters (1608 feet).

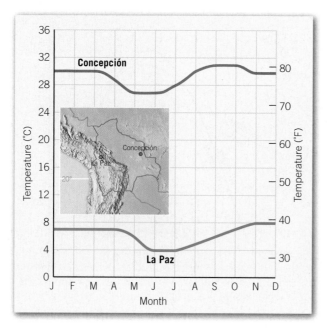

normal lapse rate results from the absorption and reradiation of solar energy by the ground surface.

Geographic Position

The geographic setting can greatly influence the temperatures experienced at a specific location. A coastal location where prevailing winds blow from the ocean onto the shore (a *windward* coast) experiences considerably different temperatures than does a coastal location where prevailing winds blow from the land toward the ocean (a *leeward* coast). The windward coast will experience the full moderating influence of the ocean—cool summers and mild winters—compared to an inland location at the same latitude.

A leeward coast, however, will have a more continental temperature regime because the winds do not carry the ocean's influence onshore. Eureka, California, and New York City, the two cities mentioned earlier, illustrate this aspect of geographic position. The annual temperature range at New York City is 19°C (34°F) greater than Eureka's (**Figure 11.34**).

Seattle and Spokane, both in Washington State, illustrate a second aspect of geographic position—mountains that act as barriers. Although Spokane is only about 360 kilometers (220 miles) east of Seattle, the towering Cascade Range separates the cities. Consequently, Seattle's temperatures show a marked marine influence, but Spokane's are more typically continental (**Figure 11.35**). Spokane is 7°C (13°F) cooler than Seattle in January and 4°C (7°F) warmer than Seattle in July. The annual range at Spokane is 11°C (20°F) greater than at Seattle. The Cascade Range effectively cuts off Spokane from the moderating influence of the Pacific Ocean.

Cloud Cover and Albedo

You may have noticed that clear days are often warmer than cloudy ones and that clear nights are usually cooler than cloudy ones. This demonstrates that cloud cover is another factor that influences temperature in the lower atmosphere. Studies using satellite images show that at any particular time, about half of our planet is covered by clouds. Cloud cover is important because many clouds have a high albedo; therefore, clouds reflect

Figure 11.35 Monthly mean temperatures for Seattle and Spokane, Washington Because the Cascade Mountains cut off Spokane from the moderating influence of the Pacific Ocean, its annual temperature range is greater than Seattle's.

Figure 11.34 Monthly mean temperatures for Eureka, California, and New York City Both cities are coastal and located at about the same latitude. Because Eureka is strongly influenced by prevailing winds from the ocean and New York City is not, the annual temperature range at Eureka is much smaller.

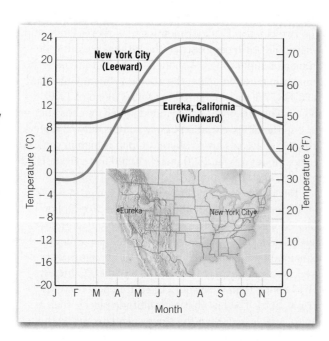

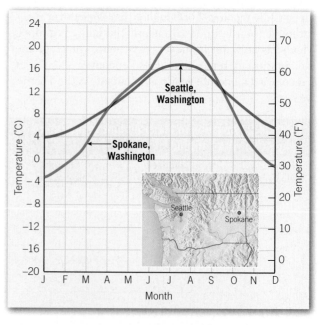

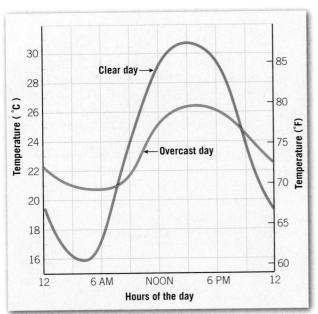

SmartFigure 11.36 *The daily cycle of temperature at Peoria, Illinois, for two July days* Clouds reduce the daily temperature range. During daylight hours, clouds reflect solar radiation back to space. Therefore, the maximum temperature is lower than if the sky were clear. At night, the minimum temperature will not fall as low because clouds retard the loss of heat. (goo.gl/909jEf)

Tutorial

back into space a significant portion of the sunlight that strikes them (see Figure 11.22). By reducing the amount of incoming solar radiation, clouds reduce daytime temperatures.

At night, clouds have the opposite effect. They act as a blanket by absorbing radiation emitted by Earth's surface and reradiating a portion of it back to the surface. Consequently, some of the heat that otherwise would have been lost remains near the ground. Thus, nighttime air temperatures do not drop as low as they would on a clear night. The effect of cloud cover is to reduce the daily temperature range by lowering the daytime maximum and raising the nighttime minimum (**Figure 11.36**).

Clouds are not the only phenomenon that increases albedo and thereby reduces air temperatures. We also recognize that snow- and ice-covered surfaces have high albedos. This is one reason why mountain glaciers do not melt away in the summer and why snow may still be present on a mild spring day. In addition, during the winter, when snow covers the ground, daytime maximums on a sunny day are lower than they otherwise would be because energy that the land would have absorbed and used to heat the air is reflected and lost.

11.9 CONCEPT CHECKS

1. List the factors that cause land and water to heat and cool differently.
2. Quito, Ecuador, is located on the equator and is *not* a coastal city. It has an average annual temperature of only 13°C (55°F). What is the likely cause of this low average temperature?
3. In what ways can geographic position be considered a control of temperature?
4. How can cloud cover influence the maximum temperature on an overcast day? How is the nighttime minimum temperature influenced by clouds?

11.10 World Distribution of Temperature

Interpret the patterns depicted on world maps of January and July temperatures.

Take a moment to study the two world maps in **Figures 11.37** and **11.38**. From hot colors near the equator to cool colors toward the poles, these maps portray sea-level temperatures in the seasonally extreme months of January and July. Temperature distribution is shown by using isotherms. On these maps, you can study global temperature patterns and the effects of the controlling factors of temperature, especially latitude, the distribution of land and water, and ocean currents. As with most other isothermal maps of large regions, all temperatures on these world maps have been reduced to sea level to eliminate the complications caused by differences in altitude.

On both maps, the isotherms generally trend east–west and show a decrease in temperatures poleward from the tropics. They illustrate one of the most fundamental aspects of world temperature distribution: that the effectiveness of incoming solar radiation in heating Earth's surface and the atmosphere above is largely a function of latitude.

Moreover, there is a latitudinal shifting of temperatures caused by the seasonal migration of the Sun's vertical rays. To see this, compare the color bands by latitude on the two maps. On the January map, the "hot spots" of 30°C are *south* of the equator, but in July they have shifted *north* of the equator.

If latitude were the only control of temperature distribution, our analysis could end here, but this is not the case. The added effect of the differential heating of land and water is also reflected on the January and July temperature maps. The warmest and coldest temperatures are found over land. Because temperatures do not fluctuate as much over water as over land, the seasonal north–south migration of isotherms is greater over the continents than over the oceans.

In addition, it is clear that the isotherms in the Southern Hemisphere, where there is little land and the oceans predominate, are much straighter and more stable

Did You Know?
A classic example of the effect of high latitude and continentality on annual temperature range is Yakutsk, a city in Siberia, approximately 60° north latitude and far from the influence of water. As a result, Yakutsk has an average annual temperature range of 62.2°C (112°F), one of the greatest in the world.

Figure 11.37 World mean sea-level temperatures in January, in Celsius (°C) and Fahrenheit (°F)

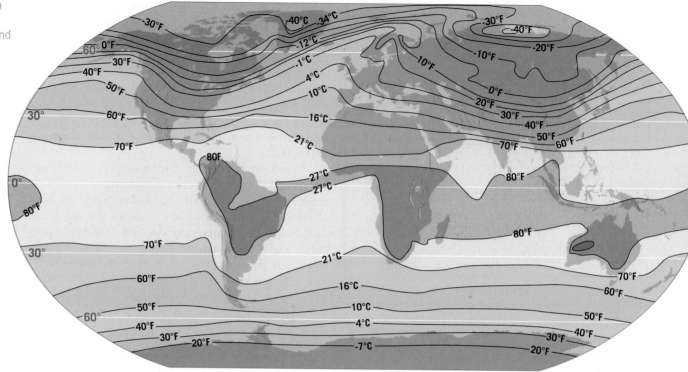

than in the Northern Hemisphere, where they bend sharply northward in July and southward in January over the continents.

Isotherms also reveal the presence of ocean currents. Warm currents cause isotherms to be deflected poleward, whereas cold currents cause an equatorward bending. The horizontal transport of water poleward warms the overlying air and results in air temperatures being higher than otherwise would be expected for the latitude. Conversely, currents moving toward the equator produce air temperatures that are cooler than expected.

Because Figures 11.37 and 11.38 show the seasonal extremes of temperature, they can be used to evaluate variations in the annual range of temperature from place

SmartFigure 11.38 World mean sea-level temperatures in July, in Celsius (°C) and Fahrenheit (°F) (goo. gl/7IWQRH)

Tutorial

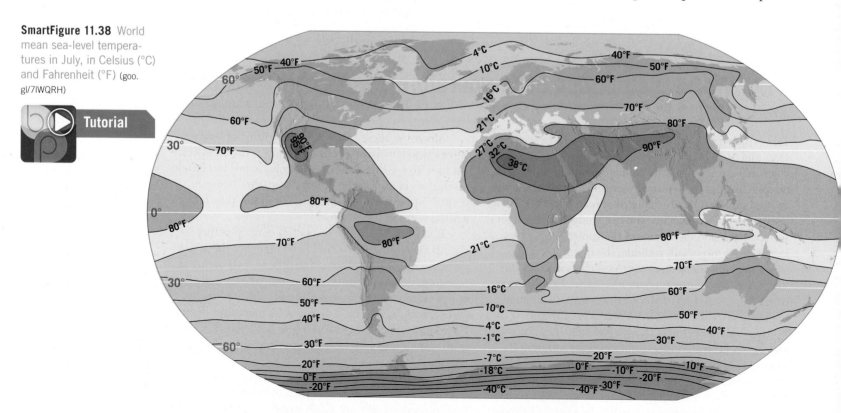

to place. A comparison of the two maps shows that a station near the equator will record a very small annual temperature range. This is true because such a station experiences little variation in the length of daylight and always has a relatively high Sun angle. By contrast, a site in the middle latitudes experiences much wider variations in Sun angle and length of daylight and thus larger temperature variations. Therefore, we can state that the annual temperature range increases with an increase in latitude.

Land and water also affect seasonal temperature variations, especially outside the tropics. A continental location must endure hotter summers and colder winters than a coastal location. Consequently, the annual range will increase with an increase in continentality.

11.10 CONCEPT CHECKS

1. Why do isotherms generally trend east–west?
2. Why do isotherms shift north and south from season to season?
3. Where do isotherms shift most—over land or over water? Explain.
4. Which area on Earth experiences the highest annual temperature range?

CONCEPTS IN REVIEW
Heating the Atmosphere

11.1 Focus on the Atmosphere

Distinguish between weather and climate and name the basic elements of weather and climate.

KEY TERMS: weather, climate, elements (of weather and climate)

- Weather is the state of the atmosphere at a particular place for a short period of time. Climate, on the other hand, is a generalization of the weather conditions of a place over a long period of time.
- The most important elements—quantities or properties that are measured regularly—of weather and climate are (1) air temperature, (2) humidity, (3) type and amount of cloudiness, (4) type and amount of precipitation, (5) air pressure, and (6) speed and direction of the wind.

❓ This is a scene on a summer day in Antarctica, showing a joint British–American research team. Write two brief statements about the locale in this image—one that relates to weather and one that relates to climate.

David Vaughn/Science Source

11.2 Composition of the Atmosphere

List the major gases composing Earth's atmosphere and identify the components that are most important to understanding weather and climate.

KEY TERMS: air, aerosols, ozone

- Air is a mixture of many discrete gases, and its composition varies from time to time and from place to place. If water vapor, dust, and other variable components of the atmosphere are removed, clean, dry air is composed almost entirely of nitrogen (N_2) and oxygen (O_2). Carbon dioxide (CO_2), although present only in minute amounts, is important because it has the ability to absorb heat radiated by Earth and thus helps keep the atmosphere warm.
- Among the variable components of air, water vapor is important because it is the source of all clouds and precipitation. Like carbon dioxide, water vapor can absorb heat emitted by Earth. When water changes from one state to another, it absorbs or releases heat. In the atmosphere, water vapor transports this latent ("hidden") heat from place to place; latent heat provides the energy that helps to drive many storms.
- Aerosols are tiny solid and liquid particles that are important because they may act as surfaces on which water vapor can condense and are also absorbers and reflectors of incoming solar radiation.
- Ozone, a form of oxygen that combines three oxygen atoms into each molecule (O_3), is a gas concentrated in the 10- to 50-kilometer (6- to 31-mile) height range in the atmosphere and is important to life because of its ability to absorb potentially harmful ultraviolet radiation from the Sun.

❓ This graph shows changes in one atmospheric component between January 2013 and September 2015. Which gas is it? How did you figure it out? Why is the line so wavy?

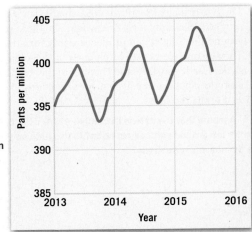

11.3 Vertical Structure of the Atmosphere

Interpret a graph that shows changes in air pressure from Earth's surface to the top of the atmosphere. Sketch and label a graph that shows atmospheric layers based on temperature.

KEY TERMS: troposphere, environmental lapse rate, radiosonde, stratosphere, mesosphere, thermosphere

- Because the atmosphere gradually thins with increasing altitude, it has no sharp upper boundary but simply blends into outer space.
- Based on temperature, the atmosphere is divided vertically into four layers. The *troposphere* is the lowermost layer. In the troposphere, temperature usually decreases with increasing altitude. This *environmental lapse rate* is variable but averages about 6.5°C per kilometer (3.5°F per 1000 feet). Essentially, all important weather phenomena occur in the troposphere.

- Beyond the troposphere is the *stratosphere*, which exhibits warming because of absorption of UV radiation by ozone. In the *mesosphere*, temperatures again decrease. Upward from the mesosphere is the *thermosphere*, a layer with only a tiny fraction of the atmosphere's mass and no well-defined upper limit.

(?) When the weather balloon in this photo was launched, the surface temperature was 17°C. The balloon is now at an altitude of 1 kilometer. What term is applied to the instrument package being carried aloft by the balloon? In what layer of the atmosphere is the balloon? If average conditions prevail, what is the air temperature at this altitude? How did you figure this out?

David R. Frazier/Science Source

11.4 Earth–Sun Relationships

Explain what causes the Sun angle and length of daylight to change during the year and describe how these changes produce the seasons.

KEY TERMS: rotation, circle of illumination, inclination of the axis, Tropic of Cancer, summer solstice, Tropic of Capricorn, winter solstice, autumnal equinox, spring equinox

- The two principal motions of Earth are (1) rotation about its axis, which produces the daily cycle of daylight and darkness; and (2) orbital motion around the Sun, which produces yearly variations.
- The seasons are caused by changes in the angle at which the Sun's rays strike Earth's surface and changes in the length of daylight at each latitude. These seasonal changes are the result of the tilt of Earth's axis as it orbits the Sun.

(?) Assume that the date is December 22. At what latitude do the Sun's rays strike the ground vertically at noon? Is this date an equinox or a solstice? What is the season in Australia?

11.5 Energy, Heat, and Temperature

Distinguish between heat and temperature. List and describe the three mechanisms of heat transfer.

KEY TERMS: heat, temperature, conduction, convection, radiation, electromagnetic radiation, visible light, infrared radiation, ultraviolet (UV) radiation

- Heat refers to the quantity of thermal energy present in a material, whereas temperature refers to intensity—how hot the material is.
- The three mechanisms of heat transfer are (1) conduction, the transfer of heat through matter by molecular activity; (2) convection, the transfer of heat by the movement of a substance from one place to another; and (3) radiation, the transfer of heat by electromagnetic waves.
- Electromagnetic radiation is energy emitted in the form of electromagnetic waves, which can transmit energy through the vacuum of space. The wavelengths of electromagnetic radiation range from very long (radio waves) to very short (gamma rays). Visible light is the only portion of the electromagnetic spectrum we can see.
- Some basic laws that relate to radiation are (1) all objects emit radiant energy; (2) for a given amount of surface area, hotter objects radiate more total energy than do colder objects; (3) the hotter the radiating body, the shorter the wavelengths of maximum radiation; and (4) objects that are good absorbers of radiation are good emitters as well.

(?) Describe how each of the three basic mechanisms of heat transfer is illustrated in this image.

Johner Images/Plattform/AGE Fotostock

11.6 Heating the Atmosphere

Sketch and label a diagram that shows the paths taken by incoming solar radiation. Summarize the greenhouse effect.

KEY TERMS: reflection, scattering, albedo, diffused light, greenhouse effect

- About 50 percent of the solar radiation that strikes the atmosphere reaches Earth's surface. About 30 percent is reflected back to space. The remaining 20 percent of the energy is absorbed by clouds and the atmosphere's gases. The fraction of radiation reflected by a surface is called the albedo of the surface.
- Radiant energy absorbed at Earth's surface is eventually radiated skyward. Because Earth has a much lower surface temperature than the Sun, its radiation is in the form of long-wave infrared radiation. Because atmospheric gases, primarily water vapor and carbon dioxide, are more efficient absorbers of long-wave than short-wave radiation, the atmosphere is heated from the ground up.
- The selective absorption of Earth's long-wave radiation by water vapor and carbon dioxide that results in Earth's average temperature being warmer than it would be otherwise is referred to as the greenhouse effect.

11.7 Human Impact on Global Climate

Summarize the nature and cause of the atmosphere's changing composition since about 1750. Describe the atmosphere's response and some possible future consequences.

- By adding carbon dioxide to the atmosphere, primarily by burning fossil fuels and through deforestation, humans are contributing significantly to global warming.
- As a result of the extra heat retention due to added CO_2, Earth's atmosphere has warmed by about 0.8°C (1.4°F) in the past 100 years, with most of this warming occurring since the 1970s. Temperatures are projected to increase by another 2° to 4.5°C (3.6° to 8.1°F) by 2100.
- Some consequences of global warming include (1) shifts in temperature and rainfall patterns, (2) a gradual rise in sea level, (3) changing storm tracks and a higher frequency and greater intensity of tropical storms, and (4) an increase in the frequency and intensity of heat waves and droughts.

11.8 For the Record: Air Temperature Data

Calculate five commonly used types of temperature data and interpret a map that depicts temperature data using isotherms.

KEY TERMS: daily mean temperature, daily range, monthly mean, annual mean, annual temperature range, isotherm, temperature gradient

- Daily mean temperature is an average of the daily maximum and daily minimum temperatures, whereas the daily range is the difference between the daily maximum and daily minimum temperatures. The monthly mean is determined by averaging the daily means for a particular month. The annual mean is an average of the 12 monthly means, whereas the annual temperature range is the difference between the highest and lowest monthly means.
- Temperature distribution is shown on a map by using isotherms, which are lines of equal temperature. Temperature gradient is the amount of temperature change per unit of distance. Closely spaced isotherms indicate a rapid rate of change.

11.9 Why Temperatures Vary: The Controls of Temperature

Discuss the principal controls of temperature and use examples to describe their effects.

KEY TERMS: temperature control, specific heat

- Controls of temperature are factors that cause temperature to vary from place to place and from time to time. Latitude (Earth–Sun relationships) is one example. Ocean currents (discussed in Chapter 10) are another example.
- Unequal heating of land and water is a temperature control. Because land and water heat and cool differently, land areas experience greater temperature extremes than water-dominated areas.
- Altitude is an easy-to-visualize control: The higher up you go, the colder it gets; therefore, mountains are cooler than adjacent lowlands.
- Geographic position as a temperature control involves such factors as mountains acting as barriers to marine influence or whether a place is on a windward or a leeward coast.

(?) **The graph shows monthly high temperatures for Urbana, Illinois, and San Francisco, California. Although both cities are located at about the same latitude, the temperatures they experience are quite different. Which line on the graph represents Urbana, and which represents San Francisco? How did you figure this out?**

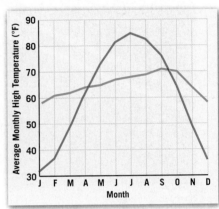

11.10 World Distribution of Temperature

Interpret the patterns depicted on world maps of January and July temperatures.

- On world maps showing January and July mean temperatures, isotherms generally trend east–west and show a decrease in temperature moving poleward from the equator. When the two maps are compared, a latitudinal shifting of temperatures is seen. Bending isotherms reveal the locations of ocean currents.
- Annual temperature range is small near the equator and increases with an increase in latitude. Outside the tropics, annual temperature range also increases as marine influence diminishes.

(?) **Refer to Figure 11.37. What causes the isotherms to bend in the North Atlantic?**

GIVE IT SOME THOUGHT

1. Determine which statements refer to weather and which refer to climate. (*Note:* One statement includes aspects of both weather and climate.)
   a. The baseball game was rained out today.
   b. January is Omaha's coldest month.
   c. North Africa is a desert.
   d. The high this afternoon was 25°C.
   e. Last evening a tornado ripped through central Oklahoma.
   f. I am moving to southern Arizona because it is warm and sunny.
   g. Thursday's low of –20°C is the coldest temperature ever recorded for that city.
   h. It is partly cloudy.

2. Refer to the graph in Figure 11.3 to answer the following questions about temperatures in New York City:
   a. What is the approximate average daily high temperature in January? In July?
   b. Approximately what are the highest and lowest temperatures ever recorded?

3. The jet in this photo is at an altitude of 10 kilometers (6.2 miles). Refer to the graph in Figure 11.7. What is the approximate air pressure where the jet is flying? About what percentage of the atmosphere is below the jet (assuming that the pressure at the surface is 1000 millibars)?

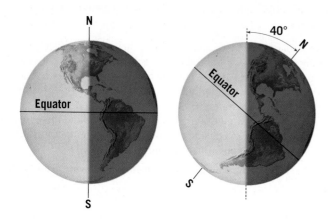

Interpixels/Shutterstock

4. Earth's axis is inclined 23½° to the plane of its orbit. What if the inclination of the axis changed? Answer the following questions that address this possibility:

   a. How would seasons be affected if Earth's axis were perpendicular to the plane of its orbit?
   b. Describe the seasons if Earth's axis were inclined 40°. Where would the Tropics of Cancer and Capricorn be located? How about the Arctic and Antarctic Circles?
5. The Sun shines continually at the North Pole for 6 months, from the spring equinox until the autumnal equinox, yet temperatures never get very warm. Explain why this is the case.

6. Imagine being at the beach in this photo on a sunny summer afternoon.
   a. Describe the temperatures you would expect if you measured the surface of the beach and at a depth of 12 inches in the sand.
   b. If you stood waist deep in the water and measured the water's surface temperature and its temperature at a depth of 12 inches, how would these measurements compare to those taken on the beach?

Tequilab/Shutterstock

7. On which of the following summer days would you expect the greatest temperature range? Which would have the smallest range in temperature? Explain your choices.
   a. Cloudy skies during the day and clear skies at night
   b. Clear skies during the day and cloudy skies at night
   c. Clear skies during the day and clear skies at night
   d. Cloudy skies during the day and cloudy skies at night
8. The accompanying sketch map represents a hypothetical continent in the Northern Hemisphere. One isotherm has been placed on the map.
   a. Is the temperature higher at City A or City B?
   b. Is the season winter or summer? How are you able to determine this?
   c. Describe (or sketch) the position of this isotherm 6 months later.

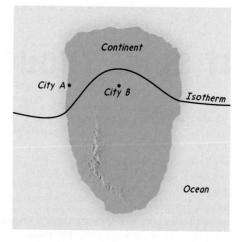

9. This photo shows a snow-covered area in the middle latitudes on a sunny day in late winter. Assume that 1 week after this photo was taken, conditions were essentially identical except that the snow was gone. Would you expect the air temperatures to be different on the two days? If so, which day would be warmer? Suggest an explanation.

CoolR/shutterstock

MasteringGeology™

www.masteringgeology.com Looking for additional review and test prep materials? With individualized coaching on the toughest topics of the course, MasteringGeology offers a wide variety of ways for you to move beyond memorization to begin thinking like a geologist. Visit the Study Area in www.masteringgeology. com to find practice quizzes, study tools, and multimedia that will improve your understanding of this chapter's content. Sign in today to enjoy the following features: **Self Study Quizzes, SmartFigure: Tutorials/Animations/Condor Videos/Mobile Field Trips, Geoscience Animation Library, GEODe, RSS Feeds, Digital Study Modules,** and an optional **Pearson eText.**

12

FOCUS ON CONCEPTS

Each statement represents the primary learning objective for the corresponding major heading within the chapter. After you complete the chapter, you should be able to:

12.1 Summarize the six processes by which water changes from one state of matter to another. For each, indicate whether energy is absorbed or released.

12.2 Write a generalization relating air temperature and the amount of water vapor needed to saturate air.

12.3 Describe adiabatic temperature changes and explain why the wet adiabatic rate of cooling is less than the dry adiabatic rate.

12.4 List and describe the four mechanisms that cause air to rise.

12.5 Describe how atmospheric stability is determined and compare conditional instability with absolute instability.

12.6 Name and describe the 10 basic cloud types, based on form and height. Contrast nimbostratus and cumulonimbus clouds and their associated weather.

12.7 Identify the basic types of fog and describe how each forms.

12.8 Describe the Bergeron process and explain how it differs from the collision–coalescence process.

12.9 Describe the atmospheric conditions that produce sleet, freezing rain (glaze), and hail.

12.10 List the advantages of using weather radar versus a standard rain gauge to measure precipitation.

Cumulonimbus clouds are often associated with thunderstorms and severe weather. (Photo by Cusp/SuperStock)

MOISTURE, CLOUDS, AND PRECIPITATION

W ater vapor is an odorless, colorless gas that mixes freely with the other gases of the atmosphere. Unlike oxygen and nitrogen—the two most abundant components of the atmosphere—water can change from one state of matter to another (solid, liquid, or gas) at the temperatures and pressures experienced on Earth (**Figure 12.1**). Because of this unique property, water leaves the oceans as a gas and returns to the oceans as a liquid.

As you observe day-to-day weather changes, you might ask: Why is it generally more humid in the summer than in the winter? Why do clouds form on some occasions but not on others? Why do some clouds look thin and harmless, whereas others form gray and ominous towers? Answers to these questions involve the role of water vapor in the atmosphere, the central theme of this chapter.

12.1 Water's Changes of State

Summarize the six processes by which water changes from one state of matter to another. For each, indicate whether energy is absorbed or released.

Water is the only substance that naturally exists on Earth as a solid (ice), liquid, and gas (water vapor). Because all forms of water are composed of hydrogen and oxygen atoms that are bonded together to form water molecules (H_2O), the primary difference among water's three phases is the arrangement of these water molecules.

Ice, Liquid Water, and Water Vapor

Ice is composed of water molecules that are held together by mutual molecular attractions. The molecules form a tight, orderly network, as shown in **Figure 12.2**. As a consequence, the water molecules in ice are not free to move relative to each other but rather vibrate about fixed sites.

When ice is heated, the molecules oscillate more rapidly. When the rate of molecular movement increases sufficiently, the bonds between the water molecules begin to break, resulting in melting.

In the liquid state, water molecules are still tightly packed but are moving fast enough that they are able to slide past one another. As a result, liquid water is fluid and takes the shape of its container.

As liquid water gains heat from its environment, some of the molecules acquire enough energy to break the remaining molecular attractions and escape from the surface, becoming water vapor.

SmartFigure 12.1 Heavy rain and hail at a ballpark in Wichita, Kansas (Photo by Fernando Salazar/AP Image) (goo.gl/ljy63o)

Tutorial

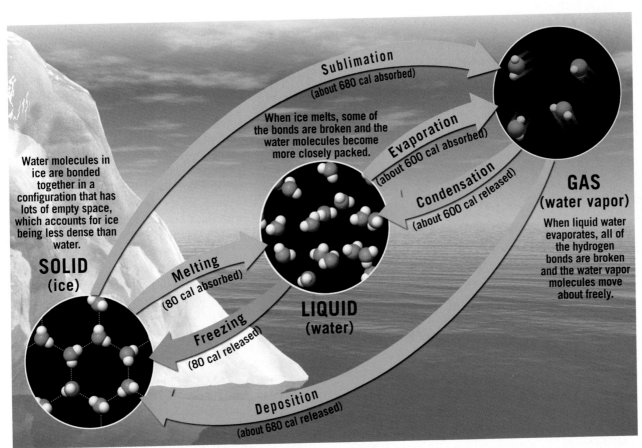

Water molecules in ice are bonded together in a configuration that has lots of empty space, which accounts for ice being less dense than water.

SOLID (ice)

Sublimation (about 680 cal absorbed)

When ice melts, some of the bonds are broken and the water molecules become more closely packed.

Evaporation (about 600 cal absorbed)

Condensation (about 600 cal released)

Melting (80 cal absorbed)

LIQUID (water)

Freezing (80 cal released)

Deposition (about 680 cal released)

GAS (water vapor)

When liquid water evaporates, all of the hydrogen bonds are broken and the water vapor molecules move about freely.

SmartFigure 12.2 Changes of state involve an exchange of heat The amounts shown here are the approximate numbers of calories absorbed or released when 1 gram of water changes from one state of matter to another. (goo.gl/5Qv8y6)

 **Animation**

Water-vapor molecules are widely spaced compared to liquid water and exhibit very energetic random motion. What distinguishes a gas from a liquid is its compressibility (and expandability). For example, you can easily put more and more air into a tire and increase its volume only slightly. However, you can't put 10 gallons of gasoline into a 5-gallon can.

Latent Heat

Whenever water changes state, heat is exchanged between water and its surroundings. When water evaporates, heat is absorbed (see Figure 12.2). Meteorologists often measure heat energy in calories. One **calorie** is the amount of heat required to raise the temperature of 1 gram of liquid water 1°C (1.8°F). Thus, when 10 calories of heat are absorbed by 1 gram of water, the molecules vibrate faster, and a 10°C (18°F) temperature rise occurs.

Under certain conditions, heat may be added to a substance without an accompanying rise in temperature. For example, when a glass of ice water is warmed, the temperature of the ice–water mixture remains a constant 0°C (32°F) until all the ice has melted. If adding heat does not raise the temperature, where does this energy go? In this case, the added energy goes into breaking the molecular attractions between the water molecules in the ice cubes.

Because the heat used to melt ice does not produce a temperature change, it is referred to as **latent heat**.

(*Latent* means "hidden," like the latent fingerprints hidden at a crime scene.) This energy can be thought of as being *stored in liquid water*, and it is not released to its surroundings as heat until the liquid returns to the solid state.

Melting 1 gram of ice requires 80 calories, an amount referred to as *latent heat of melting*. Freezing, the reverse process, releases these 80 calories per gram to its surroundings as *latent heat of freezing*.

Evaporation and Condensation We saw that heat is absorbed when ice is converted to liquid water. Heat is also absorbed during **evaporation**, the process of converting a liquid to a gas (water vapor). The energy absorbed by water molecules during evaporation is used to give them the motion needed to escape the surface of the liquid and become a gas. This energy is referred to as the *latent heat of vaporization*. During the process of evaporation, it is the higher-temperature (faster-moving) molecules that escape the surface. As a result, the average molecular motion (temperature) of the remaining water is reduced—hence the common expression "evaporation is a cooling process." You have undoubtedly experienced this cooling effect when stepping dripping wet from a swimming pool or bathtub. In this situation, the energy used to evaporate water comes from your skin, and you feel cool.

The reverse process, **condensation**, occurs when water vapor changes to the liquid state. During

Condensation of water vapor generates phenomena such as dew, clouds, and fog.

A.

Figure 12.3 Examples of condensation and deposition (Photo A by NaturePL/ SuperStock; Photo B by elen_studio/Fotolia)

Frost on a windowpane, an example of deposition.

B.

is high, this process can spur the growth of towering storm clouds.

Sublimation and Deposition You are probably least familiar with the last two processes illustrated in Figure 12.2—sublimation and deposition. **Sublimation** is the conversion of a solid directly to a gas, without passing through the liquid state. Examples you may have observed include the gradual shrinking of unused ice cubes in the freezer and the rapid conversion of dry ice (frozen carbon dioxide) to wispy clouds that quickly disappear.

Deposition refers to the reverse process, the conversion of a vapor directly to a solid. This change occurs, for example, when water vapor is deposited as ice on solid objects such as grass or windows (**Figure 12.3B**). These deposits are called *white frost* or *hoar frost*, or simply *frost*. A household example of the process of deposition is the "frost" that accumulates in a freezer. As shown in Figure 12.2, deposition releases an amount of energy equal to the total amount released by condensation and freezing.

condensation, water-vapor molecules release energy (*latent heat of condensation*) in an amount equivalent to what was absorbed during evaporation. When condensation occurs in the atmosphere, it results in the formation of such phenomena as fog and clouds (**Figure 12.3A**).

As you will see, latent heat plays an important role in many atmospheric processes. In particular, when water vapor condenses to form cloud droplets, latent heat of condensation is released, warming the surrounding air and giving it buoyancy. When the moisture content of air

 CONCEPT CHECKS

1. Summarize the processes by which water changes from one state of matter to another. Indicate whether energy is absorbed or released.
2. What is *latent heat*?
3. What is a common example of sublimation?
4. How does frost form?

12.2 Humidity: Water Vapor in the Air

Write a generalization relating air temperature and the amount of water vapor needed to saturate air.

Water vapor constitutes only a small fraction of the atmosphere, varying from as little as one-tenth of 1 percent up to about 4 percent by volume. But the importance of water in the air is far greater than these small percentages would indicate. Indeed, scientists agree that *water vapor* is the most important gas in the atmosphere when it comes to understanding atmospheric processes.

Humidity is the the amount of water vapor in air. Meteorologists employ several methods to express the water-vapor content of the air. Here we examine three: mixing ratio, relative humidity, and dew-point temperature.

Saturation

Before we consider these humidity measures further, it is important to understand the concept of **saturation**. Imagine a closed jar that contains water overlain by dry

air, both at the same temperature. As the water begins to evaporate from the water surface, a small increase in pressure can be detected in the air above. This increase is the result of the motion of the water-vapor molecules that were added to the air through evaporation. In the open atmosphere, this pressure is termed **vapor pressure** and is defined as the part of the total atmospheric pressure that can be attributed to the water-vapor content.

In the closed container, as more and more molecules escape from the water surface, the steadily increasing vapor pressure in the air above forces more and more of these molecules to return to the liquid. Eventually the number of vapor molecules returning to the surface will balance the number leaving. At that point, the air is *saturated*: It can hold no more water vapor. If we add heat to the container, thereby increasing the temperature of the water and air, more water will evaporate before a balance is reached.

Table 12.1 Amount of Water Vapor Needed to Saturate 1 Kilogram of Air at Various Temperatures

Temperature °C (°F)	Water-Vapor Content at Saturation (grams)
−40 (−40)	0.1
−30 (−22)	0.3
−20 (−4)	0.75
−10 (14)	2
0 (32)	3.5
5 (41)	5
10 (50)	7
15 (59)	10
20 (68)	14
25 (77)	20
30 (86)	26.5
35 (95)	35
40 (104)	47

Consequently, at higher temperatures, more moisture is required for saturation. The amount of water vapor required for saturation at various temperatures is shown in **Table 12.1**.

Mixing Ratio

Not all air is saturated, of course. Thus, we need ways to express how humid a parcel of air is. One method is to specify the amount of water vapor contained in a unit of air. The **mixing ratio** is the mass of water vapor in a unit of air compared to the remaining mass of dry air:

$$\text{mixing ratio} = \frac{\text{mass of water vapor (grams)}}{\text{mass of dry air (kilograms)}}$$

Table 12.1 shows the mixing ratios of saturated air at various temperatures. For example, at 25°C (77°F), 20 grams of water vapor would be needed to saturate 1 kilogram of dry air.

Because the mixing ratio is expressed in units of mass (usually in grams per kilogram), it is not affected by changes in pressure or temperature. However, measuring the mixing ratio by direct sampling is time-consuming. Thus, meteorologists commonly use other methods to express the moisture content of the air. These include relative humidity and dew-point temperature.

Relative Humidity

The most familiar and, unfortunately, the most misunderstood term used to describe the moisture content of air is relative humidity. **Relative humidity** *is the ratio of the air's actual water-vapor content to the amount of water vapor required for saturation at that temperature (and pressure).* Thus, unlike the mixing ratio, relative humidity indicates how near the air is to saturation rather than the actual quantity of water vapor in the air.

To illustrate, we see from Table 12.1 that at 25°C (77°F), air is saturated when it contains 20 grams of water vapor per kilogram of dry air. Thus, if the air contains 10 grams of water vapor per kilogram of dry air on a 25°C day, the relative humidity is expressed as 10/20, or 50 percent. If air with a temperature of 25°C has a water-vapor content of 20 grams per kilogram, the relative humidity would be expressed as 20/20, or 100 percent. When the relative humidity reaches 100 percent, the air is saturated.

Because relative humidity depends both on the air's water-vapor content and on the amount of moisture required for saturation, it can be changed in either of two ways. First, relative humidity can be changed by adding or removing water vapor. Second, because the amount of moisture required for saturation is a function of air temperature (the warmer the air, the more water is required to saturate it), relative humidity varies with temperature.

How Changes in Moisture Affect Relative Humidity

In nature, moisture is added to the air mainly via evaporation from the oceans. However, plants, soil, and smaller bodies of water also make substantial contributions.

Notice in **Figure 12.4** that when water vapor is added to a parcel of air, the relative humidity of the parcel

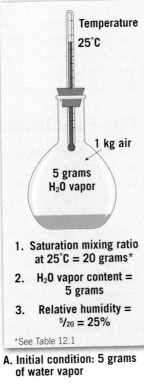

A. Initial condition: 5 grams of water vapor

1. Saturation mixing ratio at 25°C = 20 grams*
2. H_2O vapor content = 5 grams
3. Relative humidity = $^5/_{20}$ = 25%

*See Table 12.1

B. Addition of 5 grams of water vapor = 10 grams

1. Saturation mixing ratio at 25°C = 20 grams*
2. H_2O vapor content = 10 grams
3. Relative humidity = $^{10}/_{20}$ = 50%

C. Addition of 10 grams of water vapor = 20 grams

1. Saturation mixing ratio at 25°C = 20 grams*
2. H_2O vapor content = 20 grams
3. Relative humidity = $^{20}/_{20}$ = 100%

Figure 12.4 At a constant temperature the relative humidity will increase as water vapor is added to the air The saturation mixing ratio for air at 25°C is 20 g/kg (see Table 12.1). As the water-vapor content in the flask increases, the relative humidity rises from 25 percent in A to 100 percent in C.

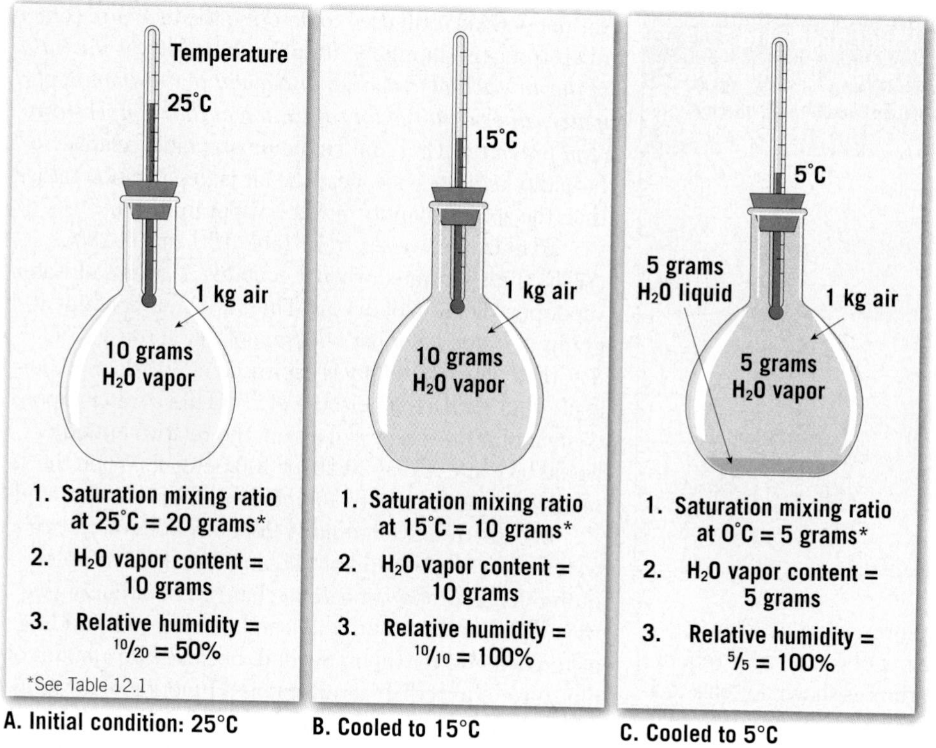

Figure 12.5 Relative humidity varies with temperature When the water-vapor content (mixing ratio) remains constant, the relative humidity will change when the air temperature either increases or decreases. In this example, when the temperature of the air in the flask was lowered from 25°C in A to 15°C in B, the relative humidity increased from 50 to 100 percent. Further cooling from 15°C in B to 5°C in C causes one-half of the water vapor to condense. In nature, when saturated air cools, it causes condensation in the form of clouds, dew, or fog.

Temperature

25°C

1 kg air

10 grams H_2O vapor

1. Saturation mixing ratio at 25°C = 20 grams*
2. H_2O vapor content = 10 grams
3. Relative humidity = $^{10}/_{20}$ = 50%

*See Table 12.1

A. Initial condition: 25°C

15°C

1 kg air

10 grams H_2O vapor

1. Saturation mixing ratio at 15°C = 10 grams*
2. H_2O vapor content = 10 grams
3. Relative humidity = $^{10}/_{10}$ = 100%

B. Cooled to 15°C

5°C

5 grams H_2O liquid

1 kg air

5 grams H_2O vapor

1. Saturation mixing ratio at 0°C = 5 grams*
2. H_2O vapor content = 5 grams
3. Relative humidity = $^5/_5$ = 100%

C. Cooled to 5°C

increases until saturation occurs (100 percent relative humidity). What if even more moisture is added to this parcel of saturated air? Does the relative humidity exceed 100 percent? Normally, this situation does not occur. Instead, the excess water vapor condenses to form liquid water.

You may have experienced such a situation while taking a hot shower. The water is composed of very energetic (hot) molecules, which means that the rate of evaporation is high. As long as you run the shower, the process of evaporation continually adds water vapor to the unsaturated air in the bathroom. If you stay in a hot shower long enough, the air eventually becomes saturated, and the excess water vapor begins to condense on the mirror, window, tile, and other cool surfaces in the room.

How Changes in Temperature Affect Relative Humidity

The second condition that affects relative humidity is air temperature. Examine **Figure 12.5** carefully. Note in Figure 12.5A that when air at 25°C contains 10 grams of water vapor per kilogram it has a relative humidity of 50 percent. This can be verified by referring to Table 12.1. Here we can see that at 25°C, air is saturated when it contains 20 grams of water vapor per kilogram of air. Because the air in Figure 12.5A contains 10 grams of water vapor, its relative humidity is 10/20, or 50 percent.

When the air in the flask is cooled from 25°C to 15°C, as shown in Figure 12.5B, the relative humidity increases from 50 to 100 percent. We can conclude that when the water-vapor content remains constant, *a*

decrease in temperature results in an increase in relative humidity.

But there is no reason to assume that cooling would cease the moment the air reached saturation. What happens when the air is cooled below the temperature at which saturation occurs? Figure 12.5C illustrates this situation. Notice from Table 12.1 that when the flask is cooled to 5°C, the air is saturated at 5 grams of water vapor per kilogram of air. Because this flask originally contained 10 grams of water vapor, 5 grams of water vapor will condense to form liquid droplets that collect on the walls of the container. In the meantime, the relative humidity of the air inside remains at 100 percent.

Similarly, when rising air reaches an elevation where it is cooled below its dew-point temperature, some of the water vapor condenses to form clouds. Because clouds are made of tiny liquid droplets (or ice crystals), this moisture is no longer part of the *water-vapor* content of the air.

We can summarize the effects of temperature on relative humidity as follows: When the water-vapor content of air remains at a constant level, a decrease in air temperature results in an increase in relative humidity, and an increase in temperature causes a decrease in relative humidity. **Figure 12.6** illustrates the variations in temperature and relative humidity during a typical day and the relationship described above.

Dew-Point Temperature

The **dew-point temperature**, or simply the **dew point**, of a given parcel of air is *the temperature at which water vapor begins to condense.* The term *dew point* stems from the fact that at night, objects near the ground often cool below the dew-point temperature and become coated with dew. You have undoubtedly seen "dew" form on an ice-cold drink on a humid summer day (**Figure 12.7**). In nature, cooling air below its dew-point temperature typically generates dew, fog, or clouds when the dew point is above freezing and frost when it is below freezing (0°C [32°F]).

Dew point can also be defined as the *temperature at which a parcel of air reaches saturation* and,

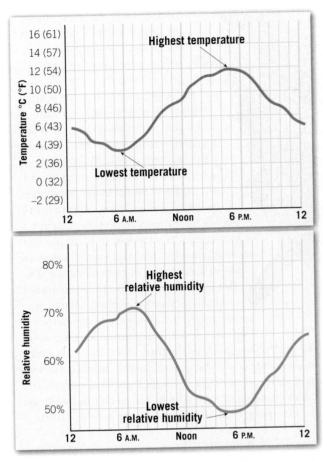

Figure 12.6 Typical daily variation in temperature and relative humidity during a spring day in Washington, DC

Table 12.2 Dew-Point Thresholds

Dew-Point Temperature	Threshold
≤ 10°F	Significant snowfall is inhibited
≥ 55°F	Minimum for severe thunderstorms to form
≥ 65°F	Considered humid by most people
≥ 70°F	Typical of the rainy tropics
≥ 75°F	Considered oppressive by most

(77°F), it contains about twice the water vapor as the air over St. Louis, Missouri, with a dew point of 15°C (59°F) and four times that of air over Tucson, Arizona, with a dew point of 5°C (41°F).

Because the dew-point temperature is a good measure of the amount of water vapor in the air, it commonly appears on weather maps. When the dew point exceeds 65°F (18°C), most people consider the air to feel humid; air with a dew point of 75°F (24°C) or higher is considered oppressive. Notice on the map in **Figure 12.8** that much of the southeastern United States has dew-point temperatures that exceed 65°F (18°C). Also notice in Figure 12.8 that although the Southeast is dominated by humid conditions, most of the remainder of the country is experiencing drier air.

How Is Humidity Measured?

Instruments called **hygrometers** (*hygro* = moisture, *metron* = measuring instrument) are used to measure the moisture content of the air.

Psychrometers One of the simplest hygrometers, a **psychrometer** (called a *sling psychrometer* when connected to a handle and spun), consists of two identical

hence, is directly related to the *actual moisture content* of that parcel. Recall that the saturation vapor pressure is temperature dependent. In fact, for every 10°C (18°F) increase in temperature, the amount of water vapor needed for saturation approximately doubles. Therefore, *saturated air* at 0°C (32°F) contains about half the water vapor of *saturated air* at 10°C (50°F) and roughly one-fourth that of *saturated air* at 20°C (68°F). Because the dew point is the temperature at which saturation occurs, we can conclude that high dew-point temperatures indicate moist air and, conversely, low dew-point temperatures indicate dry air (**Table 12.2**).

More precisely, based on what we have learned about vapor pressure and saturation, we can state that for every 10°C (18°F) increase in the dew-point temperature, air contains about twice as much water vapor. Therefore, we know that when air over Fort Myers, Florida, has a dew-point temperature of 25°C

Figure 12.7 Condensation and dew-point temperature Condensation, or "dew," occurs when a cold drinking glass chills the surrounding layer of air below the dew-point temperature. (Nitr/Fotolia)

Tutorial

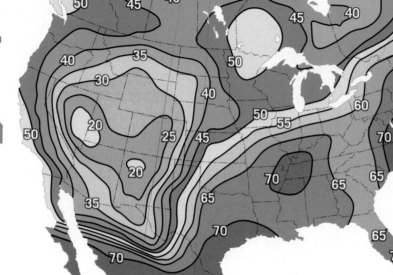

Figure 12.9 Sling psychrometer used to determine both relative humidity and dew point (Photo by E. J. Tarbuck)

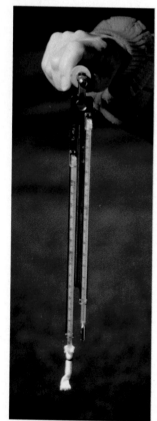

A. The dry-bulb thermometer gives the current air temperature.

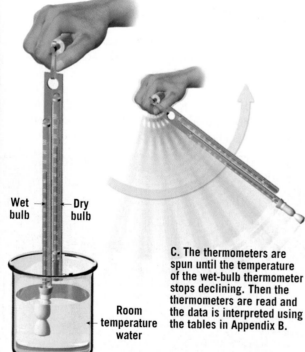

Wet bulb → ← Dry bulb

Room temperature water

B. The wet-bulb thermometer is covered with a cloth wick that is dipped in water.

C. The thermometers are spun until the temperature of the wet-bulb thermometer stops declining. Then the thermometers are read and the data is interpreted using the tables in Appendix B.

thermometers mounted side by side (**Figure 12.9A**). One thermometer, called the *dry bulb*, measures air temperature, and the other, called the *wet bulb*, has a thin cloth wick tied at the bottom. This cloth wick is saturated with water, and a continuous current of air is passed over the wick, either by swinging the psychrometer or by using an electric fan to move air past the instrument (**Figure 12.9B, C**). As a result, water evaporates from the wick, absorbing heat energy from the wet-bulb thermometer, which causes its temperature to drop. The amount of cooling that takes place is directly proportional to the dryness of the air: The drier the air, the greater the evaporation and the greater the cooling. Therefore, the larger the difference between the wet- and dry-bulb temperatures, the lower the relative humidity. By contrast, if the air is saturated, no evaporation will occur, and the two thermometers will have identical readings. By using a psychrometer and the tables provided in Appendix B, the relative humidity and the dew-point temperature can be easily determined.

Electric Hygrometers Today, a variety of *electric hygrometers* are widely used to measure humidity. The Automated Weather Observing System (AWOS) operated by the National Weather Service (NWS) employs an electric hygrometer that works on the principle of *capacitance*— a material's ability to store an electrical charge. The sensor consists of a thin hygroscopic (water-absorbent) film that is connected to an electric current. As the film absorbs or releases water, the capacitance of the sensor changes at a rate proportional to the relative humidity of the surrounding air. Thus, relative humidity can be measured by monitoring the change in the film's capacitance. Higher capacitance means higher relative humidity.

12.2 CONCEPT CHECKS

1. List three measures used to express humidity.
2. If the amount of water vapor in the air remains unchanged, how does a decrease in temperature affect relative humidity?
3. Define *dew-point temperature*.
4. Which measure of humidity, relative humidity or dew point, best describes the actual quantity of water vapor in a mass of air?
5. Briefly describe the principle of a psychrometer.

12.3 Adiabatic Temperature Changes and Cloud Formation

Describe adiabatic temperature changes and explain why the wet adiabatic rate of cooling is less than the dry adiabatic rate.

Recall that condensation occurs when sufficient water vapor is added to the air or, more commonly, when the air is cooled to its dew-point temperature. Condensation may produce dew, fog, or clouds. Heat near Earth's surface is readily exchanged between the ground and the air directly above. As the ground loses heat in the evening (radiation cooling), dew may condense on the grass, while fog may form slightly above Earth's surface. Thus, surface cooling that occurs after sunset produces some condensation. Cloud formation, however, often takes place during the warmest part of the day—an indication that another mechanism must operate aloft that cools air sufficiently to generate clouds.

Adiabatic Temperature Changes

The process that generates most clouds is easily visualized. Have you ever pumped up a bicycle tire with a hand pump and noticed that the pump barrel became very warm? When you applied energy to *compress* the air, the motion of the gas molecules increased, and the temperature of the air rose. Conversely, if you allow air to escape from a bicycle tire, the air *expands*; the gas molecules move less rapidly, and the air cools. You have probably felt the cooling effect of the propellant gas expanding as you applied hair spray or spray deodorant. The temperature changes just described, in which heat energy was neither added nor subtracted, are called **adiabatic temperature changes**. Instead, changes in pressure result in temperature changes. When air is compressed, it warms, and when air is allowed to expand, it cools.

Adiabatic Cooling and Condensation

To simplify the discussion of adiabatic cooling, imagine a volume of air enclosed in a thin balloon-like bubble. Meteorologists call this imaginary volume of air a **parcel**. Typically, we consider a parcel to be a few hundred cubic meters in volume, and we assume that it acts independently of the surrounding air. We can also assume that no heat is transferred into or out of the parcel. Although this image is highly idealized,

over short time spans, a parcel of air behaves much like a volume of air moving up or down in the atmosphere. In nature, sometimes the surrounding air infiltrates a rising or descending column of air, a process called *entrainment*. For the following discussion, however, we assume that no mixing of this type occurs.

Dry Adiabatic Rate Recall from Chapter 11 that atmospheric pressure decreases with height. Any time a parcel of air moves upward, it passes through regions of successively lower pressure. As a result, this ascending air expands and cools adiabatically. Unsaturated air cools at a constant rate of 10°C for every 1000 meters of ascent (12.5°F per 1000 feet). Conversely, descending air comes under increasing pressure and is compressed and heated 10°C for every 1000 meters of descent (**Figure 12.10**). This rate of cooling or heating applies only to *unsaturated air* and is known as the **dry adiabatic rate** ("dry" because the air is unsaturated).

Wet Adiabatic Rate If an air parcel rises high enough, it will eventually cool to its dew point and trigger the process of condensation. The altitude at which a parcel reaches saturation and cloud formation begins is called the **lifting condensation level**, or simply **condensation level**. At the lifting condensation level, an important change occurs: The *latent heat* that was absorbed by the water vapor when it evaporated is released as **sensible heat**—energy that can be measured with a thermometer. Although the parcel

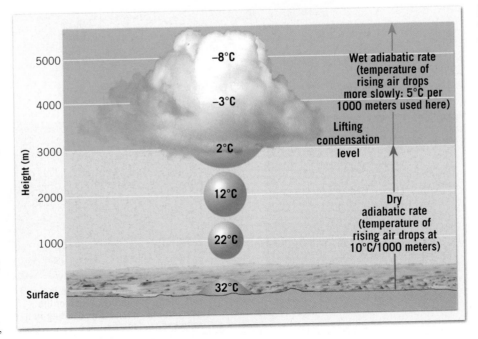

Figure 12.10 Dry versus wet adiabatic rates of cooling Rising air cools at the relatively constant dry adiabatic rate of 10°C per 1000 meters, until the air reaches the dew point and condensation (cloud formation) begins. As air continues to rise, the latent heat released by condensation reduces the rate of cooling. Because the amount of latent heat released depends on the amount of moisture present in the rising air, the wet adiabatic rate varies from about 5°C per 1000 meters for air with a high moisture content to 9°C per 1000 meters for dry air.

will continue to cool adiabatically, the release of latent heat slows the rate of cooling. In other words, when a parcel of air ascends above the lifting condensation level, the rate at which it cools is reduced. This slower rate of cooling is called the **wet adiabatic rate**, also commonly referred to as the *moist* or *saturated adiabatic rate* (see Figure 12.10).

Because the amount of latent heat released depends on the amount of moisture present in the rising air (generally between 0 and 4 percent), the wet adiabatic rate varies from about 5°C per 1000 meters for air with a high moisture content to 9°C per 1000 meters for air with a low moisture content.

To summarize, rising air cools at the dry adiabatic rate from the surface up to the lifting condensation level, at which point it cools at the slower wet adiabatic rate.

 CONCEPT CHECKS

1. What name is given to the processes whereby the temperature of air changes without the addition or subtraction of heat?
2. Why does air expand as it moves upward through the atmosphere?
3. At what rate does unsaturated air cool when it rises through the atmosphere?
4. Why does the adiabatic rate of cooling change when condensation begins?
5. Why does the wet adiabatic rate not have a constant value?

12.4 Processes That Lift Air

List and describe the four mechanisms that cause air to rise.

Why does air rise on some occasions to produce clouds, but not on others? Generally, the tendency is for air to resist vertical movement; air near the surface tends to stay near the surface, and air aloft tends to remain aloft. However, the following four processes cause air to rise, thereby generating clouds:

1. *Orographic lifting*, in which air is forced to rise over a mountainous barrier
2. *Frontal lifting*, in which warmer, less-dense air is forced over cooler, denser air
3. *Convergence*, which is a pileup of horizontal airflow that results in upward movement
4. *Localized convective lifting*, in which unequal surface heating causes localized pockets of air to rise because of their buoyancy

Orographic Lifting

Orographic lifting occurs when elevated terrain, such as a mountain range, act as a barrier to the flow of air (**Figure 12.11**). As air ascends a mountain slope, adiabatic cooling often generates clouds and copious precipitation. In fact, many of the rainiest places in the world are located on windward mountain slopes.

By the time air reaches the leeward side of a mountain, much of its moisture has been lost. If the air descends, it warms adiabatically, making condensation and precipitation even less likely. As shown in Figure 12.11, the result can be a **rainshadow desert**. The Great Basin Desert of the western United States lies only a few hundred kilometers from the Pacific Ocean, but it is effectively cut off from the ocean's moisture by the imposing Sierra

Figure 12.11 Orographic lifting and precipitation (Photo A by Shutterstock/Dean Pennala/Shutterstock; photo B by Dennis Tasa)

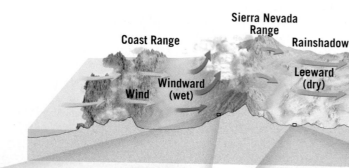

A. Orographic lifting leads to precipitation on windward slopes.

B. By the time air reaches the leeward side of the mountains, much of the moisture has been lost, resulting in a *rainshadow desert.*

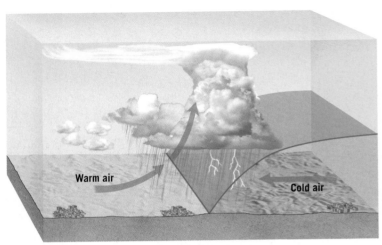

Figure 12.12 Frontal lifting Colder, denser air acts as a barrier over which warmer, less-dense air rises.

Nevada (see Figure 12.11). The Gobi Desert of Mongolia, the Takla Makan of China, and the Patagonia Desert of Argentina are other examples of deserts that exist because they are on the leeward sides of large mountain systems.

Frontal Lifting

If orographic lifting were the only mechanism that forced air aloft, the relatively flat central portion of North America would be an expansive desert rather than the area known as "the nation's breadbasket." Fortunately, this is not the case.

In central North America, warm and cold air masses often collide, producing boundaries called **fronts**. Rather than mixing, the cooler, denser air acts as a barrier over which the warmer, less-dense air rises. This process, called **frontal lifting**, also referred to as **frontal wedging**, is illustrated in **Figure 12.12**.

It should be noted that weather-producing fronts are associated with storm systems called *midlatitude cyclones*. Because these storms are responsible for producing a high percentage of the precipitation in the middle latitudes, we will examine fronts in detail in Chapter 14.

Convergence

When the wind pattern near Earth's surface is such that more air is entering an area than is leaving—a phenomenon called **convergence**—lifting occurs (**Figure 12.13**). Convergence as a mechanism of lifting is most often associated with large centers of *low pressure*, mainly midlatitude cyclones and hurricanes. The inward flow of air at the surface of these systems is balanced by rising air, cloud formation, and usually precipitation.

Convergence can also occur when an obstacle slows or restricts horizontal airflow (wind). For example, when air moves from a relatively smooth surface, such as the ocean, onto an irregular landscape, increased friction reduces its speed. The result is a pileup of air (convergence). When air converges, there is an upward flow of air molecules rather than a simple squeezing together of molecules (as happens when people are entering a crowded building).

The Florida peninsula provides an excellent example of the role that convergence can play in initiating cloud development and precipitation (see Figure 12.13). On warm days, the airflow is from the ocean to the land along both coasts of Florida. This leads to a pileup of air along the coasts and general convergence over the peninsula. This pattern of convergence and uplift is aided by intense solar heating of the land. As a result, Florida's peninsula experiences the greatest frequency of mid-afternoon thunderstorms in the United States.

Localized Convective Lifting

On warm summer days, unequal heating of Earth's surface may cause some pockets of air to be warmed

SmartFigure 12.13

Convergence over the Florida peninsula When surface air converges, it is forced to rise. Florida provides a good example. On warm days, airflow from the Atlantic Ocean and Gulf of Mexico onto the Florida peninsula generates many mid-afternoon thunderstorms. (Photo by NASA) (goo.gl/QZWct6)

Tutorial

Figure 12.14 Localized convective lifting Unequal heating of Earth's surface causes pockets of air to be warmed more than the surrounding air. These buoyant parcels of hot air (thermals) rise, and if they reach the condensation level, clouds form.

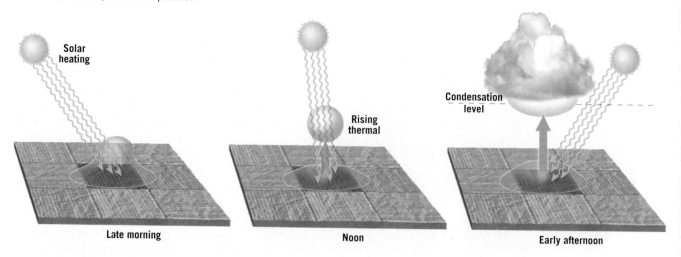

Solar heating

Late morning

Rising thermal

Noon

Condensation level

Early afternoon

more than the surrounding air (**Figure 12.14**). For instance, a plowed field will absorb more radiation and the air above it will be warmed more than the air above adjacent fields of crops. Consequently, the parcel of air above the field, which is warmer (less dense) than the surrounding air, will be buoyed upward. These rising parcels of warmer air are called *thermals*. Birds such as hawks and eagles use thermals to carry them to great heights, where they can identify unsuspecting prey. Humans take advantage of these rising parcels to use hang gliders as a way to "fly."

The phenomenon that produces rising thermals is called **localized convective lifting**, or simply **convective lifting**. When these warm parcels of air rise above the lifting condensation level, clouds form and on occasion produce mid-afternoon rain showers. The height of clouds produced in this fashion is somewhat limited

because the buoyancy caused solely by unequal surface heating is confined to, at most, the first few kilometers of the atmosphere. Also, the accompanying rains, although occasionally heavy, are of short duration and widely scattered, a phenomenon called *sun showers*.

12.4 CONCEPT CHECKS

1. Explain why the Great Basin of the western United States is dry. What term is applied to this type of desert?
2. How does frontal lifting cause air to rise?
3. Define *convergence*. Identify two types of weather systems associated with convergence in the lower atmosphere.
4. Why does Florida have abundant mid-afternoon thunderstorms?
5. Describe convective lifting.

12.5 The Critical Weathermaker: Atmospheric Stability

Describe how atmospheric stability is determined and compare conditional instability with absolute instability.

When air rises, it cools and usually produces clouds. Why do clouds vary so much in size, and why does the resulting precipitation vary so much? The answers are closely related to the *stability* of the air.

Recall that a parcel of air can be thought of as having a thin, flexible cover that allows it to expand but prevents it from mixing with the surrounding air (picture a hot-air balloon). Imagine that you find such a parcel, with some given starting temperature, and forcibly lift it to a higher location. Its temperature will decrease because of expansion; its final temperature will depend on how warm it was to begin with and how high you lift it up.

When you let go of the parcel, what happens? If it is cooler (and hence denser) than the surrounding air, it will sink back down to its original location. Air of this type, called **stable air**, resists upward movement. But if it is *warmer* than the surrounding air (and hence less dense), it will continue to rise. Specifically, it will rise

until it reaches an altitude where its temperature equals that of its surroundings. This is exactly how a hot-air balloon works, rising as long as it is warmer and less dense than the surrounding air (**Figure 12.15**). This type of air is classified as **unstable air**.

Types of Stability

Stability is a property of air that describes whether it resists rising (is stable) or may rise spontaneously (is unstable). To determine the stability of a given parcel of air, we first need to know how the temperature of the atmosphere above the parcel changes with height. Recall from Chapter 11 that this measure, determined from observations made by radiosondes and aircraft, is called the **environmental lapse rate**. It is important not to confuse this with *adiabatic temperature changes*, which are changes in the temperature of a

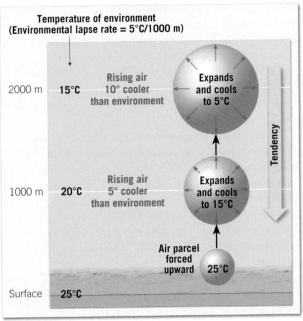

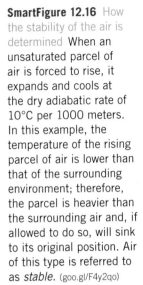

SmartFigure 12.16 How the stability of the air is determined When an unsaturated parcel of air is forced to rise, it expands and cools at the dry adiabatic rate of 10°C per 1000 meters. In this example, the temperature of the rising parcel of air is lower than that of the surrounding environment; therefore, the parcel is heavier than the surrounding air and, if allowed to do so, will sink to its original position. Air of this type is referred to as *stable*. (goo.gl/F4y2qo)

Animation

Figure 12.15 Hot air rises As long as air is warmer than its surroundings, it will rise. Hot-air balloons rise up through the atmosphere for this reason. (Photo by Steve Vidler/SuperStock)

rising or sinking parcel of air caused by expansion or compression.

To illustrate, we examine a situation in which the environmental lapse rate is 5°C per 1000 meters (**Figure 12.16**). Under this condition, when air at the surface has a temperature of 25°C, the air at 1000 meters will be 5° cooler, or 20°C, the air at 2000 meters will have a temperature of 15°C, and so forth. At first glance, it appears that the air at the surface is less dense than the air at 1000 meters because it is 5° warmer. However, if the air near the

surface were unsaturated and were to rise to 1000 meters, it would expand and cool at the dry adiabatic rate of 10°C per 1000 meters. Therefore, upon reaching 1000 meters, its temperature would have dropped 10°C. Being 5° cooler than its environment, it would be denser and tend to sink to its original position. Hence, we say that the air near the surface is potentially cooler than the air aloft and therefore will not rise on its own. The air just described is *stable* and resists vertical movement.

Absolute Stability Stated quantitatively, **absolute stability** prevails when the environmental lapse rate is less than the wet adiabatic rate. **Figure 12.17** depicts this situation using an environmental lapse rate of 5°C per

SmartFigure 12.17 Atmospheric conditions that result in absolute stability Absolute stability prevails when the environmental lapse rate is less than the wet adiabatic rate. **A.** The rising parcel of air is always cooler and heavier than the surrounding air, producing stability. **B.** Graphical representation of the conditions shown in part A. (goo.gl/gcyEoe)

Tutorial

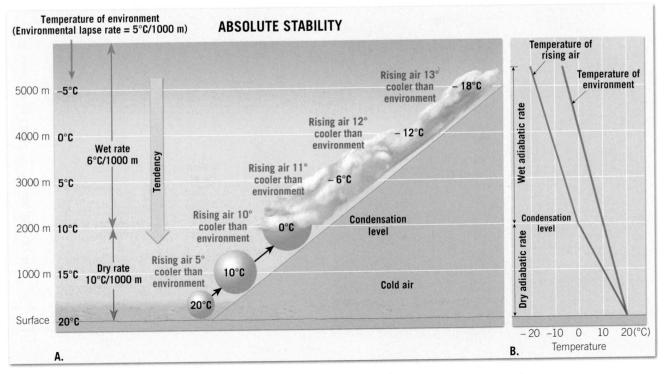

1000 meters and a wet adiabatic rate of 6°C per 1000 meters. Note that at 1000 meters, the temperature of the surrounding air is 15°C, while the rising parcel of air has cooled to 10°C and is therefore the denser air. Even if this stable air were to be forced above the condensation level, it would remain cooler and denser than its environment, and thus it would tend to return to the surface.

The most stable conditions occur when the temperature in a layer of air actually increases with altitude rather than decreases. When such a reversal occurs, a *temperature inversion* is said to exist. Temperature inversions frequently occur on clear nights as a result of radiation cooling of Earth's surface. Under these conditions, an inversion is created because the ground and the air immediately above will cool more rapidly than the air aloft. When warm air overlies cooler air, it acts as a lid and prevents appreciable vertical mixing. Because of this, temperature inversions are responsible for trapping pollutants in a narrow zone near Earth's surface.

Absolute Instability At the other extreme from absolute stability, air is said to exhibit **absolute instability** when the environmental lapse rate is greater than the dry adiabatic rate. As shown in **Figure 12.18**, the ascending parcel of air is always warmer than its environment and will continue to rise because of its own buoyancy. However, the conditions needed to render the air absolutely unstable mainly occur near Earth's surface. On hot, sunny days the air above some surfaces, such as shopping center parking lots, is heated more than the air over adjacent surfaces. These invisible pockets of more intensely heated air, being less dense than the air aloft, will rise like a hot-air balloon. This phenomenon produces the small, fluffy clouds we associate with fair weather. Occasionally, when the surface air is considerably warmer than the air aloft, clouds with considerable vertical development can form.

Conditional Instability A more common type of atmospheric instability is called **conditional instability**. This occurs when moist air has an environmental lapse rate between the dry and wet adiabatic rates (between 5°C and 10°C per 1000 meters). Simply, the atmosphere is said to be conditionally unstable when it is *stable* for an *unsaturated* parcel of air but *unstable* for a *saturated* parcel of air. Notice in **Figure 12.19** that the rising parcel of air is cooler than the surrounding air for nearly 3000 meters. With the addition of latent heat above the lifting condensation level, the parcel becomes warmer than the surrounding air. From this point along its ascent, the parcel will continue to rise because of its own buoyancy, without an outside lifting force. Thus, conditional instability depends on whether the rising air is saturated. The word *conditional* is used because the air must be forced upward, such as over mountainous terrain, before it becomes unstable and rises because of its own buoyancy.

In summary, the stability of air is determined by measuring the temperature of the atmosphere at various heights. In simple terms, a column of air is deemed unstable when the air near the bottom of the column is significantly warmer (less dense) than the air aloft, indicating a steep environmental lapse rate. Under these conditions, the air actually turns over, as the warm

Figure 12.18 Atmospheric conditions that result in absolute instability Absolute instability can develop when solar heating causes the lowermost layer of the atmosphere to be warmed to a much higher temperature than the air aloft. **A.** The result is a steep environmental lapse rate that renders the atmosphere unstable. **B.** Graphical representation of the conditions shown in part A.

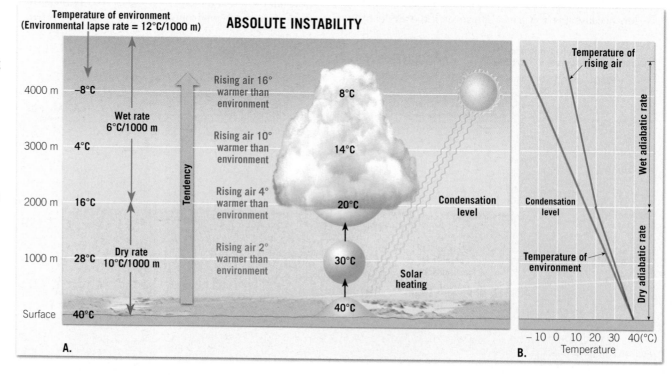

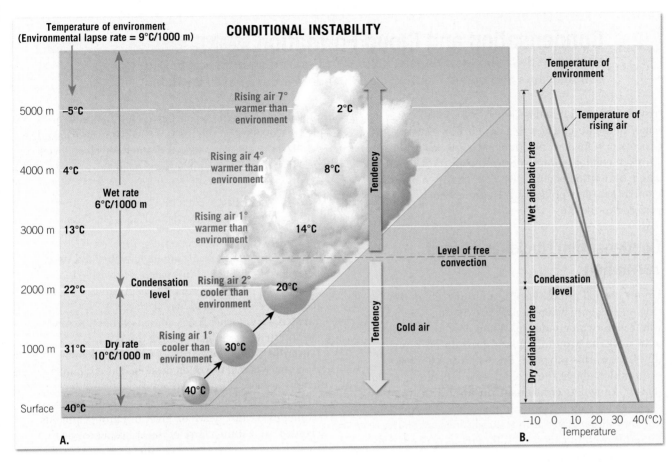

Temperature of environment
(Environmental lapse rate = 9°C/1000 m)

CONDITIONAL INSTABILITY

5000 m −5°C

Rising air 7° warmer than environment   2°C

4000 m 4°C

Rising air 4° warmer than environment   8°C

Wet rate
6°C/1000 m

3000 m 13°C

Rising air 1° warmer than environment   14°C

Level of free convection

2000 m 22°C

Condensation level

Rising air 2° cooler than environment   20°C

Cold air

1000 m 31°C

Dry rate
10°C/1000 m

Rising air 1° cooler than environment   30°C

40°C

Surface 40°C

Tendency

A.

Temperature of environment

Temperature of rising air

Wet adiabatic rate

Condensation level

Dry adiabatic rate

−10 0 10 20 30 40(°C)
Temperature

B.

Figure 12.19 Atmospheric conditions that result in conditional instability Conditional instability may result when warm air is forced to rise along a frontal boundary. Note that the environmental lapse rate of 9°C per 1000 meters lies between the dry and wet adiabatic rates. **A.** The parcel of air is cooler than the surrounding air up to nearly 3000 meters, where its tendency is to sink toward the surface (stable). Above this level, however, the parcel is warmer than its environment and will rise because of its own buoyancy (unstable). Thus, when conditionally unstable air is forced to rise, the result can be towering cumulus clouds. **B.** Graphical representation of the conditions shown in part A.

air below rises and displaces the colder air aloft. Conversely, the air is considered to be stable when the temperature drops relatively slowly with increasing altitude. The most stable conditions occur during a temperature inversion, when the temperature actually increases with height. Under these conditions, there is very little vertical air movement.

Stability and Daily Weather

From the previous discussion, we can conclude that stable air resists upward movement, whereas unstable air ascends freely because of its own buoyancy. But how do these facts manifest themselves in our daily weather?

Because stable air resists upward movement, we might conclude that clouds will not form when stable conditions prevail in the atmosphere. Although this seems reasonable, recall that processes exist that *force* air aloft. These processes include orographic lifting, frontal wedging, and convergence. When stable air is forced aloft, the clouds that form are widespread and have little vertical thickness when compared to their horizontal dimension, and precipitation, if any, is light to moderate.

By contrast, clouds associated with the lifting of unstable air are towering and often generate thunderstorms and occasionally even tornadoes. For this reason, we can conclude that on a dreary, overcast day with light drizzle, stable air has been forced aloft. On the other hand, during a day when cauliflower-shaped clouds appear to be growing as if bubbles of hot air are surging upward, we can be fairly certain that the ascending air is unstable.

In summary, stability plays an important role in determining our daily weather. To a large degree, stability determines the type of clouds that develop and whether precipitation will come as a gentle shower or a heavy downpour.

(12.5) CONCEPT CHECKS

1. Explain the difference between the environmental lapse rate and adiabatic cooling.
2. How is the stability of air determined?
3. Write a statement relating the environmental lapse rate to stability.
4. What types of clouds and precipitation, if any, form when stable air is forced aloft?
5. Describe the weather associated with unstable air.

12.6 Condensation and Cloud Formation

Name and describe the 10 basic cloud types, based on form and height. Contrast nimbostratus and cumulonimbus clouds and their associated weather.

To review briefly, condensation occurs when water vapor in the air changes to a liquid. The result of this process may be dew, fog, or clouds. For any of these forms of condensation to occur, the air must be saturated. Saturation occurs most commonly when air is cooled to its dew point, or, less often, when water vapor is added to the air.

Condensation Nuclei and Cloud Formation

Under normal atmospheric conditions, condensation occurs only when a *surface* exists on which the water vapor can condense. When dew forms, objects at or near the ground, such as grass and car windows, serve this purpose. But when condensation occurs high above the ground, tiny bits of particulate matter, known as **condensation nuclei**, serve as surfaces for water-vapor condensation. These nuclei are very important, for in their absence, a relative humidity well in excess of 100 percent is needed to produce clouds.

Condensation nuclei such as microscopic dust, smoke, pollen, and salt particles (from the ocean) are profuse in the lower atmosphere. Because of this abundance of particles, relative humidity rarely exceeds 101 percent. Some particles, such as ocean salt, are particularly good nuclei because they absorb water. These particles are termed **hygroscopic** (*hygro* = moisture, *scopic* = to seek) **nuclei<**. When condensation takes place, the tiny cloud droplets grow quickly at first, but their growth slows as they use up the excess water vapor. The result is a cloud consisting of millions upon millions of tiny water droplets, all so fine that they remain suspended in air. When cloud formation occurs at below-freezing temperatures, tiny ice crystals form. Thus, a cloud might consist of water droplets, ice crystals, or both.

Cloud Classification

Clouds are among the most conspicuous and observable aspects of the atmosphere and its weather. **Clouds** are a form of condensation best described as *visible aggregates of minute droplets of water or tiny crystals of ice*. In addition to being prominent and sometimes spectacular features in the sky, clouds are of continual interest to meteorologists because they indicate what is going on in the atmosphere.

In 1803, English naturalist Luke Howard published a cloud classification scheme that serves as the basis of our present-day system. According to Howard's system, clouds are classified on the basis of two criteria: *form* and *height* (**Figure 12.20**). We will look at the basic cloud forms (shapes) first and then examine cloud height.

Cloud Forms Clouds are classified based on how they appear when viewed from Earth's surface. The basic forms, or shapes, are:

- **Cirrus** (*cirriform*) clouds are high, white, and thin. They form delicate veil-like patches or wisplike strands and often have a feathery appearance. (*Cirrus* is Latin for "curl" or "filament.")
- **Stratus** (*stratiform*) clouds consist of sheets or layers (*strata*) that cover much or all of the sky.
- **Cumulus** (*cumuliform*) clouds consist of globular cloud masses that are often described as cottonlike in appearance. Normally cumulus clouds exhibit a flat base and appear as rising domes or towers. (*Cumulus* means "heap" or "pile" in Latin.) Cumulus clouds form within a layer of the atmosphere where there is some convection and rising air.

All clouds have at least one of these three basic forms, and some are a combination of two of them; for example, stratocumulus clouds are mostly sheetlike structures composed of long parallel rolls or broken globular patches. In addition, the term **nimbus** (Latin for "violent rain") is used in the name of a cloud that is a major producer of precipitation. Thus, *nimbostratus* denotes a flat-lying rain cloud.

Cloud Heights Four levels of cloud heights are recognized: high, middle, low, and clouds of vertical development (see Figure 12.20). **High clouds** form in the highest and coldest region of the troposphere and normally have bases above 6000 meters (20,000 feet). Temperatures at these altitudes are usually below freezing, so the high clouds are generally composed of ice crystals or supercooled water droplets. **Middle clouds** occupy heights from 2000 to 6000 meters (6500 to 20,000 feet) and may be composed of water droplets or ice crystals depending on the time of year and temperature profile of the atmosphere. **Low clouds** form nearer to Earth's surface—up to an altitude of about 2000 meters (6500 feet)—and are generally composed of water droplets. These altitudes may vary somewhat according to season of the year and latitude. For example, at high (poleward) latitudes and during cold winter months, high clouds generally occur at lower altitudes. Further, some clouds extend upward to span more than one height range and are called **clouds of vertical development**.

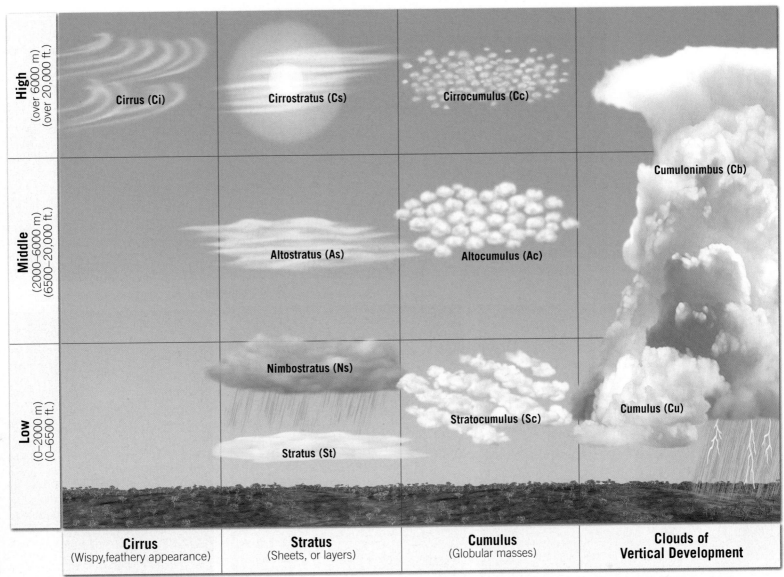

	Cirrus (Wispy, feathery appearance)	Stratus (Sheets, or layers)	Cumulus (Globular masses)	Clouds of Vertical Development
High (over 6000 m) (over 20,000 ft.)	Cirrus (Ci)	Cirrostratus (Cs)	Cirrocumulus (Cc)	
Middle (2000–6000 m) (6500–20,000 ft.)		Altostratus (As)	Altocumulus (Ac)	Cumulonimbus (Cb)
Low (0–2000 m) (0–6500 ft.)		Nimbostratus (Ns) Stratus (St)	Stratocumulus (Sc)	Cumulus (Cu)

SmartFigure 12.20
Classification of clouds
based on height and form
(goo.gl/aIQe99)

Tutorial

The internationally recognized cloud types are described in the sections that follow.

High Clouds Three cloud types make up the family of high clouds (above 6000 meters [20,000 feet]): *cirrus, cirrostratus,* and *cirrocumulus. Cirrus* (Ci) clouds are thin and delicate and sometimes appear as hooked filaments called "mares' tails" (**Figure 12.21A**). As the names suggest, **cirrocumulus** (Cc) clouds consist of fluffy masses (**Figure 12.21B**), whereas **cirrostratus** (Cs) clouds are flat layers (**Figure 12.21C**). Because of the low temperatures and small quantities of water vapor present at high altitudes, all high clouds are thin and white and are made up of ice crystals. Furthermore, these clouds are not considered precipitation makers. However, when cirrus clouds are followed by cirrocumulus clouds and increased sky coverage, they may warn of impending stormy weather.

Middle Clouds Clouds that appear in the middle range (2000–6000 meters [6500–20,000 feet]) have the prefix *alto* as part of their name. **Altocumulus** (Ac) clouds are composed of globular masses that differ from cirrocumulus clouds in that they are larger and denser (**Figure 12.21D**). **Altostratus** (As) clouds create a uniform white to grayish sheet covering the sky, with the Sun or Moon visible as a bright spot (**Figure 12.21E**). Infrequent light snow or drizzle may accompany these clouds.

Low Clouds There are three members in the family of low clouds (below 2000 meters [6500 feet]): *stratus, stratocumulus,* and *nimbostratus. Stratus* (St) are a uniform foglike layer of clouds that frequently covers much of the sky. On occasion these clouds may produce light precipitation. When stratus clouds develop a scalloped bottom that appears as long parallel rolls or broken globular patches, they are called **stratocumulus** (Sc) clouds.

Nimbostratus clouds derive their name from the Latin *nimbus,* which means "rainy cloud," and *stratus,* "to cover with a layer" (**Figure 12.21F**). Nimbostratus clouds tend to produce constant precipitation and low visibility.

Figure 12.21 Common forms of different cloud types (Photos A, B, D, E, F, and G by E. J. Tarbuck; photo C by Jung-Pang Wu/Getty Images Inc; photo H by Doug Millar/Science Source)

A. Cirrus

B. Cirrocumulus

C. Cirrostratus

D. Altocumulus

E. Altostratus

F. Nimbostratus

G. Cumulus

H. Cumulonimbus

These clouds normally form under stable conditions when air is forced to rise, as along a front (discussed in Chapter 14). Such forced ascent of stable air leads to the formation of a stratified cloud deck that is widespread and that may grow into the middle level of the troposphere. Precipitation associated with nimbostratus clouds is generally light to moderate (but can be heavy) and is usually of long duration, covering a large area.

Clouds of Vertical Development Some clouds do not fit into any one of the three height categories just mentioned. Such clouds have their bases in the low height range but often extend upward into the middle or high altitudes. Consequently, clouds in this category are called *clouds of vertical development*. They are all related to one another and are associated with unstable air. Although *cumulus* clouds are often connected with fair weather (**Figure 12.21G**), they may grow dramatically

under the proper circumstances. Once upward movement is triggered, acceleration is powerful, and clouds with great vertical extent form. The end result is often a towering cloud, called a **cumulonimbus**, which usually produces rain showers or a thunderstorm (**Figure 12.21H**).

12.6 CONCEPT CHECKS

1. Explain why a glass containing an ice-cold drink often becomes wet when it sits out at room temperature.
2. What role do condensation nuclei play in the formation of clouds?
3. What are the two criteria by which clouds are classified?
4. Why are high clouds always thin in comparison to low and middle clouds?
5. List the basic cloud types and describe each based on its form (shape) and height (altitude).

12.7 Types of Fog

Identify the basic types of fog and describe how each forms.

Fog is defined as *a cloud with its base at or very near the ground*. Physically, there are no differences between fog and a cloud; their appearances and structures are the same. The essential difference is the method and place of formation. While clouds result when air rises and cools adiabatically, fog results from cooling or when air becomes saturated through the addition of water vapor (evaporation fog).

Although fog is not inherently dangerous, it is generally considered an atmospheric hazard (**Figure 12.22**). During daylight hours, fog reduces visibility to 2 or 3 kilometers (1 or 2 miles). When the fog is particularly dense, visibility may be cut to a few dozen meters or less, making travel by any mode difficult and dangerous. Official weather stations report fog only when it is thick enough to reduce visibility to 1 kilometer (0.6 mile) or less.

Fogs Caused by Cooling

When the temperature of a layer of air in contact with the ground falls below its dew point,

condensation produces fog. Depending on the prevailing conditions, fogs formed by cooling are called either *radiation fog, advection fog,* or *upslope fog*.

Radiation Fog As the name implies, **radiation fog** results from radiation cooling of the ground and adjacent air. It is a nighttime phenomenon that requires clear skies and a high relative humidity. As the night progresses, a thin layer of air near the ground is cooled below its dew point, resulting in the formation of fog.

Because the air containing the fog is relatively cold and dense, it flows downslope in hilly terrain. As a result, radiation fog is thickest in

A.

B.

SmartFigure 12.22
Radiation fog **A.** Satellite image of dense fog in California's San Joaquin Valley on November 20, 2002. This early morning fog was caused by radiation cooling that occurred on a clear, cool evening. It was responsible for several accidents in the region, including a 14-car pileup. (Photo by NASA) **B.** Radiation fog can make the morning commute hazardous. (Photo by Tim Gainey/Alamy) (goo.gl/ZIIYpG)

Video

Figure 12.23 Advection fog rolling into San Francisco Bay This fog bank, rolling into San Francisco Bay, was generated as moist air passed over the cold California Current. (Photo by Ed Pritchard/Stone/Getty Images)

valleys, whereas the surrounding hills may remain clear (see Figure 12.22A). Normally, radiation fog dissipates within one to three hours after sunrise—and is often said to "lift." However, the fog does not actually "lift." Instead, as the Sun warms the ground, the lowest layer of air is heated first, and the fog evaporates from the bottom up.

Advection Fog When warm, moist air blows over a cold surface, it becomes chilled by contact with the cold surface below. If cooling is sufficient, the result is a blanket of fog called **advection fog**. (The term *advection* refers to air moving horizontally.) A classic example is the

frequent advection fog around San Francisco's Golden Gate Bridge (**Figure 12.23**). The fog experienced in San Francisco, California, as well as many other west coast locations, is produced when warm, moist air from the Pacific Ocean moves over the cold California Current.

Advection fog is also a common wintertime phenomenon in the Southeast and Midwest when relatively warm, moist air from the Gulf of Mexico and Atlantic moves over cold and occasionally snow-covered surfaces to produce widespread foggy conditions. This type of advection fog tends to be thick and produce hazardous driving conditions.

Figure 12.24 Steam fog rising from Sierra Lake, Blanca, Arizona (Photo by Michael Collier)

Upslope Fog When relatively humid air moves up a gradually sloping landform or, in some cases, up the steep slopes of a mountain, **upslope fog** can form. Because of the upward movement, air expands and cools adiabatically. If the dew point is reached, an extensive layer of fog will form.

It is easy to visualize how upslope fog might form in mountainous terrain. However, in the United States, upslope fog also occurs in the Great Plains, when humid air moves from the Gulf of Mexico toward the Rocky Mountains. (Recall that Denver, Colorado, is called the "mile-high city," and the Gulf of Mexico is at sea level.) Air flowing "up" the Great Plains expands and cools adiabatically by as much as 12°C (22°F), which can result in extensive upslope fog in the western plains.

Evaporation Fogs

When the saturation of air occurs primarily because of the addition of water vapor, the resulting fogs are called *evaporation fogs*. Two types of evaporation fogs are recognized: *steam fog* and *frontal (precipitation) fog*.

Steam Fog When cool, unsaturated air moves over a warm water body, enough moisture may evaporate to saturate the air directly above, generating a layer of fog. The added moisture and energy often makes the saturated air buoyant enough to cause it to rise. Because the foggy air looks like the "steam" that forms above a hot cup of coffee, the phenomenon is called **steam fog** (**Figure 12.24**). Steam fog is a fairly common occurrence over lakes and rivers on clear, crisp autumn mornings when the water is still relatively warm but the air is comparatively cold.

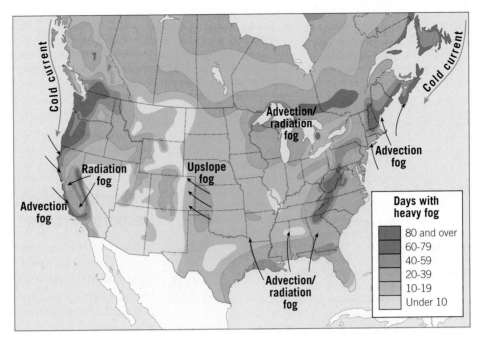

SmartFigure 12.25 Map showing the average number of days per year with heavy fog Coastal areas, particularly the Pacific Northwest and New England, where cold currents prevail, have high occurrences of dense fog. (goo.gl/pQk9sl)

Tutorial

Frontal (Precipitation) Fog Frontal boundaries where a warm, moist air mass is forced to rise over cooler, dryer air below generates **frontal (precipitation) fog**. The foggy conditions result because the raindrops falling from relatively warm air above the frontal surface evaporate in the cooler air below, causing it to become saturated. Frontal fog, which can be quite thick, is most common on cool days during extended periods of light rainfall.

Where Is Fog Most Common? The frequency of dense fog varies considerably from place to place (**Figure 12.25**). As might be expected, fog incidence is highest in coastal areas, especially where cold currents prevail, as along the Pacific and New England coasts. Relatively high frequencies are also found in the Great Lakes region and in the humid Appalachian Mountains of the Eastern United States.

12.7 CONCEPT CHECKS

1. Distinguish between clouds and fog.
2. List five main types of fog and describe how each type forms.

12.8 How Precipitation Forms

Describe the Bergeron process and explain how it differs from the collision–coalescence process.

If all clouds contain water, why do some produce precipitation while others drift placidly overhead? This seemingly simple question perplexed meteorologists for many years.

Typical cloud droplets are miniscule—0.02 millimeter (20 micrometers) in diameter (**Figure 12.26**). By comparison, a human hair is about 0.075 millimeters (75 micrometers) in diameter. Because of their small size, cloud droplets in still air fall incredibly slowly. An average cloud droplet falling from a cloud base would require several hours to reach the ground. However, it would never

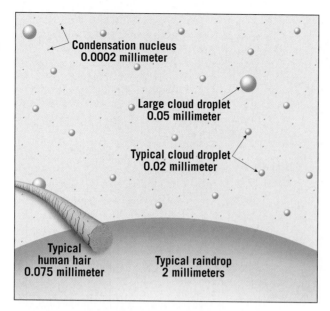

Figure 12.26 Diameters of particles involved in condensation and precipitation processes

Condensation nucleus
0.0002 millimeter

Large cloud droplet
0.05 millimeter

Typical cloud droplet
0.02 millimeter

Typical human hair
0.075 millimeter

Typical raindrop
2 millimeters

complete its journey. Instead, the cloud droplet would evaporate before it fell a few meters from the cloud base into the unsaturated air below.

How large must a cloud droplet grow in order to fall as precipitation? A typical raindrop has a diameter of about 2 millimeters, or 100 times that of the average cloud droplet (see Figure 12.26). However, the *volume* of a typical raindrop is 1 million times that of a cloud droplet. Thus, for precipitation to form, cloud droplets must grow in volume by roughly 1 million times.

Two processes are responsible for the formation of precipitation: the *Bergeron process* and the *collision–coalescence process*.

Precipitation from Cold Clouds: The Bergeron Process

You have probably watched a TV documentary in which mountain climbers have braved intense cold and ferocious snow storms to scale ice-covered peaks. Although it is hard to imagine, similar conditions exist in the upper portions of towering cumulonimbus clouds, even on sweltering summer days. It is within these cold clouds that a mechanism called the **Bergeron process** generates much of the precipitation that occurs in the middle and high latitudes.

The Bergeron process is based on the fact that cloud droplets remain liquid at temperatures as low as −40°C (−40°F). Liquid water at temperatures below freezing is termed **supercooled**, and it becomes solid, or freezes, upon impact with a surface. This explains why airplanes collect ice when they pass through a cloud composed of subzero droplets, a condition called *icing*. Supercooled water droplets also freeze upon contact with particles in the atmosphere known as **freezing nuclei**. Because freezing nuclei are relatively sparse, cold clouds primarily consist of supercooled droplets intermixed with a lesser amount of ice crystals.

When ice crystals and supercooled water droplets coexist in a cloud, the conditions are ideal for generating precipitation. Because ice crystals have a greater affinity for water vapor than does liquid water, they collect the available water vapor at a much faster rate. In turn, the water droplets evaporate to maintain saturation and replenish the diminishing water vapor, thereby providing a continual source of moisture for the growth of ice crystals. As shown in **Figure 12.27**, the result is that the ice

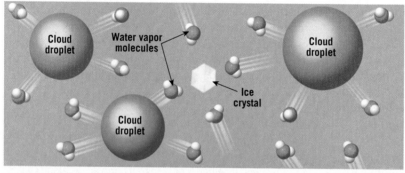

Figure 12.27 The Bergeron process Ice crystals grow at the expense of cloud droplets until they are large enough to fall.

Cloud droplet

Water vapor molecules

Ice crystal

Cloud droplet

Cloud droplet

Snow crystal

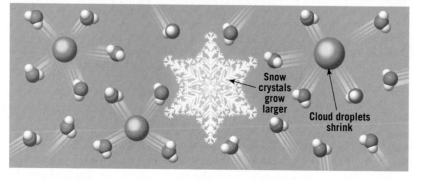

Snow crystals grow larger

Cloud droplets shrink

crystals grow larger—at the expense of the water droplets, which shrink in size.

Eventually, this process generates ice crystals large enough to fall as snowflakes. During their descent, these

ice crystals become larger as they intercept supercooled cloud droplets that freeze on them. When the surface temperature is about 4°C (39°F) or higher, snowflakes usually melt before they reach the ground and continue their descent as rain.

Precipitation from Warm Clouds: The Collision–Coalescence Process

A few decades ago, meteorologists believed that the Bergeron process was responsible for the formation of most precipitation. However, it was discovered that copious rainfall may be produced within clouds located well below the freezing level (*warm clouds*), particularly in the tropics. This led to the proposal of a second mechanism thought to produce precipitation—the **collision–coalescence process**.

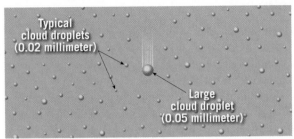

A. Because large cloud droplets fall more rapidly than smaller droplets, they are able to sweep up the smaller ones in their path and grow.

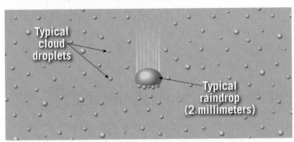

B. As drops increase in size, their fall velocity increases, resulting in increased air resistance, which causes the raindrop to flatten.

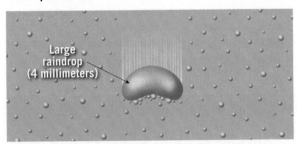

C. As the raindrop approaches 4 millimeters in size, it develops a depression in the bottom.

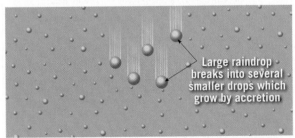

D. Finally, when the diameter exceeds about 5 millimeters, the depression grows upward almost explosively, forming a donut–like ring of water that immediately breaks into smaller drops.

Figure 12.28 The collision–coalescence process The collision–coalescence process involves multiple collisions of tiny cloud droplets that stick together (coalesce) to form raindrops large enough to reach the ground before evaporating.

Research has shown that clouds composed entirely of liquid droplets must contain some droplets larger than 20 micrometers (0.02 millimeters) for precipitation to form. These large droplets usually form when *hygroscopic particles* (particles that attract water), such as sea salt, are abundant in the atmosphere. Hygroscopic particles begin to remove water vapor from the air when the relative humidity is under 100 percent, and the cloud droplets that form on them can grow quite large. Because the rate at which drops fall is size dependent, these "giant" droplets fall most rapidly. As they plummet, they collide with smaller, slower droplets (**Figure 12.28**). After many such collisions, these droplets may grow large enough to fall to the surface without evaporating. Updrafts also aid this process because they propel the droplets to traverse the cloud repeatedly.

Raindrops can grow to a maximum size of 5 millimeters, at which point they fall at a rate of 33 kilometers (20 miles) per hour. At this size and speed, the water's surface tension, which holds the drop together, is overcome by the drag imposed by the air, causing the drops to break apart (see Figure 12.28). The resulting breakup of a large raindrop produces numerous smaller drops that begin anew the task of sweeping up cloud droplets. Drops that are less than 0.5 millimeter (0.02 inch) upon reaching the ground are termed *drizzle* and require about 10 minutes to fall from a cloud 1000 meters (3300 feet) overhead.

12.8 CONCEPT CHECKS

1. Describe the temperature conditions in clouds that are required to form precipitation by the Bergeron process.
2. Explain how snow that formed high in a towering cloud might produce rain.
3. Briefly summarize the collision–coalescence process.

12.9 Forms of Precipitation

Describe the atmospheric conditions that produce sleet, freezing rain (glaze), and hail.

Atmospheric conditions vary greatly both geographically and seasonally, resulting in several different types of precipitation. Rain and snow are the most common and familiar forms, but others, listed in **Table 12.3**, are important as well. Sleet, freezing rain (glaze), and hail often produce hazardous weather and occasionally inflict considerable damage.

Rain, Drizzle, and Mist

In meteorology, the term **rain** is restricted to drops of water that fall from a cloud and have a diameter of at least 0.5 millimeter (0.02 inch). Most rain originates either in nimbostratus clouds or in towering cumulonimbus clouds; the latter are capable of producing unusually heavy rainfalls known as *cloudbursts*. Raindrops rarely exceed about 5 millimeters (0.2 inch) in diameter. Larger drops do not survive because surface tension, which holds the drops together, is exceeded by the frictional drag of the air. Consequently, large raindrops regularly break apart into smaller ones.

Fine, uniform drops of water less than 0.5 millimeter (0.02 inch) in diameter are called **drizzle**. Drizzle can be so fine that the tiny drops appear to float, and their impact is almost imperceptible. Precipitation containing the very smallest droplets able to reach the ground is called **mist**.

Snow

Snow is precipitation in the form of ice crystals (snowflakes) or, more often, aggregates of crystals. The size, shape, and concentration of snowflakes depend to a great extent on the temperature at which they form.

Recall that at very low temperatures, the moisture content of air is low. The result is the formation of very light, fluffy snow made up of individual six-sided ice crystals. This is the "powder" that downhill skiers love so much. By contrast, at temperatures warmer than about –5°C (23°F), the ice crystals join together into larger clumps consisting of tangled aggregates of crystals. Snowfalls composed of these composite snowflakes are generally heavy and have high moisture content, which makes them ideal for making snowballs.

Sleet and Freezing Rain (Glaze)

Sleet consists of clear to translucent ice pellets. Depending on intensity and duration, sleet can cover the ground much like a thin blanket of snow. **Freezing rain**, or **glaze**, on the other hand, falls as supercooled raindrops that freeze on contact with roads, power lines, and other surfaces.

As shown in **Figure 12.29**, both sleet and freezing rain occur in the winter, and they most often form along a warm front where a mass of relatively warm air is forced over a

Table 12.3 Forms of Precipitation

Type	Approximate Size	State of Matter	Description
Mist	0.005-0.05 mm	Liquid	Droplets large enough to be felt on the face when air is moving 1 meter/second. Associated with stratus clouds.
Drizzle	0.05-0.5 mm	Liquid	Small uniform drops that fall from stratus clouds, generally for several hours.
Rain	0.5-5 mm	Liquid	Generally produced by nimbostratus or cumulonimbus clouds. When heavy, can be highly variable from one place to another.
Sleet	0.5-5 mm	Solid	Small, spherical to lumpy ice particles that form when raindrops freeze while falling through a layer of subfreezing air. Because the ice particles are small, damage, if any, is generally minor. Sleet can make travel hazardous.
Freezing rain (glaze)	Layers 1 mm-2 cm thick	Solid	Produced when supercooled raindrops freeze on contact with solid objects. Glaze can form a thick coating of ice heavy enough to seriously damage trees and power lines.
Rime	Variable accumulations	Solid	Deposits usually consisting of ice feathers that point into the wind. These delicate, frostlike accumulations form as supercooled cloud or fog droplets encounter objects and freeze on contact.
Snow	1 mm-2 cm	Solid	The crystalline nature of snow allows it to assume many shapes, including six-sided crystals, plates, and needles. Produced in supercooled clouds, where water vapor is deposited as ice crystals that remain frozen during their descent.
Hail	5 mm-10 cm or larger	Solid	Occurs as hard, rounded pellets or irregular lumps of ice. Produced in large cumulonimbus clouds, where frozen ice particles and supercooled water coexist.
Graupel	2-5 mm	Solid	"Soft hail" that forms when rime collects on snow crystals to produce irregular masses of "soft" ice. Because these particles are softer than hailstones, they normally flatten out upon impact.

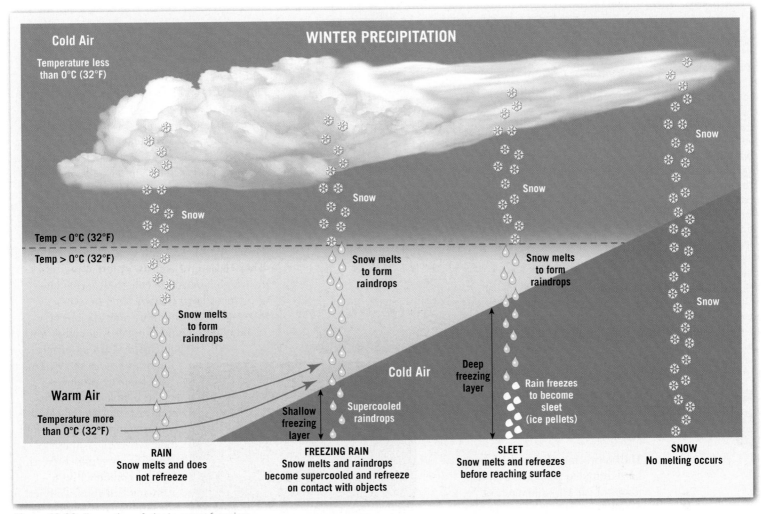

WINTER PRECIPITATION

Cold Air
Temperature less than 0°C (32°F)

Snow

Snow

Snow

Snow

Snow

Snow

Temp < 0°C (32°F)

Temp > 0°C (32°F)

Snow melts to form raindrops

Snow melts to form raindrops

Snow melts to form raindrops

Cold Air

Deep freezing layer

Rain freezes to become sleet (ice pellets)

Warm Air
Temperature more than 0°C (32°F)

Shallow freezing layer

Supercooled raindrops

RAIN
Snow melts and does not refreeze

FREEZING RAIN
Snow melts and raindrops become supercooled and refreeze on contact with objects

SLEET
Snow melts and refreezes before reaching surface

SNOW
No melting occurs

Figure 12.29 Formation of sleet versus freezing rain These forms of precipitation occur in the winter, when warm air (along a warm front) is forced over a layer of subfreezing air. When rain passes through a cold layer of air and freezes, the resulting ice pellets are called sleet. Freezing rain forms when the cold layer of air is not deep enough to refreeze the raindrops, resulting in supercooled droplets freezing on contact with objects at the surface.

layer of subfreezing air near the ground. Both begin as snow, which melts to form raindrops as it falls though the layer of warm air below. When the newly formed raindrops encounter a thick cold layer of air below the frontal boundary, *sleet* results. In this setting, as raindrops fall through the subfreezing air, they refreeze and reach the ground as small pellets of ice roughly the size of the raindrops from which they formed.

If, however, the layer of cold air near the ground is not thick enough to cause the raindrops to refreeze, they instead become supercooled—that is, they remain liquid at temperatures below freezing (see Figure 12.29). Upon striking subfreezing objects on Earth's surface, these supercooled raindrops instantly turn to ice, creating a coating of *freezing rain*. Freezing rain makes walking and driving extremely hazardous, and when it grows thick, it can break tree limbs and down power lines (**Figure 12.30**).

Figure 12.30 Freezing rain In January 1998, an ice storm of historic proportions caused enormous damage in New England and southeastern Canada. Nearly 5 days of freezing rain (glaze) left millions without electricity—some for as long as a month. (Photo by Dick Blume/Syracuse Newspapers/The Image Works)

386

SmartFigure 12.31
Formation of hailstones
A. Hailstones begin as small ice pellets that grow through the addition of supercooled water droplets as they move through a cloud. Updrafts carry stones upward, increasing the size of the hail by adding layers of ice. Eventually, the hailstones either grow too large to be supported by the updraft or else encounter a downdraft.
B. This cut hailstone, which fell over Coffeyville, Kansas, in 1970, originally weighed 766 grams (1.69 pounds). (Photo courtesy of UCAR (University Corporation for Atmospheric Research)) (goo.gl/70XFP7)

Tutorial

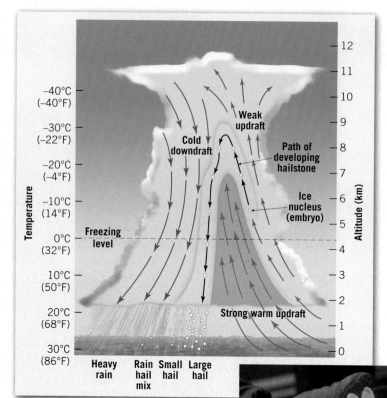

A.

B.

Hail

Hail is precipitation in the form of hard, rounded pellets or irregular lumps of ice with diameters of 5 millimeters (0.20 inches) or more. Hail is produced in the middle to upper reaches of tall cumulonimbus clouds, where updrafts can sometimes exceed speeds of 160 kilometers (100 miles) per hour and where the air temperature is below freezing. Hailstones begin as small embryonic ice pellets or graupel that coexist with supercooled droplets. The ice pellets grow by collecting supercooled water droplets and, sometimes, other small pieces of hail as they are lifted by updrafts within the cloud.

Cumulonimbus clouds that produce hail have a complex system of updrafts and downdrafts. As shown in **Figure 12.31A**, a region of intense updrafts suspends rain and hail aloft, producing a rain-free region surrounded by an area of downdrafts and heavy precipitation. The largest hailstones are generated around the core of the most intense zone of updraft, where they rise slowly enough to collect appreciable amounts of supercooled water. The process continues until a hailstone grows too heavy to be supported by the updraft or encounters a downdraft and falls to the surface.

Large hailstones often show alternating layers of clear and milky ice (**Figure 12.31B**). These layers reflect two different processes by which a hailstone can grow, termed *wet growth* and *dry growth*. Wet growth occurs in the lower and warmer regions of a cloud, where liquid droplets that collide with a hailstone wet its surface and then freeze slowly. This slow freezing produces clear, bubble-free ice. By contrast, high in the cloud, where temperatures are well below freezing, supercooled droplets immediately freeze as they collide with the growing hailstone. Trapped air bubbles are "frozen" in place, creating milky ice.

Figure 12.33 Rime Rime consists of delicate ice crystals that form when supercooled fog or cloud droplets freeze on contact with objects. (Photo by Marcus Siebert Image Broker/Age Fotostock)

Figure 12.32 Hail damage to an NOAA weather monitoring vehicle (Photo courtesy of National Weather Service)

The record for the largest hailstone ever found in the United States was set on July 23, 2010, in Vivian, South Dakota. The stone was over 20 centimeters (8 inches) in diameter and weighed nearly 900 grams (2 pounds). The stone that held the previous record of 766 grams (1.69 pounds) fell in Coffeyville, Kansas, in 1970 (see Figure 12.31B). The diameter of the stone found in South Dakota also surpassed the previous record of a 17.8-centimeter (7-inch) stone that fell in Aurora, Nebraska, in 2003. Even larger hailstones have reportedly been recorded in Bangladesh, where a 1987 hailstorm killed more than 90 people.

The destructive effects of large hailstones are well known, especially to farmers whose crops have been devastated in a few minutes and to people whose windows, roofs, and cars have been damaged (**Figure 12.32**). In the United States, annual hail damage can run into the hundreds of millions of dollars.

Rime

Rime is a deposit of ice crystals formed by the freezing of supercooled fog or cloud droplets on objects whose surface temperature is below freezing. When rime forms on trees, it adorns them with its characteristic ice feathers, which can be spectacular to observe (**Figure 12.33**). In these situations, objects such as pine needles act as freezing nuclei, causing the supercooled droplets to freeze on contact. When a wind is blowing, only the windward surfaces of objects will accumulate rime.

12.9 CONCEPT CHECKS

1. Compare and contrast rain, drizzle, and mist.
2. Describe sleet and freezing rain. Why does freezing rain result on some occasions and sleet on others?
3. How does hail form? What factors govern the ultimate size of hailstones?

12.10 Measuring Precipitation

List the advantages of using weather radar versus a standard rain gauge to measure precipitation.

The most common form of precipitation, rain, is the easiest to measure. Any open container having a consistent cross section throughout can be used as a rain gauge (**Figure 12.34A**). In general practice, however, more sophisticated devices are used to measure small amounts of rainfall more accurately and to reduce loss from evaporation.

A **standard rain gauge** has a diameter of about 20 centimeters (8 inches) at the top (**Figure 12.34B**). When the water is caught, a funnel conducts the rain into a cylindrical measuring tube that has a cross-sectional area only one-tenth as large as the receiver. Consequently, rainfall depth is magnified 10 times, which allows for accurate measurements to the nearest 0.025 centimeter (0.01 inch). When the amount of rain is less than 0.025 centimeter, it is reported as a *trace of precipitation*.

As **Figure 12.34C** illustrates, the **tipping-bucket gauge** consists of two compartments, each one capable of holding 0.025 centimeter (0.01 inch) of rain, situated at the base of a funnel. When one "bucket" fills, it tips and

A. Simple rain gauge

1 inch of rain

1 inch of rain

1 inch

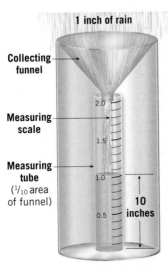

1 inch of rain

Collecting funnel

Measuring scale

Measuring tube (¹⁄₁₀ area of funnel)

2.0

1.5

1.0

0.5

10 inches

B. Standard rain gauge

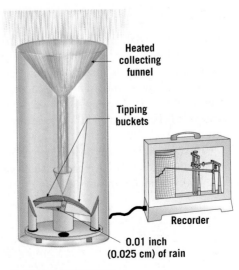

Heated collecting funnel

Tipping buckets

Recorder

0.01 inch (0.025 cm) of rain

C. Tipping–bucket gauge

Figure 12.34 Precipitation measurement **A.** The simplest gauge is any container left in the rain. **B.** The standard rain gauge increases the height of water collected by a factor of 10, allowing for accurate rainfall measurement to the nearest 0.025 centimeter (0.01 inch). **C.** The tipping-bucket rain gauge contains two "buckets," each holding the equivalent of 0.025 centimeter (0.01 inch) of liquid precipitation. When one bucket fills, it tips and the other bucket takes its place.

Figure 12.35 Doppler radar display produced by the National Weather Service Colors indicate different intensities of precipitation. Note the band of heavy precipitation (orange and red colors) along the eastern seaboard. (Courtesy of NOAA)

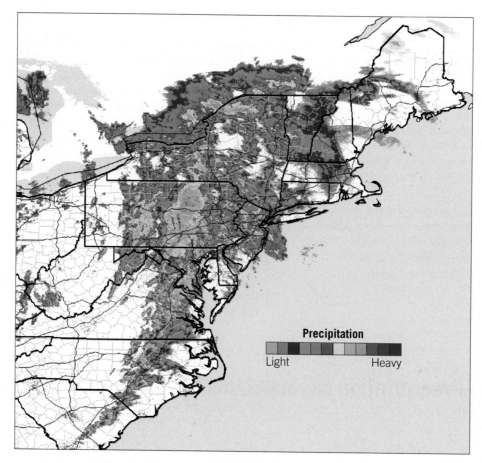

Precipitation

Light   Heavy

the actual water content of snow may deviate widely from this figure. It may take as much as 30 inches of light and fluffy dry snow (30:1) or as little as 4 inches of wet snow (4:1) to produce 1 inch of water.

Precipitation Measurement by Weather Radar

Using **weather radar**, the National Weather Service (NWS) produces maps like the one in **Figure 12.35** in which different colors illustrate precipitation intensity. The development of weather radar has given meteorologists an important tool to track storm systems and the precipitation patterns they produce, even when the storms are as far as a few hundred kilometers away.

empties its water. Meanwhile, the other "bucket" takes its place at the mouth of the funnel. Each time a compartment tips, an electrical circuit is closed, and 0.025 centimeter (0.01 inch) of precipitation is automatically recorded on a graph.

Measuring Snowfall

Snowfall is typically measured by depth and water equivalent. One way to measure the depth of snow is by using a calibrated stick. The actual measurement is not difficult, but choosing a representative spot can be. Even when winds are light or moderate, snow drifts freely. As a rule, it is best to take several measurements in an open place, away from trees and obstructions, and then average them. To obtain the water equivalent, a core taken from the snow may be weighed, or snowfall captured in a gauge may be melted and then measured as though it were rain.

The quantity of water in a given volume of snow is not constant. You may have heard media weathercasters say, "Every 10 inches of snow equals 1 inch of rain." But

Radar units have transmitters that send out short pulses of radio waves. The specific wavelengths selected depend on the objects being detected. Wavelengths between 3 and 10 centimeters are employed when monitoring precipitation. Radio waves at these wavelengths can penetrate clouds composed of small droplets, but they are reflected by larger raindrops, ice crystals, and hailstones. The reflected signal, called an *echo*, is received and displayed on a screen. Because the echo is "brighter" when the precipitation is more intense, modern weather radar is able to depict both the regional extent and the rate of precipitation. Also, because the measurements are in real time, they are particularly useful in short-term forecasting.

12.10 CONCEPT CHECKS

1. Although any open container can serve as a rain gauge, what advantages does a standard rain gauge provide?
2. Identify one advantage that weather radar has over a standard rain gauge.

CONCEPTS IN REVIEW
Moisture, Clouds, and Precipitation

12.1 Water's Changes of State

Summarize the six processes by which water changes from one state of matter to another. For each, indicate whether energy is absorbed or released.

KEY TERMS: calorie, latent heat, evaporation, condensation, sublimation, deposition,

- Water changes from one state of matter (solid, liquid, or gas) to another at the temperatures and pressures experienced near Earth's surface. The gaseous form of water is water vapor.
- The processes involved in changes of state include evaporation (liquid to gas), condensation (gas to liquid), melting (solid to liquid), freezing (liquid to solid), sublimation (solid to gas), and deposition (gas to solid). During each change, latent (hidden, or stored) heat is either absorbed or released.

(?) Label the accompanying diagram with the appropriate terms for the changes of state that are shown.

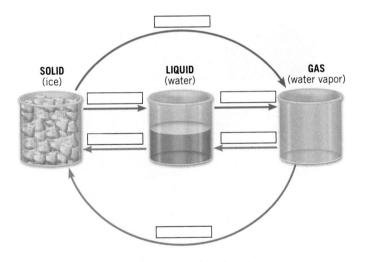

SOLID (ice)   LIQUID (water)   GAS (water vapor)

12.2 Humidity: Water Vapor in the Air

Write a generalization relating air temperature and the amount of water vapor needed to saturate air.

KEY TERMS: humidity, saturation, vapor pressure, mixing ratio, relative humidity, dew-point temperature (dew point), hygrometer, psychrometer

- Humidity is the amount of water vapor in the air. The methods used to express humidity quantitatively include (1) mixing ratio, the mass of water vapor in a unit of air compared to the remaining mass of dry air; (2) relative humidity, the ratio of the air's actual water-vapor content to the amount of water vapor required for saturation at that temperature; and (3) dew-point temperature.
- Relative humidity can be changed in two ways: by adding or subtracting water vapor or by changing the air's temperature.
- The dew-point temperature (or simply dew point) is the temperature to which a parcel of air must be cooled to reach saturation. Unlike relative humidity, dew-point temperature is a measure of the air's actual moisture content.

(?) Refer to the accompanying photo and explain how the relative humidity inside this house would compare to the relative humidity outside the house on a winter day.

Clynt Garnham Housing/Alamy

12.3 Adiabatic Temperature Changes and Cloud Formation

Describe adiabatic temperature changes and explain why the wet adiabatic rate of cooling is less than the dry adiabatic rate.

KEY TERMS: adiabatic temperature change, parcel, dry adiabatic rate, lifting condensation level (condensation level), sensible heat, wet adiabatic rate

- Cooling of air as it rises and expands due to decreasing air pressure is the basic cloud-forming process. Temperature changes that result when air is compressed or when air expands are called adiabatic temperature changes.
- Unsaturated air warms by compression and cools by expansion at the rather constant rate of 10°C per 1000 meters (5.5°F per 1000 feet) of altitude change, a quantity called the dry adiabatic rate. When air rises high enough, it cools sufficiently to cause condensation and form clouds. Air that continues to rise above the condensation level will cool at the wet adiabatic rate, which varies from 5°C to 9°C per 1000 meters of ascent. The difference in the wet and dry adiabatic rates is due to the latent heat released by condensation, which slows the rate at which air cools as it ascends.

12.4 Processes That Lift Air

List and describe the four mechanisms that cause air to rise.

KEY TERMS: orographic lifting, rainshadow desert, front, frontal lifting (wedging), convergence, localized convective lifting (convective lifting)

- Four mechanisms that cause air to rise are (1) orographic lifting, where air is forced to rise over elevated terrain such as a mountain barrier; (2) frontal lifting, where warmer, less-dense air is forced over cooler, denser air along a front; (3) convergence, a pileup of horizontal airflow resulting in an upward flow; and (4) localized convective lifting, where unequal surface heating causes localized pockets of air to rise because of their buoyancy.

12.5 The Critical Weathermaker: Atmospheric Stability

Describe how atmospheric stability is determined and compare conditional instability with absolute instability.

KEY TERMS: stable air, unstable air, environmental lapse rate, absolute stability, absolute instability, conditional instability

- Stable air resists vertical movement, whereas unstable air rises because of its buoyancy. The stability of a parcel of air is determined by the local environmental lapse rate (the temperature of the atmosphere at various heights). The three fundamental conditions of the atmosphere are (1) absolute stability, when the environmental lapse rate is less than the wet adiabatic rate; (2) absolute instability, when the environmental lapse rate is greater than the dry adiabatic rate; and (3) conditional instability, when moist air has an environmental lapse rate between the dry and wet adiabatic rates.

- In general, when stable air is forced aloft, the associated clouds have little vertical thickness, and precipitation, if any, is light. In contrast, clouds associated with unstable air are towering and can produce heavy precipitation.

(?) **Describe the atmospheric conditions that were likely associated with the development of the towering cloud shown in the accompanying photo.**

Rolf Nussbaumer/Bill Draker/Rolfnp/ Alamy

12.6 Condensation and Cloud Formation

Name and describe the 10 basic cloud types, based on form and height. Contrast nimbostratus and cumulonimbus clouds and their associated weather.

KEY TERMS: condensation nuclei, hygroscopic nuclei, cloud, cirrus, stratus, cumulus, nimbus, high clouds, middle clouds, low clouds, clouds of vertical development, cirrocumulus, cirrostratus, altocumulus, altostratus, stratocumulus, nimbostratus, cumulonimbus

- Condensation occurs when water vapor changes to liquid water. For condensation to occur aloft, the air must reach saturation, and there must be a surface on which the water vapor can condense to form liquid droplets. Condensation produces tiny cloud droplets that are held aloft by the slightest updrafts.

- Clouds, visible aggregates of minute droplets of water and/or tiny crystals of ice, are one form of condensation.

- Clouds are classified on the basis of two criteria: form and height. The three basic cloud forms are *cirrus* (high, white, and thin), *cumulus* (globular, individual cloud masses), and *stratus* (sheets or layers).

- Cloud heights can be *high*, with bases above 6000 meters (20,000 feet); *middle*, from 2000 (6500 feet) to 6000 meters; or *low*, below 2000 meters. *Clouds of vertical development* have bases in the low height range and extend upward into the middle or high range.

(?) **Which of the three basic cloud forms (cirrus, cumulus, or stratus) is illustrated by each of the accompanying images, A–C?**

A.

B.

C.   E. J. Tarbuck

12.7 Types of Fog

Identify the basic types of fog and describe how each forms.

KEY TERMS: fog, radiation fog, advection fog, upslope fog, steam fog, frontal fog (precipitation fog)

- Fog is a cloud with its base at or very near the ground. Fogs form when air is cooled below its dew point or when enough water vapor is added to the air to cause saturation.
- Fogs formed by cooling include radiation fog, advection fog, and upslope fog. Fogs formed by the addition of water vapor are steam fog and frontal fog.

(?) **Identify the fog type shown in the accompanying image and describe the mechanism that generated it.**

Pat and Chuck Blackley/Alamy

12.8 How Precipitation Forms

Describe the Bergeron process and explain how it differs from the collision–coalescence process.

KEY TERMS: Bergeron process, supercooled, freezing nuclei, collision–coalescence process

- In order for precipitation to form, millions of cloud droplets must join together into drops that are large enough to reach the ground before evaporating.
- The two mechanisms that generate precipitation are the Bergeron process, which produces precipitation from cold clouds primarily in the middle and high latitudes, and the collision–coalescence process, which occurs in warm clouds and primarily in the tropics.

12.9 Forms of Precipitation

Describe the atmospheric conditions that produce sleet, freezing rain (glaze), and hail.

KEY TERMS: rain, drizzle, mist, snow, sleet, freezing rain (glaze), hail, rime

- The two most common and familiar forms of precipitation are rain and snow. Rain can form in either warm or cold clouds. When it falls from cold clouds, it begins as snow that melts before reaching the ground.
- Sleet consists of spherical to lumpy ice particles that form when raindrops freeze while falling through a thick layer of subfreezing air. Freezing rain results when supercooled raindrops freeze upon contact with cold objects. Rime consists of delicate frostlike accumulations that form as supercooled fog droplets encounter objects and freeze on contact. Hail consists of hard, rounded pellets or irregular lumps of ice produced in towering, cumulonimbus clouds, where frozen ice particles and supercooled water coexist.

12.10 Measuring Precipitation

List the advantages of using weather radar versus a standard rain gauge to measure precipitation.

KEY TERMS: standard rain gauge, tipping-bucket gauge, weather radar

- The instruments most commonly used to measure rain are the standard rain gauge, which measures the total amount of precipitation between readings, and the tipping-bucket gauge, which is automated and records both the amount and intensity of rain. The two most common measurements of snow are depth and water equivalent.
- Modern weather radar has given meteorologists an important tool to track storm systems and precipitation patterns, even when the storms are as far as a few hundred kilometers away.

GIVE IT SOME THOUGHT

1. The accompanying photo shows a cup of hot coffee. In what state of matter is the "steam" rising from the liquid? (*Hint:* Is it possible to see water vapor?)

Dmitry Kolmakov/Shutterstock

2. Refer to Figure 12.2 to complete the following:
   a. In which state of matter is water the most dense?
   b. In which state of matter are water molecules most energetic?
   c. In which state of matter is water compressible?

3. The primary mechanism by which the human body cools itself is perspiration.
   a. Explain how perspiring cools the skin.
   b. Refer to the data for Phoenix, Arizona, and Tampa, Florida, in Table A. In which city would it be easier to stay cool by perspiring? Explain your choice.

TABLE A

City	Temperature	Dew-point Temperature
Phoenix, AZ	101°F	47°F
Tampa, FL	101°F	77°F

4. Explain why radiation fog forms mainly on clear nights rather than on cloudy nights.

5. The accompanying graph shows how air temperature and relative humidity change on a typical summer day in the Midwest. Assuming that the dew-point temperature remained constant, what would be the best time of day to water a lawn to minimize evaporation of the water spread on the grass?

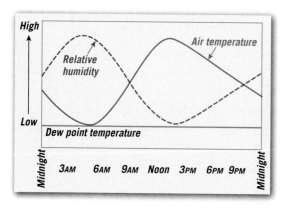

6. Refer to Table 12.1 to answer this question. How much more water is contained in saturated air at a tropical location with a temperature of 40°C compared to a polar location with a temperature of –10°C?

7. Use the data for Phoenix, Arizona, and Bismarck, North Dakota, in Table B, to complete the following:
   a. Which city has a higher relative humidity?
   b. Which city has the greater quantity of water vapor in the air?
   c. In which city is the air closest to its saturation point with respect to water vapor?

TABLE B

City	Temperature	Dew-point Temperature
Phoenix, AZ	101°F	47°F
Bismark, ND	39°F	38°F

8. Large cumulonimbus clouds like the one shown in Figure 12.21H are roughly 12 kilometers tall, 8 kilometers wide, and 8 kilometers deep. Assume that the droplets in each cubic meter of the cloud total 0.5 cubic centimeter. How much liquid would a cloud of that size contain? How many gallons is this? (*Note:* 3785 cm^3 = 1 gallon.)

9. The accompanying diagram shows air flowing from the ocean over a coastal mountain range. Assume that the dew-point temperature remains constant in dry air (air having a relative humidity less than 100 percent). If the air parcel becomes saturated, the dew-point temperature will cool at the wet adiabatic rate as it ascends, but it will not change as the air parcel descends. Use this information to complete the following:
   a. Determine the air temperature and dew-point temperature for the air parcel at each location (B–G) shown on the diagram.
   b. At what elevation will clouds begin to form (with relative humidity = 100 percent)?
   c. Compare the air temperatures at points A and G. Why are they different?
   d. How did the water vapor content of the air change as the parcel of air traversed the mountain? (*Hint:* Compare dew-point temperatures.)
   e. On which side of the mountain might you expect lush vegetation, and on which side would you expect desert-like conditions?
   f. Where in the United States might you find a situation like what is pictured here?

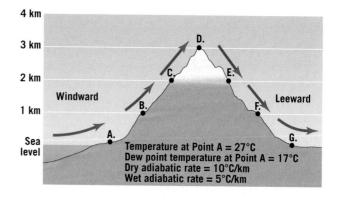

10. Weather radar provides information on the intensity of precipitation in addition to the total amount of precipitation that falls over a given time period. Table C shows the relationship between radar reflectivity (echo intensity) values and rainfall rates. If radar measured a reflectivity value of 47 dBZ for 2½ hours over a location, how much rain will have fallen there?

Table C
Conversion of radar reflectivity to rainfall rate

Radar Reflectivity (dBZ)	Rainfall Rate (inches/hr)
65	16+
60	8.0
55	4.0
52	2.5
47	1.3
41	0.5
36	0.3
30	0.1
20	trace

MasteringGeology™

www.masteringgeology.com   Looking for additional review and test prep materials? With individualized coaching on the toughest topics of the course, MasteringGeology offers a wide variety of ways for you to move beyond memorization to begin thinking like a geologist. Visit the Study Area in www.masteringgeology.com to find practice quizzes, study tools, and multimedia that will improve your understanding of this chapter's content. Sign in today to enjoy the following features: **Self Study Quizzes, SmartFigure: Tutorials/Animations/Condor Videos/Mobile Field Trips, Geoscience Animation Library, GEODe, RSS Feeds, Digital Study Modules,** and an optional **Pearson eText.**

13

FOCUS ON CONCEPTS

Each statement represents the primary learning objective for the corresponding major heading within the chapter. After you complete the chapter, you should be able to:

13.1 Define *air pressure* and describe the instruments used to measure this weather element.

13.2 Discuss the three forces that act on the atmosphere to either create or alter winds.

13.3 Contrast the weather associated with low-pressure centers (cyclones) and high-pressure centers (anticyclones).

13.4 Summarize Earth's idealized global circulation. Describe how continents and seasonal temperature changes complicate the idealized pattern.

13.5 List three types of local winds and describe their formation.

13.6 Describe the instruments used to measure wind. Explain how wind direction is expressed using compass directions.

13.7 Discuss the major factors that influence the global distribution of precipitation.

Horizontal differences in air pressure created the strong winds associated with this costal storm. (Photo by Jim Edds/Corbis)

THE ATMOSPHERE IN MOTION

O f the various elements of weather and climate, changes in air pressure are the least noticeable. When listening to a weather report, we are generally interested in moisture conditions (humidity and precipitation), temperature, and perhaps wind. Rarely do people wonder about air pressure. Although people do not generally notice the hour-to-hour and day-to-day variations in air pressure, such changes are very important factors in producing changes in our weather. Variations in air pressure from place to place cause the movement of air we call wind and are a significant factor in weather forecasting. As we will see, air pressure is closely tied to the other elements of weather in a cause-and-effect relationship.

13.1 Understanding Air Pressure

Define *air pressure* and describe the instruments used to measure this weather element.

In Chapter 11 we noted that **air pressure** is simply the pressure exerted by the weight of air above. Average air pressure at sea level is about 1 kilogram per square centimeter, or 14.7 pounds per square inch—also called 1 atmosphere. Specifically, a column of air 1 square inch in cross section, measured from sea level to the top of the atmosphere, would weigh about 14.7 pounds (**Figure 13.1**). This is roughly the same pressure that is produced by a 1-square-inch column of water 10 meters (33 feet) in height. With some simple arithmetic, you can calculate that the air pressure exerted on the top of a small (50 centimeter-by-100 centimeter [20 inch-by-40 inch]) school desk exceeds 5000 kilograms (11,000 pounds),

or about the weight of a 50-passenger school bus. Why doesn't the desk collapse under the weight of the ocean of air above? Simply, air pressure is exerted in all directions—down, up, and sideways. Thus, the air pressure pushing down on the desk exactly balances the air pressure pushing up on the desk.

Visualizing Air Pressure

Imagine a tall aquarium that has the same dimensions as the small desk mentioned in the preceding paragraph. When this aquarium is filled to a height of 10 meters (33 feet), the water pressure at the bottom equals 1 atmosphere (1 kilogram per square centimeter [14.7 pounds per square inch]). Now, imagine what will happen if this aquarium is placed on top of our student desk so that all the force is directed downward. Compare this to what results when the desk is placed inside the aquarium and allowed to sink to the bottom. In the latter example, the desk survives because the water pressure is exerted in all directions, not just downward, as in our earlier example. The desk, like your body, is "built" to withstand the pressure of 1 atmosphere. It is important to note that although we do not generally notice the pressure exerted by the ocean of air around us, except when ascending or descending in an elevator or airplane, it is nonetheless substantial. The pressurized suits that astronauts use on space walks are designed to duplicate the atmospheric pressure experienced at Earth's surface. Without these protective suits to keep body fluids from boiling away, astronauts would perish in minutes.

The concept of air pressure can also be understood if we examine the behavior of gas molecules. Gas molecules, unlike molecules of the liquid and solid phases, are not bound

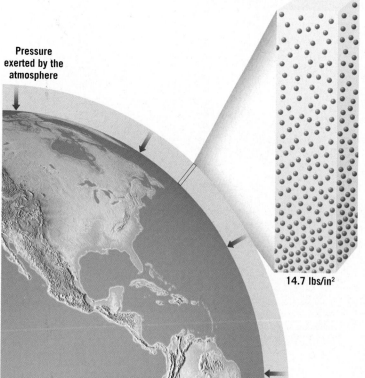

Figure 13.1 Sea-level pressure Air pressure can be thought of as the force exerted by the weight of the atmosphere above. A column of air 1 square inch in cross section extending from sea level to the top of the atmosphere would weigh about 14.7 pounds. The difference from bottom to top in the density of gas molecules is actually much greater than illustrated. In addition, the thickness of the atmosphere is exaggerated.

Pressure exerted by the atmosphere

14.7 lbs/in^2

to one another but freely move about, filling all the space available to them. When two gas molecules collide, which happens frequently under normal conditions, they bounce off each other like elastic balls. If a gas is confined to a container, this motion is restricted by its sides, much as the walls of a handball court redirect the motion of the handball. The continuous bombardment of gas molecules against the sides of the container exerts an outward push that we call air pressure. Although the atmosphere is without walls, it is confined from below by Earth's surface and effectively from above because the force of gravity prevents its escape. Here, we can define *air pressure* as the force exerted against a surface by the continuous collision of gas molecules.

Measuring Air Pressure

When meteorologists measure atmospheric pressure, they use a unit called the *millibar*. Standard sea-level pressure is 1013.25 millibars. Although the millibar has been the unit of measure on all U.S. weather maps since January 1940, the media use "inches of mercury" to describe atmospheric pressure. In the United States, the National Weather Service converts millibar values to inches of mercury for public and aviation use (**Figure 13.2**).

Inches of mercury is easy to understand. The use of mercury for measuring air pressure dates from 1643, when Torricelli, a student of the famous Italian scientist Galileo, invented the **mercury barometer**. Torricelli correctly described the atmosphere as a vast ocean of air that exerts pressure on us and all objects around us. To measure this force, he filled a glass tube, which was closed at one end, with mercury. He then inverted the tube into a dish of mercury (**Figure 13.3**). Torricelli found that the mercury flowed out of the tube until the weight of the column was balanced by the pressure that the atmosphere exerted on the surface of the mercury in the dish. In other words, the weight of mercury in the column equaled the weight of the same-diameter column of air that extended from the ground to the top of the atmosphere.

When air pressure increases, the mercury in the tube rises. Conversely, when air pressure decreases, so does the height of the mercury column. With some refinements, the mercury barometer invented by Torricelli is still the standard pressure-measuring instrument used today. Standard atmospheric pressure at sea level equals 29.92 inches of mercury.

The need for a smaller and more portable instrument for measuring air pressure led to the development of the **aneroid barometer** (*aneroid* means "without liquid"). Instead of having a mercury column held up by air pressure, an aneroid barometer uses a partially evacuated metal chamber (**Figure 13.4**). The chamber is extremely sensitive to variations in air pressure and changes shape, compressing as the pressure increases and expanding as the pressure decreases. A series of levers transmits the

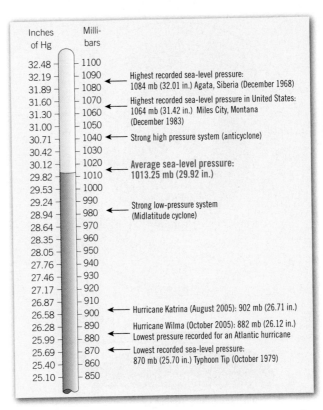

SmartFigure 13.2 Inches and millibars A comparison of two units commonly used to express air pressure. (goo.gl/QTOOmh)

Tutorial

movements of the chamber to a pointer on a dial that is calibrated to read in inches of mercury and/or millibars.

As shown in Figure 13.4, the face of an aneroid barometer intended for home use is inscribed with words such as *fair*, *change*, *rain*, and *stormy*. Notice that "fair" corresponds with high-pressure readings, whereas "rain"

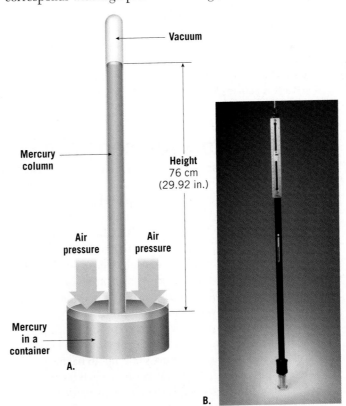

A.

B.

Figure 13.3 Mercury barometer **A.** The weight of the column of mercury is balanced by the pressure exerted on the dish of mercury by the air above. If the pressure decreases, the column of mercury falls; if the pressure increases, the column rises. **B.** Image of a mercury barometer. (Photo by Charles D. Winters/Science Source)

Figure 13.4 Aneroid barometer **A.** Illustration of an aneroid barometer. **B.** The aneroid barometer has a partially evacuated chamber that changes shape, compressing as atmospheric pressure increases and expanding as pressure decreases.

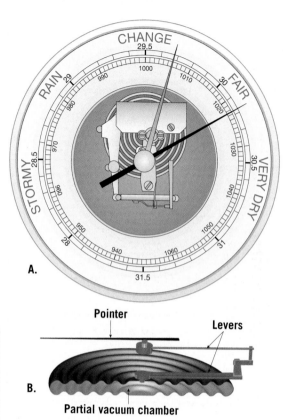

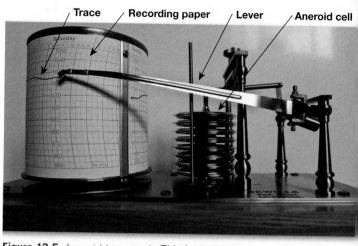

Figure 13.5 Aneroid barograph This instrument makes a continuous record of air pressure changes. The cylinder with the graph is a clock that turns once per day or once per week. (Photo by Stuart Aylmer/Alamy)

is associated with low pressures. Barometric readings, however, may not always indicate the weather. The dial may point to "fair" on a rainy day or to "rain" on a fair day. To "predict" the local weather, the change in air pressure over the past few hours is more important than the current pressure reading. Falling pressure is often associated with increasing cloudiness and the possibility of precipitation, whereas rising air pressure generally indicates clearing conditions. It is useful to remember, however, that particular barometer readings or trends do not always correspond to specific types of weather.

Another advantage of an aneroid barometer is that it can easily be connected to a recording mechanism. The resulting instrument is a **barograph**, which provides a continuous record of pressure changes with the passage of time (**Figure 13.5**). Another important adaptation of the aneroid barometer is its use to indicate altitude for aircraft, mountain climbers, and mapmakers.

13.1 CONCEPT CHECKS

1. Describe air pressure in your own words.
2. What is standard sea-level pressure in millibars, in inches of mercury, and in pounds per square inch?
3. Describe the operating principles of the mercury barometer and the aneroid barometer.
4. List two advantages of the aneroid barometer over the mercury barometer.

13.2 Factors Affecting Wind

Discuss the three forces that act on the atmosphere to either create or alter winds.

In Chapter 12, we examined the upward movement of air and its role in cloud formation. As important as vertical motion is, far more air moves horizontally, the phenomenon we call **wind**. What causes wind?

Simply stated, wind is the result of horizontal differences in air pressure. *Air flows from areas of higher pressure to areas of lower pressure.* You may have experienced this when opening something that is vacuum packed. The noise you hear is caused by air rushing from the higher pressure outside the can or jar to the lower pressure inside. Wind is nature's attempt to balance such inequalities in air pressure. Because unequal heating of Earth's surface generates these pressure differences, *solar radiation is the ultimate energy source for most wind.*

If Earth did not rotate, and if there were no friction between moving air and Earth's surface, air would flow in a straight line from areas of higher pressure to areas of lower pressure. But because Earth does rotate and friction does exist, wind is controlled by the following combination of forces: pressure gradient force, Coriolis effect, and friction.

Pressure Gradient Force

If an object experiences an unbalanced force in one direction, it will accelerate (experience a change in velocity). The force that generates winds results from horizontal pressure differences. When air is subjected to greater pressure on one side than on another, the imbalance produces a force that is directed from the region of higher pressure toward the area of lower pressure. Thus, pressure differences cause the wind to blow, and the greater these differences, the greater the wind speed.

Variations in air pressure over Earth's surface are determined from barometric readings taken at hundreds of weather stations. These pressure measurements are shown on surface weather maps using **isobars** (*iso* = equal, *bar* = pressure), or lines connecting places of equal air pressure (**Figure 13.6**). The *spacing* of the isobars indicates the amount of pressure change occurring over a given distance, which is called the **pressure gradient force**. Pressure gradient is analogous to gravity acting on a ball rolling down a hill. A steep pressure gradient, like a steep hill, causes greater acceleration of a parcel of air than does a weak pressure gradient (a gentle hill). Thus, the relationship between wind speed and the pressure gradient is straightforward: *Closely spaced isobars indicate a steep pressure gradient and strong winds; widely spaced isobars indicate a weak pressure gradient and light winds.* Figure 13.6 illustrates the relationship between the spacing of isobars and wind speed. Note also that the pressure gradient force is always directed at *right angles* to the isobars.

In order to draw isobars on a weather map to show air pressure patterns, meteorologists must compensate for the *elevation* of each station. Otherwise, all high-elevation locations, such as Denver, Colorado, would always be mapped as having low pressure. This compensation is accomplished by converting all pressure measurements to sea-level equivalents.

Figure 13.7 is a surface weather map that shows isobars (representing corrected sea-level air pressure) and

winds. Wind *direction* is shown as wind arrow shafts and *speed* as wind bars (see the key accompanying the map in the figure). Isobars, used to depict pressure patterns, are rarely straight or evenly spaced on surface maps. Consequently, wind generated by the pressure gradient force typically changes speed and direction as it flows.

The area of somewhat circular closed isobars in eastern North America represented by the red letter *L* is a *low-pressure system.* In western Canada, a *high-pressure system*, denoted by the blue letter *H*, can also be seen. We will discuss *highs* and *lows* in the next section.

In summary, the *horizontal pressure gradient is the driving force of wind.* The magnitude of the pressure gradient force is shown by the spacing of isobars. The direction of force is always from areas of higher pressure toward areas of lower pressure and at right angles to the isobars.

Coriolis Effect

Figure 13.7 shows the typical air movements associated with high- and low-pressure systems. As expected, the air

Figure 13.6 Isobars are lines connecting places of equal atmospheric pressure The spacing of isobars indicates the amount of pressure change occurring over a given distance—called the *pressure gradient force.* Closely spaced isobars indicate a strong pressure gradient and high wind speeds, whereas widely spaced isobars indicate a weak pressure gradient and low wind speeds.

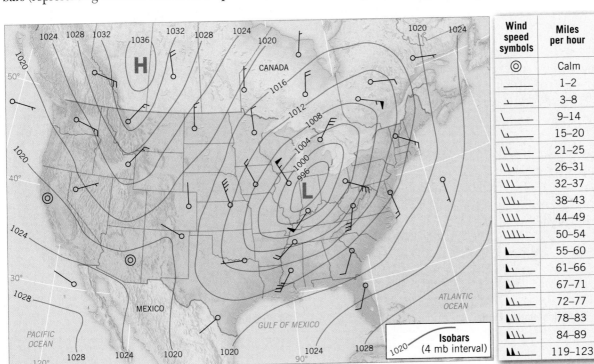

SmartFigure 13.7 Isobars show the distribution of pressure on surface weather maps Isobars are seldom straight but usually form broad curves. Concentric isobars indicate cells of high and low pressure. The "wind flags" indicate the expected airflow surrounding pressure cells and are plotted as "flying" with the wind (that is, the wind blows toward the station circle). Notice on this map that the isobars are more closely spaced and the wind speed is faster around the low-pressure center than around the high. (http://goo.gl/5D7geU)

Wind speed symbols	Miles per hour
◎	Calm
—	1–2
⊾	3–8
⊾	9–14
⊾	15–20
⊾	21–25
⊾	26–31
⊾	32–37
⊾	38–43
⊾	44–49
⊾	50–54
⊾	55–60
⊾	61–66
⊾	67–71
⊾	72–77
⊾	78–83
⊾	84–89
⊾	119–123

Tutorial

moves out of the regions of higher pressure and into the regions of lower pressure. However, the wind does not cross the isobars at right angles, as the pressure gradient force directs it to do. The direction deviates as a result of Earth's rotation. This has been named the **Coriolis effect**, after the French scientist who first thoroughly described it.

All free-moving objects or fluids, including the wind, are deflected to the *right* of their path of motion in the Northern Hemisphere and to the *left* in the Southern Hemisphere. The reason for this deflection can be illustrated by imagining the path of a rocket launched from the North Pole toward a target located on the equator (**Figure 13.8**). If the rocket took an hour to reach its target, during its flight, Earth would have rotated 15° to the east. To someone standing on Earth, it would look as if the rocket had veered off its path and hit Earth 15° west of its target. The true path of the rocket is straight and would appear so to someone out in space looking at Earth. It is Earth turning under the rocket that gives it its *apparent* deflection.

Note that the rocket is deflected to the right of its path of motion because of the counterclockwise rotation of the Northern Hemisphere. In the Southern Hemisphere, the effect is reversed. Clockwise rotation produces a similar deflection but to the *left* of the path of motion. The same deflection is experienced by wind regardless of the direction it is moving.

We attribute the apparent shift in wind direction to the Coriolis effect. This deflection (1) is always directed at right angles to the direction of airflow; (2) affects only wind direction, not wind speed; (3) is affected by wind speed (the stronger the wind, the greater the deflection);

and (4) is strongest at the poles and weakens equatorward, becoming nonexistent at the equator.

Note that any free-moving object will experience a deflection caused by the Coriolis effect. The U.S. Navy dramatically discovered this fact in World War II. During target practice, long-range guns on battleships continually missed their targets by as much as several hundred yards until ballistic corrections were made for the changing position of a seemingly stationary target. Over a short distance, however, the Coriolis effect is relatively small.

Friction with Earth's Surface

The effect of friction on wind is important only within a few kilometers of Earth's surface. Friction acts to slow air movement and, as a consequence, alters wind direction. To illustrate friction's effect on wind direction, let us look at a situation in which friction has no role. Above the friction layer, the pressure gradient force and Coriolis effect work together to direct the flow of air. Under these conditions, the pressure gradient force causes air to start moving across the isobars. As soon as the air starts to move, the Coriolis effect acts at right angles to this motion. The faster the wind speed, the greater the deflection.

Eventually, the Coriolis effect will balance the pressure gradient force, and the wind will blow parallel to the isobars (**Figure 13.9**). Upper-air winds generally take this path and are called **geostrophic winds**. The lack of friction with Earth's surface allows geostrophic winds to travel at higher speeds than do surface winds. This can be observed in **Figure 13.10** by noting the wind

SmartFigure 13.8 The Coriolis effect The deflection caused by Earth's rotation is illustrated by a rocket traveling for 1 hour from the North Pole to a location on the equator. **A.** On a nonrotating Earth, the rocket would fly straight to its target. **B.** However, Earth rotates 15° each hour. Thus, although the rocket travels in a straight line, when we plot the path of the rocket on Earth's surface, it follows a curved path that veers to the right of the target. The video illustrates the Coriolis effect by using a playground merry-go-round. (goo.gl/UT7NFg)

Video

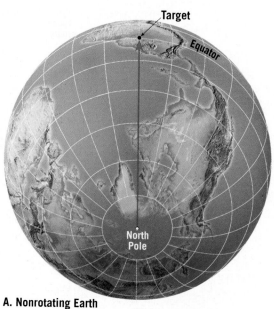

A. Nonrotating Earth

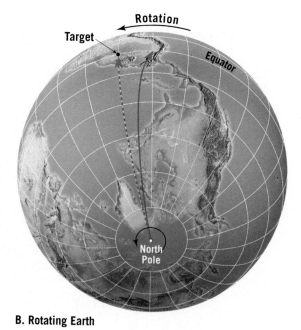

B. Rotating Earth

flags, many of which indicate winds of 50 to 100 miles per hour.

The most prominent features of upper-level flow are **jet streams**. First encountered by high-flying bombers during World War II, these fast-moving "rivers" of air travel between 120 and 240 kilometers (75 and 150 miles) per hour in a west-to-east direction. One such stream is situated over the polar front, which is the zone separating cool polar air from warm subtropical air.

Below 600 meters (2000 feet), friction complicates the airflow just described. Recall that the Coriolis effect is proportional to wind speed. Friction lowers the wind speed, so it reduces the Coriolis effect. Because the pressure gradient force is not affected by wind speed, it wins the tug of war shown in **Figure 13.11**. The result is a movement of air

at an angle across the isobars, toward the area of lower pressure.

The roughness of the terrain determines the angle of airflow across the isobars. Over the smooth ocean

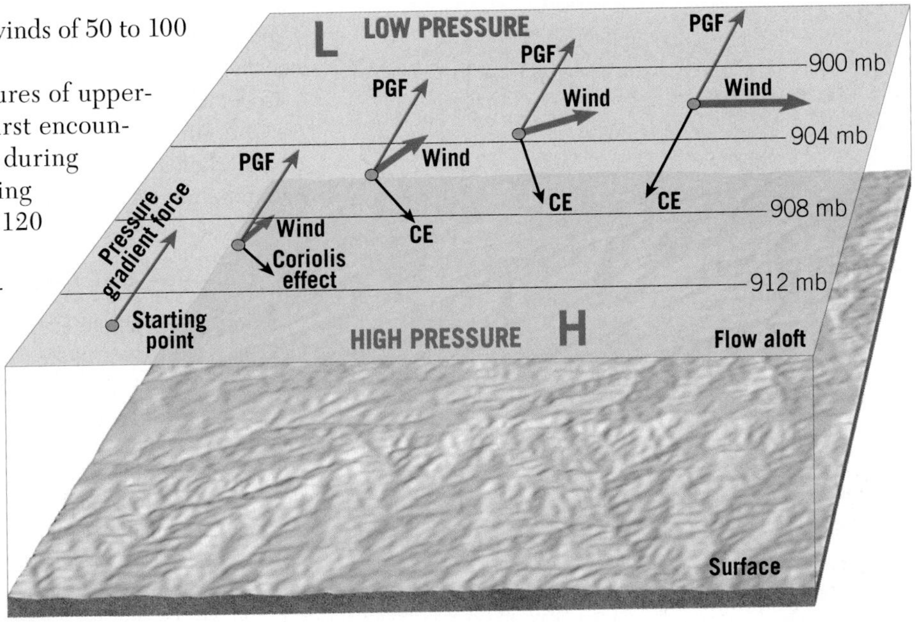

Figure 13.9 Geostrophic wind The only force acting on a stationary parcel of air is the pressure gradient force (PGF). Once the air begins to accelerate, the Coriolis effect (CE) deflects moving air to the right in the Northern Hemisphere. Greater wind speeds result in a stronger Coriolis effect (deflection) until the flow is parallel to the isobars. At this point, the pressure gradient force and Coriolis effect are in balance, and the flow is called a *geostrophic wind*. It is important to note that in the "real" atmosphere, airflow continually adjusts for variations in the pressure field. As a result, the adjustment to geostrophic equilibrium is much more irregular than shown.

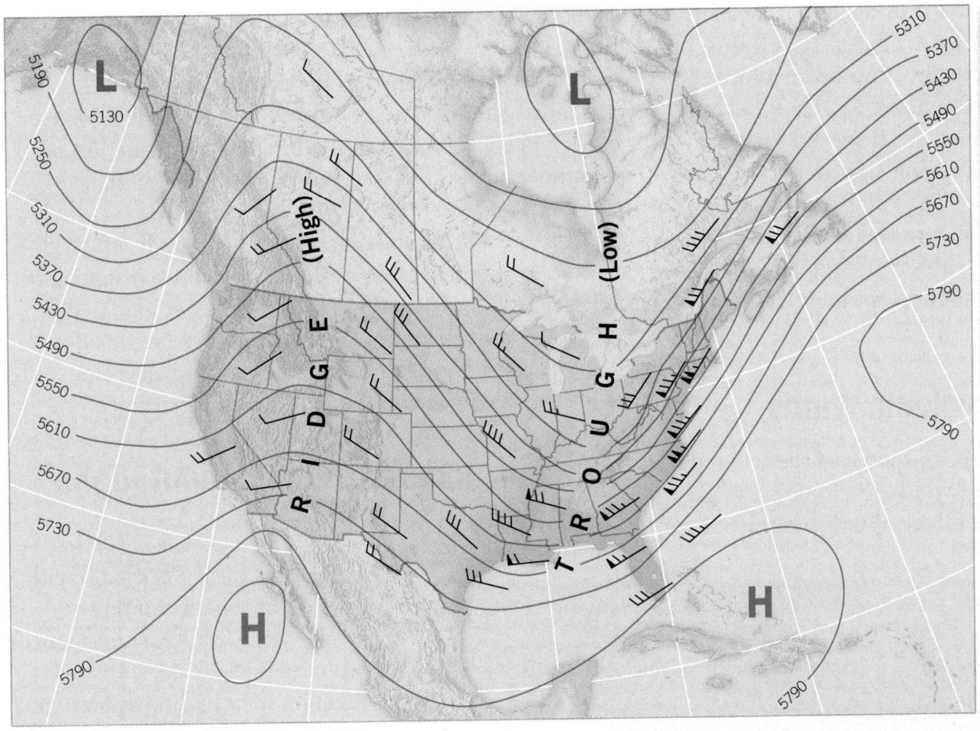

Figure 13.10 Simplified upper-air weather chart This simplified weather chart shows the direction and speed of the upper-air winds. Note from the flags that the airflow is almost parallel to the contours. Like most other upper-air charts, this one shows variations in the height (in meters) at which a selected pressure (500 millibars) is found instead of showing variations in pressure at a fixed height, like surface maps. Places experiencing 500-millibar pressure at higher altitudes (toward the south on this map) are experiencing higher pressures than places where the height contours indicate lower altitudes. Thus, *higher-elevation* contours indicate *higher* pressures, and *lower-elevation* contours indicate *lower* pressures.

Upper-level weather chart

Wind speed symbols	Miles per hour
⊚	Calm
(flag)	1–2
(flag)	3–8
(flag)	9–14
(flag)	15–20
(flag)	21–25
(flag)	26–31
(flag)	32–37
(flag)	38–43
(flag)	44–49
(flag)	50–54
(flag)	55–60
(flag)	61–66
(flag)	67–71
(flag)	72–77
(flag)	78–83
(flag)	84–89
(flag)	119–123

Representation of upper-level chart

SmartFigure 13.11 The effects of friction on wind **A.** Friction has little effect on winds aloft, so airflow is parallel to the isobars. **B.** Friction slows surface winds, which weakens the Coriolis effect, causing winds to cross the isobars and move toward the lower pressure. (https://goo.gl/PfRIPv)

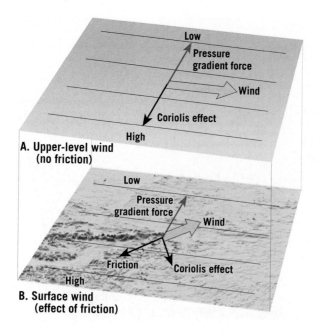

A. **Upper-level wind (no friction)**

B. **Surface wind (effect of friction)**

Animation

surface, friction is low, and the angle is small. Over rugged terrain, where friction is higher, the angle that air makes as it flows across the isobars can be as great as 45 degrees.

In summary, upper airflow is nearly parallel to the isobars, whereas the effect of friction causes the surface winds to move more slowly and cross the isobars at an angle.

13.2 CONCEPT CHECKS

1. List three factors that combine to direct horizontal airflow (wind).
2. What force is responsible for generating wind?
3. Write a generalization relating the spacing of isobars to wind speed.
4. Briefly describe how the Coriolis effect influences air movement.
5. Unlike winds aloft, which blow nearly parallel to the isobars, surface winds generally cross the isobars. Explain what causes this difference.

13.3 Highs and Lows

Contrast the weather associated with low-pressure centers (cyclones) and high-pressure centers (anticyclones).

Among the most common features on a weather map are areas designated as pressure centers. **Cyclones**, or **lows**, are centers of low pressure, and **anticyclones**, or **highs**, are high-pressure centers. As **Figure 13.12** illustrates, the pressure decreases from the outer isobars toward the center in a cyclone. In an anticyclone, just the opposite is the case: The values of the isobars increase from the outside toward the center. Knowing just a few basic facts about centers of high and low pressure greatly increases your understanding of current and forthcoming weather.

Cyclonic and Anticyclonic Winds

In the preceding section, you learned that the two most significant factors that affect wind are the pressure gradient force and the Coriolis effect. Winds move from higher pressure toward lower pressure and are deflected to the right or left by Earth's rotation. When these controls of airflow are applied to pressure centers in the Northern Hemisphere, the result is that winds blow inward and counterclockwise around a low (**Figure 13.13A**). Around a high, they blow outward and clockwise (see Figure 13.12).

In the Southern Hemisphere, the Coriolis effect deflects the winds to the left; therefore, winds around a low blow clockwise (**Figure 13.13B**), and winds around a high move counterclockwise. In either hemisphere, friction causes a net inflow (**convergence**) around a cyclone and a net outflow (**divergence**) around an anticyclone.

Weather Generalizations About Highs and Lows

Rising air is associated with cloud formation and precipitation, whereas subsidence produces clear skies. In this section we will discuss how the movement of air can itself create pressure change and generate winds. After doing so, we will examine the relationship between horizontal and vertical flow, and their effects on weather.

Let us first consider a surface low-pressure system where the air is spiraling inward. The net inward

Figure 13.12 Cyclonic and anticyclonic winds in the Northern Hemisphere Arrows show the winds blowing inward and counterclockwise around a low and outward and clockwise around a high.

Figure 13.13 Cyclonic circulation in the Northern and Southern Hemispheres The cloud patterns in these images allow us to "see" the circulation pattern in the lower atmosphere. (NASA)

A. This satellite image shows a large low-pressure center in the Gulf of Alaska. The cloud pattern clearly shows an inward and counterclockwise spiral.

B. This satellite image shows a strong cyclonic storm in the South Atlantic near the coast of Brazil. The cloud pattern shows an inward and clockwise circulation.

transport of air causes a shrinking of the area occupied by the air mass, a process that is termed *horizontal convergence.* Whenever air converges horizontally, it must pile up—that is, increase in height to allow for the decreased area it now occupies. This generates a "taller" and therefore heavier air column. Yet a surface low can exist only as long as the column of air exerts less pressure than that occurring in surrounding regions. We seem to have encountered a paradox: A low-pressure center causes a net accumulation of air, which increases its pressure, and a surface cyclone should quickly eradicate itself in a manner not unlike what happens when a vacuum-packed can is opened.

For a surface low to exist for very long, compensation must occur at some layer aloft. For example, surface convergence could be maintained if divergence (spreading out) aloft occurred at a rate equal to the inflow below. **Figure 13.14** shows the relationship between surface convergence and divergence aloft that is needed to maintain a low-pressure center.

Divergence aloft may even exceed surface convergence, thereby resulting in intensified surface inflow and accelerated vertical motion. Thus, divergence aloft can intensify storm centers as well as maintain them. On the other hand, inadequate divergence aloft permits

surface flow to "fill" and weaken the accompanying cyclone.

Note that surface convergence about a cyclone causes a net upward movement. The rate of this vertical movement is slow, generally less than 1 kilometer (0.6 mile) per day. Nevertheless, because rising air often results in cloud formation and precipitation, a low-pressure center is generally related to unstable conditions and stormy weather (**Figure 13.15A**).

As often as not, it is divergence aloft that creates a surface low. Spreading out aloft initiates upflow in the atmosphere directly below, eventually working its way to the surface, where inflow is encouraged.

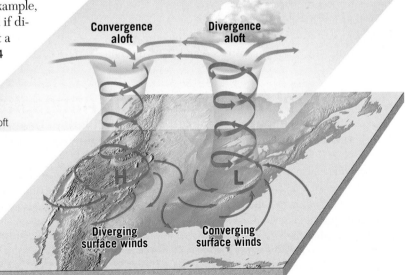

Convergence aloft

Divergence aloft

Flow aloft

H

L

Diverging surface winds

Converging surface winds

SmartFigure 13.14 Airflow associated with cyclones (L) and anticyclones (H) A low, or cyclone, has converging surface winds and rising air, resulting in cloudy conditions and often precipitation. A high, or anticyclone, has diverging surface winds and descending air, which leads to clear skies and fair weather. (goo.gl/2Fx2gM)

Animation

Figure 13.15 Weather generalizations related to pressure centers **A.** A rainy day in London. Low-pressure systems are frequently associated with cloudy conditions and precipitation. (Photo by Lourens Smak/Alamy) **B.** Clear skies and "fair" weather may be expected when an area is under the influence of high pressure. (Photo by Vidler Steve/Prisma Bildagentur/Alamy)

A.

B.

high-pressure end. By noting whether the pressure is rising, falling, or steady, we have a good indication of what the forthcoming weather will be. Such a determination, called the **pressure**, or **barometric**, **tendency**, is a useful aid in short-range weather prediction.

You should now be better able to understand why television weather reporters emphasize the positions and projected paths of cyclones and anticyclones. The "villain" on these weather programs is always the low-pressure center, which produces "bad" weather in any season. Lows move in roughly a west-to-east direction across the United States and require a few days to more than a week for the journey. Because their paths can be somewhat erratic, accurate prediction of their migration is difficult, although essential, for short-range forecasting.

Meteorologists must also determine whether the flow aloft will intensify an embryo storm or act to suppress its development. Because of the close tie between conditions at the surface and those aloft, a great deal of emphasis has been placed on the importance and understanding of the total atmospheric circulation, particularly in the midlatitudes. We will now examine the workings of Earth's general atmospheric circulation and then again consider the structure of the cyclone in light of this knowledge.

Like their cyclonic counterparts, anticyclones must be maintained from above. Outflow near the surface is accompanied by convergence aloft and general subsidence of the air column (see Figure 13.14). Because descending air is compressed and warmed, cloud formation and precipitation are unlikely in an anticyclone. Thus, fair weather can usually be expected with the approach of a high-pressure center (**Figure 13.15B**).

For reasons that should now be obvious, it has been common practice to print on household barometers the words "stormy" at the low-pressure end and "fair" on the

13.3 CONCEPT CHECKS

1. Prepare a diagram with isobars and wind arrows that shows the winds associated with surface cyclones and anticyclones in both the Northern and Southern Hemispheres.
2. For a surface low-pressure center to exist for an extended period, what condition must exist aloft?
3. What general weather conditions are to be expected when the pressure tendency is rising? When the pressure tendency is falling?

13.4 General Circulation of the Atmosphere

Summarize Earth's idealized global circulation. Describe how continents and seasonal temperature changes complicate the idealized pattern.

The underlying cause of wind is unequal heating of Earth's surface. In tropical regions, more solar radiation is received than is radiated back to space. In polar regions, the opposite is true: Less solar energy is received than is lost. Attempting to balance these differences, the atmosphere acts as a giant heat-transfer system, moving warm air poleward and cool air equatorward. On a

smaller scale, but for the same reason, ocean currents also contribute to this global heat transfer. The general circulation is complex, and a great deal has yet to be explained. We can, however, develop a general understanding by first considering the circulation that would occur on a nonrotating Earth having a uniform surface. We will then modify this system to fit observed patterns.

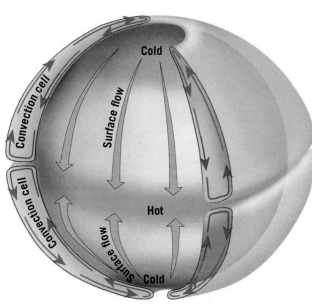

Figure 13.16 Global circulation on a nonrotating Earth A simple convection system is produced by unequal heating of the atmosphere on a nonrotating Earth.

Circulation on a Nonrotating Earth

On a hypothetical nonrotating planet with a smooth surface of either all land or all water, two large thermally produced cells would form (**Figure 13.16**). The heated equatorial air would rise until it reached the tropopause, which acts like a lid and deflects the air poleward. Eventually, this upper-level airflow would reach the poles, sink, spread out in all directions at the surface, and move back toward the equator. Once there, it would be reheated and start its journey over again. This hypothetical circulation system has upper-level air flowing poleward and surface air flowing equatorward.

If we add the effect of rotation, this simple convection system will break down into smaller cells. **Figure 13.17** illustrates the three pairs of cells proposed to carry on the task of heat redistribution on a rotating planet. The polar and tropical cells retain the characteristics of the thermally generated convection described earlier. The nature of the midlatitude circulation is more complex and will be discussed in more detail later in this chapter.

Idealized Global Circulation

Near the equator, the rising air is associated with the pressure zone known as the **equatorial low**. This region of ascending moist, hot air is marked by abundant precipitation. Because this region of low pressure is a zone where winds converge, it

is also referred to as the **intertropical convergence zone (ITCZ)**. As the upper-level flow from the equatorial low reaches 20° to 30° latitude, north or south, it sinks back toward the surface. This subsidence and associated adiabatic heating produce hot, arid conditions. The center of this zone of subsiding dry air is the **subtropical high**, which encircles the globe near 30° latitude, north and south (see Figure 13.17). The great deserts of Australia, Arabia, and Africa exist because of the stable, dry condition caused by the subtropical highs.

At the surface, airflow is outward from the center of the subtropical high. Some of the air travels equatorward and is deflected by the Coriolis effect, producing the reliable **trade winds**. The remainder travels poleward and is also deflected, generating the prevailing **westerlies** of the midlatitudes. As the westerlies move poleward, they encounter the cool **polar easterlies** in the region of the **subpolar low**. The interaction of these warm and cool winds produces the stormy belt known as the **polar front**. The source region for the variable polar easterlies is the **polar high**. Here, cold polar air is subsiding and spreading equatorward.

In summary, this simplified global circulation is dominated by four pressure zones. The subtropical and polar highs are areas of dry subsiding air that flows outward at the surface, producing the prevailing winds. The low-pressure zones of the equatorial and subpolar regions are associated with inward and upward airflow accompanied by clouds and precipitation.

Did You Know?

Approximately 0.25 percent (one-quarter of 1 percent) of the solar energy that reaches the lower atmosphere is transformed into wind. Although it is just a minuscule percentage, the absolute amount of energy is enormous. According to one estimate, North Dakota alone is theoretically capable of providing enough wind-generated power to meet more than one-third of the electricity demand of the United States.

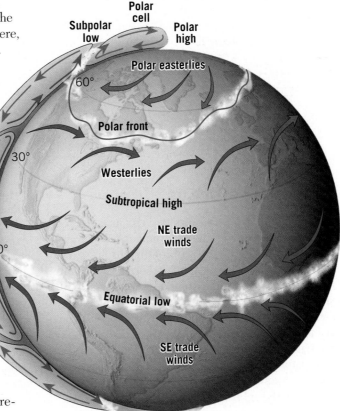

SmartFigure 13.17
Idealized global circulation for the three-cell circulation model on a rotating Earth (goo.gl/AByhGS)

Tutorial

Figure 13.18 Average surface air pressure These maps show the average surface air pressure, in millibars, for **A.** January and **B.** July, with associated winds.

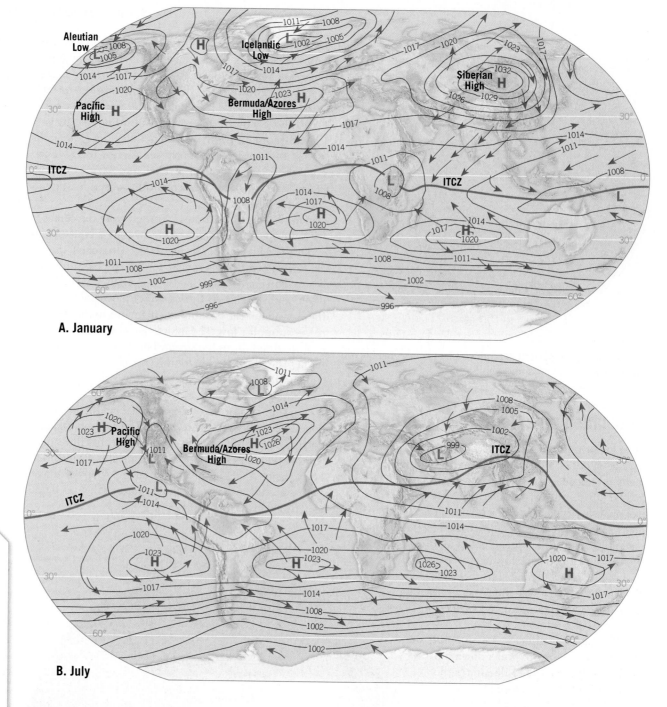

A. January

B. July

Influence of Continents

Up to this point, we have described the surface pressure and associated winds as continuous belts around Earth. However, the only truly continuous pressure belt is the subpolar low in the Southern Hemisphere, where the ocean is uninterrupted by landmasses. At other latitudes, particularly in the Northern Hemisphere, where landmasses break up the ocean surface, large seasonal temperature differences disrupt the pattern. **Figure 13.18** shows the resulting pressure and wind patterns for January and July. The circulation over the oceans is dominated by semipermanent cells of high pressure in the subtropics and cells of low pressure over the subpolar regions. The subtropical highs are

responsible for the trade winds and westerlies, as mentioned earlier.

The large landmasses, on the other hand, particularly Asia, become cold in the winter and develop a seasonal high-pressure system from which surface flow is directed off the land (see Figure 13.18). In the summer, the opposite occurs: The landmasses are heated and develop a low-pressure cell, which permits air to flow onto the land. These seasonal changes in wind direction are known as the **monsoons**. During warm months, areas such as India experience a flow of warm, water-laden air from the Indian Ocean, which produces the rainy summer monsoon. The winter monsoon is dominated by dry continental air. A similar situation exists, but to a lesser extent, over North America.

In summary, the general circulation is produced by semipermanent cells of high and low pressure over the oceans and is complicated by seasonal pressure changes over land.

The Westerlies

Circulation in the midlatitudes, the zone of the westerlies, is complex and does not fit the convection system proposed for the tropics. Between 30° and 60° latitude, the general west-to-east flow is interrupted by the migration of cyclones and anticyclones. In the Northern Hemisphere, these cells move from west to east around the globe, creating an anticyclonic (clockwise) flow or a cyclonic (counterclockwise) flow in their area of influence. A close correlation exists between the paths taken by these surface pressure systems and the position of the upper-level airflow, indicating that the upper air steers the movement of cyclonic and anticyclonic systems.

Among the most obvious features of the flow aloft are the seasonal changes. The steep temperature gradient across the middle latitudes in the winter months corresponds to a stronger flow aloft. In addition, the polar jet stream fluctuates seasonally such that its average position migrates southward with the approach of winter and northward as summer nears. By midwinter, the jet core may penetrate as far south as central Florida.

Because the paths of low-pressure centers are guided by the flow aloft, we can expect the southern tier of states to experience more of their stormy weather in the winter season. During the hot summer months, the storm track is across the northern states, and some cyclones never leave Canada. The northerly storm track associated with summer also applies to Pacific storms, which move toward Alaska during the warm months, thus producing an extended dry season for much of the West coast. The number of cyclones generated is seasonal as well, with the largest number occurring in the cooler months, when the temperature gradients are greatest. This fact is in agreement with the role of cyclonic storms in the distribution of heat across the midlatitudes.

 13.4 CONCEPT CHECKS

1. Referring to the idealized model of atmospheric circulation, in which belt of prevailing winds is most of the United States?
2. The trade winds diverge from which pressure belt?
3. Which prevailing wind belts converge in the stormy region known as the polar front?
4. Which pressure belt is associated with the equator?
5. Explain the seasonal change in winds associated with India. What term is applied to this seasonal wind shift?

13.5 Local Winds

List three types of local winds and describe their formation.

Now that we have examined Earth's large-scale circulation, let us turn briefly to winds that influence much smaller areas. Remember that all winds are produced for the same reason: pressure differences that arise because of temperature differences caused by unequal heating of Earth's surface. **Local winds** are small-scale winds produced by a locally generated pressure gradient. Those described here are caused either by topographic effects or by variations in surface composition in the immediate area.

Land and Sea Breezes

In coastal areas during the warm summer months, the land is heated more intensely during the daylight hours than is the adjacent body of water. As a result, the air above the land surface heats, expands, and rises, creating an area of lower pressure. A **sea breeze** then develops because cooler air over the water (higher pressure) moves toward the warmer land (lower pressure) (**Figure 13.19A**). The sea breeze begins to develop shortly before

SmartFigure 13.19 Sea and land breezes (goo.gl/ZQybwo)

 **Tutorial**

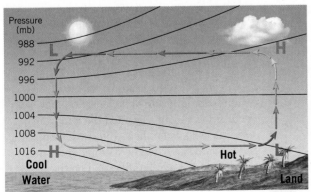

A. During daylight hours, cooler and denser air over the water moves onto the land, generating a sea breeze.

B. At night the land cools more rapidly than the sea, generating an offshore flow called a land breeze.

Figure 13.20 Valley and mountain breezes

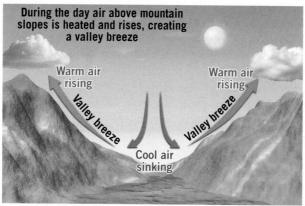

A. Valley breeze

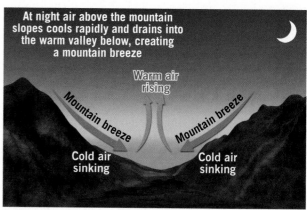

B. Mountain breeze

noon and generally reaches its greatest intensity during the mid- to late afternoon. These relatively cool winds can be a significant moderating influence on afternoon temperatures in coastal areas. Small-scale sea breezes can also develop along the shores of large lakes. People who live in a city near the Great Lakes, such as Chicago, recognize this lake effect, especially in the summer. They are reminded daily by weather reports of the cooler temperatures near the lake as compared to warmer outlying areas.

At night, the reverse may take place: The land cools more rapidly than the sea, and a **land breeze** develops (**Figure 13.19B**).

Mountain and Valley Breezes

A daily wind similar to land and sea breezes occurs in many mountainous regions. During daylight hours, the air along the slopes of the mountains is heated more intensely than the air at the same elevation over the valley floor. Because this warmer air is less dense, it glides up along the slope and generates a **valley breeze** (**Figure 13.20A**). The occurrence of these daytime upslope breezes can often be identified by the cumulus clouds that develop on adjacent mountain peaks.

After sunset, the pattern may reverse. Rapid radiation cooling along the mountain slopes produces a layer of cooler air next to the ground. Because cool air is denser than warm air, it drains downslope into the valley. This movement of air is called a **mountain breeze** (**Figure 13.20B**). The same type of cool air drainage can occur in places that have very modest slopes. The result is that the coldest pockets of air are usually found in the lowest spots. Like many other winds, mountain and valley breezes have seasonal preferences. Although valley breezes are most common during the warm season, when solar heating is most intense, mountain breezes tend to be more dominant in the cold season.

Figure 13.21 Wildfires driven by Santa Ana winds Ten large wildfires rage across southern California in this image taken on October 27, 2003, by NASA's *Aqua* satellite. The inset shows an idealized high-pressure area composed of cool, dry air that drives Santa Ana winds. As the winds move from higher elevations toward the coast, adiabatic heating causes the air temperature to increase and the relative humidity to decrease. (NASA EOS Earth Observing System)

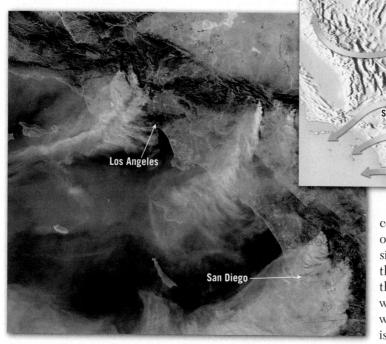

Chinook and Santa Ana Winds

Warm, dry winds sometimes move down the eastern slopes of the Rockies, where they are called **chinooks**. Such winds are often created when a strong pressure gradient develops in a mountainous region. As the air descends the leeward slopes of the mountains, it is heated adiabatically (by compression). Because condensation may have occurred as the air ascended the windward side, releasing latent heat, the air descending the leeward slope will be warmer and drier than it was at a similar elevation on the windward side. Although the temperature of these winds is generally less than 10°C (50°F), which is not particularly warm, the winds occur mostly in the winter and spring, when the affected areas may be experiencing below-freezing temperatures. Thus, by comparison, these dry, warm winds often bring a drastic change. When the ground has a snow cover, these winds are known to melt it in short order.

A chinooklike wind that occurs in southern California is the **Santa Ana**. This hot, desiccating wind greatly increases the threat of fire in this already dry area (**Figure 13.21**).

13.5 CONCEPT CHECKS

1. What is a local wind?
2. Describe the formation of a sea breeze.
3. Does a land breeze blow toward or away from the shore?
4. During what time of day would you expect to experience a well-developed valley breeze—midnight, late morning, or late afternoon?

13.6 Measuring Wind

Describe the instruments used to measure wind. Explain how wind direction is expressed using compass directions.

Two basic wind measurements, direction and speed, are particularly significant to weather observers. One simple device for determining both measurements is a *wind sock*, which is a common sight at small airports and landing strips (**Figure 13.22A**). The cone-shaped bag is open at both ends and is free to change position with shifts in wind direction. The degree to which the sock is inflated is an indication of wind speed.

Winds are always labeled by the direction from which they blow. A north wind blows *from* the north *toward* the south, an east wind *from* the east *toward* the west. The instrument most commonly used to determine wind direction is the **wind vane** (**Figure 13.22B**, upper right). This instrument, a common sight on many buildings, always points *into* the wind. The wind direction is often shown on a dial that is connected to the wind vane. The dial indicates wind direction, either by points of the compass (N, NE, E, SE, etc.) or by a scale of 0 to 360

degrees. On the latter scale, 0 degrees and 360 degrees are both north, 90 degrees is east, 180 degrees is south, and 270 degrees is west.

Wind speed is commonly measured using a **cup anemometer** (see Figure 13.22B, upper left). The wind speed is read from a dial much like the speedometer of an automobile. Places where winds are steady and speeds are relatively high are potential sites for tapping wind energy.

When the wind consistently blows more often from one direction than from any other, it is called a **prevailing wind**. You may be familiar with the prevailing westerlies that dominate the circulation in the midlatitudes. In the United States, for example, these winds consistently move the "weather" from west to east across the continent. Embedded within this general eastward flow are cells of high and low pressure, with their characteristic clockwise and counterclockwise flow. As a result, the winds associated with the

Did You Know?
The highest wind speed recorded at a surface station is 372 km (231 mi) per hour, measured April 12, 1934, at Mount Washington, New Hampshire. Located at an elevation of 1886 m (6288 ft), the observatory atop Mount Washington has an average wind speed of 56 km (35 mi) per hour. Faster wind speeds have undoubtedly occurred, but no instruments were in place to record them.

A.

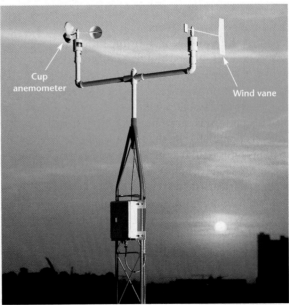

B.

Figure 13.22 Wind measurement **A.** A wind sock is a device for determining wind direction and estimating wind speed. Wind socks are common sights at small airports and landing strips. (Photo by Lourens Smak/Alamy) **B.** Wind vane (right) and cup anemometer (left). The wind vane shows wind direction, and the anemometer measures wind speed. (Photo by Belfort Instrument Company)

Cup anemometer

Wind vane

Figure 13.23 Wind roses These graphs show the percentage of time winds come from various directions.

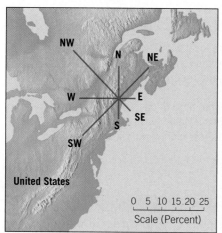

A. Wind frequency for winter in the northeastern United States.

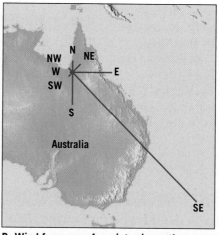

B. Wind frequency for winter in northeastern Australia. Note the reliability of the southeast trade winds in Australia as compared to the westerlies in the northeastern United States.

By knowing the locations of cyclones and anticyclones in relation to where you are, you can predict the changes in wind direction that will occur as a pressure center moves past. Because changes in wind direction often bring changes in temperature and moisture conditions, the ability to predict the winds can be very useful. In the Midwest, for example, a north wind may bring cool, dry air from Canada, whereas a south wind may bring warm, humid air from the Gulf of Mexico. Sir Francis Bacon summed it up nicely when he wrote, "Every wind has its weather."

westerlies, as measured at the surface, often vary considerably from day to day and from place to place. By contrast, the direction of airflow associated with the belt of trade winds is much more consistent, as can be seen in **Figure 13.23**.

13.6 CONCEPT CHECKS

1. What are the two basic wind measurements? What instruments are used to make these measurements?
2. *From* what direction does a northeast wind blow? *Toward* what direction does a south wind blow?

13.7 Global Distribution of Precipitation

Discuss the major factors that influence the global distribution of precipitation.

A casual glance at **Figure 13.24** shows a relatively complex pattern for the distribution of precipitation. Although the map appears to be complicated, the general features of the map can be explained by applying our knowledge of global winds and pressure systems.

In general, regions influenced by high pressure, with its associated subsidence and diverging winds, experience relatively dry conditions. On the other hand, regions under the influence of low pressure and its converging winds and ascending air receive ample precipitation. This pattern is illustrated by noting that the tropical regions dominated by the equatorial low are the rainiest regions on Earth. It is here that we find the rain forests of the Amazon basin in South America and the Congo basin in Africa. The warm, humid trade winds converge to yield abundant rainfall throughout the year. By contrast, areas dominated by subtropical high-pressure cells clearly receive much smaller amounts of precipitation. These are the regions of extensive subtropical deserts. In the Northern Hemisphere, the largest desert is the Sahara. Examples in the Southern Hemisphere

include the Kalahari in southern Africa and the dry lands of Australia.

If Earth's pressure and wind belts were the only factors controlling precipitation distribution, the pattern shown in Figure 13.24 would be simpler. The inherent nature of the air is also an important factor in determining precipitation potential. Because cold air has a low capacity for moisture compared with warm air, we would expect a latitudinal variation in precipitation, with low latitudes receiving the greatest amounts of precipitation and high latitudes receiving the smallest amounts. Figure 13.24 indeed shows heavy rainfall in equatorial regions and meager precipitation in high-latitude areas. Recall that the dry region in the warm subtropics is explained by the presence of the subtropical high.

The distribution of land and water also complicates the precipitation pattern. Large landmasses in the middle latitudes commonly experience decreased precipitation toward their interiors. For example, central North America and central Eurasia receive considerably less

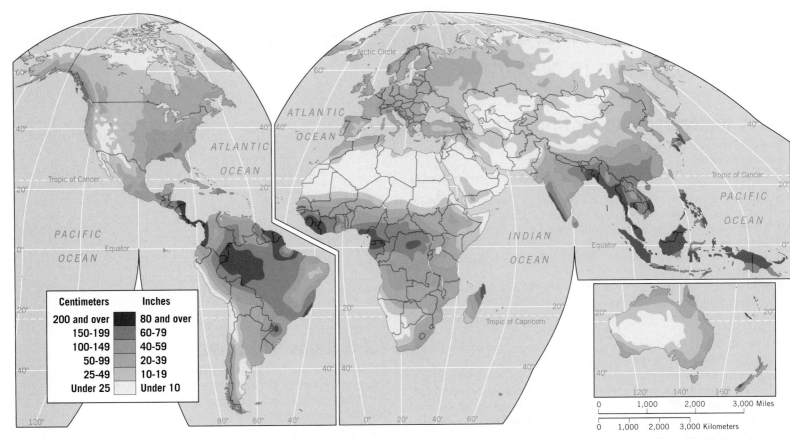

Centimeters	Inches
200 and over	80 and over
150-199	60-79
100-149	40-59
50-99	20-39
25-49	10-19
Under 25	Under 10

precipitation than do coastal regions at the same latitude. Mountain barriers also alter precipitation patterns. Windward mountain slopes receive abundant precipitation, whereas leeward slopes and adjacent lowlands are often deficient in moisture.

Figure 13.24 Global distribution of average annual precipitation

13.7 CONCEPT CHECKS

1. With which global pressure belt are the rain forests of Africa's Congo basin associated? Which pressure system is linked to the Sahara Desert?
2. What factors, in addition to the distribution of wind and pressure, influence the global distribution of precipitation?

CONCEPTS IN REVIEW
The Atmosphere in Motion

13.1 Understanding Air Pressure

Define *air pressure* and describe the instruments used to measure this weather element.

KEY TERMS: air pressure, mercury barometer, aneroid barometer, barograph

- Air has weight: At sea level, it exerts a pressure of 1 kilogram per square centimeter (14.7 pounds per square inch), or 1 atmosphere.
- Air pressure is the force exerted by the weight of air above. With increasing altitude, there is less air above to exert a force, and thus air pressure decreases with altitude—rapidly at first and then much more slowly.
- The unit meteorologists use to measure atmospheric pressure is the millibar. Standard sea-level pressure is expressed as 1013.2 millibars. Isobars are lines on a weather map that connect places of equal air pressures.
- A mercury barometer measures air pressure using a column of mercury in a glass tube sealed at one end and inverted in a dish of mercury. It measures atmospheric pressure in inches of mercury, the height of the column of mercury in the barometer. Standard atmospheric pressure at sea level equals 29.92 inches of mercury. As air pressure increases, the mercury in the tube rises, and when air pressure decreases, so does the height of the column of mercury.
- Aneroid ("without liquid") barometers consist of partially evacuated metal chambers that compress as air pressure increases and expand as pressure decreases.

13.2 Factors Affecting Wind

Discuss the three forces that act on the atmosphere to either create or alter winds.

KEY TERMS: wind, isobar, pressure gradient force, Coriolis effect, geostrophic wind, jet stream

- Wind is controlled by a combination of (1) the pressure gradient force, (2) the Coriolis effect, and (3) friction. The pressure gradient force, which results from pressure differences, is the primary force that drives wind. It is depicted by the spacing of isobars on a map. Closely spaced isobars indicate a steep pressure gradient and strong winds; widely spaced isobars indicate a weak pressure gradient and light winds.
- The Coriolis effect, which is due to Earth's rotation, produces deviation in the path of wind (to the right in the Northern Hemisphere and to the left in the Southern Hemisphere). Friction, which significantly influences airflow near Earth's surface, is negligible above a height of a few kilometers.
- Above a height of a few kilometers, the Coriolis effect is equal to and opposite the pressure gradient force, which results in geostrophic winds. Geostrophic winds flow in a nearly straight path, parallel to the isobars, with velocities proportional to the pressure gradient force.

(?) **The diagrams at the top of the right column show surface winds at two locations. All factors in both situations are identical except that one surface is land and the other is water. Which diagram represents winds over the land? Explain your choice.**

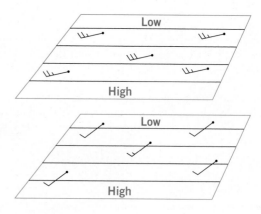

13.3 Highs and Lows

Contrast the weather associated with low-pressure centers (cyclones) and high-pressure centers (anticyclones).

KEY TERMS: cyclone (low), anticyclone (high), convergence, divergence, pressure (barometric) tendency

- The two types of pressure centers are (1) cyclones, or lows (centers of low pressure), and (2) anticyclones, or highs (centers of high pressure). In the Northern Hemisphere, winds around a low (cyclone) are counterclockwise and inward. Around a high (anticyclone), they are clockwise and outward. In the Southern Hemisphere, the Coriolis effect causes winds to be clockwise around a low and counterclockwise around a high.
- Because air rises and cools adiabatically in a low-pressure center, cloudy conditions and precipitation are often associated with their passage. In a high-pressure center, descending air is compressed and warmed; therefore, cloud formation and precipitation are unlikely in an anticyclone, and "fair" weather is usually expected.

(?) **Refer to Figure 13.4. Assume that you are an observer checking this aneroid barometer several hours after it was last checked. What is the pressure tendency? How did you figure this out? What does the tendency indicate about forthcoming weather?**

13.4 General Circulation of the Atmosphere

Summarize Earth's idealized global circulation. Describe how continents and seasonal temperature changes complicate the idealized pattern.

KEY TERMS: equatorial low, intertropical convergence zone (ITCZ), subtropical high, trade winds, westerlies, polar easterlies, subpolar low, polar front, polar high, monsoon

- If Earth's surface were uniform, four belts of pressure oriented east to west would exist in each hemisphere. Beginning at the equator, the four belts would be the (1) equatorial low, also referred to as the intertropical convergence zone (ITCZ), (2) subtropical high at about 25° to 35° on either side of the equator, (3) subpolar low, situated at about 50° to 60° latitude, and (4) polar high, near Earth's poles.

(13.4 continued)

- Particularly in the Northern Hemisphere, large seasonal temperature differences over continents disrupt the idealized, or zonal, global patterns of pressure and wind. In winter, large, cold landmasses develop a seasonal high-pressure system from which surface airflow is directed off the land. In summer, landmasses are heated, and a low-pressure system develops over them, which permits air to flow onto the land. These seasonal changes in wind direction are known as monsoons.
- In the middle latitudes, between 30° and 60° latitude, the general west-to-east flow of the westerlies is interrupted by the migration of cyclones and anticyclones. The paths taken by these cyclonic and anticyclonic systems is closely correlated to upper-level airflow and the polar jet stream. The average position of the polar jet stream, and hence the paths followed by cyclones, migrates southward with the approach of winter and northward as summer nears.

13.5 Local Winds

List three types of local winds and describe their formation.

KEY TERMS: local wind, sea breeze, land breeze, valley breeze, mountain breeze, chinook, Santa Ana

- Local winds are small-scale winds produced by a locally generated pressure gradient. Sea and land breezes form along coasts and are brought about by temperature contrasts between land and water. Valley and mountain breezes occur in mountainous areas where the air along slopes heats differently than does the air at the same elevation over the valley floor. Chinook and Santa Ana winds are warm, dry winds created when air descends the leeward side of a mountain and warms by compression.

Peter Wey/Shutterstock

(?) **With which local wind are the clouds in this photo most likely associated? Would you expect clouds such as these to form at night? Explain.**

13.6 Measuring Wind

Describe the instruments used to measure wind. Explain how wind direction is expressed using compass directions.

KEY TERMS: wind vane, cup anemometer, prevailing wind

- The two basic wind measurements are direction and speed. Winds are always labeled by the direction from which they blow. Wind direction is measured with a wind vane, and wind speed is measured using a cup anemometer.

(?) **When designing an airport, it is important to have planes take off into the wind. Refer to the accompanying wind rose and describe the orientation of the runway and the direction planes would usually travel when they took off. Where on Earth might you find a wind rose like this?**

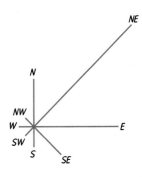

13.7 Global Distribution of Precipitation

Discuss the major factors that influence the global distribution of precipitation.

- The general features of the global distribution of precipitation can be explained by global winds and pressure systems. In general, regions influenced by high pressure, with its associated subsidence and divergent winds, experience dry conditions. Regions under the influence of low pressure, with its converging winds and ascending air, receive ample precipitation.
- Air temperature, the distribution of continents and oceans, and the location of mountains also influence the distribution of precipitation.

(?) **This satellite image was produced with data from the *Tropical Rainfall Measuring Mission* (*TRMM*). Notice the band of heavy rainfall shown in reds and yellows that extends east–west across the image. With which pressure zone is this band of rainy weather associated? Is it more likely that this image was acquired in July or January? Explain.**

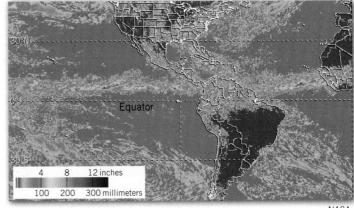

NASA

GIVE IT SOME THOUGHT

1. If divergence in the jet stream above a surface low-pressure center exceeds convergence at the surface, will surface winds likely get stronger or weaker? Explain.

2. The accompanying map is a simplified surface weather map for April 2, 2011, on which the centers of three pressure cells are numbered.

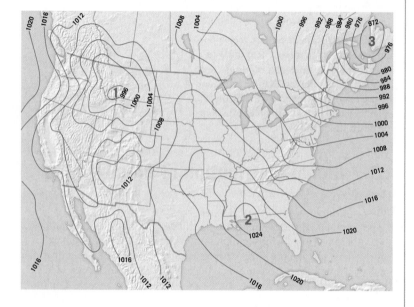

   a. Identify which of the pressure cells are anticyclones (highs) and which are cyclones (lows).
   b. Which pressure center has the steepest pressure gradient and therefore the strongest winds?
   c. Using Figure 13.2, determine whether Pressure Center 3 should be considered strong or weak.

3. You and a friend are watching TV on a rainy day, when the weather reporter says, "The barometric pressure is 28.8 inches and rising." Hearing this, you say, "It looks like fair weather is on its way." Your friend responds, "I thought air pressure had something to do with the weight of air. How does inches relate to weight? And why do you think the weather is going to improve?" How would you respond to your friend's queries?

4. If you live in the Northern Hemisphere and are directly west of the center of a cyclone, what is the probable wind direction at your location? What if you were west of an anticyclone?

5. If Earth did not rotate on its axis and if its surface were completely covered with water, what direction would a boat drift if it started its journey in the middle latitudes of the Northern Hemisphere? (*Hint:* What would the global circulation pattern be like for a nonrotating Earth?)

6. It is late afternoon on a warm summer day, and you are enjoying some time at the beach. Until the last hour or two, winds were calm. Then a breeze began to develop. Is it more likely a cool breeze from the water or a warm breeze from the adjacent land area? Explain.

7. The accompanying sketch shows a cross section of the idealized circulation in the Northern Hemisphere. Match the appropriate number on the sketch to each of the following features:
   a. Equatorial low
   b. Polar front
   c. Subtropical high
   d. Polar high

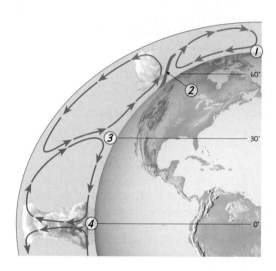

8. The accompanying maps of Africa show the distribution of precipitation for July and January. Which map represents July, and which represents January? How were you able to figure this out?

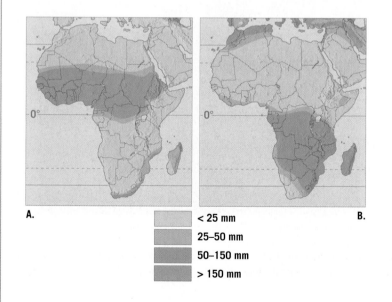

MasteringGeology™

www.masteringgeology.com Looking for additional review and test prep materials? With individualized coaching on the toughest topics of the course, MasteringGeology offers a wide variety of ways for you to move beyond memorization to begin thinking like a geologist. Visit the Study Area in www.masteringgeology.com to find practice quizzes, study tools, and multimedia that will improve your understanding of this chapter's content. Sign in today to enjoy the following features: **Self Study Quizzes, SmartFigure: Tutorials/ Animations/Condor Videos/Mobile Field Trips, Geoscience Animation Library, GEODe, RSS Feeds, Digital Study Modules,** and an optional **Pearson eText.**

14

FOCUS ON CONCEPTS

Each statement represents the primary learning objective for the corresponding major heading within the chapter. After you complete the chapter, you should be able to:

14.1 Define *air mass*. Describe the classification and weather associated with different air masses.

14.2 Compare and contrast typical weather associated with a warm front and a cold front. Describe an occluded front and a stationary front.

14.3 Summarize the weather associated with the passage of a mature midlatitude cyclone. Describe how airflow aloft is related to cyclones and anticyclones at the surface.

14.4 List the basic requirements for thunderstorm formation and locate places on a map that exhibit frequent thunderstorm activity. Describe the stages in the development of a thunderstorm.

14.5 Summarize the atmospheric conditions and locations that are favorable to the formation of tornadoes. Discuss tornado destruction and tornado forecasting.

14.6 Identify areas of hurricane formation on a world map and discuss the conditions that promote hurricane formation. List the three broad categories of hurricane destruction.

A storm is classified as a thunderstorm only after thunder is heard. Because thunder is produced by lightning, lightning must also occur. (Photo by Saverio Maria Gallotti/Alamy)

WEATHER PATTERNS AND SEVERE WEATHER

Tornadoes and hurricanes rank among nature's most destructive forces. Each spring, newspapers report the death and destruction left in the wake of "bands" of tornadoes. During late summer and fall, we hear occasional news reports about hurricanes. Storms with names such as Katrina, Rita, Sandy, and Ike make front-page headlines. Thunderstorms, although less intense and far more common than tornadoes and hurricanes, are also part of our discussion of severe weather in this chapter. Before looking at violent weather, however, we will study the atmospheric phenomena that most often affect our day-to-day weather: air masses, fronts, and traveling midlatitude cyclones. We will see the interplay of the elements of weather discussed in Chapters 11, 12, and 13.

14.1 Air Masses

Define *air mass*. Describe the classification and weather associated with different air masses.

For many people who live in the middle latitudes, including much of the United States, summer heat waves and winter cold spells are familiar experiences. In the first instance, several days of high temperatures and oppressive humidities may finally end when a series of thunderstorms pass through the area, followed by a few days of relatively cool relief. By contrast, the clear skies that often accompany a span of frigid subzero days may be replaced by thick gray clouds and a period of snow as temperatures rise to levels that seem mild compared to those that existed just a day earlier. In both examples, what was experienced was a period of generally constant weather conditions followed by a relatively short period of change and then the reestablishment of a new set of weather conditions that remained for perhaps several days before changing again.

What Is an Air Mass?

The weather patterns just described result from movements of large bodies of air, called air masses. An **air mass** is an immense body of air, typically 1600 kilometers (1000 miles) or more across and perhaps several kilometers thick, that is characterized by a similarity of temperature and moisture at any given altitude.

When this air moves out of its region of origin, it carries those temperatures and moisture conditions elsewhere, eventually affecting a large portion of a continent (**Figure 14.1**).

An excellent example of the influence of an air mass is illustrated in **Figure 14.2**, which shows a cold, dry mass from northern Canada moving southward. With a beginning temperature of –46°C (–51°F), the air mass warms to –33°C (–27°F) by the time it reaches Winnipeg. It continues to warm as it moves southward through the Great Plains and into Mexico. Throughout its southward journey, the air mass becomes warmer. But it also brings some of the coldest weather of the winter to the places in its path. Thus, the air mass is modified, but it also modifies the weather in the areas over which it moves.

The horizontal uniformity of an air mass is not perfect, of course. Because air masses extend over large areas, small differences occur in temperature and humidity from place to place. Still, the differences observed within an air mass are small compared to the rapid changes experienced across air-mass boundaries.

Since it may take several days for an air mass to move across an area, the region under its influence will probably experience fairly constant weather, a situation called **air-mass weather**. Certainly, there may be some day-to-day variations, but the events will be very unlike those in an adjacent air mass.

The air-mass concept is an important one because it is closely related to the study of atmospheric disturbances. Most disturbances in the middle latitudes originate along the boundary zones that separate different air masses.

Figure 14.1 Lake-effect snow storm This satellite image shows a cold, dry air mass moving from its source region in Canada across Lake Superior. It illustrates the process that leads to lake-effect snow storms. (NASA)

Cold, dry air mass

Lake Superior

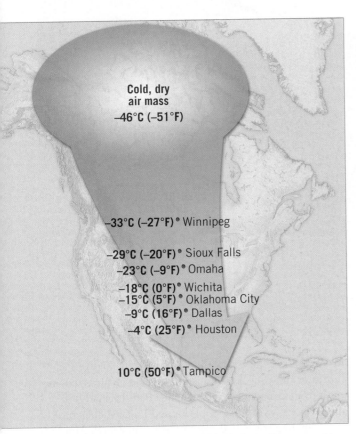

SmartFigure 14.2 An invasion of frigid air As this very cold air mass moved southward from Canada, it brought some of the coldest weather of the winter to the areas in its path. As it advanced into the United States, the air mass slowly got warmer. Thus, the air mass was gradually modified at the same time that it modified the weather in the areas over which it moved. (From *Physical Geography: A Landscape Appreciation*, 9th edition, by Tom L. McKnight and Darrell Hess, ©2008. Reprinted and electronically reproduced by permission of Pearson Education, Inc., Upper Saddle River, NJ.) (https://goo.gl/QjlzzG)

Animation

Source Regions

When a portion of the lower atmosphere moves slowly or stagnates over a relatively uniform surface, the air assumes the distinguishing features of that area, particularly with regard to temperature and moisture. The area where an air mass acquires its characteristic properties of temperature and moisture is called its **source region**. The source regions that produce air masses that influence North America are shown in **Figure 14.3**.

Air masses are classified according to their source region. **Polar (P)** and **arctic (A) air masses** originate in high latitudes toward Earth's poles, whereas those that form in low latitudes are called **tropical (T) air masses**. The designation *polar, arctic,* or *tropical* gives an indication of the temperature characteristics of an air mass. *Polar* and *arctic* indicate cold, and *tropical* indicates warm.

In addition, air masses are classified according to the nature of the surface in the source region. **Continental (c) air masses** form over land, and **maritime (m) air masses** originate over water. The designation *continental* or *maritime* suggests the moisture characteristics of the air mass. Continental air is likely to be dry, and maritime air is likely to be humid.

The basic types of air masses according to this scheme of classification are continental polar (cP), continental arctic (cA), continental tropical (cT), maritime polar (mP), and maritime tropical (mT).

Weather Associated with Air Masses

Continental polar and maritime tropical air masses influence the weather of North America most, especially east of the Rocky Mountains. Continental polar air masses originate in northern Canada, interior Alaska, and the Arctic—areas that are uniformly cold and dry in winter and cool and dry in summer. In winter, an invasion of continental polar air brings the clear skies and cold temperatures we associate with a cold wave as it moves

Figure 14.3 Air-mass source regions for North America Source regions are largely confined to subtropical and subpolar locations. The fact that the middle latitudes are where cold and warm air masses clash, often because the converging winds of a traveling cyclone draw them together, means that this zone lacks the conditions necessary to be a source region. The differences between polar and arctic are relatively small and serve to indicate the degree of coldness of the respective air masses. By comparing the winter **A.** and summer **B.** maps, it is clear that the extent and temperature characteristics of the source regions fluctuate.

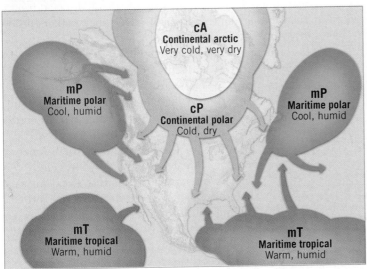

A. Winter pattern

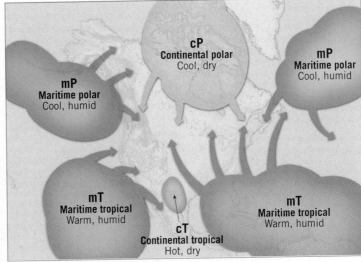

B. Summer pattern

SmartFigure 14.4 Snowfall map The snowbelts of the Great Lakes are easy to pick out on this snowfall map. (Data from NOAA) The photo was taken following a 6-day lake-effect snowstorm in November 1996 that dropped 175 centimeters (nearly 69 inches) of snow on Chardon, Ohio, setting a new state record. (Photo by Tony Dejak/AP Images) (goo.gl/jRM10n)

Tutorial

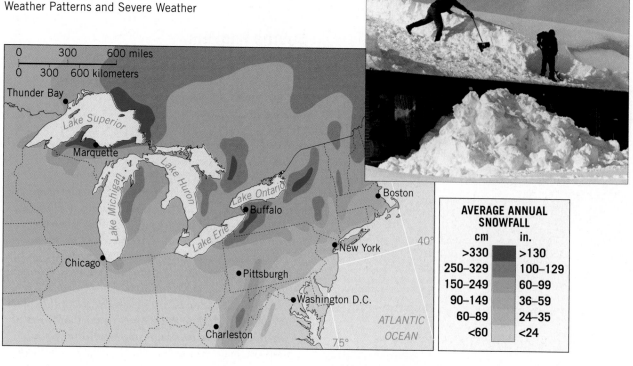

AVERAGE ANNUAL SNOWFALL

cm	in.
>330	>130
250–329	100–129
150–249	60–99
90–149	36–59
60–89	24–35
<60	<24

southward from Canada into the United States. In summer, this air mass may bring a few days of cooling relief.

Although cP air masses are not normally associated with heavy precipitation, those that cross the Great Lakes in late autumn and winter sometimes bring snow to the leeward shores. These localized storms often form when the surface weather map indicates no apparent cause for a snowstorm. These are known as **lake-effect snows**, and they make Buffalo and Rochester, New York, among the snowiest cities in the United States (**Figure 14.4**).

What causes lake-effect snow? During late autumn and early winter, the temperature contrast between the lakes and adjacent land areas can be large.° The temperature contrast can be especially great when a very cold cP air mass pushes southward across the lakes. When this occurs, the air acquires large quantities of heat and moisture from the relatively warm lake surface. By the time it reaches the opposite shore, the air mass is humid and

° Recall that land cools more rapidly and to lower temperatures than water. See the discussion "Differential Heating of Land and Water" in the section "Why Temperatures Vary: The Controls of Temperature" in Chapter 11.

Figure 14.5 Air-mass modification During winter, maritime polar (mP) air masses in the North Pacific usually begin as continental polar (cP) air masses in Siberia. The cP air is modified to mP as it slowly crosses the ocean.

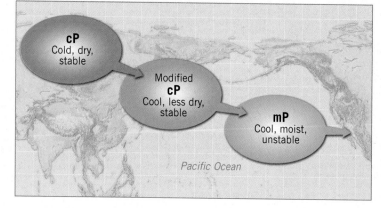

cP
Cold, dry, stable

Modified cP
Cool, less dry, stable

mP
Cool, moist, unstable

Pacific Ocean

unstable, and heavy snow showers are likely. Figure 14.1 illustrates this process. In this image, cold cP air is moving south from Canada. Notice that the cP air is cloud free until it moves over Lake Superior and Lake Michigan.

Maritime tropical air masses affecting North America most often originate over the warm waters of the Gulf of Mexico, the Caribbean Sea, or the adjacent Atlantic Ocean. As you might expect, these air masses are warm, moisture laden, and usually unstable. Maritime tropical air is the source of much, if not most, of the precipitation in the eastern two-thirds of the United States. In summer, when an mT air mass invades the central and eastern United States, and occasionally southern Canada, it brings the high temperatures and high humidity typically associated with its source region.

Of the two remaining air masses, maritime polar and continental tropical, the latter has the least influence on the weather of North America. Hot, dry continental tropical air, originating in the Southwest and Mexico during the summer, only occasionally affects the weather outside its source region.

During winter, maritime polar air masses coming from the North Pacific often originate as continental polar air masses in Siberia. The cold, dry cP air is transformed into relatively mild, humid, unstable mP air during its long journey across the North Pacific (**Figure 14.5**). As this mP air arrives at the western shore of North America, it is often accompanied by low clouds and shower activity. When this air advances inland against the western mountains, orographic uplift produces heavy rain or snow on the windward slopes of the mountains. Maritime polar air also originates in the North Atlantic, off the coast of eastern Canada, and occasionally influences the weather of

Figure 14.6 Classic nor'easter The satellite image shows a strong winter storm called a *nor'easter* along the coast of New England on January 12, 2011. In winter, a nor'easter exhibits a weather pattern in which strong northeast winds carry cold, humid mP air from the North Atlantic into New England and the middle Atlantic states. The ground-level view of the storm in Boston shows that the combination of ample moisture and strong convergence can result in heavy snow. (Satellite image by NASA; photo by Michael Dwyer/Alamy)

the northeastern United States. In winter, when New England is on the northern or northwestern side of a passing low, the counterclockwise cyclonic winds draw in maritime polar air. The result is a storm characterized by snow and cold temperatures, known locally as a **nor'easter** (**Figure 14.6**).

(14.1) CONCEPT CHECKS

1. Define *air mass*. What is air-mass weather?
2. On what basis are air masses classified?
3. Compare the temperature and moisture characteristics of the following air masses: cP, mP, mT, and cT.
4. Which air mass is associated with lake-effect snow? What causes lake-effect snow?

(14.2) Fronts

Compare and contrast typical weather associated with a warm front and a cold front. Describe an occluded front and a stationary front.

One prominent feature of middle-latitude weather is how suddenly and dramatically it can change. Most of these sudden changes are associated with the passage of weather fronts. **Fronts** are boundary surfaces that separate air masses of different densities—one of which is usually warmer and contains more moisture than the other. However, fronts can form between any two contrasting air masses. When the vast sizes of air masses are considered, the zones (fronts) that separate them are relatively narrow and are shown as lines on weather maps.

Generally, the air mass located on one side of a front moves faster than the air mass on the other side. Thus, one air mass actively advances into the region occupied by another and collides with it. During World War I, Norwegian meteorologists visualized these zones of air-mass interactions as analogous to battle lines and tagged them "fronts," as in battlefronts. It is along these zones of "conflict" that storms develop and produce much of the precipitation and severe weather in the belt of the westerlies.

As one air mass moves into a region occupied by another, minimal mixing occurs along the frontal surface. Instead, the air masses retain their identity as one is displaced upward over the other. No matter which air mass is advancing, it is always the warmer, less-dense air that is

forced aloft, whereas the cooler, denser air acts as a wedge on which lifting occurs. The process of warm air gliding up and over a cold air mass is termed **overrunning**.

Warm Fronts

When the surface position of a front moves so that warm air occupies territory formerly covered by cooler air, it is called a **warm front** (**Figure 14.7**). On a weather map, the surface position of a warm front is shown by a red line with red semicircles protruding into the cooler air. East of the Rockies, warm tropical air often enters the United States from the Gulf of Mexico and overruns receding cool air. As the cold air retreats, friction with the ground slows the advance of the surface position of the front more so than its position aloft. Stated another way, less-dense, warm air has a hard time displacing denser cold air. For this reason, the boundary separating these air masses acquires a very gradual slope. The average slope of a warm front is about 1:200, which means that if you are 200 kilometers (120 miles) ahead of the surface location of a warm front, you will find the frontal surface at a height of 1 kilometer (0.6 mile).

As warm air ascends the retreating wedge of cold air, it expands and cools adiabatically to produce clouds and,

SmartFigure 14.7 Warm front This diagram shows the idealized clouds and weather associated with a warm front. During most of the year, warm fronts produce light to moderate precipitation over a wide area. (goo.gl/eSYZBn)

Animation

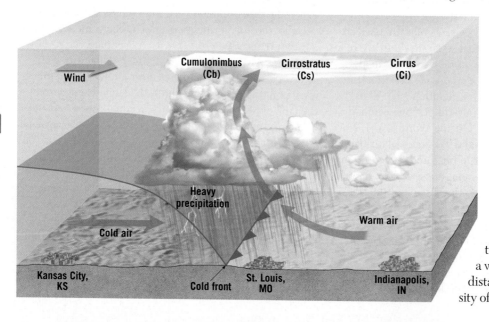

A gradual increase in temperature occurs with the passage of a warm front. The increase is most noticeable when there is a large temperature difference between the adjacent air masses. The moisture content and stability of the encroaching warm air mass largely determine when clear skies will return. During the summer, cumulus, and occasionally cumulonimbus, clouds may be embedded in the warm unstable air mass that follows the front. Precipitation from these clouds can be heavy but is usually scattered and of short duration.

frequently, precipitation. The sequence of clouds shown in Figure 14.7 typically precedes a warm front. The first sign of the approach of a warm front is the appearance of cirrus clouds overhead. These high clouds form 1000 kilometers (600 miles) or more ahead of the surface front, where the overrunning warm air has ascended high up the wedge of cold air.

As the front nears, cirrus clouds grade into cirrostratus, which blend into denser sheets of altostratus. About 300 kilometers (180 miles) ahead of the front, thicker stratus and nimbostratus clouds appear, and rain or snow begins. Because of their slow rate of advance and very low slope, warm fronts usually produce light to moderate precipitation over a large area for an extended period. Warm fronts, however, are occasionally associated with cumulonimbus clouds and thunderstorms. This occurs when the overrunning air is unstable and the temperatures on opposite sides of the front contrast sharply. At the other extreme, a warm front associated with a dry air mass could pass by without creating clouds or precipitation.

Cold Fronts

When dense cold air is actively advancing into a region occupied by warmer air, the boundary is called a **cold front** (**Figure 14.8**). As with warm fronts, friction tends to slow the surface position of a cold front more so than its position aloft. However, because of the relative positions of the adjacent air masses, the cold front steepens as it moves. On average, cold fronts are about twice as steep as warm fronts, having a slope of perhaps 1:100. In addition, cold fronts advance at speeds around 35 to 50 kilometers (20 to 35 miles) per hour compared to 25 to 35 kilometers (15 to 20 miles) per hour for warm fronts. These two differences—rate of movement and steepness of slope—largely account for the more violent nature of cold-front weather compared to the weather generally accompanying a warm front (**Figure 14.9**).

As a cold front approaches, generally from the west or northwest, towering clouds can often be seen in the distance. Near the front, a dark band of ominous clouds foretells the ensuing weather. The forceful lifting of air along a cold front is often so rapid that the latent heat released when water vapor condenses appreciably increases the air's buoyancy. The heavy downpours and vigorous wind gusts associated with mature cumulonimbus clouds frequently result. A cold front produces roughly the same amount of lifting as a warm front but over a shorter distance. As a result, the intensity of precipitation is greater, but

SmartFigure 14.8 Cold front Fast-moving cold front and cumulonimbus clouds. Thunderstorms may occur if the warm air is unstable. (goo.gl/PKYTiV)

Tutorial

Figure 14.9 Hail litters the ground Cumulonimbus clouds along a cold front produced heavy rain and damaging hail in central Mississippi in March 2013. (Photo by Rogelio V. Solis/AP Images)

the duration is shorter. In addition, a marked temperature drop and a wind shift from the south to west or northwest accompany the passage of the front. The sometimes violent weather and sharp temperature contrast along the cold front are symbolized on a weather map by a blue line with blue triangle-shaped points that extend into the warmer air mass (see Figure 14.8).

The weather behind a cold front is dominated by a subsiding and relatively cold air mass. Thus, clearing usually begins soon after the front passes. Although the compression of air due to subsidence causes some adiabatic heating, the effect on surface temperatures is minor. In winter, the long, cloudless nights that often follow the passage of a cold front allow for abundant radiation cooling that reduces surface temperatures. When a cold front moves over a relatively warm area, surface heating can produce shallow convection. This, in turn, may generate low cumulus or stratocumulus clouds behind the front.

Stationary Fronts and Occluded Fronts

Occasionally, the flow on both sides of a front is neither toward the cold air mass nor toward the warm air mass but almost parallel to the line of the front. Thus, the surface position of the front does not move. This condition is called a **stationary front**. On a weather map, stationary fronts are shown with blue triangular points on one side of the front and red semicircles on the other. At times, some overrunning occurs along a stationary front, most likely causing gentle to moderate precipitation.

The fourth type of front is an **occluded front**, an active cold front that overtakes a warm front, as shown in **Figure 14.10**. As the advancing cold air wedges the warm front upward, a new front emerges between the advancing cold air and the air over which the warm front is gliding. The weather of an occluded front is generally complex. Most precipitation is associated with the warm air being forced aloft. When conditions are suitable, however, the newly formed front is capable of initiating precipitation of its own. On a weather map, occluded fronts are shown with purple triangles and semicircles both pointing in the direction of the front's advance.

A word of caution is in order concerning the weather associated with various fronts. Although the

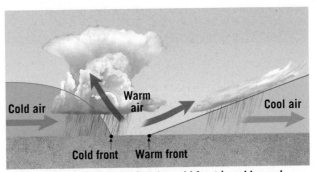

In this example, the air behind the cold front is colder and denser than the air ahead of the warm front.

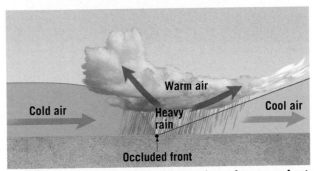

The surface cold front moves faster than the surface warm front and overtakes it to form an occluded front.

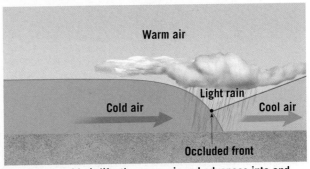

The denser cold air lifts the warm air and advances into and displaces the cool air.

Figure 14.10 Stages in the formation of an occluded front

preceding discussion will help you recognize the weather patterns associated with fronts, remember that these descriptions are generalizations. The weather generated along any individual front may or may not conform fully to this idealized picture. Fronts, like all other aspects of nature, do not lend themselves to classification as easily as we would like.

 14.2 CONCEPT CHECKS

1. Compare the weather of a typical warm front with that of a typical cold front.
2. Why is cold-front weather usually more severe than warm-front weather?
3. Describe a stationary front and an occluded front.

14.3 Midlatitude Cyclones

Summarize the weather associated with the passage of a mature midlatitude cyclone. Describe how airflow aloft is related to cyclones and anticyclones at the surface.

So far, we have examined the basic elements of weather as well as the dynamics of atmospheric motions. We are now ready to apply our knowledge of these diverse phenomena to an understanding of day-to-day weather patterns in the middle latitudes. For our purposes, *middle latitudes* refers to the region between southern Florida and Alaska. The primary weather producers here are **midlatitude**, or **middle-latitude**, **cyclones**. On weather maps they are shown by an *L*, meaning *low-pressure system*. **Figure 14.11** shows two views of a large idealized midlatitude cyclone with probable air masses, fronts, and surface wind patterns.

Midlatitude cyclones are large centers of low pressure that generally travel from west to east. Lasting from a few days to more than a week, these weather systems have a counterclockwise circulation, with an airflow inward toward their centers. Most midlatitude cyclones also have a cold front extending from the central area of low pressure, and frequently a warm front as well. Convergence and forceful lifting initiate cloud development and frequently cause abundant precipitation.

As early as the 1800s, it was known that middle-latitude cyclones were the bearers of precipitation and severe weather. But it was not until the early part of the 1900s that a model was developed to explain how cyclones form. A group of Norwegian scientists formulated and published this model in 1918. The model was created primarily from near-surface observations.

SmartFigure 14.11
Idealized structure of a large, mature midlatitude cyclone **A.** This map view shows fronts, air masses, and surface winds. **B.** The three-dimensional view is a cross section through warm and cold fronts along a line from point A to point B. (http://goo.gl/jpfHhT)

 **Tutorial**

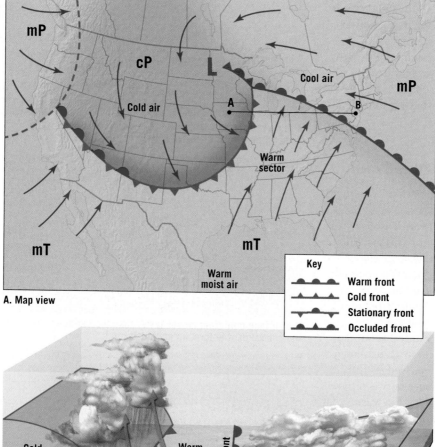

A. Map view

Key
Warm front
Cold front
Stationary front
Occluded front

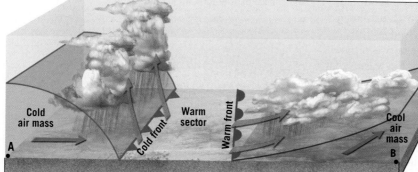

B. Three-dimensional view from points A to B

Years later, as data from the middle and upper troposphere and from satellite images became available, modifications were necessary. However, this model is still a useful working tool for interpreting the weather. If you keep this model in mind when you observe changes in the weather, the changes will no longer come as a surprise. You should begin to see some order in what once appeared to be disorder, and you might even occasionally "predict" the impending weather.

Idealized Weather of a Midlatitude Cyclone

The midlatitude cyclone model provides a useful tool for examining the weather patterns of the middle latitudes. **Figure 14.12** illustrates the distribution of clouds and thus the regions of possible precipitation associated with a mature system. Compare this drawing to the satellite image shown in **Figure 14.13**. It is easy to see why we often refer to the cloud pattern of a midlatitude cyclone as having a "comma" shape.

Guided by the westerlies aloft, cyclones generally move eastward across the United States, so we can expect the first signs of their arrival in the west. However, often in the region of the Mississippi River valley, cyclones begin a more northeasterly path and occasionally move directly northward. A midlatitude cyclone typically requires 2 to 4 days to move completely across a region. During that brief period, abrupt changes in atmospheric conditions may be experienced. This is particularly true in the winter and spring, when the largest temperature contrasts occur across the middle latitudes.

Using Figure 14.12 as a guide, we will now consider these weather producers and what we should expect from them as they move over an area. To facilitate our

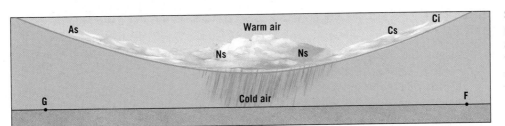

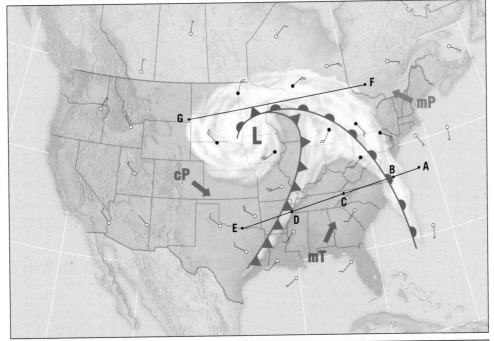

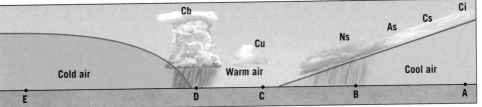

SmartFigure 14.12 Cloud patterns typically associated with a mature midlatitude cyclone The middle section is a map view. Note the cross-sectional lines (F–G, A–E). Above the map is a vertical cross section along line F–G. Below the map is a section along A–E. For cloud abbreviations, refer to Figures 14.7 and 14.8. (goo.gl/ZlJtS3)

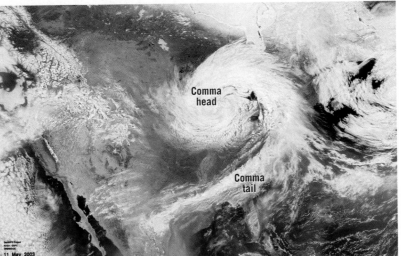

Figure 14.13 Satellite view of a mature midlatitude cyclone over the eastern half of the United States It is easy to see why we often refer to the cloud pattern of such a storm as having a "comma" shape. (NASA)

discussion, Figure 14.12 includes two profiles along lines A–E and F–G:

- Imagine the change in weather as you move along profile A–E. At point A, high cirrus clouds would be the first sign of the approaching cyclone. These high clouds can precede the surface front by 1000 kilometers (600 miles) or more, and they generally are accompanied by falling pressure. As the warm front advances, a lowering and thickening of the cloud deck is noticed.

- Usually within 12 to 24 hours after the first sighting of cirrus clouds, light precipitation begins (point B). As the front nears, the rate of precipitation increases, a rise in temperature is noticed, and winds begin to change from east or southeast to south or southwest.

- With the passage of the warm front, the area is under the influence of a maritime tropical air mass (point C). Generally, the region affected by this sector of the cyclone experiences warm to hot temperatures, southwesterly winds, fairly high humidity, and skies that may be clear or contain cumulus clouds.

- The relatively warm, humid weather of the warm sector passes quickly and is replaced by gusty winds and precipitation generated along the cold front. The approach of a rapidly advancing cold front is marked by a wall of dark clouds (point D). Severe weather accompanied by heavy precipitation, hail, and an occasional tornado is a definite possibility, especially during spring and summer. The passage of the cold front is easily detected by a wind shift; the southwest winds are replaced by winds from the west to northwest and by a pronounced drop in temperature. Also, the rising pressure hints at the subsiding cool, dry air behind the front.

- Once the front passes, skies clear as cooler air invades the region (point E). Often a day or two of almost cloudless deep blue skies occurs, unless another cyclone is edging into the region.

A very different set of weather conditions will prevail in the regions north of the storm's center, along profile F–G of Figure 14.12. In this part of the storm, temperatures remain cool. The first hints of the approaching low-pressure center are decreasing air pressure and increasingly overcast conditions that bring varying amounts of precipitation. This section of the cyclone most often generates snow during the winter months.

Once the formation of an occluded front begins, the character of the storm changes. Because occluded fronts tend to move more slowly than other fronts, the entire wishbone-shaped frontal structure of the storm rotates counterclockwise. As a result, the occluded front appears to "bend over backward." This effect adds to the misery of the region influenced by the occluded front because it lingers over the area longer than the other fronts.

The Role of Airflow Aloft

When the earliest studies of midlatitude cyclones were made, little was known about the nature of the airflow in the middle and upper troposphere. Since then, a close relationship has been established between surface disturbances and the flow aloft. Airflow aloft plays an important role in maintaining cyclonic and anticyclonic circulation. In fact, more often than not, these rotating surface wind systems are actually generated by upper-level flow.

Recall that the surface airflow around a cyclone (low-pressure system) is inward, a fact that leads to mass convergence, or coming together (**Figure 14.14**). The resulting accumulation of air must be accompanied by a corresponding increase in surface pressure. Consequently, we might expect a low-pressure system to "fill" rapidly and be eliminated. However, this does not occur. On the contrary, cyclones often exist for a week or longer. For this to happen, surface convergence must be offset by a mass outflow at some level

Figure 14.14 Flow aloft influences surface winds and pressure Idealized depiction showing the support that divergence and convergence aloft provide to cyclonic and anticyclonic circulation at the surface. Divergence aloft initiates upward air movement, reduced surface pressure, and cyclonic flow. On the other hand, convergence along the jet stream results in general subsidence of the air column, increased surface pressure, and anticyclonic surface winds.

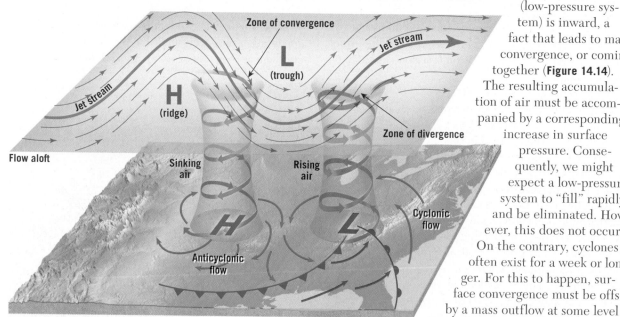

aloft (see Figure 14.14). As long as divergence (spreading out) aloft is equal to or greater than surface inflow, the low pressure and its accompanying convergence can be sustained.

Because cyclones are bearers of stormy weather, they have received far more attention than anticyclones. Nevertheless, a close relationship exists, which makes it difficult to separate any discussion of these two types of pressure systems. The surface air that feeds a cyclone, for example, generally originates as air flowing out of an anticyclone. Consequently, cyclones and anticyclones typically are found adjacent to each other. Like a cyclone, an anticyclone depends on the flow far above to maintain its circulation. Divergence at the surface is balanced by convergence aloft and general subsidence of the air column (see Figure 14.14).

(14.3) CONCEPT CHECKS

1. Briefly describe the weather associated with the passage of a mature midlatitude cyclone when the center of low pressure is about 200 to 300 kilometers (125 to 200 miles) north of your location.
2. If the midlatitude cyclone described in Question 1 took 3 days to pass your location, on which day would temperatures likely be warmest? On which day would they likely be coldest?
3. What winter weather might be expected with the passage of a mature midlatitude cyclone when the center of low pressure is located about 100 to 200 kilometers (60 to 125 miles) south of your location?
4. Briefly explain how flow aloft aids the formation of cyclones at the surface.

(14.4) Thunderstorms

List the basic requirements for thunderstorm formation and locate places on a map that exhibit frequent thunderstorm activity. Describe the stages in the development of a thunderstorm.

Thunderstorms are the first of three severe weather types we will examine in this chapter. Sections on tornadoes and hurricanes follow. All these phenomena can be related to low-pressure systems (cyclones).

Severe weather is more fascinating than everyday weather phenomena. A lightning display, such as the one in the chapter-opening photo, and the booming thunder generated by a severe thunderstorm can be a spectacular event that elicits both awe and fear. Of course, hurricanes and tornadoes also attract a great deal of much-deserved attention. A single tornado outbreak or hurricane can cause many deaths as well as billions of dollars in property damage. In a typical year, the United States experiences thousands of violent thunderstorms, hundreds of tornadoes, and several hurricanes.

What's in a Name?

Up to now we have examined midlatitude cyclones, which play an important role in causing day-to-day weather changes. Yet the use of the term *cyclone* is often confusing. To many people, the term implies only an intense storm, such as a tornado or a hurricane. When a hurricane unleashes its fury on India or Australia, for example, it is usually reported in the media as a *cyclone* (the term denoting a hurricane in that part of the world).

Similarly, tornadoes are referred to as *cyclones* in some places. This custom is particularly common in portions of the Great Plains of the United States. Recall that in *The Wizard of Oz*, Dorothy's house was carried from her Kansas farm to the land of Oz by a

cyclone. Indeed, the nickname for the athletic teams at Iowa State University is the *Cyclones* (**Figure 14.15**). Although hurricanes and tornadoes are, in fact, cyclones, the vast majority of cyclones are *not* hurricanes or tornadoes. The term *cyclone* simply refers to the circulation around any low-pressure center, no matter how large or intense it is.

Tornadoes and hurricanes are both smaller and more violent than midlatitude cyclones. Midlatitude cyclones can have a diameter of 1600 kilometers (1000 miles) or more. By contrast, hurricanes average only 600 kilometers (375 miles) across, and tornadoes, with a typical diameter of just 0.25 kilometer (0.16 mile), are much too small to show up on a weather map.

The thunderstorm, a much more familiar weather event, hardly needs to be distinguished from tornadoes, hurricanes, and midlatitude cyclones. Unlike the flow of air about these latter storms, the circulation associated with thunderstorms is characterized by strong up-and-down movements. Surface winds in the vicinity of a thunderstorm do not follow the inward spiral of a cyclone, but they are typically variable and gusty.

Thunderstorms can form "on their own," away from cyclonic storms, and they can also form in conjunction with cyclones. For instance, thunderstorms are frequently spawned along the cold front of a midlatitude cyclone, where on rare occasions a tornado may descend from the thunderstorm's cumulonimbus tower. Hurricanes also generate widespread thunderstorm activity. Thus, thunderstorms are related in some manner to all three types of cyclones mentioned here.

Figure 14.15 The term *cyclone* Sometimes the use of the term *cyclone* can be confusing. (Satellite image courtesy of NASA; logo courtesy of State University of Iowa)

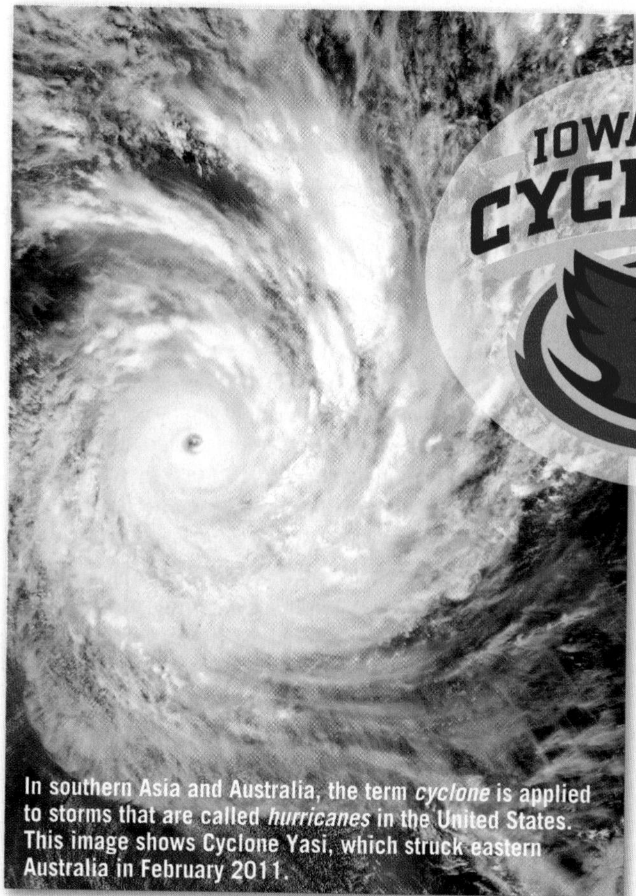

In southern Asia and Australia, the term *cyclone* is applied to storms that are called *hurricanes* in the United States. This image shows Cyclone Yasi, which struck eastern Australia in February 2011.

In parts of the Great Plains, *cyclone* is a synonym for *tornado.* The nickname for the athletic teams at Iowa State University is the *Cyclones.**

* Iowa State University is the only Division I school to use cyclones as its team name. The pictured logo was created to better communicate the school's image by combining the mascot, a cardinal bird named Cy, andthe Cyclone team name.

Did You Know?
Because lightning and thunder occur simultaneously, it is possible for you to estimate how far away a lightning stroke occurred. Lightning is seen instantaneously, but sound travels more slowly (about 1000 ft per sec) and reaches us later. If thunder is heard 5 seconds after lightning is seen, the lightning occurred about 5000 ft (1 mi) away.

Thunderstorm Occurrence

Almost everyone has observed various small-scale phenomena that result from the vertical movements of relatively warm, unstable air. Perhaps you have seen a dust devil over an open field on a hot day, whirling its dusty load to great heights. Or maybe you have noticed a bird glide effortlessly skyward on an invisible thermal of hot air. These examples illustrate the dynamic thermal instability that occurs during the development of a thunderstorm.

A **thunderstorm** is a storm that generates lightning and thunder. Thunderstorms frequently produce gusty winds, heavy rain, and hail. A thunderstorm may be produced by a single cumulonimbus cloud and influence only a small area, or it may be associated with clusters of cumulonimbus clouds covering a large area.

Thunderstorms form when warm, humid air rises in an unstable environment. Various mechanisms can trigger the upward air movement needed to create thunderstorm-producing cumulonimbus clouds. One mechanism, the unequal heating of Earth's surface, significantly contributes to the formation *of air-mass thunderstorms.* These storms are associated with the scattered puffy cumulonimbus clouds that commonly form *within* maritime tropical air masses and produce scattered thunderstorms on summer days. Such storms are usually short-lived and seldom produce strong winds or hail.

Another type of thunderstorm not only benefits from uneven surface heating but is associated with the lifting of warm air, as occurs along a front or a mountain slope. Moreover, diverging winds aloft frequently contribute to the formation of these storms because they tend to draw air from lower levels upward beneath them. Some of the thunderstorms of this type may produce high winds, damaging hail, flash floods, and tornadoes. Such storms are described as *severe.*

At any given time, an estimated 2000 thunderstorms are in progress on Earth. As we would expect, the greatest number occur in the tropics, where warmth, plentiful moisture and instability are always present. About 45,000 thunderstorms take place each day, and more than 16 million occur annually around the world. The lightning from these storms strikes Earth 100 times each second (**Figure 14.16A**). Annually, the United States experiences about 100,000 thunderstorms and millions of lightning strikes. A glance at **Figure 14.16B** shows that thunderstorms are most frequent in Florida and the eastern Gulf coast region, where such activity is recorded between 70 and 100 days each year. The region on the eastern side of the Rockies in Colorado and New Mexico is next, with thunderstorms occurring on 60 to 70 days each year. Most of the rest of the nation experiences thunderstorms on 30 to 50 days annually. The western margin of the United States has little thunderstorm activity. The same is true for the northern tier of states and for Canada, where warm, moist, unstable mT air seldom penetrates.

Stages of Thunderstorm Development

All thunderstorms require warm, moist air, which, when lifted, releases sufficient latent heat to provide the buoyancy necessary to maintain its upward flight. This instability and associated buoyancy are triggered by a number of different processes, yet most thunderstorms have a similar life history.

Because instability and buoyancy are enhanced by high surface temperatures, thunderstorms are most common in the afternoon and early evening (**Figure 14.17A**). However, surface heating alone is not sufficient for the growth of towering cumulonimbus clouds. A solitary cell of rising hot

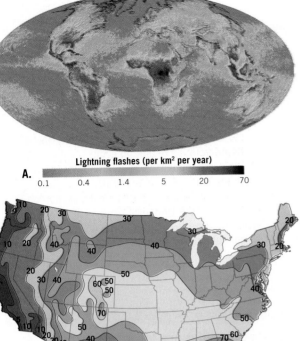

Lightning flashes (per km² per year)

0.1   0.4   1.4   5   20   70

A.

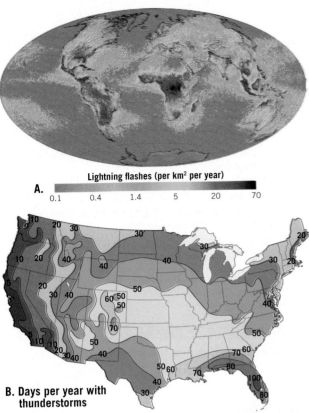

B. Days per year with thunderstorms

Figure 14.16 Occurrence of lightning and thunderstorms **A.** Data from space-based optical sensors show the worldwide distribution of lightning, with color variations indicating the average annual number of lightning flashes per square kilometer. **B.** This map shows the average number of days per year with thunderstorms in the conterminous United States. The humid subtropical climate that dominates the southeastern United States receives much of its precipitation in the form of thunderstorms. Most of the Southeast averages 50 or more days each year with thunderstorms.

air produced by surface heating could, at best, produce a small cumulus cloud, which would evaporate within 10 to 15 minutes.

A.

B.

Figure 14.17 Cumulus development **A.** Buoyant thermals produce fair-weather cumulus clouds that soon evaporate into the surrounding air, making the air more humid. As this process of cumulus development and evaporation continues, the air eventually becomes sufficiently humid so that newly forming clouds do not evaporate but continue to grow. **B.** This developing cumulonimbus cloud became a towering August thunderstorm over central Illinois. (Photos by E.J. Tarbuck)

The development of 12,000-meter (40,000-foot) (or, on rare occasions, 18,000-meter [60,000-foot]) cumulonimbus towers requires a continual supply of moist air (**Figure 14.17B**). Each new surge of warm air rises higher than the last, adding to the height of the cloud (**Figure 14.18**). These updrafts occasionally reach speeds greater

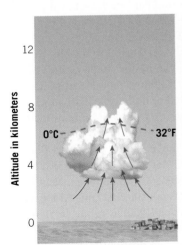

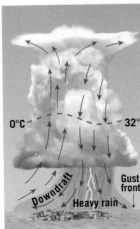

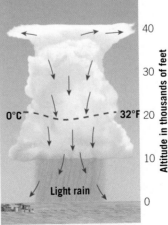

During the cumulus stage, strong updrafts provide moisture that condenses and builds the cloud.

The mature stage is marked by heavy precipitation. Updrafts exist side by side with downdrafts and continue to enlarge the cloud.

When updrafts disappear, precipitation becomes light and then stops. Without a supply of moisture from updrafts, the cloud evaporates.

SmartFigure 14.18 Thunderstorm development Once a cloud passes beyond the freezing level, the Bergeron process begins producing precipitation. Eventually, the accumulation of precipitation in the cloud is too great for the updraft to support. The falling precipitation causes drag on the air and initiates a downdraft. Once downdrafts dominate, rainfall diminishes, and the cloud starts to dissipate. (goo.gl/vOoRcH)

 Tutorial

than 100 kilometers (60 miles) per hour based on the size of hailstones they are capable of carrying upward. Usually within an hour, the amount and size of precipitation that has accumulated is too much for the updrafts to support, and consequently downdrafts develop in one part of the cloud, releasing heavy precipitation. This is the most active stage of the thunderstorm. Gusty winds, lightning, heavy precipitation, and sometimes hail are experienced.

Eventually the warm, moist air supplied by updrafts ceases as downdrafts dominate throughout the cloud. The cooling effect of falling precipitation, coupled with the influx of colder air aloft, marks the end of the thunderstorm activity. The life span of a typical cumulonimbus cell within a thunderstorm complex is only about an

hour, but as the storm moves, fresh supplies of warm, water-laden air generate new cells to replace those that are dissipating.

14.4 CONCEPT CHECKS

1. Briefly compare and contrast midlatitude cyclones, hurricanes, and tornadoes. How are thunderstorms related to each?
2. What are the basic requirements for the formation of a thunderstorm?
3. Where are thunderstorms most common on Earth? In the United States?
4. Summarize the stages in the development of a thunderstorm.

14.5 Tornadoes

Summarize the atmospheric conditions and locations that are favorable to the formation of tornadoes. Discuss tornado destruction and tornado forecasting.

Figure 14.19 Condensation and debris make tornadoes visible A tornado is a violently rotating column of air in contact with the ground. The air column is visible when it contains condensation or when it contains dust and debris. Often the appearance is a result of both. When the column of air is aloft and does not produce damage, the visible portion is properly called a *funnel cloud*. (Photo by Jason Persoff Stormdoctor/Cultura RM Exclusive/Getty Images)

Tornadoes are local storms of short duration that rank high among nature's most destructive forces (**Figure 14.19**). Their sporadic occurrence and violent winds cause many deaths each year. The nearly total destruction in some stricken areas has led many to liken their passage to bombing raids during war.

Such was the case during a very stormy period in late May 2013 in central Oklahoma. On May 20, an EF-5 tornado, the most severe category (see Table 14.1), struck the city of Moore. Peak winds were estimated at 340 kilometers (210 miles) per hour. The storm took 25 lives and injured more than 350 people. Entire neighborhoods were destroyed; at least 13,000 structures were demolished or damaged. Estimated damages exceeded $2 billion (**Figure 14.20**). The tornado was one of many that occurred across the Great Plains over a 2-day span, including five that struck central Oklahoma on May 19. This was not the first time Moore had experienced such a storm. Thirteen years earlier, in May 1999, an even stronger and deadlier tornado hit this community.

Tornadoes, sometimes called *twisters* or *cyclones*, are violent

SmartFigure 14.20 Tornado destruction at Moore, Oklahoma On May 20, 2013, central Oklahoma was devastated by an EF-5 tornado, the most severe category. At its peak, the tornado was 2.1 kilometers (1.3 miles) wide and had winds of 340 kilometers (210 miles) per hour. (Photo by Jewel Samad/AFP/Getty Images) (goo.gl/1qe1EM)

 Video

Did You Know?
According to the National Weather Service, during the 15-year span 2000–2014, tornadoes took an average of 88 lives each year. The number of deaths each year varied greatly, from a low of 22 in 2009 to a high of 553 in 2011.

windstorms that take the form of a rotating column of air, or *vortex*. Pressures within some tornadoes have been estimated to be as much as 10 percent lower than pressures immediately outside the storm. Drawn by the much lower pressure in the center of the vortex, air near the ground rushes into the tornado from all directions. As the air streams inward, it is spiraled upward around the core until it eventually merges with the airflow of the parent thunderstorm deep in the cumulonimbus tower. Because of the tremendous pressure gradient associated with a strong tornado, maximum winds can sometimes approach 480 kilometers (300 miles) per hour.

A tornado may consist of a single vortex, but within many stronger tornadoes are smaller whirls called *suction vortices* that rotate within the main vortex (**Figure 14.21**). Suction vortices have diameters of only about 10 meters (33 feet) and rotate very rapidly. This structure accounts for occasional observations of virtually total destruction of one building while another one, just 10 meters (33 feet) away, suffers little damage.

Tornado Occurrence and Development

Tornadoes form in association with severe thunderstorms that produce high winds, heavy (sometimes torrential) rainfall, and often damaging hail. Fortunately, fewer than 1 percent of all thunderstorms produce tornadoes. Nevertheless, a much higher number of thunderstorms must be monitored as potential tornado producers. A tornado is the product of the interaction between strong updrafts in a thunderstorm and the winds in the troposphere.

Tornadoes can form in any situation that produces severe weather, including cold fronts and tropical cyclones (hurricanes). The most intense tornadoes are usually those that form in association with huge thunderstorms called *supercells*. An important precondition linked to tornado formation in severe thunderstorms is the development of a *mesocyclone*—a vertical cylinder of rotating air, typically about 3 to 10 kilometers (2 to

SmartFigure 14.21
Multiple-vortex tornado
Some tornadoes have multiple suction vortices. These small and very intense vortices are roughly 10 meters (30 feet) across and move in a counterclockwise path around the tornado center. Because of this multiple-vortex structure, one building might be heavily damaged and another one, just 10 meters away, might suffer little damage. (goo.gl/8zfeuJ)

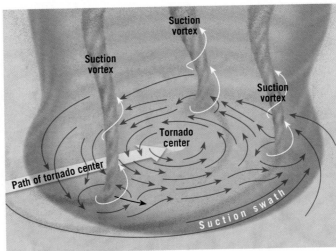

Suction vortex

Suction vortex

Suction vortex

Tornado center

Path of tornado center

Suction swath

 Animation

SmartFigure 14.22 The formation of a mesocyclone often precedes tornado formation **A.** Winds are stronger aloft than at the surface (called speed wind shear), producing a rolling motion about a horizontal axis. **B.** Strong thunderstorm updrafts tilt the horizontally rotating air to a nearly vertical alignment. **C.** The mesocyclone, a vertical cylinder of rotating air, is established. **D.** If a tornado develops, it will descend from a slowly rotating wall cloud in the lower portion of the mesocyclone. (Photo by Gene Rhoden/Weatherpix/Getty Images) (goo.gl/DpFk9S)

Tutorial

6 miles) across, that develops in the updraft of a severe thunderstorm (**Figure 14.22**). The formation of this large vortex often precedes tornado formation by 30 minutes or so.

The formation of a mesocyclone does not necessarily mean that tornado formation will follow. Only about half of all mesocyclones produce tornadoes. Forecasters cannot determine in advance which mesocyclones will spawn tornadoes.

General Atmospheric Conditions

Severe thunderstorms—and, hence, tornadoes—are most often spawned along the cold front of a midlatitude cyclone or in association with a supercell thunderstorm such as the one pictured in Figure 14.22D. Throughout spring, air masses associated with middle-latitude cyclones are most likely to have greatly contrasting conditions. Continental polar air from Canada may still be very cold and dry, whereas maritime tropical air from the Gulf of Mexico is warm, humid, and unstable. The greater the contrast when these air masses meet, the more intense the storm. The two contrasting air masses are most likely to meet in the central United States because there is no significant natural barrier separating the center of the country from the Arctic or the Gulf of Mexico. Consequently, this region generates more tornadoes than any other area of the country or, in fact, the world. The map in **Figure 14.23**, which depicts tornado incidence in the United States for a 27-year period, readily substantiates this fact.

Tornado Climatology

An average of 1297 tornadoes were reported annually in the United States between 2000 and 2014. Still, the actual number that occurs from one year to the next varies greatly. During this

Figure 14.23 Tornado occurrence The map shows average annual tornado incidence per 26,000 square kilometers (10,000 square miles) for a 27-year period. The graph shows average number of tornadoes and tornado days each month in the United States for the same period.

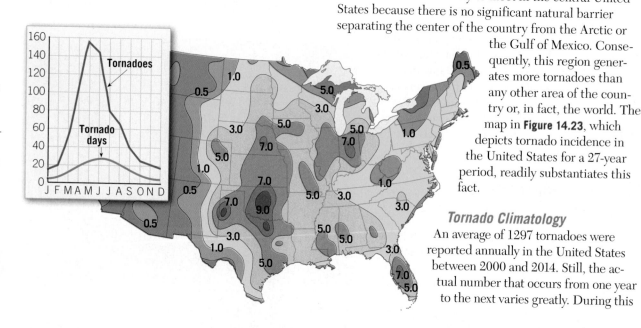

15-year span, for example, yearly totals ranged from a low of 938 in 2002 to a high of 1894 in 2011.

Tornadoes occur during every month of the year. April through June is the period of greatest tornado frequency in the United States; the number is lowest during December and January (see Figure 14.23, graph inset). Of the 40,522 confirmed tornadoes reported over the contiguous 48 states during the 50-year period from 1950 through 1999, an average of almost 6 per day occurred during May. At the other extreme, a tornado was reported only about every other day in December and January.

Profile of a Tornado

An average tornado has a diameter of between 150 and 600 meters (500 and 2000 feet), travels across the landscape at approximately 45 kilometers (30 miles) per hour, and cuts a path about 10 kilometers (6 miles) long.* Because tornadoes usually occur slightly ahead of a cold front, in the zone of southwest winds, most move toward the northeast.

Of the hundreds of tornadoes reported in the United States annually, more than half are comparatively weak and short-lived. Most of these small tornadoes have lifetimes of 3 minutes or less and paths that seldom exceed 1 kilometer (0.6 mile) in length and 100 meters (330 feet) in width. Typical wind speeds are on the order of 150 kilometers (90 miles) per hour or less. On the other end of the tornado spectrum are the infrequent and often long-lived violent tornadoes. Although

---

* The 10-kilometer (6-mile) figure applies to documented tornadoes. Because many small tornadoes may go undocumented, the real average path of all tornadoes is unknown but is shorter than 10 kilometers.

large tornadoes constitute only a small percentage of the total reported, their effects are often devastating. Such tornadoes may exist for periods in excess of 3 hours and produce an essentially continuous damage path more than 150 kilometers (90 miles) long and perhaps a kilometer or more wide. Maximum winds may approach 500 kilometers (310 miles) per hour.

Tornado Destruction and Loss of Life

The potential for tornado destruction depends largely on the strength of the winds generated by the storm. Because tornadoes generate the strongest winds in nature, they have accomplished many seemingly impossible tasks, such as the ones shown in **Figure 14.24**. Although it may seem impossible for winds to cause some of the extensive damage attributed to tornadoes, tests in engineering facilities have repeatedly demonstrated that winds in excess of 320 kilometers (200 miles) per hour are capable of incredible feats.

Most tornado losses are associated with the rare storms that strike urban areas or devastate entire small communities. The amount of destruction caused by such storms depends to a significant degree (but not completely) on the strength of the winds. A wide spectrum of tornado strengths, sizes, and lifetimes are observed. The commonly used guide to tornado intensity is the **Enhanced Fujita intensity scale**, or **EF scale** for short (**Table 14.1**). Because tornado winds are rarely measured directly, a rating on the EF scale is determined by assessing the worst damage produced by a storm. Although widely used, the EF scale is not perfect. Estimating tornado intensity based on damage alone does not take into

Figure 14.24 Tornado winds: The strongest in nature **A.** The force of the wind during a tornado near Wichita, Kansas, in April 1991 was enough to drive this piece of metal into a utility pole. (Photo by NOAA) **B.** The remains of a truck wrapped around a tree in Bridge Creek, Oklahoma, on May 4, 1999, following a major tornado outbreak. (LM Otero/ AP Images)

A.

B.

TABLE 14.1 Enhanced Fujita Intensity Scale*

Scale	Wind Speed		Damage
	Km/Hr	Mi/Hr	
EF-0	105–137	65–85	*Light.* Some damage to siding and shingles.
EF-1	138–177	86–110	*Moderate.* Considerable roof damage. Winds can uproot trees and overturn single-wide mobile homes. Flagpoles bend.
EF-2	178–217	111–135	*Considerable.* Most single-wide mobile homes destroyed. Permanent homes can shift off foundations. Flagpoles collapse. Softwood trees debarked.
EF-3	218–265	136–165	*Severe.* Hardwood trees debarked. All but small portions of houses destroyed.
EF-4	266–322	166–200	*Devastating.* Complete destruction of well-built residences, large sections of school buildings.
EF-5	>322	>200	*Incredible.* Significant structural deformation of mid- and high-rise buildings.

* The original Fujita scale was developed by T. Theodore Fujita in 1971 and put into use in 1973. The Enhanced Fujita intensity scale is a revision that was put into use in February 2007. Wind speeds are estimates (not measurements) based on damage, and represent 3-second gusts at the point of damage.

Did You Know?

One tornado easily ranks as the single most dangerous and destructive. Known as the Tri-State Tornado, it occurred on March 18, 1925. Starting in southeastern Missouri, the tornado remained on the ground for 352 km (219 mi), traveling across southern Illinois and finally ending in Indiana. The losses included 695 dead and 2027 injured. Property losses were also great, with several small towns almost totally destroyed.

account the structural integrity of the objects hit by a tornado. A well-constructed building can withstand very high winds, whereas a poorly built structure can suffer devastating damage from the same or weaker winds.

Although the greatest part of tornado damage is caused by violent winds, most tornado injuries and deaths result from flying debris. The proportion of tornadoes that result in the loss of life is small. In most years, fewer than 2 percent of all reported tornadoes in the United States are "killers." Although the percentage of tornadoes resulting in death is small, each tornado is potentially lethal. When tornado fatalities and storm intensities are compared, the results are quite interesting: The majority (63 percent) of tornadoes are weak (EF-0 and EF-1), and the number of storms decreases as tornado intensity increases. The distribution of tornado fatalities, however, is just the opposite. Although only 2 percent of tornadoes are classified as violent (EF-4 and EF-5), they account for nearly 70 percent of the deaths.

Tornado Forecasting

Because severe thunderstorms and tornadoes are small and relatively short-lived phenomena, they are among the most difficult weather features to forecast precisely. Nevertheless, the prediction, detection, and monitoring of such storms are some of the most important services provided by professional meteorologists. Both the timely issuance and dissemination of watches and warnings are critical to the protection of life and property.

The Storm Prediction Center (SPC) located in Norman, Oklahoma, is part of the National Weather Service (NWS) and the National Centers for Environmental Prediction (NCEP). The mission of the SPC is to provide timely and accurate forecasts and watches for severe thunderstorms and tornadoes.

Severe thunderstorm outlooks are issued several times daily. *Day 1* outlooks identify the areas that are likely to be affected by severe thunderstorms during the next 6 to 30 hours, and *day 2* outlooks extend the forecast through the following day. Both outlooks describe the type, coverage, and intensity of the severe weather expected. Many local NWS field offices also issue severe weather outlooks that provide more local descriptions of the severe weather potential for the next 12 to 24 hours.

Tornado Watches and Warnings

Tornado watches alert the public to the possibility of tornadoes over a specified area for a particular time interval. Watches serve to fine-tune forecast areas already identified in severe weather outlooks. A typical watch covers an area of about 65,000 square kilometers (25,000 square miles) for a 4- to 6-hour period. A tornado watch is an important part of the tornado alert system because it sets in motion the procedures necessary to deal adequately with detection, tracking, warning, and response. Watches are generally reserved for organized severe weather events where the tornado threat will affect at least 26,000 square kilometers (10,000 square miles) and/or persist for at least 3 hours. Watches typically are not issued when the threat is thought to be isolated and/or short-lived.

Whereas a tornado watch is designed to alert people to the possibility of tornadoes, a **tornado warning** is issued by local offices of the National Weather Service when a tornado has actually been sighted in an area or is indicated by weather radar. It warns of a high probability of imminent danger. Warnings are issued for much smaller areas than are watches, usually covering portions of a county or counties. In addition, they are in effect for much shorter periods, typically 30 to 60 minutes. Because a tornado warning may be based on an actual sighting, warnings are occasionally issued after a tornado has already developed. However, most warnings are issued prior to tornado formation, sometimes by several tens of minutes, based on Doppler radar data and/or spotter reports of funnel clouds.

If the direction and the approximate speed of a storm are known, an estimate of its most probable path can be made. Because tornadoes often move erratically, the warning area is fan-shaped downwind from the point where the tornado has been spotted. Improved forecasts and advances in technology have contributed to a significant decline in tornado deaths over the past 50 years.

A.   Alaska   Hawaii   Puerto Rico   Guam   B.

SmartFigure 14.25
Doppler radar **A.** Doppler radar sites in the United States. If you go to http://radar.weather.gov, you will see a similar map. You can click on any site to see the current National Weather Service Doppler radar display. **B.** Doppler on Wheels is a portable unit that researchers use in field studies of severe weather events. (Photo by University Corporation for Atmospheric Research) (goo.gl/cGCnXY)

Video

Doppler Radar

Many of the difficulties that once limited the accuracy of tornado warnings have been reduced or eliminated by an advancement in radar technology called **Doppler radar** (**Figure 14.25**). Doppler radar not only performs the same tasks as conventional radar but also has the ability to detect motion directly. Doppler radar can detect the initial formation and subsequent development of a *mesocyclone*, the intense rotating wind system in the lower part of a thunderstorm that frequently precedes tornado development. Almost all mesocyclones produce damaging hail, severe winds, or tornadoes. Those that produce tornadoes (about 50 percent) can sometimes be distinguished by their stronger wind speeds and their sharper gradients of wind speeds.

It should also be pointed out that not all tornado-bearing storms have clear-cut radar signatures and that other storms can give false signatures. Detection, therefore, is sometimes a subjective process, and a given display could be interpreted in several ways. Consequently, trained observers continue to be an important part of the warning system.

The benefits of Doppler radar are many. As a research tool, it not only provides data on the formation of tornadoes but also helps meteorologists gain new insights into thunderstorm development, the structure and dynamics of hurricanes, and air-turbulence hazards that plague aircraft. As a practical tool for tornado detection, Doppler radar has significantly improved our ability to track thunderstorms and issue warnings.

14.5 CONCEPT CHECKS

1. Why do tornadoes have such high wind speeds?
2. What general atmospheric conditions are most conducive to the formation of tornadoes?
3. During what months is tornado activity most pronounced in the United States?
4. Name the scale commonly used to rate tornado intensity. How is a rating on this scale determined?
5. Distinguish between a tornado watch and a tornado warning.

14.6 Hurricanes

Identify areas of hurricane formation on a world map and discuss the conditions that promote hurricane formation. List the three broad categories of hurricane destruction.

Most of us view the weather in the tropics with favor. Places such as the islands of the Caribbean are known for their lack of significant day-to-day variations. Warm breezes, steady temperatures, and rains that come as heavy but brief tropical showers are often the rule. It is ironic that these relatively tranquil regions sometimes produce the most violent storms on Earth.

Hurricanes are intense centers of low pressure that form over tropical oceans and are characterized by intense convective (thunderstorm) activity and strong cyclonic circulation (**Figure 14.26**). Sustained winds must equal or exceed 119 kilometers (74 miles) per hour. Unlike midlatitude cyclones, hurricanes lack contrasting air masses and fronts. Rather, the source of energy that produces and maintains hurricane-force winds is the huge quantity of latent heat liberated during the formation of the storm's cumulonimbus towers.

The vast majority of hurricane-related deaths and damage are caused by relatively infrequent, yet powerful, storms. The storm that pounded an unsuspecting Galveston, Texas, in 1900 was not just the deadliest U.S. hurricane ever but the deadliest natural disaster of

Figure 14.26 Hurricane Patricia This satellite image from October 23, 2015, shows the hurricane over the eastern Pacific Ocean near the west coast of Mexico. With sustained winds of 352 kilometers (200 miles) per hour, it was the strongest hurricane ever recorded in the Western Hemisphere. Fortunately, the storm quickly weakened after making landfall. The counterclockwise spiral of the clouds indicates that it is a Northern Hemisphere storm. In the Southern Hemisphere, the spiral is clockwise. (NASA)

The eye wall surrounds the eye and is the most intense part of the storm

Well-developed eye

N

Profile of a Hurricane

Most hurricanes form between the latitudes of 5° and 20° over all the tropical oceans except the South Atlantic and the eastern South Pacific (**Figure 14.27**). The North Pacific has the greatest number of storms, averaging 20 each year. Fortunately for those living in the coastal regions of the southern and eastern United States, fewer than 5 hurricanes, on average, develop annually in the warm sector of the North Atlantic.

These intense tropical storms are known in various parts of the world by different names. In the western Pacific, they are called *typhoons*, and in the Indian Ocean, including the Bay of Bengal and Arabian Sea, they are simply called *cyclones*. In the following discussion, these storms will be referred to as hurricanes. The term *hurricane* is derived from the name Huracan, a Carib god of evil.

Although many tropical disturbances develop each year, only a few reach hurricane status. By international agreement, a hurricane has wind speeds in excess of 119 kilometers (74 miles) per hour and a rotary circulation. Mature hurricanes average 600 kilometers (375 miles) across, although they can range in diameter from 100 kilometers (60 miles) up to about 1500 kilometers (930 miles). From the outer edge to the center, the barometric pressure has on occasion dropped 60 millibars, from 1010 millibars to 950 millibars. The lowest pressures ever recorded in the Western Hemisphere are associated with these storms.

A steep pressure gradient generates the rapid, inward-spiraling winds of a hurricane (**Figure 14.28**). As the air rushes toward the center of the storm, its velocity increases. This occurs for the same reason that skaters with their arms extended spin faster as they pull their arms in close to their bodies.

As the inward rush of warm, moist surface air approaches the core of the storm, it turns upward and ascends

any kind to affect the United States. More recently, an extremely deadly and costly storm occurred in August 2005, when Hurricane Katrina devastated the Gulf Coast of Louisiana, Mississippi, and Alabama and took an estimated 1800 lives. Although hundreds of thousands fled before the storm made landfall, thousands of others were caught by the storm. In addition to the human suffering and tragic loss of life that were left in the wake of Hurricane Katrina, the financial losses caused by the storm are practically incalculable.

SmartFigure 14.27 The where and when of hurricanes **A.** The world map shows the regions where most hurricanes form as well as their principal months of occurrence and the tracks they most commonly follow. Hurricanes do not develop within about 5° of the equator because the Coriolis effect (a force related to Earth's rotation that gives storms their "spin") there is too weak. Because warm ocean-surface temperatures are necessary for hurricane formation, hurricanes seldom form poleward of 20° latitude or over the cool waters of the south Atlantic and the eastern south Pacific. **B.** The graph shows the frequency of tropical storms and hurricanes from May 1 through December 31 in the Atlantic basin. It shows the number of storms to be expected over a span of 100 years. The period from late August through October is clearly the most active. (Data from National Hurricane Center/NOAA) (goo.gl/oijOe2)

Video

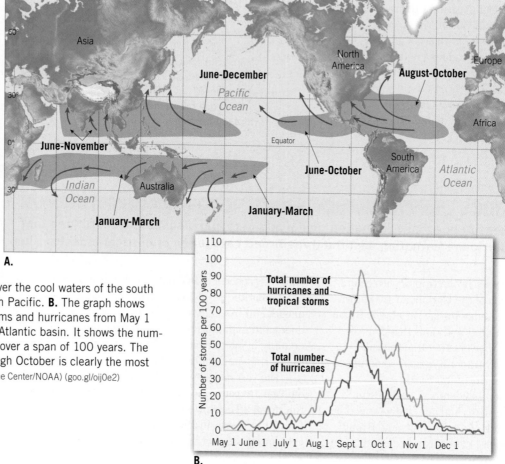

A.

B.

in a ring of cumulonimbus towers. This doughnut-shaped wall of intense convective activity surrounding the center of the storm is called the **eye wall**. It is here that the greatest wind speeds and heaviest rainfall occur. Surrounding the eye wall are curved bands of clouds that trail away in a spiral fashion. Near the top of the hurricane, the airflow is outward, carrying the rising air away from the storm center, thereby providing room for more inward flow at the surface.

At the very center of the storm is the **eye** of the hurricane. This well-known feature is a zone about 20 kilometers (12.5 miles) in diameter where precipitation ceases and winds subside. It offers a brief but deceptive break from the extreme weather in the enormous curving wall clouds that surround it. The air within the eye gradually descends and heats by compression, making it the warmest part of the storm. Although many people believe that the eye is characterized by clear blue skies, this is usually not the case because the subsidence in the eye is seldom strong enough to produce cloudless conditions. Although the sky appears much brighter in this region, scattered clouds at various levels are common.

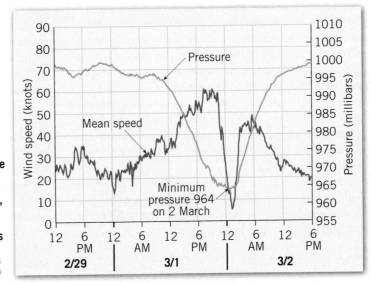

Outflow of air at the top of the hurricane is important because it prevents the convergent flow at lower levels from "filling in" the storm.

Sinking air in the eye warms by compression.

Eye

Eye wall, the zone where winds and rain are most intense.

Tropical moisture spiraling inward creates rain bands that pinwheel around the storm center.

Cross section of a hurricane. Note that the vertical dimension is greatly exaggerated. (After NOAA)

SmartFigure 14.28
Conditions inside a hurricane (Data from World Meteorological Organization) (http://goo.gl/yT2d1q)

Video

Measurements of surface pressure and wind speed during the passage of Cyclone Monty at Mardie Station, Western Australia, between February 29 and March 2, 2004. (Hurricanes are called "cyclones" in this part of the world.)

Pressure

Mean speed

Minimum pressure 964 on 2 March

Hurricane Formation and Decay

A hurricane is a heat engine that is fueled by the latent heat liberated when huge quantities of water vapor condense. The amount of energy produced by a typical hurricane in just a single day is truly immense. The release of latent heat warms the air and provides buoyancy for its upward flight. The result is to reduce the pressure near the surface, which encourages a more rapid inward flow of air. To get this engine started, a large quantity of warm, moisture-laden air is required, and a continual supply is needed to keep it going.

Hurricane Formation

Hurricanes develop most often in the late summer, when ocean waters have reached temperatures of 27°C (80°F) or higher and thus are able to provide the necessary heat and moisture to the air (**Figure 14.29**). This ocean-water temperature requirement accounts for the fact that

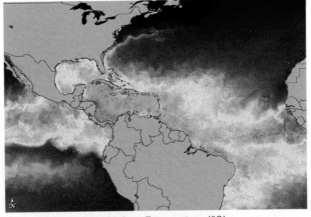

Sea Surface Temperature (°C)

−2   16.5   27.8   35

Figure 14.29 Sea-surface temperatures Among the necessary ingredients for a hurricane is warm ocean temperatures above 27°C (80°F). This color-coded satellite image from June 1, 2010, shows sea-surface temperatures at the beginning of hurricane season. (NASA)

hurricanes do not form over the relatively cool waters of the South Atlantic and the eastern South Pacific. For the same reason, few hurricanes form poleward of 20° latitude. Although water temperatures are sufficiently high, hurricanes do not form within 5° of the equator because the Coriolis effect is too weak to initiate the necessary rotary motion.

Many tropical storms begin as disorganized arrays of clouds and thunderstorms that develop weak pressure gradients but exhibit little or no rotation. Such areas of low-level convergence and lifting are called *tropical disturbances*. Most of the time, these zones of convective activity die out. However, tropical disturbances occasionally grow larger and develop a strong cyclonic rotation.

What happens on occasions when conditions favor hurricane development? As latent heat is released from the clusters of thunderstorms that make up the tropical disturbance, areas within the disturbance get warmer. As a result, air density lowers and surface pressure drops, creating a region of weak low pressure and cyclonic circulation. As pressure drops at the storm center, the pressure gradient steepens. If you were watching an animated weather map of the storm, you would see the isobars get closer together. In response, surface wind speeds increase and bring additional supplies of moisture to nurture storm growth. The water vapor condenses, releasing latent heat, and the heated air rises. Adiabatic cooling of rising air triggers more condensation and the release of more latent heat, which causes a further increase in buoyancy. And so it goes.

Meanwhile, at the top of the storm, air is diverging. Without this outward flow up top, the inflow at lower levels would soon raise surface pressures (that is, fill in the low) and thwart storm development.

Other Tropical Storms

Many tropical disturbances occur each year, but only a few develop into full-fledged hurricanes. By international agreement, lesser tropical cyclones are placed in different categories, based on wind strength. When a cyclone's strongest winds do not exceed 61 kilometers (38 miles) per hour, it is called a **tropical depression**. When winds are between 61 and 119 kilometers (38 and 74 miles) per hour, the cyclone is termed a **tropical storm**. It is during this phase that a name is given (Andrew, Katrina, Sandy, etc.). If the tropical storm becomes a hurricane, the name

remains the same. Each year, between 80 and 100 tropical storms develop around the world. Of these, usually half or more eventually become hurricanes.

Hurricane Decay

Hurricanes diminish in intensity whenever they (1) move over ocean waters that cannot supply warm, moist tropical air; (2) move onto land; or (3) reach a location where the large-scale flow aloft is unfavorable. When a hurricane moves onto land, it loses its punch rapidly. The most important reason for this rapid demise is the fact that the storm's source of warm, moist air is cut off. When an adequate supply of water vapor does not exist, condensation and the release of latent heat must diminish. In addition, friction from the increased roughness of the land surface rapidly slows surface wind speeds. This factor causes the winds to move more directly into the center of the low, thus helping to eliminate the large pressure differences.

Hurricane Destruction

A location only a few hundred kilometers from a hurricane—just 1 day's striking distance away—may experience clear skies and virtually no wind. Prior to the age of weather satellites, this situation made the task of warning people of impending storms very difficult.

The amount of damage caused by a hurricane depends on several factors, including the size and population density of the area affected and the shape of the ocean bottom near the shore. The most significant factor, of course, is the strength of the storm itself. By studying past storms, a scale has been established to rank the relative intensities of hurricanes. As **Table 14.2** indicates, a *category 5* storm is the worst possible, whereas a *category 1* hurricane is least severe.

During hurricane season, it is common to hear scientists and reporters use the numbers from the **Saffir–Simpson hurricane scale**. When Hurricane Katrina made landfall, sustained winds were 225 kilometers (140 miles) per hour, making it a strong category 4 storm. Storms that fall into category 5 are rare. Damage caused by hurricanes can be divided into three categories: (1) storm surge, (2) wind damage, and (3) heavy rains and inland flooding.

Storm Surge

The most devastating damage in the coastal zone is usually caused by storm surge (**Figure 14.30**). It not only accounts for a large share of coastal property losses but also is responsible for a high percentage of all hurricane-caused deaths. A **storm surge** is a dome of water 65 to 80 kilometers (40 to 50 miles) wide that sweeps across the coast near the point where the eye makes landfall. If all wave

TABLE 14.2 Saffir–Simpson Hurricane Scale				
Scale Number (category)	Central Pressure (millibars)	Winds (km/hr)	Storm Surge (meters)	Damage
1	≥980	119–153	1.2–1.5	Minimal
2	965–979	154–177	1.6–2.4	Moderate
3	945–964	178–209	2.5–3.6	Extensive
4	920–944	210–250	3.7–5.4	Extreme
5	<920	>250	>5.4	Catastrophic

Figure 14.30 Storm surge destruction This is Crystal Beach, Texas, on September 16, 2008, 3 days after Hurricane Ike came ashore. At landfall the storm had sustained winds of 165 kilometers (105 miles) per hour. The extraordinary storm surge caused most of the damage shown here. (Photo by Smiley N. Pool/Rapport Press/Newscom)

activity were smoothed out, the storm surge would be the height of the water above normal tide level. In addition, tremendous wave activity is superimposed on the surge. The worst surges occur in places like the Gulf of Mexico, where the continental shelf is very shallow and gently sloping. In addition, local features such as bays and rivers can cause the surge height to double and increase in speed.

As a hurricane advances toward the coast in the Northern Hemisphere, storm surge is always most intense on the right side of the eye (viewed from the ocean), where winds are blowing *toward* the shore. In addition, on this side of the storm, the forward movement of the hurricane contributes to the storm surge. In **Figure 14.31**, assume that a hurricane with peak winds of 175 kilometers (109 miles) per hour is moving toward the shore at 50 kilometers (31 miles) per hour. In this case, the net wind speed on the right side of the advancing storm is 225 kilometers (140 miles) per hour. On the left side, the hurricane's winds are blowing opposite the direction of storm movement, so the net winds are *away* from the coast at 125 kilometers (78 miles) per hour. Along the shore facing the left side of the oncoming hurricane, the water level may actually decrease as the storm makes landfall.

Wind Damage

Destruction caused by wind is perhaps the most obvious of the classes of hurricane damage. Debris such as signs, roofing materials, and small items left outside become dangerous flying missiles in hurricanes. For some structures, the force of the wind is sufficient to cause total ruin. Mobile homes are particularly vulnerable. High-rise buildings are also susceptible to hurricane-force winds. Upper floors are most vulnerable because wind speeds usually increase with height. Recent research suggests that people should stay below the tenth floor but

remain above any floors at risk for flooding. In regions with good building codes, wind damage is usually not as catastrophic as storm-surge damage. However, hurricane-force winds affect a much larger area than storm surge and can cause huge economic losses. For example, in 1992 it was largely the winds associated with Hurricane Andrew that produced more than $25 billion of damage in southern Florida and Louisiana.

Hurricanes sometimes produce tornadoes that contribute to the storm's destructive power. Studies have shown that more than half of the hurricanes that make landfall produce at least one tornado. In 2004 the number of tornadoes associated with tropical storms and hurricanes was extraordinary. Tropical Storm Bonnie and five landfalling hurricanes—Charley, Frances, Gaston, Ivan, and Jeanne—produced nearly 300 tornadoes that affected the southeastern and mid-Atlantic states.

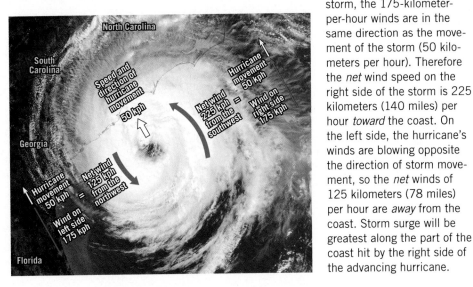

Figure 14.31 An approaching hurricane This hypothetical Northern Hemisphere hurricane, with peak winds of 175 kilometers (109 miles) per hour, is moving toward the coast at 50 kilometers (31 miles) per hour. On the right side of the advancing storm, the 175-kilometer-per-hour winds are in the same direction as the movement of the storm (50 kilometers per hour). Therefore the *net* wind speed on the right side of the storm is 225 kilometers (140 miles) per hour *toward* the coast. On the left side, the hurricane's winds are blowing opposite the direction of storm movement, so the *net* winds of 125 kilometers (78 miles) per hour are *away* from the coast. Storm surge will be greatest along the part of the coast hit by the right side of the advancing hurricane.

SmartFigure 14.32 Five-day track forecast for Tropical Storm Gonzalo, issued at 5 A.M. EDT, Monday, October 13, 2014 When a hurricane track forecast is issued by the National Hurricane Center, it is termed a *forecast cone*. The cone represents the probable track of the center of the storm and is formed by enclosing the area swept out by a set of circles along the forecast track (at 12 hours, 24 hours, 36 hours, etc.). The circles get larger as they extend further into the future. The entire track of an Atlantic tropical cyclone can be expected to remain entirely within the cone roughly 60 to 70 percent of the time. (National Weather Service/National Hurricane Center) (http://goo.gl/Dn8nOI)

Video

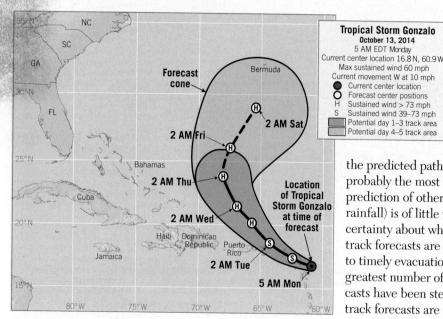

Heavy Rains and Inland Flooding

The torrential rains that accompany most hurricanes pose a third significant threat: flooding. Whereas the effects of storm surge and strong winds are concentrated in coastal areas, heavy rains may affect places hundreds of kilometers from the coast for up to several days after the storm has lost its hurricane-force winds.

In September 1999, Hurricane Floyd brought flooding rains, high winds, and rough seas to a large portion of the Atlantic seaboard. More than 2.5 million people evacuated their homes from Florida north to the Carolinas and beyond. It was the largest peacetime evacuation in U.S. history up to that time. Torrential rains falling on already saturated ground created devastating inland flooding. Altogether Floyd dumped more than 48 centimeters (19 inches) of rain on Wilmington, North Carolina, 33.98 centimeters (13.38 inches) in a single 24-hour span.

Tracking Hurricanes

Today we have the benefit of numerous observational tools for tracking tropical storms and hurricanes. Using input from satellites, aircraft reconnaissance, coastal radar, and remote data buoys in conjunction with sophisticated computer models, meteorologists monitor and forecast storm movements and intensity. The goal is to issue timely watches and warnings.

An important part of this process is the *track forecast*—the predicted path of the storm. The track forecast is probably the most basic information because accurate prediction of other storm characteristics (winds and rainfall) is of little value if there is significant uncertainty about where the storm is going. Accurate track forecasts are important because they can lead to timely evacuations from the surge zone, where the greatest number of deaths usually occur. Track forecasts have been steadily improving. Current 5-day track forecasts are now as accurate as the 3-day forecasts of 20 years ago (**Figure 14.32**). Despite improvements in accuracy, forecast uncertainty still requires that hurricane warnings be issued for relatively large coastal areas. Only about one-quarter of an average warning area experiences hurricane conditions.

14.6 CONCEPT CHECKS

1. Define *hurricane*. What other names are used for this type of storm?
2. In what latitude zone do hurricanes develop?
3. Distinguish between the eye and the eye wall of a hurricane. How do conditions differ in these zones?
4. What is the source of energy that drives a hurricane?
5. Why do hurricanes *not* form near the equator? Explain the lack of hurricanes in the South Atlantic and eastern South Pacific.
6. When do most hurricanes in the North Atlantic and Caribbean occur? Why are these months the most common times for hurricanes?
7. Why does the intensity of a hurricane diminish rapidly when it moves over land?
8. What are the three broad categories of hurricane damage?

CONCEPTS IN REVIEW
Weather Patterns and Severe Weather

14.1 Air Masses

Define *air mass*. Describe the classification and weather associated with different air masses.

KEY TERMS: air mass, air-mass weather, source region, polar (P) air mass, arctic (A) air mass, tropical (T) air mass, continental (c) air mass, maritime (m) air mass, lake-effect snow, nor'easter

- An *air mass* is a large body of air, usually 1600 kilometers (1000 miles) or more across, that is characterized by a sameness of temperature and moisture at any given altitude. When this air moves out of its region of origin, called the source region, it carries these temperatures and moisture conditions elsewhere, perhaps eventually affecting a large portion of a continent.
- Air masses are classified according to the nature of the surface in the source region and the latitude of the source region. Continental (c) designates an air mass of land origin, with the air likely to be dry; a maritime (m) air mass originates over water and, therefore, will be relatively humid. Polar (P) and arctic (A) air masses originate in high latitudes and are cold. Tropical (T) air masses form in low latitudes and are warm. According to this classification scheme, the four main types of air masses are continental polar (cP), continental tropical (cT), maritime polar (mP), and maritime tropical (mT).
- Continental polar (cP) and maritime tropical (mT) air masses influence the weather of North America most, especially east of the Rocky Mountains. Maritime tropical air is the source of much, if not most, of the precipitation received in the eastern two-thirds of the United States.

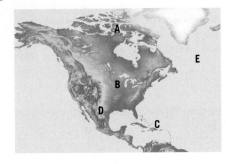

(?) Identify the source region associated with each letter on this map. One letter *is not* associated with a source region. Which one is it?

14.2 Fronts

Compare and contrast typical weather associated with a warm front and a cold front. Describe an occluded front and a stationary front.

KEY TERMS: front, overrunning, warm front, cold front, stationary front, occluded front

- Fronts are boundary surfaces that separate air masses of different densities, one usually warmer and more humid than the other. As one air mass moves into another, the warmer, less dense air mass is forced aloft in a process called overrunning.
- Along a warm front, a warm air mass overrides a retreating mass of cooler air. As the warm air ascends, it cools adiabatically to produce clouds and, frequently, light to moderate precipitation over a large area.
- A cold front forms where cold air is actively advancing into a region occupied by warmer air. Cold fronts are about twice as steep as and move more rapidly than do warm fronts. Because of these two differences, precipitation along a cold front is generally more intense and of shorter duration than precipitation associated with a warm front.

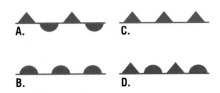

A.   C.

B.   D.

(?) Identify each of these symbols used to designate fronts. On which side of each symbol are the warm air and the cool air?

14.3 Midlatitude Cyclones

Summarize the weather associated with the passage of a mature midlatitude cyclone. Describe how airflow aloft is related to cyclones and anticyclones at the surface.

KEY TERM: midlatitude (middle-latitude) cyclone

- The primary weather producers in the middle latitudes are large centers of low pressure that generally travel from west to east, called middle-latitude cyclones. These bearers of stormy weather, which last from a few days to a week, have a counterclockwise circulation pattern in the Northern Hemisphere, with an inward flow of air toward their centers.
- Most middle-latitude cyclones have a cold front and frequently a warm front extending from the central area of low pressure. Convergence and forceful lifting along the fronts initiate cloud development and frequently cause precipitation. The particular weather experienced by an area depends on the path of the cyclone.
- Guided by west-to-east-moving jet streams, cyclones generally move eastward across the United States. Airflow aloft (divergence and convergence) plays an important role in maintaining cyclonic and anticyclonic circulation. In cyclones, divergence aloft supports the inward flow at the surface.

14.4 Thunderstorms

List the basic requirements for thunderstorm formation and locate places on a map that exhibit frequent thunderstorm activity. Describe the stages in the development of a thunderstorm.

KEY TERM: thunderstorm

- Thunderstorms are caused by the upward movement of warm, moist, unstable air. They are associated with cumulonimbus clouds that generate heavy rainfall, lightning, thunder, and occasionally hail and tornadoes.
- Air-mass thunderstorms frequently occur in maritime tropical (mT) air during spring and summer in the middle latitudes. Generally, three stages are involved in the development of these storms: the cumulus stage, mature stage, and dissipating stage.

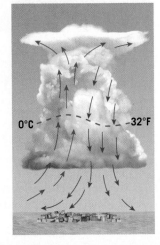

0°C   -32°F

(?) Which stage in the development of a thunderstorm is shown in this sketch? Describe what is occurring. Is there a stage that follows this one? If so, describe what occurs during that stage.

14.5 Tornadoes

Summarize the atmospheric conditions and locations that are favorable to the formation of tornadoes. Discuss tornado destruction and tornado forecasting.

KEY TERMS: tornado, Enhanced Fujita intensity scale (EF scale), tornado watch, tornado warning, Doppler radar

- Tornadoes are violent windstorms that take the form of a rotating column of air called a vortex that extends downward from a cumulonimbus cloud. Many strong tornadoes contain smaller internal vortices. Because of the tremendous pressure gradient associated with a strong tornado, maximum winds can approach 480 kilometers (300 miles) per hour.
- Tornadoes are most often spawned along the cold front of a midlatitude cyclone or in association with a supercell thunderstorm. Tornadoes also form in association with tropical cyclones (hurricanes). In the United States, April through June is the period of greatest tornado activity, but tornadoes can occur during any month of the year.
- Most tornado damage is caused by the tremendously strong winds. One commonly used guide to tornado intensity is the Enhanced Fujita intensity scale (EF scale). A rating on the EF scale is determined by assessing damages produced by the storm.
- Because severe thunderstorms and tornadoes are small and short-lived phenomena, they are among the most difficult weather features to forecast precisely. When weather conditions favor the formation of tornadoes, a tornado watch is issued. The National Weather Service issues a tornado warning when a tornado has been sighted in an area or is indicated by Doppler radar.

14.6 Hurricanes

Identify areas of hurricane formation on a world map and discuss the conditions that promote hurricane formation. List the three broad categories of hurricane destruction.

KEY TERMS: hurricane, eye wall, eye, tropical depression, tropical storm, Saffir–Simpson hurricane scale, storm surge

- Hurricanes, the greatest storms on Earth, are tropical cyclones with wind speeds in excess of 119 kilometers (74 miles) per hour. These complex disturbances develop over tropical ocean waters and are fueled by the latent heat that is liberated when huge quantities of water vapor condense.
- Hurricanes form most often in late summer, when ocean-surface temperatures reach 27°C (80°F) or higher and thus are able to provide the necessary heat and moisture to the air. Hurricanes diminish in intensity when they move over cool ocean water that cannot supply adequate heat and moisture, move onto land, or reach a location where large-scale flow aloft is unfavorable.
- The Saffir–Simpson scale ranks the relative intensities of hurricanes. A 5 on the scale represents the strongest storm possible, and a 1 indicates the lowest severity. Damage caused by hurricanes has three main causes: (1) storm surge, (2) wind, and (3) heavy rains and inland flooding.

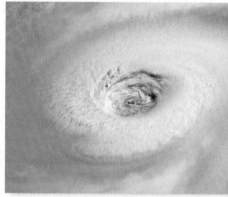

(?) **This image taken from the International Space Station shows the inner portion of Hurricane Igor in September 2010. Identify the eye and eye wall. In which of these zones are winds and rainfall most intense?**

NASA

GIVE IT SOME THOUGHT

1. We know that during the winter, all polar (P) air masses are cold. Which is likely to be colder: a wintertime mP air mass or a wintertime cP air mass? Briefly explain. We expect tropical (T) air masses to be warm, but some are warmer than others. Which should be warmer: a summertime cT air mass or a summertime mT air mass? How did you figure this out?

2. Refer to Figure 14.4 to answer these questions.
   a. Thunder Bay and Marquette are both on the shore of Lake Superior, yet Marquette gets much more snow than Thunder Bay. Why is this the case?
   b. Notice the narrow, north–south zone of relatively heavy snow east of Pittsburgh and Charleston. This area is not affected by the Great Lakes. Speculate on the likely reason for the higher snowfall here. Does your answer explain the shape of this snowy area?

3. If you hear that a cyclone is approaching, should you immediately seek shelter? Why or why not?

4. Refer to the accompanying weather map to answer the following questions:
   a. What is a likely wind direction at each city?
   b. Identify the likely air mass that is influencing each city.
   c. Identify the cold front, warm front, and occluded front.
   d. What is the barometric tendency at City A and City C?
   e. Which one of the three cities is probably coldest? Which one is probably warmest?

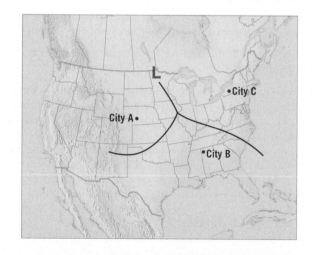

5. The accompanying diagrams show surface temperatures with isotherms labeled in degrees Fahrenheit for noon and 6 P.M. on January 29, 2008. On this day, a powerful front moved through Missouri and Illinois.
   a. What type of front passed through the Midwest?
   b. Describe how the temperature changed in St. Louis, Missouri, over the 6-hour period.
   c. Describe the likely shift in wind direction in St. Louis during this time span.

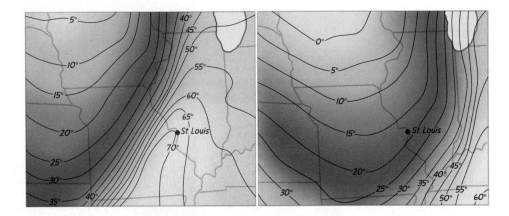

6. The accompanying table lists the number of tornadoes reported in the United States by decade. Propose a reason to explain why the totals for the 1990s and 2000s are so much higher than for the 1950s and 1960s.

Number of U.S. Tornadoes Reported, by Decade	
Decade	Number of Tornadoes Reported
1950–1959	4796
1960–1969	6613
1970–1979	8579
1980–1989	8196
1990–1999	12,138
2000–2009	12,914

7. The number of tornado deaths in the United States in the 2000s was less than 40 percent the number that occurred in the 1950s, even though there was a significant increase in population. Suggest a likely reason for the decline in the death toll.

8. A television meteorologist is able to inform viewers about the intensity of an approaching hurricane. However, the meteorologist can report the intensity of a tornado only after it has occurred. Why is this the case?

9. Refer to the graph in Figure 14.28. Explain why wind speeds are greatest when the slope of the pressure curve is steepest.

10. This world map shows the tracks and intensities of thousands of hurricanes and other tropical cyclones. It was put together by the National Hurricane Center and the Joint Typhoon Warning Center.
    a. What area has experienced the greatest number of category 4 and 5 storms?
    b. Why do hurricanes not form in the very heart of the tropics, astride the equator?
    c. Explain the absence of storms in the South Atlantic and the eastern South Pacific.

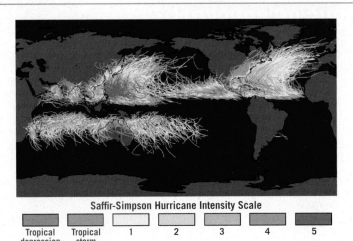

Saffir-Simpson Hurricane Intensity Scale

Tropical depression	Tropical storm	1	2	3	4	5

11. Assume that it is late September 2018, and that the eye of Hurricane Gordon, a category 5 storm, is projected to follow the path shown on the accompanying map of Texas. Answer the following questions:
    a. Name the stages of development that Gordon must have gone through to become a hurricane. At what stage did the storm receive its name?
    b. If the storm follows the projected path, will the city of Houston experience Gordon's fastest winds and greatest storm surge? Explain why or why not.
    c. What is the greatest threat to life and property if this storm approaches the Dallas-Fort Worth area?

MasteringGeology™

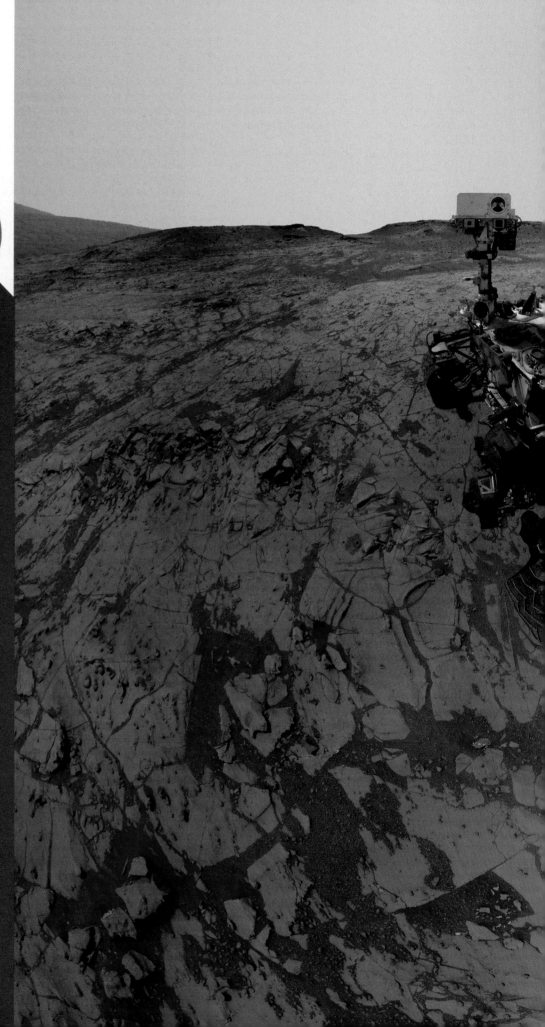

15

FOCUS ON CONCEPTS

Each statement represents the primary learning objective for the corresponding major heading within the chapter. After you complete the chapter, you should be able to:

15.1 Explain the geocentric view of the solar system and describe how it differs from the heliocentric view.

15.2 List and describe the contributions to modern astronomy of Nicolaus Copernicus, Tycho Brahe, Johannes Kepler, Galileo Galilei, and Isaac Newton.

15.3 Describe the formation of the solar system according to the nebular theory. Compare and contrast the terrestrial and Jovian planets.

15.4 List and describe the major features of Earth's Moon and explain how maria basins were formed.

15.5 Outline the principal characteristics of Mercury, Venus, and Mars. Describe their similarities to and differences from Earth.

15.6 Summarize and compare the features of Jupiter, Saturn, Uranus, and Neptune, including their ring systems.

15.7 List and describe the principal characteristics of the small bodies that inhabit the solar system.

Selfie of *Curiosity* rover as it moves up the slopes of Mount Sharp on Mars. This is a composite of several images captured by a camera located at the end of the rover's robotic arm. (Photo courtesy of NASA)

THE NATURE OF THE SOLAR SYSTEM*

*This chapter was revised with the assistance of Professors Teresa Tarbuck and Mark Watry.

The science of astronomy provides a rational way of knowing and understanding the origins of Earth, the solar system, and the universe. Earth was once thought to be unique, different in every way from everything else in the universe. However, through the science of astronomy, we have discovered that our planet is similar to other objects in the universe and that the physical laws that apply on Earth also apply everywhere else in the universe.

In this chapter we examine the transformation from the ancient view of the universe, which focused on the *positions and movements of celestial objects* (mainly planets), to the modern perspective, which focuses on *understanding how* the planets came to be and *why* they differ from one another.

15.1 Ancient Astronomy

Explain the geocentric view of the solar system and describe how it differs from the heliocentric view.

Long before recorded history, people were aware of the close relationship between events on Earth and the positions of heavenly bodies. They realized that changes in the seasons and floods of great rivers such as the Nile in Egypt occurred when certain celestial bodies, including the Sun, Moon, planets, and stars, reached particular places in the heavens. Early agrarian cultures, whose survival depended on seasonal change, believed that if these heavenly objects could control the seasons, they could also strongly influence all Earthly events. These beliefs undoubtedly encouraged early civilizations to begin keeping records of the positions of celestial objects.

The ancient Chinese, Egyptians, and Babylonians are well known for their record keeping. These cultures recorded the locations of the Sun, the Moon, and the five visible planets as these objects moved slowly against the background of "fixed" stars. Eventually, it was not enough to track the motions of celestial objects; predicting their future positions (to avoid getting married at an unfavorable time, for example) became important.

The Chinese recorded every appearance of the famous Halley's Comet for at least 10 centuries. However, because this comet appears only once every 76 years, they were unable to link these appearances to establish that what they saw was the same object multiple times. Like most other ancients, the Chinese considered comets to be mystical. Generally, comets were seen as bad omens and were blamed for a variety of disasters, from wars to plagues (**Figure 15.1**).

The ancient Chinese also kept quite accurate records of "guest stars." Today we know that a "guest star" is a normal star, usually too faint to be visible, which increases in brightness as it explosively ejects gases from its surface. We call this phenomenon a *nova* (*novus* = new) or *supernova* (**Figure 15.2**).

Figure 15.1 The Bayeux Tapestry This tapestry, which hangs in Bayeux, France, shows the apprehension caused by Halley's Comet in A.D. 1066. This event preceded King Harold's defeat by William the Conqueror. (Gianni Dagli Orti/The Art Archive/Art Resource)

The Golden Age of Astronomy

The "Golden Age" of early astronomy (600 B.C.–A.D. 150) was centered in Greece. Although the early Greeks have been criticized

Figure 15.2 The sudden appearance of a "guest star" The Chinese recorded the sudden appearance of a "guest star" in A.D. 1054. The scattered remains of that supernova is the Crab Nebula in the constellation Taurus. This image comes from the Hubble Space Telescope. (J Hester/A Loll/NASA)

for using purely philosophical arguments to explain natural phenomena, they employed observational data as well. The Greeks developed geometry and trigonometry and used the basics of these disciplines to measure the sizes of and distances to the largest-appearing bodies in the heavens—the Sun and the Moon.

The early Greeks held the incorrect **geocentric** (*geo* = Earth, *centric* = centered) view of the universe—which positioned Earth as a motionless sphere at the center of the universe. Orbiting Earth were the Moon, Sun, and known planets—Mercury, Venus, Mars, Jupiter, and Saturn. The Sun and Moon were thought to be perfect crystal spheres.

Beyond the planets was a transparent, hollow **celestial sphere** to which the stars were attached and that traveled daily around Earth. Some early Greeks realized that the motion of the stars could be explained just as easily by a rotating Earth, but they rejected that idea because Earth exhibits no sense of motion and was considered too large to be movable. In fact, proof that Earth rotates on its axis was not demonstrated until 1851.

The first Greek to propose a *Sun-centered*, or **heliocentric** (*helios* = Sun, *centric* = centered), universe was Aristarchus (312–230 B.C.), who used simple geometric relationships to calculate the relative distances from Earth to the Sun and the Moon. He later used these data to calculate their sizes. As a result of an observational error beyond his control, he came up with measurements that were much too small. However, he did discover that the Sun was many times more distant than the Moon and many times larger than Earth. The latter fact may have prompted him to suggest a Sun-centered universe.

Ptolemy's Model

Much of our knowledge of Greek astronomy comes from a 13-volume treatise, *Almagest* (Greek for "the great work"), compiled by Claudius Ptolemy in A.D. 141. In addition to presenting a summary of Greek astronomical knowledge, Ptolemy is credited with developing a model of the universe, known as the **Ptolemaic system** (Figure 15.3).

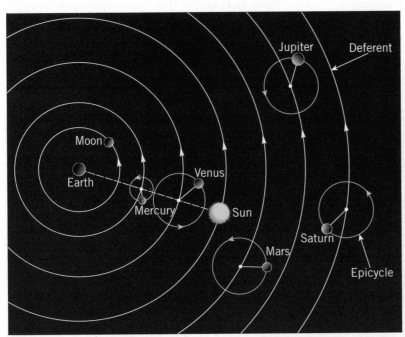

A.

B.

Figure 15.3 The universe according to Ptolemy, second century A.D. **A.** Ptolemy believed that a star-studded celestial sphere made a daily trip around a motionless Earth. In addition, he proposed that the Sun, Moon, and planets made trips of various lengths along individual orbits. **B.** A three-dimensional model of an Earth-centered system. Ptolemy likely utilized something similar to this to calculate the motions of the heavens. (Photo by Science Museum, London UK/Bridgeman Images)

Figure 15.4 Retrograde motion of Mars, as seen against the background of distant stars When viewed from Earth, Mars moves eastward among the stars each day and then periodically appears to stop and reverse direction. This apparent westward drift is a result of the fact that Earth has a faster orbital speed than Mars and overtakes it. As this occurs, Mars appears to be moving backward—that is, it exhibits retrograde motion.

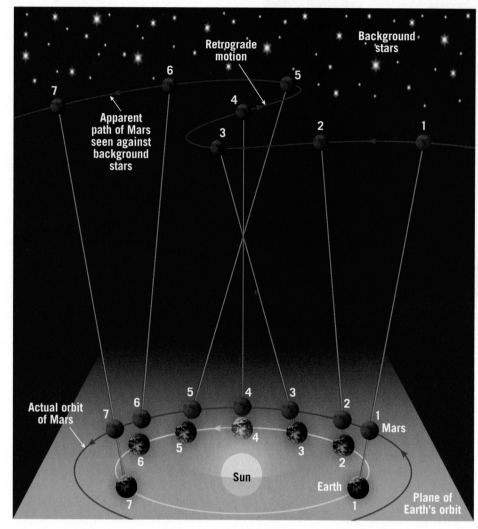

each planet appears to stop, reverse direction for a period of time, and then resume an eastward motion. The apparent westward drift is called **retrograde** (*retro* = to go back, *gradus* = walking) **motion**. This rather odd apparent motion results from the combination of the motion of Earth and the planet's own motion around the Sun.

The retrograde motion of Mars is shown in **Figure 15.4**. Because Earth has a faster orbital speed than Mars, it overtakes its neighbor. While that happens, Mars *appears* to be moving backward, in retrograde motion. This is analogous to what a driver sees out the side window when passing a slower car. The slower planet, like the slower car, appears to be going backward, although its actual motion is in the same direction as the faster-moving body.

Although it is difficult to accurately represent retrograde motion using the incorrect Earth-centered model, Ptolemy did so successfully (**Figure 15.5**). Rather than use a single circle for each planet's orbit, he proposed that the planets orbited on small circles (*epicycles*), revolving along large circles (*deferents*). By trial and error, he found the right combination of circles to produce the amount of retrograde motion observed for each planet.

In the Greek tradition, the Ptolemaic model had the planets moving in perfect circular orbits around a motionless Earth. (The Greeks considered the circle to be a pure and perfect shape.) However, the motion of the planets, as seen against the background of stars, is not so simple. Each planet, if watched night after night, moves slightly eastward among the stars. Periodically,

SmartFigure 15.5 Ptolemy's explanation of retrograde motion Retrograde motion is the apparent backward motion of planets against the background of fixed stars. In Ptolemy's model, the planets move on small circles (epicycles) while they orbit Earth on larger circles (deferents). Through trial and error, Ptolemy discovered the right combination of circles to produce the retrograde motion observed for each planet. (goo.gl/amES3s)

Tutorial

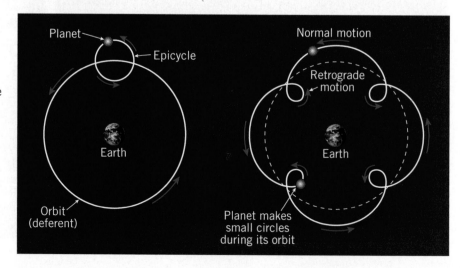

It is a tribute to Ptolemy's genius that he was able to account for the planets' motions as well as he did, considering that his model was incorrect. When Ptolemy's predicted positions for the planets became out of step with the observed positions (which took 100 years or more), his model was simply recalibrated, using the new observed positions as a starting point.

Much accumulated knowledge disappeared as libraries were destroyed after the decline of Greek and Roman civilizations, and the center of astronomical study moved east, to Baghdad where, fortunately, Ptolemy's work was translated into Arabic. It wasn't until sometime after the tenth century that the ancient Greeks' contributions to astronomy were reintroduced to Europe. The Ptolemaic model soon dominated European thought as the correct representation of the heavens, which created problems for anyone who found errors in it.

15.1 CONCEPT CHECKS

1. Why did the ancients believe that celestial objects had some influence over their lives?
2. What is the modern explanation of the "guest stars" that suddenly appear in the night sky?
3. Explain the *geocentric* view of the universe.
4. Describe what produces the retrograde motion of Mars. What geometric arrangements did Ptolemy use to explain this motion?

15.2 The Birth of Modern Astronomy

List and describe the contributions to modern astronomy of Nicolaus Copernicus, Tycho Brahe, Johannes Kepler, Galileo Galilei, and Isaac Newton.

Modern astronomy's development was more than a scientific endeavor; it required a break from deeply entrenched Western philosophical and religious views that had been held for thousands of years. This transition was brought about by the discovery of a new and much larger universe, governed by discernible laws.

We examine the work of five noted scientists involved in this progression from an astronomy that merely describes what is observed to one that tries to explain what is observed and, more importantly, why the universe behaves the way it does: Nicolaus Copernicus, Tycho Brahe, Johannes Kepler, Galileo Galilei, and Sir Isaac Newton.

necessary to add smaller circles (epicycles) like those Ptolemy had used. The discovery that the planets actually have *elliptical* orbits occurred a century later and is credited to Johannes Kepler.

Copernicus's monumental work *De Revolutionibus Orbium Coelestium* (*On the Revolution of the Heavenly Spheres*), which set forth his controversial Sun-centered solar system, was published as he lay on his deathbed. Hence, he never suffered the criticisms that fell on many of his followers. Although Copernicus's model represented a breakthrough, it did not attempt to explain how planetary motions occurred or why.

Nicolaus Copernicus

For almost 13 centuries after the time of Ptolemy, very few astronomical advances were made in Europe. Some were even lost, including the notion of a spherical Earth. The first great astronomer to emerge after the Middle Ages was Nicolaus Copernicus (1473–1543) from Poland (**Figure 15.6**). After discovering Aristarchus's writings, Copernicus became convinced that Earth is a planet, just like the other five then-known planets. The daily motions of the heavens, he reasoned, could be more simply explained by a rotating Earth.

Having concluded that Earth is a planet, Copernicus constructed a *heliocentric* model for the solar system, with the Sun at the center and the planets Mercury, Venus, Earth, Mars, Jupiter, and Saturn orbiting it. This was a major break from the ancient and prevailing idea that a motionless Earth lies at the center of all movement in the universe. However, Copernicus retained a link to the past and used circles to represent the orbits of the planets. Because of this, he was unable to accurately predict the future locations of the planets and found it

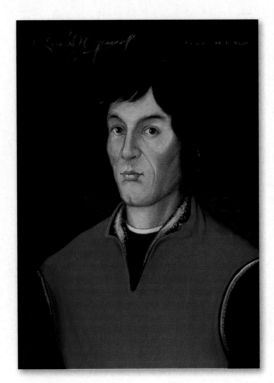

Figure 15.6 Painting of Polish astronomer Nicolaus Copernicus (1473–1543) Copernicus's proclamation that Earth was just another planet was very controversial for more than 100 years after his death. (Detlev van Ravenswaay/Science Source)

The greatest contribution of the Copernican system to modern science is its challenge of the primacy of Earth in the universe. At the time, many Europeans considered this heretical.

Tycho Brahe

Tycho Brahe (1546–1601), a Danish nobleman, was born 3 years after the death of Copernicus. Tycho reportedly became interested in astronomy while viewing a solar eclipse that astronomers had predicted. He persuaded King Frederick II to establish an observatory near Copenhagen, which Tycho headed. There he designed and built pointers (the telescope would not be invented for a few more decades), which he used for 20 years to systematically measure the locations of the heavenly bodies in an effort to disprove the Copernican theory (**Figure 15.7**). His observations, particularly of Mars, were far more precise than any made previously and are his legacy to astronomy.

Tycho did not believe in the Copernican model because he was unable to observe an apparent shift in the position of stars that should result if Earth traveled around the Sun. He argued that if Earth orbits the Sun, the position of a nearby star, when observed from two locations in Earth's orbit 6 months apart, should shift with respect to the more distant stars. Tycho was correct, but his measurements were not precise enough to show any displacement. This apparent shift of the stars is called *stellar parallax* (discussed in Appendix C, page 507) and is used today to measure distances to the nearest stars.

The principle of parallax is easy to visualize: Close one eye, and with your index finger vertical, use your eye to line up your finger with some distant object. Now, without moving your finger, view the object with your other eye and notice that the object's position appears to change. The farther away you hold your finger, the less the object's position seems to shift. Herein lay the flaw in Tycho's argument. He was right about parallax, but the distance to even the nearest stars is enormous compared to the width of Earth's orbit. Consequently, the shift that Tycho was looking for is too small to be detected without the aid of a telescope—an instrument that had not yet been invented.

With the death of his patron, the king of Denmark, Tycho was forced to leave his observatory. Known for his arrogance and extravagant nature, Tycho was unable to continue his work under Denmark's new ruler and moved to Prague, in the present-day Czech Republic. There, in the last year of his life, he acquired an able assistant, Johannes Kepler. Kepler retained most of Tycho's observations and put them to exceptional use. Ironically, the data Tycho collected to refute the Copernican view of the solar system would later be used by Kepler to support it.

Johannes Kepler

If Copernicus ushered out the old astronomy, Johannes Kepler (1571–1630) ushered in the new (**Figure 15.8**). Armed with Tycho's data, a good mathematical mind, and, of much importance, a strong belief in the accuracy of Tycho's work, Kepler derived three basic laws of planetary motion. The first two laws resulted from his inability to fit Tycho's observations of Mars to a circular orbit. Unwilling to concede that the discrepancies were due to observational error, he searched for another solution. This endeavor led him to discover that the orbit of Mars is not a perfect circle but is slightly elliptical (**Figure 15.9**). Kepler also realized that the orbital speed of Mars varies in a predictable way; Mars speeds up as it approaches the Sun and slows down as it moves away.

In 1609, after nearly a decade of work, Kepler proposed his first two laws of planetary motion:

1. The path of each planet around the Sun is actually an ellipse, with the Sun at one focus (see Figure 15.9).

2. Each planet revolves so that an imaginary line connecting it to the Sun sweeps over equal areas in

Figure 15.7 Painting of Tycho Brahe (1546–1601) in his observatory in Uraniborg, on the Danish island of Hveen Tycho (central figure) and the background are painted on the wall of the observatory within the arc of the sighting instrument called a quadrant. In the far right, Tycho can be seen "sighting" a celestial object through the "hole" in the wall. His accurate measurements of Mars enabled Johannes Kepler to formulate his three laws of planetary motion.
(Courtesy of Royal Geographic Society, London UK/Bridgeman Images)

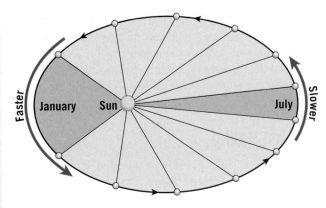

Figure 15.10 Kepler's law of equal areas A line connecting a planet (Earth) to the Sun sweeps out an area in such a manner that equal areas are swept out in equal times. Thus, Earth revolves slower when it is farther from the Sun and faster when it is closer. The eccentricity of Earth's orbit is greatly exaggerated in this diagram.

Figure 15.8 German astronomer Johannes Kepler (1571–1630) Kepler's contribution to modern astronomy was the derivation of his three laws of planetary motion. (Photo by Imagno/Getty Images)

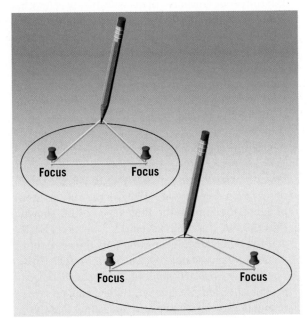

Figure 15.9 Drawing ellipses with various eccentricities Using two straight pins for foci and a loop of string, trace out a curve while keeping the string taut, and you will have drawn an ellipse. The farther apart the pins (the foci) are moved, the more flattened (more eccentric) is the resulting ellipse.

equal intervals of time (**Figure 15.10**). This *law of equal areas* geometrically expresses the variations in orbital speeds of the planets.

Figure 15.10 illustrates the second law. Note that in order for a planet to sweep equal areas in the same amount of time, it must travel more rapidly when it is nearer the Sun and more slowly when it is farther from the Sun.

Kepler was devout and believed that the Creator had made an orderly universe and that this order would be reflected in the positions and motions of the planets. The uniformity he tried to find eluded him for nearly a decade. Then, in 1619, Kepler published his third law in *The Harmony of the Worlds*:

3. The orbital periods of the planets and their distances to the Sun are proportional.

In the simplest form of this law, the orbital period is measured in Earth years, and the planet's distance to the Sun is expressed in terms of Earth's mean distance to the Sun. The latter "yardstick" is called the **astronomical unit (AU)** and is equal to about 150 million kilometers (93 million miles). Using these units, Kepler's third law states that the planet's orbital period squared is equal to its mean solar distance cubed. Consequently, the solar distances of the planets can be calculated when their periods of revolution are known. For example, Mars has an orbital period of 1.88 years, and 1.88 squared equals 3.54. The cube root of 3.54 is 1.52, and that is the mean distance from Mars to the Sun, in astronomical units (**Table 15.1**).

Table 15.1 Period of Revolution and Solar Distances of Planets			
Planet	Mean Solar Distance (AU)*	Period (years)	Eccentricity (0 = circle)
Mercury	0.39	0.24	0.205
Venus	0.72	0.62	0.007
Earth	1.00	1.00	0.017
Mars	1.52	1.88	0.094
Jupiter	5.20	11.86	0.049
Saturn	9.54	29.46	0.057
Uranus	19.18	84.01	0.046
Neptune	30.06	164.80	0.011
* AU = astronomical unit			

Kepler's findings provide strong support for the Copernican theory that the planets revolve around the Sun. Kepler, however, did not determine the *forces* that act to produce the planetary motion he had so ably described. That task would remain for Galileo Galilei and Sir Isaac Newton.

Galileo Galilei

Galileo Galilei (1564–1642) was the greatest Italian scientist of the Renaissance (**Figure 15.11**). A contemporary of Kepler, Galileo also strongly supported the Copernican theory of a heliocentric system. Galileo's greatest contributions to science were his descriptions of the behavior of moving objects, which he derived from experimentation. The method of using experiments to determine natural laws had essentially been lost since the time of the early Greeks.

All astronomical discoveries before Galileo's time were made without the aid of a telescope. In 1609, Galileo heard that a Dutch lens maker had devised a system of lenses that magnified objects. Apparently without ever having seen a telescope, Galileo constructed his own, which magnified distant objects to three times the size seen by the unaided eye. He immediately made others, the best having a magnification of about 30 (**Figure 15.12**).

With the telescope, Galileo was able to view the universe in a new way. He made many important discoveries that supported the Copernican view of the universe, including the following:

Figure 15.12 One of Galileo's telescopes Although Galileo did not invent the telescope, he built several—the largest of which had a magnification of 30. (Photo by Gianni Tortoli/Science Source)

1. *He observed Jupiter's four largest satellites, or moons* (**Figure 15.13**). This finding dispelled the old idea that Earth was the sole center of motion in the universe, for here, plainly visible, was another center of motion—Jupiter. This finding also countered the frequently used argument that the Moon would be left behind if Earth revolved around the Sun.

2. *Planets are circular disks rather than just points of light, as had previously been thought.* This indicated that the planets must be Earth-like as opposed to star-like.

3. *Venus exhibits phases just as the Moon does, and Venus appears smallest when it is in full phase and thus is farthest from Earth* (**Figure 15.14B, C**). This observation demonstrates that Venus orbits its source of light—the Sun. In the Ptolemaic system, shown in **Figure 15.14A**, the orbit of Venus lies between Earth and the Sun, which means that only the crescent phases of Venus should ever be seen from Earth.

4. *The Moon's surface is not a smooth glass sphere, as the ancients had proclaimed.* Rather, Galileo saw mountains, craters, and plains, indicating that the Moon is Earth-like. He thought the plains might be bodies of water, and this idea was strongly promoted by others, as we can tell from the names given to these features (Sea of Tranquility, Sea of Storms, etc.).

Figure 15.11 Italian scientist Galileo Galilei (1564–1642) Galileo was the first scientist to use a new invention, the telescope, to observe the Sun, Moon, and planets in more detail than ever before. (Nimatallah/Art Resource)

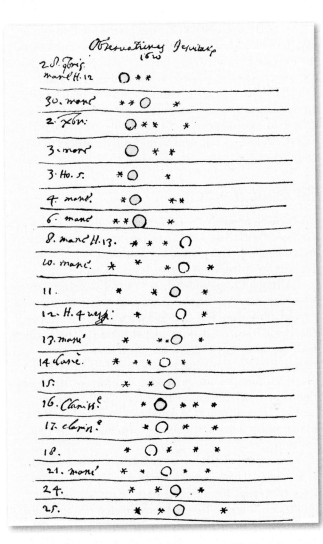

Figure 15.13 Sketch by Galileo of Jupiter and its four largest satellites Using a telescope, Galileo discovered Jupiter's four largest Moons (drawn as stars) and noted that their positions change nightly. You can observe these same changes with binoculars. (Yerkes Observatory Photograph/University of Chicago)

5. *The Sun has sunspots—dark regions caused by slightly lower temperatures.* Galileo tracked the movement of these spots (which may have caused the eye damage that later blinded him) and estimated the rotational period of the Sun as just under a month. Hence, another heavenly body was found to have both "blemishes" and rotational motion.

Each of these observations eroded a bedrock principle held by the prevailing view on the nature of the universe.

In 1616, the Roman Catholic Church condemned Copernican theory as contrary to biblical scripture because it did not put humans in their rightful place at the center of Creation, and Galileo was told to abandon this theory. Undeterred, Galileo began writing his most famous work, *Dialogue of the Great World Systems*, and in 1630 he went to Rome, seeking permission from Pope Urban VIII to publish it. Because the book was a

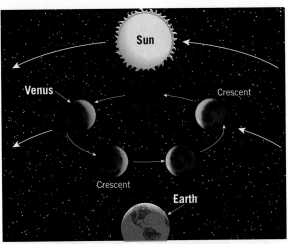

A. In the Ptolemaic (Earth-centered) system, the orbit of Venus lies between the Sun and Earth, as shown here. Thus, in an Earth-centered solar system, only the crescent phase of Venus would be visible from Earth.

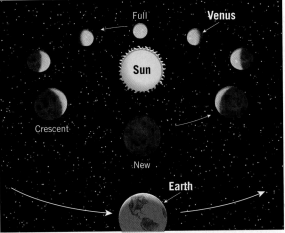

B. In the Copernican (Sun-centered) system, Venus orbits the Sun and hence all of the phases of Venus should be visible from Earth.

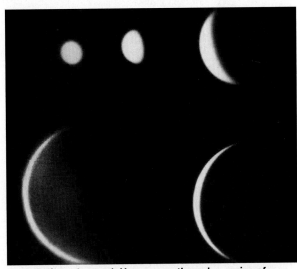

C. As Galileo observed, Venus goes through a series of Moonlike phases. Venus appears smallest during the full phase when it is farthest from Earth and largest in the crescent phase when it is closest to Earth. This led Galileo to conclude that the Sun was the center of the solar system.

SmartFigure 15.14 Using a telescope, Galileo discovered that Venus, just like the Moon, has phases **A.** In the Ptolemaic (geocentric) system, the orbit of Venus lies between the Sun and Earth, as shown in Figure 15.4A. Thus, in an Earth-centered solar system, only the crescent phase of Venus would be visible from Earth. **B.** In the Copernican (heliocentric) system, Venus orbits the Sun, and hence all the phases of Venus should be visible from Earth. **C.** As Galileo observed, Venus goes through a series of Moonlike phases. Venus appears smallest during the full phase, when it is farthest from Earth, and largest in the crescent phase, when it is closest to Earth. This verified Galileo's belief that the Sun is the center of the solar system. (Photo courtesy of Lowell Observatory Archives) (goo.gl/L2JcLm)

Tutorial

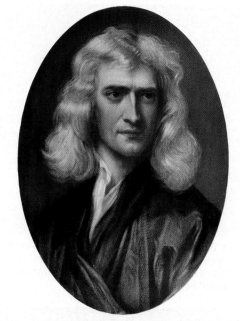

Figure 15.15 Prominent English scientist Sir Isaac Newton (1642–1727) Newton discovered that gravity is the force that holds planets in orbit around the Sun. (Photo by G. Nimatallah/Dea/ Getty Images)

dialogue that expounded both the Ptolemaic and Copernican systems, publication was allowed. However, Galileo's detractors were quick to realize that he was promoting the Copernican view. Sale of the book was quickly halted, and Galileo was called before the Inquisition. Convicted of proclaiming doctrines contrary to religious teachings, he was sentenced to permanent house arrest, under which he remained for the last 10 years of his life.

SmartFigure 15.16 Orbital motion of the planets (goo.gl/iDSiEo)

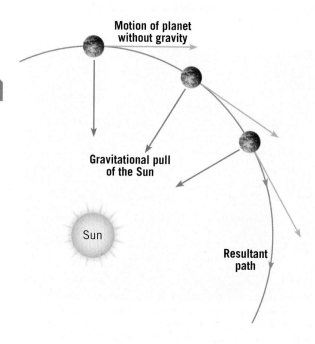

Despite this restriction, and his grief following the death of his eldest daughter, Galileo continued to work. In 1637 he became totally blind yet completed his finest scientific work, a book on the study of motion in which he stated that the natural tendency of an object in motion is to remain in motion. Later, as more scientific evidence supporting the Copernican system was discovered, the Church allowed Galileo's works to be published.

Sir Isaac Newton

Sir Isaac Newton (1642–1727) was born in the year of Galileo's death (**Figure 15.15**). His many accomplishments in mathematics and physics led a successor to say, "Newton was the greatest genius that ever existed."

Although Kepler and those who followed attempted to explain the forces involved in planetary motion, their explanations were less than satisfactory. Kepler believed that some force pushed the planets along in their orbits. Galileo, however, correctly reasoned that no force is required to keep an object in motion. Instead, Galileo proposed that the natural tendency for a moving object that is unaffected by an outside force is to continue moving at a uniform speed and in a straight line. Newton later formalized this concept, **inertia**, as his first law of motion.

The problem, then, was not to explain the force that keeps the planets moving but rather to determine the force that *keeps them from going in a straight line out into space.* It was to this end that Newton conceptualized the force of gravity. At the early age of 23, he envisioned a force that extends from Earth into space and holds the Moon in orbit around Earth. Although others had theorized the existence of such a force, he was the first to formulate and test the **law of universal gravitation**. It states:

> *Every body in the universe attracts every other body with a force that is directly proportional to their masses and inversely proportional to the square of the distance between them.*

Thus, gravitational force decreases with distance so that two objects 3 kilometers apart have 3^2, or 9, times less gravitational attraction than if the same objects were 1 kilometer apart.

The law of gravitation also states that the greater the mass of an object, the greater the gravitational force it exerts on other objects. For example, the large mass of the Moon has a gravitational force strong enough to cause ocean tides on Earth, whereas the tiny mass of a communications satellite has very little effect on Earth.

With his laws of motion, Newton proved that the force of gravity—combined with the tendency of a

Motion of planet without gravity

Gravitational pull of the Sun

Sun

Resultant path

planet to remain in straight-line motion—would result in a planet having an elliptical orbit, as established by Kepler. Earth, for example, moves forward in its orbit about 30 kilometers (18.5 miles) each second, and during the same second, the force of gravity pulls it toward the Sun about 0.5 centimeter (1/8 inch). Therefore, as Newton concluded, it is the combination of Earth's forward motion and its "falling" motion that defines its orbit (**Figure 15.16**). If gravity were somehow eliminated, Earth would move in a straight line out into space. Conversely, if Earth's forward motion suddenly stopped, gravity would pull it until it crashed into the Sun.

15.2 CONCEPT CHECKS

1. What major change did Copernicus make to the Ptolemaic system? Why was this change philosophically different?
2. What data did Tycho Brahe collect that was useful to Johannes Kepler in his quest to describe planetary motion?
3. Who discovered that planetary orbits are ellipses rather than circles?
4. Explain how Galileo's discovery of the phases of Venus supported the Copernican view of a Sun-centered universe.
5. In your own words, describe the law of universal gravitation. Who discovered this important concept?

15.3 Our Solar System: An Overview

Describe the formation of the solar system according to the nebular theory. Compare and contrast the terrestrial and Jovian planets.

The Sun is the center of a revolving system, trillions of miles wide, consisting of eight planets, their satellites, and numerous smaller bodies—dwarf planets, asteroids, comets, and meteoroids. An estimated 99.85 percent of the mass of our solar system is contained within the Sun. Collectively, the planets account for most of the remaining 0.15 percent. Starting from the Sun, the planets are Mercury, Venus, Earth, Mars, Jupiter, Saturn, Uranus, and Neptune (**Figure 15.17**).

Tethered to the Sun by gravity, the planets travel in the same direction, on slightly elliptical orbits; those nearest the Sun travel fastest (**Table 15.2**). Therefore, Mercury has the highest orbital velocity, 48 kilometers (30 miles) per second, and the shortest period of revolution around the Sun, 88 Earth days. By contrast, Neptune has an orbital speed of just 5.3 kilometers (3.3 miles) per second and requires 165 Earth-years to complete one revolution. Most large bodies orbit the Sun approximately in the same plane. The planets' inclination with respect to the Earth–Sun orbital plane, known as the *ecliptic*, is shown in Table 15.2.

Nebular Theory: Formation of the Solar System

The **nebular theory**, which explains the formation of the solar system, proposes that the Sun and planets formed from a rotating cloud of interstellar gases (mainly hydrogen and helium) and dust called the **solar nebula**. As gravity contracted the solar nebula, most of the material collected in the center to form the hot *protosun*. The remaining materials formed a thick, flattened,

rotating disk, within which matter gradually cooled and condensed into grains and clumps of icy, rocky material. Repeated collisions resulted in most of the material clumping together into increasingly larger chunks that eventually became asteroid-sized objects called **planetesimals**.

The composition of planetesimals was largely determined by their proximity to the protosun. As you might expect, temperatures were highest in the inner solar system and decreased toward the outer edge of the disk. Between the present orbits of Mercury and Mars, the planetesimals were composed of materials with high melting temperatures—metals and rocky substances. Then, through repeated collisions and accretion, these asteroid-sized rocky bodies combined to form the four **protoplanets** that eventually became Mercury, Venus, Earth, and Mars.

The planetesimals that formed beyond the orbit of Mars, where temperatures were low, contained high percentages of ices—water, carbon dioxide, ammonia, and methane—as well as small amounts of rocky and metallic debris. It was mainly from these planetesimals that the four outer planets eventually formed. The accumulation of ices accounts, in part, for the large sizes and low densities of these outer planets. The two most massive planets, Jupiter and Saturn, had surface gravities sufficient to attract and retain large quantities of hydrogen and helium, the lightest elements.

It took roughly 1 billion years after the protoplanets formed for the planets to gravitationally accumulate most of the interplanetary debris. This was a period of intense bombardment as the planets cleared their orbits

Planetary orbits **A.** Artistic view of the solar system; the planets are not drawn to scale. **B.** The distances of the planets from the Sun and relative sizes of the planets shown using two different scales. The distances are given in astronomical units (AU), where 1 AU is equal to the mean distance from Earth to the Sun—150 million kilometers (93 million miles). If the Sun and planets were shown at the same scale as the distances, they would be about 5000 times smaller than illustrated here. (goo.gl/ocqiug)

Tutorial

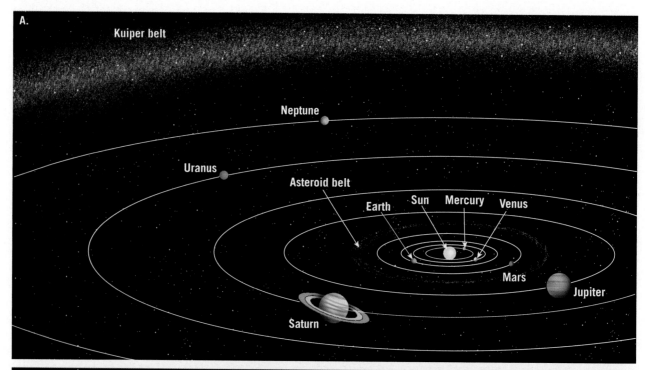

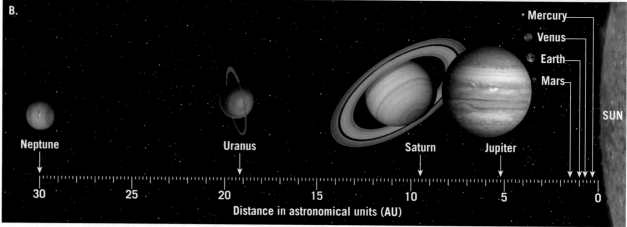

The Planets: Internal Structures and Atmospheres

The planets fall into two groups, based on location, size, and density: the **terrestrial (Earth-like) planets** (Mercury, Venus, Earth, and Mars) and the **Jovian (Jupiter-like) planets** (Jupiter, Saturn, Uranus, and Neptune). Because of their locations relative to the Sun, the four terrestrial planets are also known as *inner planets*, and the four Jovian planets are known as *outer planets*. A correlation exists between planetary locations and sizes: The inner planets are substantially smaller than the outer

by collecting much of the leftover material. The "scars" of this period are still evident on the Moon's surface. Because of the gravitational effect of the planets, particularly Jupiter, small bodies were flung into planet-crossing orbits or into interstellar space. The small fraction of interplanetary matter that escaped this violent period became either asteroids or comets.

planets, also called *gas giants*. For example, the diameter of Neptune (the smallest Jovian planet) is nearly 4 times larger than the diameter of Earth. Furthermore, Neptune's mass is 17 times greater than that of Earth or Venus.

Other properties that differ among the planets include densities, chemical compositions, orbital periods, and numbers of satellites (see Table 15.2). Variations in the chemical composition of planets are largely responsible for their density differences. Specifically, the average density of the terrestrial planets is about 5 times the density of water, whereas the average density of the Jovian planets is only 1.5 times that of water. In fact, Saturn has a density only 0.7 times that of water, which means that it would float in a sufficiently large tank of water. The outer planets are also characterized by long orbital periods and numerous satellites.

Recall from Chapter 6 that shortly after Earth formed, segregation of material formed three major layers, defined by their chemical composition—the crust, mantle, and core. This type of chemical separation also occurred

Table 15.2 Planetary Data

Planet	Symbol	Mean Distance from Sun			Orbital Period	Inclination of Orbit	Orbital Velocity	
		AU*	Millions of Miles	Millions of Kilometers			mi/s	km/s
Mercury	☿	0.39	36	58	88 days	7°00′	29.5	47.5
Venus	♀	0.72	67	108	245 days	3°24′	21.8	35.0
Earth	⊕	1.00	93	150	365.25 days	0°00′	18.5	29.8
Mars	♂	1.52	142	248	687 days	1°51′	14.9	24.1
Jupiter	♃	5.20	483	778	12 years	1°18′	8.1	13.1
Saturn	♄	9.54	886	1427	30 years	2°29′	6.0	9.6
Uranus	♅	19.18	1783	2870	84 years	0°46′	4.2	6.8
Neptune	♆	30.06	2794	4497	165 years	1°46′	3.3	5.3

Planet	Period of Rotation Around Axis	Diameter		Relative Mass (Earth = 1)	Average Density (g/cm³)	Polar Flattening (%)	Eccentricity†	Number of Known Satellites††
		Miles	Kilometers					
Mercury	59 days	3015	4878	0.06	5.4	0.0	0.206	0
Venus	243 days	7526	12,104	0.82	5.2	0.0	0.007	0
Earth	23ʰ56ᵐ04ˢ	7920	12,756	1.00	5.5	0.3	0.017	1
Mars	24ʰ37ᵐ23ˢ	4216	6794	0.11	3.9	0.5	0.093	2
Jupiter	9ʰ56ᵐ	88,700	143,884	317.87	1.3	6.7	0.048	67
Saturn	10ʰ30ᵐ	75,000	120,536	95.14	0.7	10.4	0.056	62
Uranus	17ʰ14ᵐ	29,000	51,118	14.56	1.2	2.3	0.047	27
Neptune	16ʰ07ᵐ	28,900	50,530	17.21	1.7	1.8	0.009	14

*AU = astronomical unit, Earth's mean distance from the Sun.

†Eccentricity is a measure of the amount an orbit deviates from a circular shape. The larger the number, the less circular the orbit.

††Includes all satellites discovered as of July 2015. Satellites are celestial bodies that orbit a planet rather than orbiting a star like the Sun.

in the other planets, but because the terrestrial planets are compositionally different from the Jovian planets, the nature of these layers differs as well (**Figure 15.18**).

Figure 15.18 Comparing internal structures of the planets

Interiors of the Terrestrial Planets

The terrestrial planets are dense, having relatively large cores composed mainly of iron and nickel. Based on seismological evidence, we know that Earth's outer core is molten. We also know that Earth's strong magnetic field is generated by convection within the molten outer core, aided by moderately rapid planetary rotation.

Mars's core is thought to be partially molten but not hot enough to support convection; as a result, Mars lacks a magnetic field. Venus also lacks a magnetic field, even though its core is thought to have a molten metallic layer

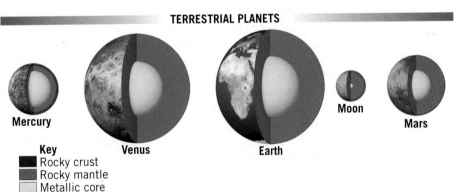

TERRESTRIAL PLANETS

Mercury
Venus
Earth
Moon
Mars

Key
- Rocky crust
- Rocky mantle
- Metallic core

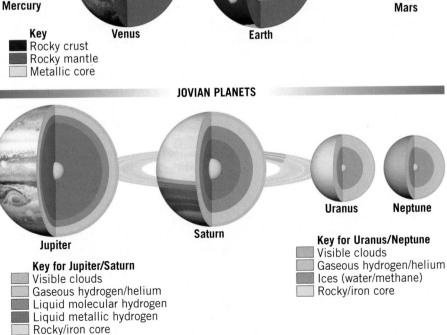

JOVIAN PLANETS

Jupiter
Saturn
Uranus
Neptune

Key for Jupiter/Saturn
- Visible clouds
- Gaseous hydrogen/helium
- Liquid molecular hydrogen
- Liquid metallic hydrogen
- Rocky/iron core

Key for Uranus/Neptune
- Visible clouds
- Gaseous hydrogen/helium
- Ices (water/methane)
- Rocky/iron core

SmartFigure 15.19 Bodies with atmospheres versus airless bodies Airless worlds have relatively warm surface temperatures and/or weak gravities. Bodies with significant atmospheres have low surface temperatures and/or strong gravities. (goo.gl/NgQuw4)

Tutorial

Airless worlds have relatively warm surface temperatures and/or weak gravities.

Airless bodies

Mercury

Venus

Moon

Earth

Mars

Jupiter

Asteroids

Galilean moons

Europa

Saturn

Titan

Uranus

Charon

Triton

Neptune

Pluto

Bodies with an atmosphere

Solar heating (temperature)

Bodies with significant atmospheres have low surface temperatures and/or strong gravities.

Gravity

metallic hydrogen. Above this metallic layer, both Jupiter and Saturn are thought to be composed of molecular liquid hydrogen intermixed with helium. The outermost layers are gases of hydrogen and helium, as well as ices of water, ammonia, and methane—which mainly account for the low densities of these giants.

Uranus and Neptune also have small iron-rich, rocky cores, but their mantles are likely hot, dense water and ammonia. Above their mantles, the amount of hydrogen and helium increases, but these gases exist in much smaller amounts than in Jupiter and Saturn.

The Atmospheres of the Planets The Jovian planets have very thick atmospheres composed mainly of hydrogen and helium, with lesser amounts of water, methane, ammonia, and other hydrocarbons. By contrast, the terrestrial planets, including Earth, have relatively meager atmospheres that are typically dominated by carbon dioxide or nitrogen, and most small solar system bodies are airless. Two factors explain these significant differences: solar heating (temperature) and gravity (**Figure 15.19**). These variables determine what planetary gases, if any, were captured by the planets during the formation of the solar system and which were ultimately retained.

During planetary formation, the inner regions of the developing solar system were too hot for ices and gases to condense, but the outer planets formed where temperatures were low and solar heating of planetesimals was minimal. This allowed water vapor, ammonia, and methane to condense into ices. Hence, the Jovian planets contain large amounts of these volatiles. As the planets grew, the largest Jovian planets, Jupiter and Saturn, also attracted large quantities of the lightest gases, hydrogen and helium, due to their strong gravitational fields. This explains why the Jovian planets have thick atmospheres and why Saturn's moon Titan, which is smaller than Earth but much further from the Sun, retains an atmosphere.

How did Earth acquire water and other volatile gases? It seems that early in the history of the solar system, gravitational tugs by the developing protoplanets sent planetesimals into very eccentric orbits. As a result, Earth was bombarded with icy objects (comets and asteroids) that brought in water and other elements. This was a fortuitous event for organisms that currently inhabit our planet. Mercury, our Moon, and numerous other small bodies lack significant atmospheres, although they certainly would have been bombarded by icy bodies early in their development.

Airless bodies develop where solar heating is strong and/or gravities are weak. Simply stated, *small warm*

much like Earth's. Presumably, it lacks a magnetic field either because the core is not hot enough to drive convection or because the planet's 243-day period of rotation is too slow to generate a magnetic field. Researchers were surprised to discover that Mercury, despite its small size and slow 59-day period of rotation, possess a measurable magnetic field, albeit only 1 percent the strength of Earth's. This may be the result of Mercury's partially molten metallic core, which is unusually large compared to the size of the planet.

Silicate minerals and other lighter compounds make up the mantles of the terrestrial planets. Finally, the silicate crusts of terrestrial planets are relatively thin compared to their mantles.

Interiors of the Jovian Planets The two largest Jovian planets, Jupiter and Saturn, likely have small, solid cores consisting of iron compounds, like the cores of the terrestrial planets, and rocky material similar to Earth's mantle. Progressing outward, the layer above the core consists of liquid hydrogen that is under extremely high temperatures and pressures. There is substantial evidence that under these conditions, hydrogen behaves like a metal in that its electrons move freely about and are efficient conductors of both heat and electricity. Jupiter's intense magnetic field is thought to be the result of electric currents flowing within a spinning layer of liquid metallic hydrogen. Saturn's magnetic field is much weaker than Jupiter's, due to its smaller shell of liquid

bodies have a better chance of losing their atmospheres: Gas molecules are more energetic (and hence faster-moving) on a warm body, and they need less speed to escape the weak gravity of a small body. Mercury, the smallest and least massive of the eight planets, has a low surface gravity, which makes holding on to an atmosphere challenging. In addition, because it is the planet closest to the Sun, it is constantly bombarded by solar wind. Mercury therefore has the thinnest atmosphere of all the planets.

The somewhat larger terrestrial planets, Earth, Venus, and Mars, retain some heavy gases, including nitrogen, carbon dioxide, oxygen, and even water vapor. However, their atmospheres are miniscule compared to their total mass. Early in their development, the terrestrial planets probably had much thicker atmospheres. Over time, however, these primitive atmospheres gradually changed as light gases trickled away into space. For example, Earth's atmosphere continues to leak hydrogen and helium (the two lightest gases) into space. This phenomenon occurs near the top of Earth's atmosphere, where air is so tenuous that nothing stops the fastest-moving particles from flying off into space. The speed required to escape a planet's gravity is called **escape velocity**. Because hydrogen is the lightest gas, it most easily reaches the speed needed to overcome Earth's gravity.

Planetary Impacts

Impacts between solar system bodies have occurred throughout the history of the solar system. On bodies that have little or no atmosphere (like the Moon) and, therefore, no air resistance, even the smallest pieces of interplanetary debris (meteorites) can reach the surface. At high enough velocities, this debris can produce microscopic cavities on individual mineral grains. By contrast, large **impact craters** result from collisions with massive bodies, such as asteroids and comets.

Planetary impacts were considerably more common in the early history of the solar system than they are today. Following that early period of intense bombardment, the rate of cratering diminished dramatically and now remains essentially constant. Because weathering and erosion are almost nonexistent on the Moon and Mercury, evidence of their cratered past is clearly evident.

On larger bodies, the presence of an atmosphere may cause the impacting objects to break up and/or decelerate. For example, Earth's atmosphere causes meteoroids with masses of less than 10 kilograms (22 pounds) to lose up to 90 percent of their speed as they penetrate the atmosphere. Therefore, impacts of low-mass bodies produce only small craters on Earth. Our atmosphere is much less effective in slowing large bodies; fortunately, they make very rare appearances.

The formation of a large impact crater is illustrated in **Figure 15.20**. The meteoroid's high-speed impact compresses the material it strikes, causing an almost instantaneous rebound, which ejects material from the surface. On Earth, impacts can occur at speeds that exceed 50

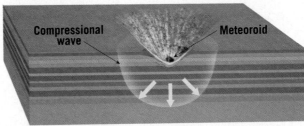

Figure 15.20 Formation of an impact crater

A. The energy of a rapidly moving body is transformed into heat and shock waves.

B. The rebound of over-compressed rock causes debris to be explosively ejected from the crater. Some of this material may melt and be deposited as glass beads.

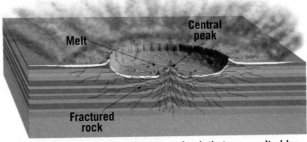

C. Large craters may contain areas of ock that was melted by the impact and a rebounded central peak.

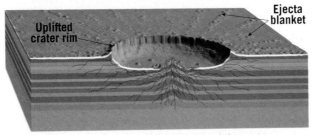

D. Ejected material forms a "blanket" around the crater.

kilometers (30 miles) per second. Impacts at such high speeds produce shock waves that compress both the impactor and the material being impacted. Almost instantaneously, the over-compressed material rebounds and explosively ejects material out of the newly formed crater. In addition, large craters often exhibit a central peak, such as the one shown in **Figure 15.21**. These central peaks also result from crustal rebound.

Much of the material expelled, called *ejecta*, lands in or near the crater, where it accumulates to form a rim. The remaining material forms a blanket around the crater. Upon impact, large meteoroids may eject large projectiles that strike the surrounding landscape to generate smaller structures called *secondary craters*. Large meteoroids also generate sufficient heat to melt and then eject some of the impacted rock as glass beads. Specimens of glass beads

Figure 15.21 Lunar crater Euler This 20-kilometer-wide (12-mile-wide) crater is located in the southwestern part of Mare Imbrium. Clearly visible are the bright rays, central peak, secondary craters, and large accumulation of ejecta near the crater rim. (Courtesy of NASA)

produced in this manner, as well as melt breccia consisting of broken fragments welded by the heat of impact, have been collected on Earth and on the Moon.

15.3 CONCEPT CHECKS

1. Briefly outline the steps in the formation of our solar system, according to the nebular theory.
2. By what criteria are planets considered either terrestrial or Jovian?
3. What accounts for the large density differences between the terrestrial and Jovian planets?
4. Explain why the terrestrial planets have meager atmospheres compared to the Jovian planets.

15.4 Earth's Moon: A Chip Off the Old Block

List and describe the major features of Earth's Moon and explain how maria basins were formed.

The Earth–Moon system is unique partly because of the Moon's large size relative to other bodies in the inner solar system. The Moon's diameter is 3475 kilometers (2160 miles), about one-fourth of Earth's 12,756 kilometers (7926 miles), and its surface temperature averages about 107°C (225°F) during daylight hours and –153°C (–243°F) at night. Because its period of rotation on its axis equals its period of revolution around Earth, the same lunar hemisphere always faces Earth. All of the landings of staffed *Apollo* missions were confined to the side of the Moon that faces Earth.

The Moon's density is 3.3 times that of water, comparable to that of mantle rocks on Earth but considerably less than Earth's average density (5.5 times that of water). The Moon's low mass relative to Earth results in a lunar gravitational attraction that is one-sixth that of Earth. The Moon's small mass (and low gravity) is the primary reason it was not able to retain an atmosphere.

How Did the Moon Form?

Current models show that Earth is too small to have formed with a moon, particularly one so large. Furthermore, a captured moon would likely have an eccentric orbit similar to the captured moons that orbit the Jovian planets.

Astronomers generally agree that the Moon formed as a result of a collision between a Mars-sized body and a youthful, semimolten Earth about 4.5 billion years ago. During this collision, some of the ejected debris was thrown into orbit around Earth and gradually coalesced to form the Moon. Computer simulations show that most of the ejected material would have come from the rocky mantle of the impactor, while its core was assimilated into the growing Earth. This *impact model* is consistent with the Moon having

a proportionately smaller core than Earth's and, hence, a lower density.

The Lunar Surface

When Galileo first pointed his telescope toward the Moon, he observed two different types of terrain: dark lowlands and brighter, highly cratered highlands (**Figure 15.22**). Because the dark regions appeared to be smooth,

Figure 15.22 Telescopic view of the lunar surface The major features are the dark maria and the light, highly cratered highlands. (Lick Observatory Publications Office)

resembling seas on Earth, they were called **maria** (*mar* = sea, singular *mare*). The *Apollo 11* mission showed conclusively that the maria are exceedingly smooth plains composed of basaltic lavas. These vast plains are strongly concentrated on the side of the Moon facing Earth and cover about 16 percent of the lunar surface. The lack of large volcanic cones on these surfaces is evidence of high eruption rates of very fluid basaltic lavas similar to the Columbia Plateau flood basalts on Earth.

By contrast, the Moon's light-colored areas resemble Earth's continents, so the first observers dubbed them *terrae* (Latin for "lands"). These areas are now called **lunar highlands** because they are elevated several kilometers above the maria. Rocks retrieved from the highlands are mainly breccias, pulverized by massive bombardment early in the Moon's history. The arrangement of terrae and maria has resulted in the legendary "face" of the "man in the Moon."

Some of the most obvious lunar features are impact craters. A meteoroid 3 meters (10 feet) in diameter can blast out a crater 50 times larger, or about 150 meters (500 feet) in diameter. The larger craters shown in Figure 15.22, such as Kepler and Copernicus (32 and 93 kilometers [20 and 58 miles] in diameter, respectively), were created from bombardment by bodies 1 kilometer (0.62 mile) or more in diameter.

History of the Lunar Surface The evidence used to unravel the history of the lunar surface comes primarily from radiometric dating of rocks returned from *Apollo* missions and studies of crater densities—counting the number of craters per unit area. The greater the crater density, the older the feature is inferred to be. Such evidence suggests that, after the Moon coalesced, it passed through four phases: (1) formation of the original crust, (2) excavation of the large impact basins, (3) filling of maria basins, and (4) formation of rayed craters.

During the late stages of its accretion, the Moon's outer shell was most likely completely melted—literally a magma ocean. Then, about 4.4 billion years ago, the magma ocean began to cool and underwent magmatic differentiation (see Chapter 2). Most of the dense minerals, olivine and pyroxene, sank, while less-dense silicate minerals floated to form the Moon's crust. The highlands are made of these igneous rocks, which rose buoyantly like "scum" from the crystallizing magma. The most common highland rock type is *anorthosite*, composed mainly of calcium-rich plagioclase feldspar.

Once formed, the lunar crust was continually impacted as the Moon swept up debris from the solar nebula. During this time, several large impact basins were created. Then, about 3.8 billion years ago, the Moon, as well as the rest of the solar system, experienced a sudden drop in the rate of meteoritic bombardment.

The Moon's next major event was the filling of the large impact basins that were created at least 300 million years earlier (**Figure 15.23**). Radiometric dating of

 Tutorial

Impact of an asteroid-size body produced a huge crater hundreds of kilometers in diameter and disturbed the lunar crust far beyond the crater.

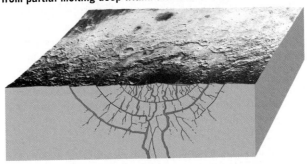

Filling of the impact crater with fluid basalts, perhaps derived from partial melting deep within the lunar mantle.

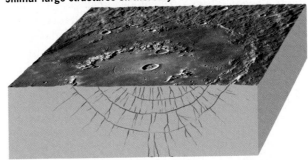

Today these lava-filled basins make up the lunar maria and similar large structures on Mercury.

the maria basalts puts their age between 3.0 billion and 3.5 billion years, considerably younger than the initial lunar crust.

The maria basalts are thought to have originated at depths between 200 and 400 kilometers (125 and 250 miles). They were likely generated by a slow rise in temperature attributed to the decay of radioactive elements. Partial melting probably occurred in several isolated pockets, as indicated by the diverse chemical makeup of the rocks retrieved during the *Apollo* missions. Recent evidence suggests that some mare-forming eruptions may have occurred as recently as 1 billion years ago.

Figure 15.24 Astronaut Harrison Schmitt, sampling the lunar surface Notice the footprint (inset) in the lunar "soil," called regolith, which lacks organic material and is therefore not a true soil. (Courtesy of NASA)

Other lunar surface features related to this period of volcanism include small shield volcanoes (8 to 12 kilometers [5 to 7.5 miles] in diameter), evidence of pyroclastic eruptions, narrow winding valleys (*rills*) thought to be collapsed lava tubes, and long linear depressions similar to down-faulted valleys (*grabens*) on Earth.

The last prominent features to form were rayed craters, as exemplified by the roughly 93-kilometer-wide (58-mile-wide) Copernicus crater shown in Figure 15.22. Light-colored material ejected from these craters, called *rays* because they radiate outward, blankets the maria surfaces and many older, rayless craters. The relatively young Copernicus crater is thought to be about 1 billion years old. Had it formed on Earth, weathering and erosion would have long since obliterated it.

Today's Lunar Surface The Moon's small mass and low gravity account for its lack of atmosphere and flowing water; therefore, the processes of weathering and erosion that continually modify Earth's surface are absent. In addition, tectonic forces are no longer active on the Moon, so quakes and volcanic eruptions have ceased. Because the Moon is unprotected by an atmosphere, erosion is

dominated by the impact of tiny particles from space (*micrometeorites*) that continually bombard its surface and gradually smooth the landscape. This activity has crushed and repeatedly mixed the upper portions of the lunar crust.

Both the maria and terrae are mantled with a layer of gray, unconsolidated debris derived from a few billion years of meteoric bombardment (**Figure 15.24**). This soil-like layer, properly called **lunar regolith** (*rhegos* = blanket, *lithos* = stone), is anywhere from 2 to 20 meters (6.6 to 66 feet) thick, composed of igneous rocks, breccia, glass beads, and fine *lunar dust*.

15.4 CONCEPT CHECKS

1. Briefly describe the origin of the Moon.
2. Compare and contrast the Moon's maria and highlands.
3. How are maria on the Moon similar to the Columbia Plateau in the Pacific Northwest?
4. How is crater density used in the relative dating of the Moon's surface features?
5. Summarize the major stages in the development of the modern lunar surface.

15.5 Terrestrial Planets

Outline the principal characteristics of Mercury, Venus, and Mars. Describe their similarities to and differences from Earth.

The terrestrial planets, in order from the Sun, are Mercury, Venus, Earth, and Mars. Here we consider the three other Earth-like planets and compare their features to those of Earth.

Mercury: The Innermost Planet

Mercury, the innermost and smallest planet, revolves around the Sun quickly (88 days) but rotates slowly

on its axis. Mercury's day–night cycle, which lasts 176 Earth days, is very long compared to Earth's 24-hour cycle. One "night" on Mercury is roughly equivalent to 3 months on Earth and is followed by the same duration of daylight. Mercury has the greatest temperature extremes, from as low as –173°C (–280°F) at night to noontime temperatures exceeding 427°C (800°F), hot enough to melt tin and lead and making life as we know it impossible.

Mercury absorbs most of the sunlight that strikes it, reflecting only 6 percent into space, a characteristic of terrestrial bodies with little or no atmosphere. The minuscule amount of gas that is present on Mercury may have originated from several sources, including ionized gas from the Sun, ices that vaporized during a relatively recent comet impact, and outgassing of the planet's interior.

Although Mercury is small and scientists expected the planet's interior to have already cooled, NASA's *Messenger* spacecraft found in 2012 that Mercury has a magnetic field, although it is about 100 times weaker than Earth's. This suggests that Mercury has a large core that remains hot and fluid—a requirement for generating a magnetic field.

Mercury resembles Earth's Moon in that it has very low reflectivity, no sustained atmosphere, numerous volcanic features, and a heavily cratered terrain (**Figure 15.25**). The largest-known impact crater on Mercury is Caloris Basin (1300 kilometers [800 miles] in diameter). Like our Moon, Mercury has extensive smooth plains; they cover nearly 40 percent of the area imaged by *Mariner 10*. Most of these smooth areas are associated with large impact basins, including Caloris Basin, where lava partially filled the basins and surrounding lowlands. Consequently, they appear to be similar in origin to lunar maria. Recently, *Messenger* found evidence of other types of volcanism on Mercury, including a huge flood basalt province reminiscent of, but much larger than, Earth's Columbia Plateau. Researchers also confirmed the presence of substantial deposits of ice within perpetually shadowed polar craters.

Venus: The Veiled Planet

Venus, second only to the Moon in brilliance in the night sky, is named for the Roman goddess of love and beauty. It orbits the Sun in a nearly perfect circle once every 225 Earth days. However, Venus rotates in the opposite direction of the other planets (*retrograde motion*) at an agonizingly slow pace: 1 Venus day is equivalent to about 243 Earth days. Venus has the densest atmosphere of the terrestrial planets, consisting mostly of carbon dioxide (97 percent)—and it is the prototype for an extreme *greenhouse effect*. As a consequence, the surface temperature of Venus averages more than 450°C (900°F) day and night. Temperature variations at the surface are generally minimal because of the intense mixing within the planet's dense atmosphere. Investigations of the planet's extreme and uniform surface temperature led scientists to more fully understand how the greenhouse effect operates on Earth.

The composition of the Venusian interior is probably similar to Earth's. However, Venus's weak magnetic field means its internal dynamics must be very different from Earth's. Scientists think that mantle convection operates on Venus, but the processes of plate tectonics do not appear to have contributed to the present Venusian topography.

The surface of Venus is completely hidden from view by a thick cloud layer composed mainly of tiny sulfuric acid droplets. Between 1961 and 1984, despite extreme temperatures and pressures, 10 Russian spacecraft landed successfully and transmitted data including surface images. As expected, however, all the probes were

Figure 15.25 Two views of Mercury On the left is a monochromatic image, while the image on the right is color enhanced. These are high-resolution mosaics constructed from thousands of images obtained by the *Messenger* orbiter. (Courtesy of NASA)

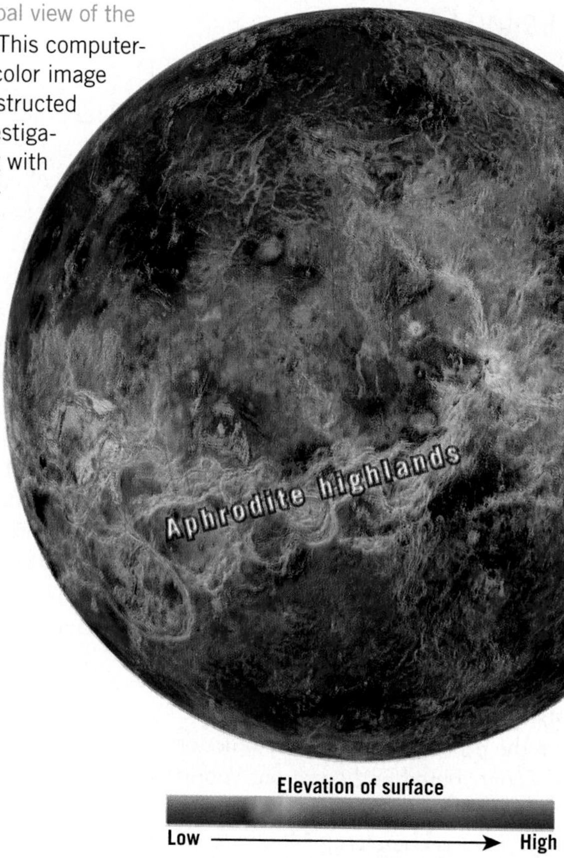

Figure 15.26 Global view of the surface of Venus This computer-generated, false-color image of Venus was constructed from years of investigations, culminating with the *Magellan* mission. The twisting bright features that cross the globe are highly fractured ridges and canyons of the eastern Aphrodite highland. (Courtesy of NASA)

Aphrodite highlands

Elevation of surface

Low ⟶ High

crushed by the planet's immense atmospheric pressure, approximately 90 times that on Earth, within an hour of landing. Using radar imaging, the unstaffed spacecraft *Magellan* mapped Venus's surface in stunning detail (**Figure 15.26**).

A few thousand impact craters have been identified on Venus—far fewer than on Mercury and Mars but more than on Earth. Researchers expected that Venus

would show evidence of extensive cratering from the heavy bombardment period but found instead that a period of extensive volcanism was responsible for resurfacing Venus. The planet's thick atmosphere also limits the number of impacts by breaking up large incoming meteoroids and incinerating most of the small debris.

About 80 percent of the Venusian surface consists of low-lying plains covered by lava flows, some of which traveled along lava channels that extend hundreds of kilometers. Venus's Baltis Vallis, the longest-known lava channel in the solar system, meanders 6800 kilometers (4255 miles) across the planet. More than 1000 volcanoes with diameters greater than 20 kilometers (12 miles) have been identified on Venus. However, high surface pressures keep the gaseous components in lava from escaping and limits production of pyroclastic material and lava fountaining, phenomena that tend to steepen volcanic cones. In addition, Venus's high temperatures allow lava to remain mobile longer and, thus, flow far from the vent. Both of these factors result in volcanoes that tend to be flatter and wider than those on Earth or Mars (**Figure 15.27**). Maat Mons, the largest volcano on Venus, is about 8.5 kilometers (5 miles) high and 400 kilometers (250 miles) wide. By comparison, Mauna Loa, Earth's largest volcano, is about 9 kilometers high (5.5 miles) and only 120 kilometers (75 miles) wide.

Venus also has major highlands consisting of plateaus, ridges, and topographic rises that stand above the plains. The rises are thought to have formed where hot mantle plumes encountered the base of the planet's crust, causing uplift. Much like mantle plumes on Earth, abundant volcanism is associated with mantle upwelling on Venus. Recent data collected by the European Space Agency's *Venus Express* suggest that Venus's highlands contain silica-rich granitic rock. These elevated landmasses resemble Earth's continents, albeit on a much smaller scale.

Mars: The Red Planet

Mars, approximately one-half the diameter of Earth, revolves around the Sun in 687 Earth days. Mean surface temperatures range from lows of –140°C (–220°F) at the poles in the winter to highs of 20°C (68°F) at the equator in the summer. Although seasonal temperature variations are similar to Earth's, daily temperature variations are greater due to the very thin atmosphere of Mars (only 1 percent as dense as Earth's). The tenuous Martian atmosphere consists primarily of carbon dioxide (95 percent), with small amounts of nitrogen, oxygen, and water vapor.

Figure 15.27 Volcanoes on Venus Sapas Mons is a broad volcano, 400 kilometers (250 miles) wide. The bright areas in the foreground are lava flows. Another large volcano, Maat Mons, is in the background. (This image has considerable vertical exaggeration.) (Courtesy of NASA)

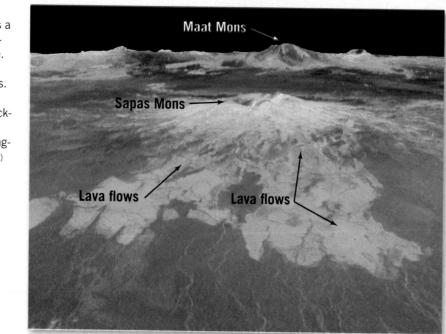

Maat Mons

Sapas Mons

Lava flows

Lava flows

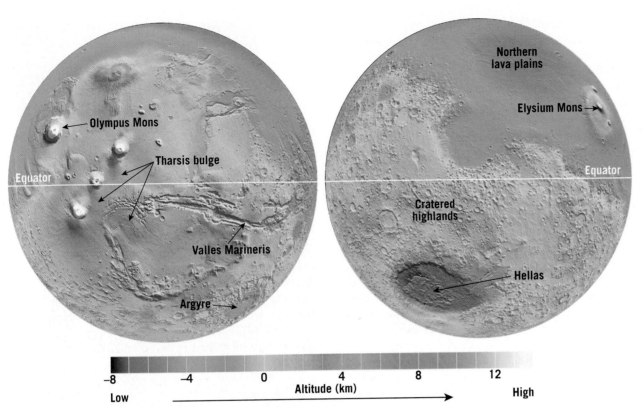

Figure 15.28 Two hemispheres of Mars Color represents height above (or below) the mean planetary radius: White is about 12 kilometers above average, and dark blue is 8 kilometers below average. (Courtesy of NASA)

Martian Topography Mars, like the Moon, is pitted with impact craters. The smaller craters are usually filled with wind-blown dust—confirming that Mars is a dry, desert world. The reddish color of the Martian landscape is due to iron oxide (rust). Large impact craters provide information about the nature of the Martian surface. For example, where the Martian surface is composed of dry rocky debris, ejecta similar in size and shape to that surrounding lunar craters is found. However, some Martian craters feature ejecta that looks like muddy slurry was splashed from the crater. Planetary geologists infer that a layer of permafrost (frozen, icy soil) lies below portions of the Martian surface and that the heat of impacts melted the ice to produce the fluid-like appearance of these ejecta.

About two-thirds of the surface of Mars consists of heavily cratered highlands, concentrated mostly in its southern hemisphere (**Figure 15.28**). The period of extreme cratering occurred early in the planet's history and ended about 3.8 billion years ago, as it did in the rest of the solar system. Thus, Martian highlands are similar in age to the lunar highlands.

The remaining one-third of the planet, located in the northern hemisphere, is covered by low plains. Based on their relatively low crater counts, these northern plains are younger than the highlands. Their flat topography, possibly the smoothest surface in the solar system, is consistent with vast outpourings of fluid basaltic lavas. Visible on the plains are volcanic cones, some with summit pits (craters) and lava flows with wrinkled edges. If Mars once had abundant water, it would have flowed to the north, which is lower in elevation, possibly forming an expansive ocean.

Located along the Martian equator is an enormous elevated region about the size of North America, called the *Tharsis bulge*. This feature, about 10 kilometers (6 miles) high, appears to have been uplifted and capped with a massive accumulation of volcanic rock that includes the solar system's largest volcanoes.

The tectonic forces that created the Tharsis region also produced fractures that radiate from its center, like spokes on a bicycle wheel. Along the eastern flanks of the bulge, a series of vast canyons called *Valles Marineris* (Mariner Valleys) developed (see Figure 15.28). This canyon network was largely created by down-faulting rather than the stream erosion that carved Arizona's Grand Canyon. Thus, it consists of graben-like valleys similar to the East African Rift valleys. Once formed, Valles Marineris grew thanks to water erosion and collapse of the rift walls. The main canyon is more than 5000 kilometers (3000 miles) long, 7 kilometers (4 miles) deep, and 100 kilometers (60 miles) wide.

Other prominent features on the Martian landscape are large impact basins. Hellas, the largest visible impact basin on the planet, is about 2300 kilometers (1400 miles) in diameter and has the planet's lowest elevation. Debris ejected from this basin contributed to the elevation of the adjacent highlands. Other buried crater basins even larger than Hellas exist, including Utopia Basin, where *Viking 2* landed.

Volcanoes on Mars Volcanism has been prevalent on Mars during most of its history. The scarcity of impact craters on some volcanic surfaces suggests that the planet is still active. Mars has several of the solar system's largest known volcanoes, including the largest, Olympus Mons, which is about the size of Arizona and stands nearly three times higher than Mount Everest. This enormous volcano was active as recently as a few million years ago and resembles Earth's Hawaiian shield volcanoes (**Figure 15.29**).

How did the volcanoes on Mars grow so much larger than similar structures on Earth? The largest volcanoes on the terrestrial planets tend to form where plumes of hot rock rise from deep within their interiors. On Earth, moving plates keep the crust in constant motion. Consequently, mantle plumes tend to produce a chain of volcanic structures, like the Hawaiian Islands. By contrast, plate movement on Mars is absent, so successive eruptions accumulate in the same location, creating enormous volcanoes rather than a string of smaller ones.

Wind Erosion on Mars The dominant force currently shaping the Martian surface is wind erosion. Extensive dust storms with winds up to 270 kilometers (170 miles) per hour can persist for weeks. Dust devils have also been photographed. Most of the Martian landscape resembles Earth's rocky deserts, with abundant dunes and low areas partially filled with dust.

Water on Mars in the Past Considerable evidence indicates that in the first 1 billion years of the planet's history, liquid water flowed on Mars's surface, creating stream valleys and related features. One location where running water was involved in carving valleys can be seen in the *Mars Reconnaissance Orbiter* image

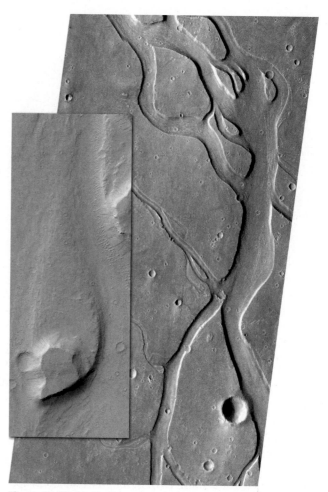

Figure 15.30 Earth-like stream channels are strong evidence that Mars once had flowing water Inset shows a close-up of a streamlined island where running water encountered resistant material along its channel. (Courtesy of NASA)

SmartFigure 15.29
Olympus Mons This massive inactive shield volcano on Mars covers an area about the size of the state of Arizona. (Courtesy of NASA) (goo.gl/ejZphx)

in **Figure 15.30**. Notice the stream-like banks that contain numerous teardrop-shaped islands. These valleys appear to have been cut by catastrophic floods with discharge rates that were more than 1000 times greater than those of the Mississippi River. Most of these large flood channels emerge from areas of chaotic topography that appear to have formed when the surface collapsed. The most likely source of water for these flood-created valleys was the melting of subsurface ice. However, not all Martian valleys were generated from water released in this manner. Some exhibit branching, tree-like patterns resembling dendritic stream drainage networks on Earth.

On August 6, 2012, the Mars rover *Curiosity* landed in Gale Crater, an impact crater that contains a 5-kilometer-high (3-mile-high) accumulation of sediment called Mount Sharp. As of October 2015, *Curiosity* had traveled over 10 kilometers (6 miles), examining the lower slopes of this layered mountain. At a target zone NASA calls "Big Arm," the Mars Hand Lens Imager (MAHLI) took images of sediment containing rounded grains, indicating that the grains traveled long distances before being deposited. Analysis of these sediments indicates that Gale Crater periodically filled with water, forming a lake that lasted hundreds or possibly even thousands of years. This is strong evidence that Mars must have had a much thicker atmosphere that supported a hydrologic cycle similar to Earth's. It also means that other craters most likely supported long-lived lakes that may have provided suitable habitat for microbial life.

Does Liquid Water Exist on Present-Day Mars? We have learned from NASA's *Phoenix Mars Lander*, which dug into the Martian surface, that poleward of about 30° latitude, ice can be found within 1 meter (3 feet) of the surface. Furthermore, Mars's permanent polar ice caps are composed of mainly pure water ice, blanketed by a thin layer of carbon dioxide ice during the cold season. Current estimates place the maximum amount of water ice held by the Martian polar ice caps at about 1.5 times the amount covering Greenland. However, this water appears to remain frozen year-round.

Recent images from a high-resolution camera aboard NASA's *Mars Reconnaissance Orbiter* show dark streaks on Mars, called *recurring slope lineae* (**Figure 15.31**). Researchers believe that these streaks, which appear seasonally on steep, relatively warm Martian slopes, may be caused by the flow of briny (salty) liquid water. Although these dark streaks are just 0.5 to 5 meters (1.6 to 16 feet) wide, they can extend for hundreds of meters downslope. In addition, these features appear during warm weather but fade away when the temperatures drop, providing further evidence that liquid water is involved in their formation. Salts, which are widespread on the Martian surface, would lower the freezing point of water from 0°C (32°F) to–70°C (–94°F). The discovery of these dark streaks has implications for future human exploration of Mars. In the late 2030s, when NASA plans to

Figure 15.31 These dark streaks on Mars are thought to be caused by the flow of briny (salty) water These streaks, called *recurring slope lineae*, are found on steep, warm Martian slopes and disappear during the cold season. (Photo courtesy of NASA)

Dark streaks
(Recurring slope lineae)

send astronauts to the red planet, the presence of liquid water—even very salty water—would aid that ambitious effort.

15.5 CONCEPT CHECKS

1. What body in our solar system is most like Mercury?
2. Venus was once referred to as "Earth's twin." How are these two planets similar? How do they differ from one another?
3. What surface features do Mars and Earth have in common?
4. Why are the largest volcanoes on Earth so much smaller than the largest ones on Mars?
5. What evidence suggests that Mars had an active hydrologic cycle in the past?

15.6 Jovian Planets

Summarize and compare the features of Jupiter, Saturn, Uranus, and Neptune, including their ring systems.

The four Jovian planets, in order from the Sun, are Jupiter, Saturn, Uranus, and Neptune. Because of their location within the solar system and their size and composition, they are also commonly called the *outer planets* and the *gas giants*.

Jupiter: Lord of the Heavens

The giant among planets, Jupiter has a mass 2.5 times greater than the combined mass of all other planets, satellites, and asteroids in the solar system. However, it pales in comparison to the Sun, with only 1/800 of the Sun's mass.

Jupiter orbits the Sun once every 12 Earth years, and it rotates more rapidly than any other planet, completing one rotation in slightly less than 10 hours. When viewed telescopically, the effect of this fast spin is noticeable. The bulge of the equatorial region and the slight flattening at the poles are evident (see the "Polar Flattening" column in Table 15.2).

Jupiter's appearance is mainly attributable to the colors of light reflected from its three main cloud layers

Figure 15.32 The structure of Jupiter's atmosphere The areas of light clouds (*zones*) are regions where gases are ascending and cooling. Sinking and warming dominate the flow in the darker cloud layers (*belts*). This convective circulation, along with the planet's rapid rotation, generates the high-speed winds observed between the belts and zones.

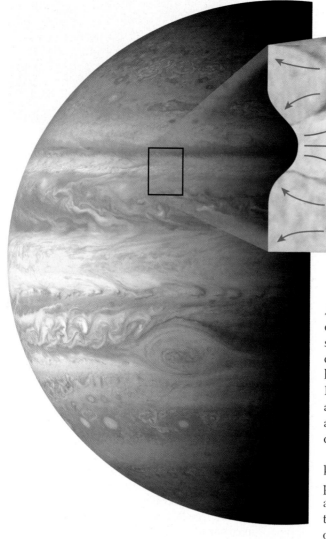

The largest storm on the planet is the Great Red Spot. This enormous anticyclonic storm, twice the size of Earth, has been known for 300 years. In addition to the Great Red Spot, there are various white and brown oval-shaped storms. The white ovals are the cold cloud tops of huge storms many times larger than hurricanes on Earth. The brown storm clouds reside at lower levels in the atmosphere. Lightning in various white oval storms has been photographed by the *Cassini* spacecraft, but the strikes appear to be less frequent than on Earth.

Jupiter's Moons Jupiter's satellite system, consisting of 67 moons discovered thus far, resembles a miniature solar system. Galileo discovered the 4 largest satellites, called Galilean satellites, in 1610 (**Figure 15.33**). The 2 largest, Ganymede and Callisto, are roughly the size of Mercury, whereas the two smaller ones, Europa and Io, are about the size of Earth's Moon. The 8 largest moons appear to have formed around Jupiter as the solar system condensed.

Jupiter also has many very small satellites (about 20 kilometers [12 miles] in diameter) that revolve in the opposite direction (retrograde motion) of the largest moons and have eccentric (elongated) orbits steeply inclined to the Jovian equator. These satellites appear to be asteroids or comets that passed near enough to be either gravitationally captured by Jupiter or remnants of the collisions of larger bodies.

The Galilean moons can be observed with binoculars or a small telescope and are interesting in their own right. Images from *Voyagers 1* and *2* revealed, to the surprise of most geoscientists, that each of the four Galilean satellites is a unique world (Figure 15.33). The *Galileo* mission also unexpectedly revealed that the composition of each satellite is strikingly different, implying a different evolution for each. For example, Ganymede has a dynamic core that generates a strong magnetic field not observed in other satellites.

The innermost Galilean moon, Io, is perhaps the most volcanically active body in our solar system. More than 80 active, sulfurous volcanic centers have been discovered. Umbrella-shaped plumes have been observed rising from Io's surface to heights exceeding 100 kilometers (60 miles) (**Figure 15.34A**). The heat source for volcanic activity is tidal energy generated by a relentless "tug of war" between Jupiter and the other Galilean satellites—with Io as the rope. The gravitational field of Jupiter and the other nearby satellites pull and push on Io's tidal bulge as its slightly eccentric orbit takes it

(**Figure 15.32**). The warmest, and lowest, layer is composed mainly of water ice and appears blue-gray, while the middle layer, where temperatures are lower, consists of brown to orange-brown clouds of ammonium hydrosulfide droplets. These colors are thought to be by-products of chemical reactions occurring in Jupiter's atmosphere. Near the top of its atmosphere lie white wispy clouds of ammonia ice.

Because of its immense gravity, Jupiter is shrinking a few centimeters each year. This contraction generates most of the heat that drives Jupiter's atmospheric circulation. Thus, unlike winds on Earth, which are driven by solar energy, the heat emanating from Jupiter's interior produces the huge convection currents observed in its atmosphere.

Jupiter's convective flow produces alternating dark-colored *belts* and light-colored *zones*, as shown in Figure 15.32. The light clouds (*zones*) are regions where warm material is ascending and cooling, whereas the dark belts represent cool material that is sinking and warming. This convective circulation, along with Jupiter's rapid rotation, generates the high-speed, east–west flow observed between the belts and zones.

alternately closer to and farther from Jupiter. This gravitational flexing of Io is transformed into heat (similar to the back-and-forth bending of a piece of sheet metal) and results in Io's spectacular sulfurous volcanic eruptions. Lava, thought to be mainly composed of silicate minerals, regularly erupts on its surface (**Figure 15.34B**).

One of the best prospects of finding liquid water within our solar system lies beneath the icy surfaces of some of Jupiter's moons. For example, detailed images from *Galileo* have revealed that Europa's icy surface is quite young and exhibits cracks apparently filled with dark fluid from below. This suggests that under its icy shell, Europa must have a warm, mobile interior—perhaps an ocean. Because liquid water is a necessity for life as we know it, there is considerable interest in sending an orbiter to Europa—and, eventually, a lander capable of launching a robotic submarine—to determine whether it harbors life.

Jupiter's Rings One of the surprising aspects of the *Voyager 1* mission was the discovery of Jupiter's ring system. More recently, the ring system was thoroughly investigated by the *Galileo* mission. By analyzing how these rings scatter light, researchers determined that the rings are composed of fine, dark particles similar in size to smoke particles. Furthermore, the faint nature of the rings indicates that these minute particles are widely dispersed. The main ring is composed of particles believed to be fragments blasted from the surfaces of Metis and Adrastea, two small moons of Jupiter. Impacts on Jupiter's moons Amalthea and Thebe are believed to be the source of the debris from which the outer gossamer ring formed.

Saturn: The Elegant Planet

Requiring more than 29 Earth years to make one revolution, Saturn is almost twice as far from the Sun as Jupiter,

A. Io, perhaps the most volcanically active body in our solar system, has more than 80 active, sulfurous volcanic structures .

B. Europa's icy surface is quite flat and thought to cover a vast ocean composed of briny water.

C. Ganymede, the largest of the Jovian satellites, contains both smooth as well as cratered regions, which suggest this body is still active.

D. Callisto, the outermost of the Galilean satellites, is densely cratered, much like Earth's Moon.

Figure 15.33 Jupiter's four largest moons
These moons are often referred to as the Galilean moons because Galileo discovered them. (Courtesy of NASA)

yet their atmospheres, compositions, and internal structures are remarkably similar. The most striking feature of Saturn is its system of rings, first observed by Galileo in 1610 (**Figure 15.35**). Through his primitive telescope, the rings appeared as two small bodies adjacent to the planet. Their ring nature was determined 50 years later by Dutch astronomer Christiaan Huygens.

Saturn's atmosphere, like Jupiter's, is dynamic. Although the bands of clouds are fainter and wider near the equator, rotating "storms" similar to Jupiter's Great Red Spot occur in Saturn's atmosphere, as does intense lightning. Although the atmosphere is about 93 percent hydrogen and 3 percent helium by volume, the clouds

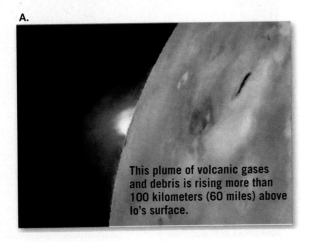

A.

This plume of volcanic gases and debris is rising more than 100 kilometers (60 miles) above Io's surface.

B.

The bright red area on the left side of the image (see arrow) is newly erupted lava.

Figure 15.34 A volcanic eruption on Jupiter's moon Io (Courtesy of NASA; Jet Propulsion Laboratory/University of Arizona/NASA)

Figure 15.35 Saturn's major rings The two bright rings, called A ring (outer) and B ring (inner), are separated by the Cassini division. A second small gap (Encke gap) is also visible as a thin line in the outer portion of the A ring. (Courtesy of NASA)

(or condensed gases) are composed mainly of ammonia, ammonium hydrosulfide, and water, each segregated by temperature. Also, much like Jupiter, Saturn emits roughly twice as much energy as it receives from the Sun. This implies that it must have an internal heat source, which may come from chemical differentiation in its interior.

Saturn's Moons The Saturnian satellite system consists of 62 known moons, of which 53 have been named. The moons vary significantly in size, shape, surface age, and origin. Twenty-three of the moons are "original" satellites that formed in tandem with their parent planet. Many of Saturn's smallest moons have irregular shapes and are only a few tens of kilometers in diameter.

Saturn's largest moon, Titan, is larger than Mercury and is the second-largest satellite in the solar system. Titan and Neptune's Triton are the only satellites in the solar system known to have substantial atmospheres. Titan was visited and photographed by the *Cassini-Huygens* probe in 2005. The atmospheric pressure at Titan's surface is about 1.5 times that at Earth's surface, and the atmospheric composition is about 98 percent nitrogen and 2 percent methane, with trace organic compounds. Titan has Earth-like landforms and geologic processes, such as dune formation and stream-like erosion caused by methane "rain." In addition, the northern latitudes appear to have lakes of liquid methane.

Enceladus is another unique satellite of Saturn—one of a few icy moons that erupt "fluid" ice containing minor amounts of other debris. This amazing manifestation of volcanism, called **cryovolcanism** (from the Greek *kryos*, meaning "frost") describes the eruption of magmas derived from the partial melting of ice instead of silicate rocks (**Figure 15.36**). This outgassing occurs in areas called "tiger stripes" that consist of large fractures with ridges on either side. The material ejected by these eruptions

is thought to be the source of material that replenishes Saturn's E ring.

Saturn's Ring System In the early 1980s, the nuclear-powered *Voyagers 1* and *2* explored Saturn within 160,000 kilometers (100,000 miles) of its surface. More information was collected about Saturn in that short time than had been acquired since Galileo first viewed this "elegant planet" in the early 1600s. More recently,

Figure 15.36 Enceladus, one of Saturn's tectonically active, icy satellites Enceladus has active linear features, called tiger stripes, which are a source region for cryovolcanic activity. Inset image shows jets spurting ice particles, water, and organic compounds from the area of the tiger stripes. (Courtesy of NASA)

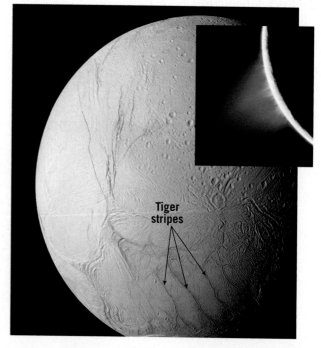

observations from ground-based telescopes, the Hubble Space Telescope, and the *Cassini-Huygens* spacecraft, have added to our knowledge of Saturn's ring system. In 1995 and 1996, when the positions of Earth and Saturn allowed the rings to be viewed edge-on, Saturn's faintest rings and satellites became visible. (The rings were visible edge-on again in 2009.)

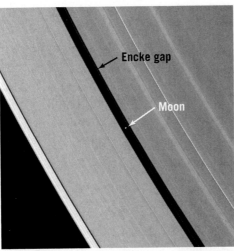 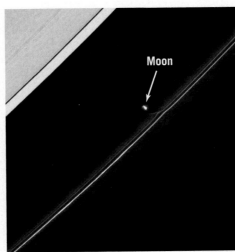

Figure 15.37 Two of Saturn's ring moons (Courtesy of NASA)

A. Pan is a small moon about 30 kilometers in diameter that orbits in the Encke gap, located in the A ring. It is responsible for keeping the Encke gap open by sweeping up any stray material that may enter.

B. Prometheus, a potato-shaped moon, acts as a ring shepherd. Its gravity helps confine the particles that make up Saturn's thin Fring.

Saturn's ring system is more like a large rotating disk of varying density and brightness than a series of independent ringlets. Each ring is composed of individual particles—mainly water ice, with lesser amounts of rocky debris—that circle the planet while regularly impacting one another. There are only a few gaps; most of the areas that look like empty space either contain fine dust particles or coated ice particles that are inefficient light reflectors.

Most of Saturn's rings fall into one of two categories, based on density. Saturn's main (bright) rings, designated A and B, are tightly packed and contain particles ranging in size from a few centimeters (pebble-size) to tens of meters (house-size), with most of the particles being roughly the size of a large snowball (see Figure 15.35). In these dense rings, particles collide frequently as they orbit the planet. Although Saturn's main rings (A and B) are 40,000 kilometers (25,000 miles) wide, they are very thin, only 10–30 meters (30–100 feet) from top to bottom.

At the other extreme are Saturn's faint rings. Saturn's outermost ring (E ring), not visible in Figure 15.21, is composed of widely dispersed, tiny particles. Recall that cryovolcanism (eruption of water-ice mixture) on Saturn's satellite Enceladus is thought to be the source of material for the E ring.

Studies have shown that the gravitational tugs of nearby moons tend to "shepherd" the ring particles by gravitationally altering their orbits (**Figure 15.37**). For example, the F ring, which is very narrow, appears to be the work of satellites located on either side that confine the ring by pulling back particles that try to escape. On the other hand, the Cassini division, a clearly visible gap in Figure 15.35, arises from the gravitational pull of Mimas, one of Saturn's moons.

Some of the ring particles are believed to be debris ejected from the moons embedded in them. It is also possible that material is continually recycled between the rings and the ring moons. The ring moons gradually sweep up particles, which are subsequently ejected by collisions with large chunks of ring material, or perhaps by energetic collisions with other moons. It seems, then, that planetary rings are not the timeless features that we once thought; rather, they are continually recycled.

The origin of planetary ring systems is still being debated. Saturn's rings probably formed when objects like comets, asteroids, or perhaps even moons were pulled apart by Saturn's strong gravity. Pieces of these objects would have collided with each other, breaking into even smaller pieces. Collisions among these fragments would tend to jostle one another and cause them to spread out to form the flat, thin ring system we observe today. Saturn's rings, as well as those of the other planets, are thought to be short-lived compared to the age of our solar system. This means that Saturn probably lacked rings early in its history, and its existing ring system will likely dissipate in the distant future.

Uranus and Neptune: Twins

Although Earth and Venus have many similar traits, Uranus and Neptune are perhaps more deserving of being called "twins." They are nearly equal in diameter (both about four times the size of Earth), and they are both bluish in appearance, as a result of methane in their atmospheres. Their days are nearly the same length, and their cores are made of rocky silicates and iron—similar to the other Jovian planets. Their mantles, made mainly of water, ammonia, and methane, are thought to be very different from those of Jupiter and Saturn. One of the most pronounced differences between Uranus and Neptune is the time they take to complete one revolution around the Sun—84 and 165 Earth years, respectively.

Figure 15.38 Uranus, surrounded by its major rings and a few of its known moons Also visible in this image are cloud patterns and several oval storm systems. This false-color image was generated from data obtained by Hubble's Near Infrared Camera. (Courtesy of NASA)

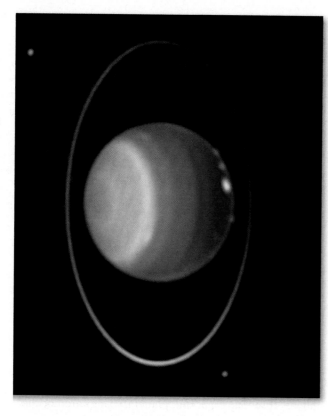

Uranus: The Sideways Planet

Unique to Uranus is the orientation of its axis of rotation. Whereas the other planets resemble spinning toy tops as they circle the Sun, Uranus is like a top that has been knocked on its side but remains spinning (**Figure 15.38**). This unusual characteristic of Uranus is likely due to one or more impacts that essentially knocked the planet sideways from its original orientation early in its evolution.

Uranus shows evidence of huge storm systems larger than Earth's continents. Recent photographs from the Hubble Space Telescope also reveal banded clouds composed mainly of ammonia and methane ice—similar to the cloud systems of the other Jovian planets.

Uranus's Moons

Spectacular views from *Voyager 2* showed that Uranus's five largest moons have varied terrains. Some have long, deep canyons and linear scars, whereas others possess large, smooth areas on otherwise crater-riddled surfaces. Studies conducted at NASA's Jet Propulsion Laboratory suggest that Miranda, the innermost of the five largest moons, was recently geologically active—most likely driven by gravitational heating, as occurs on Io.

Uranus's Rings

A surprise discovery in 1977 showed that Uranus has a ring system. The discovery was made as Uranus passed in front of a distant star and blocked its view, a process called *occultation* (*occult* = hidden). Observers saw the star "wink" briefly five times (meaning five rings) before the primary occultation and again five

times afterward. More recent ground- and space-based observations indicate that Uranus has at least 10 sharp-edged, distinct rings orbiting its equatorial region. Interspersed among these distinct structures are broad sheets of dust.

Neptune: The Windy Planet

Because of Neptune's great distance from Earth, astronomers knew very little about this planet until 1989. Twelve years and nearly 3 billion miles of *Voyager 2* travel provided investigators an amazing opportunity to view the outermost planet.

Neptune has a dynamic atmosphere, much like that of the other Jovian planets (**Figure 15.39**). Record wind speeds exceeding 2400 kilometers (1500 miles) per hour encircle the planet, making Neptune one of the windiest places in the solar system. Neptune also exhibits large dark spots, thought to be rotating storms similar to Jupiter's Great Red Spot. However, Neptune's storms appear to have comparatively short life spans—usually only a few years. Another feature that Neptune has in common with the other Jovian planets is layers of white, cirrus-like clouds (probably frozen methane) about 50 kilometers (30 miles) above the main cloud deck.

Neptune's Moons

Neptune has 14 known satellites, the largest of which is Triton; the remaining 13 are small, irregularly shaped bodies. Recall that Triton and a few other icy moons erupt "fluid" ices—a manifestation of volcanism called *cryovolcanism*. Triton's icy magma is a mixture of water ice, methane, and probably ammonia. When partially melted, this mixture behaves as molten rock does on Earth. In fact, upon reaching the surface, these magmas can generate quiet outpourings of ice lavas

Figure 15.39 Neptune's dynamic atmosphere (Courtesy of NASA)

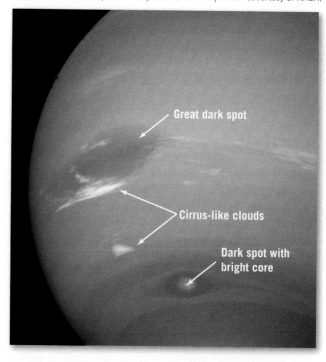

Great dark spot

Cirrus-like clouds

Dark spot with bright core

that can flow great distances from their source—similar to the fluid basaltic flows on Hawaii. They also occasionally produce explosive eruptions that can generate the ice equivalent of volcanic ash. In 1989, *Voyager 2* detected active plumes on Triton that rose 8 kilometers (5 miles) above the surface and were blown downwind for more than 100 kilometers (60 miles).

Neptune's Rings Neptune has five named rings; two of them are broad, and three are narrow, perhaps no more than 100 kilometers (60 miles) wide. The outermost ring appears to be partially confined by the satellite Galatea. Neptune's rings, like Jupiter's, appear faint, which suggests that they are composed mostly of dust-size particles. Neptune's rings also display red colors, indicating that the dust is composed of organic compounds.

 15.6 **CONCEPT CHECKS**

1. What is the nature of Jupiter's Great Red Spot?
2. What is distinctive about Jupiter's satellite Io?
3. How are Jupiter and Saturn similar to one another?
4. What two roles do ring moons play in the nature of planetary ring systems?
5. How are Saturn's satellite Titan and Neptune's satellite Triton similar to one another?
6. Name three bodies in the solar system that exhibit active volcanism.

15.7 Small Solar System Bodies

List and describe the principal characteristics of the small bodies that inhabit the solar system.

There are countless chunks of debris in the vast spaces separating the eight planets and in the outer reaches of the solar system. In 2006, the International Astronomical Union organized solar system objects not classified as planets or moons into two broad categories: (1) **small solar system bodies**, including *asteroids, comets,* and *meteoroids*, and (2) **dwarf planets**. The newest grouping, dwarf planets, includes Ceres, a body about 1000 kilometers (600 miles) in diameter and the largest known object in the asteroid belt, and Pluto, formerly considered a planet.

Asteroids and meteoroids are compositionally quite similar, both being composed of rocky and/or metallic material similar to that which makes up the terrestrial planets. They are usually distinguished by size, with asteroids being much larger than meteoroids, although the exact size difference is not well defined. Comets, on the other hand, are collections of ices, with lesser amounts of dust and small rocky particles. Comets mainly inhabit the outer reaches of the solar system.

Asteroids: Leftover Planetesimals

Asteroids are small bodies (planetesimals) that remain from the formation of the solar system, which means they are about 4.6 billion years old. The orbits of more than 100,000 asteroids have been accurately measured, and thousands more have orbits that are incompletely known, keeping them off the "official" list.

Most asteroids orbit the Sun between Mars and Jupiter, in the region known as the **asteroid belt** (**Figure 15.40**). There are only about 2 dozen asteroids that are more than 200 kilometers (125 miles) across. However, our solar system hosts an estimated 1 to 2 million asteroids larger than 1 kilometer (0.6 mile) across and many millions that are smaller.

A smaller number of asteroids travel along eccentric orbits that take them near the Sun, and about 1300 of these, called *Earth-crossing asteroids*, will eventually collide with Earth. Many of the recent large impact craters discovered on Earth resulted from collisions with asteroids. Although these events are rare, they merit our attention because of their potential destruction. As a result, observational initiatives that aim to measure asteroid orbits with great accuracy are ongoing.

Asteroid Structure and Composition The largest asteroids are roughly spherical because, as for planets and large moons, gravity determines their shape. Indirect evidence from meteorites suggests that the largest asteroids were heated by impact events early in their history, which caused them to melt. This resulted in an early period of chemical differentiation that produced their dense iron-rich cores and rocky mantles.

Figure 15.40 The asteroid belt The orbits of most asteroids lie between Mars and Jupiter. Also shown in red are the orbits of a few known near-Earth asteroids.

Figure 15.41 Asteroid Itokawa The barren rocky surface of asteroid Itokawa appears to be a pile of rubble held together by the asteroid's weak gravitational field. This potato-shaped asteroid orbits between Mars and Jupiter and is only about 0.5 kilometers across—the size of about five football fields. (Courtesy of Japan Aerospace Exploration Agency)

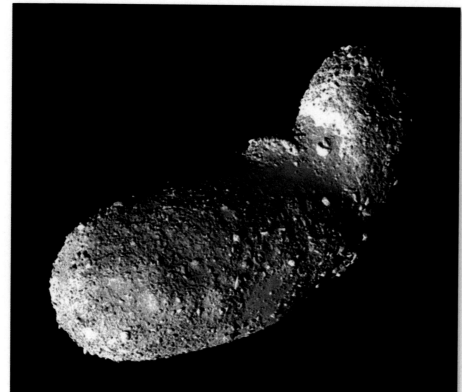

Most asteroids, however, are small and have irregular shapes, which led planetary geologists to conclude that they are leftover debris from the solar nebula. In addition, these small asteroids have densities lower than scientists originally predicted, indicating that they are relatively porous bodies, like "piles of rubble," loosely bound together by their weak gravitational fields (**Figure 15.41**).

In February 2001, an American spacecraft became the first visitor to an asteroid. Although it was not designed for landing, *NEAR Shoemaker* landed successfully on Eros and collected information that has planetary geologists both intrigued and perplexed. Images obtained as the spacecraft drifted toward the surface of Eros revealed a barren, rocky surface composed of particles ranging in size from fine dust to boulders up to 10 meters (30 feet) across. Researchers unexpectedly discovered that fine debris tends to concentrate in the low areas, where it forms flat deposits resembling ponds. Surrounding the low areas, the landscape is marked by an abundance of large boulders.

One of several hypotheses to explain the boulder-strewn topography is seismic shaking, which would cause the boulders to move upward as the finer materials sink. This is analogous to what happens when a jar of sand and various-sized pebbles is shaken: The larger pebbles rise to the top, while the smaller sand grains settle to the bottom (sometimes referred to as the Brazil nut effect).

Exploring Asteroids In November 2005, the Japanese probe *Hayabusa* made a soft landing on a small near-Earth asteroid named Itokawa, and picked up some rocky debris before returning to Earth in June 2010 (see Figure 15.41). Chemical analyses of samples from this mission show that the surface of this asteroid is nearly identical in composition to rocky meteorites. This finding strongly supports the idea that asteroids are the source of most meteoroids large enough to reach Earth's surface.

Hayabusa 2, launched in 2014, is expected to land on its target (asteroid 1999 JU3) in July 2018. After exploring this asteroid by digging into its surface to extract fresh samples, it is scheduled to return to Earth in 2020.

Comets: Dirty Snowballs

Comets, like asteroids, are leftover material from the formation of the solar system. They are loose collections of rocky material, dust, water ice, and frozen gases (ammonia, methane, and carbon dioxide), thus the nickname "dirty snowballs." Recent space missions to comets have shown their surfaces to be dry and dusty, which indicates that their ices are hidden beneath a layer of rocky debris.

Most comets reside in the outer reaches of the solar system and take hundreds of thousands of years to complete a single orbit around the Sun. However, a smaller number of *short-period comets* (those having orbital periods of less than 200 years), such as the famous Halley's Comet, make regular encounters with the inner solar system (**Figure 15.42**). The shortest-period comet (Encke's Comet) orbits the Sun once every 3 years.

Structure and Composition of Comets The phenomena associated with comets come from a small central body called the **nucleus**. These structures are typically 1 to 10 kilometers in diameter, but comet nuclei 40 kilometers across have been observed. When a comet comes within about 5 AU from the Sun, solar energy heats its surface sufficiently to cause its icy components to begin to vaporize into gas. The escaping gases carry dust from the comet's surface, producing a huge dusty atmosphere called a **coma** (**Figure 15.43**). As a comet approaches the inner solar system, the coma grows, and some of the dust and gas is pushed away from the Sun to form the comet's tails, which can grow to hundreds of millions of kilometers in length.

Bright comets have two visible tails, a dark blue tail that points straight away from the Sun and a brighter tail

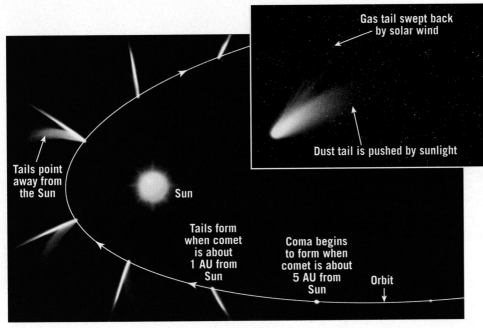

Figure 15.42 Changing orientation of a comet's tail as it orbits the Sun (Photo by Dan Schechter/Science Source)

that points away from the Sun but also curves slightly in the direction from which the comet came. Scientists have determined the mechanisms that account for the formation of these tails. The faint, straight *gas tail* consists of ionized cometary gases that are pushed away from the coma by the pressure of the *solar wind* of charged particles emitted by the Sun. The brighter, curved *dust tail*

Figure 15.43 Coma and nucleus of Comet Holmes The nucleus of Comet Holmes is the bright yellow spot within the reddish-orange coma. (Courtesy of NASA)

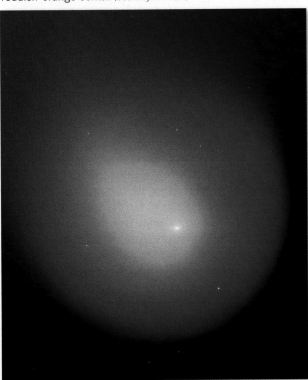

consists of dust particles that are pushed away from the coma by the much weaker pressure of sunlight (*radiation pressure*). The dust tail is curved because its bright, relatively slow-moving particles record how the direction toward the Sun changes as the comet moves along its orbit.

As a comet's orbit carries it away from the Sun, the gases forming the coma dissipate, the tails disappear, and the comet returns to cold storage. Material that was blown from the coma to form the tails is lost forever. When all the gases are expelled, the inactive comet continues its orbit without a coma or tail. Sometimes the comet literally disintegrates into small fragments that continue to orbit the Sun. Scientists believe that few comets remain active for more than a few hundred close orbits of the Sun.

In 2015 the *Rosetta* spacecraft, launched by the European Space Agency, obtained a close-up image of the nucleus of Comet 67P/Churyumov–Gerasimenko that offered a new perspective on the extent of the comet's activity. **Figure 15.44** shows jets of gas and dust emanating

Figure 15.44 Jets of gas and dust erupting from the nucleus of Comet 67P This comet's full name is Comet 67P/Churyumov–Gerasimenko; like all other comets, it is named after its discoverers. It is a regular visitor to the inner solar system, orbiting the Sun every 6.5 years. (Courtesy of European Space Agency)

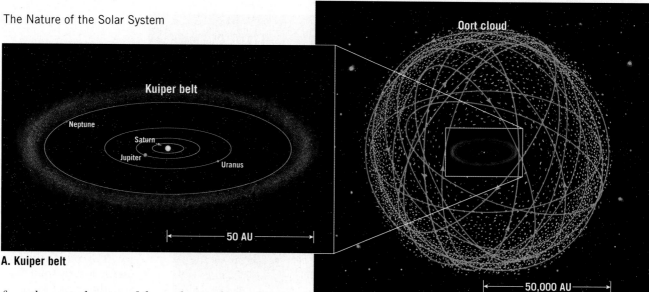

A. Kuiper belt

B. Oort cloud

from the central region of the nucleus and extending toward the upper right of the image. This image also shows the nebulous glow of material escaping the bright sunlit surface of the comet.

The Realm of Comets: The Kuiper Belt and Oort Cloud

Most comets originate in one of two regions: the *Kuiper belt* or the *Oort cloud*. Named in honor of astronomer Gerald Kuiper, who predicted its existence, the **Kuiper belt** hosts a large group of icy objects that reside in the outer solar system, beyond the orbit of Neptune. Pluto's orbit lies within the Kuiper belt, and in 2005 scientists discovered Eris, a more massive Kuiper belt body than Pluto.

Like most asteroids, Kuiper belt comets orbit the Sun in the same direction and along nearly the same plane as the planets (**Figure 15.45A**). This disc-shaped structure is thought to contain about 100,000 bodies more than 100 kilometers (60 miles) across, as well as many smaller objects. The largest Kuiper belt objects are much larger than the comets we observe in the inner solar system, probably because relatively small objects are more likely to have their orbits altered than are larger bodies.

Kuiper belt objects are thought to be leftover planetesimals that formed and remain in the frigid outer reaches of the solar system. Interactions with other nearby bodies or the gravitational influence of one of the Jovian planets occasionally alters their orbits sufficiently to send them into our view.

Named for Dutch astronomer Jan Oort, the **Oort cloud** consists of icy planetesimals that form a roughly spherical shell around the outer reaches of the solar system (**Figure 15.45B**). Oort cloud objects have random orbits at distances often greater than 50,000 times the Earth–Sun distance. This places them at nearly half the distance to Proxima Centauri, the nearest star to the Sun. These icy objects are so loosely bound to the solar system that the gravitational effect of a distant passing star may send an occasional Oort cloud body into a highly eccentric orbit that carries it toward the Sun. Because most comets that enter the inner solar system have random orbits, these are assumed to have come from the Oort cloud.

The Oort cloud is estimated to contain at least a trillion comets. How did so many comets end up in the outer reaches of the solar system? The most widely accepted hypothesis is that these bodies formed early in our solar system's history, in the region occupied by the still-developing Jovian planets. Rather than being captured by one of these growing planets, the comets were gravitationally flung in all directions.

Meteors, Meteoroids, and Meteorites

Nearly everyone has seen **meteors**, commonly (but inaccurately) called "shooting stars." These streaks of light can be observed in as little as the blink of an eye or can last as "long" as a few seconds. They occur when a small solid particle, a **meteoroid**, enters Earth's atmosphere from interplanetary space. Heat, created by friction between the meteoroid and Earth's atmosphere, produces the streak of light we see trailing across the sky. Most meteoroids originate from one of three sources: (1) interplanetary debris missed by the gravitational sweep of the planets during solar system formation, (2) material ejected from the asteroid belt, or (3) the rocky and/or metallic remains of comets that once passed through Earth's orbit.

Meteoroids less than about 1 meter (3 feet) in diameter generally vaporize before reaching Earth's surface. Some, called *micrometeorites*, are so tiny and their rate of fall so slow that they drift to Earth continually as space dust. Researchers estimate that thousands of meteoroids enter Earth's atmosphere every day. After sunset on a clear, dark night, many are bright enough to be seen with the naked eye.

Meteor Showers

Occasionally, meteor sightings increase dramatically to 60 or more per hour. Such displays, called **meteor showers**, result when Earth encounters a swarm of meteoroids traveling in the same direction at nearly the same speed as Earth. The close association of these swarms to the orbits of some short-term

Table 15.3 Major Meteor Showers

Shower	Approximate Dates	Associated Comet
Quadrantids	January 4–6	Unknown
Lyrids	April 20–23	Comet Thatcher
Eta Aquarids	May 3–5	Halley's Comet
Delta Aquarids	July 30	Unknown
Perseids	August 12	Comet Swift–Tuttle
Draconids	October 7–10	Comet Giacobini–Zinner
Orionids	October 20	Halley's Comet
Taurids	November 3–13	Comet Encke
Andromedids	November 14	Comet Biela
Leonids	November 18	Comet Tempel–Tuttle
Geminids	December 4–16	Unknown

comets strongly suggests that they represent material lost by these comets (**Table 15.3**). Some swarms, not associated with the orbits of known comets, are probably the scattered remains of the nucleus of a long-defunct comet. The notable *Perseid meteor shower* that occurs each year around August 12 is likely material ejected from the comet Swift–Tuttle on previous approaches to the Sun.

Meteorites: Visitors to Earth The remains of meteoroids that have impacted Earth's surface are called **meteorites** (**Figure 15.46**). Most meteoroids large enough to survive passage through the atmosphere likely started out as asteroids; a chance collision or gravitational interactions with Jupiter modifies the orbit and sends the asteroid toward Earth. Earth's gravity does the rest. A few meteoroids are fragments of the Moon, Mars, or possibly even Mercury, ejected by a violent asteroid impact. Before *Apollo* astronauts brought Moon rocks back to Earth, meteorites were the only extraterrestrial materials that could be studied in laboratories.

A few large meteorites have blasted craters on Earth's surface that strongly resemble craters on our Moon. More than 40 terrestrial craters with diameters larger than 20 kilometers (12 miles) exist. These craters exhibit features that could only have been produced by an explosive impact of an asteroid or perhaps even a comet nucleus. More than 250 smaller craters are also thought to have an impact origin. Notable among them is Arizona's Meteor Crater, a cavity more than 1 kilometer (0.6 mile) wide and 170 meters (560 feet) deep, with an upturned rim that rises above the surrounding countryside (**Figure 15.47**). More than 30 tons of iron fragments have been

found in the immediate area, but attempts to locate the main body have been unsuccessful. Based on the amount of erosion observed on the crater rim, the impact likely occurred within the past 50,000 years.

Types of Meteorites Classified by their composition, meteorites are either (1) *irons*, mostly aggregates of iron with 5–20 percent nickel; (2) *stony*, silicate minerals with inclusions of other minerals; or (3) *stony–irons*, mixtures of the two. Although stony meteorites are the most common, irons are found in large numbers because metallic meteorites withstand impacts better, weather more slowly, and are easily distinguished from terrestrial rocks. Iron meteorites are probably fragments of once-molten cores of large asteroids or small planets.

Data from meteorites have been used to ascertain the internal structure of Earth and the age of the solar system. If meteorites represent the composition of the terrestrial planets, as some planetary geologists suggest, our planet must contain a much larger percentage of iron than is indicated by surface rocks. This is one reason that geologists think Earth's core is mostly iron and nickel. In addition, radiometric dating of meteorites indicates that the age of our solar system is about 4.6 billion years. This "old age" has been confirmed by data obtained from lunar samples.

Dwarf Planets

Pluto was discovered in 1930 by astronomers who were searching for another planet in order to explain irregularities in the orbit of Uranus. However, Pluto's discovery turned out to be pure coincidence; astronomers quickly realized that it was too small to account for the "irregularities," which ultimately were explained by recalculating the mass of Neptune. Pluto has a diameter of about 2370 kilometers (1470 miles), or about one-fifth that of Earth and less than half that of Mercury (long considered the solar system's "runt").

SmartFigure 15.46 Iron meteorite found near Meteor Crater, Arizona (Courtesy of M2 Photography/Alamy) (https://goo.gl/bbT16J)

Tutorial

Mobile
Field Trip

More attention was given to Pluto's status as a planet when astronomers began to discover other large Kuiper belt bodies. Clearly, Pluto was like this new category of objects and completely different from either the terrestrial planets or the Jovian planets.

In 2006, the International Astronomical Union, the group responsible for naming and classifying celestial objects, voted to designate a new class of solar system objects called *dwarf planets*. These celestial bodies orbit the Sun and are essentially spherical due to their own gravity but are not large enough to sweep their orbits clear of other debris. By this definition, Pluto is recognized as a dwarf planet and the prototype for this new category of planetary objects. Other dwarf planets include Eris, a Kuiper belt object, and Ceres, the largest-known asteroid.

In July 2015, as this chapter was being finalized, NASA's *New Horizons* spacecraft shot past Pluto, after a 9-year journey that brought it within 12,500 kilometers (7800 miles) of Pluto's surface. Images transmitted from *New Horizons* showed Pluto to be a complex body with several distinct terrains, including mountainous areas, ice plains, and rugged areas indicating a history of impacts (**Figure 15.48**). One of the most interesting regions, named Sputnik Planum (*planum* = plain, or flat area), is found in Pluto's southern hemisphere. Sputnik Planum is presumed to be a large ice field that has tongues of ice flowing from its edges, similar to glaciers found on Earth. With surface temperatures of about −235°C (just shy of −400°F), Pluto is too cold for these ice flows to be made of water. Instead, Pluto's surface ice is most likely frozen nitrogen, which will flow at these frigid temperatures.

Because of Pluto's distance from Earth (about 32 times the distance between Earth and the Sun), a complete set of data cannot be retrieved until autumn 2016. After its flyby of Pluto, *New Horizons* changed course and was aimed at another icy Kuiper belt object about 40 kilometers (25 miles) in diameter.

Figure 15.48 This enhanced color image is used to detect differences in the composition and texture of Pluto's surface The bright area in the lower central region, sometimes referred to as "the heart of the heart," is formally named Sputnik Planum and is thought to be the source of exotic ices that flowed to produce the two bluish lobes near the bottom of the image. (Courtesy of NASA)

Sputnik
Planum

(15.7) **CONCEPT CHECKS**

1. Compare and contrast asteroids and comets. Where are most asteroids found?
2. Where are most comets thought to reside? What eventually becomes of comets that orbit close to the Sun?
3. Differentiate among the following solar system bodies: meteoroid, meteor, and meteorite.
4. What are the three main sources of meteoroids?
5. Why was Pluto reclassified as a dwarf planet?

CONCEPTS IN REVIEW
The Nature of the Solar System

15.1 Ancient Astronomy

Explain the geocentric view of the solar system and describe how it differs from the heliocentric view.

KEY TERMS: geocentric, celestial sphere, heliocentric, Ptolemaic system, retrograde motion

- The early Greeks held a geocentric ("Earth-centered") view of the universe, in which Earth is a motionless sphere at the center of the universe, orbited by the Moon, the Sun, and the known planets—Mercury, Venus, Mars, Jupiter, and Saturn.
- The early Greeks believed that the stars traveled daily around Earth on a transparent, hollow celestial sphere. In A.D. 141, Claudius Ptolemy documented this geocentric view, now called the Ptolemaic system, which became the dominant view of the solar system for over 15 centuries.

15.2 The Birth of Modern Astronomy

List and describe the contributions to modern astronomy of Nicolaus Copernicus, Tycho Brahe, Johannes Kepler, Galileo Galilei, and Isaac Newton.

KEY TERMS: astronomical unit (AU), inertia, law of universal gravitation

- Modern astronomy evolved during the 1500s and 1600s, facilitated by scientists going beyond merely describing what is observed to explaining why the universe behaves the way it does.
- Nicolaus Copernicus (1473–1543) reconstructed the solar system with the Sun at the center and the planets orbiting around it, but he erroneously continued to use circles to represent the orbits of planets. His Sun-centered view was rejected by the establishment of his day.
- Tycho Brahe's (1546–1601) observations of the planets were far more precise than any made previously and are his legacy to astronomy.
- Johannes Kepler (1571–1630) used Tycho Brahe's observations to usher in a new astronomy with the formulation of his three laws of planetary motion.
- After constructing his own telescope, Galileo Galilei (1564–1642) made many important discoveries that supported the Copernican view of a Sun-centered solar system. This included charting the movement of Jupiter's four largest moons, proving that Earth was not the center of all planetary motion.
- Sir Isaac Newton (1642–1727) demonstrated that the orbit of a planet is a result of the planet's inertia (its tendency to move in a straight line) and the Sun's gravitational attraction, which bends the planet's path into an elliptical orbit.

15.3 Our Solar System: An Overview

Describe the formation of the solar system according to the nebular theory. Compare and contrast the terrestrial and Jovian planets.

KEY TERMS: nebular theory, solar nebula, planetesimal, protoplanet, terrestrial (Earth-like) planet, Jovian (Jupiter-like) planet, escape velocity, impact crater

- Our Sun is the most massive body in our solar system, which includes planets, dwarf planets, moons, and other small bodies. The planets orbit in the same direction and at speeds proportional to their distance from the Sun, with inner planets moving faster and outer planets moving more slowly.
- The solar system began as a solar nebula before condensing due to gravity. While most of the matter ended up in the Sun, some material formed a thick disc around the early Sun and later clumped together into larger and larger bodies. Planetesimals collided to form protoplanets, and protoplanets grew into planets.
- The four terrestrial planets are enriched in rocky and metallic materials, whereas the Jovian planets have a higher proportion of ice and gas. The terrestrial planets are relatively dense, with thin atmospheres, while the Jovian planets are less dense and have thick atmospheres.
- Smaller planets have less gravity to retain gases in their atmosphere. Lightweight gases such as hydrogen and helium more easily reach escape velocity, so the atmospheres of the terrestrial planets tend to be enriched in heavier gases, such as water vapor, carbon dioxide, and nitrogen.

15.4 Earth's Moon: A Chip Off the Old Block

List and describe the major features of Earth's Moon and explain how maria basins were formed.

KEY TERMS: maria, lunar highlands, lunar regolith

- The Moon has a composition that is approximately the same as that of Earth's mantle. The Moon likely formed from a collision between a Mars-sized protoplanet and the early Earth.
- The lunar surface is dominated by light-colored lunar highlands (or terrae) and darker lowlands called maria, the latter formed primarily from flood basalts. Both terrae and maria are partially covered by lunar regolith produced by micrometeorite bombardment.

(?) **Briefly describe how our Moon formed and how its formation accounts for its low density compared to that of Earth.**

15.5 Terrestrial Planets

Outline the principal characteristics of Mercury, Venus, and Mars. Describe their similarities to and differences from Earth.

- Mercury has a very thin atmosphere and a weak magnetic field. Like Earth's moon, Mercury has both heavily cratered areas and smooth plains; the smooth plains are similar to lunar maria.
- Venus has a very dense atmosphere, dominated by carbon dioxide. The resulting extreme greenhouse effect produces surface temperatures around 450°C (900°F). The topography of Venus has been resurfaced by active volcanism.
- Mars has about 1 percent as much atmosphere as Earth, so it is relatively cold (–140°C to 20°C [–220°F to 68°F]). Mars appears to be the closest planetary analog to Earth, showing surface evidence of rifting, volcanism, and modification by flowing water. Volcanoes on Mars are much bigger than volcanoes on Earth because of the lack of plate motion on Mars.

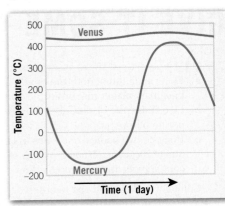

(?) **As you can see from this graph, Mercury's temperature varies significantly from "day" to "night," but Venus's temperature is relatively constant "around the clock." Suggest a reason for this difference.**

15.6 Jovian Planets

Summarize and compare the features of Jupiter, Saturn, Uranus, and Neptune, including their ring systems.

KEY TERMS: cryovolcanism

- Jupiter's mass is several times larger than the combined mass of everything else in the solar system except for the Sun. Convective flow, combined with its three cloud layers, produces its characteristic banded appearance. Persistent, giant rotating storms exist between these bands. Many moons orbit Jupiter, including Io, which shows active volcanism, and Europa, which is believed to have a liquid ocean under its icy shell.
- Saturn, like Jupiter, is big, gaseous, and endowed with dozens of moons. Some moons show evidence of tectonics, while Titan has its own atmosphere. Saturn's well-developed rings are made of many particles of water ice and rocky debris.
- Uranus, like its "twin" Neptune, has a blue atmosphere dominated by methane, and its diameter is about four times greater than Earth's. Uranus rotates sideways relative to the plane of the solar system. It has a relatively thin ring system and at least five moons.
- Neptune has an active atmosphere, with fierce wind speeds and giant storms. It has 1 large moon, Triton, which shows evidence of cryovolcanism, as well as a 13 smaller moons and a ring system.

15.7 Small Solar System Bodies

List and describe the principal characteristics of the small bodies that inhabit the solar system.

KEY TERMS: small solar system body, dwarf planet, asteroid, asteroid belt, comet, nucleus, coma, Kuiper belt, Oort cloud, meteor, meteoroid, meteor shower, meteorite

- Small solar system bodies include rocky asteroids and icy comets. Both are basically scraps left over from the formation of the solar system or fragments from later impacts.
- Most asteroids are concentrated in a wide belt between the orbits of Mars and Jupiter. Some are rocky, some are metallic, and some are basically "piles of rubble," loosely held together by their own weak gravity.

- Comets are dominated by ices, "dirtied" by rocky material and dust. Most originate in either the Kuiper belt beyond Neptune or the Oort cloud. When a comet's orbit brings it through the inner solar system, solar radiation causes its ices to vaporize, generating the coma and its characteristic "tail."
- A meteoroid is a small rocky or metallic body traveling through space. When it enters Earth's atmosphere, it flares briefly as a meteor before either burning up or striking Earth's surface to become a meteorite. Asteroids and material lost from comets as they travel through the inner solar system are the most common sources of meteoroids.
- Bodies massive enough to have a spherical shape but not so massive as to have cleared their orbits of debris are classified as dwarf planets. They include the rocky asteroid Ceres as well as the icy worlds Pluto and Eris, which are located in the Kuiper belt.

❓ **Shown here are four small solar system bodies. Identify each and explain the differences among them.**

GIVE IT SOME THOUGHT

1. Use Kepler's third law to answer the following questions:
   a. Determine the period (the time it takes to orbit the Sun) of an imaginary planet that orbits the Sun at a distance of 10 AU.
   b. Determine the distance between the Sun and a planet with a period of 5 years.
   c. Imagine two bodies, one twice as large as the other, orbiting the Sun at the same distance. Which of the bodies, if either, would complete its orbit around the Sun in less time? Explain.

2. Galileo used his telescope to observe the planets and moons in our solar system. These observations allowed him to determine the positions and relative motions of the Sun, Earth, and other objects in the solar system. Refer to Figure 15.14A, which shows an Earth-centered solar system, and Figure 15.14B, which shows a Sun-centered solar system, to complete the following:
   a. Describe the phases of Venus that an observer on Earth would see for the Earth-centered model of the solar system.
   b. Describe the phases of Venus that an observer on Earth would see for the Sun-centered model of the solar system.
   c. Explain how Galileo used observations of the phases of Venus to determine the correct positions of the Sun, Earth, and Venus.

3. The accompanying diagram shows three asteroids (A, B, and C) that are being pulled by the gravitational force exerted on them by their

partner asteroid. How will the strength of the gravitational force felt by each asteroid (A, B, and C) compare? (Assume that all these asteroids are composed of the same material.)

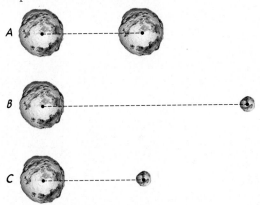

4. Assume that a solar system has been discovered in a nearby region of the Milky Way Galaxy. The accompanying table shows data that have been gathered about three of the planets orbiting the central star of this newly discovered solar system. Using Table 15.2 as a guide, classify each planet as Jovian, terrestrial, or neither. Explain your reasoning.

	Planet 1	Planet 2	Planet 3
Relative Mass (Earth = 1)	1.2	15	0.1
Diameter (km)	15,000	52,000	5000
Mean Distance from Star (AU)	1.4	17	35
Density (g/cm3)	4.8	1.22	5.3
Orbital Eccentricity	0.01	0.05	0.23

5. In order to conceptualize the size and scale of Earth and Moon as they relate to the rest of the solar system, complete the following:
   a. Approximately how many Moons (diameter 3475 kilometers [2160 miles]) would fit side by side across the diameter of Earth (diameter 12,756 kilometers [7926 miles])?
   b. Given that the Moon's orbital radius is 384,798 kilometers, approximately how many Earths would fit side by side between Earth and the Moon?
   c. Approximately how many Earths would fit side by side across the Sun, whose diameter is about 1,390,000 kilometers?
   d. Approximately how many Suns would fit side by side between Earth and the Sun, a distance of about 150,000,000 kilometers?

6. The graph in the upper right shows the temperatures at various distances from the Sun during the formation of our solar system. (You can assume that the planets formed at roughly their current distances from the Sun, although that may not be strictly true.) Use it to answer the following:
   a. Which planet(s) formed at locations in the solar system where the temperature was hotter than the boiling point of water?
   b. Which planet(s) formed at locations in the solar system where the temperature was cooler than the freezing point of water?

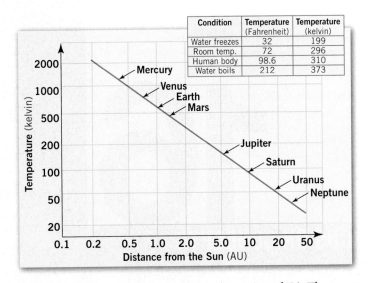

Condition	Temperature (Fahrenheit)	Temperature (kelvin)
Water freezes	32	199
Room temp.	72	296
Human body	98.6	310
Water boils	212	373

7. This sketch shows four primary craters (A, B, C, and D). The impact that produced crater A produced two secondary craters (labeled a) and three rays. Crater D has one secondary crater (labeled d). Rank the four primary craters from oldest to youngest and explain your ranking.

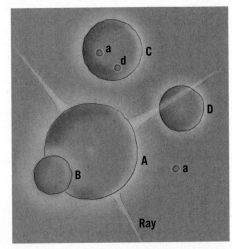

8. Halley's Comet has a mass estimated at 100 billion tons. Furthermore, it is estimated to lose about 100 million tons of material when its orbit brings it close to the Sun. With an orbital period of 76 years, calculate the maximum remaining life span of Halley's Comet.

9. Assume that three irregularly shaped planet-like objects, each smaller than our Moon, have just been discovered orbiting the Sun at a distance of 35 AU. One of your friends argues that the objects should be classified as planets because they are large and orbit the Sun. Another friend argues that the objects should be classified as dwarf planets, like Pluto. State whether you agree or disagree with either or both of your friends. Explain your reasoning.

MasteringGeology™

16

FOCUS ON CONCEPTS

Each statement represents the primary learning objective for the corresponding major heading within the chapter. After you complete the chapter, you should be able to:

16.1 Define *cosmology* and describe Edwin Hubble's most significant discovery about the universe.

16.2 Define *main-sequence star*. Explain the criteria used to classify stars as giants.

16.3 List and describe the stages in the evolution of a typical Sun-like star.

16.4 Compare and contrast the final state of Sun-like stars to the remnants of the most massive stars.

16.5 List the three major types of galaxies. Explain the formation of large elliptical galaxies.

16.6 Describe the big bang theory. Explain what it tells us about the universe.

The Whirlpool Galaxy (M51) has excited astronomers for nearly three centuries. Its majestic spiral arms are star-forming factories. (NASA)

BEYOND OUR SOLAR SYSTEM*

*Revised with the assistance of Professors Teresa Tarbuck and Mark Watry.

Astronomers and cosmologists study the nature of our vast universe, attempting to answer questions such as these: Is our Sun a typical star? Do other stars have solar systems with planets similar to Earth? Are galaxies distributed randomly, or are they organized into groups? How do stars form? What happens when stars expend their fuel? If the early universe consisted mostly of hydrogen and helium, how did other elements come into existence? How large is the universe? Did it have a beginning? Will it have an end? This chapter explores these as well as other questions.

16.1 The Universe

Define *cosmology* and describe Edwin Hubble's most significant discovery about the universe.

Figure 16.1 The Trifid Nebula, in the constellation Sagittarius This colorful nebula is a cloud consisting mostly of hydrogen and helium gases. These gases are excited by light emitted by hot, young stars within and produce a reddish glow. (Courtesy of National Optical Astronomy Observatory [NOAO])

The universe is more than a collection of dust clouds, stars, stellar remnants, and galaxies (**Figure 16.1**). It is an entity with its own properties. **Cosmology** is the study of the universe, including its properties, structure, and evolution. Over the years, cosmologists have developed a comprehensive theory that describes the structure and evolution of the universe in order to answer questions such as: How did the universe evolve to its present state? How long has it existed, and how will it end? Modern cosmology addresses these important issues and helps us understand the universe we inhabit.

How Large Is It?

For most of human existence, our universe was thought to be Earth centered, containing only the Sun, Moon, five planets, and the roughly 6000 stars visible to the naked eye. Even after the Copernican view of a

Sun-centered universe became widely accepted, the entire universe was believed to consist of a single galaxy—the Milky Way, composed of innumerable stars, along with many faint "fuzzy patches," thought to be clouds of dust and gases.

In the mid-1700s, German philosopher Immanuel Kant proposed that many of the telescopically visible fuzzy patches of light scattered among the stars were actually distant galaxies similar to the Milky Way. Kant described them as "island universes." Each galaxy, he believed, contained billions of stars and was a universe in itself. In Kant's time, however, the weight of opinion favored the hypothesis that the faint patches of light occurred within our galaxy. Admitting otherwise would have implied a vastly larger universe, thereby diminishing the status of Earth and, likewise, humankind.

In 1919 Edwin Hubble arrived at the observatory at Mount Wilson, California, to conduct research using a 2.5-meter (100-inch) telescope, then the world's largest and most advanced astronomical instrument. Armed with this modern tool, Hubble set out to solve the mystery of the "fuzzy patches." At that time, the debate was still raging as to whether the fuzzy patches were "island universes," as Kant had proposed more than 150 years earlier, or clouds of dust and gases (nebulae). To accomplish this task, Hubble studied a group of pulsating stars known as *Cepheid variables*—extremely bright variable stars that increase and decrease in brightness in a repetitive cycle. This group of stars is significant because their "true" brightness, called **absolute magnitude** (see Appendix C), can be determined by knowing the rate at which they pulsate. When the absolute magnitude of a star is compared to its observed brightness, a reliable approximation of distance can be established. (This is similar to how we judge the distance of an oncoming vehicle when driving at night.) Thus, Cepheid variables are important because they can be used to measure large astronomical distances.

Using the telescope at Mount Wilson, Hubble found several Cepheid variables embedded in one of the fuzzy patches. However, because these intrinsically bright stars appeared faint, Hubble concluded that they must lie outside the Milky Way. Indeed, he concluded that this fuzzy

A.

B.

patch lies more than 2 million light-years away. It is now known as the *Andromeda Galaxy* (**Figure 16.2**).

Figure 16.2 Andromeda: A nearby galaxy that is larger than the Milky Way **A.** Photo of Andromeda Galaxy taken by a telescope aboard GALEX Orbiter. (NASA) **B.** Andromeda Galaxy, as it appears under low magnification. When viewed with the naked eye, Andromeda appears as a fuzzy patch surrounded by stars. (European Southern Observatory/ESO)

Based on his observations, Hubble determined that the universe extended far beyond the limits of our imagination. Today, we know that there are hundreds of billions of galaxies, each containing hundreds of billions of stars. For example, researchers estimate that a million galaxies exist in the portion of the sky bounded by the cup of the Big Dipper. There are literally more stars in the heavens than grains of sand in all the beaches on Earth.

A Brief History of the Universe

Large telescopes can actually "look back in time," which accounts for much of the knowledge astronomers have acquired about the history of the universe. Light from celestial objects that are great distances from Earth require millions or even billions of years to reach Earth. The distance light travels in 1 year is called a **light-year** (slightly less than 10 trillion kilometers [6 trillion miles]). Therefore, the farther out telescopes can "see," the farther back in time astronomers are able to study. Even the closest large galaxy, the Andromeda Galaxy, is a staggering 2.5 million light-years away. Light that left the Andromeda Galaxy 2.5 million years ago is just now reaching Earth, allowing scientists to observe this galaxy as it was 2.5 million years ago. We have now observed the almost

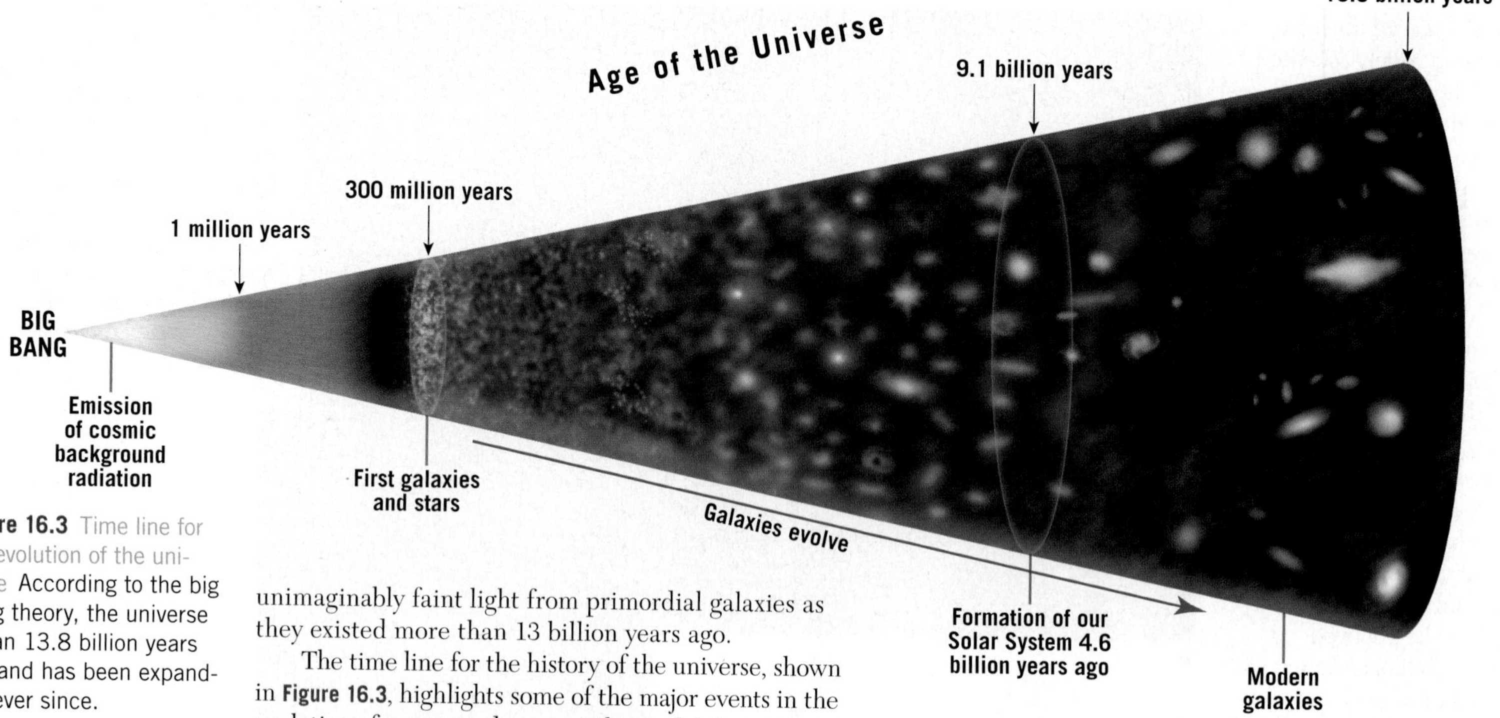

Age of the Universe

Present day
13.8 billion years

9.1 billion years

300 million years

1 million years

BIG
BANG

Emission
of cosmic
background
radiation

First galaxies
and stars

Galaxies evolve

Formation of our
Solar System 4.6
billion years ago

Modern
galaxies

Figure 16.3 Time line for the evolution of the universe According to the big bang theory, the universe began 13.8 billion years ago and has been expanding ever since.

unimaginably faint light from primordial galaxies as they existed more than 13 billion years ago.

The time line for the history of the universe, shown in **Figure 16.3**, highlights some of the major events in the evolution of matter and energy. The model that most accurately describes the birth and current state of the universe is the **big bang theory**. According to this theory, all of the energy and matter of the universe originally existed in an incomprehensibly hot and dense state. About 13.8 billion years ago, our universe began as a cataclysmic explosion, which continued to expand, cool, and evolve to its current state.

In the earliest moments of this expansion, only energy and quarks (subatomic particles that are the building blocks of protons and neutrons) existed. Not until 380,000 years after the initial expansion did the universe cool sufficiently for electrons and protons to combine to form hydrogen and helium atoms—the lightest elements in the universe. Eventually, temperatures

decreased sufficiently to allow this primordial gas to condense gravitationally into clumps, which quickly evolved into the first stars and galaxies. Our Sun and planetary system, having formed about 5 billion years ago (nearly 9 billion years after the big bang), is a latecomer to the universe.

16.1 CONCEPT CHECKS

1. What is cosmology?
2. Explain how Edwin Hubble used Cepheid variables to change our view of the structure of the universe.
3. When do cosmologists think the universe began?
4. What two elements were the first to form?

16.2 Classifying Stars: Hertzsprung–Russell Diagrams (H-R Diagrams)

Define *main-sequence star*. Explain the criteria used to classify stars as giants.

Early in the twentieth century, Einar Hertzsprung and Henry Russell independently studied the relationship between the true brightness (absolute magnitude) of stars and their respective temperatures. Their work resulted in the development of a graph, called a **Hertzsprung–Russell diagram (H-R diagram)**. By studying H-R diagrams, we can learn a great deal about the relationships among the sizes, colors, and temperatures of stars (see Appendix C). For example, we learned that the hottest stars are blue in color and the coolest are red.

To produce an H-R diagram, astronomers survey a portion of the sky and plot each star according to its absolute magnitude (stellar brightness) and temperature (**Figure 16.4**). Notice that the stars in Figure 16.4 are not uniformly distributed. Rather, about 90 percent of all stars fall along a band that runs from the upper-left corner to the lower-right corner of the H-R diagram. These "ordinary" stars are called **main-sequence stars**. As you can see in Figure 16.4, the hottest main-sequence stars are intrinsically the brightest, and, conversely, the coolest are the dimmest.

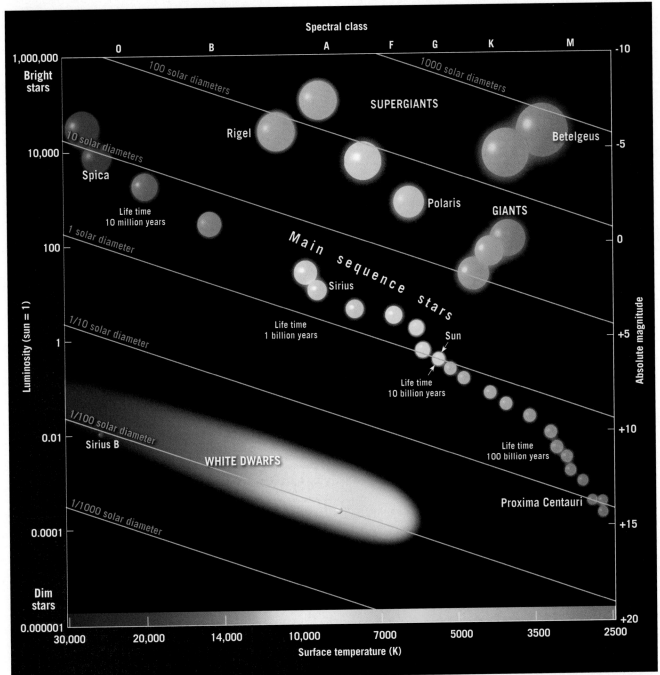

SmartFigure 16.4
Hertzsprung–Russell
diagram Astronomers use
these diagrams to study
stellar evolution by plotted
stars according to their
temperatures and luminosi-
ties (absolute magnitudes).
Stellar temperatures are
given in kelvins (K); to
convert to degrees Celsius
(°C), subtract 273. (goo
.gl/5Auwf6)

The absolute magnitude of main-sequence stars is also related to their mass. The hottest (blue) stars can be 50 or more times more massive than the Sun, whereas the coolest (red) stars can be less than 1/10 as massive. Therefore, on the H-R diagram, the main-sequence stars appear in decreasing order, from hotter, more massive blue stars to cooler, less massive red stars.

Note the location of the Sun in Figure 16.4. The Sun is a yellow main-sequence star with an absolute magnitude, or "true" brightness, of about 5 (see Appendix C). Because the vast majority of main-sequence stars have magnitudes between –10 and 20, the Sun's midpoint

position in this range results in its classification as an "average star." (Note that stellar magnitudes are measured so that the lower the number, the brighter the star and vice versa.)

While most stars fall along the main sequence, some lie well off it. Above and to the right of the main-sequence stars lies a group of very luminous stars called **giants**, or, on the basis of their color, **red giants** (see Figure 16.4). The size of a giant can be estimated by comparing it with stars of known size that have the same surface temperature. Scientists have discovered that objects having equal surface temperatures radiate the same amount of energy per unit area. Any difference in the brightness of two stars having the same surface temperature can be attributed to their relative sizes. Therefore, if one red star is 100 times more luminous than another red star, it must have a surface area that is 100 times larger. Thus, stars with large radiating surfaces that appear in the upper-right position of an H-R diagram are appropriately called *giants*.

Some stars are so immense that they are called **supergiants**. Betelgeuse, a bright red supergiant in the constellation Orion, has a radius about 800 times that of the Sun. If this star were at the center of our solar system, it would extend beyond the orbit of Mars, and Earth would find itself buried inside this supergiant!

In the lower portion of the H-R diagram, opposite conditions are observed. These stars are much fainter than main-sequence stars of the same temperature and likewise are much smaller. Some approximate Earth in size (see Figure 16.4). These stars are called **white dwarfs**.

Despite their name, the hottest white dwarfs are blue in color, but as they cool, they gradually become white and then reddish in color, before eventually going dark.

H-R diagrams have proved to be an important tool for interpreting stellar evolution—the stages in which stars, similar to living things, are born, age, and die. Considering that almost 90 percent of stars lie on the main sequence, we can be relatively certain that most stars spend their active years as main-sequence stars. Only a few percent are giants, and perhaps 10 percent are white dwarfs.

16.2 CONCEPT CHECKS

1. On an H-R diagram, where do stars spend most of their life span?
2. How does the Sun compare in size and brightness to other main-sequence stars?
3. Describe how the H-R diagram is used to determine which stars are "giants."

16.3 Stellar Evolution

List and describe the stages in the evolution of a typical Sun-like star.

The idea of describing how a star is born, ages, and dies may seem a bit presumptuous, for most stars have life spans that exceed billions of years. However, by studying stars of different ages, at different points in their life cycles, astronomers have been able to assemble a model for stellar evolution.

The method used to create this model is analogous to how an alien, upon reaching Earth, might determine the developmental stages of human life. By observing large numbers of humans, this stranger would witness the onset of life, the progression of life in children and adults, and the death of the elderly. From this information, the alien could put the stages of human development into their natural sequence. Based on the relative abundance of humans in each stage of development, it would even be possible to conclude that humans spend more of their lives as adults than as toddlers. Similarly, astronomers have pieced together the life story of stars.

The first stars probably formed about 300 million years after the big bang, when clouds of gas became dense enough to collapse under their own gravitation. Like the primordial gas from which they formed, these *first-generation stars* consisted of hydrogen and a small amount of helium. Massive stars have relatively short lifetimes followed by violent, explosive deaths. During their lives and particularly during their explosive deaths, these stars synthesized heavier elements and expelled them into space. Some of this ejected matter is incorporated into subsequent generations of stars such as our Sun.

Every stage of a star's life is ruled by gravity. The mutual gravitational attraction of particles in a thin, gaseous cloud causes the cloud to collapse on itself. As

the cloud's center is squeezed to unimaginable pressures, its temperature increases; eventually its nuclear furnace ignites, and a star is born. A star is a ball of very hot gases, caught between the opposing forces of gravity trying to contract it and thermal nuclear energy trying to expand it. Eventually, all of a star's nuclear fuel will be exhausted, and gravity will prevail, collapsing the stellar remnant into a small, dense body.

The stages in the evolution of a typical Sun-like star are illustrated in **Figure 16.5**.

Stellar Birth

The birthplaces of stars are interstellar clumps of matter, rich in dust and gases, called **nebulae** (meaning "clouds"; the singular of *nebulae* is *nebula*). In the Milky Way, these gaseous clouds consist of about 92 percent hydrogen and 7 percent helium. The remainder consists of heavier atoms, molecules such as carbon monoxide, and tiny grains of *interstellar dust*. Although most nebulae are vast in size and their total mass is many times that of the Sun, their particles are widely dispersed. Nevertheless, some nebulae contain star-forming regions such as the "Pillars of Creation" in the Eagle Nebula (**Figure 16.6**). In these regions, gas and dust clump together to form denser masses, which attract more matter until they are massive enough to contract gravitationally to form stars.

Protostar Stage

As these massive clouds of dust and gases collapse, gravitational energy is converted into energy of motion, or

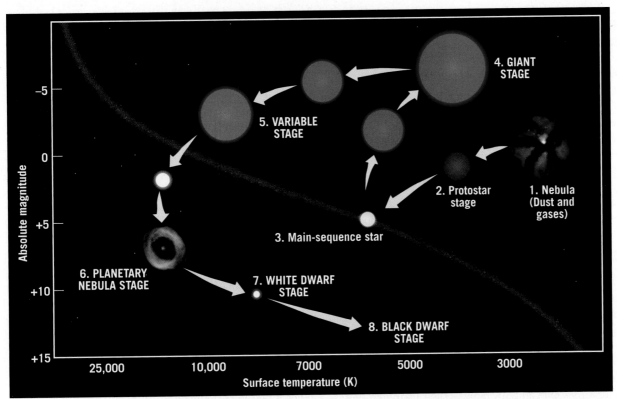

Figure 16.5 H-R diagram illustrating the evolution of a Sun-like star Stellar temperatures are given in kelvins (K); to convert to degrees Celsius (°C), subtract 273.

thermal energy, causing the contracting gases to gradually increase in temperature. When the temperature of these gaseous bodies increases sufficiently, they begin to radiate energy in the form of long-wavelength red light. Because these large red objects are not hot enough to engage in nuclear fusion, they are not yet stars. The name **protostar** is applied to these bodies.

During the protostar stage, gravitational contraction continues, slowly at first and then much more rapidly. This collapse causes the core of the developing star to heat much faster than its outer envelope. (Stellar temperatures are expressed in kelvins [K]; see Appendix A.) When the core reaches a temperature of 10 million K, the pressure within is so intense that groups of four hydrogen nuclei (through a several-step process)

Figure 16.6 Star-forming regions in Eagle Nebula These dark, eerie pillar-like structures are columns of cool interstellar dust, hydrogen, and helium, which are incubators for new stars. They protrude from the wall of the Eagle Nebula, also referred to as M16. (Courtesy of NASA)

fuse into single helium nuclei. Astronomers refer to this nuclear reaction, in which hydrogen nuclei are fused into helium, as **hydrogen fusion**.

The immense amount of heat released by hydrogen fusion causes the gases inside stars to move with increased vigor, raising the internal gas pressure. At some point, the increased atomic motion produces an outward force (gas pressure) that balances the inward-directed force of gravity. Upon reaching this balance, the stars become *stable main-sequence stars* (see Figure 16.5). In other words, a main-sequence star is one in which the force of gravity, in an effort to squeeze the star into the smallest possible ball, is precisely balanced by the gas pressure created by hydrogen fusion in the star's interior.

Main-Sequence Stage

During the main-sequence stage, stars experience minimal changes in size or energy output. Hydrogen is continually being converted into helium, and the energy released maintains the gas pressure sufficiently high to prevent gravitational collapse.

How long can stars maintain this balance? Hot, massive blue stars radiate energy at such an enormous rate that they substantially deplete their hydrogen fuel in only a few million years, approaching the end of their main-sequence stage rapidly. By contrast, the smallest (red) main-sequence stars may take hundreds of billions of years to burn their hydrogen; they live practically forever. A yellow star, such as the Sun, typically remains a main-sequence star for about 10 billion years. Because the solar system is about 5 billion years old, the Sun is expected to remain a stable main-sequence star for another 5 billion years.

An average star spends 90 percent of its life as a hydrogen-burning main-sequence star. Once the hydrogen fuel in a star's core is depleted, it evolves rapidly and dies. However, with the exception of the least massive stars, death is delayed when another type of nuclear reaction is triggered, and the star becomes a *red giant* (see Figure 16.5).

Red Giant Stage

Evolution to the red giant stage begins when the usable hydrogen in a star's interior is consumed, leaving a helium-rich core. Although hydrogen fusion is still progressing in the star's outer shell, no fusion is taking place in the core. Without a source of energy, the core no longer has the gas pressure necessary to support itself against the inward force of gravity. As a result, the core begins to contract.

The collapse of a star's interior causes its temperature to rise rapidly as gravitational energy is converted into thermal energy. Some of this energy is radiated outward, initiating a more vigorous level of hydrogen fusion in the shell surrounding the star's core. The additional

heat from the accelerated rate of hydrogen fusion expands the star's outer gaseous shell enormously. Sun-like stars become bloated *red giants*, while the most massive stars become *supergiants*, which can be thousands of times larger than their main-sequence size.

As a star expands, its surface cools, which explains the star's color: Relatively cool objects radiate more energy as long-wavelength radiation (nearer the red end of the visible spectrum). Eventually the star's gravitational force stops this outward expansion, and the two opposing forces—gravity and gas pressure—once again achieve balance. The star enters a stable state but is much larger in size. Some red giants overshoot the equilibrium point and instead rebound like an overextended spring. These stars, which alternately expand and contract, and never reach equilibrium, are known as **variable stars**.

While the outer envelope of a red giant expands, the core continues to collapse, and the internal temperature eventually reaches 100 million K. This astonishing temperature triggers another nuclear reaction, in which helium is converted to carbon. At this point, a red giant consumes both hydrogen and helium to produce energy. In stars more than eight times as massive as the Sun, other thermonuclear reactions occur that generate the elements on the periodic table, up to and including number 26, iron.

Burnout and Death

What happens to stars in the final phase of their lives? We know that stars, regardless of their size, must eventually exhaust their usable nuclear fuel and collapse in response to their immense gravity. Because the gravitational field of a star is dependent on its mass, low-mass stars and high-mass stars have different fates. To simplify the discussion of the final stage in stellar lives, we divide stars into three groups based on mass: low-mass stars, intermediate-mass (Sun-like) stars, and high-mass stars.

Death of Low-Mass Stars Stars less than one-half the mass of the Sun (0.5 solar mass) consume their fuel at relatively low rates (**Figure 16.7A**). Consequently, there are many small, *cool red stars* that may remain stable for longer than 100 billion years. Because the interiors of low-mass stars never attain sufficiently high temperatures and pressures to fuse helium, their only energy source is hydrogen fusion. Thus, a low-mass star never becomes a bloated red giant. Rather, it remains a stable main-sequence star until it consumes its usable hydrogen fuel and collapses into a hot, dense white dwarf.

Death of Intermediate-Mass (Sun-Like) Stars Stars with masses ranging between one-half and eight times that of the Sun have a similar evolutionary history (**Figure 16.7B**). During their red giant phase, Sun-like stars fuse hydrogen and helium fuel at accelerated rates. Once this fuel is exhausted, these stars (like low-mass stars)

Tutorial

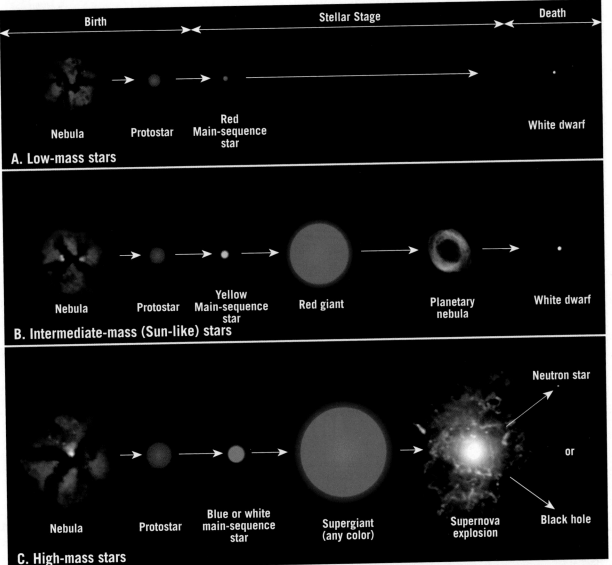

Birth   **Stellar Stage**   **Death**

Nebula   Protostar   Red Main-sequence star   White dwarf

A. Low-mass stars

Nebula   Protostar   Yellow Main-sequence star   Red giant   Planetary nebula   White dwarf

B. Intermediate-mass (Sun-like) stars

Nebula   Protostar   Blue or white main-sequence star   Supergiant (any color)   Supernova explosion   Neutron star   or   Black hole

C. High-mass stars

collapse into Earth-size bodies of great density—white dwarfs. Without a source of nuclear energy, white dwarfs become cooler and dimmer as they continually radiate thermal energy into space.

During their collapse from red giants to white dwarfs, intermediate-mass stars cast off their bloated outer atmosphere, creating an expanding cloud of gas. The remaining hot, central white dwarf heats the gas cloud, causing it to glow. These spectacular clouds are called **planetary nebulae** (**Figure 16.8**).

Death of Massive Stars In contrast to Sun-like stars, which expire nonviolently, stars that are more than eight times the mass of the Sun have

Figure 16.8 Planetary nebula The Helix Nebula, the nearest planetary nebula to our solar system. A planetary nebula is the ejected outer envelope of a Sun-like star that formed during the star's collapse from a red giant to a white dwarf. (Courtesy of NASA/JPL-Caltech)

Figure 16.9 Crab Nebula in the constellation Taurus This spectacular nebula is thought to be the remains of the supernova of A.D. 1054. (Courtesy of NASA)

recorded an even brighter supernova in A.D. 1054. Today, the remnant of that great outburst is the Crab Nebula, shown in **Figure 16.9**.

Supernova events can be triggered when a massive star has consumed most of its nuclear fuel. Without a heat engine to generate the gas pressure required to balance its immense gravitational field, it collapses. This implosion is enormous, resulting in a shock wave that rebounds out from the star's interior. This energetic shock wave blasts the star's outer shell into space, generating the fiery supernova event.

Theoretical work predicts that during this type of supernova, the star's interior condenses into an incredibly hot object, possibly no larger than 20 kilometers (12 miles) in diameter. These incomprehensibly dense bodies have been named *neutron stars*. Some supernova events are thought to produce even smaller and more intriguing objects called *black holes*. We consider the nature of neutron stars and black holes in the following section.

relatively short life spans and terminate in brilliant explosions called **supernovas** (**Figure 16.7C**). During supernova events, these stars become millions of times brighter than they were in prenova stages. If a star located near Earth produced such an outburst, its brilliance would surpass that of the Sun. Fortunately for us, supernovae are relatively rare events; none have been observed in our galaxy since the advent of the telescope, although Tycho Brahe and Johannes Kepler each recorded one, about 30 years apart, late in the sixteenth century. Chinese astronomers

16.3 CONCEPT CHECKS

1. What element is the fuel for main-sequence stars?
2. Describe how main-sequence stars become giants.
3. Why are less massive stars thought to age more slowly than more massive stars, despite the fact they have much less "fuel"?
4. List the steps thought to be involved in the evolution of Sun-like stars.

16.4 Stellar Remnants

Compare and contrast the final state of Sun-like stars to the remnants of the most massive stars.

Eventually, all stars consume their nuclear fuel and collapse into one of three types of celestial object: *white dwarfs*, *neutron stars*, or *black holes*. How a star's life ends and what final form it takes depend largely on the star's mass during its main-sequence stage (**Table 16.1**).

White Dwarfs

All that is left of a low- to medium-mass star after its outer layers are ejected into space is its gaseous core. Recall that without a source of fuel, the remnant core gravitationally collapses into a small hot star called a

white dwarf. These objects, which are typically the size of Earth and have masses roughly equal to or greater than that of the Sun, are composed of super-compressed gases called **degenerate matter**. A spoonful of such matter would weigh several tons on Earth. Densities of this magnitude are the result of electrons being displaced inward toward an atom's nucleus, so that each atom takes up less space.

Degenerate matter has been squeezed together so tightly that the gas particles have no "elbow room" and, unlike ordinary gases, they resist being compressed further. Thus, degenerate matter behaves more like a solid

TABLE 16.1 Summary of Evolution for Stars of Various Masses

Initial Mass of Main-Sequence Star (Sun = 1)*	Main-Sequence Stage	Giant Phase	Evolution After Giant Phase	Terminal State (Final Mass)
0.001	None (Planet)	No	Not applicable	Planet (0.001)
0.1	Red	No	Not applicable	White dwarf (0.1)
1–3	Yellow	Yes	Planetary nebula	White dwarf (<1.4)
8	White	Yes	Supernova	Neutron star (1.4–3)
25	Blue	Yes (Supergiant)	Supernova	Black hole (>3)

* These masses are estimates.

than a gas. However, the electrical repulsion that occurs between the negatively charged electrons supports the star's core against further gravitational collapse.

As main-sequence stars contract into white dwarfs, their surfaces become extremely hot, sometimes exceeding 25,000 K. Without an internal source of energy, white dwarfs slowly cool and eventually become small, cold, burned-out embers called **black dwarfs**. However, current estimates of the cooling rates of white dwarfs indicates that our galaxy is not yet old enough for any black dwarfs to have formed.

Neutron Stars

A study of white dwarfs produced a surprising conclusion: The *smallest white dwarfs* are the *most massive*, and the *largest* are the *least massive* (**Figure 16.10**). This occurs because when more massive stars collapse, their stronger gravitational fields cause them to be squeezed into smaller, more densely packed objects than less massive stars. Thus, the smallest white dwarfs were produced from the collapse of larger, more massive main-sequence stars than are the largest white dwarfs.

This conclusion led to the prediction that stellar remnants even *smaller* and *more massive* than white dwarfs must exist. Named **neutron stars**, these objects are the remnants (cores) of massive stars (originally more than 8 times as massive as the Sun) and are generated during an explosive supernova event. As the outer envelope of a star is violently ejected, it leaves behind a small, dense core that collapses into a very hot star less than 20 kilometers (12 miles) in diameter.

Recall that white dwarfs are composed of particles in which the electrons are pulled close to the nucleus. By contrast, in neutron stars, the electrons can be thought of as combining with the protons located *inside* the nucleus to produce neutrons (hence the name *neutron star*). A pea-size sample of this matter would weigh 100 million tons. This is approximately the density of an atomic nucleus; thus, a neutron star can be thought of as a large nucleus composed mainly of neutrons.

Although neutron stars have high surface temperatures, their small size greatly limits their luminosity, making them difficult to locate visually. Neutron stars also have a very strong magnetic field and a high rate of rotation. As stars collapse, they rotate faster—similar to the way ice skaters rotate faster as they pull their arms in as they spin. Radio waves generated by the rapidly rotating magnetic fields of neutron stars are concentrated into two narrow zones that align with the star's magnetic poles. Consequently, these stars resemble rapidly rotating beacons emitting strong radio waves. If Earth happened to be in the path of these beacons, the star would appear to blink on and off, or pulsate, as the waves swept past.

In the early 1970s, a source that radiates short pulses of radio energy, named a **pulsar** (*pulsating radio source*), was discovered in the Crab Nebula (**Figure 16.11**). Visual inspection of this radio source indicated that it was coming from a small star centered in the nebula. The pulsar found in the Crab Nebula is most likely the remains of the supernova of A.D. 1054 (see Figure 16.9).

Black Holes

Although neutron stars are extremely dense, they are not the densest objects in the universe. Stellar evolutionary theory predicts that neutron stars cannot exceed three times the mass of the Sun. Above this mass, not even tightly packed neutrons can withstand the star's gravitational pull. Following supernova explosions, if the core of a remaining star exceeds three solar masses, gravity prevails and the stellar remnant collapses into an object more dense than a neutron star. (The pre-supernova

Figure 16.10 Comparing the sizes of white dwarfs of different masses to Earth Researchers unexpectedly discovered that smaller white dwarfs are more dense than larger white dwarfs.

Earth

White dwarf having a mass equal to the Sun

White dwarf having a mass about 30% greater than the Sun

Figure 16.11 Crab Pulsar:
A young neutron star centered in the Crab Nebula
This is the first pulsar to
have been associated with
a supernova. The energy
emitted from this star illuminates the Crab Nebula.
(Courtesy of NASA EOS Earth
Observing System)

mass of such a star was likely 25 times or more than that of the Sun.) The incredible objects, or celestial phenomena, created by such a collapse are called **black holes**.

Einstein's theory of general relativity predicts that even though black holes are extremely hot, their surface gravity is so immense that even light cannot escape. Consequently, they literally disappear from sight. Anything that moves too close to a black hole can be swept in and devoured by its immense gravitational field.

How do astronomers find objects whose gravitational field prevents the escape of all matter and energy? Theory predicts that as matter is pulled into a black hole, it should become extremely hot and emit a flood of x-rays before it is engulfed. Because *isolated* black holes do not have a source of matter to engulf, astronomers decided to look at binary-star systems for evidence of matter emitting x-rays while being rapidly swept into a region of apparent nothingness.

X-rays cannot penetrate our atmosphere; therefore, the existence of black holes was not confirmed until the advent of orbiting observatories. The first black hole to be identified, Cygnus X-1, orbits a massive supergiant companion once every 5.6 days. The gases that are pulled from the companion form an *accretion disk* that spirals around a "void" while emitting a steady stream of x-rays (**Figure 16.12**). Recent observations have determined that pairs of jets extend outward from these accretion disks and are thought to return some of this material back to space (see Figure 16.12).

Cygnus X-1, which is about 8 or 9 times as massive as our Sun, probably formed from a star of approximately 40 solar masses. Since the discovery of Cygnus X-1, many other x-ray sources have been discovered that are assumed to be black holes.

Astronomers have established that black holes are common objects in the universe and vary considerably in size. Small black holes have masses approximately 10 times that of our Sun but are only about 32 kilometers (20 miles) across, less than the distance of a marathon course. Intermediate black holes have masses 1000 times our Sun, and the largest black holes (*supermassive black holes*), found in the centers of galaxies, are estimated to be millions of solar masses. Because the earliest stars were thought to be massive, their deaths could have provided the seeds that eventually formed the supermassive black holes at the centers of galaxies.

(16.4) CONCEPT CHECKS

1. Describe degenerate matter.
2. What is the final state of a medium-mass (Sun-like) star?
3. How do the "lives" of the most massive stars end? What are the two possible products of this event?
4. Explain how it is possible for the *smallest* white dwarfs to be the *most* massive.
5. Black holes are thought to be abundant, yet they are hard to find. Explain why.

Figure 16.12 Artist's view of a black hole and a giant companion star Note the accretion disk surrounding the black hole.
(Courtesy of European Southern Observatory/L. Calcada/M. Kornmesser)

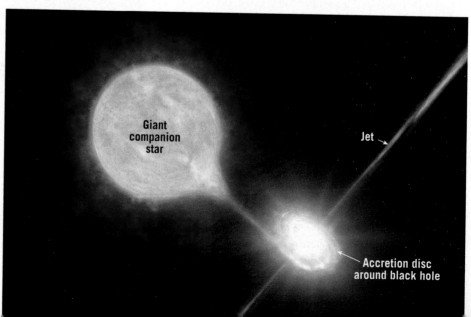

Giant
companion
star

Jet

Accretion disc
around black hole

16.5 Galaxies and Galactic Clusters

List the three major types of galaxies. Explain the formation of large elliptical galaxies.

On a clear and moonless night away from city lights, you can see a truly marvelous sight—a band of light stretching from horizon to horizon. With his telescope, Galileo discovered that this band of light was composed of countless individual stars. Today, we realize that the Sun is actually a part of this vast system of stars—the Milky Way Galaxy (**Figure 16.13**).

Galaxies (*galaxias* = milky), including the Milky Way, are collections of interstellar matter, stars, and stellar remnants that are gravitationally bound (**Figure 16.14A**). Recent observational data indicate that supermassive black holes may exist at the centers of most galaxies. In addition, spherical halos of very tenuous gas and numerous star clusters (*globular clusters*) appear to surround many of the largest galaxies (**Figure 16.14B, C**).

The first galaxies were small and composed mainly of massive stars and abundant interstellar matter. These galaxies grew quickly by accreting nearby interstellar matter and colliding and merging with other galaxies. In fact, our galaxy is currently absorbing at least two tiny satellite galaxies.

Types of Galaxies

Among the hundreds of billions of galaxies, three basic types have been identified: *spiral, elliptical,* and *irregular*. Within each of these categories are many variations, the causes of which are still in part a mystery.

Spiral Galaxies Our Milky Way Galaxy is an example of a large **spiral galaxy** (**Figure 16.15**). Spiral galaxies are flat, disk-shaped objects that range from 20,000 to about

Figure 16.13 View of our Milky Way Galaxy at sunset The dark patches in the "milky" band of light are caused by the presence of dark nebulae. (Courtesy of European Southern Observatory)

125,000 light-years in diameter. Typically, spiral galaxies have a greater concentration of stars near their centers, but there are numerous variations. As shown in Figure

A. Face-on view

B. Edge-on view

C. Globular cluster

SmartFigure 16.14 Galaxies are collections of stars and interstellar matter that are gravitationally bound **A.** Face-on view of a large spiral galaxy. Spiral galaxies typically have a greater concentration of older stars near their center, which gives the central bulge its yellowish color. By contrast, the arms of spiral galaxies contain numerous hot, young stars that give these structures a bluish or violet tint. **B.** Edge-on view showing the central bulge. **C.** Surrounding most large galaxies is a spherical halo that includes groups of stars called globular clusters. (Image A courtesy of NASA; images B and C courtesy of European Southern Observatory) (goo.gl/eK72Uz)

Tutorial

Figure 16.15 Dramatic image of the spiral galaxy Messier 83 Although smaller, Messier 83 is thought to be very similar to the Milky Way Galaxy. (European Space Observatory)

16.15, spiral galaxies have arms (usually two or more) extending from the central nucleus. Generally, the central bulge contains older stars that give the galaxy's center a yellowish color, while younger hot stars are located in the arms. The young, hot stars in the arms are found in large groups that appear as bright patches of blue and violet light, as shown in Figure 16.15.

Many spiral galaxies have a band of stars extending outward from the central bulge that merges with the spiral arms. These are called **barred spiral galaxies** (**Figure 16.16**). Recent investigations have found evidence that our galaxy probably has a bar structure. What produces these bar-shaped structures is a matter of ongoing research.

Figure 16.16 Barred spiral galaxy (Courtesy of NASA)

Elliptical Galaxies As the name implies, **elliptical galaxies** have ellipsoidal shapes; they can be nearly spherical, and they have no arms (**Figure 16.17**). Some of the largest and the smallest galaxies are elliptical. All small galaxies, regardless of type, are known as **dwarf galaxies**. The two small companions of Andromeda shown in Figure 16.2 are dwarf galaxies.

The very largest known galaxies (1 million light-years in diameter) are elliptical. For comparison, the Milky Way, a large spiral galaxy, about one-tenth the diameter of a large elliptical galaxy. Most large elliptical galaxies are believed to result from the merger of two or more smaller galaxies.

Large elliptical galaxies tend to be composed of older, low-mass stars (red) and have minimal amounts of interstellar matter. Thus, unlike the arms of spiral galaxies, they have low rates of star formation. As a result, elliptical galaxies appear yellow to red in color, as compared to the bluish tint emanating from the young, hot stars in the arms of spiral galaxies.

Irregular Galaxies Approximately 25 percent of known galaxies show no symmetry and are classified as **irregular galaxies**. Some were once spiral or elliptical galaxies that were subsequently distorted by the gravity of a neighbor. Two well-known irregular galaxies, the *Large* and *Small Magellanic Clouds*, are named for explorer Ferdinand Magellan, who observed them when he circumnavigated the globe in 1520. They are among our nearest galactic neighbors.

Recent images of the Large Magellanic Cloud reveal a central bar-shaped structure. Thus, the Large Magellanic Cloud was once a barred spiral galaxy that was subsequently distorted by gravity exerted by another galaxy as it passed by.

Galactic Clusters

Once astronomers discovered that stars occur in groups (galaxies), they set out to determine whether galaxies also occur in groups or whether they are randomly distributed. They found that galaxies are grouped into gravitationally bound clusters (**Figure 16.18**). Some large **galactic clusters** contain thousands of galaxies. Our own galactic cluster, called the **Local Group**, consists of more than 40 galaxies and may contain many undiscovered dwarf galaxies. Of the Local Group galaxies, three are large

Figure 16.17 Large elliptical galaxy This large elliptical galaxy belongs to the Fornax Cluster. Dark clouds of interstellar matter are visible within the central nucleus of this galaxy. Some of the star-like objects in this image are large groups of stars (globular clusters) that belong to the galaxy. (Courtesy of European Southern Observatory)

Figure 16.18 The Fornax Galaxy Cluster This is one of the nearest groupings of galaxies to our Local Group. Although many of the galaxies shown are elliptical, an elegant barred spiral galaxy is visible in the lower right. (Courtesy of European Southern Observatory/J. Emerson/VISTA)

Figure 16.19 The collision of the Antennae Galaxies When two galaxies collide, the stars generally do not. However, the clouds of dust and gas that are common to both collide. During these galactic encounters, there is a rapid birth of millions of stars, shown as the bright regions in this image. (Courtesy of NASA)

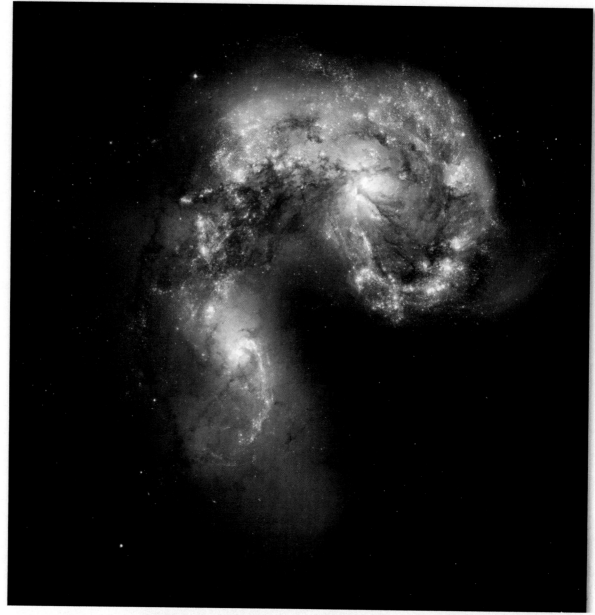

our Local Group is found in the Laniakea Supercluster. From visual observations, it appears that superclusters are the largest entities in the universe.

Galactic Collisions

Within galactic clusters, interactions between galaxies, often driven by one galaxy's gravity disturbing another, are common. For example, a large galaxy may engulf a dwarf satellite galaxy. In this case, the larger galaxy will retain its form, while the smaller galaxy will be torn apart and assimilated into the larger galaxy. Recall that two dwarf satellite galaxies are currently merging with the Milky Way.

Galactic interactions may also involve two galaxies of similar size passing through one another without merging. It is unlikely that the individual stars within these

galaxies will collide because they are so widely dispersed. However, the interstellar matter will likely interact, triggering an intense period of star formation.

In an extreme case, two large galaxies may collide and merge into a single large galaxy (**Figure 16.19**). Many of the largest elliptical galaxies are thought to have been produced by the merger of two large spiral galaxies. Some studies have predicted that in 2 to 4 billion years, there is a 50 percent probability that the Milky Way and Andromeda Galaxies will collide and merge.

16.5 CONCEPT CHECKS

1. Compare the three main types of galaxies.
2. What type of galaxy is our Milky Way?
3. Describe a possible scenario for the formation of a large elliptical galaxy.

16.6 The Big Bang Theory

Describe the big bang theory. Explain what it tells us about the universe.

The *big bang theory* describes the birth, evolution, and fate of the universe. According to the big bang theory, the universe was originally in an extremely hot, super-massive state that expanded rapidly in all directions. Based on astronomers' best calculations, this expansion began about 13.8 million years ago. What scientific evidence exists to support this theory?

Evidence for an Expanding Universe

In 1912 Vesto Slipher, while working at the Lowell Observatory in Flagstaff, Arizona, was the first to discover that galaxies exhibit motion. The motions he detected were twofold: Galaxies rotate, and galaxies move relative to each other. Slipher's efforts focused on the shifts in the spectra of the light emanating from galaxies (see the section "The Doppler Effect" in Appendix C, page 509). When a source of light is moving away from an observer, the spectral lines shift toward the red end of the spectrum (longer wavelengths). Conversely, when celestial objects approach the observer, the spectral lines shift to the blue end of the spectrum (shorter wavelengths).

In 1929, Edwin Hubble's study of galaxies expanded the groundwork established by Slipher. Hubble noticed that most galaxies have spectral shifts toward the red end of the spectrum—which occurs when an object emitting light is receding from an observer (**Figure 16.20**). Therefore, all galaxies (except those in the Local Group) appear to be moving away from the Milky Way. These patterns were later named **cosmological red shifts** because the movement they revealed resulted from the expansion of the universe.

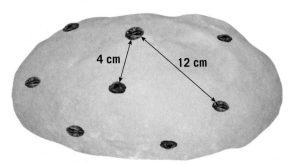
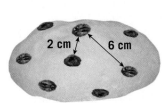

A. Raisin bread dough before it rises.

B. Raisin bread dough a few hours later.

SmartFigure 16.21 Raisin bread analogy for an expanding universe As the dough rises, raisins that were originally farthest apart travel greater distances than those located closer together. Thus, in an expanding universe (as with the raisins), more space is created between two objects that are farther apart than between two objects that are closer together. (goo.gl/ZnaSVv)

Tutorial

Recall that Hubble had also found a way to measure galactic distances. By comparing the distance to a galaxy with Vesto Slipher's measurements of its red shift, Hubble made an unexpected discovery: He discovered that the red shifts of galaxies increase with distance and that the most distant galaxies are receding from the Milky Way at the fastest rate. This concept, now called **Hubble's law**, states that *galaxies recede at speeds proportional to their distances from the observer.*

This discovery surprised Hubble because conventional wisdom was that the universe was unchanging and would likely remain unchanged. What cosmological theory could explain Hubble's findings? Researchers concluded that an expanding universe accounts for the observed red shifts.

To help visualize why Hubble's law implies an expanding universe, imagine a batch of raisin bread dough that has been set out to rise for a few hours (**Figure 16.21**). In this analogy, the raisins represent galaxies, and the dough represents space. As the dough doubles in size, so does the distance between all of the raisins. The distance between raisins that were originally 2 centimeters apart will become 4 centimeters, while the distance between raisins originally 6 centimeters apart will increase to 12 centimeters. The raisins that were originally farthest apart traveled greater distances than those located closer together. Therefore, in an expanding universe, as in our analogy, more space is created between two objects located farther apart than between two objects that are closer together.

Another feature of the expanding universe can be demonstrated using the same bread analogy. Regardless of which raisin you look at, it will move away from all the other raisins. Likewise, at any point in the universe, all other galaxies (except those in the same cluster) are receding from an observer at that location. Hubble's law implies a centerless universe that is expanding uniformly. The Hubble Space Telescope is named in honor of Edwin

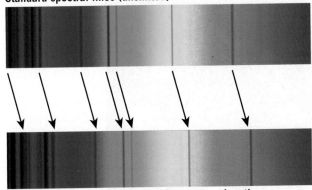

Standard spectral lines (unshifted)

Red shift moves spectral lines to longer wavelengths

Figure 16.20 Redshift Illustration of the shift in spectral lines toward the red end of the spectrum, which occurs when an object emitting light is receding from an observer.

Hubble's invaluable contributions to the scientific understanding of the universe.

Predictions of the Big Bang Theory

Recall from "The Nature of Scientific Inquiry" in this textbook's Introduction that in order for a hypothesis to become an accepted component of scientific knowledge (a theory), it must incorporate predictions that can be tested. One prediction of the big bang model is that if the universe was initially unimaginably hot, then researchers should be able to detect the remnant of that heat. The electromagnetic radiation (light) emitted by a white-hot universe would have extremely high energy and short wavelengths. However, according to the big bang theory, the continued expansion of the universe would have stretched the waves so that by now they should be detectable as long-wavelength radio waves called microwave radiation. Scientists began to search for this "missing" radiation, which they named *cosmic microwave background radiation*. As predicted, in 1965, this microwave radiation was detected and found to fill the entire visible universe.

Detailed observations of the cosmic microwave background radiation since its original discovery have confirmed many theoretical details of the big bang theory, including the order and timing of important events in the early history of the universe.

What Is the Fate of the Universe?

Cosmologists have developed different scenarios for the ultimate fate of the universe (**Figure 16.22**). In one scenario, the stars will slowly burn out and be replaced by invisible degenerate matter and black holes that travel outward through an endless, dark, cold universe. This scenario is sometimes called the "big chill" because the universe will slowly cool as it expands to the point that it is unable to sustain life. Another possibility is that the outward flight of the galaxies will slow and eventually stop. Gravitational contraction would follow, causing all matter to eventually collide and coalesce into the

Figure 16.22 Cosmic tug of war Illustration showing two possible fates of the universe. The gravity of dark matter tries to pull the universe together, while dark energy tries to push it apart. There is growing consensus that dark energy will prevail and produce an ever-expanding universe.

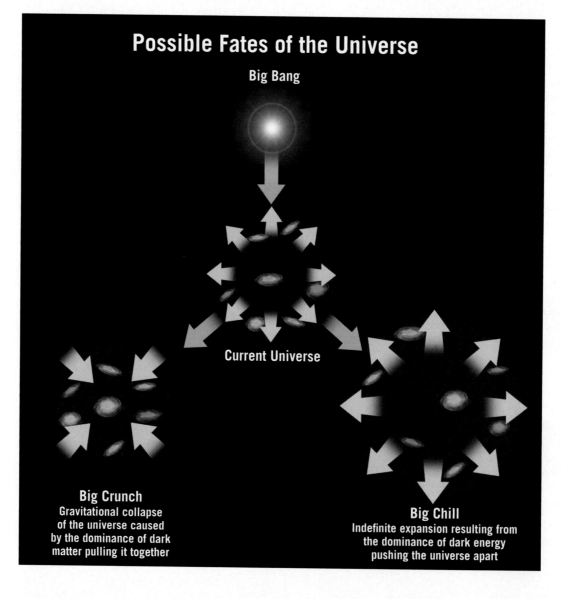

Possible Fates of the Universe

Big Bang

Current Universe

Big Crunch
Gravitational collapse
of the universe caused
by the dominance of dark
matter pulling it together

Big Chill
Indefinite expansion resulting from
the dominance of dark energy
pushing the universe apart

high-energy, high-density state, from which the universe began. This fiery death of the universe, the big bang operating in reverse, has been called the "big crunch."

Whether the universe will expand forever or eventually collapse upon itself is partially contingent on its density. If the average density of the universe is greater than an amount known as its *critical density* (about five atoms for every cubic meter), gravitational attraction should be sufficient to stop the outward expansion and cause the universe to collapse. On the other hand, if the density of the universe is less than the critical value, the universe will expand forever. Complicating the possibilities for the fate of the universe are two other constituents of the universe that are thought to exist: *dark matter* and *dark energy*.

Dark Matter The universe contains perhaps 100 billion galaxies, each with billions of stars, massive clouds of gas and dust, and large numbers of planets, moons, and other debris. Yet everything we see is like the tip of the cosmic iceberg; it accounts for less than 5 percent of the total mass of the universe. Astronomers came to this conclusion when studying the rotational periods of stars as they orbit the center of the Milky Way Galaxy. The law of gravity states that the stars closest to galactic center should travel faster than those near the galaxy's outer edge. (This is the reason Mercury travels around the Sun at a much faster speed than Neptune.) Yet these researchers found that all stars orbit the galactic center at roughly the same speed. This implies that something surrounding our galaxy is tugging on the stars. This yet undetected material was named **dark matter**.

Approximately one-quarter of the universe consists of dark matter, which produces no detectable light energy but exerts a force much like gravity that pulls on all "visible" matter in the universe. Thus, dark matter exerts a force that helps hold our galaxy together and at the same time works to slow the expansion of the universe as a whole.

Although the concept of *dark matter* may sound foreboding, it simply allows for the possibility that matter exists that does not interact with electromagnetic radiation. Recall that most of our knowledge about the universe comes to us via light. If there is a form of matter that does not interact with light, we will not be able to "see" it—hence the term *dark matter*.

Dark Energy In the early 1990s most cosmologists held the view that gravity was certain to slow the expansion of the universe over time, resulting eventually in the "big crunch." However, in 1998 observations of very distant galaxies by the Hubble Space Telescope showed that the universe is actually expanding faster today than it was early in its history. Therefore, the expansion of the universe was not slowing due to gravity as scientists had thought but instead was accelerating, or speeding up. To explain this unexpected result, researchers concluded that some unusual material, generally referred to as **dark energy**, must exist. Unlike dark matter, which works to slow the expansion of the universe, dark energy exerts a force that pushes matter outward, causing the expansion to speed up.

It has not been determined whether dark matter and dark energy are related, or exactly what they are. Most researchers think that dark matter consists of a type of subatomic particle that has not yet been detected. Dark energy may have its own particle, but there is no evidence of the existence of this particle.

There is growing consensus among cosmologists that dark energy, which is propelling the universe outward, is the dominant force. If dark energy is, in fact, the driving force behind the fate of the universe, the universe will expand forever (see Figure 16.22). Consider the following as astronomers search for dark energy: "Absence of evidence is not evidence of absence."

(16.6) **CONCEPT CHECKS**

1. In your own words, explain how astronomers determined that the universe is expanding.
2. What did the big bang theory predict that was finally confirmed years after it was formulated?
3. Which view of the fate of the universe is currently favored: a "big crunch" or an endless, ever-expanding universe?
4. What property does the universe possess that will determine its final state?

CONCEPTS IN REVIEW
Beyond Our Solar System

16.1 The Universe

Define *cosmology* and describe Edwin Hubble's most significant discovery about the universe.

KEY TERMS: cosmology, absolute magnitude, light-year, big bang theory

- Cosmology is the study of the universe, including its properties, structure, and evolution.

- The universe consists of hundreds of billions of galaxies, most containing billions of stars.
- The model that most accurately describes the birth and current state of the universe is the big bang theory. According to this model, the universe began about 13.8 billion years ago, in a cataclysmic explosion, and then it continued to expand, cool, and evolve to its current state.

16.2 Classifying Stars: Hertzsprung–Russell Diagrams (H-R Diagrams)

Define *main-sequence star*. Explain the criteria used to classify stars as giants.

KEY TERMS: Hertzsprung–Russell diagram (H-R diagram), main-sequence star, giant, red giant, supergiant, white dwarf

- Hertzsprung–Russell diagrams are constructed by plotting the absolute magnitudes and temperatures of stars on graphs. Considerable information about stars and stellar evolution has been discovered through the use of H-R diagrams.
- Stars are positioned within H-R diagrams as follows: (1) Main-sequence stars, 90 percent of all stars, are in the band that runs from the upper-left corner (massive, hot blue stars) to the lower-right corner (low-mass, red stars); (2) red giants and supergiants, very luminous stars with large radii, are located in the upper-right position; and (3) white dwarfs, which are small, dense stars, are located in the lower portion.

(?) **Identify the type of stars located in the positions labeled A, B, and C on the accompanying H-R diagram.**

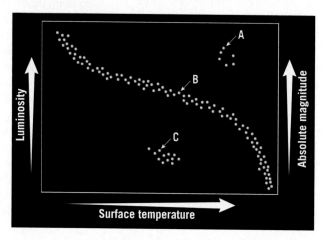

16.3 Stellar Evolution

List and describe the stages in the evolution of a typical Sun-like star.

KEY TERMS: nebula, protostar, hydrogen fusion, variable star, planetary nebula, supernova

- Stars are born when their nuclear furnaces are ignited by the unimaginable pressures and temperatures generated during the collapse of nebulae.
- Red star-like objects not yet hot enough for nuclear fusion are called protostars. When the core of a protostar reaches a temperature of about 10 million K, a process called hydrogen fusion begins, marking the birth of the star. Hydrogen fusion involves the conversion of four hydrogen nuclei into a single helium nucleus and the release of thermal nuclear energy.
- Two opposing forces act on a star: gravity, which tries to contract it into the smallest possible ball, and gas pressure (created by thermal nuclear energy), which tries to blow it apart. When the forces are balanced, the star becomes a stable main-sequence star.
- Medium- and high-mass stars experience another type of nuclear fusion that causes their outer envelopes to expand enormously (hundreds to thousands of times larger), making them red giants or supergiants. When a star exhausts all of its usable nuclear fuel, gravity takes over, and the stellar remnant collapses into a small, dense body.

16.4 Stellar Remnants

Compare and contrast the final state of Sun-like stars to the remnants of the most massive stars.

KEY TERMS: degenerate matter, black dwarf, neutron star, pulsar, black hole

- The final fate of a star is determined by its mass.
- Stars with less than one-half the mass of the Sun collapse into hot, dense white dwarfs.
- Medium-mass stars become red giants, collapse, and end up as white dwarfs, often surrounded by expanding clouds of glowing gas called planetary nebulae. Our Sun is in this category.
- Massive stars terminate in a brilliant explosion called a supernova. Supernova events can produce small, extremely dense neutron stars, composed entirely of neutrons, or smaller, even denser black holes—objects that have such immense gravity that light cannot escape their surface.

16.5 Galaxies and Galactic Clusters

List the three major types of galaxies. Explain the formation of large elliptical galaxies.

KEY TERMS: galaxy, spiral galaxy, barred spiral galaxy, elliptical galaxy, dwarf galaxy, irregular galaxy, galactic cluster, Local Group

- The various types of galaxies include (1) irregular galaxies, which lack symmetry and account for about 25 percent of the known galaxies; (2) spiral galaxies, which are disk shaped and have a greater concentration of stars near their centers and arms extending from their central nucleus; and (3) elliptical galaxies, which have an ellipsoidal shape and may be nearly spherical.
- Galaxies are grouped in galactic clusters, some containing thousands of galaxies. Our own, called the Local Group, contains at least 40 galaxies.

(?) **What type of galaxy is shown in the accompanying image?**

NASA

16.6 The Big Bang Theory

Describe the big bang theory. Explain what it tells us about the universe.

KEY TERMS: cosmological red shift, Hubble's law, dark matter, dark energy

- Evidence for an expanding universe came from the study of red shifts in the spectra of galaxies. Edwin Hubble concluded that the observed red shifts,

(16.6 continued)

called cosmological red shifts, result from the expansion of space. This evidence strongly supports the big bang model of an expanded universe.

• One question that remains is whether the universe will expand forever or gravitationally contract in a "big crunch." Dark matter works to slow the expansion of the universe, while dark energy exerts a force that pushes matter outward and causes the expansion to speed up. Most cosmologists favor an endless, ever-expanding universe.

GIVE IT SOME THOUGHT

1. Assume that NASA is sending space probes to each of the locations listed below. Arrange the order in which each probe will encounter its destination, *from nearest Earth to farthest*.
   a. Polaris (the North Star)
   b. A comet near the outer edge of our solar system
   c. Jupiter
   d. The far edge of the Milky Way Galaxy
   e. The Andromeda Galaxy
   f. The Sun

2. How a star evolves is closely related to its mass as a main-sequence star. Complete the accompanying diagram by labeling the evolutionary stages for the three groups of main-sequence stars shown.

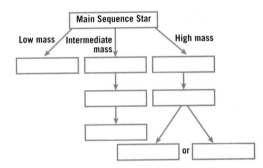

3. Use the information provided below about three main-sequence stars (A, B, and C) and Figure 16.4 to complete and explain your reasoning for the following:
   • Star A has a main-sequence life span of 5 billion years.
   • Star B has the same luminosity (absolute magnitude) as the Sun.
   • Star C has a surface temperature of 5000 K.
   a. Rank the mass of these stars from *greatest to least*.
   b. Rank the energy output of these stars from *greatest to least*.
   c. Rank the main-sequence life span of these stars from *longest to shortest*.

4. The masses of three clouds of gas and dust (nebulae) are provided below:
   • Cloud A is 60 times the mass of the Sun.
   • Cloud B is 7 times the mass of the Sun.
   • Cloud C is 2 times the mass of the Sun.

Imagine that all the material in each cloud will collapse to form a single star. Use this information to answer the questions and explain your reasoning.
   a. Which cloud or clouds, if any, will evolve into a red main-sequence star?
   b. Which of the stars that will form from these clouds, if any, will reach the giant stage?
   c. Which of the stars that will form from these clouds, if any, will go through the supernova stage?

5. Refer to the accompanying photos of an elliptical galaxy and a spiral galaxy to answer the following:
   a. Which image (A or B) is an elliptical galaxy?
   b. Which of these galaxies appears to contain more young, hot massive stars? How did you determine your answer?
   c. When stars are born from a cloud of dust and gases, large and small stars form at about the same time. Which group of stars, the large or the small, will die out first? Over time, how will this affect the color of the light we observe coming from this group of stars?

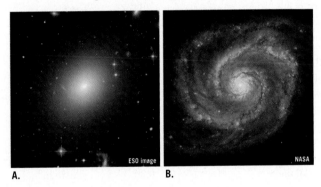

A.   **B.**

6. Consider these three characteristics of the universe:
   • It does not have a center.
   • It does not have edges.
   • Its galaxies are all moving away from each other.
   a. Which of the three characteristics of the universe does the raisin bread dough analogy depict (see Figure 16.21)?
   b. Which of the three characteristics of the universe is not accurately depicted in Figure 16.21?

MasteringGeology™

www.masteringgeology.com Looking for additional review and test prep materials? With individualized coaching on the toughest topics of the course, MasteringGeology offers a wide variety of ways for you to move beyond memorization to begin thinking like a geologist. Visit the Study Area in www.masteringgeology.com to find practice quizzes, study tools, and multimedia that will improve your understanding of this chapter's content. Sign in today to enjoy the following features: **Self Study Quizzes, SmartFigure: Tutorials/Animations/Condor Videos/Mobile Field Trips, Geoscience Animation Library, GEODe, RSS Feeds, Digital Study Modules,** and an optional **Pearson eText.**

Appendix A Metric and English Units Compared

Units

1 kilometer (km)	= 1000 meters (m)
1 meter (m)	= 100 centimeters
1 centimeter (cm)	= 0.39 inch (in.)
1 mile (mi)	= 5280 feet (ft)
1 foot (ft)	= 12 inches (in.)
1 inch (in.)	= 2.54 centimeters (cm)
1 square mile (mi^2)	= 640 acres (a)
1 kilogram (kg)	= 1000 grams (g)
1 pound (lb)	= 16 ounces (oz)
1 fathom	= 6 feet (ft)

Conversions

Length

When you want to convert:	multiply by:	to find:
inches	2.54	centimeters
centimeters	0.39	inches
feet	0.30	meters
meters	3.28	feet
yards	0.91	meters
meters	1.09	yards
miles	1.61	kilometers
kilometers	0.62	miles

Area

When you want to convert:	multiply by:	to find:
square inches	6.45	square centimeters
square centimeters	0.15	square inches
square feet	0.09	square meters
square meters	10.76	square feet
square miles	2.59	square kilometers
square kilometers	0.39	square miles

Volume

When you want to convert:	multiply by:	to find:
cubic inches	16.38	cubic centimeters
cubic centimeters	0.06	cubic inches
cubic feet	0.028	cubic meters
cubic meters	35.3	cubic feet
cubic miles	4.17	cubic kilometers
cubic kilometers	0.24	cubic miles
liters	1.06	quarts
liters	0.26	gallons
gallons	3.78	liters

Masses and Weights

When you want to convert:	multiply by:	to find:
ounces	28.35	grams
grams	0.035	ounces
pounds	0.45	kilograms
kilograms	2.205	pounds

Temperature

When you want to convert degrees Fahrenheit (°F) to degrees Celsius (°C), subtract 32 degrees and divide by 1.8.

When you want to convert degrees Celsius (°C) to degrees Fahrenheit (°F), multiply by 1.8 and add 32 degrees.

When you want to convert degrees Celsius (°C) to kelvins (K), delete the degree symbol and add 273. When you want to convert kelvins (K) to degrees Celsius (°C), add the degree symbol and subtract 273.

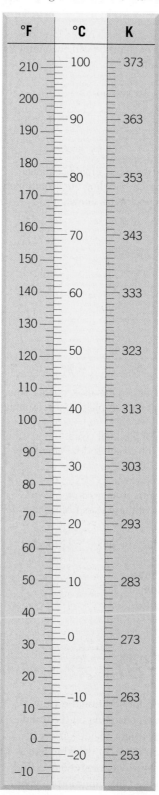

Figure A.1 Temperature scales.

Appendix B   Relative Humidity and Dew-Point Tables

TABLE B.1 Relative Humidity (percent)

Dry bulb (°C)	Depression of Wet-Bulb Temperature (Dry-Bulb Temperature Minus Wet-Bulb Temperature = Depression of the Wet Bulb)																					
	1	2	3	4	5	6	7	8	9	10	11	12	13	14	15	16	17	18	19	20	21	22
−20	28																					
−18	40																					
−16	48	0																				
−14	55	11																				
−12	61	23																				
−10	66	33	0																			
−8	71	41	13																			
−6	73	48	20	0																		
−4	77	54	32	11																		
−2	79	58	37	20	1																	
0	81	63	45	28	11																	
2	83	67	51	36	20	6																
4	85	70	56	42	27	14																
6	86	72	59	46	35	22	10	0														
8	87	74	62	51	39	28	17	6														
10	88	76	65	54	43	33	24	13	4													
12	88	78	67	57	48	38	28	19	10	2												
14	89	79	69	60	50	41	33	25	16	8	1											
16	90	80	77	62	54	45	37	29	21	74	7	1										
18	91	81	72	64	56	48	40	33	26	19	12	6	0									
20	91	82	74	66	58	51	44	36	30	23	17	11	5									
22	92	83	75	68	60	53	46	40	33	27	21	15	10	4	0							
24	92	84	76	69	62	55	49	42	36	30	25	20	14	9	4	0						
26	92	85	77	70	64	57	51	45	39	34	28	23	18	13	9	5						
28	93	86	78	71	65	59	53	45	42	36	31	26	21	17	12	8	4					
30	93	86	79	72	66	61	55	49	44	39	34	29	25	20	16	12	8	4				
32	93	86	80	73	68	62	56	51	46	41	36	32	27	22	19	14	11	8	4			
34	93	86	81	74	69	63	58	52	48	43	38	34	30	26	22	18	14	11	8	5		
36	94	87	81	75	69	64	59	54	50	44	40	36	32	28	24	21	17	13	10	7	4	
38	94	87	82	76	70	66	60	55	51	46	42	38	34	30	26	23	20	16	13	10	7	5
40	94	89	82	76	71	67	61	57	52	48	44	40	36	33	29	25	22	19	16	13	10	7

*To determine the relative humidity, or dew point, find the air (dry-bulb) temperature on the vertical axis (far left) and the depression of the wet bulb on the horizontal axis (top). Where the two meet, the relative humidity (TABLE B.1) or dew point (TABLE B.2) is found. For example, when the dry-bulb temperature is 20°C and a wet-bulb temperature is 14°C, the depression of the wet bulb is 6°C (20°–14°C). From TABLE B.1, the relative -humidity is 51 percent, and from TABLE B.2, the dew point is 10°C.

TABLE B.2 Dew-Point Temperature (°C)*

Dry bulb (°C)	1	2	3	4	5	6	7	8	9	10	11	12	13	14	15	16	17	18	19	20	21	22
						(Dry-Bulb Temperature Minus Wet-Bulb Temperature = Depression of the Wet Bulb)																
−20	−33																					
−18	−28																					
−16	−24																					
−14	−21	−36																				
−12	−18	−28																				
−10	−14	−22																				
−8	−12	−18	−29																			
−6	−10	−14	−22																			
−4	−7	−12	−17	−29																		
−2	−5	−8	−13	−20																		
0	−3	−6	−9	−15	−24																	
2	−1	−3	−6	−11	−17																	
4	1	−1	−4	−7	−11	−19																
6	4	1	−1	−4	−7	−13	−21															
8	6	3	1	−2	−5	−9	−14															
10	8	6	4	1	−2	−5	−9	−14	−18													
12	10	8	6	4	1	−2	−5	−9	−16													
14	12	11	9	6	4	1	−2	−5	−10	−17												
16	14	13	11	9	7	4	1	−1	−6	−10	−17											
18	16	15	13	11	9	7	4	2	−2	−5	−10	−19										
20	19	17	15	14	12	10	7	4	2	−2	−5	−10	−19									
22	21	19	17	16	74	12	10	8	5	3	−1	−5	−10	−19								
24	23	21	20	18	16	14	12	10	8	6	2	−1	−5	−10	−18							
26	25	23	22	20	18	17	15	13	11	9	6	3	0	−4	−9	−18						
28	27	25	24	22	27	19	17	16	14	11	9	7	4	1	−3	−9	−16					
30	29	27	26	24	23	21	19	18	16	14	12	70	8	5	1	−2	−8	−15				
32	31	29	28	27	25	24	22	21	19	17	15	13	11	8	5	2	−2	−7	−14			
34	33	31	30	29	27	26	24	23	21	20	18	16	14	12	9	6	3	−1	−5	−12	−29	
36	35	33	32	31	29	28	27	25	24	22	20	19	17	15	13	10	7	4	0	−4	−10	
38	37	35	34	33	32	30	29	28	26	25	23	21	19	17	15	13	11	8	5	1	−3	−9
40	39	37	36	35	34	32	31	30	28	37	25	24	22	20	18	16	14	12	9	6	2	−2

Dew-Point Values

Dry-Bulb (Air) Temperature

Appendix C Stellar Properties

Measuring Distances to the Closest Stars

Measuring the distance to stars is difficult. Nevertheless, astronomers have developed some direct as well as indirect methods to measure stellar distances. One simple measurement, called *stellar parallax*, is effective in determining the distances to only the closest stars.

Stellar parallax is the slight back-and-forth shift of the apparent position of a nearby star due to the orbital motion of Earth around the Sun. The principle of parallax is easy to visualize. Close one eye, and with your index finger in a vertical position, use your open eye to align your finger with some distant object. Without moving your finger, view the object with your other eye and notice that its position appears to have changed. Now repeat the exercise, this time holding your finger farther away, and notice that the farther away you hold your finger, the less its position seems to shift. This method of measuring stellar distances is elementary and was practiced by ancient Greek astronomers.

Modern cosmologists determine parallax by photographing a nearby star against the background of distant stars. Then, when Earth has moved halfway around its orbit 6 months later, the same star is photographed again. When these two photographs are compared, the position of the nearby star appears to have shifted with respect to the background stars. **Figure C.1** illustrates this shift, and the parallax angle that is determined from it. The nearest stars have the largest parallax angles, whereas the parallax angles of distant stars are much too small to measure.

In practice, conducting parallax measurements is complex because of the miniscule angles being measured. The process is further complicated because both the Sun and the star being measured are moving relative to each other. The first accurate stellar parallax was not determined until 1838. Even today, parallax angles for only a few thousand of the nearest stars are known with certainty; nearly all others have such small parallax shifts that accurate measurements are not possible.

Fortunately, other methods have been developed to estimate distances to more distant stars. In addition, the Hubble Space Telescope, which is not hindered by Earth's light-distorting atmosphere, has obtained accurate parallax distances for many more stars.

Stellar Brightness

The oldest means of classifying stars is based on their *brightness*, also called *luminosity* or *magnitude*. Three factors control the brightness of a star as seen from Earth: *how large* it is, *how hot* it is, and its *distance* from Earth. The stars in the night sky come in a grand assortment of sizes, temperatures, and distances, so their apparent brightness varies widely.

Apparent Magnitude

Stars have been classified according to their apparent brightness since at least the second century B.C., when Hipparchus placed about 850 stars into six categories, based on his ability to see differences in brightness. Because he could only reliably see six different brightness levels, he created six categories. These categories were later called *magnitudes*, with first magnitude being the brightest and sixth magnitude the dimmest. Because some stars may appear dimmer than others only because they are farther away, a star's brightness, *as it appears when viewed from Earth*, is called its *apparent magnitude*. With the invention of the telescope, many stars fainter than the sixth magnitude were discovered.

In the mid-1800s, a method was developed to standardize the magnitude scale. An absolute comparison was made between the light coming from stars of the first magnitude and from stars of the sixth magnitude. It was determined that a first-magnitude star was about 100 times brighter than a sixth-magnitude star. On the scale that was devised, any two stars that differ by five magnitudes have a ratio in

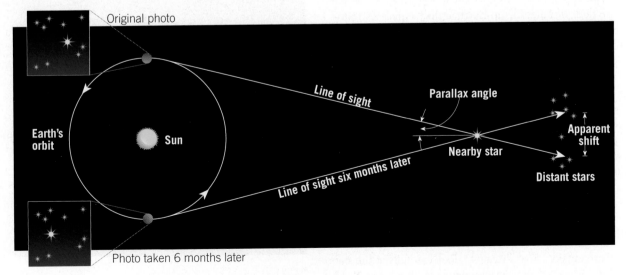

Figure C.1 Geometry of stellar parallax The parallax angle shown here is enormously exaggerated to illustrate the principle. Because distances to even the nearest stars are thousands of times greater than the Earth–Sun distance, the triangles that astronomers work with are extremely long and narrow, making the angles that are measured very small.

TABLE C.1	Ratios of Star Brightness
Difference in Magnitude	Brightness Ratio
0.5	1.6:1
1	2.5:1
2	6.3:1
3	16:1
4	40:1
5	100:1
10	10,000:1
20	100,000,000:1

Calculations: 2.512 × 2.512 × 2.512 × 2.512 × 2.512, or 2.512 raised to the fifth power, equals 100.

TABLE C.2	Distance, Apparent Magnitude, and Absolute Magnitude of Some Stars		
Name	Distance (light-years)	Apparent Magnitude	Absolute Magnitude*
Sun	NA	−26.7	5.0
Alpha Centauri	4.27	0.0	4.4
Sirius	8.70	−1.4	1.5
Arcturus	36	−0.1	−0.3
Betelgeuse	520	0.8	−5.5
Deneb	1600	1.3	−6.9

* The more negative, the brighter; the more positive, the dimmer.

brightness of 100 to 1. Hence, a third-magnitude star is 100 times brighter than an eighth-magnitude star. It follows, then, that the brightness ratio of two stars differing by only one magnitude is about 2.5.* A star of the first magnitude is about 2.5 times brighter than a star of the second magnitude. **Table C.1** shows how differences in magnitude correspond to brightness ratios.

Because some celestial bodies are brighter than first-magnitude stars, zero and negative magnitudes were introduced. On this scale, the Sun has an apparent magnitude of −26.7. At its brightest, Venus has a magnitude of −4.3. At the other end of the scale, the Hubble Space Telescope can view stars with an apparent magnitude of 30, more than 1 billion times dimmer than stars that are visible to the unaided eye.

Absolute Magnitude

Apparent magnitudes were good approximations of the true brightness of stars when astronomers thought that the universe was very small—containing no more than a few thousand stars, all at very similar distances from Earth. However, we now know that the universe is unimaginably large and contains innumerable stars at wildly varying distances. Because astronomers are interested in the "true" brightness of stars, they devised a measure called *absolute magnitude.*

Stars of the same apparent magnitude usually do not have the same brightness because their distances from Earth are not equal. Astronomers correct for distance by determining what brightness (magnitude) the stars would have if they were at a standard distance—about 32.6 light-years. For example, if the Sun, which has an apparent magnitude of −26.7, were located 32.6 light-years from Earth, it would have an absolute magnitude of about +5. Thus, stars with absolute magnitudes greater than 5 (smaller numerical value) are intrinsically brighter than the Sun but appear much dimmer because of their distance from Earth. **Table C.2** lists the absolute and apparent magnitudes of some stars as well as their distances from Earth. Most stars have an absolute magnitude between −5 (very bright) and 15 (very dim). The Sun is near the midpoint of this range.

Stellar Color and Temperature

The next time you are outside on a clear night, look carefully at the stars and note their colors (**Figure C.2**). Because human eyes do not respond

*The more negative, the brighter; the more positive, the dimmer.

well to color in low-intensity light (when it is very dark, we see in only black and white), we tend to look at the brightest stars. Some stars that are quite colorful can be found in the constellation Orion. Of the two brightest stars in Orion, Rigel (β Orionis) appears blue, and Betelgeuse (α Orionis) is definitely red.

Very hot stars with surface temperatures above 30,000 K emit most of their energy in the form of short-wavelength light and therefore appear blue. On the other hand, cooler red stars, with surface temperatures generally less than 3,000 K, emit most of their energy as longer-wavelength red light. Stars, such as the Sun, with surface temperatures between 5,000 and 6,000 K, appear yellow. Because color is primarily a manifestation of a star's surface temperature, this characteristic provides astronomers with useful information. As you will see, combining temperature data with stellar magnitude tells us a great deal about the size and mass of stars.

Binary Stars and Stellar Mass

One of the night sky's best-known constellations, the Big Dipper, appears to consist of seven stars. But those with good eyesight can see that the

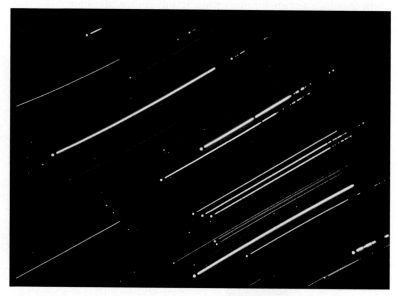

Figure C.2 Time-lapse photograph of stars in the constellation Orion These star trails show some of the various star colors. It is important to note that the human eye sees color somewhat differently than photographic film. (Courtesy of National Optical Astronomy Observatories)

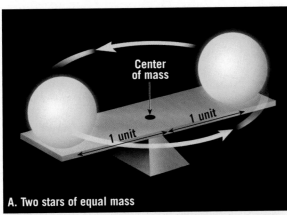

A. Two stars of equal mass

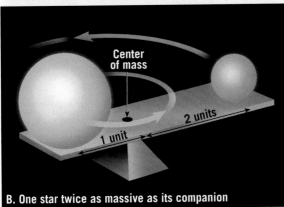

B. One star twice as massive as its companion

Figure C.3 Binary stars orbit each other around their common center of mass **A**. For stars of equal mass, the center of mass lies exactly halfway between them. **B**. If one star is twice as massive as its companion, it is twice as close to the two stars' common center of mass. Therefore, more massive stars have proportionately smaller orbits than their less massive companions.

second star in the handle is actually two stars. In the early nineteenth century, careful examination of numerous star pairs by William Herschel showed that many stars found in pairs actually orbit one another. In such cases, the two stars are in fact united by their mutual gravitation. These pairs, in which the members are far enough apart to be telescopically identified as two stars, are called *visual binaries* (*binaries* = double). The idea of one star orbiting another may seem unusual, but many stars in the universe exist in pairs or multiples.

Binary stars can be used to determine the star property that is most difficult to calculate—its *mass*. The mass of a body can be

established if it is gravitationally attached to a partner. Binary stars orbit each other around a common point called the *center of mass* (**Figure C.3**). For stars of equal mass, the center of mass lies exactly halfway between them. When one star is more massive than its partner, their common center will be located closer to the more massive one. Thus, if the sizes of their orbits can be observed, their individual masses can be determined. You can experience this relationship on a seesaw by trying to balance a person who has a much greater (or smaller) mass.

For illustration, when one star has an orbit half the size (radius) of its companion, it is twice as massive as its companion. If their combined masses are equal to three solar masses, then the larger will be twice as massive as the Sun, and the smaller will have a mass equal to that of the Sun. Most stars have a mass that falls in a range between 1/10 and 50 times the mass of the Sun.

The Doppler Effect

The positions of the bright and dark lines in the spectra shift when the source of energy moves relative to the observer. This effect is observed for all types of waves. Think about the change in pitch you hear in a car horn or an ambulance siren as it passes by. When it is approaching, the sound seems to have a higher-than-normal pitch, and when it is moving away, the pitch sounds lower than normal. This effect, first explained by Christian Doppler in 1842, is called the *Doppler effect*. The reason for the difference in pitch is that it takes time for the wave to be emitted. If the source is moving away, the beginning of the wave is emitted nearer to you than the end, which stretches the wave—that is, gives it a longer wavelength (**Figure C.4**). The opposite is true for an approaching source.

In the case of light, when a source is moving away, its light appears redder than it actually is because its waves are lengthened. Objects approaching have their light waves shifted toward the blue (shorter-wavelength) end of the spectrum. Thus, if a source of red light approached you at a very high speed (near the speed of light), it would actually appear blue. The same effect is produced if you are moving and the light remains stationary.

The Doppler effect is important because it reveals whether Earth is approaching or receding from a star or another celestial body. In addition, the amount of shift allows us to calculate the rate at which the relative movement is occurring. Large Doppler shifts indicate high velocities; small Doppler shifts indicate low velocities.

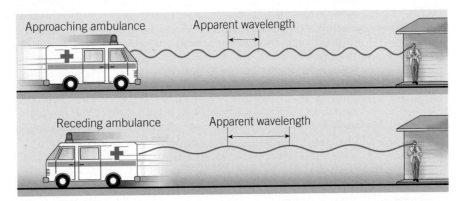

Figure C.4 The Doppler effect This image illustrates the apparent lengthening and shortening of wavelengths caused by the relative motion between a source and an observer.

GLOSSARY

Aa flow A type of lava flow that has a jagged, blocky surface.

Abrasion The grinding and scraping of a rock surface by the friction and impact of rock particles carried by water, wind, or ice.

Absolute instability The condition of air that has an environmental lapse rate that is greater than the dry adiabatic rate (1°C per 100 meters).

Absolute magnitude The apparent brightness of a star if it were viewed from a distance of 10 parsecs (32.6 light-years). Used to compare the true brightness of stars.

Absolute stability The condition of air that has an environmental lapse rate that is less than the wet adiabatic rate.

Abyssal plain A very level area of the deep-ocean floor, usually lying at the foot of the continental rise.

Accretionary wedge A large wedge-shaped mass of sediment that accumulates in subduction zones. Here sediment is scraped from the subducting oceanic plate and accreted to the overriding crustal block.

Active continental margin A margin that is usually narrow and consists of highly deformed sediments. Such margins occur where oceanic lithosphere is being subducted beneath the margin of a continent.

Adiabatic temperature change Cooling or warming of air that occurs because air is allowed to expand or is compressed, not because heat is added or subtracted.

Advection fog A fog formed when warm, moist air is blown over a cool surface.

Aerosols Tiny solid and liquid particles suspended in the atmosphere.

Air The mixture of gases and particles that make up Earth's atmosphere. Nitrogen and oxygen are most abundant.

Air mass A large body of air that is characterized by a sameness of temperature and humidity.

Air-mass weather The conditions experienced in an area as an air mass passes over it. Because air masses are large and fairly homogeneous, air-mass weather will be fairly constant and may last for several days.

Air pressure Force exerted by the weight of the air above.

Albedo The reflectivity of a substance, usually expressed as a percentage of the incident radiation reflected.

Alluvial fan A fan-shaped deposit of sediment formed when a stream's slope is abruptly reduced.

Alluvium Unconsolidated sediment deposited by a stream.

Alpine glacier See *valley glacier*.

Altocumulus White to gray clouds, often made up of separate globules; "sheepback" clouds.

Altostratus Stratified veil of clouds that is generally thin and may produce very light precipitation. When thin, the Sun or Moon may be visible as a "bright spot," but no halos are produced.

Ambiguous property A property of a mineral that is not diagnostic because it varies among different specimens of the mineral.

Andesite A gray, fine-grained igneous rock, primarily of volcanic origin and commonly exhibiting a porphyritic texture.

Andesitic (intermediate) composition A compositional group of igneous rocks, in which the rock contains at least 25 percent dark silicate minerals. The other dominant mineral is plagioclase feldspar.

Aneroid barometer An instrument for measuring air pressure that consists of evacuated metal chambers that are very sensitive to variations in air pressure.

Angle of repose The steepest angle at which loose material remains stationary without sliding downslope.

Angular unconformity An unconformity in which the strata below dip at an angle different from that of the beds above.

Annual mean An average of the 12 monthly means (for instance, of temperature).

Annual temperature range The difference between the highest and lowest monthly means.

Anticline A fold in sedimentary strata that resembles an arch.

Anticyclone (high) A high-pressure center characterized by a clockwise flow of air in the Northern Hemisphere.

Apparent magnitude The brightness of a star when viewed from Earth.

Aquifer Rock, sediment, or soil through which groundwater moves easily.

Aquitard An impermeable bed that hinders or prevents groundwater movement.

Arctic (A) air mass Air mass that originates at high polar latitudes. Generally very cold and dry.

Arête A narrow knifelike ridge separating two adjacent glaciated valleys.

Artesian system A situation in which groundwater under pressure rises above the level of the aquifer

Asteroid One of thousands of small planetlike bodies, ranging in size from a few hundred kilometers to less than 1 kilometer, whose orbits lie mainly between those of Mars and Jupiter.

Asteroid belt The region in which most asteroids orbit the Sun between Mars and Jupiter.

Asthenosphere A subdivision of the mantle situated below the lithosphere. This zone of weak material exists below a depth of about 100 kilometers and in some regions extends as deep as 700 kilometers. The rock within this zone is easily deformed.

Astronomical unit (AU) The average distance from Earth to the Sun; 1.5×10^8 kilometers (93×10^6 miles).

Astronomy The scientific study of the universe; it includes the observation and interpretation of celestial bodies and phenomena.

Atmosphere The gaseous portion of a planet; the planet's envelope of air. One of the traditional subdivisions of Earth's physical environment.

Atoll A continuous or broken ring of coral reef surrounding a central lagoon.

Atom The smallest particle that exists as an element.

Atomic number The number of protons in the nucleus of an atom.

Augite A black, opaque silicate mineral of the pyroxene group that is a dominant component of basalt.

Autumnal equinox See *Equinox*.

Back swamp A poorly drained area on a floodplain that results when natural levees are present.

Bajada An apron of sediment along a mountain front created by the coalescence of alluvial fans.

Bar *In a stream channel*, a common term for deposits of sand and gravel. *In meteorology*, a unit of air pressure, where standard sea level pressure is defined as 1 bar.

Barograph A recording barometer.

Barometer An instrument that measures atmospheric pressure.

Barometric tendency See *Pressure tendency*.

Barred spiral galaxy A galaxy that has straight arms extending from its nucleus.

Barrier island A low, elongated ridge of sand that parallels the coast.

Basalt A fine-grained igneous rock of mafic composition.

Basalt plateau Vast accumulations of basaltic lava resulting in a flat, broad plain of great thickness. See also *Flood basalts*.

Basaltic (mafic) composition A compositional group of igneous rocks in which the rock contains substantial dark silicate minerals and calcium-rich plagioclase feldspar.

Base level The level below which a stream cannot erode.

Basin A circular downfolded structure.

Batholith A large mass of igneous rock that formed when magma was emplaced at depth, crystallized, and subsequently exposed by erosion.

Bathymetry The measurement of ocean depths and the charting of the topography of the ocean floor.

Baymouth bar A sandbar that completely crosses a bay, sealing it off from the open ocean.

Beach An accumulation of sediment found along the landward margin of the ocean or a lake.

Beach drift The transport of sediment in a zigzag pattern along a beach caused by the uprush of water from obliquely breaking waves.

Beach nourishment The addition of large quantities of sand to a beach system to offset losses caused by wave erosion.

Bed load Sediment that is carried by a stream along the bottom of its channel.

Bergeron process A theory that relates the formation of precipitation to supercooled clouds, freezing nuclei, and the different saturation levels of ice and liquid water.

Big bang theory A theory which proposes that the universe originated as a single mass, which subsequently exploded.

Biochemical sedimentary rock A sedimentary rock composed of material that was extracted from water by organisms to create hard parts such as shells.

Biogenous sediment Seafloor sediment consisting of material of marine-organic origin.

Biosphere The totality of life on Earth; the parts of the lithosphere, hydrosphere, and atmosphere in which living organisms can be found.

Biotite A dark, iron-rich mineral and a member of the mica family with excellent cleavage.

Black dwarf A final state of evolution for a star, in which all of its energy sources are exhausted and it no longer emits radiation.

Black hole A massive star that has collapsed to such a small volume that its gravity prevents the escape of all radiation.

Blowout A depression excavated by the wind in easily eroded deposits.

Body waves Seismic waves that travel through Earth's interior.

Bowen's reaction series A concept proposed by N. L. Bowen that illustrates the relationships between magma and the minerals crystallizing from it during the formation of igneous rocks.

Braided channel Type of alluvial channel consisting of a broad network of diverging and converging channels. Forms where the stream's load includes abundant coarse material and the discharge is highly variable.

Breakwater A structure that protects a near-shore area from breaking waves.

Breccia A sedimentary rock composed of angular fragments that were lithified.

Brittle deformation The loss of strength by a material, usually in the form of sudden fracturing.

Calcite Calcium carbonate ($CaCO_3$), one of the two most common carbonate minerals.

Caldera A large depression typically caused by collapse or ejection of the summit area of a volcano.

Calorie The amount of energy needed to raise the temperature of 1 gram of water through 1 degree Centigrade.

Capacity The total amount of sediment a stream is able to transport.

Carbonic acid A weak acid formed when carbon dioxide is dissolved in water. It plays an important role in chemical weathering.

Catastrophism The concept that Earth was shaped by catastrophic events of a short-term nature.

Cavern A naturally formed underground chamber or series of chambers most commonly produced by solution activity in limestone.

Celestial sphere An imaginary hollow sphere on which the ancients believed the stars were hung and were carried around Earth.

Cementation One way in which sedimentary rocks are lithified. As material precipitates from water that percolates through the sediment, open spaces are filled and particles are joined into a solid mass.

Cenozoic era A time span on the geologic calendar beginning about 66 million years ago, following the Mesozoic era.

Chemical bond A strong attractive force that exists between atoms in a substance. It involves the transfer or sharing of electrons that allows each atom to attain a full valence shell.

Chemical compound A substance formed by the chemical combination of two or more elements in definite proportions and usually having properties different from those of its constituent elements.

Chemical sedimentary rock Sedimentary rock consisting of material that was precipitated from water by either inorganic or organic means.

Chemical weathering The processes by which the internal structure of a mineral is altered by the removal and/or addition of elements.

Chinook A wind that blows down the leeward side of a mountain and that warms by compression.

Cinder cone (scoria cone) A rather small volcano built primarily of pyroclastics ejected from a single vent.

Circle of illumination The great circle that separates daylight from darkness.

Circum-Pacific belt The zone of intense seismic activity that encompasses the coastal regions of Chile, Central America, Indonesia, Japan, and Alaska, including the Aleutian Islands.

Cirque An amphitheater-shaped basin at the head of a glaciated valley produced by frost wedging and plucking.

Cirrocumulus Thin, white, ice-crystal clouds that take the form of ripples, waves, or globular masses all in a row. May produce a "mackerel sky." Least common of high clouds.

Cirrostratus Thin sheet of white, ice-crystal clouds that may give the sky a milky look. Sometimes produce halos around the Sun and Moon.

Cirrus *As one of the three basic cloud forms*, refers to high, thin ice-crystal clouds. *As one of the three high cloud types*, refers to thin, delicate, fibrous, ice-crystal clouds, which sometimes take the form of hooked filaments called "mares' tails."

Clay mineral A group of light-colored silicates that typically form as products of chemical weathering of igneous rocks. Major components of soil and sedimentary rocks. Kaolinite is a common clay mineral derived from the weathering of feldspar.

Cleavage The tendency of a mineral to break along planes of weak bonding.

Climate A description of aggregate weather conditions; the sum of all statistical weather information that helps describe a place or region.

Cloud A form of condensation best described as a dense concentration of suspended water droplets or tiny ice crystals.

Clouds of vertical development Clouds having bases in the low height range but extending upward into the middle or high altitudes.

Coal A sedimentary rock consisting primarily of organic matter, formed in stages from accumulations of large quantities of undecayed plant material. Used as a fossil fuel.

Coarse-grained texture An igneous rock texture in which the crystals are roughly equal in size and large enough that individual minerals can be identified with the unaided eye.

Coastal upwelling The process by which deep, cold, nutrient-rich water is brought to the surface, usually by coastal currents that move surface water away from the coast.

Cold front A front along which a cold air mass thrusts beneath a warmer air mass.

Collision–coalescence process A theory of raindrop formation in warm clouds (above 0°C [32°F]) in which large cloud droplets (giants) collide and join together with smaller droplets to form raindrops. Opposite electrical charges may bind the cloud droplets together.

Color An obvious mineral characteristic that is often unreliable as a diagnostic property.

Columnar jointing A pattern of cracks that form during cooling of molten rock, resulting in columns that are generally six sided.

Coma The fuzzy, gaseous component of a comet's head.

Comet A small body that generally revolves around the Sun in an elongated orbit.

Compaction A type of lithification in which the weight of overlying material compresses more deeply buried sediment. It is most important in the fine-grained sedimentary rocks such as shale.

Competence A measure of the largest particle a stream can transport; a factor dependent on velocity.

Composite volcano (stratovolcano) A volcano composed of both lava flows and pyroclastic material.

Compound A substance formed by the chemical combination of two or more elements in definite proportions and usually having properties different from those of its constituent elements.

Compressional mountains Mountains in which great horizontal forces have shortened and thickened the crust. Most major mountain belts are of this type.

Concordant A term used to describe intrusive igneous masses that form parallel to the bedding of the surrounding rock.

Condensation The change of state from a gas to a liquid.

Condensation nuclei Tiny bits of particulate matter that serve as surfaces on which water vapor condenses.

Conditional instability The condition of moist air with an environmental lapse rate between the dry and wet adiabatic rates.

Conduction The transfer of heat through matter by molecular activity. Energy is transferred through collisions from one molecule to another.

Conduit A pipelike opening through which magma moves toward Earth's surface. It terminates at a surface opening called a vent.

Cone of depression A cone-shaped depression in the water table immediately surrounding a well.

Confined aquifer An aquifer that has impermeable layers (aquitards) both above and below.

Confining pressure Stress that is applied uniformly in all directions.

Conformable Having layers of rock deposited without interruption.

Conglomerate A sedimentary rock composed of rounded, gravel-size particles.

Contact metamorphism Changes in rock caused by the heat from a nearby magma body. Also known as *thermal metamorphism*.

Continental (c) air mass An air mass that forms over land; it is normally relatively dry.

Continental drift A theory which originally proposed that the continents are rafted about. It has essentially been replaced by the plate tectonics theory.

Continental margin The portion of the seafloor adjacent to the continents. It may include the continental shelf, continental slope, and continental rise.

Continental rift A linear zone along which continental lithosphere stretches and pulls apart. Its creation may mark the beginning of a new ocean basin.

Continental rise The gently sloping surface at the base of the continental slope.

Continental shelf The gently sloping submerged portion of the continental margin that extends from the shoreline to the continental slope.

Continental slope The steep gradient that leads to the deep-ocean floor and marks the seaward edge of the continental shelf.

Continental volcanic arc Mountains formed in part by igneous activity associated with the subduction of oceanic lithosphere beneath a continent. Examples include the Andes and the Cascades.

Convection The transfer of heat by the movement of a mass or substance. It can take place only in fluids.

Convergence The condition that exists when the distribution of winds within a given area results in a net horizontal inflow of air into the area. Since convergence at lower levels is associated with an upward movement of air, areas of convergent winds are regions favorable to cloud formation and precipitation.

Convergent plate boundary (subduction zone) A boundary in which two plates move together, causing one of the slabs of lithosphere to be consumed into the mantle as it descends beneath an overriding plate.

Coquina A coarse rock composed of loosely cemented shells and shell fragments.

Core The innermost layer of Earth, based on composition. It is thought to be largely an iron–nickel alloy, with minor amounts of oxygen, silicon, and sulfur.

Coriolis effect The deflective force of Earth's rotation on all free-moving objects, including the atmosphere and oceans. Deflection is to the right in the Northern Hemisphere and to the left in the Southern Hemisphere. Also called Coriolis force.

Correlation The establishment of the equivalence of rocks of similar age in different areas.

Cosmological red shift Changes in the spectra of galaxies which indicate that they are moving away from the Milky Way as a result of the expansion of space.

Cosmology The study of the universe, including its properties, structure, and evolution.

Covalent bond A chemical bond produced by the sharing of electrons.

Crater The depression at the summit of a volcano, or a depression that is produced by a meteorite impact.

Creep The slow downhill movement of soil and regolith.

Crevasse A deep crack in the brittle surface of a glacier.

Cross bedding Relatively thin layers that are inclined at an angle to the main bedding. Formed by currents of wind or water.

Cross-cutting relationship A principle of relative dating which states that a rock or fault is younger than any rock or fault through which it cuts.

Crust The very thin outermost layer of Earth.

Cryovolcanism The eruption of magmas derived from the partial melting of frozen volatiles (ices) instead of silicate rocks.

Crystal settling The downward movement of minerals during the crystallization of magma. The earlier-formed minerals are denser than the liquid portion and settle to the bottom of the magma chamber.

Crystal shape See *Habit*.

Crystallization The formation and growth of a crystalline solid from a liquid or gas.

Cumulonimbus Towering clouds, sometimes spreading out on top to form an "anvil head." Associated with heavy rainfall, thunder, lightning, hail, and tornadoes.

Cumulus *As one of the three basic cloud forms*, refers to clouds that consist of globular masses that are often described as cottonlike in appearance. *As one of the two types of clouds of vertical development*, refers to dense, billowy clouds often characterized by flat bases, which may occur as isolated clouds or may be closely packed.

Cup anemometer An instrument used to determine wind speed.

Curie point The temperature above which a material loses its magnetization.

Cut bank The area of active erosion on the outside of a meander.

Cutoff A short channel segment created when a river erodes through the narrow neck of land between meanders.

Cyclone (low) A low-pressure center characterized by a counterclockwise flow of air in the Northern Hemisphere.

Daily mean temperature A statistic that is determined by averaging the hourly readings or, more commonly, by averaging the maximum and minimum temperatures for a day.

Daily range The difference between the maximum and minimum for a day (for instance, of temperature).

Dark energy A hypothetical form of energy that produces a force that opposes gravity and is thought to be the cause of the accelerating expansion of the universe.

Dark matter Undetected matter that is thought to exist in great quantities in the universe.

Dark silicate mineral A silicate mineral that contains ions of iron and/or magnesium in its structure. It is dark in color and has a higher specific gravity than a light silicate mineral.

Decompression melting Melting that occurs as rock ascends due to a drop in confining pressure.

Deep-ocean basin The portion of the seafloor that lies between the continental margin and the oceanic ridge. This region comprises almost 30 percent of Earth's surface.

Deep-ocean trench A narrow, elongated depression on the floor of the ocean.

Deep-sea fan A cone-shaped deposit at the base of the continental slope. The sediment is transported to the fan by turbidity currents that follow submarine canyons.

Deflation The lifting and removal of loose material by wind.

Deformation General term for the processes of folding, faulting, shearing, compression, or extension of rocks.

Degenerate matter Incomprehensibly dense material formed when stars collapse and form a white dwarf.

Delta An accumulation of sediment formed where a stream enters a lake or an ocean.

Density The weight per unit volume of a particular material.

Deposition The process by which water vapor is changed directly to a solid, without passing through the liquid state.

Desert One of the two types of dry climate; the driest of the dry climates.

Desert pavement A layer of coarse pebbles and gravel created when wind removes the finer material.

Detrital sedimentary rock Rock formed from the accumulation of material that originated and was transported in the form of solid particles derived from both mechanical and chemical weathering.

Dew-point temperature (dew point) The temperature to which air has to be cooled in order to reach saturation. Often shortened to *dew point*.

Diagnostic property A property of a mineral that aids in mineral identification. Taste or feel, crystal shape, and streak are examples of diagnostic properties.

Differential stress Forces that are unequal in different directions.

Diffused light Solar energy is scattered and reflected in the atmosphere and reaches Earth's surface in the form of diffuse blue light from the sky.

Dike A tabular-shaped intrusive igneous feature that cuts through the surrounding rock.

Diorite A coarse-grained, intrusive igneous rock primarily composed of plagioclase feldspar and amphibole minerals.

Dip-slip fault A fault in which the movement is parallel to the dip of the fault.

Discharge The quantity of water in a stream that passes a given point in a period of time.

Disconformity A type of unconformity in which the beds above and below are parallel.

Discordant A term used to describe plutons that cut across existing rock structures, such as bedding planes.

Dissolved load The portion of a stream's load that is carried in solution.

Distributary A section of a stream that leaves the main flow.

Diurnal tidal pattern A tidal pattern exhibiting one high tide and one low tide during a tidal day; a daily tide.

Divergence The condition that exists when the distribution of winds within a given area results in a net horizontal outflow of air from the region. In divergence at lower levels, the resulting deficit is compensated for by a downward movement of air from aloft; hence, areas of divergent winds are unfavorable to cloud formation and precipitation.

Divergent plate boundary (spreading center) A region where the rigid plates are moving apart, typified by the mid-ocean ridges.

Divide An imaginary line that separates the drainage of two streams; often found along a ridge.

Dolomite Calcium/magnesium carbonate, $CaMg(CO_3)_2$, one of the two most common carbonate minerals.

Dome A roughly circular upfolded structure similar to an anticline.

Doppler effect The apparent change in wavelength of radiation caused by the relative motions of the source and the observer.

Doppler radar A new generation of radar that can handle the tasks performed by conventional radar and also detect motion directly. It greatly improves tornado and severe storm warnings.

Drainage basin The land area that contributes water to a stream. Also called a *watershed*.

Drawdown The difference in height between the bottom of a cone of depression and the original height of the water table.

Drizzle Precipitation from stratus clouds consisting of tiny droplets.

Drumlin A streamlined asymmetrical hill composed of glacial till. The steep side of the hill faces the direction from which the ice advanced.

Dry adiabatic rate The rate of adiabatic cooling or warming in unsaturated air. The rate of temperature change is 1°C per 100 meters.

Dry climate A climate in which yearly precipitation is less than the potential loss of water by evaporation.

Ductile deformation A type of solid-state flow that produces a change in the size and shape of a rock body without fracturing. Occurs at depths where temperatures and confining pressures are high.

Dune A hill or ridge of wind-deposited sand.

Dwarf galaxy A very small galaxy, usually elliptical and lacking spiral arms.

Dwarf planet A celestial body that orbits stars, massive enough to be spherical but has not cleared its neighboring region of planetesimals.

Earth science The name for all the sciences that collectively seek to understand Earth. It includes geology, oceanography, meteorology, and astronomy.

Earth system Earth viewed as a dynamic system of interacting parts and processes, including the geosphere, hydrosphere, biosphere, and atmosphere.

Earth system science An interdisciplinary study that seeks to examine Earth as a system composed of numerous interacting parts or subsystems.

Earthquake Vibration of Earth produced by the rapid release of energy.

Echo sounder An instrument used to determine the depth of water by measuring the time interval between emission of a sound signal and the return of its echo from the bottom.

Economic mineral A mineral used extensively in the manufacture of products.

Elastic deformation Rock deformation in which the rock will return to nearly its original size and shape when the stress is removed.

Elastic rebound The sudden release of stored strain in rocks that results in movement along a fault.

Electromagnetic radiation Transfer of energy in the form of light and related types of radiation, including gamma rays, x rays, ultraviolet light, infrared light, microwaves, and radio waves.

Electron A negatively charged subatomic particle that has a negligible mass and is found outside an atom's nucleus.

Element A substance that cannot be decomposed into simpler substances by ordinary chemical or physical means.

Elements (of weather and climate) Quantities or properties of the atmosphere that are measured regularly and that are used to express the nature of weather and climate.

Elliptical galaxy A galaxy that is round or elliptical in outline. It contains little gas and dust, no disk or spiral arms, and few hot, bright stars.

Emergent coast A coast where land that was formerly below sea level has been exposed either because of crustal uplift or a drop in sea level or both.

End moraine A ridge of till that marks a former position on the front of a glacier.

Enhanced Fujita intensity scale (EF-scale) A scale originally developed by Theodore Fujita for classifying the severity of a tornado, based on the correlation of wind speed and the degree of destruction.

Entrenched meander A meander cut into bedrock when uplifting rejuvenated a meandering stream.

Environmental lapse rate The rate of temperature decrease with increasing height in the troposphere.

Eon The largest time unit on the geologic time scale, next in order of magnitude above era.

Ephemeral stream A stream that is usually dry because it carries water only in response to specific episodes of rainfall. Most desert streams are of this type.

Epicenter The location on Earth's surface that lies directly above the forces of an earthquake.

Epoch A unit of the geologic calendar that is a subdivision of a period.

Equatorial low A belt of low pressure that lies near the equator and between the subtropical highs.

Equinox (spring or autumnal) The time when the vertical rays of the Sun are striking the equator. At equinox, the length of daylight and darkness is equal at all latitudes.

Era A major division on the geologic calendar; eras are divided into shorter units called periods.

Erosion The incorporation and transportation of material by a mobile agent, such as water, wind, or ice.

Eruption column A buoyant plume of hot, ash-laden gases that can extend thousands of meters into the atmosphere.

Escape velocity The initial velocity an object needs to escape from the surface of a celestial body.

Esker A sinuous ridge composed largely of sand and gravel deposited by a stream flowing in a tunnel within or beneath a glacier near its terminus.

Estuary A funnel-shaped inlet of the sea that formed when a rise in sea level or subsidence of land caused the mouth of a river to be flooded.

Evaporation The process of converting a liquid to a gas.

Evaporite deposit A sedimentary rock formed of material deposited from solution by evaporation of the water.

Evapotranspiration The combined effect of evaporation and transpiration.

Exfoliation dome A large, dome-shaped structure, usually composed of granite, formed by sheeting.

External process A process such as weathering, mass wasting, or erosion that is powered by the Sun and transforms solid rock into sediment.

Extrusive (volcanic) rock Igneous rock formed when magma solidifies at Earth's surface.

Eye A zone of scattered clouds and calm averaging about 20 kilometers in diameter at the center of a hurricane.

Eye wall The doughnut-shaped area of intense cumulonimbus development and very strong winds that surrounds the eye of a hurricane.

Fault A break in a rock mass along which movement has occurred.

Fault-block mountain A mountain formed by the displacement of rock along a fault.

Fault creep Gradual displacement along a fault. Such activity occurs relatively smoothly and with little noticeable seismic activity.

Fault scarp A cliff created by movement along a fault. It represents the exposed surface of the fault prior to modification by weathering and erosion.

Felsic composition See *Granitic composition.*

Fetch The distance that the wind has traveled across the open water.

Fine-grained texture A texture of igneous rocks in which the crystals are too small for individual minerals to be distinguished with the unaided eye.

Fiord A steep-sided inlet of the sea that formed when a glacial trough was partially submerged.

Fissure A crack in rock along which there is a distinct separation.

Fissure eruption An eruption in which lava is extruded from narrow fractures or cracks in the crust.

Flood The overflow of a stream channel that occurs when discharge exceeds the channel's capacity. A flood is the most common and destructive geologic hazard.

Flood basalts Flows of basaltic lava that issue from numerous cracks or fissures and that commonly cover extensive areas to thicknesses of hundreds of meters.

Floodplain A flat, low-lying portion of a stream valley that is subject to periodic inundation.

Focus (earthquake) The zone within Earth where rock displacement produces an earthquake. Also known as a *hypocenter.*

Fog A cloud with its base at or very near Earth's surface.

Fold A bent rock layer or series of layers that were originally horizontal and subsequently deformed.

Foliation A texture of metamorphic rocks that gives the rock a linear or layered appearance.

Footwall block The rock surface below a fault.

Foreshocks Small earthquakes that often precede a major earthquake.

Fossil The remains or traces of an organism preserved from the geologic past.

Fossil assemblage A set of fossil organisms found together in a given layer; used in relative dating.

Fossil magnetism See *Paleomagnetism.*

Fracture Any break or rupture in rock along which no appreciable movement has taken place.

Fracture zone A linear zone of irregular topography on the deep-ocean floor that follows transform faults and their inactive extensions.

Freezing The change of state from a liquid to a solid.

Freezing nuclei Solid particles that serve as cores for the formation of ice crystals.

Freezing rain (glaze) A coating of ice on objects formed when supercooled rain freezes on contact.

Front The boundary between two adjoining air masses that have contrasting characteristics.

Frontal fog Fog formed when rain evaporates as it falls through a layer of cool air.

Frontal lifting (wedging) Lifting of air that results when cool air acts as a barrier over which warmer, lighter air will rise.

Frost wedging The mechanical breakup of rock caused by the expansion of freezing water in cracks and crevices.

Fumarole A vent in a volcanic area from which fumes or gases escape.

Gabbro A dark-green to black intrusive igneous rock composed of dark silicate minerals. Gabbro makes up a significant percentage of oceanic crust.

Galactic cluster A system of galaxies that contains from several to thousands of member galaxies.

Galaxy A collection of interstellar matter, stars, and star remains that are gravitationally bound to one another.

Garnet A dark silicate mineral with a glassy luster, lacking cleavage and of varying colors; can be used as a gemstone.

Geocentric The concept of an Earth-centered universe.

Geologic time The span of time since the formation of Earth.

Geologic time scale The division of Earth history into blocks of time, such as eons, eras, periods, and epochs. The time scale was originally created using relative dating principles.

Geology The science that examines Earth, its form and composition, and the changes it has undergone and is undergoing.

Geosphere The solid Earth; one of Earth's four basic spheres.

Geostrophic wind A wind, usually above a height of 600 meters (2000 feet), that blows parallel to the isobars.

Geothermal gradient The gradual increase in temperature with depth in the crust. The average is 30°C per kilometer in the upper crust.

Geyser A fountain of hot water ejected periodically.

Giant In astronomy: a luminous star of large radius.

Glacial budget The balance, or lack of balance, between ice formation at the upper end of a glacier and ice loss in the zone of wastage.

Glacial drift An all-embracing term for sediments of glacial origin, no matter how, where, or in what shape they were deposited. Also known as *drift*.

Glacial erratic An ice-transported boulder that was not derived from bedrock near its present site.

Glacial striation A scratch or groove on bedrock caused by glacial abrasion.

Glacial trough A mountain valley that has been widened, deepened, and straightened by a glacier.

Glacier A thick mass of ice that originates on land from the compaction and recrystallization of snow and that shows evidence of past or present flow.

Glassy texture A term used to describe the texture of certain igneous rocks, such as obsidian, that contain no crystals.

Glaze *See* Freezing rain (glaze).

Gneiss Medium- to coarse-grained banded metamorphic rocks in which granular and elongated minerals dominate.

Graben A valley formed by the downward displacement of a fault-bounded block.

Gradient The slope of a stream; generally measured in feet per mile.

Granite A coarse-grained igneous rock of approximately 10–20 percent quartz and 50 percent potassium feldspar that forms where large masses of magma solidify at depth. Used as a building material.

Granitic (felsic) composition A compositional group of igneous rocks in which the rock is made up almost entirely of light-colored silicates.

Greenhouse effect The transmission of short-wave solar radiation by the atmosphere coupled with the selective absorption of longer-wavelength terrestrial radiation, especially by water vapor and carbon dioxide.

Groin A short wall built at a right angle to the shore to trap moving sand.

Ground moraine An undulating layer of till deposited as the ice front retreats.

Groundmass The matrix of smaller crystals within an igneous rock that has porphyritic texture.

Groundwater Water in the zone of saturation.

Guyot A submerged flat-topped seamount. Also known as a *tablemount*.

Gypsum Nonsilicate mineral commonly found in sedimentary rocks used to manufacture plaster and similar building materials.

Gyre The large, circular surface-current pattern found in each ocean.

Habit The common or characteristic shape of a crystal or aggregate of crystals. Also known as *crystal shape*.

Hail Nearly spherical ice pellets having concentric layers and formed by the successive freezing of layers of water.

Half-life The time required for one-half of the atoms of a radioactive substance to decay.

Halite Mineral name for common table salt (NaCl); a nonsilicate mineral commonly found in sedimentary rocks.

Hanging valley A tributary valley that enters a glacial trough at a considerable height above its floor.

Hanging wall block The rock surface immediately above a fault.

Hard stabilization An artificial structure built to protect a coast or to prevent the movement of sand along a beach. Examples include groins, jetties, breakwaters, and seawalls.

Hardness The resistance a mineral offers to scratching.

Heat The kinetic energy of random molecular motion.

Heliocentric The view that the Sun is at the center of the solar system.

Hertzsprung-Russell (H-R) diagram A plot of stars according to their absolute magnitudes and spectral types.

High clouds Clouds that normally have their base above 6000 meters; the base may be lower in winter and at high-latitude locations.

High-pressure center See *Anticyclone*.

Horn A pyramid-like peak formed by glacial action in three or more cirques surrounding a mountain summit.

Hornblende A dark green to black mineral of the amphibole group, often found in igneous rocks.

Horst An elongated, uplifted block of crust bounded by faults.

Host (country) rock Pre-existing crustal rocks intruded by magma. Host rock may be displaced or assimilated by magmas.

Hot spot A concentration of heat in the mantle that is capable of producing magma, which in turn extrudes onto Earth's surface. The intraplate volcanism that produced the Hawaiian islands is one example.

Hot-spot track A chain of volcanic structures produced as a lithospheric plate moves over a mantle plume.

Hot spring A spring in which the water is 6° to 9°C (10° to 15°F) warmer than the mean annual air temperature of its locality.

Hubble's law A law that relates the distance to a galaxy and its velocity.

Humidity A general term referring to water vapor in the air but not to liquid droplets of fog, cloud, or rain.

Hurricane A tropical cyclonic storm that has winds in excess of 119 kilometers (74 miles) per hour.

Hydrogen fusion A nuclear reaction in which hydrogen nuclei are fused into helium nuclei.

Hydrogenous sediment Seafloor sediment consisting of minerals that crystallize from seawater. An important example is manganese nodules.

Hydrologic cycle The unending circulation of Earth's water supply. The cycle is powered by energy from the Sun and is characterized by continuous exchanges of water among the oceans, the atmosphere, the geosphere, and the biosphere.

Hydrosphere The water portion of our planet; one of the traditional subdivisions of Earth's physical environment.

Hygrometer An instrument designed to measure relative humidity.

Hygroscopic nuclei Condensation nuclei that have a high affinity for water, such as salt particles.

Hypocenter (focus) The zone within Earth where rock displacement produces an earthquake.

Hypothesis A tentative explanation that is tested to determine whether it is valid.

Ice cap A mass of glacial ice covering a high upland or plateau and spreading out radially.

Ice sheet A very large, thick mass of glacial ice flowing outward in all directions from one or more accumulation centers.

Iceberg A mostly submerged mass of floating ice created by the calving of a glacier.

Igneous rock A rock formed by the crystallization of molten magma.

Impact crater A depression resulting from collisions with bodies such as asteroids and comets.

Incised meander A meandering channel that flows in a steep, narrow valley. It forms either when an area is uplifted or when the base level drops.

Inclination of the axis The tilt of Earth's axis from the perpendicular to the plane of Earth's orbit.

Inclusion A piece of one rock unit contained within another. Inclusions are useful in relative dating.

Index fossil A fossil that is associated with a particular span of geologic time.

Inertia A property by which objects at rest tend to remain at rest, and objects in motion tend to stay in motion unless either is acted upon by an outside force.

Infiltration The movement of surface water into rock or soil through cracks and pore spaces.

Infrared radiation Radiation with a wavelength from 0.7 to 200 micrometers.

Inner core The solid innermost layer of Earth, about 1300 kilometers (800 miles) in radius.

Intensity A measure of the degree of earthquake shaking at a given locale, based on the amount of damage.

Interface A common boundary where different parts of a system interact.

Interior drainage A discontinuous pattern of intermittent streams that do not flow to the ocean.

Internal process A process such as volcanic activity, earthquake, and mountain building that derives its energy from Earth's interior.

Intertropical convergence zone (ITCZ) The zone of general convergence between the Northern and Southern Hemisphere trade winds.

Intraplate volcanism Igneous activity that occurs within a tectonic plate away from plate boundaries.

Intrusion (pluton) A structure that results from the emplacement and crystallization of magma beneath the surface of Earth. See *Pluton*.

Intrusive (plutonic) rock Igneous rock that formed below Earth's surface.

Ion An atom or molecule that possesses an electrical charge.

Ionic bond A chemical bond between two oppositely charged ions formed by the transfer of valence electrons from one atom to the other.

Iron meteorite One of the three main categories of meteorites. This group is composed largely of iron, with varying amounts of nickel (5–20 percent). Most meteorite finds are irons.

Irregular galaxy A galaxy that lacks symmetry.

Island arc See *Volcanic island arc.*

Isobar A line drawn on a map that connects points of equal atmospheric pressure, usually corrected to sea level.

Isotherm A line on a diagram that connect points of equal temperature.

Isotopes Varieties of the same element that have different mass numbers; their nuclei contain the same number of protons but different numbers of neutrons.

Jet stream Swift, high-altitude winds (120–240 kilometers per hour).

Joint A fracture in rock along which there has been no movement.

Jovian (Jupiter-like) planet The Jupiter-like planets Jupiter, Saturn, Uranus, and Neptune. These planets have relatively low densities. Also known as the *outer planets.*

Kame A steep-sided hill composed of sand and gravel that originates when sediment collects in openings in stagnant glacial ice.

Karst topography A topography that consists of numerous depressions called sinkholes.

Kettle A depression created by a block or blocks of ice becoming lodged in a glacial deposit and subsequently melting.

Kuiper belt A region outside the orbit of Neptune where most short-period comets are thought to originate.

Laccolith A massive igneous body intruded between preexisting strata.

Lahar Mudflows on the slopes of volcanoes that result when unstable layers of ash and debris become saturated and flow downslope, usually following stream channels.

Lake-effect snow Snow showers associated with a cP air mass to which moisture and heat are added from below as it traverses a large and relatively warm lake (such as one of the Great Lakes), rendering the air mass humid and unstable.

Laminar flow The movement of water particles in straight-line paths that are parallel to the channel. The water particles move downstream without mixing.

Land breeze A local wind that blows from land toward the water during the night in coastal areas.

Latent heat The energy absorbed or released during a change in state.

Lateral moraine A ridge of till along the sides of an alpine glacier composed primarily of debris that fell to the glacier from the valley walls.

Lava Magma that reaches Earth's surface.

Lava tube A tunnel in hardened lava that acts as a horizontal conduit for lava flowing from a volcanic vent. Lava tubes allow fluid lavas to advance great distances.

Law of universal gravitation Newton's law stating that any two bodies in the universe attract each other with a force that is directly proportional to the product of their masses and inversely proportional to the square of the distance between them.

Lifting condensation level The height at which rising air that is cooling at the dry adiabatic rate becomes saturated and condensation begins.

Light silicate mineral A silicate mineral that lacks iron and/or magnesium. It is generally lighter in color and has a lower specific gravity than a dark silicate mineral.

Light-year The distance light travels in a year; about 6 trillion miles.

Limestone A chemical sedimentary rock composed chiefly of calcite. Limestone can form by inorganic means or from biochemical processes.

Liquefaction A phenomenon, sometimes associated with earthquakes, in which soils and other unconsolidated materials containing abundant water are turned into a fluidlike mass that is not capable of supporting buildings.

Lithification The process, generally cementation and/or compaction, of converting sediments to solid rock.

Lithosphere The rigid outer layer of Earth, including the crust and upper mantle.

Lithospheric plate (plate) A coherent unit of Earth's rigid outer layer that includes the crust and uppermost mantle. Also known as a *tectonic plate.*

Local Group The cluster of 20 or so galaxies to which our galaxy belongs.

Local wind A small-scale wind produced by a locally generated pressure gradient.

Localized convective lifting Unequal surface heating that causes localized pockets of air (thermals) to rise because of their buoyancy.

Loess Deposits of windblown silt that are lacking visible layers, generally buff-colored, and capable of maintaining a nearly vertical cliff.

Longitudinal profile A cross section of a stream channel along its descending course from the head to the mouth.

Longshore current A near-shore current that flows parallel to the shore.

Low clouds Clouds that form below a height of about 2000 meters.

Low-pressure center See *Cyclone.*

Lower mantle The part of the mantle that extends from the core–mantle boundary to a depth of 660 kilometers.

Lunar highlands The extensively cratered highland areas of the Moon. Also known as *terrae.*

Lunar regolith A thin, gray layer on the surface of the Moon, consisting of loosely compacted, fragmented material believed to have been formed by repeated meteoric impacts.

Luster The appearance or quality of light reflected from the surface of a mineral.

Mafic composition See *Basaltic composition.*

Magma A body of molten rock found at depth, including any dissolved gases and crystals.

Magmatic differentiation The process of generating more than one rock type from a single magma.

Magnetic reversal A change in the polarity of Earth's magnetic field that occurs over time intervals of roughly 200,000 years.

Magnetic time scale The history of magnetic reversals through geologic time.

Magnetometer A sensitive instrument used to measure the intensity of Earth's magnetic field.

Magnitude In seismology, the total amount of energy released during an earthquake.

Magnitude (stellar) A number given to a celestial object to express its relative brightness.

Main-sequence star A sequence of stars on the Hertzsprung-Russell diagram, containing the majority of stars, that runs diagonally from the upper left to the lower right.

Mantle The 2900-kilometer- (1800-mile-) thick layer of Earth located below the crust.

Mantle plume Structures that originate at great depth and, upon reaching the crust, spread laterally, creating a localized volcanic zone called a hot spot. A mantle plume is a source of some intraplate basaltic magma.

Marble A soft metamorphic rock formed from limestone or dolostone. Marble of various colors is used for building stones and monuments.

Maria The Latin name for the smooth areas of the Moon formerly thought to be seas.

Marine terrace A wave-cut platform that has been exposed above sea level.

Maritime (m) air mass An air mass that originates over the ocean. Maritime air masses are relatively humid.

Mass number The number of neutrons and protons in the nucleus of an atom.

Mass wasting The downslope movement of rock, regolith, and soil under the direct influence of gravity.

Massive Descriptive term for an igneous pluton that is not tabular in shape.

Meander A looplike bend in the course of a stream.

Mechanical weathering The physical disintegration of rock, resulting in smaller fragments.

Medial moraine A ridge of till formed when lateral moraines from two coalescing alpine glaciers join.

Megathrust fault The plate boundary separating a subducting slab of oceanic lithosphere and the overlying plate.

Melting The change of state from a solid to a liquid.

Mercury barometer A mercury-filled glass tube in which the height of the mercury column is a measure of air pressure.

Mesosphere The layer of the atmosphere immediately above the stratosphere and characterized by decreasing temperatures with height.

Mesozoic era A time span on the geologic calendar between the Paleozoic and Cenozoic eras—from about 245 to 66.4 million years ago.

Metallic bond A chemical bond that is present in all metals that may be characterized as an extreme type of electron sharing in which the electrons move freely from atom to atom.

Metamorphic rock Rock formed by the alteration of preexisting rock deep within Earth (but still in the solid state) by heat, pressure, and/or chemically active fluids.

Metamorphism The changes in mineral composition and texture of a rock subjected to high temperatures and pressures within Earth.

Meteor The luminous phenomenon observed when a meteoroid enters Earth's atmosphere and burns up; popularly called a "shooting star."

Meteor shower Many meteors appearing in the sky that occurs when Earth intercepts a swarm of meteoritic particles.

Meteorite Any portion of a meteoroid that survives its traverse through Earth's atmosphere and strikes Earth's surface.

Meteoroid Small, solid particles that have orbits in the solar system.

Meteorology The scientific study of the atmosphere and atmospheric phenomena; the study of weather and climate.

Microcontinent A relatively small fragment of continental crust that may lie above sea level, such as the island of Madagascar, or that may be submerged, as exemplified by the Campbell Plateau located near New Zealand.

Middle clouds Clouds that occupy the height range from 2000 to 6000 meters.

Middle-latitude (midlatitude) cyclone A large low-pressure center with a diameter often exceeding 1000 kilometers (600 miles) that moves from west to east and may last from a few days to more than a week and usually has a cold front and a warm front extending from the central area of low pressure.

Mid-ocean ridge A continuous elevated zone on the floor of all the major ocean basins representing divergent plate boundaries. See *Oceanic ridge (rise)*.

Mineral A naturally occurring, inorganic crystalline material that has a unique chemical composition.

Mineral resources All discovered and undiscovered deposits of a useful mineral that can be extracted now or at some time in the future.

Mineralogy The study of minerals.

Mist A cloud of water droplets suspended in the atmosphere at or near Earth's surface.

Mixed tidal pattern A tidal pattern exhibiting two high tides and two low tides per tidal day, with a large inequality in high water heights, low water heights, or both. Coastal locations that experience such a tidal pattern may also show alternating periods of diurnal and semidiurnal tidal patterns. Also called *mixed semidiurnal*.

Mixing ratio The mass of water vapor in a unit mass of dry air; commonly expressed as grams of water vapor per kilogram of dry air.

Modified Mercalli Intensity scale A 12-point scale developed to evaluate earthquake intensity based on the amount of damage to various structures.

Mohs scale A series of 10 minerals used as a standard in determining mineral hardness.

Moment magnitude A more precise measure of earthquake magnitude than the Richter scale that is derived from the amount of displacement that occurs along a fault zone.

Monocline A one-limbed flexure in strata. The strata are usually flat-lying or very gently dipping on both sides of the monocline.

Monsoon A seasonal reversal of wind direction associated with large continents, especially Asia. In winter, the wind blows from land to sea; in summer, from sea to land.

Monthly mean The mean (average) for a month that is calculated by averaging the daily means of measurable phenomena for the month.

Mountain breeze The nightly downslope winds commonly encountered in mountain valleys.

Muscovite A common member of the mica family of minerals, with excellent cleavage.

Natural hazard A natural phenomenon that represents a potential danger to people, such as an earthquake, a volcanic eruption, or a hurricane.

Natural levee An elevated landform that parallels some streams and acts to confine their waters, except during flood stage.

Neap tide The lowest tidal range, occurring near the times of the first and third quarters of the Moon.

Nebula A cloud of interstellar gas and/or dust. See also *Solar nebula*.

Nebular theory A model for the origin of the solar system that supposes a rotating nebula of dust and gases that contracted to form the Sun and planets.

Neutron A subatomic particle found in the nucleus of an atom. A neutron is electronically neutral and has a mass approximately that of a proton.

Neutron star A star of extremely high density that is composed entirely of neutrons.

Nimbostratus Amorphous layer of dark gray clouds. One of the primary precipitation-producing clouds.

Nimbus A cloud that is a major producer of precipitation.

Nonconformity An unconformity in which older metamorphic or intrusive igneous rocks are overlain by younger sedimentary strata.

Nonfoliated Descriptive term for metamorphic rocks that do not exhibit foliation.

Nonrenewable resource A resource that forms or accumulates over such long time spans that it must be considered as fixed in total quantity.

Nonsilicate A mineral group that lacks silicas in their mineral structures; accounts for less than 10 percent of Earth's crust.

Nor'easter The weather associated with an incursion of mP air into the Northeast from the North Atlantic; strong northeast winds, freezing or near-freezing temperatures, and possible precipitation make this an unwelcome weather event.

Normal fault A fault in which the rock above the fault plane has moved down relative to the rock below.

Normal polarity A magnetic field that is the same as that which exists at present.

Nucleus (atomic) The small, heavy core of an atom that contains all of its positive charge and most of its mass.

Nucleus (comet) The small central body of a comet, typically 1 to 10 kilometers in diameter.

Nuée ardente See *Pyroclastic flow*.

Numerical date A date that specifies the actual number of years that have passed since an event occurred.

Obsidian A volcanic glass of felsic composition.

Occluded front A front formed when a cold front overtakes a warm front.

Oceanic plateau An extensive region on the ocean floor that is composed of thick accumulations of pillow basalts and other mafic rocks that, in some cases, exceed 30 kilometers (20 miles) in thickness.

Oceanic ridge (rise) A continuous elevated zone on the floor of all the major ocean basins and varying in width from 500 to 5000 kilometers (300 to 3000 miles). The rifts at the crests of these ridges represent divergent plate boundaries. Also known as a *mid-ocean ridge*.

Oceanic ridge system A continuous elevated zone on the floor of all the major ocean basins and varying in width from 500 to 5000 kilometers (300–3000 miles). The rifts at the crests of ridges represent divergent plate boundaries.

Oceanography The scientific study of the oceans and oceanic phenomena.

Octet rule A rule which states that atoms combine in order that each may have the electron arrangement of a noble gas—that is, so the outer energy level contains eight neutrons.

Olivine A high temperature, dark silicate mineral typically found in basalt.

Oort cloud A spherical shell composed of comets that orbit the Sun at distances generally greater than 10,000 times the Earth–Sun distance.

Ore Usually a useful metallic mineral that can be mined at a profit. The term is also applied to certain nonmetallic minerals such as fluorite and sulfur.

Organic matter Carbon-rich substances derived from living things that are in various stages of decomposition.

Organic sedimentary rock Sedimentary rock composed of organic carbon from the remains of plants that died and accumulated on the floor of a swamp. Coal is the primary example.

Orogenesis The processes that collectively result in the formation of mountains.

Orographic lifting Mountains acting as barriers to the flow of air and forcing the air to ascend. The air cools adiabatically, and clouds and precipitation may result.

Outer core A layer beneath the mantle that is about 2200 kilometers (1364 miles) thick and has the properties of a liquid.

Outgassing The escape of gases that were once dissolved in magma.

Outlet glacier A tongue of ice normally flowing rapidly outward from an ice cap or ice sheet, usually through mountainous terrain to the sea.

Outwash plain A relatively flat, gently sloping plain consisting of materials deposited by meltwater streams in front of the margin of an ice sheet.

Overrunning Warm air gliding up a retreating cold air mass.

Oxbow lake A curved lake produced when a stream cuts off a meander.

Ozone A molecule of oxygen that contains three oxygen atoms.

P waves A type of seismic wave that involves alternating compression and expansion of the material through which it passes. Also called *primary waves*.

Pahoehoe flow A lava flow with a smooth-to-ropey surface.

Paleomagnetism (fossil magnetism) The natural remnant magnetism in rock bodies. The permanent magnetization acquired by rock that can be used to determine the location of the magnetic poles and the latitude of the rock at the time it became magnetized.

Paleontology The systematic study of fossils and the history of life on Earth.

Paleozoic era A time span on the geologic calendar between the Precambrian and Mesozoic eras—from about 570 million to 245 million years ago.

Pangaea A proposed supercontinent that 200 million years ago began to break apart and form the present landmasses.

Parasitic cone A volcanic cone that forms on the flank of a larger volcano.

Parcel An imaginary volume of air enclosed in a thin elastic cover. Typically it is considered to be a few hundred cubic meters in volume and is assumed to act independently of the surrounding air.

Partial melting The process by which most igneous rocks melt. Since individual minerals have different melting points, most igneous rocks melt over a temperature range of a few hundred degrees. If the liquid is squeezed out after some melting has occurred, a melt with a higher silica content results.

Passive continental margin A margin that consists of a continental shelf, continental slope, and continental rise. These margins are not associated with plate boundaries and therefore experience little volcanism and few earthquakes.

Perched water table A localized zone of saturation above the main water table that is created by an impermeable layer (aquitard).

Peridotite An igneous rock of ultramafic composition thought to be abundant in the upper mantle.

Period A basic unit of the geologic calendar that is a subdivision of an era. Periods may be divided into smaller units called epochs.

Periodic table An arrangement of the elements in which atomic number increases from the left to right and elements with similar properties appear in columns called families or groups.

Permeability A measure of a material's ability to transmit water.

Perturbation The gravitational disturbance of the orbit of one celestial body by another.

Phanerozoic eon The part of geologic time represented by rocks containing abundant fossil evidence. The eon extending from the end of the Proterozoic eon (570 million years ago) to the present.

Phenocryst A conspicuously large crystal embedded in a matrix of finer-grained crystals.

Phyllite A metamorphic rock composed mainly of fine crystals of muscovite, chlorite, or both.

Physical environment The part of the environment that encompasses water, air, soil, and rock, as well as conditions such as temperature, humidity, and sunlight.

Piedmont glacier A glacier that forms when one or more alpine glaciers emerge from the confining walls of mountain valleys and spread out to create a broad sheet in the lowlands at the base of the mountains.

Pillow lava Basaltic lava that solidifies in an underwater environment and develops a structure that resembles a pile of pillows.

Plagioclase feldspar A relatively hard light silicate mineral containing both sodium and calcium ions that freely substitute for one another depending on the crystallization environment.

Planetary nebula A shell of incandescent gas expanding from a star.

Planetesimal A solid celestial body that accumulated during the first stages of planetary formation. Planetesimals aggregated into increasingly larger bodies, ultimately forming the planets.

Plate One of numerous rigid sections of the lithosphere that move as a unit over the material of the asthenosphere. Also called lithospheric plate, tectonic plate.

Playa lake A temporary lake in a playa.

Pleistocene epoch An epoch of the Quaternary period beginning about 1.6 million years ago and ending about 10,000 years ago. Best known as a time of extensive continental glaciation.

Plucking The process by which pieces of bedrock are lifted out of place by a glacier. Sometimes called *quarrying*.

Pluton A structure that results from the emplacement and crystallization of magma beneath the surface of Earth. Also known as an *intrusion*.

Plutonic rock See *Intrusive (plutonic) rock*.

Pluvial lake A lake formed during a period of increased rainfall. During the Pleistocene epoch, this occurred in some nonglaciated regions during periods of ice advance elsewhere.

Point bar A crescent-shaped accumulation of sand and gravel deposited on the inside of a meander.

Polar (P) air mass A cold air mass that forms in a high-latitude source region.

Polar easterlies In the global pattern of prevailing winds, winds that blow from the polar high toward the subpolar low. Unlike the trade winds, however, these winds should not be thought of as persistent winds.

Polar front A stormy frontal zone that separates air masses of polar origin from air masses of tropical origin.

Polar high An anticyclone that is assumed to occupy the inner polar regions and is believed to be thermally induced, at least in part.

Polar wandering A theory based on paleomagnetic studies in the 1950s, in which researchers proposed that either the magnetic poles migrated greatly through time or the continents had gradually shifted their positions.

Porosity The volume of open spaces in rock or soil.

Porphyritic texture An igneous texture consisting of large crystals embedded in a matrix of much smaller crystals.

Positive-feedback mechanism As used in climatic change, any effect that acts to reinforce an initial change.

Potassium feldspar An abundant, relatively hard light silicate mineral containing potassium ions in its structure.

Pothole A depression formed in a stream channel by the abrasive action of the water's sediment load.

Precambrian All geologic time prior to the Paleozoic era.

Precipitation fog Fog formed when rain evaporates as it falls through a layer of cool air.

Pressure (barometric) tendency The nature of the change in atmospheric pressure over a period of several hours. It can be a useful aid in short-range weather prediction.

Pressure gradient force The force that results from a difference in atmospheric pressure between two locations. Horizontal pressure gradient forces are what cause winds to blow.

Prevailing wind A wind that consistently blows from one direction more than from any other.

Primary (P) waves A type of seismic wave that involves alternating compression and expansion of the material through which it passes. Also called *P waves*.

Principle of cross-cutting relationships A principle of relative dating which states that a rock or fault is younger than any rock or fault through which it cuts.

Principle of fossil succession A principle by which fossil organisms succeed one another in a definite and determinable order, and any time period can be recognized by its fossil content.

Principle of inclusions A principle that uses pieces of rock contained within another to determine a relative date. According to the principle of inclusions, the rock mass that provided the inclusion is older than the rock mass containing the inclusion.

Principle of lateral continuity A principle by which sedimentary beds originate as continuous layers until they grade into another rock type or thin out. Used to correlate outcrops that are now isolated from each other.

Principle of original horizontality A principle by which layers of sediment are generally deposited in a horizontal or nearly horizontal position.

Principle of superposition A principle which states that in any undeformed sequence of sedimentary rocks, each bed is older than the one above and younger than the one below.

Proglacial lake A lake created when a glacier acts as a dam, blocking the flow of a river or trapping glacial meltwater. The term refers to the position of such lakes just beyond the outer limits of a glacier.

Proton A positively charged subatomic particle found in the nucleus of an atom.

Proton–proton chain A chain of thermonuclear reactions by which nuclei of hydrogen are built up into nuclei of helium.

Protoplanet A developing planetary body that grows by the accumulation of planetesimals.

Protostar A collapsing cloud of gas and dust that is destined to become a star.

Psychrometer A device consisting of two thermometers (wet-bulb and dry-bulb) that is rapidly whirled and, with the use of tables, yields the relative humidity and dew point.

Ptolemaic system An Earth-centered system of the universe.

Pulsar A variable radio source of small size that emits radio pulses in very regular periods.

Pumice A light-colored, glassy vesicular rock that commonly has a granitic composition.

Pycnocline A layer of water in which there is a rapid change of density with depth.

Pyroclastic flow (nuée ardente) A highly heated mixture, largely of ash and pumice fragments, traveling down the flanks of a volcano or along the surface of the ground.

Pyroclastic (fragmental) texture An igneous rock texture resulting from the consolidation of individual rock fragments that are ejected during a violent eruption.

Pyroclastic materials The volcanic rock ejected during an eruption, including ash, bombs, and blocks; also called *tephra*.

Quartz A common silicate mineral consisting entirely of silicon and oxygen that resists weathering.

Quartzite A hard metamorphic rock formed from quartz sandstone.

Quaternary period The most recent period on the geologic time scale. It began about 2.6 million years ago and extends to the present.

Radiation (electromagnetic radiation) The transfer of energy (heat) through space by electromagnetic waves.

Radiation fog Fog resulting from radiation heat loss by Earth.

Radioactive decay The spontaneous decay of certain unstable atomic nuclei.

Radioactivity The process by which atomic nuclei spontaneously break apart; also called *radioactive decay*. Used to calculate the ages of rocks and minerals that contain radioactive isotopes.

Radiocarbon dating Dating of events from the very recent geologic past (the past few tens of thousands of years) based on the fact that the radioactive isotope of carbon is produced continuously in the atmosphere.

Radiometric dating The procedure of calculating the absolute ages of rocks and minerals that contain radioactive isotopes.

Radiosonde A lightweight package of weather instruments fitted with a radio transmitter and carried aloft by a balloon.

Rain Drops of water that fall from a cloud and have a diameter of at least 0.5 millimeter.

Rainshadow desert A dry area on the lee side of a mountain range. Many middle-latitude deserts are of this type.

Red giant A large, cool star of high luminosity; a star occupying the upper-right portion of the Hertzsprung-Russell diagram.

Reflection The process by which light, sound, or energy is returned without being absorbed or scattered.

Refraction The process by which the portion of a wave in shallow water slows, causing the wave to bend and tend to align itself with the underwater contours. Also known as *wave refraction*.

Regional metamorphism Metamorphism associated with large-scale mountain-building processes.

Regolith The layer of rock and mineral fragments that nearly everywhere covers Earth's land surface.

Rejuvenation A change, often caused by regional uplift, that causes the force of erosion to intensify.

Relative date Rocks placed in their proper sequence or order of formation based on geologic principles.

Relative humidity The ratio of the air's water-vapor content to its water-vapor capacity.

Renewable resource A resource that is virtually inexhaustible or that can be replenished over relatively short time spans.

Reserve Identified deposits from which minerals can be extracted profitably.

Retrograde motion The apparent westward motion of the planets with respect to the stars.

Reverse fault A fault in which the material above the fault plane moves up in relation to the material below.

Reverse polarity A magnetic field opposite that which exists at present.

Revolution The motion of one body about another, as Earth about the Sun.

Rhyolite The fine-grained equivalent of the igneous rock granite, composed primarily of the light-colored silicates.

Richter scale A scale of earthquake magnitude based on the motion of a seismograph.

Ridge push A mechanism that may contribute to plate motion. It involves the oceanic lithosphere sliding down the oceanic ridge under the pull of gravity.

Rift valley A region of Earth's crust along which divergence is taking place.

Rime A thin coating of ice on objects that is produced when supercooled fog or cloud droplets freeze on contact.

Ring of Fire The zone of active volcanoes surrounding the Pacific Ocean.

Rock A consolidated mixture of minerals.

Rock cycle A model that illustrates the origin of the three basic rock types and the interrelatedness of Earth materials and processes.

Rock flour Ground-up rock produced by the grinding effect of a glacier.

Rock-forming mineral One of the few dozen most abundant minerals that make up most common rocks. Minerals rich in oxygen and silicon (silicate minerals) dominate this group.

Rotation The spinning of a body, such as Earth, about its axis.

Runoff Water that flows over land rather than infiltrating into the ground.

S waves A seismic wave that involves oscillation perpendicular to the direction of propagation. Also called *Secondary waves*.

Saffir–Simpson hurricane scale A scale from 1 to 5 that is used to rank relative intensities of hurricanes.

Salinity The proportion of dissolved salts to pure water, usually expressed in parts per thousand (‰).

Sandstone An abundant, durable sedimentary rock primarily composed of sand-size grains.

Santa Ana The local name given to a chinook wind in southern California.

Saturation The maximum possible quantity of water vapor that the air can hold at any given temperature and pressure.

Scattering The process in which electromagnetic radiation or particles are deflected or diffused.

Schist Medium- to coarse-grained metamorphic rocks having a foliated texture, in which platy minerals dominate.

Scientific method The process by which researchers raise questions, gather data, and formulate and test scientific hypotheses.

Scoria Vesicular ejecta that is the product of basaltic magma.

Scoria cone See *Cinder cone*.

Sea arch An arch formed by wave erosion when caves on opposite sides of a headland unite.

Sea breeze A local wind blowing from the sea during the afternoon in coastal areas.

Sea ice Frozen seawater that is associated with polar regions. The area covered by sea ice expands in winter and shrinks in summer.

Sea stack An isolated mass of rock standing just offshore, produced by wave erosion of a headland.

Seafloor spreading The process of producing new seafloor between two diverging plates.

Seamount An isolated volcanic peak that rises at least 1000 meters (3000 feet) above the deep-ocean floor.

Seawall A barrier constructed to prevent waves from reaching the area behind the wall. Its purpose is to defend property from the force of breaking waves.

Secondary (S) waves A seismic wave that involves oscillation perpendicular to the direction of propagation. Also called *S waves*.

Sediment Unconsolidated particles created by the weathering and erosion of rock, by chemical precipitation from solution in water, or from the secretions of organisms and transported by water, wind, or glaciers.

Sedimentary rock Rock formed from the weathered products of preexisting rocks that have been transported, deposited, and lithified.

Seismic sea wave A rapidly moving ocean wave generated by earthquake activity that is capable of inflicting heavy damage in coastal regions.

Seismic wave A form of elastic energy released during an earthquake that causes vibrations in the materials that transmit them.

Seismogram A record made by a seismograph.

Seismograph An instrument that records earthquake waves. Also known as a *seismometer*.

Seismology The study of earthquakes and seismic waves.

Seismometer See *Seismograph*.

Semidiurnal tidal pattern A tidal pattern exhibiting two high tides and two low tides per tidal day, with small inequalities between successive highs and successive lows; a semidaily tide.

Sensible heat The heat we can feel and measure with a thermometer.

Shale The most common sedimentary rock, consisting of silt- and clay-size particles.

Sheeting A mechanical weathering process that is characterized by the splitting off of slablike sheets of rock.

Shield volcano A broad, gently sloping volcano built from fluid basaltic lavas.

Silicate mineral Any one of numerous minerals that have the silicon–oxygen tetrahedron as their basic structure.

Silicon–oxygen tetrahedron A structure composed of four oxygen atoms surrounding a silicon atom that constitutes the basic building block of silicate minerals.

Sill A tabular igneous body that was intruded parallel to the layering of preexisting rock.

Siltstone A fine-grained sedimentary rock composed of clay-sized sediment mixed with silt-sized grains.

Sink hole (sink) A depression produced in a region where soluble rock has been removed by groundwater.

Slab pull A mechanism that contributes to plate motion in which cool, dense oceanic crust sinks into the mantle and "pulls" the trailing lithosphere along.

Slate A very fine-grained metamorphic rock containing platy minerals and having excellent rock cleavage.

Sleet Frozen or semifrozen rain that forms when raindrops freeze as they pass through a layer of cold air.

Slip face The steep, leeward slope of a sand dune; it maintains an angle of about 34 degrees.

Small solar system body An object in the solar system that is not a planet or dwarf planet. Examples include comets and asteroids.

Snow A solid form of precipitation produced by sublimation of water vapor.

Solar nebula A cloud of interstellar gas and/or dust from which the bodies of our solar system formed.

Solstice (summer or winter) The time when the vertical rays of the Sun are striking either the Tropic of Cancer or the Tropic of Capricorn. Solstice represents the longest or shortest day (length of daylight) of the year.

Sonar An instrument that uses acoustic signals (sound energy) to measure water depths. Sonar is an acronym for *so*und *na*vigation and *r*anging.

Sorting The process by which solid particles of various sizes are separated by moving water or wind. Also the degree of similarity in particle size in sediment or sedimentary rock.

Source region The area where an air mass acquires its characteristic properties of temperature and moisture.

Specific gravity The ratio of a substance's weight to the weight of an equal volume of water.

Specific heat The amount of heat needed to raise 1 gram of a substance 1°C at sea level atmospheric pressure.

Spiral galaxy A flattened, rotating galaxy with pinwheel-like arms of interstellar material and young stars winding out from its nucleus.

Spit An elongated ridge of sand that projects from the land into the mouth of an adjacent bay.

Spreading center See *Divergent plate boundary*.

Spring A flow of groundwater that emerges naturally at the ground surface.

Spring equinox See *Equinox*.

Spring tide The highest tidal range. Occurs near the times of the new and full moons.

Stable air Air that resists vertical displacement. If it is lifted, adiabatic cooling will cause its temperature to be lower than the surrounding environment; if it is allowed to do so, it will sink to its original position.

Stalactite An icicle-like structure that hangs from the ceiling of a cavern.

Stalagmite A columnlike rock formation that grows upward from the floor of a cavern.

Standard rain gauge A gauge that has a diameter of about 20 centimeters and funnels rain into a cylinder that magnifies precipitation amounts by a factor of 10, allowing for accurate measurement of small amounts.

Stationary front A situation in which the surface position of a front does not move; the flow on either side of such a boundary is nearly parallel to the position of the front.

Steam fog Fog that has the appearance of steam; produced by evaporation from a warm-water surface into the cool air above.

Stellar parallax A measure of stellar distance.

Steppe One of the two types of dry climate. A marginal and more humid variant of the desert that separates it from bordering humid climates.

Stock A pluton similar to, but smaller than, a batholith.

Stony meteorite One of the three main categories of meteorites. Such meteorites are composed largely of silicate minerals, with inclusions of other minerals.

Stony–iron meteorite One of the three main categories of meteorites. This group is a mixture of iron and silicate minerals.

Storm surge The abnormal rise of the sea along a shore as a result of strong winds.

Strata (beds) Parallel layers of sedimentary rock.

Stratified drift Sediments deposited by glacial meltwater.

Stratocumulus Soft, gray clouds in globular patches or rolls. Rolls may join together to make a continuous cloud.

Stratosphere The layer of the atmosphere immediately above the troposphere, characterized by increasing temperatures with height due to the concentration of ozone.

Stratovolcano See *Composite volcano*.

Stratus *As one of the three basic cloud forms*, refers to sheets or layers (strata) of cloud that cover much or all of the sky. *As one of the three low cloud types*, refers to a low, uniform cloud layer that resembles fog but does not rest on the ground and that may produce drizzle.

Streak The color of a mineral in powdered form.

Stream terrace A flat, benchlike structure produced by a stream, which was left elevated as the stream cut downward.

Stream valley The channel, valley floor, and sloping valley walls of a stream.

Strike-slip fault A fault along which the movement is horizontal.

Subduction erosion A process at subduction zones in which sediment and rock are scraped off the bottom of the overriding plate and transported into the mantle.

Subduction zone A long, narrow zone where one lithospheric plate descends beneath another. See also *Convergent plate boundary*.

Sublimation The conversion of a solid directly to a gas, without passing through the liquid state.

Submarine canyon A canyon carved into the outer continental shelf, slope, and rise by turbidity currents.

Submergent coast A coast with a form that is largely a result of the partial drowning of a former land surface either because of a rise of sea level or subsidence of the crust or both.

Subpolar low Low pressure located at about the latitudes of the Arctic and Antarctic Circles. In the Northern Hemisphere, the low takes the form of individual oceanic cells; in the Southern Hemisphere, there is a deep and continuous trough of low pressure.

Subtropical high Not a continuous belt of high pressure but rather several semipermanent, anticyclonic centers characterized by subsidence and divergence located roughly between latitudes 25° and 35°.

Summer solstice See *Solstice*.

Supercontinent A large landmass that contains all, or nearly all, of the existing continents.

Supercooled The condition of water droplets that remain in the liquid state at temperatures well below 0°C.

Supergiant A very large star with high luminosity.

Supernova An exploding star that increases in brightness many thousands of times.

Superplume A large mantle plume thought to be responsible for bursts of volcanic activity that emit the vast outpourings of lava forming large basaltic provinces.

Surf A collective term for breakers; also the wave activity in the area between the shoreline and the outer limit of breakers.

Surface waves Seismic waves that travel along the outer layer of Earth.

Suspended load The fine sediment carried within the body of flowing water.

Syncline A linear downfold in sedimentary strata; the opposite of anticline.

System A group of interacting or interdependent parts that form a complex whole.

Tabular A term used to describe a feature such as an igneous pluton having two dimensions that are much longer than the third; a pluton that is thin in one dimension.

Temperature A measure of the degree of hotness or coldness of a substance.

Temperature control A factor that causes temperature to vary from place to place or from time to time.

Temperature gradient The amount of temperature change per unit of distance.

Tenacity Describes a mineral's toughness or resistance to breaking or deforming.

Tephra The volcanic rock ejected during an eruption, also called *pyroclastic materials*.

Terrae See *Lunar highlands*.

Terrane A crustal block bounded by faults, whose geologic history is distinct from the histories of adjoining crustal blocks.

Terrestrial (Earth-like) planet Any of the Earthlike planets, including Mercury, Venus, Mars, and Earth. Also known as the *inner planets*.

Terrigenous sediment Seafloor sediment derived from weathering and erosion on land.

Texture The size, shape, and distribution of the particles that collectively constitute a rock.

Theory A well-tested and widely accepted view that explains certain observable facts.

Theory of plate tectonics Tested theory proposing that Earth's outer shell consists of individual plates that interact in various ways and thereby produce earthquakes, volcanoes, mountains, and the crust itself.

Thermocline A layer of water in which there is a rapid change in temperature in the vertical dimension.

Thermohaline circulation Movements of ocean water caused by density differences brought about by variations in temperature and salinity.

Thermosphere The region of the atmosphere immediately above the mesosphere, which is characterized by increasing temperatures due to absorption of very shortwave solar energy by oxygen.

Thrust fault A low-angle reverse fault.

Thunderstorm A storm produced by a cumulonimbus cloud and always accompanied by lightning and thunder. It is of relatively short duration and usually accompanied by strong wind gusts, heavy rain, and sometimes hail.

Tidal current The alternating horizontal movement of water associated with the rise and fall of the tide.

Tidal delta A deltalike feature created when a rapidly moving tidal current emerges from a narrow inlet and slows, depositing its load of sediment.

Tidal flat A marshy or muddy area that is covered and uncovered by the rise and fall of the tide.

Tide Periodic change in the elevation of the ocean surface.

Till Unsorted sediment deposited by a glacier.

Tipping-bucket gauge A recording rain gauge consisting of two compartments ("buckets"), each capable of holding 0.025 centimeter of water. When one compartment fills, it tips, and the other compartment takes its place.

Tombolo A ridge of sand that connects an island to the mainland or to another island.

Tornado A small, very intense cyclonic storm with exceedingly high winds, most often produced along cold fronts in conjunction with severe thunderstorms.

Tornado warning A warning issued when a tornado has actually been sighted in an area or is indicated by radar.

Tornado watch A forecast issued for areas of about 65,000 square kilometers (25,000 square miles), indicating that conditions are such that tornadoes may develop; it is intended to alert people to the possibility of tornadoes.

Trade winds Two belts of winds that blow almost constantly from easterly directions and are located on the equatorward sides of the subtropical highs.

Transform fault A major strike-slip fault that cuts through the lithosphere and accommodates motion between two plates. Also called a *transform plate boundary*.

Transform plate boundary A boundary in which two plates slide past one another without creating or destroying lithosphere. Also called a *transform fault* or transform boundary.

Transpiration The release of water vapor to the atmosphere by plants.

Travertine A form of limestone ($CaCO_3$) that is deposited by hot springs or as a cave deposit.

Trench An elongated depression in the seafloor produced by bending of oceanic crust during subduction.

Trigger An event, such as an earthquake or heavy rainfall, that initiates a mass wasting process.

Tropic of Cancer The parallel of latitude, 23½° north latitude, marking the northern limit of the Sun's vertical rays.

Tropic of Capricorn The parallel of latitude, 23½° south latitude, marking the southern limit of the Sun's vertical rays.

Tropical (T) air mass A warm-to-hot air mass that forms in the subtropics.

Tropical depression By international agreement, a tropical cyclone with maximum winds that do not exceed 61 kilometers (38 miles) per hour.

Tropical storm By international agreement, a tropical cyclone with maximum winds between 61 and 119 kilometers (38 and 74 miles) per hour.

Troposphere The lowermost layer of the atmosphere. It is generally characterized by a decrease in temperature with height.

Tsunami A rapidly moving ocean wave generated by earthquake activity that is capable of inflicting heavy damage in coastal regions.

Turbidity current A downslope movement of dense, sediment-laden water created when sand and mud on the continental shelf and slope are dislodged and thrown into suspension.

Turbulent flow Erratic movement of water often characterized by swirling, whirlpool-like eddies. Most streamflow is of this type.

Ultramafic Term for igneous rocks that consist mostly of olivine and pyroxene.

Ultraviolet (UV) radiation Radiation with a wavelength from 0.2 to 0.4 micrometer.

Unconformity A surface that represents a break in the rock record, caused by erosion or nondeposition.

Uniformitarianism The concept that the processes that have shaped Earth in the geologic past are essentially the same as those operating today.

Unsaturated zone The area above the water table where openings in soil, sediment, and rock are not saturated with water but filled mainly with air.

Unstable air Air that does not resist vertical displacement. If it is lifted, its temperature will not cool as rapidly as the surrounding environment, and so it will continue to rise on its own.

Upslope fog Fog created when air moves up a slope and cools adiabatically.

Upwelling See *Coastal upwelling*.

Valence electron The electrons involved in the bonding process; the electrons occupying the highest principal energy level of an atom.

Valley breeze The daily upslope winds commonly encountered in a mountain valley.

Valley glacier (Alpine glacier) A glacier confined to a mountain valley, which in most instances was previously a stream valley.

Valley train A relatively narrow body of stratified drift deposited on a valley floor by meltwater streams that issue from a valley glacier.

Vapor pressure The part of the total atmospheric pressure that is attributable to water-vapor content.

Variable star A red giant that overshoots equilibrium and then alternately expands and contracts.

Vent A conduit that connects a magma chamber to a volcanic crater.

Vesicular texture A term applied to fine-grained igneous rocks that contain many small cavities called vesicles, which are openings on the outer portion of a lava flow that were created by escaping gases.

Viscosity A measure of a fluid's resistance to flow.

Visible light Radiation with a wavelength from 0.4 to 0.7 micrometer.

Volatiles Gaseous components of magma dissolved in the melt. Volatiles readily vaporize (form a gas) at surface pressures.

Volcanic See *Extrusive*.

Volcanic bomb A streamlined pyroclastic fragment ejected from a volcano while molten.

Volcanic cone A cone-shaped structure built by successive eruptions of lava and/or pyroclastic materials.

Volcanic island arc A chain of volcanic islands generally located a few hundred kilometers from a trench where there is active subduction of one oceanic plate beneath another. Also known simply as an *island arc*.

Volcanic neck (plug) An isolated, steep-sided, erosional remnant consisting of lava that once occupied the vent of a volcano.

Volcano A mountain formed of lava and/or pyroclastics.

Warm front A front along which a warm air mass overrides a retreating mass of cooler air.

Water table The upper level of the saturated zone of groundwater.

Watershed See *Drainage basin*.

Wave height The vertical distance between the trough and crest of a wave.

Wave period The time interval between the passage of successive crests at a stationary point.

Wave refraction See *Refraction*.

Wave-cut cliff A seawater-facing cliff along a steep shoreline formed by wave erosion at its base and mass wasting.

Wave-cut platform A bench or shelf in the bedrock at sea level, cut by wave erosion.

Wavelength The horizontal distance separating successive crests or troughs.

Weather The state of the atmosphere at any given time.

Weathering The disintegration and decomposition of rock at or near the surface of Earth.

Weather radar Instruments that utilize transmitters to send out radio waves at wavelengths that can penetrate clouds, to produce a reflected signal called an echo that can be displayed on a monitor to show the location and intensity of precipitation.

Well An opening bored into the zone of saturation.

Westerlies The dominant west-to-east motion of the atmosphere that characterizes the regions on the poleward side of the subtropical highs.

Wet adiabatic rate The rate of adiabatic temperature change in saturated air. The rate of temperature change is variable, but it is always less than the dry adiabatic rate.

White dwarf A star that has exhausted most or all of its nuclear fuel and has collapsed to a very small size; believed to be near its final stage of evolution.

Wind Air flowing horizontally with respect to Earth's surface.

Wind vane An instrument used to determine wind direction.

Winter solstice See *Solstice*.

Xenolith An inclusion of unmelted country rock in an igneous pluton.

Yazoo tributary A tributary that flows parallel to the main stream because a natural levee is present.

Zone of accumulation The part of a glacier characterized by snow accumulation and ice formation. Its outer limit is the snowline.

Zone of saturation The zone where all open spaces in sediment and rock are completely filled with water.

Zone of wastage The part of a glacier beyond the zone of accumulation, where all the snow from the previous winter melts, as does some of the glacial ice.

INDEX

Note: Boldfaced entries are key terms. References to figures are marked *f*, to tables marked *t*.